On the Origin of Extraterrestrial Industrial Civilizations

Steven Suan Zhu

April 29, 2021

Abstract

Recent discovery of billions of habitable planets within the Milky Way alone and a practical route to nuclear fusion using Project PACER approach, suggesting that any habitable planet with intelligent life should be able to expand beyond their home planet and colonize the galaxy within a relatively short time. Given the absence of detection by SETI for the past few decades, we take this result for granted that no other industrial civilization exists within the galaxy and validated the rare earth and rare intelligence hypothesis by using rigorous astronomical and geological filter to reduce the potential candidate pool to host civilization < 1 per galaxy. So that, the total number of habitable extraterrestrial planets within the Milky Way capable of supporting advanced, intelligent life within the next 500 Myr is < 969. Most of which are earth-like orbiting around a single star with mass ranges from 0.712 to 1 solar mass. No exomoons are capable of supporting advanced life, and a negligible number of low mass binary systems (<0.712 solar mass) are habitable. Among these habitable, the emergence of intelligence is still rare and must be a relatively recent phenomena.

By specifying species as a combination and permutation of traits acquired through evolutionary time, multi-nominal distribution profile of species can be constructed. Those with fewer traits are the most common. A particular multi-nominal distribution is build to model the emergence of civilization by specifying homo sapiens as an outlier. The deviation is calculated based on known cranial capacity of homo sapiens and the explosive growth of angiosperm. The multi-nominal distribution is then transformed/approximated into a more manipulative, generalized *multivariate time-dependent exponential lognormal distribution* to model biological evolution from the perspective of man.

Most surprisingly, given that the emergence chance of civilization decreases exponentially into the past, predicted by the distribution model, a wall of semi-invisibility exists due to relativistic time delay of signal arrival at cosmological distance so that the universe appears empty even if a significant portion of the space could have already been occupied. The nearest extraterrestrial industrial civilization lies at least 51.85 million light years away, and possibly at least 100 million light years or beyond. Based on the starting model, no extraterrestrial civilization arises before 119 Mya within the observable universe, and no extraterrestrial civilization arises before 138 Mya within the universe by co-moving distance. Despite great distances between the nearest civilizations and the low probability of emergence within our vicinity, given the sheer size of the universe, the total number of intelligent extraterrestrial civilizations likely approaches infinity or $\left(\frac{1}{4.4 \cdot 10^7}\right)^3 \cdot 3.621 \cdot 10^6 \cdot 10^{10^{10^{10^{122}}}}$ if the universe is finitely bounded. Based on incentives for economic growth and assuming wormhole shortens cosmic distances, all civilizations tend to expands near the speed of light and will eventually universally connect with each other via wormhole networks. Within such a network, the farthermost distances traversable from earth can be either infinite or $3.621 \cdot 10^6 \cdot 10^{10^{10^{10^{122}}}}$ light years in radius if the universe is finitely bounded.

This work distinguishes from and enhances previous works on SETI by focusing on the biological and statistical aspect of the evolution of intelligence, statistical distributions can serve as indispensable tools for SETI to model the pattern and behavior of civilization's emergence and development and bridging the inter-disciplinary gap between astrophysical, biological, and social aspects of extraterrestrial study.

This book constitutes the second print version of the same text published dated Feb 26th, 2018.

Out of this edition, significant changes were made. The original book with 340 pages has increased to 600 pages.

April 2021 Update: Added proofs and comparisons to section 8.7.3.5 Conclusion and 3 Laws of Evolution, added proofs in section 8.8.2 k Bounds for Weighted Deviation.

Feb 2021 Update: Added a new Section 7.2.2 Permutation attempts and Generalized Biocomplexity, completely revised Section 7.2.3 Best Fit, Section 7.2.4 Multinominal Distribution In-Depth Analysis, Section 7.2.5 Transforming Multinomial Distribution to Lognormal Distribution, added content to Section 8.7.5 The Rate of Civilization Emergence.

Dec 2020 Update: Added a new section 8.9 The Self Indication Assumption for the Assessment Earliest Window and Nearest Civilization, section 8.12 Overall Landscape Analysis.

The following major changes are made:

December 2019: added Section 8.8 "Constraining the Model Using Observations", Section 8.9 "Earliest Window Revisited and the Outer Wall", Section 8.10 "Other Types of Wall of Semi-Invisibility".

July 2019: added Section 10.6 "Temporal Convergence",

April 2019: Modified Chapter 7's counting on Deviation and YAABER.

Feb 2019: Modified Chapter 8's model and Generalized model section extensively.

Nov 2018: Completed the revision of the section "Habitability of Binaries and Multiples systems", redone entire "Rotational speed section", pages expanded to 490.

Aug 2018: Introduced Generalized model section

July 2018: Universal Non-Intentional Exclusion Revised the section "Habitability of Binaries and Multiples systems".

June 2018: Almost redone calculation for entire Chapter 2, 3, most of 4, 5. Separated Chapter 5 into two chapters due to increasing details and volumes. Redone entire lognormal distribution on Chapter 8 (formerly Chapter 7). Word usage, equation corrections, mistakes, typos, and clarifications were updated throughout the chapters. Major revisions are done to Chapter 2 Section 2.8 "Red Dwarves' Habitability", Chapter 4 Section 4.5 "Speed of Multicellular Evolution" (redid equations) 4.7 "Continent Cycle" (added equation derivation steps), Chapter 5 Section 5.3 "Expected Ice Age Interval" (added substantial new content and equations), Section 5.4 "Supercontinent Cycle and Ice Age" (added substantial new content and equations), 5.5 "The Probability of the Hominid Lineage" (added equations and corrected and added tables) Chapter 7 Section 7.1 "Number of Habitable Earth" (updates on the probability on Number of Civilizations), Section 7.3 "The Wall of Semi-Invisibility" (corrected mistakes and added substantial new content and equations), Section 7.4 "7.4 Complexity Equivalence" (added content), Section 7.9 "Observational Equations" (corrected mistakes), Chapter 8 Section 8.2 "Earthbound Democracy" (corrected mistakes added new content and equations), Section 8.5 "Earthbound Ruling Class" (corrected mistakes on equations and graphs), Section 8.6 "Shipbound with Energy Gathering Case" (corrected mistakes on equations and graphs), Section 8.11 "Worm Hole Maintenance Cost" (corrected mistakes on calculations and graphs), Chapter 9 Section 9.1 "E(d, v) Derivation and the Limit of Our Reach" (clarification on previous equations and explanations), Section 9.2 "Connected/Disconnected" (developed more robust explanations), and Section 9.4 "Looking Back in Time" (clarification on previous equations).

Prologue

One hundred and sixty years have passed since the publication of the *Origin of Species*, which provided convincing evidence that evolution took place on our planet which has existed billion of years before man. Many of the early objections such as Lord Kelvins disagreement on earth's age and the mechanism through which genes are inherited have been settled. The discovery of DNA as the carrier of genetic information and the sequencing of human genome helped to buttress the theory with solid authentication and to establish it as an experimental scientific discipline. The theory has stood the test of time, but increasingly a more knowledgeable, complex society urgently demands questions beyond the original author's imagination and capacity to follow and answer. First of all, Darwin had shown that evolution is goal-less and direction-less process. As a result, it offers little explanatory power to the human society which was rooted and originally evolved from nature. He has left such difficult, arduous tasks to those of the sociologists, economists, and anthropologists, who have embraced them with full dedication. Social sciences have provided enormously helpful theories to explain and predict the intricate, evolving society, yet they disagree on the final outcome of human society's progress. Each stood from their own stance and perspective. Each has their own merits and faults. Not to mention those took and possibly misunderstood their teachings and brought horrible consequences on the world as whole during the 20th century.

As a result, the dispute lead to the second problem. Is it possible to accommodate economics, social sciences, and anthropology into the natural sciences? Much like Boyle's Law, the behavior of each individual gas molecules can hardly be predictable, yet the overall behavior of gas itself follows simple and elegant equations? If there exists simple equations that can describe the ultimate pattern and behavior of human, or any intelligent species social trajectory? Thirdly, much like the Copernican revolution dethrone earth's unique position within the solar system, the discovery of immeasurable number of exoplanets have again put ourselves through another round of revolution, shaking the foundation in our thinking that life and even intelligent life is unique in the cosmos. The earthquake in our thinking has just barely started, yet those most curious and the deepest seeker of the truth are already excited by the roar and rumbling from the foundation below. How can we build upon the *Origin's* author's perspicacity and serendipity for life on earth and expand it to take into account all possible scenarios across different habitable planets? How can we provide simple, and elegant mathematical equations to encompass all possible evolutionary tales? Furthermore, our understanding in physics, cosmology, astronomy, economics, and information science since Darwin gave us additional tools and thinking models helped to pave numerous paths that potentially can lead to our goal. In fact, we are standing on shoulders of giants and peeking through the binoculars to see the farthermost possible horizon that is coming into our view and focus.

If these questions can be resolved and addressed cogently, a hypothetically back to life Darwin, our contemporaries, and future generations would be thankful to the contributors and weave these answers as critical bridges joining hitherto impassable domains of knowledge into the ever-increasing web of human rationality frontier. This author, due to his youth, has bravely came forth and attempted to untangle these difficulties, though somewhat maverick, but as if young David in the fight with the Goliath without fear but with whole-hearted confidence, has attempted to address these three questions fully, and has been deeply humbled and thankful for by the amount others have already contributed to each of the domains of knowledge, and is awed by the breadth, and deepness of the problems. Finally, the author is receptive to constructive criticism at: suan.zhu@wustl.edu

Acknowledgments

It has been almost 17 years since I first read about the Fermi Paradox. It has always remained a fascination for me over the possibility of extraterrestrials. I have also been a keen follower of the Kepler mission and the related topics of exoplanets. However, I have never dreamed and imagined to undertake a project like this all by myself from start to finish. What really prompted me to start working on this project was my accidental rediscovery of nuclear fusion using the Project Pacer approach in July 2014. Prior to this date, I always believed that the Fermi Paradox implies a pessimistic future for the humanity. That is, every civilization eventually exhausts its limited fossil fuel stock and dies out. My rediscovery of nuclear fusion gives me hope for the sustainability of the modern industrial infrastructure. Most importantly, being the youngest and an industry outsider I personally knew to have discovered such approach, I become confident enough to believe that I may have the talent and potential to tackle the even greater Fermi Paradox. Over the last few years, I used all of my free time besides work to contemplate over many different issues across many disciplines. Notes have been piled up. As I solidified problems and calculated precision up to my ability, I am finally presenting my results to all. My research serves as a starting framework, and I am eager to collaborate with others across all disciplines to make the calculation more precise and proofs more rigorous. I am thankful for those that helped to shape my academic career including but not limited to: Kenneth J. Goldman, Richard J. Smith, Paul Rothstein, Weixiong Zhang, Bill Smart, Chenyang Lu, David C. Butler, Jon Turner, Lihao Xu, and Xinwen Zhang. This book is also dedicated to my mother and my deceased father.

Contents

1 Introduction   9

    1.1 New Understanding Sharpens the Fermi Paradox 9

    1.2 Assumptions in the Solution to the Fermi Paradox 10

    1.3 Reconciliation of the Principles of Mediocrity with the Rare Human Hypothesis 13

    1.4 Our limited Observation Window 14

2 Number of Terrestrial Planets in Habitable Zone   17

    2.1 Temporal Window and Galactic Habitable Zone 17

    2.2 Definition of Stellar Habitable Zone 19

    2.3 Definition of Sun-like Stars 21

    2.4 Number of Stars 24

    2.5 Peak of Terrestrial Planet formation 31

    2.6 Excluding Over-counted Binaries and Multiples 32

       2.6.1 Habitability of Binaries 33

       2.6.2 Habitability of Ternaries 41

       2.6.3 Habitability of Quaternaries 50

       2.6.4 Habitability of All Systems 69

    2.7 Habitability of Low Mass Binaries 75

    2.8 Red Dwarves' Habitability 78

    2.9 Habitability of Exomoon 91

3 Number of Earths   94

    3.1 Orbital Eccentricity 94

    3.2 Orbital Period 95

    3.3 Earth & Moon Separation 96

       3.3.1 Final Separation Distance 96

       3.3.2 Earth's Locking Time to the Moon 99

    3.4 Earth-Moon Collision Probability Explanation 106

    3.5 The Right Rotational Speed 113

    3.6 Non-Locked Moons 121

    3.7 Moon's Orbital Obliquity Evolution 127

    3.8 Earth like Planet Size Requirement 129

    3.9 The Chance of Getting Watered 130

    3.10 Total Water Budget of Earth 134

    3.11 Right Ocean and Land Mix 137

    3.12 Plate Tectonics 145

4 Evolution   **147**

  4.1   Water vs. Other Solvents 147

  4.2   Biocomplexity Explanation 148

  4.3   Probability on the Emergence of Prokaryotes from Amino Acids 154

  4.4   Probability on the Emergence of Eukaryotes, Sex, and Multicellularity . . 157

  4.5   Speed of Multicellular Evolution 160

  4.6   BCS, BER, and Evolutionary Speed 168

      4.6.1   BCS 168

      4.6.2   Evolutionary Speed 170

      4.6.3   BER 170

      4.6.4   Altering the Speed of Evolution 171

      4.6.5   Measuring BCS and BER 173

  4.7   Continent Cycle 174

  4.8   Supercontinent and Island continent biodiversity in depth-analysis 177

  4.9   Continental Movement Speed 182

5 Glaciation and Super Continent Cycles   **185**

  5.1   Ice Age as an Accelerator and Its Causes 185

  5.2   Expected Ice Age Interval 188

      5.2.1   Proportion of island to supercontinent - lower bound biodiversity calculation 193

      5.2.2   Upper bound biodiversity calculation 196

  5.3   Supercontinent Cycle and Ice Age Cycle 202

  5.4   Weighted Emergence Rate across All Dryland Ranges: Intro 207

  5.5   The Permissible Range Factor 209

  5.6   Weighted Emergence Rate across All Dryland Ranges: Detailed Analysis . 214

      5.6.1   Chance within a breaking up phase 214

      5.6.2   Chance of rejoining 215

      5.6.3   Glaciation chance 215

      5.6.4   Partial chance of emergence 216

  5.7   Generalized Emergence Curve across All Temporal Periods 222

  5.8   Chance of Human Emergence Recalibration 227

6 Homo Sapiens Emergence Probability   **230**

  6.1   Why Human Did not Appear Earlier 230

  6.2   Why intelligent species can not emerge from Arthropods 231

  6.3   The Probability of the Hominid Lineage 233

      6.3.1   Binocular Vision 234

      6.3.2   Large Cranial Capacity 234

      6.3.3   Language 234

      6.3.4   Bipedal 235

      6.3.5   Thumbs 235

      6.3.6   Social 235

      6.3.7   Omnivorous Feeding 236

  6.4   The Probability of Alternative Intelligence 247

  6.5   Probability of the Emergence of Homo Sapiens within the Genus Homo . 250

  6.6   Probability of Fruit Trees 251

  6.7   Probability of Crop Plants 252

  6.8   Probability of Angiosperm 253

7 The Distribution Model   **256**

  7.1 Mathematical Model for Human Evolution 256

  7.2 Background Rate 258

     7.2.1 Sample Data 258

     7.2.2 Permutation attempts and Generalized Biocomplexity 261

     7.2.3 Best Fit 267

     7.2.4 Multinominal Distribution In-Depth Analysis 279

     7.2.5 Transforming Multinomial Distribution to Lognormal Distribution 283

     7.2.6 Threshold Test 303

     7.2.7 Projection onto 3D Space 307

     7.2.8 Conclusion 310

  7.3 Counting Deviation and YAABER 311

  7.4 Deviation and YAABER for Evolution of Homo Sapiens 316

  7.5 YAABER for Hunter Gatherer 320

  7.6 YAABER for Feudal Society 323

     7.6.1 City states period 325

     7.6.2 Middle Ages 326

     7.6.3 Age of Exploration 327

  7.7 YAABER for Industrial Society 331

8 Model Predictions   **340**

  8.1 Number of Habitable Earth 340

  8.2 The Model 349

  8.3 Space Occupancy Constraint 356

  8.4 The Wall of Semi-Invisibility: Introduction 361

  8.5 The Wall of Semi-Invisibility: Proof 367

     8.5.1 Base Case: 368

     8.5.2 Inductive Step: 370

  8.6 The Wall of Semi-Invisibility: Detailed Analysis with the Theoretical Upper Bound 374

  8.7 Generalized Model 383

     8.7.1 Different values of BCS and BER 383

     8.7.2 Distribution Placement and Variable k 385

     8.7.3 BER's corresponding speed and the selection factor: 389

        8.7.3.1 First Interpretation 389

        8.7.3.2 Second Interpretation 397

        8.7.3.3 Test Cases 397

        8.7.3.4 The Rate of Civilization Emergence 411

        8.7.3.5 Conclusion and 3 Laws of Evolution 417

  8.8 Constraining the Model Using Observations 433

     8.8.1 Using Major Events and Genomic Complexity to define BER 433

     8.8.2 k Bounds for Weighted Deviation 434

     8.8.3 k Bounds for Unweighted Deviation 446

     8.8.4 Self-similarity of the Distribution 448

     8.8.5 BCS, k and Selection Factor Relationship 450

     8.8.6 The Lower and Upper Bound on Deviation 453

     8.8.7 Step by Step Diagram for Constructing the Generalized Model and a Real Case Instantiation 455

  8.9 The Self Indication Assumption for the Assessment Earliest Window and Nearest Civilization . . 459

  8.10 Earliest Window Revisited and the Outer Wall 464

  8.11 Other Types of Wall of Semi-Invisibility 465

      8.11.1 Civilization Becoming More Frequent in Space but not in Time 466

      8.11.2 Civilization Becoming More Frequent in Time but not in Space 468

  8.12 Overall Landscape Analysis 470

  8.13 Complexity Equivalence 473

  8.14 Complexity Transformation 476

  8.15 Darwin's Great-Great Grandson's Cosmic Voyage 478

  8.16 Upper Bound & Lower Bound 481

  8.17 Subluminal Expansion 482

  8.18 Observational Equations 483

9 Relativistic Economics   **487**

  9.1 Overview 487

  9.2 Earthbound Democracy 497

  9.3 Earthbound Investing Nearest Galaxy 498

  9.4 Galaxy Bound Investing Nearest Galaxy 500

  9.5 Earthbound Ruling Class 501

  9.6 Shipbound with Energy Gathering Case 502

  9.7 Shipbound as Lottery Winners Case 503

  9.8 Post-Singularity 504

  9.9 Expansion Speed from Outsider's Perspective 506

  9.10 Worm Hole 508

  9.11 Worm Hole Maintenance Cost 516

10 The Principle of Universal Contacts   **521**

  10.1 E(d, v) Derivation and the Limit of Our Reach 521

  10.2 Connected/Disconnected 529

  10.3 Cosmic Nash Equilibrium 534

  10.4 Looking Back in Time 535

  10.5 Universal Non-intentional Exclusion 539

  10.6 Temporal Convergence 540

11 Conclusion   **543**

  11.1 Extra-terrestrials vs. Time 543

  11.2 Final Thoughts 548

12 Special Chapter: Gravitational Effect on the Final Stellar to Planetary Mass Ratio   **550**

  12.1 Overview 550

  12.2 Empirical Law Derivation and Proof 550

  12.3 Stellar Data and Derivation 564

  12.4 Conclusion 569

  12.5 Appendix 571

A Proof for $P_{df}(t,x)$ represents the cumulative emergence chance of all previous periods   **580**

B Review and Response to Milan Cirkovic's The Great Silence: Science and Philosophy of Fermi's Paradox   **581**

  B.1 Intro 581

  B.2 My Theory 582

Chapter 1

Introduction

1.1 New Understanding Sharpens the Fermi Paradox

The recent discovery by Kepler Exoplanet hunting mission revealed that Earth-like planets within the habitable zones of main sequence stars in the Milky Way galaxy might well be within the range of billions.[63] On the other hand, a practical technique to achieve nuclear fusion using the project PACER approach, which was further refined upon using small yield hydrogen detonation device with minimal fissioning plutonium underground concrete cavity[104][20] has guaranteed a cheap, sustainable energy budget for industrial use into the indefinite future.[83] Therefore, a Hubbert like peak of resource exhaustion due to falling EROEI for fossil fuel based industrial civilization can be avoided technically speaking.[83][107][70][69] Though energy conservation must be enacted in the future for such scenario to hold and to avoid Jevon's Paradox as a planet based civilization.[64] Furthermore, city-sized spaceship based on nuclear fusion power plants based on Project PACER model, also enable human interstellar travel in less than geological timescale and magnitudes lower than astronomical scales.[22] Based on the calculation, nuclear fusion powered vessels are capable of carrying the total population of human race to any predetermined destinations in a few generations with fully furnished and self-sustaining life quarters at up to a small fraction the speed of light c. If nuclear fusion inter-stellar vessels project is initiated, populations magnitudes higher than the current population can be migrated to predetermined destinations. Assuming every extraterrestrial industrial civilization arise in the Milky Way follow a similar developmental pattern of earth based on the Principle of Mediocrity, we should expect them to discover at least project PACER model of nuclear fusion. They should also be able to confirm the existence of billions of habitable planets scattered in their galaxy. Then, a strong case is presented for their preference to expand and explore the galaxies less than geologic time and bounded by the speed of light from stationary observers on earth at a maximum of 10^5 years for the diameter of the entire galaxy disk. (10^5 yrs $<$ x$<10^9$ yrs) However, no overwhelming scientific evidence since the inception of SETI project has proved that any star in the Milky Way host an industrial civilization. Given the age of Milky Way Galaxy at 10^{11} yrs and the Principle of Mediocrity applying to the temporal aspect of cosmic biological evolution, then the chance that all extraterrestrial industrial civilization evolved at around the margin of 10^4 yrs so that they currently remain undetectable is $\frac{10^4 \text{ Yrs}}{10^{11} \text{ Yrs}} = 10^{-7}$ for each habitable planet. With 10^9 habitable planets within the galaxy, the chance is 10^{-16} in Milky Way alone. This chance can also be argued as the chance of success of extraterrestrial industrial civilization arising in our galaxy yet we failed to observe them so far. Indeed, this hope crushing number can be interpreted as such that our chance to colonize the galaxy is not much greater than 0. Assuming intelligent life can be evolved easily and transforms into an industrial civilization and all intelligent life destroys themselves eventually regardless their worldview and culture. However, I have just shown that cosmic expansion is the predicted outcome of industrial civilization. Furthermore, the recent analysis that an average earth-like habitable planet has a median age of 78 Gyr, that is 25 Gyr ahead of earth in development. Paradoxically, advocates of Technological Singularity

9

argues for accelerating return generalized from the Moore's Law and expect machine intelligence overtaking human intelligence around mid-century. [105][77] Robotic successors to biological humans, capable of greater endurance in all types of environments deemed hostile to biological life, can accelerate industrial civilization expansion faster and by more economical means. This is shown as rover opportunity still roaming on Mars at the temperature and atmospheric pressure significantly lower than earth after a decade following the assumed lifespan for its exploratory mission. On one hand, we have shown the predicted expansion trajectory for the future of human or human-machine industrial civilization. On the other hand, an eerily silent sky filled with billions of Earth-like planets in the Milky Way galaxy yet no detectable sign of industrial civilization comparable to our current epoch and earlier. [54] The immense discrepancy between two sides sharpens the Fermi Paradox more than at any time before. Even more disturbingly, billions of galaxies exist within the observable universe, assuming the universe is isotropic, then each of these galaxies should hold billions of Earth-like planets as well. A study done in late 2014 to detect infrared emission proposed by Freeman Dyson and hunting for galaxy wide Dyson sphere energy harvesting civilization from the most promising and suspicious 100,000 galaxies within a light cone of $2 \cdot 10^8$ light years found nothing unusual. [114]

It is then of paramount importance and urgency to resolve this paradox with every tool and piece of knowledge we currently have, though probably still incomplete or never will be truly complete. The most significant central concern of Fermi Paradox is to address the problem of sustainability of industrial civilization; therefore, the resolution of an astrophysical-biological question has an immediate terrestrial interest.[89]

1.2   Assumptions in the Solution to the Fermi Paradox

We shall make some assumptions before our further discussion.

First, any intelligent civilization, no matter how efficient they are at energy manipulation, even down to Planck scale or beyond, still expands until it meets another extra-terrestrial civilization's sphere of influence or had grabbed all possible resources reachable. After that point in time, whether the extraterrestrial civilization will continue to consume its energy in an exponential growth based model and collapse or resort to energy conservation is not a concern for this assumption to hold. British economist Jevon in 1865 made a strong case that more efficient use of coal in Britain led to more rapid and greater consumption of the same resource. This is called the Jevon's Paradox.[23] This is further elaborated in scenarios of post-singularity civilization. Other compelling reasons for expansion will be discussed in later sections but should not be mentioned here, because these reasons are derived from this basic assumption.

Secondly, no overwhelming evidence of any extraterrestrial industrial civilization based on our current observation of the Milky Way and beyond implies the non-existence of human comparable or beyond at that point in time where the information transmitted from that original source point. That is, we do not see evidence of extraterrestrial civilization from a star 2,000 light years away implies that there is no emerged extraterrestrial industrial civilization from that star 2,000 years ago or earlier. A natural looking galaxy lying 20 million light years way implies that no emerged extraterrestrial industrial civilization from that galaxy 20 million years ago or earlier based on the speed of light to a stationary observer on earth.

Thirdly, our current knowledge and understanding regarding harnessing nuclear fusion is sufficient evidence or a counterexample to suggest that physics and nature do not impose a ceiling on the amount of available energy a civilization can achieve before it reaches an interplanetary and a galactic scale. Once it does, it can expand at sub geological time scale and bounded by the speed of light. Our local derivations of physical laws apply universally throughout the universe. It is still possible that the civilization destroys itself through nuclear wars, biological warfare or self-decay. However, such scenarios are social and culturally specific, it does not universally apply to all possible emerging civilizations given the sheer number of Earth-like planets in the entire observable universe.

Lastly, all potentially habitable planets within the habitable zones of their parent star contain a hospitable environment for life. The low eccentricities of planet orbit, the presence of moon is not considered to be essential to the emergence of intelligent life. (Furthermore, it is assumed earth and moon type of binary planetary system are common. The reasoning behind this will be further explained in section Reconciliation of Principles of Mediocrity with Rare Human Hypothesis. Calculation is presented to explain why moons are common in Chapter 2) This is a possible overstatement since Mars, also lie on the outer edge of the habitable zone is currently inhabitable. The motive behind the assumption **is** that all potentially habitable planet with at most one magnitude within the range of earth's condition such as water presence, atmospheric pressure, oxygen abundance is to accommodate the principle of Mediocrity, showing mathematically an upper bound on the shortest possible temporal time span and spatial distance to find nearest extra-terrestrial industrial civilization.

These assumptions led us to case scenarios. If we consider pq=w. Where w is the probability that expanding industrial civilization excluding earth exists in a given region, and p is the probability that intelligent life evolves in somewhere other than earth and q is the probability that laws of physics permits such intelligent species eventually expands. Then, from assumption 3), we know that q=1. Yet we know from assumption 2), that w $\leq 10^{-16}$ for region occupying the Milky Way galaxy with a temporal range between $13.2 \cdot 10^9 > x > 10^5$ years ago. This implies that $p \leq 10^{-16}$ for region occupying the Milky Way galaxy. This concludes that there is an extremely high probability $(1- 10^{-16})$ that intelligent life does not exist somewhere other than earth within the Milky Way region with a temporal range between $13.2 \cdot 10^9 > x > 10^5$ years ago. We do not know the more recent 10^5 years because signals from the outermost regions of the galaxies take 10^5 light years to reach observers on earth. However, given the sheer size of the observable universe, which makes up with $> 10^{11}$ galaxies, then $p \leq 10^{-5}$ for region occupying the observable universe. This concludes that there is a good probability $(1- 10^{-5})$ that intelligent life does not exist somewhere other than earth within the observable universe region with a temporal range between $13.2 \cdot 10^9 > x > 10^5$ years ago. It will take a stationary observer on earth at most $13.2 \cdot 10^9$ more years to verify the validity of the prediction because signals from the outermost edges of the observable universe take $13.2 \cdot 10^9$ light years to reach observers on earth. Furthermore, given the sheer size of the entire universe, which can be infinite in size, then $p \leq 1$ for region occupying the universe is possible but it can take a stationary observer infinite amount time to verify apart from observational limitation due to the expansion of the universe and redshifts.

This all suggests that intelligent life evolved on earth is rare both in space and time since the Big Bang.

However, further extrapolation leads into 4 case scenarios.

First, the earth continues to be rare or even rarer in space and time. In this scenario, events observed on earth are almost never repeated anywhere in the universe for an infinite amount of time into the future. Then, the chance of encountering any intelligent civilization infinitely approaches 0 from now to infinitely long future periods.

Secondly, earth-like cases become more frequent in space but not in time. This case is slightly optimistic but similar to the first case, where intelligent life forms evolve on Earth-like planets can repeatedly happen in the same galaxy or even neighboring stars many times, but each is separated by billions and trillions of years apart. In this case, communication can be unidirectional where an ancient civilization communicates to the younger through ancient remains and assuming symbols can endure such long epochs, but such communication is irrelevant to bidirectional communications shall be concerned in this paper.

Thirdly, earth-like cases become more frequent in time but not in space. This case is similar to the second and also slightly more optimistic than the first. Whereas each extraterrestrial civilization is emerging only by 10^5 years or less apart from each other but given the sheer size of the universe, each is separated from each other by billions or trillion light years apart. Whereas it is possible that every galaxy can guarantee the emergence of intelligent industrial civilization but only at an infinite amount of time assuming the infinite size of the universe.

Finally and most interestingly, earth-like cases become both frequent in space and time. This scenario requires particular attention because it implies encountering with extraterrestrial civilization is bounded by a limited temporal range and limited spatial distance. We shall model and abstract all scenarios of encountering behaviors based on this case. Since the first case is trivial to show or derive (it simply states that earth is a sporadic event and never repeated again), then we can show that scenario 2 and scenario 3 falls between scenario 1 and scenario 4. Models from scenario 4 can also be applied to scenario 2 and 3. If scenario 4 is true where life is assumed to exist and abundant starting now and into the future yet its evolutionary trajectory has to satisfy our current understanding of biology, chemistry, and physical constraints. Then *these constraints create an illusion that we are alone as far as our current observation dictates yet it can be shown mathematically that they must exist.* Any intelligent life in the universe arising should be well-aware of their "neighbors" well before they ever detect them through physical means.

Fortunately, existing literature suggests that scenario 4 is possible. Two camps, the catastrophic camp and bio-complexity camp supports scenario 4. The great concerns and differences have been raised by two different camps, but both arrived at the same conclusion. That is, complex, intelligent, and technological life has only recently begun to appear in the universe.

The catastrophic camp argues that a cosmic phase transition is taking place because cosmic regulating mechanisms such as Gamma Ray Bursts, and Quasar-like super black holes were dominant in the cosmic past since the Big Bang. Prior to the formation of galaxy spiral arms, stars are much closely packed and Supernovae explosion causes great havoc on potentially life-bearing planets. If only the simplest and the sturdiest survives from each holocaust, each event disrupts biological evolution and systematically resets regional and global fauna complexity and diversity. The frequency of cataclysmic global events guarantees the universe remains silent. In the recent epoch, the evolution of spiral arms in galaxies and an exponential decrease in the large energetic release of cosmic events contributes to the growth of biological complexity and diversity guaranteeing the emergence of industrial civilization.

On the other hand, the bio-complexity camp argues that the biological complexity encoded in DNA, RNA, and epigenetics, and functional modularity increases over geological time.[106] If all genome complexity is considered and follows the Law of Accelerating Return of complexity doubling over a period $3.75 \cdot 10^8$ yrs as observed on earth from the prokaryotes to mammalian lineage, then, prokaryotes dispersed to earth by panspermia must be formed before the formation of the sun. Reaching prokaryotes level of biological complexity from a single pair of DNA requires another 5 Gyr prior to the formation of the sun. Thus, life evolved before earth. This camp also frequently cites the contradiction between rapid emergence of Prokaryote life following the Late Heavy Bombardment, yet it is never synthetically replicated in laboratory settings as an evidence supporting their hypothesis. A weaker argument can also be used to support their view given the metallicity of older stars belong to generation II are about 1~2 magnitudes lower than the sun, it has been shown that metal-poor stars offer a lower chance of planet formation and life on earth demonstrates the necessity of elements higher than Helium. However, a careful analysis of host star age and its metallicity shows a weak positive correlation, where metal-rich stars exist among the very old, and metallicity buildup of stars following the first 10^9 yrs is slow compares to the earliest times. Therefore, enough metallicity is sufficient to sustain life 10^9 yrs after the big bang, which nevertheless is consistent with their hypothesis that life evolves early.

Both camps agree that the universe is undergoing a cosmic phase transition where a lifeless universe is transitioning toward one filled with life.[32][60] However, a crucial difference exists where the catastrophic camp believes that life is easy to evolve at any time and at anywhere as long as global cataclysmic regulatory mechanisms cease. The biological complexity camp holds that evolution of life was a slow and hard process, complexity doubling was slow from the very start; therefore, even if the universe were conducive to the evolution of life from very early on, the universe would still be lifeless until very recently. In this paper, the author is inclined toward the catastrophic camp, mainly persuaded by the mathematical models (including derivations from molecular biology, confirming punctuated equilibrium which formed as one of the pillars of the entire model) build in order to resolve this paradox.

1.3 Reconciliation of the Principles of Mediocrity with the Rare Human Hypothesis

If extraterrestrial life is now becoming abundant in all possible earth-like exoplanet candidates given in scenario 4, we still need to analyze the median and mean average level of bio-complexity of the highest form of life currently attained for each of these planets. In order to solve this problem, we need to reconcile the rare human hypothesis (that human-like intelligent life with opposable thumbs, bipedal locomotion, big brain, omnivorous diet, binocular vision, land residing, and complex language to grow knowledge in successive generations) to a weaker form of rare earth hypothesis with the Principle of Mediocrity. It basically states that if an item is drawn at random from one of several sets or categories, it's likelier to come from the most numerous category than from any one of the less numerous categories.[15] The principle has been taken to suggest that there is nothing unusual about the evolution of the Solar System, Earth's history, the evolution of biological complexity, human evolution, or the developmental path of civilization leading to expanding cosmic civilization. It is a heuristic in the vein of the Copernican principle and is sometimes used as a philosophical statement about the place of humanity. The idea is to assume mediocrity, rather than starting with the assumption that a phenomenon is special, privileged, exceptional, or even superior.[33, 95]

Before the discovery of exoplanets and Kepler's mission for exoplanets, a rare earth hypothesis generally assumed that Earth-like planets and possibly even planets are rare in the galaxy and the universe. However, it is now believed that as many as 40 billion Earth-sized planets orbiting in the habitable zones of sun-like stars and red dwarf stars within the Milky Way Galaxy alone, based on Kepler space mission data.[63] 11 billion of these estimated planets may be orbiting sun-like stars.[43] The nearest such planet may be 12 light-years away, according to the scientists. However, it is equally erroneous to assume that all life-bearing planets underwent exact or very similar evolution happened here on earth where intelligent creatures similar to human in almost every aspect. This over-generalization is in direct contradiction with the evidence obtained from our local neighborhood with a high confidence. For the delay in light signal transmission between $0 < x < 10^5$ years, there is a low probability that all intelligent life evolved recently within a window of $0 < x < 10^5$ years in the Milky Way and not earlier. Even under the same solar radiation and environmental factors, human civilization diverged since the great exodus from Africa 100,000 years ago have evolved socially at different rates, where hunter-gatherers in Amazon forests are $10^2 < x < 10^3$ years apart in complexity from the advanced industrial nations. Therefore, the standard deviation is likely higher than $> 10^3$ years for different intelligent civilization within the Milky Way at different stages of development even if global regulating mechanisms persisted in the past suppressing the early emergence of complex life or genomic complexity and functional complexity of life and just recently enabled human level observers to exist. It is much more likely, that all extra-terrestrial beings, humans included are nearly identical in the physical, chemical, even biological aspects, but with differences mainly in subtle details and variations. These variations, then, prevented extra-terrestrial from cosmic expansion.

To understand this argument better, we need to introduce the following thought experiment. Imagine you are Nicolas Copernicus in the early 16th century living in celibacy. For some reason, the Pope, through prophetic dream knowing that you are going to disrupt the Catholic church when you grow up, imprisoned you since you are a baby and food is delivered to your prison cell. Separating yourself from any other human since you are born, except you are able to view the night sky from your cell. (the Pope does not know prior what kind of idea you comes up to disrupt the church, so he did not know your disruptive idea stems from astronomy) You are oblivious and unaware of your surroundings and people around you. While you are in the cell, you viewed the sky, made observations, and come up with the Heliocentric model of the universe. You, for a long time, thought that the Pope, the prison guard, and yourself are the only three people in Europe, which makes you 1 in 3 chance as the only one comes up with Heliocentrism if you follow the Principle of Mediocrity. Curiously enough, you then ask the prison guard if there are as many people in Europe as there are the number of visible stars. When the prison guard replied that there are tens of millions, your jaw dropped because you never expected these many people. Then, you start to ask yourself, if I come up with this theory, based on the Principle of Mediocrity, tens of millions must have done the same in Europe, so the chance that I am the first to write about it is 1 out of tens of millions. Because there are tens of millions of Europeans, then each of them, viewed the sky will have similar thoughts and worldviews. However, any outsider observer or us from the future will laugh at his conclusion. So how was his reasoning leading himself to the conclusion and where did the reasoning go wrong?

First of all, he assumed that everywhere in Europe, people use the same mathematics principle and equations, which is true. He assumed everyone lives on land just like himself and subjected to the laws of physics and electromagnetism, which is true. Then, he assumed that everyone is made of chemical composition with Carbon, hydrogen, oxygen atoms, which is also true. Then, he further narrows down his assumptions that everyone is made of flesh and blood if the biological condition is universal among all humans: they all subject to the laws of biology, they follow specific developmental patterns, they grow hair, has two pairs of eyes, ears, legs, and arms. They are capable of self-cognition, comprehension, and language communication. This is again also true. Then, he even narrowed his assumption further and reasoned that everyone in Europe drinks beer and lives in a temperate climate like that of Poland if ecological conditions are same throughout Europe. This is only partially true because Italians, Greeks, and Spaniards enjoy a Mediterranean climate and frequently drinks wine. Then, he further reasoned that all European speak the Polish language if cultural is universal. This is assumption starts to deviate from the objective truth because only Polish people speak the Polish Language. Then, he further assumes that everyone in Europe bears the last name Copernicus if everyone descended from the same family. This over-generalization does not even apply to most Polish people. Finally, he reasoned that all Europeans have an unquenchable curiosity for astronomy and did extensive analysis of motions of heavenly bodies and came up with the Heliocentric model. In reality, only himself came up with the theory.

The conclusion from the thought experiment is significant because one should not be alarmed by the discovery of billions of Earth-like planets in the Milky Way galaxy. This does not imply every planet currently hosts an earth-like expanding industrial civilization. However, based on the evolutionary paradigm hierarchy, it is certain without a doubt these planets use the same mathematical formula just like on earth, and are subject to the same physical laws with more or less gravitational effect. Their biological body made up of oxygen, hydrogen, carbon, and maybe few different atoms not abundant here on earth. Even possibly we all share many similar biological characteristics such as DNA blueprint, sexual reproduction, and multi-cellularity. However, it is likely that they evolved bipedal locomotion, opposable thumbs but no big brain. Or Having all the attributes of human but lived on an exclusively carnivorous diet so that agricultural revolution is not possible and civilization can only be maintained at a low level of complexity. Or they are omnivorous like human but lacks the essential crop plants to usher themselves into an agriculture revolution and creates a civilization with high population density. Therefore, we could even refine our hypothesis as *Many-earth to fewer habitable to fewer hunter-gatherer to rarer agricultural to industrial to an exceptionally rare cosmic expanding human hypothesis*, which truly captures the essence of the Fermi Paradox. It is then not surprising, to predict that intelligent life forms may already present, if not abundant in the Milky way under scenario 4 complied with the Principle of Mediocrity but lacks some trivial yet essential ingredients to usher themselves into a full-blown cosmic expanding industrial civilization.

1.4 Our limited Observation Window

To put our current lack of observation of extra-terrestrial civilization into perspective, the temporal window that is visible currently to us is very limited, yet we are looking for a resolution. By plotting the graph of all possible signals ever reachable from earth, we can compute the fraction of all possible signals we are receiving.

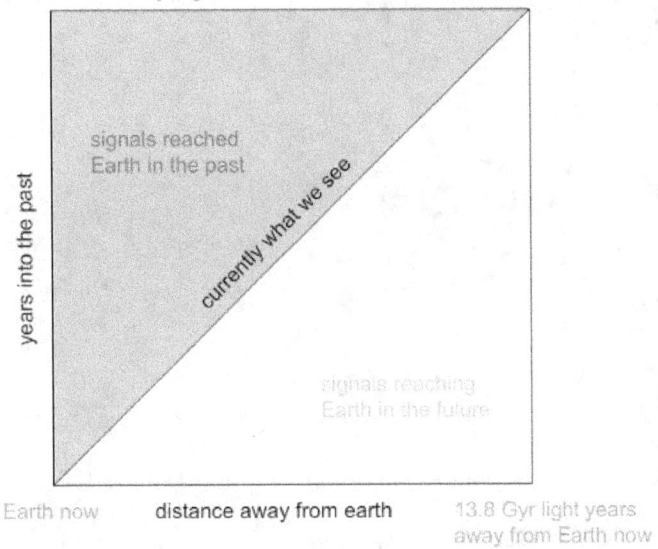

Figure 1.4.1: Signal detection landscape of time vs distance

From the graph, one can see that the vertical axis indicates the passage of time where the past 13.8 Gyr ago starts from the top and reaches the bottom which is the current present time. The horizontal axis represents the distance away from earth in units of light years, whereas the leftmost point represents 0 light years away from earth, the rightmost point represents 13.8 billion light-years away. The diagonal line represents the signals we are currently experiencing. That is, signals from 13.8 Gya from 13.8 billion light years away had just had enough time to reach us and all remaining signals from 13.8 billion years ago up to today had yet to be received. Somewhere along the diagonal, say 6.5 billion light years away, any signals sent from earlier than 6.5 billion years ago had passed earth and signals emitted from 6.5 billion years ago is reaching earth right now, and signals emitted later than 6.5 billion years ago yet to reach us. At the leftmost point along the diagonal, which is 0 light years away from earth, all signals emitted before the present time had passed earth except the current signals and only signals yet to be emitted from the future will be captured.

Based on this graph, we can make the following logical deduction:

$$wh \geq \frac{wh}{2} \geq \sqrt{w^2 + h^2} \tag{1.4.1}$$

where the signals we are receiving from different types and distances away from earth is a subset of signals ever received by earth, which in turn is half all of the signals ever emitted. From a purely observational point of view, seeing the probability of extraterrestrial civilization is tiny because we can only observe a very narrow window of the range of all possible signals.

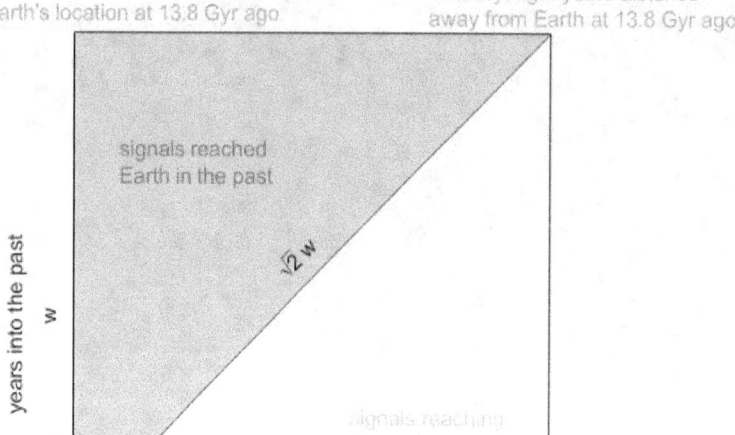

Figure 1.4.2: Signal detection landscape of time vs distance quantified

If we simplify the equation and assume that both the height and width of the box are of the same length, which is a reasonable assumption excluding the expansion of the universe, then the equation can be simplified as:

$$\frac{\sqrt{2}w}{\frac{w^2}{2}} = \frac{2\sqrt{2}}{w} \Rightarrow \lim_{w \to \infty} \frac{2\sqrt{2}}{w} = 0 \tag{1.4.2}$$

and taking the limit, we found that the observation window at the current time is diminishingly small. Of course, one can argue if extraterrestrial civilization ever expanded to include the earth from the past then earth will be transformed. The lack of evidence of artificial alteration of earth, the solar system, and the Milky Way, is an indication, that at least half the signals ever reachable since the Big Bang can be excluded from our search space, and one should focus on the remaining half. Nevertheless, the narrowness of the observational window at current time does give some solace to SETI which is complaining the eerie silence.

Chapter 2

Number of Terrestrial Planets in Habitable Zone

2.1 Temporal Window and Galactic Habitable Zone

Out of the total number of earth like terrestrial planets within the habitable zones of stars with the mass range from 0.712 to 1 solar mass, the planets have to be formed between 5 Gyr ago and 4 Gyr ago. Prior to 5 Gyr, episodes of short Gamma-Ray Bursts, at the earlier epoch of galaxy formation, are dominant and sterilize the emergence of life from those planets.[97][11][31] The formation of terrestrial planets is correlated in both space and time. Earlier generations of planets formed closer to the galactic core, where numerous supernovae and other energetic cosmic events, as well as excessive cometary impacts caused by perturbations of the host star's Oort cloud persisted. Therefore, even if Gamma Ray Bursts becomes less frequent as the Galactic metallicity increases (which still only reaches -0.2 near the center), earlier generations of terrestrial planets are vulnerable. Milky Way's spiral arms, constituting the Galactic Goldilocks zone with less density and higher metallicity to minimize life-threatening conditions, only start to form 8 Gya. Therefore, life favoring conditions in the galaxy only start after the start of the formation of spiral arms. It cannot be later than 4 Gyr ago because merely 500 Myr head start on earth will give the planet a head start of 500 million years. We have shown in our chapter 4 that at early stages of evolution it takes billions of years for the cyanobacteria to fill all the major oxygen sinks on earth before free oxygen will become available throughout the ocean and the atmosphere. Therefore, having an additional 500 million years head start means cyanobacteria have extra half billion years to transform the planet, implying a 500 Myr head start into every other later stage of evolution of life. Moreover, the oxygen concentration can only reach an even higher level and enables the emergence of multicellular eukaryotes after another billion years of geoengineering by biologic life. Furthermore, a head start of 500 million years can be translated into 2.78^5, or 166 fold increase in biological complexity in the evolutionary history of multicellular life, as evidenced by earth's history. Therefore, everything being equal, the ground chance [1] of intelligent life emergence at the current epoch on a planet formed 4.4 Gyr ago is 36% of earth formed 4.5 Gyr ago. The ground chance of intelligent life emergence at the current epoch on a planet formed 4.3 Gyr ago is 12.9% of earth formed 4.5 Gyr ago. The ground chance of intelligent life emergence at the current epoch on a planet formed 4.2 Gyr ago is 4.65% of earth formed 4.5 Gyr ago. On the other hand, the ground chance of intelligent life emergence at the current epoch on a planet formed 4.6 Gyr ago is not 278% of earth formed 4.5 Gyr ago because the average metallicity of the systems of an earlier epoch is lower. Nevertheless, the biocomplexity and diversity

---

[1] The ground chance refers to the size of BCS (biocomplexity search space, the number of species alive on a planet at any given period). The chance of civilization emergence follows the growth rate of BCS if it is one of complete passive evolution with BER = 1 and k = ∞ We later demonstrated that for BER > 1 and k < ∞, the chance of civilization emergence grows faster than the growth rate of BCS. In that case, emergence of civilization's curve is steeper than the BCS size curve, indicating much rare chance of civilization formed on planet younger than earth, and much greater chance formed on planet older than earth.

grows exponentially larger for older planets than earth. It is demonstrated in Chapter 8, that depending on the type of evolution (progressive vs. passive), the earliest possible time window for formed planets that host intelligent life can be derived based on observational constraints (such as we have 70% confidence we are the only civilization within the nearest 3 galaxies) coupled with the emergence chance of civilization currently observed on earth and the formation rate of earth like planet from the past. We basically exclude those planet formed earlier than 5 Gya and treat those planets formed between 5 Gya and 4.5 Gya (earth's formation date) on a case by case basis, depending on the initial input values. Detailed analysis is followed in Chapter 8.

Finally, we compute the total probability of planets within so-called the galactic habitable zone. The thickness of the galaxy disk averaged 2,000 light years. However, the thickness is non-uniform, the disk is thicker near the galactic center and tapers off near the galactic edge. Therefore, we modeled the galactic height by a decreasing linear function which decreases to 0 at unit 165 (representing 165,000 ly), which is the estimated edge of the galactic disk. It is speculated that 50,000 to 90,000 light years from the galactic center lies the galactic habitable zone. The entire zone is hospitable to life except the deepest layer, where the stellar density rivals the galactic center.

$$y_0(x) = -0.02212(x - 165) \tag{2.1.1}$$

$$\frac{2\pi \int_{50}^{90} y_0(x)\, x\, dx}{2\pi \int_{0}^{165} y_0(x)\, x\, dx} \cdot 0.74 \tag{2.1.2}$$

$$\approx 25.76\% \tag{2.1.3}$$

The integration sums up each value of $y_0(x)$ for $50 \leq x \leq 90$ as the height of the disc multiplied with each successive perimeter size $2\pi x$ to yield the total volume of the galactic habitable zone. It is multiplied by 2 because the disc is symmetric above and below the disc central plane. The volume of the galactic habitable zone is then divided by the volume of the entire galaxy to yield the probabilistic percentage. This gave 25.76% as the upper bound for the habitability of the galaxy.

Alternatively, Gowanlock [49] give the lower bound estimate. According to Gowanlock, the concept of galactic habitable zone needs to be revised, it is found that no particular region of the galaxy is inhabitable for life, and the overall 0.3% of all stars may be capable of supporting complex life on a non-tidally locked planet. In a clear contrast to our earlier calculation showing that 25.76% falls within the galactic habitable zone (Chapter 2), the calculation seem to suggest that one reduce the number of habitable planet by a factor of $\frac{25.76}{0.3} = 85.867$. However, the original study already taking planetary habitable zone and metallicity selection into effect, so we need to remove them from our calculation to yield the percentage we are interested. We have done calculation showing that 9.20% of planets falls within the planetary habitable zone and 25% of stars falls within the metallicity selection criteria. As a result, the adjusted percentage indicates that Gowanlock predicts that $\frac{0.3}{0.25 \cdot 0.092} = 13.04\%$ of all stars may be capable of supporting complex life. We also shown that 36.8% of planets falls within the planetary habitable zone according to Nyambuya's generalized Titius Bode's law. Then, the adjusted percentage indicates that Gowanlock predicts that $\frac{0.3}{0.25 \cdot 0.368} = 3.261\%$ of all stars may be capable of supporting complex life.

We adopt the calculation using both approaches and use the geometric mean of the upper and lower bound to estimate the chance of galactic habitability ranges from 9.165% to 18.327%.

$$\sqrt{25.76 \cdot 3.261} = 9.165335 \tag{2.1.4}$$

$$\sqrt{25.76 \cdot 13.04} = 18.327859 \tag{2.1.5}$$

2.2 Definition of Stellar Habitable Zone

Various definitions of the stellar habitable zone have been proposed, we shall define habitable zone more rigorously. It is known that the increasing luminosity of the Sun will render earth to be inhabitable. The rate of weathering of silicate minerals will increase as rising temperatures speed up chemical processes. This in turn will decrease the level of carbon dioxide in the atmosphere, as these weathering processes convert carbon dioxide gas into solid carbonates. Within the next 600 million years from the present, the concentration of CO_2 will fall below the critical threshold needed to sustain C_3 photosynthesis: about 50 parts per million. At this point, trees and forests in their current forms will no longer be able to survive. C_4 carbon fixation can continue at much lower concentrations, down to above 10 parts per million. Thus plants using C_4 photosynthesis may be able to survive for at least 0.8 billion years and possibly as long as 1.2 billion years from now, after which rising temperatures will make the biosphere unsustainable. Therefore, we set the distance at which earth currently will receive the equivalent of Sun's luminosity in 1 billion years into the future as the threshold of the inner edge of the habitable zone. Since luminosity increased in a nearly linear fashion to the present, rising by 1% per 110 myr, then, the total solar output experienced by the inner edge should be 9.091% more than at earth at its current location. Since the Sun's radiant power decrease by the inverse squared as celestial objects move away from them, *the inner edge must occur at 0.840278 AU,* about the midpoint between Venus and earth's current orbit.[50] The outer edge, on the other hand, is that earth's temperature does not fall below the freezing point of water at 273.15 K. Currently, with moderate greenhouse effect, earth's temperature is at 285 K, slightly warmer than predicted by the solar radiation model alone at 279 K. We use 285 K as our reference and find that *1.0887 AU as the outer edge of the habitable zone.* From these edges, we found the midpoint at 0.964489 AU, earth's current position lies somewhat on the colder side and it is expected to cross into the warmer side 166 Myr into the future. If one computes the effective habitable window within the habitable zone, then the habitable zone of the parent star can only support 2.3 Gyr of habitability. Beyond the definition of the habitable zone, there is also the definition of a continuously habitable zone. Since the host star increases its luminosity throughout its lifespan, in the most stringent case, a planet has to be continuously within the habitable zone in order to nurture life. If this requirement is necessary, then, the earth was beyond the outer edge of the habitable zone 1.3 Gya. The earth was nevertheless active and liquid water was present from the very beginning. Now it is commonly understood that early earth has as much as three times the heat budget as it is today with completely different atmospheric composition composing mostly greenhouse gasses such as CO_2 and Methane. A thicker atmosphere also enabled the warming of the earth beyond the outer edge of the habitable zone. Prior to 2.5 Gya, the earth was also covered by ocean, a lower albedo also helped maintain the absorption of solar insolation. Most importantly, the moon was much closer to earth at the earlier times and heat generated from tidal heating was much more significant. We will see in our later chapters that the aforementioned condition applies to many terrestrial planets early in its evolution. Therefore, it is typical that as terrestrial planets formed, can lie outside the outer edge of the habitable zone and as its heat budget dwindled the increasingly warmer sun continues to maintain a stable temperature for the planet by including the planet into its edge of habitability. However, the planet cannot lie too far from the host star, as the case of Mars, which was geologically active but turned dead and frozen as its internal heat budget is exhausted. Under such scenario, the planet can only be warmed up again when sun increases its luminosity in a few billion years, but then life has to restart on such planet after billions of years of hiatus. In general, the more massive planet can retain its heat for longer than its smaller cousins so they can lie further beyond the outer habitable edge to start with. Since all planetary system's total mass budget is distributed from a central median value, earth's value can be used as the weighted average over all possible scenarios and combinations, with planets more massive and less massive included into consideration as well as its stable distance from the host star.

At the same time, for stars at solar mass, a planet lies at the outer edge or inside the habitable zone when the planet is first formed will experience higher heat budget every point in its evolutionary history. It will less likely enter an ice age, and most importantly shorter effective habitable period of no more than 2.3 Gyr. Whereas the habitability extendable by its own internal heat complemented by solar heating completely overlapped the host stars'. For earth, the continuous habitability spans 5.5 Gyr (the first 3.2 Gyr of habitability is maintained by earth's itself in addition to insufficient solar insolation, and the later 2.3 Gyr in which we are currently residing is maintained by sufficient solar insolation with limited internal heat budget and tidal heating. With minimal overlapping between the two stages).

Solar mass stars increase its luminosity quickly in contrasts to stars with smaller mass and habitable zone shifts relatively quickly, and the effective continuous habitable zone is much smaller than the definition of the habitable zone. Luckily, our potential candidate pool includes stars ranges in mass from 0.712 to 1 solar mass. Therefore, the habitable zone shifts up to 3.2 times slower, or 7.36 Gyr. If one sets the definition of maximum temporal range for the continuously habitable zone for solar mass stars to be 500 Myr longer than earth's, then the planet experience a period of cool temperature lower than earth's geologic past with almost no to very minimal overlapping between the two stages. On the other hand, if one sets the minimum requirement for continuous habitability window to be 500 Myr shorter than earth's, then a planet experience at most only 500 million additional years for all life's habitability with more overlapping between the two stages. Then, only 40% of the currently defined habitable zone falls into the continuously habitable zone. Stars with 0.712 to 0.73 solar mass will guarantee a continuously habitable zone of 5 Gyr or longer by stellar insolation alone; therefore, 100% of the currently defined habitable zone falls into the continuously habitable zone. Stars with 0.73 to 1 solar mass will have continuously habitable zone from 5 Gyr to 2.3 Gyr; therefore, 100% to 40% of the currently defined habitable zone falls into the continuously habitable zone. Taking the weighted average of all cases, 75.73% of the currently defined habitable zone falls into the continuously habitable zone. However, stars with lower mass still allow all terrestrial planets to form beyond the outer edge of habitability. As a result, the continuous habitability zone miraculously matches the currently defined habitable zone 100% if not a bit more. Therefore, the habitable zone restricts planets to be between 0.840278 AU and 1.0887 AU, out of a total radius of 2.7 AU radius (inside the snow line) in which terrestrial planet can form. The chance is therefore 9.20%.

Very recently, Nyambuya has shown that a generalized Titius Bode's law, which states that the placement of planet are actually non-random and follows a pattern, can be derived based on the third solution to a four Poisson-Laplace differential equation with a time dependent term making it Lorentz invariant within the context of gravitomagnetic theory, whereas the first solution is simply Newton's universal gravitation.[18] Based on this theory, not only one able to find all potential "wells" or ring nodes between planetary orbit spaces, a generalized law is shown to be only dependent on the stellar mass and radius. If planetary orbits indeed follows an exponentially spaced gaps as theorized, then the probability of planet formed within the habitable zone of any star has to be adjusted to taking into this fact.

Below is the table taking Nyambuya's analysis of 25 stellar systems including the solar system. The node range is computed based on Nyambuya's best fit for each system that falls within the frost line, where terrestrial planet forms. The frost line is computed based on the rough relationship between the stellar mass and its luminosity. $m^{4.5}$. In reality, many system's luminosity under investigation does not follow such relation. It is likely due to selection bias. Only the dimmer star's planets gives a higher signal to noise ratio. The starting node is usually assumed to be the first planet's node occupation within the system, which is not necessarily the first one. In the case of system lacking close orbiting planet around the host star, such as the solar system, an arbitrary distance of 0.03 AU is assumed and the node closest to this distance is selected as the starting node. Nyambuya assumed the impossibility of planet formation between 0.03 AU to the orbit of Mercury within the solar system. However, many system detected shown reasonable accommodation of even Jupiter sized gas giants within very tight orbits around their host stars. Therefore, sun's node ranges is much larger than his prediction. If the most distant planet detected located beyond the frost line, then nodes up to the most distant planet is included. The number of confirmed occupations are simply the number of planets detected. The habitable nodes are those nodes that falls within the habitable zone of the host star. The habitable confirmed indicates the number of habitable planets detection within the habitable zone. Almost every star system contains one node ring within its habitable zone and a weighted average of 1.04. It is shown that 35.38% of all nodes are occupied within the frost line. The percent of detection within the habitable zone is 16%, which is significantly lower than the overall occupancy rate. This discrepancy occurs due to selection bias. Nearly half of the planet detected are terrestrial planets orbiting much closer to the host star than the habitable zone though many of the planets including some gas giants are found at or beyond the habitable zone. Overall, the results are inconclusive. On one hand, this gives us confidence that the true occupancy rate should be somewhat higher if not significantly higher as more detection is possible around habitable zone. On the other hand, all systems we picked could be fall within selection bias. That is, the majority of systems occupancy rate is lower and close orbiting planet are less common. If one tentatively assumes that planet falls within the habitable zone gives a chance of $1.04 \cdot 35.38$, then it is 4 times higher than our previous estimate. We will adopt the second value **as** the more accurate estimate. *The chance is therefore 36.8%.*

Name	Node range	Confirmed	Ratio	Habitable nodes	Habitable confirmed
Sun	23.00	8	0.35	1	1
55 Cancri	17.50	5	0.29	1	0
HD 10180	14.80	7	0.47	1	0
HD 40307	15.00	6	0.40	1	1
HR 8799	17.00	4	0.24	1	0
HR 8832	15.80	5	0.32	1	0
Gliese 876	29.70	4	0.13	2	1
Kepler 11	13.85	6	0.43	1	0
Kepler 20	14.00	6	0.43	1	0
Kepler 24	12.20	4	0.33	0	0
Kepler 26	19.00	4	0.21	1	0
Kepler 32	25.00	5	0.20	2	0
Kepler 33	8.60	5	0.58	0	0
Kepler 37	9.60	4	0.42	1	0
Kepler 62	13.20	5	0.38	1	0
Kepler 80	19.00	6	0.32	1	0
Kepler 89	9.00	4	0.44	0	0
Kepler 90	14.30	8	0.56	1	0
Kepler 102	16.00	5	0.31	1	0
Kepler 169	17.00	5	0.29	2	0
Kepler 186	18.00	5	0.28	2	1
Kepler 292	15.70	6	0.38	1	0
Kepler 444	22.00	5	0.23	2	0
Mu Arae	8.50	4	0.47	0	0
Upsilon-Andromedae	10.40	4	0.38	1	0
Average	-	- 0.3538		**1.04**	**0.16**

Table 2.2.1: Number of nodal detection vs expected nodes and number of habitable nodes per each system

2.3 Definition of Sun-like Stars

By sunlike stars, one meant stars that are not too massive so that it provides ample time for evolution to take place for the emergence of intelligent species. The sun is a boundary case because it has shown that in just another 500 million years the increasing luminosity of the sun will accelerate the carbon cycle, and just another 1 billion years will increase the surface temperature high enough rendering the planet inhabitable. Since the evolution of intelligent life on earth, as it is demonstrated in this paper, is currently ahead most of the earth like planets, therefore, setting our selection criteria stringent, 1 solar mass is the upper bound chosen for sunlike stars. Then, how low can the stellar mass go to be hospitable to intelligent life? We have shown in our section discussing red dwarfs with 0.35 solar mass or smaller is unsuitable candidates due to strong magnetic fields. Then, it leaves us mass between 0.35 solar mass to 1 solar mass for our current investigation. The greatest challenge facing intelligent life is then tidally locking to their host star. Once tidal locking completes, the planet faces one hemisphere around its star in its daylight and another half in perpetual darkness. The weather would be extreme on such a planet. In addition, much slower rotation rate reduces the magnetic field strength of the planet, though the interference from the cosmic stellar magnetic storm is not as catastrophic as those observed on red dwarves, it is nevertheless grave compared to earth. We define a planet is tidally locked if 4.5 Gyr after its formation, it currently completes a day and night cycle every 7 days or longer. The first step is to find the habitable zone for a given stellar mass, and then assuming a habitable planet of 1 earth mass revolves at such distance from their star and the time it takes to tidally lock to their star. Finally, based on the remaining time to lock to their host star, deriving the current number of rotational days of their planet.

Tidal locking is defined by the classical equation:[26]

$$t_{lock} \approx \frac{\omega a^6 I Q}{3 G m_s^2 k_2 R^5} \tag{2.3.1}$$

Whereas ω is the initial spin rate expressed in radians per second, we substitute with $\frac{2\pi}{24 \cdot 60 \cdot 60} = \frac{1}{13750}$. We found in Chapter 3 that the average initial spin rate for terrestrial planet formation is 0.6079 days. Therefore, one can add $\frac{1}{0.6079}$ factor to the initial spin rate.

a is the semi-major axis of the motion of the planet around the star.

$I \approx 0.3307 \cdot m_p R^2$ is the moment of inertia of the planet, where m_p is the mass of the satellite and R is the mean radius of the planet, and the moment inertia factor for earth is 0.3307 and we assumed all habitable planet has a similar factor.

Q is the dissipation function of earth, which is estimated to be 280 based on existing literature.

G is the gravitational constant.

m_s is the mass of the star.

R is the radius of the planet.

k_2 is the tidal love number of the satellite, with various estimates for earth between $0.302 \sim 0.82$. 0.302 is based on existing literature, 0.82 is derived based on equation given for love number for earth:

$$k_2 = \frac{1.5}{1 + \frac{19(3 \times 10^{10})}{2 \cdot 5.514 \cdot 10^3 \cdot 9.807 \cdot (6371 \cdot 1000)}} = 0.821 \tag{2.3.2}$$

We are uncertain regarding the discrepancy between the prediction and observation. We shall adopt the value from the existing literature as the more accurate one. The combined $\frac{Q}{k_2} = 927.15$

One can derive the total time required for earth tidally lock to any stellar mass given current spin rate. When $x=1$, the locking time is 230 billion years. One can observe that it takes half of the total locking time to reduce the rotation rate by half to 2 earth days, and $\frac{3}{4}$ of the locking time to reduce the rotation rate to $\frac{1}{4}$ or 4 earth days.

$$j(x) = \sqrt{x^{4.5}} \tag{2.3.3}$$

$$T_{earthlockingtosun}(x) = \frac{\left(\frac{1}{13750}\right) \cdot (j(x) \cdot a_{earth})^6 \cdot I_{earth} \cdot M_{earth} \cdot 927.15}{3G(x \cdot M_{sun})^2 \cdot R_{earth}^5 \cdot 60 \cdot 60 \cdot 24 \cdot 365} = 2.3040519541 \times 10^{11} \text{ yrs} \tag{2.3.4}$$

Whereas $j(x)$ is the habitable zone radius given the stellar mass. The cumulative tidal locking calculation across all stellar mass can then be computed through recursive method, the tidal locking time depends on each step of change of the angular spin and semi-major axis. Unlike tidal locking for the moon, the semi-major axis will grow at most 1.5 km during the course of earth tidally locked to the sun. So one can assume it is largely fixed.

$$\frac{dt}{d(\omega, x)} = \frac{\omega (j(x) \cdot a_{earth})^6 \cdot I_{earth} \cdot M_{earth} \cdot 927.15}{3 \cdot G \cdot (x M_{sun})^2 \cdot R_{earth}^3} \tag{2.3.5}$$

Given $\frac{dt}{d(\omega,x)}$, we now define the recursive function that plots the rotational spin vs time as pairs of points:

$$S(0) = (t_1 = 0, \quad \omega)$$

$$S(1) = \left(t_1 = t_1 + \frac{dt}{d(\omega,x)} \cdot \frac{1}{d}, \quad \omega(1+d)\right)$$

$$S(2) = \left(t_1 = t_1 + \frac{dt}{d(\omega(1+d),x)} \cdot \frac{1}{d}, \omega(1+d)(1+d)\right)$$

$$\ldots$$

$$S(n) = \left(t_1 = t_1 + \frac{dt}{d\left(\omega(1+d)^{n-1},x\right)} \cdot \frac{1}{d}, \omega(1+d)^n\right)$$

Whereas each step such as step $S(1)$ is defined as the time required to decrease the angular spin per hr of earth from ω to $\omega(1+d)$. $\frac{dt}{d(\omega(1+d),x)}$ is the locking time required for an initial spin of $\omega(1+d)$. $\frac{dt}{d(\omega,x)}$ is the locking time for required for initial spin of ω. d can be made arbitrarily small in simulation to increase precision. For our simulation, we chosen $d = 0.001$. The time required to decrease the angular spin varies by each period depending on the current values of ω as the parameter for the tidal locking equation.

The initial value of ω is chosen to be at 4 hr. Based on possible initial spin rates generated by oligarchic protoplanet merging process (see Section 3.5), for stellar mass between 0.8 and 1 solar mass, only spin < 8 hr can provide a lunar mass satellite and host star tidal-locking free planet within 4.5 Gyr. For stellar mass of 0.8 solar mass or below, only spin < 6 hr can provide a lunar mass satellite and host star tidal-locking free planet within 4.5 Gyr. That is, the addition of natural satellite forms a more stringent constraint criterion on initial rotational spin requirement to avoid tidal locking than just with the star alone. Initial spin less than 6 hr generates an average spin rate of 4 hr.

$$\frac{\int_0^6 x R_{otation}(x)\, dx}{\int_0^6 R_{otation}(x)\, dx} = 4 \tag{2.3.6}$$

Spin rates less than 6 hr represents 13.64% of all initial spin rates.

$$\frac{\int_0^8 R_{otation}(x)\, dx}{\int_0^\infty R_{otation}(x)\, dx} = 0.1364 \tag{2.3.7}$$

Based on these assumptions, a planet within the habitable zone of star with 0.712 solar mass or smaller will have a rotational period of 7 days or longer assuming earth's 1 day rotation is typical for any habitable planet 4.5 Gyr after its formation, mimicking tidal locking condition. As a result, we will set star from 0.712 solar mass to 1 solar mass as what we call a sun-like star.

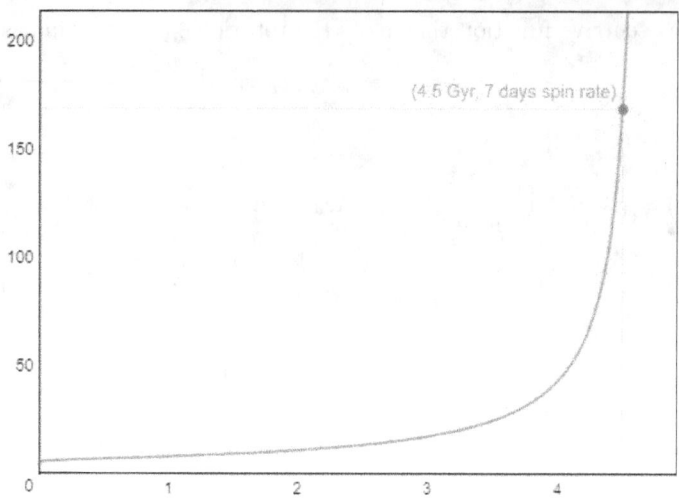

Figure 2.3.1: For 0.712 solar mass star, tidal forces of both the host star and satellite slow downs initial spin at 4 hr to 7 days per rotation by 4.5 Gyr of evolution

One could set an even lower mass limit for the host star. Along with its natural satellite of at least 1 lunar mass, stars with mass between 0.695 and 0.712 solar mass can be tidally locking free if the initial spin rates were between 1 and 4 hr (avg of 3.1 hr, 4.2121% of all possible initial spin rate). Stars with mass between 0.663 and 0.695 solar mass can be tidally locking free if the initial spin rates is between 1 and 2 hr (avg of 1.653 hr, 0.3373% of all possible initial spin rate). Both cases does not nullify the limit of 0.712 solar mass because the number of additional stars added is negligible and does not affect the general conclusion. Both cases require an initial spin rate constitute $< 5\%$ of all initial spin speeds. Furthermore, only marginally larger satellite than lunar mass met the tidal locking requirements even if their planet fall within $< 5\%$ initial spin speeds. On the other hand, by including these lower limits, however, the results are distorted, since $> 90\%$ cases these stars lead to non-habitable condition. Finally, the average stellar characteristics does not differ much even if stellar characteristics within this ranges is added if one uses the weighted average approach. Its analysis are nevertheless included in the rotational speed section of Chapter 3.5 to provide a robust and complete picture. As a result, the absolute minimum stellar mass for tidal locking is at 0.663 solar mass, but we chose instead 0.712 as the minimum stellar mass due to the cost of inclusion outweigh the benefit.

2.4   Number of Stars

To compute the total number of stars in the Milky Way, we first obtain the mass of the Milky Way galaxy which ranges from $0.8 \cdot 10^{12}$ solar mass to $1.5 \cdot 10^{12}$ solar mass. We take the median mass value at $1.15 \cdot 10^{12}$ solar mass. Then, we multiply by 0.1546 because visible matter occupies 15.46% of matter (The universe contains 26.8% dark matter and 4.9% ordinary matter). We end up with $1.7779 \cdot 10^{11}$ solar mass.

$$M = 1.15 \cdot 10^{12} \cdot 0.1546 = 1.7779 \times 10^{11} \ M_{sol} \tag{2.4.1}$$

Some have argued that the total mass of all stars in the Milky way is between 4.6×10^{10} and 6.43×10^{10} solar mass. If it were true, based on the proportion of dark matter, the mass of the galaxy is between 2.9754×10^{11} and 4.159×10^{11} solar mass. This is only between 25.87% and 36.17% of galaxy's mass. The alternative explanation would be the ratio of matter to dark matter is exceptionally high within the Milky Way than typical. This is unlikely since it is in violation of the Copernican principle. Therefore, we are content to settle on our solution of $1.7779 \times 10^{11} \ M_{sol}$. This is mass that made up all the stars, nebulae, and the interstellar medium. The mass of interstellar medium and nebulae occupies roughly 1% of the total mass of ordinary matter, so we can just ignore it. Then, we compute the number of stars in the stellar mass range from 0.712 to 1 solar mass based on initial mass function [94][56]

The initial mass function is a step function and we find the best fit by considering the percentage of the number of stars falling into each respective spectral class based on observation (whereas stellar mass less than 0.3843 solar mass is derived based on Chabrier):

$$I_{mf}(x) = \begin{cases} 0.254 \left(\frac{1}{\ln 10x}\right) \cdot \exp\left(\frac{-(\log(x)-\log(0.08))^2}{(2 \cdot 0.69^2)}\right) & 0.08 \leq x \leq 0.3843 \\ -6.78043x^4 + 13.9936x^3 - 9.17486x^2 + 1.59331x + 0.27259 & 0.3837 \leq x \leq 0.7662 \\ 2.23991x^4 - 9.30733x^3 + 13.6611x^2 - 8.53514x + 1.999 & 0.7662 \leq x \leq 1.1287 \\ 6\exp(-5x) & 1.1287 \leq x \end{cases} \quad (2.4.2)$$

so that the percentage constraints are satisfied:

$$M_{class} = \frac{100 \int_{0.08}^{0.45} I_{mf}(x)\,dx}{\int_{0.08}^{\infty} I_{mf}(x)\,dx} = 76.43\% \qquad K_{class} = \frac{100 \int_{0.45}^{0.8} I_{mf}(x)\,dx}{\int_{0.08}^{\infty} I_{mf}(x)\,dx} = 12.88\% \qquad (2.4.3)$$

$$G_{class} = \frac{100 \int_{.8}^{1.04} I_{mf}(x)\,dx}{\int_{0.08}^{\infty} I_{mf}(x)\,dx} = 7.1095\% \qquad F_{class} = \frac{100 \int_{1.04}^{1.4} I_{mf}(x)\,dx}{\int_{0.08}^{200} I_{mf}(x)\,dx} = 3.018\% \qquad (2.4.4)$$

$$A_{class} = \frac{100 \int_{1.4}^{2.1} I_{mf}(x)\,dx}{\int_{0.08}^{\infty} I_{mf}(x)\,dx} = 0.5337\% \qquad B_{class} = \frac{100 \int_{2.1}^{16} I_{mf}(x)\,dx}{\int_{0.08}^{200} I_{mf}(x)\,dx} = 0.0166\% \qquad (2.4.5)$$

$$O_{class} = \frac{100 \int_{16}^{\infty} I_{mf}(x)\,dx}{\int_{0.08}^{\infty} I_{mf}(x)\,dx} = 0\% \qquad (2.4.6)$$

and plotted below:

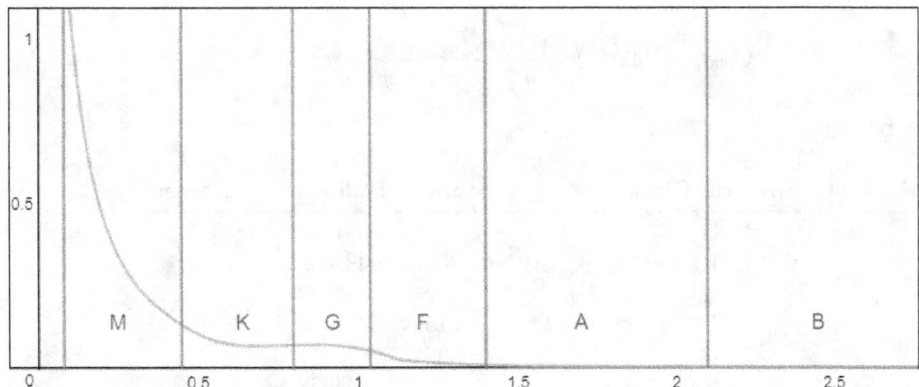

Figure 2.4.1: Initial mass function (PDF) by different stellar mass separated by their respective spectral class, whereas class O representing stellar mass > 16 solar mass is off the chart to the right

and the initial mass function combined with its own mass yields the PDF for the initial mass function for mass distribution:

$$xI_{mf}(x) \qquad (2.4.7)$$

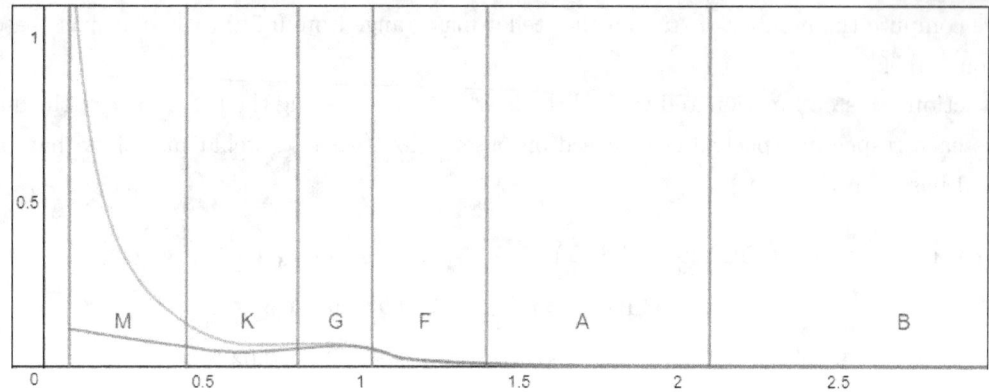

Figure 2.4.2: Initial mass function (PDF) by different stellar mass vs. PDF for the initial mass function for mass distribution

By knowing the weighted percentage of mass occupied by a particular range of stellar mass out of the entire galaxy and taking the particular range to infinitesimally small, one can find the number of stars within each spectral class. Whereas M is the mass of the Milky way responsible for stellar formation and we multiplied by a factor of 99% to exclude interstellar medium.

$$M_{\text{class}} = .99M \int_{.08}^{.45} \frac{y I_{mf}(y)}{y \int_{0.08}^{\infty} x I_{mf}(x)\,dx}\,dy \qquad K_{\text{class}} = .99M \int_{.45}^{.8} \frac{y I_{mf}(y)}{y \int_{0.08}^{\infty} x I_{mf}(x)\,dx}\,dy \qquad (2.4.8)$$

$$G_{\text{class}} = .99M \int_{.8}^{1.04} \frac{y I_{mf}(y)}{y \int_{0.08}^{\infty} x I_{mf}(x)\,dx}\,dy \qquad F_{\text{class}} = .99M \int_{1.04}^{1.4} \frac{y I_{mf}(y)}{y \int_{0.08}^{\infty} x I_{mf}(x)\,dx}\,dy \qquad (2.4.9)$$

$$A_{\text{class}} = .99M \int_{1.4}^{2.1} \frac{y I_{mf}(y)}{y \int_{0.08}^{\infty} x I_{mf}(x)\,dx}\,dy \qquad B_{\text{class}} = .99M \int_{2.1}^{16} \frac{y I_{mf}(y)}{y \int_{0.08}^{\infty} x I_{mf}(x)\,dx}\,dy \qquad (2.4.10)$$

$$O_{\text{class}} = .99M \int_{16}^{\infty} \frac{y I_{mf}(y)}{y \int_{0.08}^{\infty} x I_{mf}(x)\,dx}\,dy \qquad (2.4.11)$$

and given in the table below:

Percentage by Mass	Spectral Class	Stars in Billions	Percentage by Number
0.00%	O	0.0001433	0.00%
0.1%	B	0.08239286099	0.02%
2.37%	A	2.64608396883	0.51%
10.43%	F	15.8896758915	3.06%
19.77%	G	38.0236373562	7.32%
23.13%	K	67.2832444038	12.95%
44.185%	M	395.492488388	76.14%
100.00%	**Total**	**519.417522869**	**100.00%**

Table 2.4.1: Star distributions of Milky Way

For stars within our definition of habitable mass range, we have:

$$.99M \int_{.712}^{1} \frac{y I_{mf}(y)}{y \int_{0.08}^{\infty} x I_{mf}(x)\,dx}\,dy = 4.6937 \times 10^{10} \qquad (2.4.12)$$

The total number of stars in the Milky Way galaxy is 519.418 billion. The total number of stars that is between 0.712 to 1 solar mass is then 46.9 billion. However, only 1 in 4 sun-like stars can host terrestrial planets with their metallicity higher than -1. Many of the earlier stars are born when the cosmic neighborhood is deficient in elements heavier than helium (Population II and Population III stars). Even at the era when the Sun was born from 5 Gya to 4 Gya, the average metallicity of the sun like stars is -0.3 and only 1 in 4 sunlike (derived based on the stellar formation and earth formation rate of Lineweaver's approach) stars can host terrestrial planets. One then needs to find the number of stars formed between 5 Gya to 4 Gya out of the total number of stars formed within the mass range of 0.712 to 1 solar mass. The total number of stars ever formed within this range is > 46.9 billion since the more massive stars born earlier has already died out. The lifespan of stars is proportional to its mass, the equation is given as (will be mentioned multiple times in the future):

$$T_{MS} \approx 10^{10} \left[\frac{M}{M_{sol}} \right] \left[\frac{L_{sol}}{L} \right] = 10^{10} \left[\frac{M}{M_{sol}} \right]^{-2.5} \tag{2.4.13}$$

simplifying the equation into:

$$x^{-2.5} \tag{2.4.14}$$

and the further one looks back in time, the smaller mass stars and a higher proportion of stars have died out. Since the oldest age of the universe is at 13.8 Gyr, we modify our equation into:

$$13.8 - 10x^{-2.5} \tag{2.4.15}$$

and then we take the inverse the equation to find the highest stellar mass can still remain on the main sequence given its current age:

$$D_{eath}(t) = \left(\frac{-t + 13.8}{10} \right)^{-\frac{1}{2.5}} \tag{2.4.16}$$

At the start of the big bang, only stars with mass of 0.879 solar mass or lower still remain on the main sequence, as the time approaches the current time, more massive stars can stay on the main sequence since there is not enough time have passed since their birth. We combine this information with the initial mass function (the percentage of stars with the threshold mass or above) within our selected range (0.712~1 solar mass) and the stellar formation rate (the absolute number of stars formed at any given time) to yield the percentage of stars ever formed but died as a function of time:

$$J(t) = \frac{\int_{D_{eath}(t)}^{1} I_{mf}(x) \, dx}{\int_{0.712}^{1} I_{mf}(x) \, dx} \tag{2.4.17}$$

and it can be shown that

$$\frac{\int_{0}^{13.799} (1 - J(x)) S_{tellar}(x) \, dx}{\int_{0}^{13.799} S_{tellar}(x) \, dx} = 0.91748 \tag{2.4.18}$$

Over 91.7% stars ever born are still alive. So one needs to multiply by a factor of $\frac{1}{0.91748}$ to scale to the total number of stars ever created within the mass range between 0.712 and 1 solar mass.

The ratio of the total number of stars formed between 5 Gya to 4 Gya out of all time period is found using Lineweaver's approach based on earth formation rate, which will be discussed immediately, and the total number of stars is 6.024 billion.

$$N_{Earthbetween5Gyato4Gya} = \frac{\int_{9.199-0.5}^{9.199+0.5} S_{tellar}(x) \, dx}{\int_{0}^{13.799} S_{tellar}(x) \, dx} \cdot \frac{1}{0.91748} \cdot 4.693 \times 10^{10} \cdot 0.247 \tag{2.4.19}$$

$$= 602,413,516$$

We shall derive the lower bound using Lineweaver's approach.

Lineweaver[38]has computed the earth formation rate through time based on the stellar formation rate of the Milky Way and using metallicity as the selection criterion. He has shown that the rate of terrestrial planet formation increases sublinearly as the metallicity of the stellar system increases from -1. A system with metallicity below -1 lack substantial heavy elements heavier than helium to form terrestrial planets. A system with a metallicity of 0 or higher has an increasing chance of hosting hot Jupiters (which swept through terrestrial planets' orbit and destroy them). Therefore, the rate of terrestrial planet formation is the probability of creating earth times the probability of destroying the earth, and one found that terrestrial planet formation is greatest at metallicity of 0.1, slightly higher than the sun.

The following equations are a faithful duplication originally obtained by Lineweaver:

The probability of producing earth:

$$P_{PE}(x) = \begin{cases} (0.625x + 0.625)^{1.68} & -1 \le x \le 0.6 \\ 1 & 0.6 \le x \le 1 \end{cases} \tag{2.4.20}$$

The probability of destroying earth:

$$P_{DE}(x) = \begin{cases} 0.0279081\,(47010.3)^x & x \le 0.2741 \\ \frac{1}{2}\tanh 9.2\,(x - 0.267) + 0.5 & 0.2741 \le x \end{cases} \tag{2.4.21}$$

The probability of harboring earth:

$$P_{HE}(x) = P_{PE}(x)\,(1 - P_{DE}(x)) \tag{2.4.22}$$

The normal distribution of stellar metallicity at 4.5 Gya:

$$P(x) = \frac{1}{0.3\sqrt{2\pi}} \exp\left(-\frac{(x + 0.3)^2}{2\,(0.3)^2}\right) \tag{2.4.23}$$

The PDF for earth formation at 4.5 Gya:

$$P_E = P(x) \cdot P_{HE} \tag{2.4.24}$$

which is only

$$\frac{\int_{-1}^{1} P_E\,dx}{\int_{-2}^{2} P(x)\,dx} = 0.247 \tag{2.4.25}$$

24.7% of all possible stars formed at the time.

Lineweaver only produced the final earth formation rate curve but did not include the equation for the curve. As a result, carefully matched curves are produced as the starting point of our derivation.

$$f_1(t) = -1.38 \cdot 10^{-6}t^4 + 3.83 \cdot 10^{-5}t^3 - 3.93 \cdot 10^{-4}t^2 + 1.741 \cdot 10^{-3}t - 0.0024 \tag{2.4.26}$$

$$f_2(t) = -7.07 \cdot 10^{-8}t^4 + 3.248 \cdot 10^{-6}t^3 - 5.17 \cdot 10^{-5}t^2 + 2.974 \cdot 10^{-4}t - 1.56 \cdot 10^{-4} \tag{2.4.27}$$

$$f_3(t) = 0.00194272\,(0.790599)^t \tag{2.4.28}$$

$$f_{earth}(t) = \begin{cases} f_1(t) & 2.3851 \le t \le 4.504 \\ f_2(t) & 4.504 \le t \le 11.653 \\ f_3(t) & 11.653 \le t \end{cases} \tag{2.4.29}$$

Alternatively, one could use a fit of:

$$f_{earth}(t) = \left(17.374t^{-5.33} + 0.5386t^{-0.198} + 0.964(1.0195)^t\right)^{-20.7738} \tag{2.4.30}$$

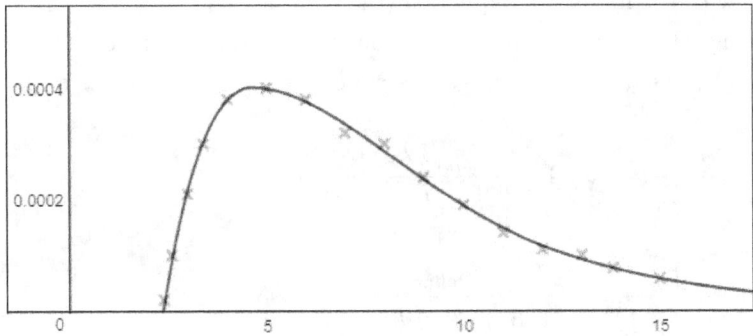

Figure 2.4.3: Non-logarithmic fit for earth formation

Using this equation above, one can derive the number of terrestrial planets formed from 5 Gya to 4 Gya. However, nowhere did Lineweaver mentioned about the habitability of these planets. Therefore, a factor is later added as a selection criterion for those planets that formed within the habitable zone of their host star. Moreover, we also reproduced the equation for the stellar formation also given by Lineweaver.

$$S_{tellar}(t) = \begin{cases} 0.00380914\,(13)^{(t-1.2)} - 0.00018 & 0 \leq t \leq 2.27 \\ 0.035\tanh 4.9\,(t-2.1) + 0.0352 & 2.27 \leq t \leq 2.948 \\ 0.138746\,(0.791754)^{.99t} & 2.948 \leq t \end{cases} \tag{2.4.31}$$

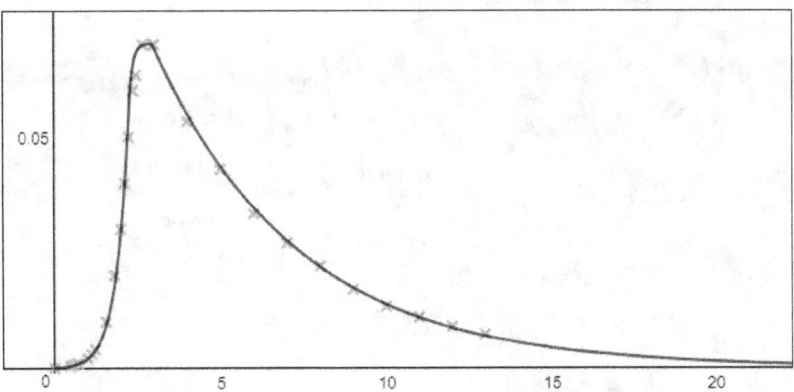

Figure 2.4.4: Non-logarithmic fit for stellar formation

The original earth and stellar formation curve is expressed in production density rate units of millions per million cubic parsecs per year. We determined the spatial volume of the galaxy by its average radius at 87,500 ly and a disc average height of 3,000 ly. Although the galaxy's volumetric size have been dynamically increasing as a function of time, we treat each satellite galaxy that gaves birth to stars and migrated to the galaxy at the later time as part of the overall volume we observe today. Therefore, the computation is justified by assuming that the volumetric size of the galaxy remain constant throughout all times. Integrating the equation over the whole range of stellar formation time window from the big bang until today gives us the total number of stars ever produced within the galaxy given its volumetric size. However, some of the stars have already died during the course of cosmic evolution, in order to find the percentage of stars that are dead we use the same equation that finds the highest stellar mass can still remain on the main sequence given its current age we used earlier:

$$D_{eath}\left(t\right) = \left(\frac{-t + 13.8}{10}\right)^{-\frac{1}{2.5}}$$ (2.4.32)

We combine this information with the initial mass function (the percentage of stars with the threshold mass or above) within our selected range (0.08~21 solar mass) and the stellar formation rate (the absolute number of stars formed at any given time) to yield the percentage of stars ever formed but died as a function of time:

$$J\left(t\right) = \frac{\int_{D_{eath}(t)}^{1} I_{mf}\left(x\right) dx}{\int_{0.08}^{\infty} I_{mf}\left(x\right) dx}$$ (2.4.33)

and it can be shown that:

$$\frac{\int_{0}^{13.799} \left(1 - J\left(x\right)\right) S_{tellar}\left(x\right) dx}{\int_{0}^{13.799} \left(1\right) S_{tellar}\left(x\right) dx} = 0.956717$$ (2.4.34)

Over 95.6% stars ever born are still alive. So one needs to multiply by a factor of $\frac{1}{0.956717}$ to scale to the total number of stars ever created within the mass range between 0.08 and 1 solar mass.

However, this rescale factor is non-applicable to our selected mass range (0.712~1) and the selected temporal range (5 Gya to 4 Gya). Within this range, it is shown that:

$$\frac{\int_{9.199-0.5}^{9.199+0.5} \left(1 - J\left(x\right)\right) S_{tellar}\left(x\right) dx}{\int_{9.199-0.5}^{9.199+0.5} S_{tellar}\left(x\right) dx} = 1$$ (2.4.35)

100% of all stars ever born are still alive. So no additional rescaling is needed. We simply assumed every other stars outside our range of consideration behaves similarly. We also need to rescale the number of years to 1 billion 10^9 since each unit in our integration represents a time scale of 1 billion years. Finally, a factor of 20 is required because Lineweaver's final plot for stellar formation rate denotes the star formation rate of sun like stars, and it is assumed that 5% of all stars are sunlike.

$$N_{stars} = 20 \cdot \frac{10^6}{\left(3.262 \cdot 10^6\right)^3} \cdot \left(\pi R_{milky}^2 H_{milky}\right) \cdot 10^9 \cdot 35.9096 \int_{0}^{13.799} S_{tellar}\left(x\right) dx$$ (2.4.36)

$$N_{earth} = 20 \cdot \frac{10^6}{\left(3.262 \cdot 10^6\right)^3} \cdot \left(\pi R_{milky}^2 H_{milky}\right) \cdot 10^9 \cdot \frac{9.0363}{5} \cdot 35.9096 \int_{9.199-0.5}^{9.199+0.5} f_{earth}\left(x\right) dx$$ (2.4.37)

$$= 611,741,790$$

By parameterization, we found that, in order to fit the actual number of stars observed, in fact, 35.9096 times higher than the star formation curve given. We are unsure why this factor is required. A possible justification for this factor is that the original star formation rate assumed the galaxy size to be 3.6 times larger by radius so the formation rate density has to be decreased accordingly. We multiply the same factor to the earth formation rate by 35.91. That is, 35.91 times higher than the earth formation curve given other than the scale conversion. The $\frac{9.0363}{5}$ factor can be explained as we further taking into consideration those stars with stellar mass from 0.712 to 0.8, part of the more massive Spectral K class. We have to revise the mass range of stars represented in the selection criteria, instead of 0.8 solar mass to 1.2 solar mass, we specified 0.712 solar mass to 1 solar mass instead. The final result is the number of terrestrial planets within the habitable zone of their host star arose from 5 Gya to 4 Gya. The cross-examined results with our earliest derivation show that these two numbers closely match each other.

We compare our earlier number of earth using the initial mass function method with the Lineweaver's method one find that the number closely match each other:

$$\frac{L_{ineweaver}}{N_{Earthbetween5Gyato4Gya}} = 1.01548 \tag{2.4.38}$$

They still mismatch each other by 1.5%. The discrepancy could be explained by the difference between metallicity selection criteria for selecting earth like planets from the sunlike stars. Although we faithfully trying to duplicate Lineweaver's result, we found that the integration of stellar formation curve, representing the accumulation of metallicity does no sub-monotonically increasing 5 Gyr after the big bang contrary to Lineweaver's original plot. The final earth formation also shifted further to the left, so that terrestrial planet formation started at 1 Gyr post Big bang instead of 2.5 Gyr.

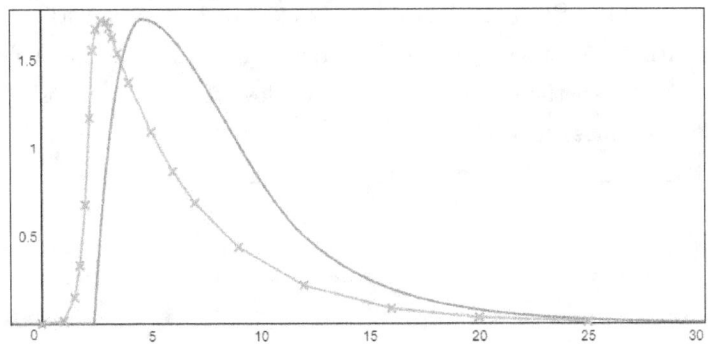

Figure 2.4.5: Our duplication of Lineweaver's SFR with metallicity as its selection criterion and shows that earth formation occurs earlier than the model predicts, as result, the total number of earth formation within a selected time period can differ.

Another possible explanation is that the initial mass function we adopted may not match the initial mass function assumed in Lineweaver's model, so that the numerator of the $\frac{9.0363}{5}$ factor requires adjustment. This does not mean we are wrong, it is simply suggests that our current observation does not have the resolution to determine which of the model is the more accurate.

Therefore, both approaches are well justified, we take the average of both approaches and compromise on the mean value as our final tally for the number of earths:

$$\frac{L_{ineweaver} + N_{Earthbetween5Gyato4Gya}}{2} = 607,077,653 \tag{2.4.39}$$

In conclusion, *the total number of terrestrial planets formed between 5 Gya to 4 Gya is 0.607 billion.*

2.5   Peak of Terrestrial Planet formation

Follow up from the previous calculation regarding the number of terrestrial planets between the temporal window of 5 Gya to 4 Gya, the earth was formed at the time when the total number of terrestrial planets are just falling off the peak. Once the star enters its main sequence lifespan, the terrestrial planet enters its window of habitability, but its habitability is increasingly being eroded by the increasing luminosity of the host star. Because the sun is a more massive yellow dwarf of the spectral class G, its main sequence is considerably shorter than most of the stars within spectral class G and K, the yellow and orange dwarves respectively. The amount of fuel available for nuclear fusion is proportional to the mass of the star. Thus, the lifetime of a star on the main sequence can be estimated by comparing it to solar evolutionary models. The Sun has been a main-sequence star for about 4.5 billion years, and it will become a red giant in 6.5 billion years, for a total main sequence lifetime of roughly 10^{10} years. Hence:

$$T_{MS} \approx 10^{10} \left[\frac{M}{M_{sol}}\right]\left[\frac{L_{sol}}{L}\right] = 10^{10} \left[\frac{M}{M_{sol}}\right]^{-2.5} \tag{2.5.1}$$

where M and L are the mass and luminosity of the star, respectively, M_{sol} is a solar mass, L_{sol} is the solar luminosity, and T_{MS} is the star's estimated main sequence lifetime. It then can be estimated that most of the orange dwarves with less than 0.74895 solar mass are still in their main sequence since their formation at the earliest times. Their window of habitability are still open because the current age of universe subtracted from the age of first possible terrestrial planet formation is still too short compares to the lower mass stars' habitability window, which ranges from 11 Gyr years up to 40 Gyr years. In fact, an orange dwarf with 0.5 solar mass will remain on the main sequence for $5.65 \cdot 10^{10}$ years, about five and half times longer than the sun. Taking the window of habitability into considerations and assuming a weighted stellar mass of 1 M_{sol} and the window of habitability of 5.3 Gyr, we have come up with a model which indicates that a peak of a number of habitable terrestrial planets 9.723 Gyr after the Big Bang. Weighted stellar mass of 1 M_{sol} is justified, because at earlier epoch, we are more concerned with the synthesis of complex organic molecules or the evolution of single cell life. As a result, stars more massive than the sun with shorter window of habitability is also considered and, overall, increased the weighted stellar mass to 1 M_{sol}.

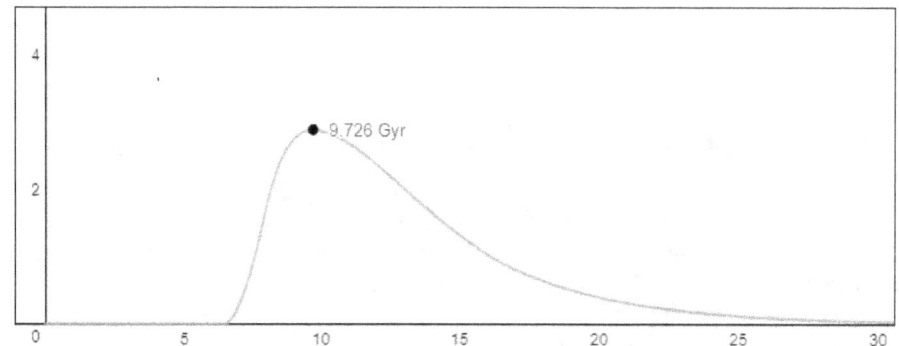

Figure 2.5.1: The total number of terrestrial planet through time

This result has strengthened both the panspermia theory and the catastrophic theory on the origin of life on earth. Since the formation of the first terrestrial planet occurred around 2.5 gyr after the Big Bang, or 11.3 Gyr ago, this closely matches Sharov and Gordon's back-extrapolation of DNA complexity's exponential doubling with a starting point around 9.7 Gyr ago. It further strengthened their arguments that life arise before earth's formation and the prokaryotes may have evolved much earlier on other terrestrial planets. They possibly formed long before earth, and the bacteria themselves may have survived and scattered to earth since the formation stage of the solar system but was only able to get strong hold after the cooling and the presence of water followed the late heavy bombardment 3.8 Gyr ago. Both the catastrophic theory and the panspermia model is strengthened by the fact that the total number of terrestrial planets reaches a maximum at the time of earth's formation. Regardless whether the simplicity of life back then was constrained by cataclysmic cosmic events such as Gamma Ray bursts or simply a matter of slow information doubling rate at the early stages of evolution or low metallicity in general, it suggests that panspermia reaches its greatest chance of success in terms of contamination during the formation age of the earth. It has been decreasing quasi exponentially as the star formation rate slows down. This contamination could be as complex as prokaryotes surviving interstellar journey as the panspermia theory suggests or could be much more modest as an enrichment of interstellar medium with complex organic molecules which facilitated the emergence of life based on the catastrophic theory once the persistence of life becomes possible at 5 Gya.

2.6   Excluding Over-counted Binaries and Multiples

Although binary stars are not the majority of all stars within the Milky Way galaxy, they are more likely to occur around sun-like stars (GFK spectral class). The probability of binary stars forming increases with stellar mass. Between 0.7 to 1.3 solar mass, 44 percent of all stars are binaries. The original paper does not intend to formulate further detailed model for correlation between stellar mass and multiplicity due to inadequate data. Therefore, we take 44 percent as the weighted average value for all stellar mass ranging from 0.7 to 1.3 where stellar mass at the lower mass end tends to formulate multiples with **a** lower chance closer to 35% while stars at the higher mass end tends to formulate multiples with 47% or more. Based on existing literature, binaries stars, in most cases, do not impede life formation. The general formula for multiples follow the formula [12]:

$$N(n) \propto 2.5^{-n} \tag{2.6.1}$$

where triple and higher-order systems represent 25% of all solar-type multiple systems, with a distribution of systems with n components that roughly follows a geometric distribution. Knowing the frequency of multiplicity among solar mass stars and using this equation, one finds that binary represents 75% of all multiple star systems.

2.6.1 Habitability of Binaries

The habitability of binaries strongly depends on the configurations of the system. In general, four cases of binary configurations are possible. All cases depends on the separation distance between the primary and the companion.

We derived a probability distribution function that closely matches the empirical data from a catalog of 2,728 binaries and shall use this equation for the remainder of the binary and multiple system habitability calculations.[90] We find the peak at 2.199 on the log-scale plot, indicating the most frequent value of 158 AU.

$$D\left(x\right) = \begin{cases} 0.0107982\left(3.02356\right)^{x} + 5.90068 \cdot 10^{-4} & x \leq 2.199 \\ 0.07\tanh\left(1.8\left(2.7 - x\right)\right) + 0.0734 & 2.199 \leq x \leq 2.886 \\ 605.211x^{-8.85514} & 2.886 \leq x \end{cases} \tag{2.6.2}$$

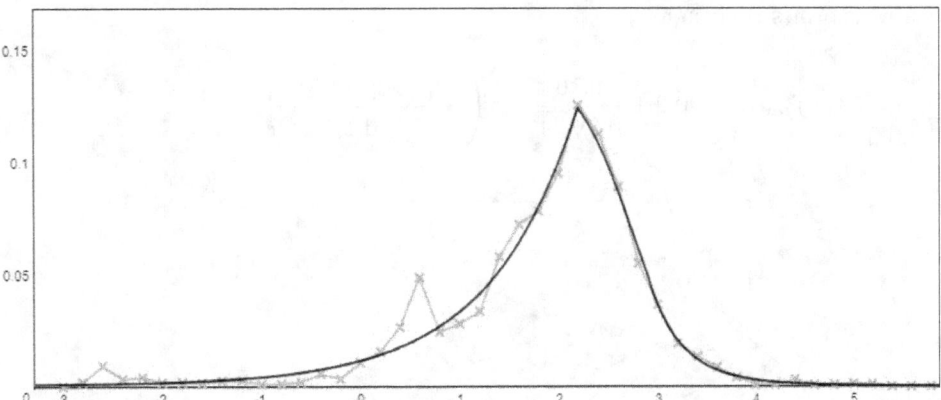

Figure 2.6.1: Probability Distribution Function of binary stars by separation distance based on log scale

In the first case, the average separation between the stars has to be small enough so that the minimum distance required for a stable orbit for any planets revolving around the binary pairs can fall on the combined stars' luminosity's habitable zone. Research has shown that planets orbiting binaries have to be between 3 to 5 times the distance between the pairs in order to be stable and we take the mid value at 4.

For close orbiting binaries, one also has to find the percentage of the companion that are 0.712 solar mass or above. For lower-mass companion are not included in our original tally of stars and are deemed inhabitable. Therefore, we need to exclude those companions with over 0.712 solar mass from our tally for those stars that are over-counted as habitable.

Before we proceed, we find the median mass for stars with our selected stellar range $0.712 \leq x \leq 1$ to be 0.855 solar mass using the initial mass function.

$$\frac{\int_{0.712}^{0.855} I_{mf}(x)\, dx}{\int_{0.712}^{1} I_{mf}(x)\, dx} = 0.5 \tag{2.6.3}$$

The mass distribution PDF between binary primary and companion is modeled by the binary companion mass distribution function (which is not the initial mass function observed among single stars). The distribution is faithfully duplicated based on existing literature.[12] Binary companion mass distribution functions are distinguished as three separate ones based on the proximity of the companion star.

Whereas binaries with tight orbits is given as:

$$f_{nearpeak}(x) = \frac{0.012}{0.05\sqrt{2\pi}} \exp\left(-\frac{\ln(-x+2)^2}{2(0.05)^2}\right) \tag{2.6.4}$$

$$f_{near}(x) = \left(x^{0.1} - 0.1\right) + f_{nearpeak}(x) \tag{2.6.5}$$

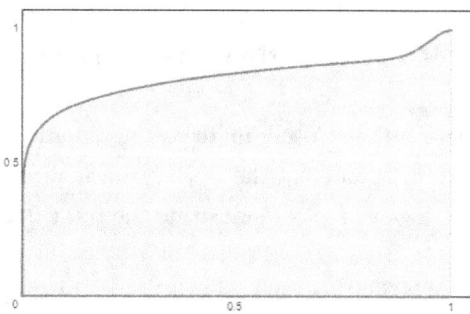

Figure 2.6.2: The PDF function of the secondary stellar mass to the primary stellar mass ratio for tight orbiting binaries, whereas horizontal axis represents companion mass ratio relative to the primary.

Whereas binaries with wide orbits is given as:

$$f_{far}(x) = 0.2 + \frac{0.16}{0.1\sqrt{2\pi}} \exp\left(-\frac{\ln(-x+1.2)^2}{2(0.1)^2}\right) \tag{2.6.6}$$

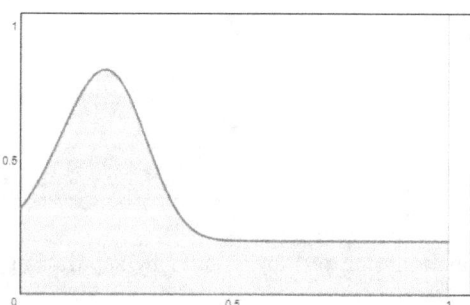

Figure 2.6.3: The PDF function of the secondary stellar mass to the primary stellar mass ratio for wide orbiting binaries

The binaries with intermediate separation is given as the average of the two:

$$f_{mid}(x) = \frac{1}{3} \cdot f_{far}(x) + \frac{2}{3} \cdot f_{near}(x) \tag{2.6.7}$$

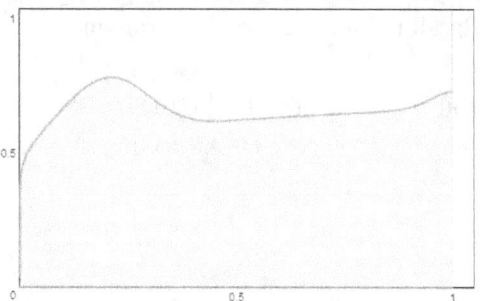

Figure 2.6.4: The PDF function of the secondary stellar mass to the primary stellar mass ratio with intermediate separating binaries

We found the percentage to be 19.045%. That is, for closely orbiting binaries 19 percent of which hosts a companion with a mass greater than 0.712 solar mass. 0.712 is divided by 0.855 because the average primary mass within our selected range is 0.855 solar mass. Therefore, 0.712 solar mass is rescaled relative to 0.855 solar mass.

$$P_{near} = \frac{\int_{\frac{0.712}{0.855}}^{1} f_{near}(x)\,dx}{\int_0^1 f_{near}(x)\,dx} = 0.19045 \tag{2.6.8}$$

We also found for inter-mediate orbiting binaries 17.33% which hosts a companion with a mass greater than 0.712 solar mass.

$$P_{mid} = \frac{\int_{\frac{0.712}{0.855}}^{1} f_{mid}(x)\,dx}{\int_0^1 f_{mid}(x)\,dx} = 0.17333 \tag{2.6.9}$$

Case 1: 0 ~ 0.1812 AU
We then compute the probability of tightly orbiting binaries with a separation between 0 AU to 0.1812 AU. This is determined by first assuming that the combined luminosity of tightly orbiting binaries is approximately equal to the combined luminosity of 2 stars at the same exact location and its habitable zone radius is given as:

$$\sqrt{2x^{4.5}} \tag{2.6.10}$$

However, the mass of the secondary companion, on average, has a lower mass than the primary with $0 \leq q \leq 1$. Evaluation shows a median value of 0.5402 solar mass.

$$\frac{\int_0^{0.5402} f_{near}(x)\,dx}{\int_0^1 f_{near}(x)\,dx} = 0.5 \tag{2.6.11}$$

So we modify our earlier habitable zone radius:

$$\sqrt{x^{4.5} + (0.5402x)^{4.5}} \tag{2.6.12}$$

The close orbiting binaries' semi-major axis must be less than 1/4 of the habitable zone radius to guarantee a stable orbit around the habitable zone, therefore, the maximum allowable separation is given by:

$$\frac{1}{4}\sqrt{x^{4.5} + (0.5402x)^{4.5}} \tag{2.6.13}$$

We are only interested in the range within $0.712 \leq x \leq 1$. If we take the upper limit of 0.2577 AU for 1 solar mass, the maximum stable orbit of all stars less than a solar mass within our range becomes larger than their habitable zone radius, grossly over-estimating the number of habitable systems. If we take the lower limit of 0.12 AU for 0.712 solar mass, the maximum stable orbit of all stars within our range would be significantly smaller than the habitable zone radius, grossly under-estimating the number of habitable systems. As a compromise, we use the median mass value instead. Recall the median mass for stars mass with our selected stellar range is 0.855 solar mass.

As a result, we settle on the minimum orbit requirement to be $\frac{1}{4}\sqrt{(0.855)^{4.5} + (0.5402\,(0.855))^{4.5}} \approx 0.1812$ AU. Therefore, the first case requires binary separation between 0 AU to 0.1812 AU.

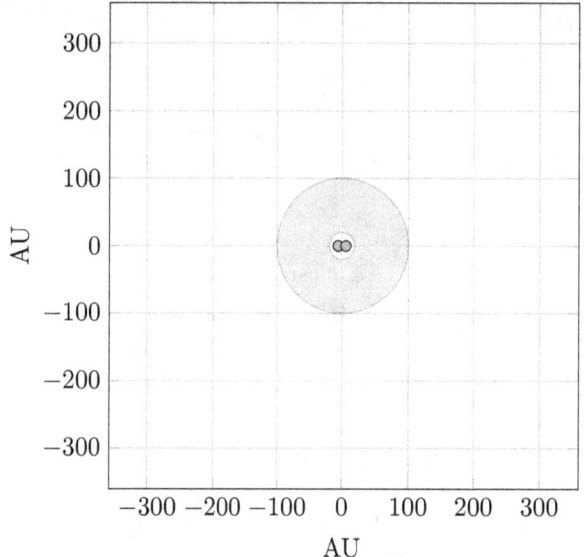

Figure 2.6.5: Binary pair with separation < 0.1812 AU

$$S_1 = \frac{\int_{-\infty}^{\log(0.1812)} D\,(x)\,dx}{\int_{-\infty}^{\infty} D\,(x)\,dx} = 0.0461 \tag{2.6.14}$$

$$S_{1\,voidzone1} = P_{near} \cdot S_1 = 0.00878 \tag{2.6.15}$$

The final results show that 4.61% of all binaries have tight orbits and 0.878% of them have companions of $0.712 \leq x \leq 1$ solar mass.

Case 2: 0.1812 ~ 100 AU

In the second case, the separation between the pairs is between 0.1812 AU to 100 AU. Both stars in the pair in this category can not host habitable planets. As the pair separation increases, the stable orbit moves beyond the habitable zone of the pairs so no habitable planet can revolve around both stars. As the separation widens, one may speculate that planetary system can revolve around one of the two stars. However, we set a very rigorous standard for planetary habitability. We define a planetary system to be habitable if no major astronomical objects comparable to solar mass lies within 100 AU from the habitable planet. Although Pluto, Neptune lies around 30 AU from the sun, the outermost dwarf planet Sedna within the solar system lies at 506 AU, Eris, and Makemake lies at 50 AU on average. If a solar mass object lies close to the planetary system, then, over the course of millions of years, a significant number of additional comets and disturbances can be brought to the inner planets from the outer rings. Therefore, binaries with separation between 0.1812 AU to 100 AU constitute the total dead zone of habitability.

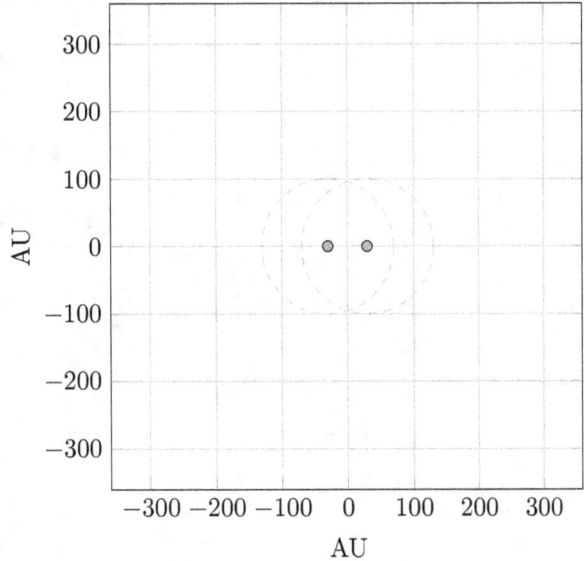

Figure 2.6.6: Binary pair with separation 0.1812 AU < x < 100 AU

Because binaries within the dead zone can host companions with different mass, we have to exclude those binaries in which the central star has a mass above 0.712 solar mass and those binaries in which both stars have a mass above 0.712 solar mass. Using the previous binary mass distribution equation, we found that 17.33% of all pairs have companions with mass 0.712 or greater.

$$P_{mid} = \frac{\int_{\frac{0.712}{0.855}}^{1} f_{mid}(x)\,dx}{\int_{0}^{1} f_{mid}(x)\,dx} = 0.17333 \tag{2.6.16}$$

Using the binary probability distribution based on separation distance, we found that 43.53% of all binaries fall into this dead zone. Therefore, we over-counted 7.546% of these binaries by both primary and companion and over counted 35.99% of these binaries by the primary only.

$$S_2 = \frac{\int_{\log(0.1626)}^{\log(100)} D(x)\,dx}{\int_{-\infty}^{\infty} D(x)\,dx} = 0.43533 \tag{2.6.17}$$

$$S_{2voidzone1} = S_2 - S_{2voidzone2} = 0.35987 \tag{2.6.18}$$

$$S_{2voidzone2} = P_{mid} \cdot S_2 = 0.07546 \tag{2.6.19}$$

Case 3: 100 ~ 200 AU

In the third case, binaries are separated by a distance between 100 AU to 200 AU and are capable of hosting one planetary system around one of the stars. We compute the probability of binaries within this range of separation and compute the percentage of companions with a mass greater than 0.712 solar mass and found that we over-counted the number of stars habitable by 3%.

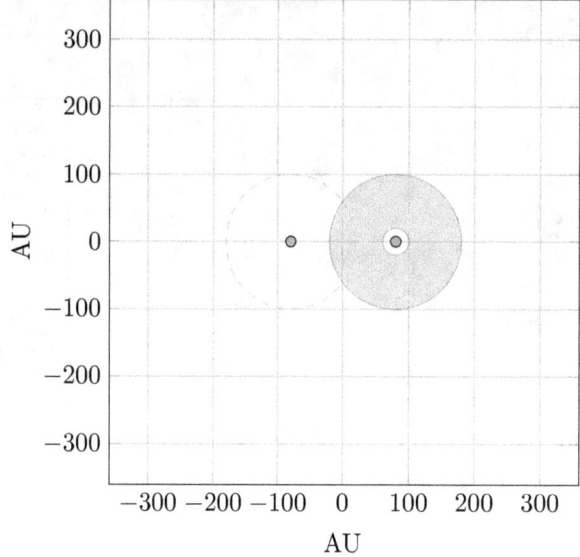

Figure 2.6.7: Binary pair with separation 100 AU < x < 200 AU

$$S_3 = \frac{\int_{\log(100)}^{\log(200)} D(x)\, dx}{\int_{-\infty}^{\infty} D(x)\, dx} = 0.1728 \tag{2.6.20}$$

$$S_{3voizone1} = P_{mid} \cdot S_3 = 0.02996 \tag{2.6.21}$$

Case 4: > 200 AU

Finally, in the fourth case, binaries are separated by a distance of 200 AU or greater. Under this scenario, both stars can host habitable planetary systems. Since both are capable hosting life, we just treat them, conceptually, as two separate life hosting stars and no star is over-counted from the number of our existing habitable systems.

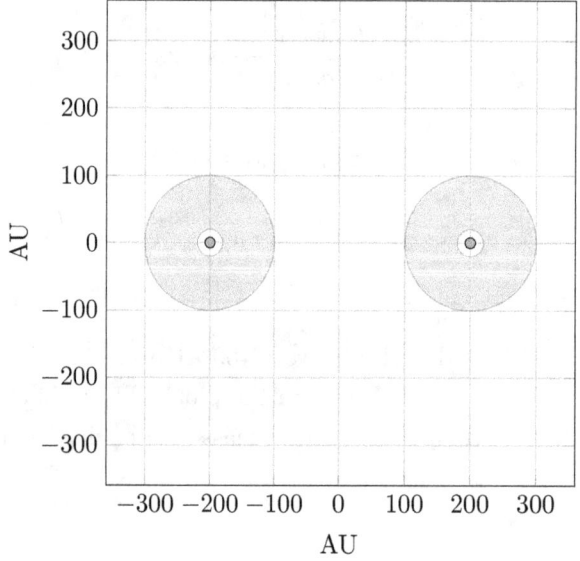

Figure 2.6.8: Binary pair with separation > 200 AU

$$S_4 = \frac{\int_{\log(200)}^{\infty} D(x)\, dx}{\int_{-\infty}^{\infty} D(x)\, dx} = 0.3457 \tag{2.6.22}$$

38

Our calculation is simplified by excluding taking binary pairs' eccentricity into account. The eccentricity of binaries is minimal in tightly orbiting pairs due to mutual tidal interaction but can become significant in binaries with greater separation. Luckily, binaries with eccentricity up to 0.5 do not significantly alter our previous derived results and likely to reduce the number of habitable systems by 10% at most. Star eccentricity mostly affect those binary systems with separation between 100 AU to 250 AU, where the slightest deviation from perfect circular orbits can result in diminishing chance of life. We excluded this calculation from our binaries computation to ease our later computation on ternary, quaternary, quintenary, sextenary, and higher order systems.

By deriving all possible cases for binaries, we can now sum up our results and obtain the following conclusion. This shows that 54.52% of binaries can host potential life-bearing planets with the remaining 45.47% can not. Since our original assumption for habitable star computation requires every star to be habitable, $\frac{S_{1voidzone1}}{2S_1}$ ratio is further decreased by half because for tightly orbiting habitable binaries, one of the stars have to be discounted as non-habitable despite the fact that the whole binary pair is deemed habitable as a system. The same logic applies to $\frac{S_{3voizone1}}{2S_3}$, which is decreased by half since one of the stars have to be discounted as non-habitable. S_2 requires no further modification since both stars of the binary system is deemed inhabitable, so any percentage of distribution fallen within this range is deducted from the overall habitability.

$$A_1 = \left(\frac{S_{1voidzone1}}{2S_1}\right)S_1 + S_2 + \left(\frac{S_{3voizone1}}{2S_3}\right)S_3 \tag{2.6.23}$$

$$= 0.45471$$

$$T = S_1 + S_2 + S_3 + S_4 = 1 \tag{2.6.24}$$

$$1 - \frac{A_1}{T} = 0.545295 \tag{2.6.25}$$

However, things get more complicated. It is possible that out of the binary pair, the habitable star with mass (0.712~1) solar mass happens to be the companion instead of the primary. This brings serious consequences. We have shown earlier that more massive stars leaves main sequence faster at the rate of $x^{-2.5}$ relative to solar mass. As result, an increasingly luminous massive primary raises the background temperature of the companion star within the habitable range. One may speculate that a more massive companion brings relatively mild consequences to background temperature. However, study points to serious consequences even for massive primary thousands of AU away. The expected surface temperature in Kelvins of different planets at their distance away from the sun is governed by the equation: [50, 51]

$$T(a) = \frac{(P_s)^{\frac{1}{4}}}{(16\pi k)^{\frac{1}{4}} a^{\frac{1}{2}}} \tag{2.6.26}$$

Whereas P_s is the power output of the sun in watts, and k is the Stefan-Boltzmann constant. a is the semi-major axis of the planet. This equation fits well with observed surface temperature of planets minus the greenhouse effect. The equation can be transformed into:

$$T(i, R) = \frac{i \cdot 1.08 \cdot 10^8}{R^{\frac{1}{2}}} \tag{2.6.27}$$

Whereas i stands for the increase in solar luminosity output and R stands for distance to the planet in AU. 66.2% of the stars with mass greater than 0.712 solar has mass greater than 1 solar mass. More massive stars has a higher chance hosting binaries but such an increase is negligible over the range from 1 to 5 solar mass (from 44% to 50%) whereas only 16 solar mass stars has 80% chance being binaries but almost no binary system hosts such massive star. So we assumed that there is almost equal chance of forming binaries within our range of interests:

$$P_{within1} = 1 - \frac{\int_1^{15} I_{mf}(x)\, dx}{\int_{0.712}^{15} I_{mf}(x)\, dx} = 0.662015 \tag{2.6.28}$$

and out of those stars with greater than 1 solar mass the median mass is 1.11 solar mass:

$$\frac{\int_1^{1.11} I_{mf}(x)\, dx}{\int_1^{15} I_{mf}(x)\, dx} \approx 0.5 \tag{2.6.29}$$

and mean mass of 1.176 solar mass:

$$\frac{\int_1^{15} x I_{mf}(x)\, dx}{\int_1^{15} I_{mf}(x)\, dx} = 1.176 \tag{2.6.30}$$

Knowing that typical mass of larger companion has 1.11 solar mass implies that the star moving off main sequence $\left(1.11^{-2.5}\right)^{-1} = 1.298$ times faster than the sun, so that after 4.5 Gyr, its luminosity has increased 1.44 times greater than the sun, which translates to greater surface temperature of surrounding planets.[61] This relationship can be computed based on the stellar age and luminosity ratio with best fit as:

$$L = \left(ax^b + x^d + f\right)^v = \left(0.00000582 x^{4.30} + x^{0.234} - 0.29\right)^{4.258} \tag{2.6.31}$$

and other non solar mass can be derived by adding a factor F to exponents b and d as:

$$L = \left(ax^{Fb} + x^{Fd} + f\right)^v \tag{2.6.32}$$

Ignasi Ribas

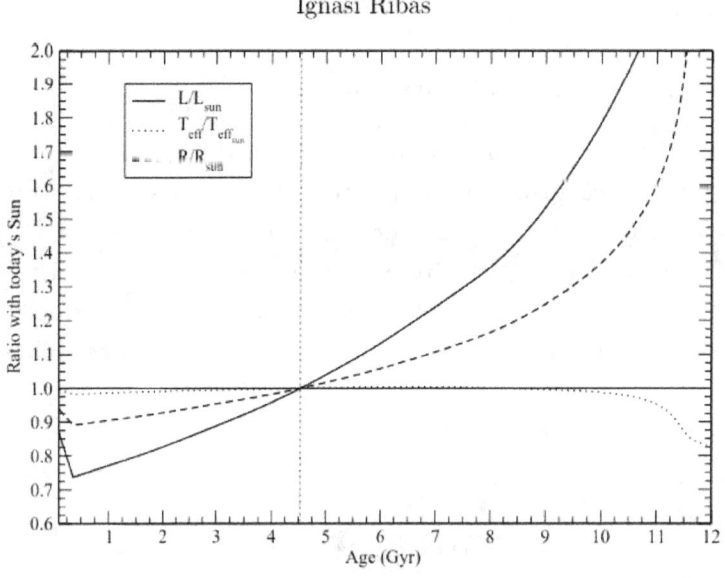

Figure 2.6.9: Solar mass star and luminosity across its age profile

This is the lower bound. If habitability window is defined by orange dwarfs with longer lifespan than the sun, the luminosity exceeds 1.44 times of solar luminosity. It is shown that it increases temperature by at least 5.644 K even at 1,000 AU away. Since most binaries occur within 1,000 AU, it effectively shows that all binary system hosting companion larger than 1 solar mass is inhospitable to complex life.

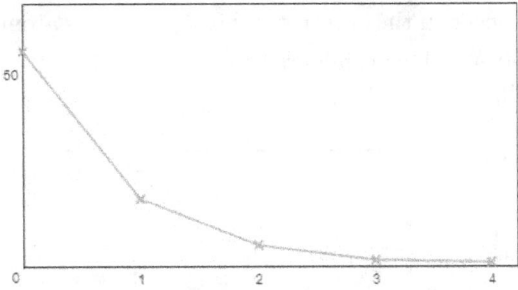

Figure 2.6.10: The lower bound log-plot of temperature increase in K vs distance for a 1.1 solar mass star

As a result, we add $P_{within1}$ as a factor to our existing calculation, we exclude those cases in which the primary exceeds 1 solar mass.

The final results lowers the habitability of binary system to 36.10%:

$$A_1 = \left(\frac{S_{1voidzone1}}{2S_1}\right) S_1 \cdot P_{within1} + S_1\left(1 - P_{within1}\right) + S_2 + \left(\frac{S_{3voizone1}}{2S_3}\right) S_3 \cdot P_{within1}$$
$$+ S_3\left(1 - P_{within1}\right) + S_4\left(1 - P_{within1}\right) \quad (2.6.33)$$

$$= 0.63900$$

$$C_2 = 1 - \frac{A_1}{T} = 0.36099 \quad (2.6.34)$$

Having derived our results for the binaries, we move to ternary systems consisting of three stars.

2.6.2 Habitability of Ternaries

For ternary star systems, all possible cases can be reduced to two independent scenarios interacting with each other. The first independent scenario determines the characteristics of the inner two stars. If the inner two stars are in a tight orbit, then they are circling each other. Otherwise, one star orbits another at some distance away. The innermost two stars among the triples and their probabilistic distribution can be modeled based on the binary probability distribution over separation distance. Conceptually, we treat the inner two stars as a pair of binary. We can then treat the inner two stars system as a single star, and its separation between the third stars circling around them constitute another pair of binary.

Now, a prudent reader may point out that there exists a case in which a pair of binary circles around a central star. This case, however, is quantitatively symmetrical to the condition we just described above, except that we would start our computation around the outer pair of stars, but the final probability stays the same as if the binary pair stayed at the center. There are just a few exceptions due to structural asymmetry which we will deal with on a case by case basis. Therefore, this overall extra step of computation is not necessary.

Case 1: 0 ~ 0.1812 AU

For the first case, the inner pair have a separation between 0 to 0.1812 AU. The probability of forming such pair is multiplied by the chance of having a binary separation between $\frac{0.1812}{2} \cdot 4$ AU or greater for the third star circling the first two. $\frac{0.1812}{2}$ because we take the weighted average of all possible distance ranges from 0 to 0.1812 AU and times a factor of 4 because only stars circling around the inner two stars with 4 times the pair separation distance away are stable.

$$S_1 = \frac{\int_{-\infty}^{\log(0.1812)} D(x)\,dx}{\int_{-\infty}^{\infty} D(x)\,dx} \cdot \frac{\int_{\log\left(\frac{0.1812}{2} \cdot 4\right)}^{\infty} D(x)\,dx}{\int_{-\infty}^{\infty} D(x)\,dx} \quad (2.6.35)$$

Within this scenario, a total dead zone occurs if the outermost third-star circles around the first two between the minimal stable orbit distance up to 100 AU. This configuration does not allow either habitable planet circling around the inner pair nor circling around the third star.

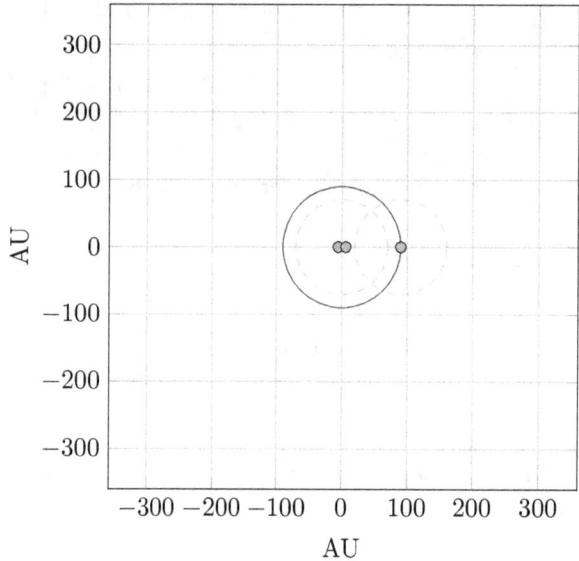

Figure 2.6.11: Ternary system with the inner pair < 0.1812 AU and the outer third star < 100 AU

$$S_{1deadzone} = \frac{\int_{-\infty}^{\log(0.1812)} D(x)\, dx}{\int_{-\infty}^{\infty} D(x)\, dx} \cdot \frac{\int_{\log\left(\frac{0.1812}{2} \cdot 4\right)}^{\log(100)} D(x)\, dx}{\int_{-\infty}^{\infty} D(x)\, dx} \tag{2.6.36}$$

There can be 1, 2, or 3 stars all found within the same ternary system satisfies the condition $0.712 < x < 1$. In every case, they are deemed inhabitable.

Beyond the dead zone, when the outermost third star circles around the inner pair between 100 AU and 200 AU, one planetary system can exist around either the inner pair or around the outermost star.

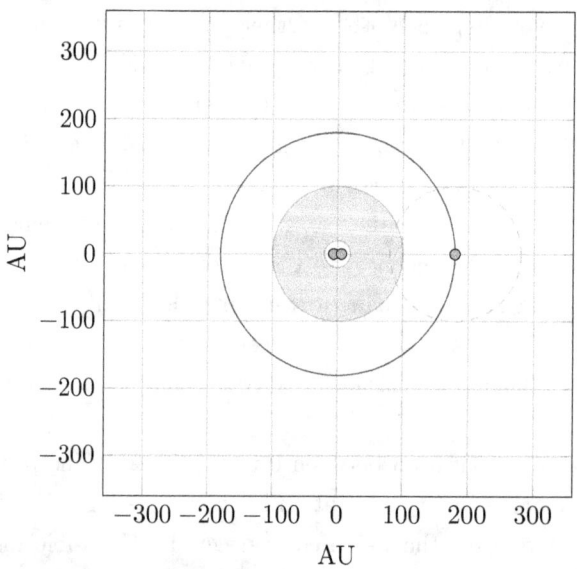

Figure 2.6.12: Ternary system with the inner pair < 0.1812 AU and the outer third star 100 AU < x < 200 AU

$$S_{1o} = \frac{\int_{-\infty}^{\log(0.1812)} D(x)\, dx}{\int_{-\infty}^{\infty} D(x)\, dx} \cdot \frac{\int_{\log(100)}^{\log(200)} D(x)\, dx}{\int_{-\infty}^{\infty} D(x)\, dx} \tag{2.6.37}$$

The chance of both the inner pair are > 0.712 solar mass and is circled by the 3rd star < 0.712 solar mass is given by:

$$\left(\frac{S_{1o}}{S_1} \cdot P_{near} \right)(1 - P_{mid}) \tag{2.6.38}$$

then one of the habitable star is over-counted as:

$$\left(\frac{S_{1o}}{2S_1} \cdot P_{near} \right)(1 - P_{mid}) \tag{2.6.39}$$

Here a case of structural asymmetry leads quantitative asymmetry. The chance of a central star with > 0.712 solar mass surrounded by a pair of binaries in which the pair < 0.712 solar mass yields:

$$\left(\frac{S_{1o}}{S_1} \right)(1 - P_{mid}) \tag{2.6.40}$$

If one were to take the average of the 2 cases, the percentage needed to be discounted is then:

$$\frac{1}{2} \left(\frac{S_{1o}}{2S_1} \cdot P_{near} \right)(1 - P_{mid}) \tag{2.6.41}$$

Alternatively, one can be more accurately expressed it as (because there is 3 possible placements: placements within a central pair or single star in the center):

$$\left(\frac{S_{1o}}{(2+1)S_1} \cdot P_{near} \right)(1 - P_{mid}) \tag{2.6.42}$$

Neither are the exact solutions to the percentage of over-counted. It is evident from calculation shown below.

	binary at the center	single star at the center
P_{near}	2	1
$1 - P_{near}$	1	1

The table depicts the number of slots available for each possible cases of over-count concerning *only star placement at the center*. When the binary pair sits at the center of the configuration, there can be either 2 slots (when both primary and companion > 0.712 solar mass). When the third star sits at the center, there always remains only 1 slot regardless if the circling binary primary or companion > 0.712 solar mass.

The total possible slot placement is simply:

$$2P_{near} + P_{near} + (1 - P_{near}) + (1 - P_{near}) \tag{2.6.43}$$

$$= (P_{near} + 1) + 1 \tag{2.6.44}$$

The over counts assuming binary pair placement at the center constitutes a dead zone (0.1812 AU < x < 100 AU):

$$2P_{near} + (1 - P_{near}) = P_{near} + 1 \tag{2.6.45}$$

and the ratio becomes:

$$D_{eadzone3}(P_{near}) = \frac{P_{near} + 1}{(P_{near} + 1) + 1} \tag{2.6.46}$$

43

and it can be compared with our approximation:

$$D_{eadzone3}\left(P_{near}\right) = \left(1 - P_{near}\right)\left(\frac{1}{1+1}\right) + P_{near}\left(\frac{2}{2+1}\right) \tag{2.6.47}$$

The over counts if assuming binary pair placement at the center constitutes a semi-habitable zone (100 AU $<$ x $<$ 200 AU):

$$P_{near} + 0 \cdot \left(1 - P_{near}\right) = P_{near} \tag{2.6.48}$$

and the ratio becomes:

$$H_{alfzone3}\left(P_{near}\right) = \frac{P_{near}}{\left(P_{near}+1\right)+1} \tag{2.6.49}$$

and it can be compared with our approximation:

$$H_{alfzone3}\left(P_{near}\right) = \left(1 - P_{near}\right)(0) + P_{near}\left(\frac{1}{2+1}\right) \tag{2.6.50}$$

It can be shown that depending on the chosen P_{near} value, the final over-count over total slots available ratio changes and our approximation is a fairly accurate one.

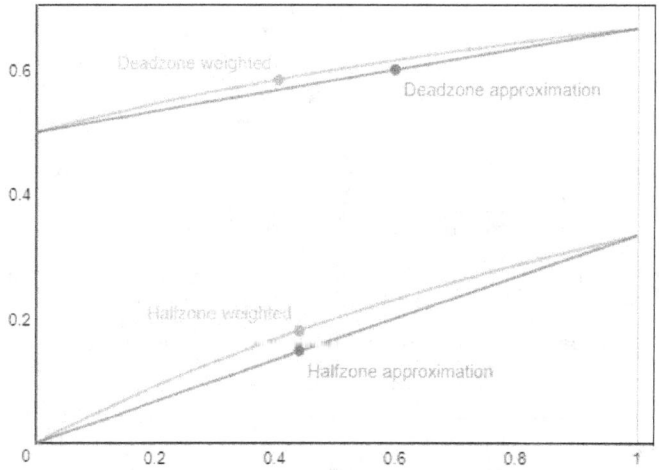

Figure 2.6.13: Weighted vs. approximation

The model is applicable to binary system as:

$$D_{eadzone2}\left(P_{near}\right) = \frac{P_{near}+1}{\left(P_{near}+1\right)} \tag{2.6.51}$$

$$H_{alfzone2}\left(P_{near}\right) = \frac{P_{near}}{\left(P_{near}+1\right)} \tag{2.6.52}$$

Whereas the total possible slot count for a single star placement becomes impossible and decrease by 1. Likewise, for higher order systems, the total slots placement increases. Therefore, it can be generalized as:

$$D_{eadzone}\left(P_{near}, n\right) = \frac{P_{near}+1}{\left(P_{near}+1\right)+\left(n-2\right)} \tag{2.6.53}$$

$$H_{alfzone}\left(P_{near}, n\right) = \frac{P_{near}}{\left(P_{near}+1\right)+\left(n-2\right)} \tag{2.6.54}$$

Whereas n stands for the number of stars in the system.
Nevertheless, we utilize the second approximate approach to ease the computation on higher order systems.

The chance of only one of the inner pair is greater than 0.712 solar mass and is circled by the 3rd star below 0.712 solar mass is given by:

$$\left(\frac{S_{1o}}{S_1} \cdot (1 - P_{near})\right)(1 - P_{mid}) \tag{2.6.55}$$

This case is excluded since only one habitable star satisfies the requirement and there is no over-count.

When the 3rd > 0.712 solar mass and the companion of the inner primary < 0.712 solar mass, over-counted is expressed as (structural asymmetry does not affect quantitative symmetry):

$$\left(\frac{S_{1o}}{2S_1}(1 - P_{near})\right)P_{mid} \tag{2.6.56}$$

When both the 3rd and the companion of the inner primary > 0.712 solar mass, over-counted is expressed as (structural asymmetry does not affect quantitative symmetry):

$$\left(\frac{2S_{1o}}{3S_1} \cdot P_{near}\right)P_{mid} \tag{2.6.57}$$

Combining all cases yields:

$$F_1 = \left(\frac{S_{1o}}{(2+1)S_1} \cdot P_{near}\right)(1 - P_{mid}) + \left(\frac{S_{1o}}{2S_1}(1 - P_{near}) + \frac{2S_{1o}}{3S_1} \cdot P_{near}\right)P_{mid} \tag{2.6.58}$$

When neither the companion of the inner pair nor the third star satisfies $0.712 < x < 1$ solar mass, no star is over-counted. If planets circle around the inner pair, the companion is already excluded from the original computation and the planets revolves around the primary (satisfied) and companion (non-satisfied). If planets circle around the third (non-satisfied), the star is already excluded from habitability.

When the outermost third-star circles around the first two with separation 200 AU or greater, planetary systems can exist around both the inner pair and the outermost star. Therefore, one of the inner pair is over-counted from the number of our existing habitable systems.

$$S_{2o} = \frac{\int_{-\infty}^{\log(0.1812)} D(x)\, dx}{\int_{-\infty}^{\infty} D(x)\, dx} \cdot \frac{\int_{\log(200)}^{\infty} D(x)\, dx}{\int_{-\infty}^{\infty} D(x)\, dx} \tag{2.6.59}$$

We apply the same approach as before, and the over-counted is expressed as:

$$F_2 = \left(\frac{S_{2o}}{(2+1)S_1} \cdot P_{near}\right)(1 - P_{far}) + \left(\frac{S_{2o}}{3S_1}P_{near}\right)P_{far} \tag{2.6.60}$$

and can be further simplified to:

$$F_2 = \frac{S_{2o}}{(2+1)S_1} \cdot P_{near} \tag{2.6.61}$$

since both the 1st and 2nd terms contains equal value.

There is no over-count for the case $\left(\frac{S_{2o}}{S_1} \cdot (1 - P_{near})\right)(1 - P_{far})$ since only 1 star satisfies 2 habitability placement requirement.

and there is no over-count for the case $\left(\frac{S_{2o}}{3S_1}(1 - P_{near})\right)P_{far}$ since only 2 stars satisfy 2 habitability placement requirement.

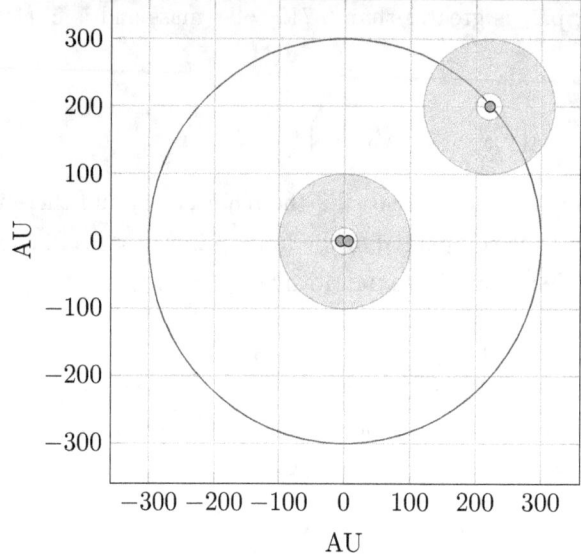

Figure 2.6.14: Ternary system with the inner pair < 0.1812 AU and the outer third star > 200 AU

Case 2: 0.1812 ~ 25 AU

For our second case, the inner pair with separation between 0.1812 AU to 25 AU, the outer ring can only be stable if it lies at $\frac{(25+0.1812)}{2}$ (the average distance between the inner pair fall within this range) times 4 (in order for the orbit to be stable). First of all, the inner pair is always non-habitable.

$$S_3 = \frac{\int_{\log(0.1812)}^{\log(25)} D(x)\, dx}{\int_{-\infty}^{\infty} D(x)\, dx} \cdot \frac{\int_{\log\left(\frac{(25+0.1812)}{2}\cdot 4\right)}^{\infty} D(x)\, dx}{\int_{-\infty}^{\infty} D(x)\, dx} \tag{2.6.62}$$

A total dead zone occurs if the outermost third-star circles around the first two between the minimal stable orbit distance up to 100 AU. This configuration does not allow either habitable planet circling around the inner pair nor circling around the third star.

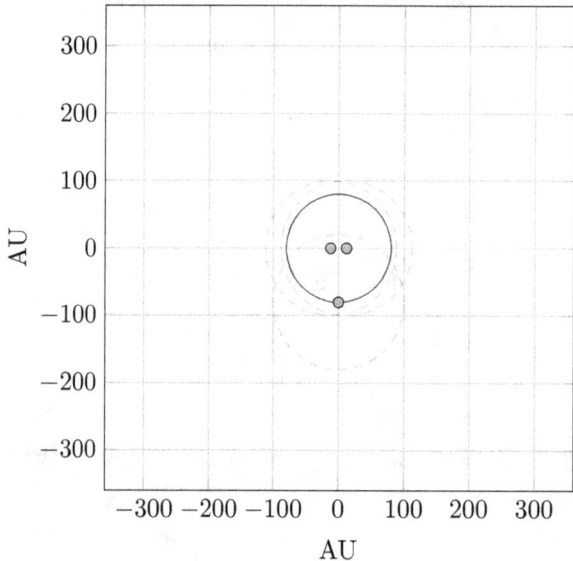

Figure 2.6.15: Ternary system with the inner pair 0.1812 AU < x < 25 AU and the outer third star < 100 AU

$$S_{3deadzone} = \frac{\int_{\log(0.1812)}^{\log(25)} D\left(x\right) dx}{\int_{-\infty}^{\infty} D\left(x\right) dx} \cdot \frac{\int_{\log\left(\frac{(25+0.1812)}{2}\cdot 4\right)}^{\log(100)} D\left(x\right) dx}{\int_{-\infty}^{\infty} D\left(x\right) dx} \qquad (2.6.63)$$

Beyond the dead zone, when the outermost third-star circles around the first two between 100 AU and 200 AU, one planetary system can exist only around the outermost star. The inner pair is over-counted from the number of our existing habitable systems.

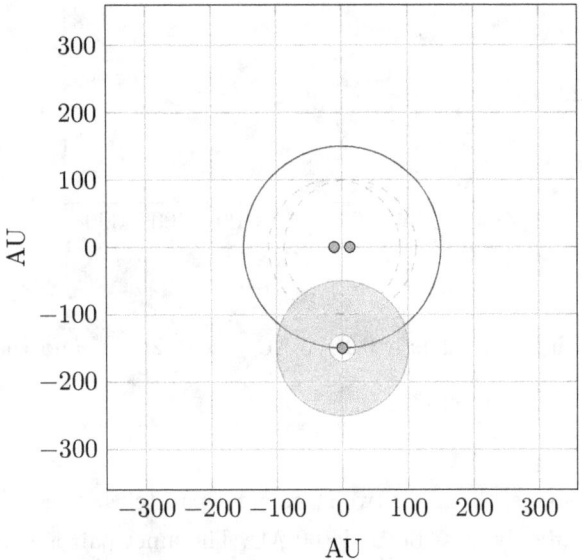

Figure 2.6.16: Ternary system with the inner pair 0.1812 AU < x < 25 AU and the outer third star 100 AU < x < 200 AU

$$S_{3o} = \frac{\int_{\log(0.1812)}^{\log(25)} D\left(x\right) dx}{\int_{-\infty}^{\infty} D\left(x\right) dx} \cdot \frac{\int_{\log(100)}^{\infty} D\left(x\right) dx}{\int_{-\infty}^{\infty} D\left(x\right) dx} \qquad (2.6.64)$$

The over-counted is expressed as:

$$F_3 = \left(\frac{1S_{3o}}{(1+1)S_3}(1 - P_{mid}) + \frac{2S_{3o}}{(2+1)S_3}P_{mid}\right)(1 - P_{mid}) + \left(\frac{S_{3o}}{2S_3}(1 - P_{mid}) + \frac{2S_{3o}}{3S_3}P_{mid}\right)P_{mid} \quad (2.6.65)$$

and can be simplified to:

$$F_3 = \frac{1S_{3o}}{(1+1)S_3}(1 - P_{mid}) + \frac{2S_{3o}}{(2+1)S_3}P_{mid} \qquad (2.6.66)$$

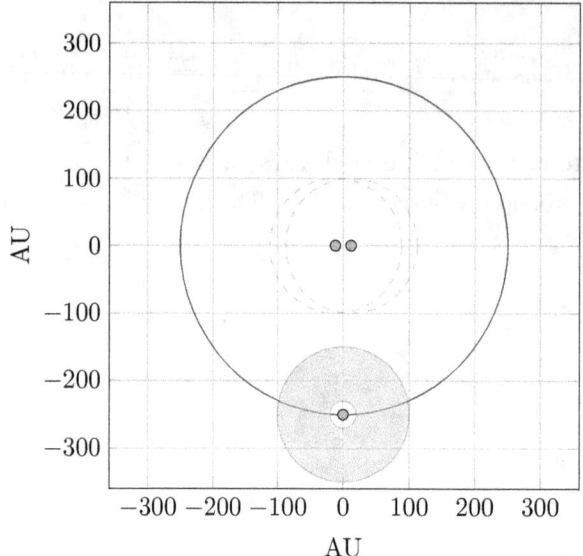

Figure 2.6.17: Ternary system with the inner pair 0.1812 AU < x < 25 AU and the outer third star > 200 AU

Case 3: 25 ~ 100 AU

In the third case for ternary star systems, the inner two stars are separated between 25 AU to 100 AU, and a stable orbit for the outermost star can only exist beyond 100 AU. The inner pair is non-habitable, but no total dead zone exists for such configurations for the third star. Therefore, the inner pair is over-counted from the number of our existing habitable systems.

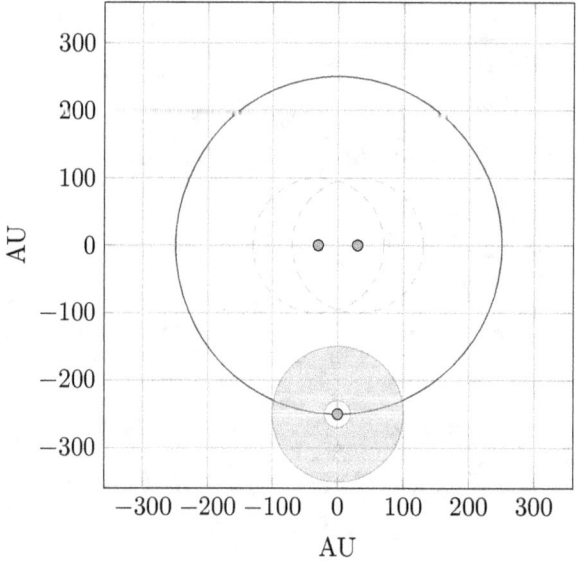

Figure 2.6.18: Ternary system with the inner pair 25 AU < x < 100 AU and the outer third star > **250 AU**

$$S_4 = \frac{\int_{\log(25)}^{\log(100)} D\left(x\right) dx}{\int_{-\infty}^{\infty} D\left(x\right) dx} \cdot \frac{\int_{\log\left(\frac{125}{2} \cdot 4\right)}^{\infty} D\left(x\right) dx}{\int_{-\infty}^{\infty} D\left(x\right) dx} \qquad (2.6.67)$$

The over-counted is expressed as:

$$F_4 = \left(\frac{1 S_4}{(1+1) S_4} \left(1 - P_{mid}\right) + \frac{2 S_4}{(2+1) S_4} P_{mid} \right) \left(1 - P_{far}\right) + \left(\frac{S_4}{2 S_4} \left(1 - P_{mid}\right) + \frac{2 S_4}{3 S_4} \left(P_{mid}\right) \right) P_{far} \qquad (2.6.68)$$

and can be simplified to:

$$F_4 = \frac{1 S_4}{(1+1) S_4} (1 - P_{mid}) + \frac{2 S_4}{(2+1) S_4} P_{mid} \tag{2.6.69}$$

Case 4: 100 ~ 200 AU

In the fourth case, the inner two stars are separated between 100 AU to 200 AU, and a stable orbit for the outermost star can only exist beyond 400 AU. As a result, one of the inner pair is over-counted from the number of our existing habitable systems, and no dead zone exists for such configurations for the third star.

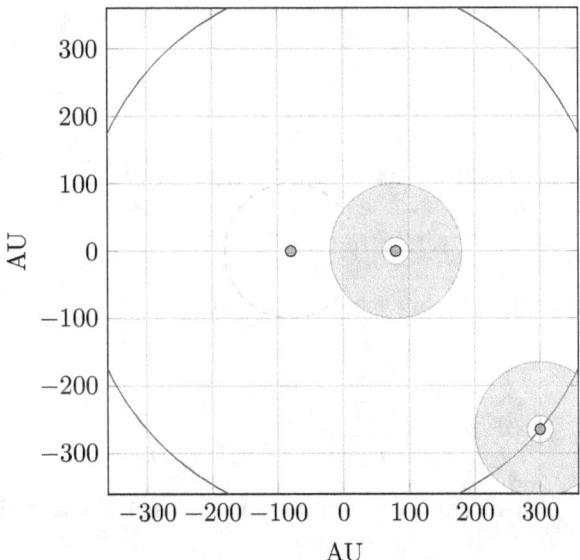

Figure 2.6.19: Ternary system with the inner pair 100 AU < x < 200 AU and the outer third star > 400 AU

$$S_5 = \frac{\int_{\log(100)}^{\log(200)} D(x) \, dx}{\int_{-\infty}^{\infty} D(x) \, dx} \cdot \frac{\int_{\log(\frac{300}{2} \cdot 4)}^{\infty} D(x) \, dx}{\int_{-\infty}^{\infty} D(x) \, dx} \tag{2.6.70}$$

The over-counted is expressed as:

$$F_5 = \left(\frac{S_5}{(2+1) S_5} P_{mid} + 0 \cdot \frac{S_5}{S_5} (1 - P_{mid}) \right) (1 - P_{far}) + \left(\frac{S_5}{3 S_5} P_{mid} + 0 \cdot \frac{S_5}{S_5} (1 - P_{mid}) \right) P_{far} \tag{2.6.71}$$

and can be simplified to:

$$F_5 = \frac{S_5}{(2+1) S_5} P_{mid} + 0 \cdot \frac{S_5}{S_5} (1 - P_{mid}) \tag{2.6.72}$$

Case 5: > 200 AU

In the final case, the inner two stars are separated by greater than 200 AU, and the stable orbit for the outer circling star can only exist beyond 800 AU. As a result, no dead zone exists for such configurations. The great separation distance between the inner pair allows habitable planetary system circling around both inner stars as well as the outermost one. As a result, the final case is the only case where all three stars are not over-counted from the number of our existing habitable systems. Since all three are capable hosting life, we just treat them, conceptually, as three separate life-capable hosting stars.

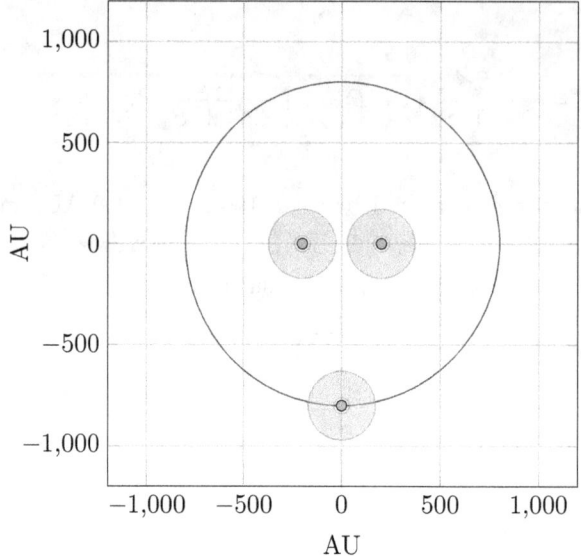

Figure 2.6.20: Ternary system with the inner pair > 200 AU and the outer third star > 800 AU

$$S_6 = \frac{\int_{\log(200)}^{\infty} D(x)\, dx}{\int_{-\infty}^{\infty} D(x)\, dx} \cdot \frac{\int_{\log(200\cdot 4)}^{\infty} D(x)\, dx}{\int_{-\infty}^{\infty} D(x)\, dx} \tag{2.6.73}$$

By deriving all possible cases for ternary, we can now sum up our results and obtain the following conclusion.

$$Q_0 = \left(\frac{S_{1deadzone}}{S_1} + F_1 + F_2\right) S_1 P_{within1} + S_1 (1 - P_{within1}) \tag{2.6.74}$$

$$Q_1 = \left(\frac{S_{3deadzone}}{S_3} + F_3\right) S_3 P_{within1} + S_3 (1 - P_{within1}) \tag{2.6.75}$$

$$Q_2 = F_4 S_4 P_{within1} + S_4 (1 - P_{within1}) \tag{2.6.76}$$

$$Q_3 = F_5 S_5 P_{within1} + S_5 (1 - P_{within1}) \tag{2.6.77}$$

$$Q_4 = S_6 (1 - P_{within1}) \tag{2.6.78}$$

$$T = S_1 + S_3 + S_4 + S_5 + S_6 \tag{2.6.79}$$

$$C_3 = 1 - \frac{Q_0 + Q_1 + Q_2 + Q_3 + Q_4}{T} = 0.345836386634 \tag{2.6.80}$$

It shows that 34.58% of ternaries can host potentially life-bearing planets with the remaining 65.42% can not. This decreasing trend is not hard to interpret because as more bodies are capable of disturbing the zone of habitability, the chance of hosting habitable planets decreases.

2.6.3 Habitability of Quaternaries

We then shift our attention to Quaternary star systems, the quaternary system is more complicated because not only we have to consider cases where two innermost stars can be treated as a binary pair while the two remaining outermost stars can be treated as two separate stars circling the inner pair conceptually as a single star. We also have to consider that a pair of binary can circle another pair of binary. Therefore, we have two major cases to cover.

The first major case is basically all the cases covered under the ternary star system scenario with an additional star circling beyond the orbit of the third star. We can then treat the inner three stars system as a single star, and its separation between the fourth circling around them constitute another pair of binary. No new dead zone for habitability arises other than the ones existed from the ternary system. Because in order for the fourth star to be stable, it has to circle the other three stars in a wide orbit 4 times the separation distance between the innermost pair and the third, resulting in even the closest orbiting fourth star to be 200 AU away.

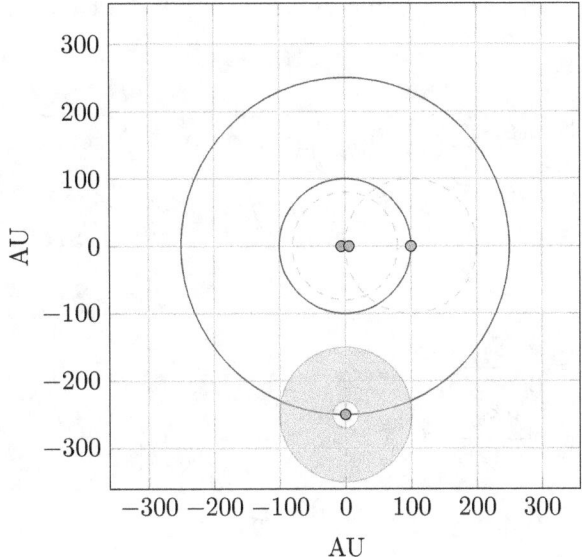

Figure 2.6.21: Quadruple system with one possible ternary system and the outer fourth star > 250 AU

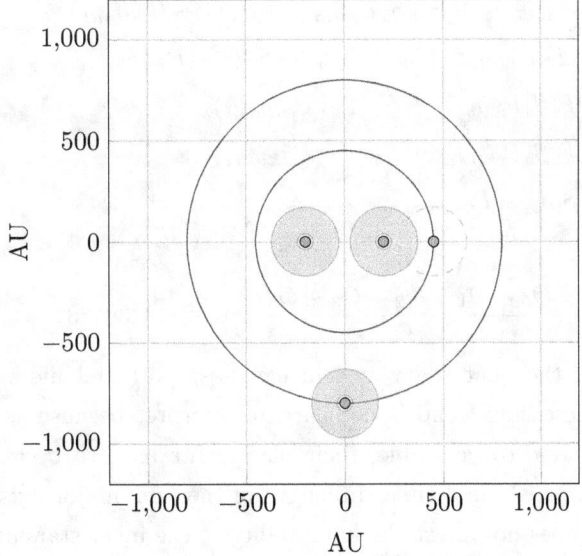

Figure 2.6.22: Quadruple system with one possible ternary system and the outer fourth star > 800 AU

$$F_0 = \left(\frac{2S_{1deadzone}}{(2+1)\,S_1} \cdot P_{near} + \frac{1S_{1deadzone}}{(1+1)\,S_1} \cdot (1 - P_{near}) \right)(1 - P_{mid})$$
$$+ \left(\frac{3S_{1deadzone}}{(3+1)\,S_1} \cdot P_{near} + \frac{2S_{1deadzone}}{(2+1)\,S_1}\,(1 - P_{near}) \right) P_{mid} \quad (2.6.81)$$

$$F_1 = \left(\frac{S_{1o}}{(2+2)\, S_1} \cdot P_{near} \right) (1 - P_{mid})$$

$$+ \left(\frac{S_{1o}}{(2+1)\, S_1} (1 - P_{near}) + \frac{2S_{1o}}{(3+1)\, S_1} \cdot P_{near} \right) P_{mid} \quad (2.6.82)$$

$$F_{3dead} = \left(\frac{2S_{3deadzone}}{(2+1)\, S_3} \cdot P_{mid} + \frac{1S_{3deadzone}}{(1+1)\, S_3} \cdot (1 - P_{mid}) \right) (1 - P_{mid})$$

$$+ \left(\frac{3S_{3deadzone}}{(3+1)\, S_3} \cdot P_{mid} + \frac{2S_{3deadzone}}{(2+1)\, S_3} (1 - P_{mid}) \right) P_{mid} \quad (2.6.83)$$

$$F_2 = \frac{S_{2o}}{(2+2)\, S_1} \cdot P_{near}$$

$$F_3 = \frac{1S_{3o}}{(1+2)\, S_3} (1 - P_{mid}) + \frac{2S_{3o}}{(2+2)\, S_3} P_{mid}$$

$$F_4 = \frac{1S_4}{(1+2)\, S_4} (1 - P_{mid}) + \frac{2S_4}{(2+2)\, S_4} P_{mid}$$

$$F_5 = \frac{S_5}{(2+2)\, S_5} P_{mid}$$

$$Q_0 = (F_0 + F_1 + F_2)\, S_1 P_{within1} + S_1 (1 - P_{within1}) \quad (2.6.84)$$

$$Q_1 = (F_{3dead} + F_3)\, S_3 P_{within1} + S_3 (1 - P_{within1}) \quad (2.6.85)$$

$$Q_2 = F_4 S_4 P_{within1} + S_4 (1 - P_{within1}) \quad (2.6.86)$$

$$Q_3 = F_5 S_5 P_{within1} + S_5 (1 - P_{within1}) \quad (2.6.87)$$

$$Q_4 = S_6 (1 - P_{within1}) \quad (2.6.88)$$

$$P_0 = 1 - \frac{Q_0 + Q_1 + Q_2 + Q_3 + Q_4}{T} = 0.528434394332 \quad (2.6.89)$$

We obtained a result of 52.84% of the quaternary system can host potential life-bearing planets with the remaining 47.16% can not. This increasing trend is not hard to interpret because as more bodies are added to the system. In order for the system to be stable, their placements have to be at a distance of 16 times the distance between the innermost pair, and 4 times in length of the semi-major axis of the third star. As a result, the additional star virtually does not affect the habitability of the inner stars at all, then the chance of habitability of planets increases. From this point onward, the habitability for multiple systems only increases. The second major case consists of all possible combinations of two pairs of binaries where each has separation ranges from 0 AU to beyond 200 AU.

Case 1: inner pair < 0.1812 AU & outer pair < 0.1812 AU

In the first case, one computes the probability of two binary pairs each separating less than 0.1812 AU apart in tight orbits hosting habitable planets.

The total possible configurations range from 4 times the weighted average separation between the binary pairs in order to form stable orbits, which turns out to be 0.1812·4 = 0.7248 AU, up to the theoretical maximum separation between any pair.

$$P_{case1all} = S_1 \cdot S_1 \frac{\int_{\log(0.1812 \cdot 4)}^{\infty} D(x)\, dx}{\int_{-\infty}^{\infty} D(x)\, dx} \quad (2.6.90)$$

Where S_1, S_3, S_4, S_5, and S_6 is defined as:

$$S_1 = \frac{\int_{\log(0.001)}^{\log(0.1812)} D(x)\,dx}{\int_{-\infty}^{\infty} D(x)\,dx} \qquad S_3 = \frac{\int_{\log(0.1812)}^{\log(25)} D(x)\,dx}{\int_{-\infty}^{\infty} D(x)\,dx} \qquad (2.6.91)$$

$$S_4 = \frac{\int_{\log(25)}^{\log(100)} D(x)\,dx}{\int_{-\infty}^{\infty} D(x)\,dx} \qquad S_5 = \frac{\int_{\log(100)}^{\log(200)} D(x)\,dx}{\int_{-\infty}^{\infty} D(x)\,dx} \qquad (2.6.92)$$

$$S_6 = \frac{\int_{\log(200)}^{\infty} D(x)\,dx}{\int_{-\infty}^{\infty} D(x)\,dx} \qquad (2.6.93)$$

$$S_1 = 0.0265313228252 \qquad S_3 = 0.215290475175$$

$$S_4 = 0.220044988625 \qquad S_5 = 0.17282150494$$

$$S_6 = 0.345728721312$$

With $P_{case1w1pair}$, the binary pairs can only host a single planetary system when they are separated between 100 AU and 200 AU apart.

$$P_{case1w1pair} = S_1 \cdot S_1 \cdot S_5 \qquad (2.6.94)$$

The over-counts for symmetric cases for only the inner pair > 0.712 solar mass:

$$F_{111inner} = \frac{2}{2}\left(0 \cdot (1 - P_{near}) + \frac{1}{2}P_{near}\right) \qquad (2.6.95)$$

The over-counts for symmetric cases for both the inner and the outer pair > 0.712 solar mass:

$$F_{111outer} = \frac{1}{2}(1 - P_{near})(1 - P_{near}) + \frac{2}{3}P_{near}(1 - P_{near}) + \frac{2}{3}(1 - P_{near})P_{near} + \frac{3}{4}P_{near} \cdot P_{near} \qquad (2.6.96)$$

The combined cases:

$$F_{111} = \frac{P_{case1w1pair}}{P_{case1all}}\left(F_{111inner} \cdot (1 - P_{mid}) + F_{111outer} \cdot P_{mid}\right) \qquad (2.6.97)$$

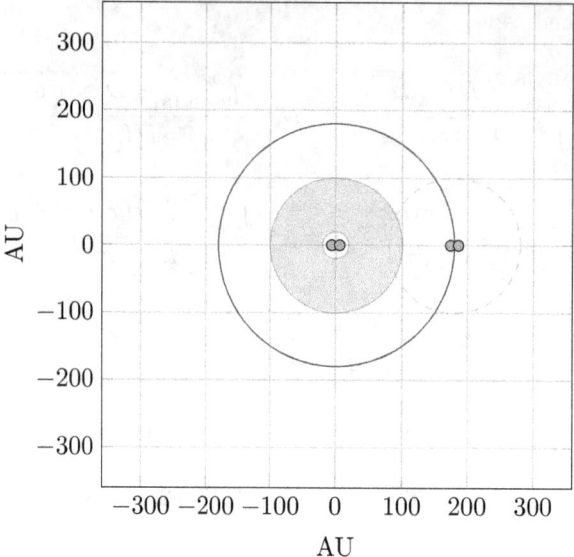

Figure 2.6.23: Quadruple system with the inner pair < 0.1812 AU and the outer pair < 0.1812 AU and a separation 100 AU $< x < 200$ AU

With $P_{case1w2pair}$, the binary pairs can host two planetary systems when they are separated by greater than 200 AU.

$$P_{case1w2pair} = S_1 \cdot S_1 \cdot S_6 \tag{2.6.98}$$

The over-counts for symmetric cases for only the inner pair > 0.712 solar mass:

$$F_{112inner} = \frac{2}{2}\left(0 \cdot (1 - P_{near}) + \frac{1}{2}P_{near}\right) \tag{2.6.99}$$

The over-counts for symmetric cases for both the inner and the outer pair > 0.712 solar mass:

$$F_{112outer} = 0\left(1 - P_{near}\right)\left(1 - P_{near}\right) + \frac{1}{3}P_{near}\left(1 - P_{near}\right) + \frac{1}{3}\left(1 - P_{near}\right)P_{near} + \frac{2}{4}P_{near} \cdot P_{near} \tag{2.6.100}$$

The combined cases:

$$F_{112} = \frac{P_{case1w2pair}}{P_{case1all}}\left(F_{112inner} \cdot (1 - P_{far}) + F_{112outer} \cdot P_{far}\right) \tag{2.6.101}$$

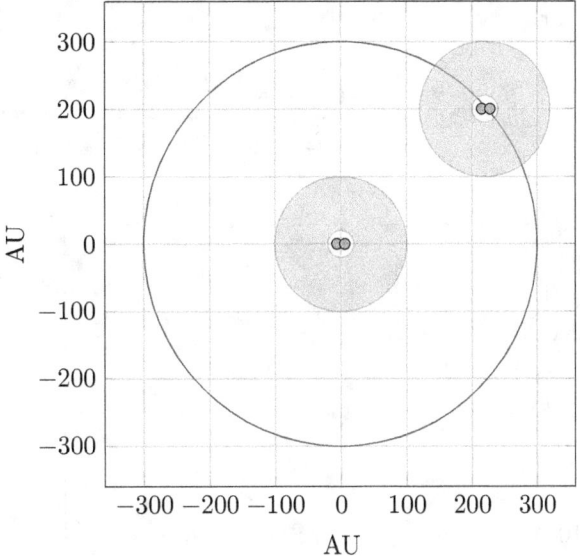

Figure 2.6.24: Quadruple system with the inner pair < 0.1812 AU and the outer pair < 0.1812 AU and a separation > 200 AU

In addition, a dead zone of habitability occurs from the distance of minimum stable orbit of 0.7248 AU up to 100 AU.

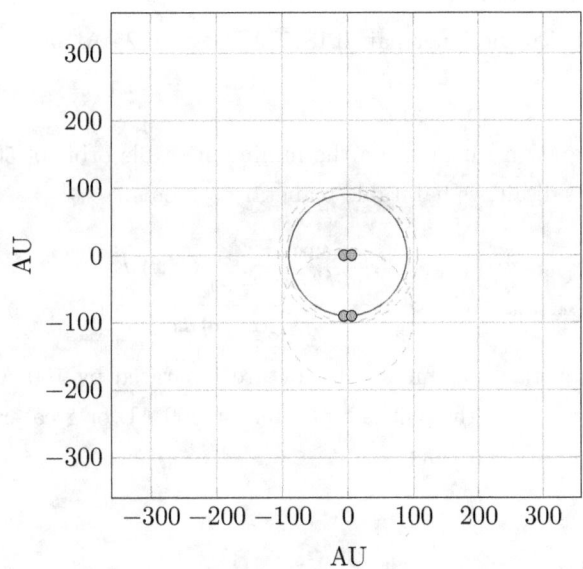

Figure 2.6.25: Quadruple system with the inner pair < 0.1812 AU and the outer pair < 0.1812 AU and a separation < 100 AU

$$P_{case1dead} = S_1 \cdot S_1 \frac{\int_{\log(0.1812\cdot4)}^{\log(100)} D\left(x\right) dx}{\int_{-\infty}^{\infty} D\left(x\right) dx} \tag{2.6.102}$$

Case 2: inner pair 0.1812~25 AU & outer pair < 0.1812 AU

In the second case, the inner pair has a separation between 0.1812 to 25 AU, and the outer pair has a separation < 0.1812 AU. The total possible configurations range from 4 times the weighted average separation between the binary pairs in order to form stable orbits, which turns out to be 12.59·4 = 50.36 AU, up to the theoretical maximum separation between any pair.

$$P_{case2all} = S_3 \cdot S_1 \frac{\int_{\log\left(\frac{0.1812+25}{2} \cdot 4\right)}^{\infty} D(x)\, dx}{\int_{-\infty}^{\infty} D(x)\, dx}$$ (2.6.103)

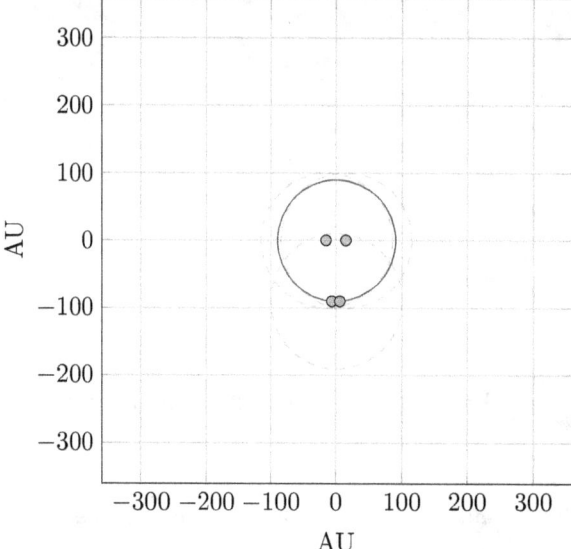

Figure 2.6.26: Quadruple system with the inner pair 0.1812 AU < x < 25 AU and the outer pair < 0.1812 AU and a separation < 100 AU

A dead zone occurs when pairs separation ranges from the minimum stable orbit of 50.36 AU up to 100 AU, where the distance between the pairs disallows habitable planetary systems.

$$P_{case2dead} = S_3 \cdot S_1 \cdot \frac{\int_{\log\left(\frac{0.1812+25}{2} \cdot 4\right)}^{\log(100)} D(x)\, dx}{\int_{-\infty}^{\infty} D(x)\, dx}$$ (2.6.104)

The pairs can only host a single planetary system when they are separated by 100 AU or greater. It is not possible to host two planetary systems when the pair is separated by 200 AU or greater because the separation distance of the inner pair itself creates a dead zone.

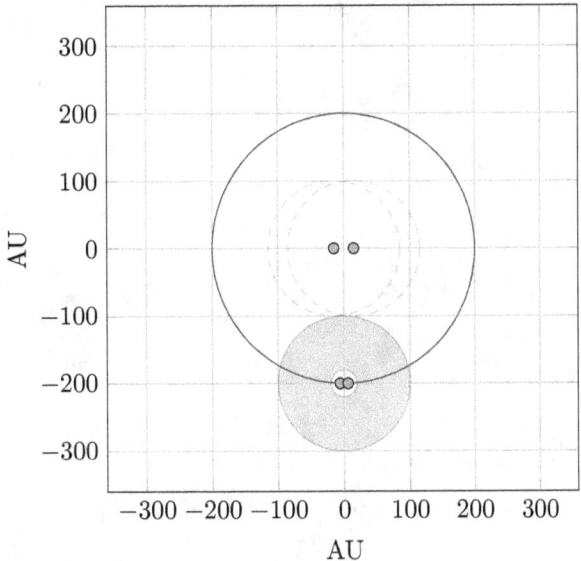

Figure 2.6.27: Quadruple system with the inner pair 0.1812 AU < x < 25 AU and the outer pair < 0.1812 AU and a separation > 100 AU

$$P_{case2w1pair} = S_3 \cdot S_1 \cdot \frac{\int_{\log(100)}^{\infty} D(x)\,dx}{\int_{-\infty}^{\infty} D(x)\,dx} \tag{2.6.105}$$

The over-counts for symmetric cases for only the inner pair > 0.712 solar mass (the average of the current inner/outer pairs switch places):

$$F_{31inner} = \frac{1}{2}\left(\left(\frac{1}{1}(1 - P_{mid}) + \frac{2}{2}P_{mid}\right) + \left(0 \cdot (1 - P_{near}) + \frac{1}{2}P_{near}\right)\right) \tag{2.6.106}$$

The over-counts for symmetric cases for both the inner and the outer pair > 0.712 solar mass:

$$F_{31outer} = \frac{1}{2}(1 - P_{mid})(1 - P_{near}) + \frac{2}{3}P_{mid}(1 - P_{near}) + \frac{2}{3}(1 - P_{mid})P_{near} + \frac{3}{4}P_{mid} \cdot P_{near} \tag{2.6.107}$$

The combined cases:

$$F_{31} = \frac{P_{case2w1pair}}{P_{case2all}}\left(F_{31inner} \cdot (1 - P_{far}) + F_{31outer} \cdot P_{far}\right) \tag{2.6.108}$$

Case 3: inner pair 0.1812~25 AU & outer pair 0.1812~25 AU

When the inner pair has a separation between 0.1812 to 25 AU, and the outer pair also has a separation between 0.1812 to 25 AU, all possible cases constitute dead zones because both pairs neither permits habitable planets circling around any one of the stars within the pairs nor permits habitable planets circling around the pairs.

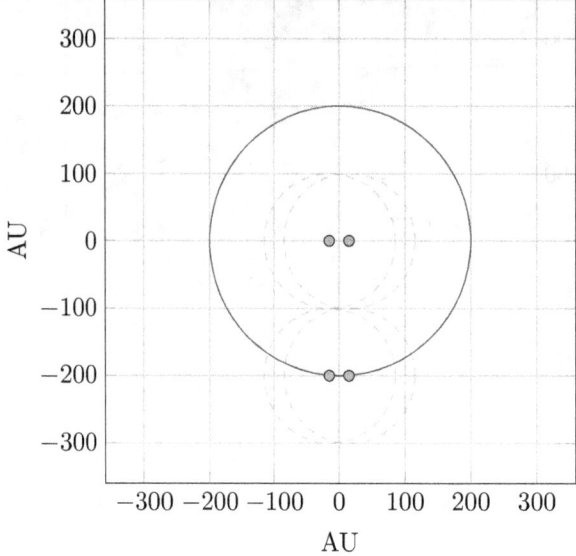

Figure 2.6.28: Quadruple system with the inner pair 0.1812 AU $< x < 25$ AU and the outer pair 0.1812 AU $< x < 25$ AU and a separation > 100 AU

$$P_{case3all} = P_{case3dead} = S_3 \cdot S_3 \frac{\int_{\log\left(\frac{0.1812+25}{2}\cdot 4\right)}^{\infty} D(x)\,dx}{\int_{-\infty}^{\infty} D(x)\,dx} \tag{2.6.109}$$

Case 4: inner pair 25~100 AU & outer pair < 0.1812 AU

The next case composes cases where the inner pair has a separation between 25 AU to 100 AU. If the outer pair has a separation less than 0.1812 AU, then the total possible configurations range from 4 times the weighted average separation between the binary pairs in order to form stable orbits, which turns out to be 250 AU, up to the theoretical maximum separation between any pair.

$$P_{case4w1pair} = P_{case4all} = S_4 \cdot S_1 \frac{\int_{\log\left(\frac{25+100}{2}\cdot 4\right)}^{\infty} D(x)\,dx}{\int_{-\infty}^{\infty} D(x)\,dx} \tag{2.6.110}$$

Because the minimum stable orbit lies beyond the 100 AU space requirement for the habitable planetary system, no dead zone is observed under this configuration. When they are separated by 100 AU or more, one habitable system is possible to orbit around the outer pair.

The over-counts for symmetric cases for only the inner pair > 0.712 solar mass (the average of the current inner/outer pairs switch places):

$$F_{41inner} = \frac{1}{2}\left(\left(\frac{1}{1}(1-P_{mid}) + \frac{2}{2}P_{mid}\right) + \left(0 \cdot (1-P_{near}) + \frac{1}{2}P_{near}\right)\right) \tag{2.6.111}$$

The over-counts for symmetric cases for both the inner and the outer pair > 0.712 solar mass:

$$F_{41outer} = \frac{1}{2}(1-P_{mid})(1-P_{near}) + \frac{2}{3}P_{mid}(1-P_{near}) + \frac{2}{3}(1-P_{mid})(P_{near}) + \frac{3}{4}P_{mid} \cdot P_{near} \tag{2.6.112}$$

The combined cases:

$$F_{41} = \frac{P_{case4w1pair}}{P_{case4all}}\left(F_{41inner} \cdot (1-P_{far}) + F_{41outer} \cdot P_{far}\right) \tag{2.6.113}$$

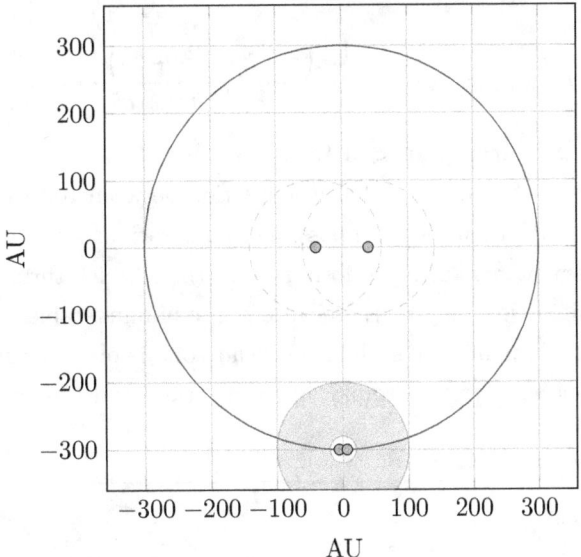

Figure 2.6.29: Quadruple system with the inner pair 25 AU < x < 100 AU and the outer pair < 0.1812 AU and a separation > 250 AU

Case 5: inner pair 25~100 AU & outer pair 0.1812~25 AU

If the outer pair has a separation between 0.1812 AU to 25 AU, then all possible configurations lead to dead zones because the separation distance for both binary pairs alone results in dead zones for habitability.

$$P_{case5all} = P_{case5dead} = S_4 \cdot S_3 \frac{\int_{\log\left(\frac{25+100}{2}\cdot 4 + \left(\frac{25+0.1812}{2}\right)\right)}^{\infty} D(x)\,dx}{\int_{-\infty}^{\infty} D(x)\,dx} \qquad (2.6.114)$$

Case 6: inner pair 25~100 AU & outer pair 25~100 AU

If the outer pair has a separation between 25 AU to 100 AU, again, all possible configurations lead to dead zones because the separation distance for both binary pairs alone results in dead zones for habitability.

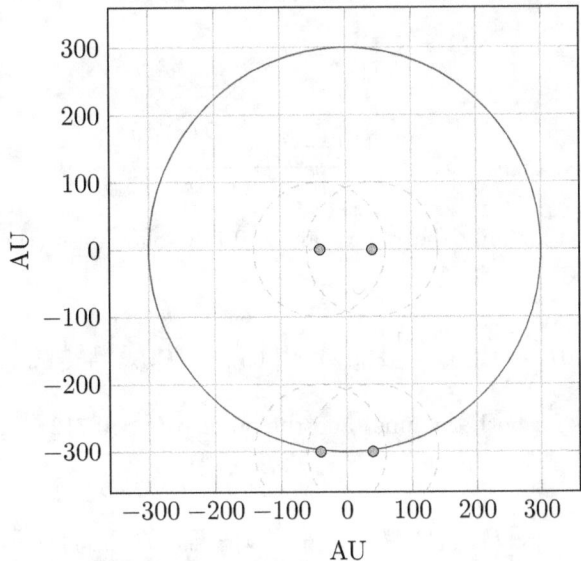

Figure 2.6.30: Quadruple system with the inner pair 25 AU < x < 100 AU and the outer pair 25 AU < x < 100 AU and a separation > 250 AU

$$P_{case6all} = P_{case6dead} = S_4 \cdot S_4 \frac{\int_{\log\left(\frac{25+100}{2}\cdot 4 + \left(\frac{25+100}{2}\right)\right)}^{\infty} D\left(x\right) dx}{\int_{-\infty}^{\infty} D\left(x\right) dx} \qquad (2.6.115)$$

Case 7: inner pair 100~200 AU & outer pair < 0.1812 AU

The 7th case composes scenarios where the inner pair has a separation between 100 AU to 200 AU. If the outer pair has a separation less than 0.1812 AU, then the total possible configurations range from 4 times the weighted average separation between the binary pairs in order to form stable orbits, which turns out to be 600 AU, up to the theoretical maximum separation between any pair. Because the minimum stable orbit lies beyond 100 AU, the space requirement for the habitable planetary system, no dead zone is observed under this configuration, and the inner pair can host one planetary system around one of its stars, and the outer pair can host planets around the pair.

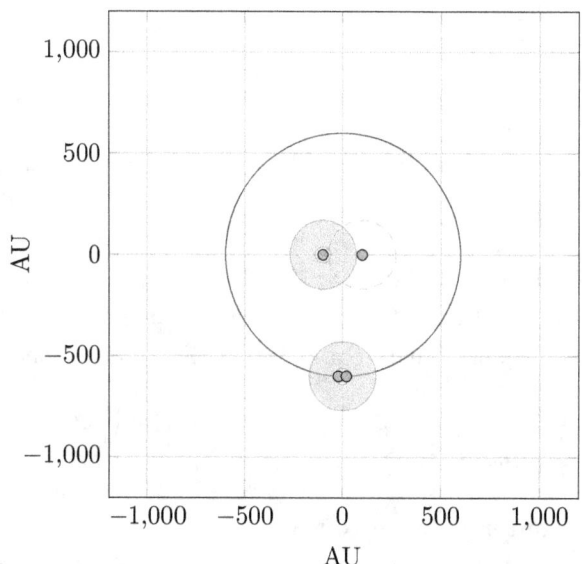

Figure 2.6.31: Quadruple system with the inner pair 100 AU < x < 200 AU and the outer pair < 0.1812 AU and a separation > 400 AU

$$P_{case7all} = S_5 \cdot S_1 \frac{\int_{\log\left(\frac{100+200}{2}\cdot 4\right)}^{\infty} D\left(x\right) dx}{\int_{-\infty}^{\infty} D\left(x\right) dx} \qquad (2.6.116)$$

The over-counts for symmetric cases for only the inner pair > 0.712 solar mass (the average of the current inner/outer pairs switch places):

$$F_{51inner} = \frac{1}{2}\left(\left(0\left(1 - P_{mid}\right) + \frac{1}{2}P_{mid}\right) + \left(0 \cdot \left(1 - P_{near}\right) + \frac{1}{2}P_{near}\right)\right) \qquad (2.6.117)$$

The over-counts for symmetric cases for both the inner and the outer pair > 0.712 solar mass:

$$F_{51outer} = 0\left(1 - P_{mid}\right)\left(1 - P_{near}\right) + \frac{1}{3}P_{mid}\left(1 - P_{near}\right) + \frac{1}{3}\left(1 - P_{mid}\right)\left(P_{near}\right) + \frac{2}{4}P_{mid} \cdot P_{near} \qquad (2.6.118)$$

The combined cases:

$$F_{51} = \frac{P_{case7all}}{P_{case7all}}\left(F_{51inner} \cdot \left(1 - P_{far}\right) + F_{51outer} \cdot P_{far}\right) \qquad (2.6.119)$$

Case 8: inner pair 100~200 AU & outer pair 0.1812~25 AU

If the outer pair has a separation between 0.1812 AU and 25 AU, again there is no dead zone between inner and outer pair, but only the inner pair can host one planetary system around one of its stars because the outer pair with such a separation falls under the binary dead zone list.

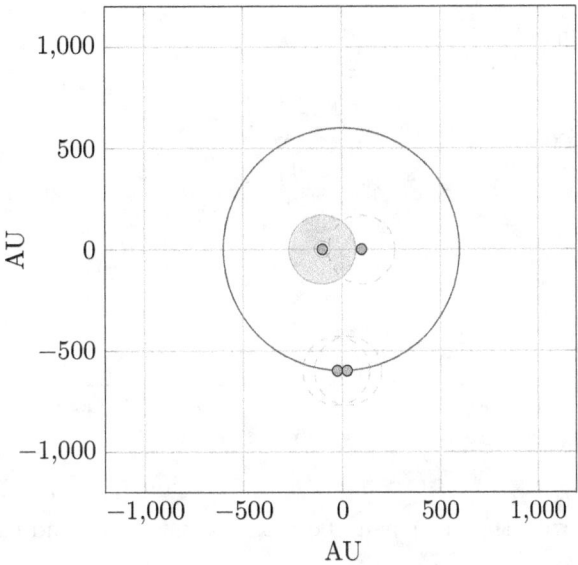

Figure 2.6.32: Quadruple system with the inner pair 100 AU < x < 200 AU and the outer pair 0.1812 AU < x < 25 AU and a separation > 400 AU

$$P_{case8all} = S_5 \cdot S_3 \frac{\int_{\log\left(\frac{100+200}{2} \cdot 4 + \left(\frac{0.1812+25}{2}\right)\right)}^{\infty} D(x)\, dx}{\int_{-\infty}^{\infty} D(x)\, dx} \qquad (2.6.120)$$

The over-counts for symmetric cases for only the inner pair > 0.712 solar mass (the average of the current inner/outer pairs switch places):

$$F_{53inner} = \frac{1}{2} \left(\left(0\left(1 - P_{mid}\right) + \frac{1}{2} P_{mid} \right) + \left(\frac{1}{1} \cdot \left(1 - P_{mid}\right) + \frac{2}{2} P_{mid} \right) \right) \qquad (2.6.121)$$

The over-counts for symmetric cases for both the inner and the outer pair > 0.712 solar mass:

$$F_{53outer} = \frac{1}{2} \left(1 - P_{mid}\right)\left(1 - P_{mid}\right) + \frac{2}{3} P_{mid}\left(1 - P_{mid}\right) + \frac{2}{3} \left(1 - P_{mid}\right)\left(P_{mid}\right) + \frac{3}{4} P_{mid} \cdot P_{mid} \qquad (2.6.122)$$

The combined cases:

$$F_{53} = \frac{P_{case8all}}{P_{case8all}} \left(F_{53inner} \cdot \left(1 - P_{far}\right) + F_{53outer} \cdot P_{far} \right) \qquad (2.6.123)$$

Case 9: inner pair 100~200 AU & outer pair 25~100 AU

If the outer pair has a separation between 25 AU and 100 AU, again there is no dead zone, but only the inner pair can host one planetary system around one of its stars because the outer pair with such a separation falls under the binary dead zone list.

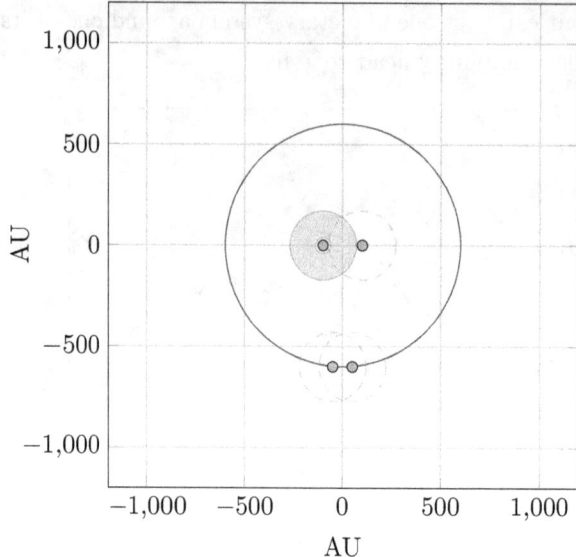

Figure 2.6.33: Quadruple system with the inner pair 100 AU < x < 200 AU and the outer pair 25 AU < x < 100 AU and a separation > 400 AU

$$P_{case9all} = S_5 \cdot S_4 \frac{\int_{\log\left(\frac{100+200}{2}\cdot 4 + \left(\frac{25+100}{2}\right)\right)}^{\infty} D\left(x\right) dx}{\int_{-\infty}^{\infty} D\left(x\right) dx} \tag{2.6.124}$$

The over-counts for symmetric cases for only the inner pair > 0.712 solar mass (the average of the current inner/outer pairs switch places):

$$F_{54inner} = \frac{1}{2}\left(\left(0\left(1 - P_{mid}\right) + \frac{1}{2}P_{mid}\right) + \left(\frac{1}{1}\cdot\left(1 - P_{mid}\right) + \frac{2}{2}P_{mid}\right)\right) \tag{2.6.125}$$

The over-counts for symmetric cases for both the inner and the outer pair > 0.712 solar mass:

$$F_{54outer} = \frac{1}{2}\left(1 - P_{mid}\right)\left(1 - P_{mid}\right) + \frac{2}{3}P_{mid}\left(1 - P_{mid}\right) + \frac{2}{3}\left(1 - P_{mid}\right)\left(P_{mid}\right) + \frac{3}{4}P_{mid}\cdot P_{mid} \tag{2.6.126}$$

The combined cases:

$$F_{54} = \frac{P_{case9all}}{P_{case9all}}\left(F_{54inner}\cdot\left(1 - P_{far}\right) + F_{54outer}\cdot P_{far}\right) \tag{2.6.127}$$

Case 10: inner pair 100~200 AU & outer pair 100~200 AU

If the outer pair has a separation between 100 AU and 200 AU, again there is no dead zone, and the inner pair can host one planetary system around one of its stars, and the outer pair can host one planetary system around one of its stars.

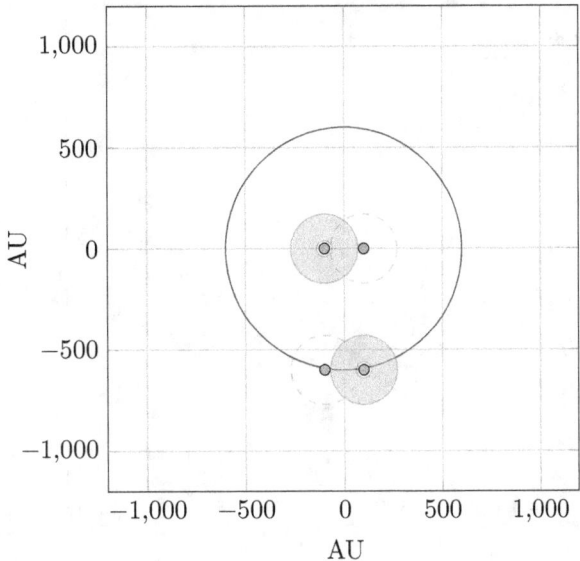

Figure 2.6.34: Quadruple system with the inner pair 100 AU < x < 200 AU and the outer pair 100 AU < x < 200 AU and a separation > 400 AU

$$P_{case10all} = S_5 \cdot S_5 \frac{\int_{\log\left(\frac{100+200}{2}\cdot 4 + \left(\frac{100+200}{2}\right)\right)}^{\infty} D\left(x\right) dx}{\int_{-\infty}^{\infty} D\left(x\right) dx} \tag{2.6.128}$$

The over-counts for symmetric cases for only the inner pair > 0.712 solar mass (the average of the current inner/outer pairs switch places):

$$F_{55inner} = \frac{1}{2}\left(\left(0\left(1 - P_{mid}\right) + \frac{1}{2}P_{mid}\right) + \left(0\left(1 - P_{mid}\right) + \frac{1}{2}P_{mid}\right)\right) \tag{2.6.129}$$

The over-counts for symmetric cases for both the inner and the outer pair > 0.712 solar mass:

$$F_{55outer} = 0\left(1 - P_{mid}\right)\left(1 - P_{mid}\right) + \frac{1}{3}P_{mid}\left(1 - P_{mid}\right) + \frac{1}{3}\left(1 - P_{mid}\right)\left(P_{mid}\right) + \frac{2}{4}P_{mid} \cdot P_{mid} \tag{2.6.130}$$

The combined cases:

$$F_{55} = \frac{P_{case10all}}{P_{case10all}}\left(F_{55inner} \cdot \left(1 - P_{far}\right) + F_{55outer} \cdot P_{far}\right) \tag{2.6.131}$$

Case 11: inner pair > 200 AU & outer pair < 0.1812 AU

The 11th case composes scenarios where the inner pair has a separation greater than 200 AU. If the outer pair has a separation less than 0.1812 AU, then the total possible configurations range from 4 times the weighted average separation between the binary pairs in order to form stable orbits, which turns out to be 800 AU, up to the theoretical maximum separation between any pair. Because the minimum stable orbit lies beyond 100 AU, the space requirement for the habitable planetary system, no dead zone is observed under this configuration, and the inner pair can host two planetary systems around both of its stars, and the outer pair can host planets around the pair.

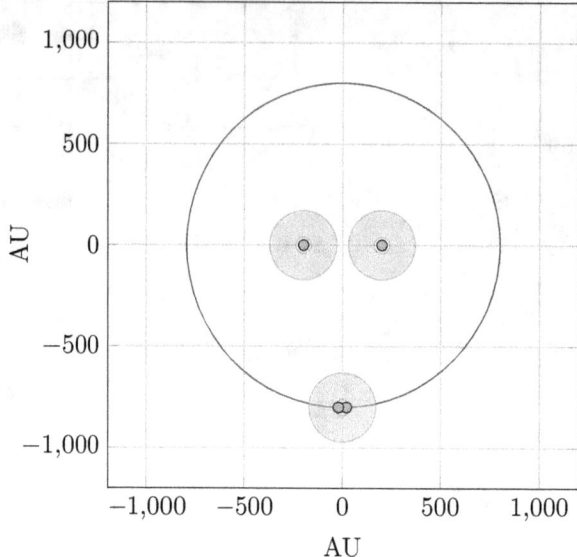

Figure 2.6.35: Quadruple system with the inner pair > 200 AU and the outer pair < 0.1812 AU and a separation > 800 AU

$$P_{case11all} = S_6 \cdot S_1 \frac{\int_{\log(200 \cdot 4)}^{\infty} D(x)\, dx}{\int_{-\infty}^{\infty} D(x)\, dx} \tag{2.6.132}$$

The over-counts for symmetric cases for only the inner pair > 0.712 solar mass (the average of the current inner/outer pairs switch places):

$$F_{61inner} = \frac{1}{2}\left(\left(0\left(1 - P_{far}\right) + 0 P_{far}\right) + \left(0\left(1 - P_{near}\right) + \frac{1}{2} P_{near}\right)\right) \tag{2.6.133}$$

The over-counts for symmetric cases for both the inner and the outer pair > 0.712 solar mass:

$$F_{61outer} = 0\left(1 - P_{far}\right)\left(1 - P_{near}\right) + 0 P_{far}\left(1 - P_{near}\right) + \frac{1}{3}\left(1 - P_{far}\right)\left(P_{near}\right) + \frac{1}{4} P_{far} \cdot P_{near} \tag{2.6.134}$$

The combined cases:

$$F_{61} = \frac{P_{case11all}}{P_{case11all}} \left(F_{61inner} \cdot \left(1 - P_{far}\right) + F_{61outer} \cdot P_{far}\right) \tag{2.6.135}$$

Case 12: inner pair > 200 AU & outer pair 0.1812~25 AU

If the outer pair has a separation between 0.1812 AU and 25 AU, there is no dead zone but only the inner pair can host two planetary systems around both of its stars because the outer pair with such separation falls under the binary dead zone list.

64

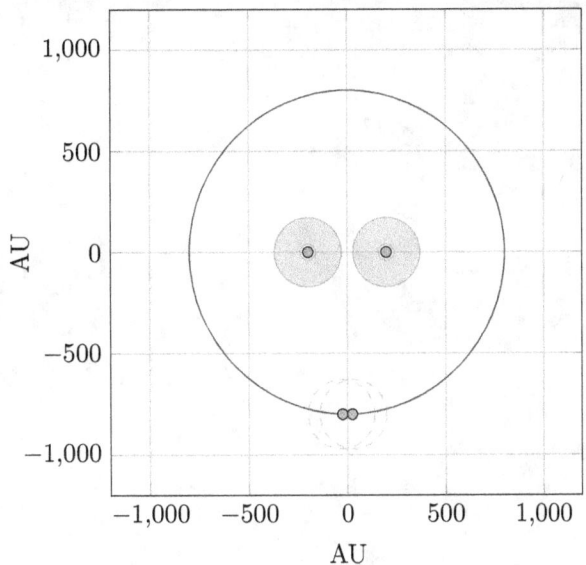

Figure 2.6.36: Quadruple system with the inner pair > 200 AU and the outer pair 0.1812 AU < x < 25 AU and a separation > 800 AU

$$P_{case12all} = S_6 \cdot S_3 \frac{\int_{\log\left(200 \cdot 4 + \left(\frac{0.1812 + 25}{2}\right)\right)}^{\infty} D\left(x\right) dx}{\int_{-\infty}^{\infty} D\left(x\right) dx} \tag{2.6.136}$$

The over-counts for symmetric cases for only the inner pair > 0.712 solar mass (the average of the current inner/outer pairs switch places):

$$F_{63inner} = \frac{1}{2}\left(\left(0\left(1 - P_{far}\right) + 0 P_{far}\right) + \left(\frac{1}{1}\left(1 - P_{mid}\right) + \frac{2}{2} P_{mid}\right)\right) \tag{2.6.137}$$

The over-counts for symmetric cases for both the inner and the outer pair > 0.712 solar mass:

$$F_{63outer} = \frac{1}{2}\left(1 - P_{far}\right)\left(1 - P_{mid}\right) + \frac{1}{3} P_{far}\left(1 - P_{mid}\right) + \frac{2}{3}\left(1 - P_{far}\right)\left(P_{mid}\right) + \frac{2}{4} P_{far} \cdot P_{mid} \tag{2.6.138}$$

The combined cases:

$$F_{63} = \frac{P_{case12all}}{P_{case12all}}\left(F_{63inner} \cdot \left(1 - P_{far}\right) + F_{63outer} \cdot P_{far}\right) \tag{2.6.139}$$

Case 13: inner pair > 200 AU & outer pair 25~100 AU

If the outer pair has a separation between 25 AU and 100 AU, again there is no dead zone but only the inner pair can host two planetary systems around both of its stars because outer pair with such separation falls under the binary dead zone list.

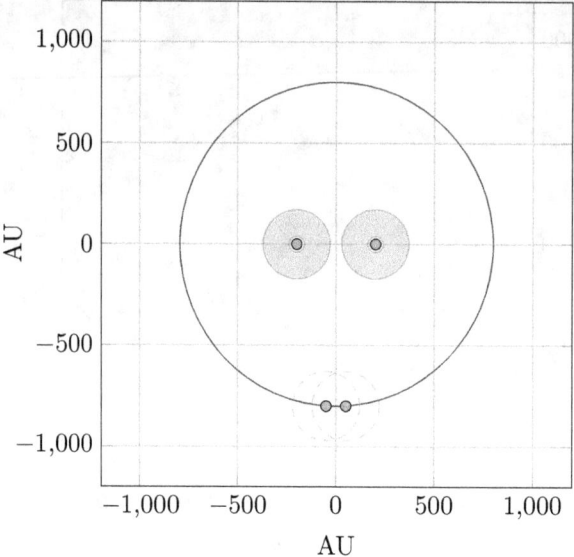

Figure 2.6.37: Quadruple system with the inner pair > 200 AU and the outer pair 25 AU < x < 100 AU and a separation > 800 AU

$$P_{case13all} = S_6 \cdot S_4 \frac{\int_{\log\left(200 \cdot 4 + \left(\frac{25+100}{2}\right)\right)}^{\infty} D(x)\, dx}{\int_{-\infty}^{\infty} D(x)\, dx} \tag{2.6.140}$$

The over-counts for symmetric cases for only the inner pair > 0.712 solar mass (the average of the current inner/outer pairs switch places):

$$F_{64inner} = \frac{1}{2}\left(\left(0\left(1 - P_{far}\right) + 0 P_{far}\right) + \left(\frac{1}{1}\left(1 - P_{mid}\right) + \frac{2}{2}P_{mid}\right)\right) \tag{2.6.141}$$

The over-counts for symmetric cases for both the inner and the outer pair > 0.712 solar mass:

$$F_{64outer} = \frac{1}{2}\left(1 - P_{far}\right)\left(1 - P_{near}\right) + \frac{1}{3}P_{far}\left(1 - P_{mid}\right) + \frac{2}{3}\left(1 - P_{far}\right)\left(P_{mid}\right) + \frac{2}{4}P_{far} \cdot P_{mid} \tag{2.6.142}$$

The combined cases:

$$F_{64} = \frac{P_{case13all}}{P_{case13all}}\left(F_{64inner} \cdot \left(1 - P_{far}\right) + F_{64outer} \cdot P_{far}\right) \tag{2.6.143}$$

Case 14: inner pair > 200 AU & outer pair 100~200 AU

If the outer pair has a separation between 100 AU and 200 AU, there is no dead zone and the inner pair can host two planetary systems around both of its stars, and the outer pair can host one planetary system around one of its stars.

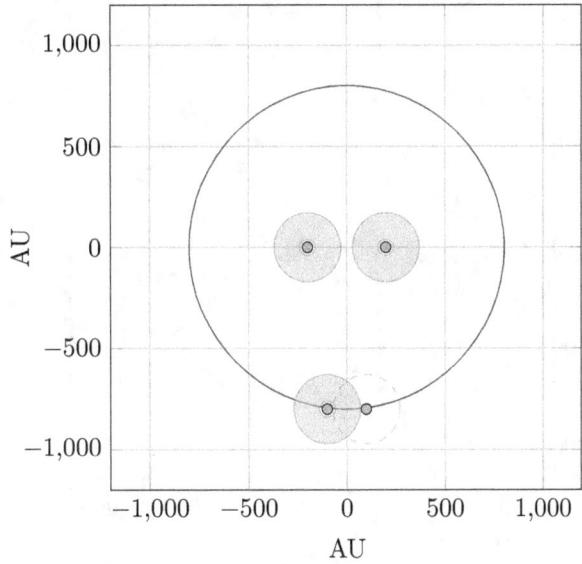

Figure 2.6.38: Quadruple system with the inner pair > 200 AU and the outer pair 100 AU < x < 200 AU and a separation > 800 AU

$$P_{case14all} = S_6 \cdot S_5 \frac{\int_{\log\left(200 \cdot 4 + \left(\frac{100+200}{2}\right)\right)}^{\infty} D\left(x\right) dx}{\int_{-\infty}^{\infty} D\left(x\right) dx} \tag{2.6.144}$$

The over-counts for symmetric cases for only the inner pair > 0.712 solar mass (the average of the current inner/outer pairs switch places):

$$F_{65inner} = \frac{1}{2}\left(\left(0\left(1 - P_{far}\right) + 0P_{far}\right) + \left(0\left(1 - P_{mid}\right) + \frac{1}{2}P_{mid}\right)\right) \tag{2.6.145}$$

The over-counts for symmetric cases for both the inner and the outer pair > 0.712 solar mass:

$$F_{65outer} = 0\left(1 - P_{far}\right)\left(1 - P_{mid}\right) + 0P_{far}\left(1 - P_{mid}\right) + \frac{1}{3}\left(1 - P_{far}\right)\left(P_{mid}\right) + \frac{1}{4}P_{far} \cdot P_{mid} \tag{2.6.146}$$

The combined cases:

$$F_{65} = \frac{P_{case14all}}{P_{case14all}}\left(F_{65inner} \cdot \left(1 - P_{far}\right) + F_{65outer} \cdot P_{far}\right) \tag{2.6.147}$$

Case 15: inner pair > 200 AU & outer pair > 200 AU

If the outer pair has a separation greater than 200 AU, there is no dead zone, and planetary systems can orbit around all four-star system. This is the only case where all four stars are not over-counted from the number of our existing habitable systems. Since all four are capable of hosting life, we just treat them, conceptually, as four separate life-capable hosting stars.

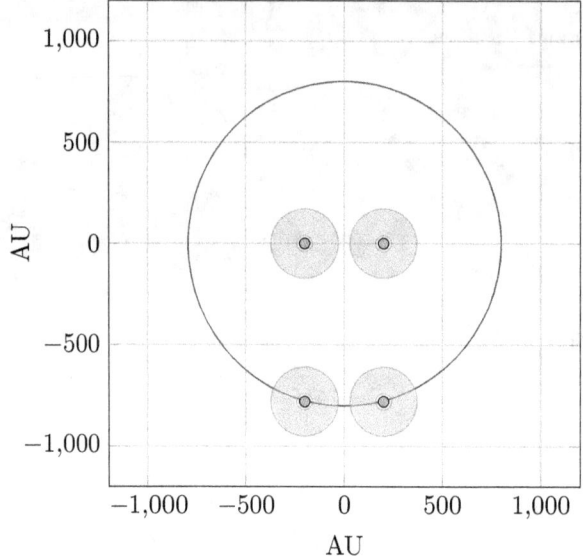

Figure 2.6.39: Quadruple system with the inner pair > 200 AU and the outer pair > 200 AU and a separation > 800 AU

$$P_{case15all} = S_6 \cdot S_6 \frac{\int_{\log(200 \cdot 4 + 200)}^{\infty} D(x)\, dx}{\int_{-\infty}^{\infty} D(x)\, dx} \tag{2.6.148}$$

By deriving all possible cases for pairs of binaries, we can now sum up our results and obtain the following conclusion. This shows that 17.83% of the quaternary system with binary pairs can host potentially life-bearing planets with the remaining 82.16% can not.

$$Q_1 = \left(\frac{P_{case1dead}}{P_{case1all}} + F_{111} + F_{112} \right) P_{case1all} \cdot P_{within1} + P_{case1all} \left(1 - P_{within1} \right) \tag{2.6.149}$$

$$Q_2 = \left(\frac{P_{case2dead}}{P_{case2all}} + F_{31} \right) P_{case2all} \cdot P_{within1} + P_{case2all} \left(1 - P_{within1} \right) \tag{2.6.150}$$

$$Q_3 = \left(\frac{P_{case3dead}}{P_{case3all}} \right) P_{case3all} \tag{2.6.151}$$

$$Q_4 = \left(F_{41} \right) P_{case4all} \cdot P_{within1} + P_{case4all} \left(1 - P_{within1} \right) \tag{2.6.152}$$

$$Q_5 = \left(\frac{P_{case5dead}}{P_{case5all}} \right) P_{case5all} \tag{2.6.153}$$

$$Q_6 = \left(\frac{P_{case6dead}}{P_{case6all}} \right) P_{case6all} \tag{2.6.154}$$

$$Q_7 = \left(F_{51} \right) P_{case7all} \cdot P_{within1} + P_{case7all} \left(1 - P_{within1} \right) \tag{2.6.155}$$

$$Q_8 = \left(F_{53} \right) P_{case8all} \cdot P_{within1} + P_{case8all} \left(1 - P_{within1} \right) \tag{2.6.156}$$

$$Q_9 = \left(F_{54} \right) P_{case9all} \cdot P_{within1} + P_{case9all} \left(1 - P_{within1} \right) \tag{2.6.157}$$

$$Q_{10} = \left(F_{55} \right) P_{case10all} \cdot P_{within1} + P_{case10all} \left(1 - P_{within1} \right) \tag{2.6.158}$$

$$Q_{11} = \left(F_{61} \right) P_{case11all} \cdot P_{within1} + P_{case11all} \left(1 - P_{within1} \right) \tag{2.6.159}$$

$$Q_{12} = \left(F_{63} \right) P_{case12all} \cdot P_{within1} + P_{case12all} \left(1 - P_{within1} \right) \tag{2.6.160}$$

$$Q_{13} = \left(F_{64} \right) P_{case13all} \cdot P_{within1} + P_{case13all} \left(1 - P_{within1} \right) \tag{2.6.161}$$

$$Q_{14} = \left(F_{65} \right) P_{case14all} \cdot P_{within1} + P_{case14all} \left(1 - P_{within1} \right) \tag{2.6.162}$$

$$Q_{15} = \left(F_{66} \right) P_{case15all} \cdot P_{within1} + P_{case15all} \left(1 - P_{within1} \right) \tag{2.6.163}$$

$$T_2 = \sum_{n=1}^{15} P_{case(n)all} \tag{2.6.164}$$

$$P_0 = 1 - \frac{\sum_{n=1}^{15} Q_n}{T_2} = 0.17835168856 \tag{2.6.165}$$

Finally, we take the weighted average of the two parts and arrive at the final habitability of the quaternary system at 33.34%.

$$C_4 = 1 - \frac{\left(\sum_{n=1}^{15} Q_n + \sum_{n=0}^{4} Q_n\right)}{T_2 + T_1} = 0.333400924927 \tag{2.6.166}$$

2.6.4   Habitability of All Systems

Once we have derived the probability for the ternary and the quaternary system, one starts to see a pattern. All higher order systems composed of systems of binary, ternary, and quaternary systems. To compute a given higher order systems, one has to enumerate all possible combinations (symmetry guarantees all possible permutation) of the subsystems within the entire system and taking the weighted average of habitability of all combinations to yield the final habitability of the given system.

Multiple System	Combinations	Multiple System	Combinations
2	2	6	5+1
3	2+1		4+2
4	3+1		3+3
	2+2	7	6+1
5	4+1		5+2
	3+2		4+3

Table 2.6.1: Higher systems' combinations

For any system with n stars, The habitability for the combination of $(n\text{-}1)+1$ is the simplest. It can be expressed by the formula:

$$j_1(n) = d \cdot (1 - P_{far}) + \frac{n-1}{n} \cdot d \cdot P_{far} \tag{2.6.167}$$

$$j_2(n) = (1 - P_{within1}) \cdot (1 - P_{far}) + \frac{n-1}{n} \cdot d \cdot P_{far} \tag{2.6.168}$$

$$C_{case1}(n) = 1 - \frac{1}{2}(j_1 + j_2) \tag{2.6.169}$$

Whereas d is the over-count on the habitability of n-1 star system. $j_1(n)$ is the case of the probability of the over-count with n-1 star system at the center. It is further divided into sub cases of $(1 - P_{far})$ and P_{far}. For the additional star circling around n-1 system and falling within the 0.712~1 solar mass range, the over-count is reduced by a factor of $\frac{n-1}{n}$. $j_2(n)$ is the case of the probability of the over-count with a single star at the center circled by n-1 star system. The single central star is over-counted by $(1 - P_{within1})$, equals to the percentage of stars with mass > 1 solar mass. For quintary system's case of 4+1, we have:

$$C_{case1}(5) = 0.49477982603 \tag{2.6.170}$$

For other cases, one has to work case by case. For quintary system's case of 3+2, one first tabulate the possible configuration as the follows: (we assumed that 2+3 leads to similar results and is nearly symmetric. one can think of 2+3 as a binary weighted by a ternary system)

	2nd (avg)	3rd	Gap	Permitted	Outer Binary Pair	Description
	< 0.1812 AU	< 33 AU	< 133 AU	< 100 AU	< 25 AU	mostly dead
p_1	< 0.1812 AU	33 ~ 100 AU	133 ~ 400 AU	100 ~ 300 AU	25 ~ 75 AU	dead
	< 0.1812 AU	100 ~ 200 AU	400 ~ 800 AU	300 ~ 600 AU	75 ~ 150 AU	mostly dead
p_2	< 0.1812 AU	> 200 AU	> 800 AU	> 600 AU	> 150 AU	habitable
p_3	12.6 AU	50 ~ 100 AU	200 ~ 400 AU	150 ~ 300 AU	37.5 ~ 75 AU	dead
	12.6 AU	100 ~ 200 AU	400 ~ 800 AU	300 ~ 600 AU	75 ~ 150 AU	mostly dead
p_4	12.6 AU	> 200 AU	> 800 AU	> 600 AU	> 150 AU	
p_5	62.5 AU	> 250 AU	> 1000 AU	> 750 AU	> 187.5 AU	habitable
p_6	150 AU	> 600 AU	> 2400 AU	> 1800 AU	> 450 AU	
p_7	> 200 AU	> 800 AU	> 3200 AU	> 2700 AU	> 675 AU	

Table 2.6.2: Case for 3+2

Notice that each layer is assumed to be 4 times farther in distance than the immediate inner layers. This number can be easily expanded > 4 times. However, increasingly greater separation leads to smaller chance of existence in the first place, so it is justified to assume each layer is kept stable by minimum stable distances and the calculation is done as follows:

Whereas D is defined as:

$$D = \int_{-\infty}^{\infty} D(x)\, dx \tag{2.6.171}$$

$$p_1 = S_1 \cdot \int_{\log(0.1626\cdot4)}^{\log(200)} D(x)\, dx \cdot \int_{\log\left(\frac{200+0.1626}{2}\cdot4\right)}^{\infty} D(x)\, dx \cdot \left(\frac{1}{D}\right)^2 \tag{2.6.172}$$

$$p_2 = S_1 \cdot \int_{\log(200)}^{\infty} D(x)\, dx \cdot \int_{\log(200\cdot4)}^{\infty} D(x)\, dx \cdot \left(\frac{1}{D}\right)^2 \tag{2.6.173}$$

$$p_3 = S_3 \cdot \int_{\log(50)}^{\log(200)} D(x)\, dx \cdot \int_{\log\left(\frac{50+200}{2}\cdot4\right)}^{\log(200\cdot4)} D(x)\, dx \cdot \left(\frac{1}{D}\right)^2 \tag{2.6.174}$$

$$p_4 = S_3 \cdot \int_{\log(200)}^{\infty} D(x)\, dx \cdot \int_{\log(200\cdot4)}^{\infty} D(x)\, dx \cdot \left(\frac{1}{D}\right)^2 \tag{2.6.175}$$

$$p_5 = S_4 \cdot \int_{\log(250)}^{\infty} D(x)\, dx \cdot \int_{\log(250\cdot4)}^{\infty} D(x)\, dx \cdot \left(\frac{1}{D}\right)^2 \tag{2.6.176}$$

$$p_6 = S_5 \cdot \int_{\log(600)}^{\infty} D(x)\, dx \cdot \int_{\log(600\cdot4)}^{\infty} D(x)\, dx \cdot \left(\frac{1}{D}\right)^2 \tag{2.6.177}$$

$$p_7 = S_6 \cdot \int_{\log(800)}^{\infty} D(x)\, dx \cdot \int_{\log(800\cdot4)}^{\infty} D(x)\, dx \cdot \left(\frac{1}{D}\right)^2 \tag{2.6.178}$$

$$T_1 = p_1 + p_2 + p_3 + p_4 + p_5 + p_6 + p_7 \tag{2.6.179}$$

$$C_{ase2} = \left(\frac{p_1 + p_3}{T_1}\right)\left(\frac{3}{5}C_3 + 0\right) + \left(\frac{p_2 + p_4 + p_5 + p_6 + p_7}{T_1}\right)\left(\frac{3}{5}C_3 + \frac{2}{5}C_2\right) \tag{2.6.180}$$

Whereas C_2 is the habitability of all binary systems and C_3 is the habitability of all ternary systems as we have obtained before.

The total habitability of quintary system is given by combination cases (4+1) and (3+2).

$$C_5 = C_{ase1} \cdot \frac{T_0}{T_0 + T_1} + C_{ase2} \cdot \frac{T_1}{T_0 + T_1} = 0.48971 \tag{2.6.181}$$

However, this result is only partially correct. T_0 actually stands for the probability of 3+1 case for the inner 4 star system. For case 1 there actually exists two cases of inner 4 star system (2+2 and 3+1). We could

70

refine our computation by finding the probability of (2+2)+1. However, later steps will become unnecessarily complicated. To simplify computation, we simply assume that each of the cases has an equal chance of existence. Inner ternary system occurs more frequently than any cases of quaternaries. However, ternary is surrounded by binaries which is rarer and less habitable than single circling star. Therefore, we modify our equation to:

$$C_5 = C_{ase1} \cdot \left(\frac{2}{2+1} \right) + C_{ase2} \cdot \left(\frac{1}{2+1} \right) = 0.43056 \tag{2.6.182}$$

For sextuple system's case of 5+1, we have:

$$C_{ase1}(6) = 0.546113755908 \tag{2.6.183}$$

For sextuple system's case of 4+2, one first tabulate the possible configuration as the follows:

	Pair 1	Pair 2	Gap	Permitted	Outer Binary Pair	Description
s_{11}	< 0.1812 AU	< 0.1812 AU	> 3.44 AU	> 2.58 AU	> 0.65 AU	dead
s_{31}	12.6 AU	< 0.1812 AU	> 200 AU	> 150 AU	> 37.5 AU	mostly dead
s_{33}	12.6 AU	0.1812 ~ 25 AU	> 250 AU	> 187.5 AU	> 46.88 AU	mostly dead
s_{41}	62.5 AU	< 0.1812 AU	> 1000 AU	> 750 AU	> 187.5 AU	
s_{43}	62.5 AU	0.1812 ~ 25 AU	> 1048 AU	> 786 AU	> 196.5 AU	
s_{44}	62.5 AU	25 ~ 100 AU	> 1248 AU	> 936 AU	> 234 AU	
s_{51}	100 ~ 200 AU	< 0.1812 AU	> 2400 AU	> 1800 AU	> 450 AU	
s_{53}	100 ~ 200 AU	0.1812 ~ 25 AU				
s_{54}	100 ~ 200 AU	25 ~ 100 AU				habitable
s_{55}	100 ~ 200 AU	100 ~ 200 AU	> 3000 AU	> 2250 AU	> 562 AU	
s_{61}	> 200 AU	< 0.1812 AU	> 3200 AU	> 2400 AU	> 600 AU	
s_{63}	> 200 AU	0.1812 ~ 25 AU				
s_{64}	> 200 AU	25 ~ 100 AU				
s_{65}	> 200 AU	100 ~ 200 AU				
s_{66}	> 200 AU	> 200 AU	> 4000 AU	> 3000 AU	> 750 AU	

Table 2.6.3: Case for 4+2

To compute for each case, we use:

$$s_{11} = S_1 S_1 \left(\int_{\log\left(\frac{0.1812}{2}\cdot 4 + \frac{0.1812}{2}\right)}^{\infty} D\left(x\right) dx \right) \cdot \left(\int_{\log(0.453\cdot 4)}^{\infty} D\left(x\right) dx \right) \cdot \left(\frac{1}{D}\right)^2 \tag{2.6.184}$$

$$s_{31} = S_3 S_1 \left(\int_{\log(50)}^{\infty} D\left(x\right) dx \right) \cdot \left(\int_{\log(50\cdot 4)}^{\infty} D\left(x\right) dx \right) \cdot \left(\frac{1}{D}\right)^2 \tag{2.6.185}$$

$$s_{33} = S_3 S_3 \left(\int_{\log(62.5)}^{\infty} D\left(x\right) dx \right) \cdot \left(\int_{\log(62.5\cdot 4)}^{\infty} D\left(x\right) dx \right) \cdot \left(\frac{1}{D}\right)^2 \tag{2.6.186}$$

$$s_{41} = S_4 S_1 \left(\int_{\log(250)}^{\infty} D\left(x\right) dx \right) \cdot \left(\int_{\log(250\cdot 4)}^{\infty} D\left(x\right) dx \right) \cdot \left(\frac{1}{D}\right)^2 \tag{2.6.187}$$

$$s_{43} = S_4 S_3 \left(\int_{\log(262)}^{\infty} D\left(x\right) dx \right) \cdot \left(\int_{\log(262\cdot 4)}^{\infty} D\left(x\right) dx \right) \cdot \left(\frac{1}{D}\right)^2 \tag{2.6.188}$$

$$s_{44} = S_4 S_4 \left(\int_{\log(312)}^{\infty} D\left(x\right) dx \right) \cdot \left(\int_{\log(312\cdot 4)}^{\infty} D\left(x\right) dx \right) \cdot \left(\frac{1}{D}\right)^2 \tag{2.6.189}$$

$$s_{51} = S_5 S_1 \left(\int_{\log(600)}^{\infty} D\left(x\right) dx \right) \cdot \left(\int_{\log(600\cdot 4)}^{\infty} D\left(x\right) dx \right) \cdot \left(\frac{1}{D}\right)^2 \tag{2.6.190}$$

$$s_{53} = S_5 S_3 \left(\int_{\log(612)}^{\infty} D\left(x\right) dx \right) \cdot \left(\int_{\log(612\cdot 4)}^{\infty} D\left(x\right) dx \right) \cdot \left(\frac{1}{D}\right)^2 \tag{2.6.191}$$

$$s_{54} = S_5 S_4 \left(\int_{\log(662)}^{\infty} D\left(x\right) dx \right) \cdot \left(\int_{\log(662\cdot 4)}^{\infty} D\left(x\right) dx \right) \cdot \left(\frac{1}{D}\right)^2 \tag{2.6.192}$$

$$s_{55} = S_5 S_5 \left(\int_{\log(750)}^{\infty} D\left(x\right) dx \right) \cdot \left(\int_{\log(750\cdot 4)}^{\infty} D\left(x\right) dx \right) \cdot \left(\frac{1}{D}\right)^2 \tag{2.6.193}$$

$$s_{61} = S_6 S_1 \left(\int_{\log(800)}^{\infty} D\left(x\right) dx \right) \cdot \left(\int_{\log(800\cdot 4)}^{\infty} D\left(x\right) dx \right) \cdot \left(\frac{1}{D}\right)^2 \tag{2.6.194}$$

$$s_{63} = S_6 S_3 \left(\int_{\log(812)}^{\infty} D\left(x\right) dx \right) \cdot \left(\int_{\log(812\cdot 4)}^{\infty} D\left(x\right) dx \right) \cdot \left(\frac{1}{D}\right)^2 \tag{2.6.195}$$

$$s_{64} = S_6 S_4 \left(\int_{\log(860)}^{\infty} D\left(x\right) dx \right) \cdot \left(\int_{\log(860\cdot 4)}^{\infty} D\left(x\right) dx \right) \cdot \left(\frac{1}{D}\right)^2 \tag{2.6.196}$$

$$s_{65} = S_6 S_5 \left(\int_{\log(950)}^{\infty} D\left(x\right) dx \right) \cdot \left(\int_{\log(950\cdot 4)}^{\infty} D\left(x\right) dx \right) \cdot \left(\frac{1}{D}\right)^2 \tag{2.6.197}$$

$$s_{66} = S_6 S_6 \left(\int_{\log(1000)}^{\infty} D\left(x\right) dx \right) \cdot \left(\int_{\log(1000\cdot 4)}^{\infty} D\left(x\right) dx \right) \cdot \left(\frac{1}{D}\right)^2 \tag{2.6.198}$$

$$T_2 = s_{11} + s_{31} + ... + s_{65} + s_{66} \tag{2.6.199}$$

$$C_{ase2} = \frac{(s_{11} + s_{31} + s_{33})}{T_2}\left(\frac{4}{6}C_4 + 0\right) + \frac{(s_{41} + ... + s_{66})}{T_2}\left(\frac{4}{6}C_4 + \frac{2}{6}C_2\right) \tag{2.6.200}$$

Whereas C_2 is the habitability of all binary systems and C_4 is the habitability of all quaternary systems as we have obtained before.

For sextuple system's case of 3+3, one first tabulate the possible configuration as the follows:

Pair (avg)	3rd	Gap	Permitted	3rd	Pair	Description
< 0.1812 AU	< 33 AU	< 133 AU	< 100 AU	25 AU	6.25 AU	dead
p_1 < 0.1812 AU	33 ~ 100 AU	133 ~ 400 AU	100 ~ 300 AU	25 ~ 75 AU	6~18 AU	dead
< 0.1812 AU	100 ~ 200 AU	400 ~ 800 AU	300 ~ 600 AU	75 ~ 150 AU	18 ~ 37 AU	mostly dead
p_2 < 0.1812 AU	> 200 AU	> 800 AU	> 600 AU	> 150 AU	> 37.5 AU	partially dead
p_3 12.6 AU	50 ~ 100 AU	200 ~ 400 AU	150 ~ 300 AU	37.5 ~ 75 AU	9.3 ~ 18.7 AU	dead
12.6 AU	100 ~ 200 AU	400 ~ 800 AU	300 ~ 600 AU	75 ~ 150 AU	18 ~ 37.5 AU	mostly dead
p_4 12.6 AU	> 200 AU	> 800 AU	> 600 AU	> 150 AU	> 37.5 AU	partially dead
p_5 62.5 AU	> 250 AU	> 1000 AU	> 750 AU	> 187.5 AU	> 46.875 AU	partially dead
p_6 150 AU	> 600 AU	> 2400 AU	> 1800 AU	> 450 AU	> 112.5 AU	mostly habitable
p_7 > 200 AU	> 800 AU	> 3200 AU	> 2700 AU	> 675 AU	> 168.75 AU	habitable

Table 2.6.4: Case for 3+3

To compute for each case, we use:

$$p_1 = S_1 \cdot \left(\int_{\log\left(\frac{0.1812}{2}\cdot 4\right)}^{\log(200)} D(x)\, dx \right) \cdot \left(\int_{\log\left(\frac{200+0.1812}{2}\cdot 4\right)}^{\infty} D(x)\, dx \right) \cdot \left(\frac{1}{D}\right)^2$$

$$p_2 = S_1 \cdot \left(\int_{\log(200)}^{\log(500)} D(x)\, dx \right) \cdot \left(\int_{\log(200\cdot 4)}^{\infty} D(x)\, dx \right) \cdot \left(\frac{1}{D}\right)^2$$

$$p_3 = S_3 \cdot \left(\int_{\log(50)}^{\log(200)} D(x)\, dx \right) \cdot \left(\int_{\log\left(\frac{50+200}{2}\cdot 4\right)}^{\log(200\cdot 4)} D(x)\, dx \right) \cdot \left(\frac{1}{D}\right)^2$$

$$p_4 = S_3 \cdot \left(\int_{\log(200)}^{\infty} D(x)\, dx \right) \cdot \left(\int_{\log(200\cdot 4)}^{\infty} D(x)\, dx \right) \cdot \left(\frac{1}{D}\right)^2$$

$$p_5 = S_4 \cdot \left(\int_{\log(250)}^{\log(550)} D(x)\, dx \right) \cdot \left(\int_{\log(250\cdot 4)}^{\infty} D(x)\, dx \right) \cdot \left(\frac{1}{D}\right)^2$$

$$p_6 = S_5 \cdot \left(\int_{\log(600)}^{\log(1100)} D(x)\, dx \right) \cdot \left(\int_{\log(600\cdot 4)}^{\infty} D(x)\, dx \right) \cdot \left(\frac{1}{D}\right)^2$$

$$p_7 = S_6 \cdot \left(\int_{\log(800)}^{\infty} D(x)\, dx \right) \cdot \left(\int_{\log(800\cdot 4)}^{\infty} D(x)\, dx \right) \cdot \left(\frac{1}{D}\right)^2$$

$$T_3 = p_1 + p_2 + p_3 + p_4 + p_5 + p_6 + p_7 \tag{2.6.201}$$

$$C_{ase3} = \left(\frac{p_1 + p_3}{T_3}\right)\left(\frac{3}{6}C_3 + 0\right) + \left(\frac{p_2 + p_4}{T_3}\right)\left(\frac{3}{6}C_3 + \frac{3}{6}P_1\right) +$$
$$\left(\frac{p_7}{T_3}\right)\left(\frac{3}{6}C_3 + \frac{3}{6}C_3\right) + \left(\frac{p_5}{T_3}\right)\left(\frac{3}{6}C_3 + \frac{3}{6}P_1\right) + \left(\frac{p_6}{T_3}\right)\left(\frac{3}{6}C_3 + \frac{3}{6}P_3\right) = 0.26457 \tag{2.6.202}$$

Whereas C_3 is the habitability of all ternary systems and P_1 and P_3 is the habitability of restricted ternary system ranges. The total habitability of quintary system is given by combination case (5+1) and (4+2) and (3+3):

$$C_{ase1}\left(\frac{2}{2+2+1}\right) + C_{ase2}\left(\frac{2}{2+2+1}\right) + C_{ase3}\left(\frac{1}{2+2+1}\right) = 0.38569 \tag{2.6.203}$$

Based on the results of pair of 1+1, 2+2, and 3+3, one can extrapolate habitability of pair n+n:

$$y_{n+n} = 0.0908867x^{0.972583} \tag{2.6.204}$$

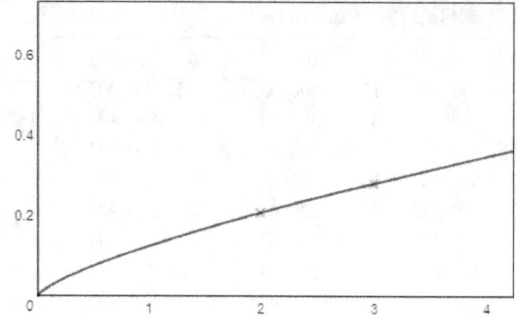

Figure 2.6.40: 2+2, 3+3, to n+n extrapolation

Based on calculation on these results, one can approximate the higher order systems by:

7 star system:

$$C_7 \approx \frac{3}{7} \cdot C_{ase1}(7) + \frac{2}{7} \cdot \left(\frac{5}{7} C_5 + \frac{2}{7} C_2 \right) + \frac{2}{7} \cdot \left(\frac{4}{7} C_4 + \frac{3}{7} C_3 \right) \qquad (2.6.205)$$

8 star system:

$$C_8 \approx \frac{3}{10} \cdot C_{ase1}(8) + \frac{3}{10} \cdot \left(\frac{6}{8} C_6 + \frac{2}{8} C_2 \right) + \frac{2}{10} \cdot \left(\frac{5}{8} C_5 + \frac{3}{8} C_3 \right) + \frac{2}{10} \cdot C_{ase}(4+4) \qquad (2.6.206)$$

9 star system:

$$C_9 \approx \frac{4}{12} \cdot C_{ase1}(9) + \frac{3}{12} \cdot \left(\frac{7}{9} C_7 + \frac{2}{9} C_2 \right) + \frac{3}{12} \cdot \left(\frac{6}{9} C_6 + \frac{3}{9} C_3 \right) + \frac{2}{12} \cdot \left(\frac{5}{9} C_5 + \frac{4}{9} C_4 \right) \qquad (2.6.207)$$

10 star system:

$$C_{10} \approx \frac{4}{16} \cdot C_{ase1}(10) + \frac{4}{16} \cdot \left(\frac{8}{10} C_8 + \frac{2}{10} C_2 \right) + \frac{3}{16} \cdot \left(\frac{7}{10} C_7 + \frac{3}{10} C_3 \right) +$$
$$\frac{3}{16} \cdot \left(\frac{6}{10} C_6 + \frac{4}{10} C_4 \right) + \frac{2}{16} \cdot C_{ase}(5+5) \qquad (2.6.208)$$

11 star system:

$$C_{11} \approx \frac{5}{19} \cdot C_{ase1}(11) + \frac{4}{19} \cdot \left(\frac{9}{11} C_9 + \frac{2}{11} C_2 \right) + \frac{4}{19} \cdot \left(\frac{8}{11} C_8 + \frac{3}{11} C_3 \right) +$$
$$\frac{3}{19} \cdot \left(\frac{7}{11} C_7 + \frac{4}{11} C_4 \right) + \frac{3}{19} \cdot \left(\frac{6}{11} C_6 + \frac{5}{11} C_5 \right) \qquad (2.6.209)$$

There is a general trend of increasing habitability as the number of stars in the system goes up as more stars circling around a group of inner stars in stable orbits. As the system hosting number of stars increases, all additional stars orbiting around an inner system can be treated, conceptually, as separate life hosting stars or systems. For higher order system with an additional binary, ternary surrounds the inner system, the outermost pair form stable orbit beyond the 200 AU limit for planetary habitability for both the circling and the circled. As a result, all additional system orbiting around inner system can be treated, conceptually, as separate cases of star systems which would be already computed in our earlier calculation. The final results is derived from the weighted average of the habitability of the circled and the habitability of the circling. Consequently, the probability of hosting habitable planets gradually increases. The exception occurs for even numbered systems such as 4, 6, 8 stars. For such system, the case for the habitability of (2+2), (3+3), and (4+4) is always lower since the circling system is as complex as the circled. As a result, more dead zone can form around such case for circling system. A system with greater than 11 stars are possible but are not counted toward the final calculation because the results do not alter the conclusion to several digits of precision. Based on the computed table, *one can see that 35.95% of all multiple systems are habitable.*

Multiple System	Percentage	Habitability
2*	75.00%	36.10%
3*	15.00%	34.58%
4*	6.00%	33.34%
5	2.40%	≈43.06%
6	0.96%	≈38.57%
7	0.38%	≈44.59%
8	0.15%	≈43.53%
9	0.06%	≈45.29%
10	0.02%	≈44.95%
11	0.01%	≈45.69%
Total		**35.95%**

Table 2.6.5: Multiple stars systems habitability break down

2.7   Habitability of Low Mass Binaries

There remains the case where binary pairs with lower masses can match and provide enough heating for planets in their habitable zone and avoiding tidally lock at the same time with their combined luminosity. Multiple star system can be simply ruled out because all stable multiple systems come in a hierarchy, that is, a stable pair of binary revolves around another pair of binaries or single stars and their separation have to be at least 3 times the separation distance from the binary pairs they revolve. In essence, the habitable zone is an applicable definition only to single or binaries. Multiple star system can host habitable planet, but it must be revolving around one of its subunit consisting of a single or a binary star.

Since luminosity increases to the 4.5th power of the stellar mass between 0.43 to 2 solar mass, luminosity drops to the inverse of 4.5th power as the stellar mass decreases. This indicates that only stars from 0.6643 solar mass to 0.712 solar mass have enough mass to shed enough luminosity and when combined in pairs (assuming the secondary has a mass 80% of the primary) gives enough heat that provides a tidal locking free habitable zone.

$$j_{lumiositybinary} = (0.8 \cdot x)^{4.5} + x^{4.5} \tag{2.7.1}$$

$$j_{lumiosity} = x^{4.5} \tag{2.7.2}$$

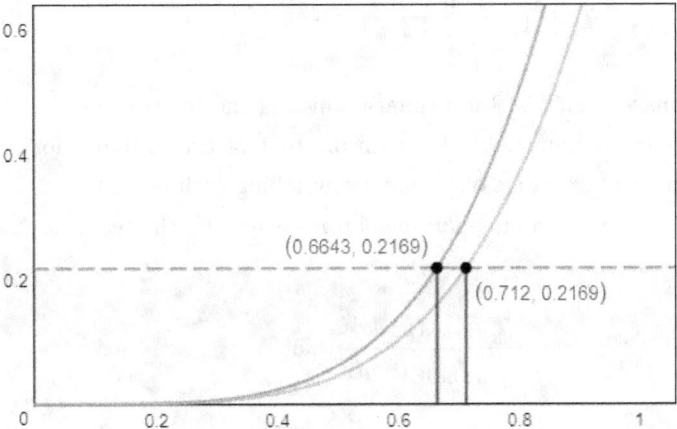

Figure 2.7.1: The minimum binary stars' primary stellar mass threshold requirement for the pair's combined luminosity output matches the luminosity of the smallest single star

Stars with 0.712 solar mass or greater alone, without pairs, can provide a tidal locking free habitable zone and it is already counted toward our total number of habitable planets. However, by stricter definition, the chance of two equal mass closely paired binaries are rare. In fact, the primary to the secondary mass ratio in closely paired binaries do not follow the stellar mass power law distribution. There is an observed little spike in the probability distribution for secondary mass and the q number closely matching 1, but in general, the secondary companion can have all possible mass from 1 primary mass to 0.1 primary mass. An approximate distribution is given below (as stated earlier):

$$f_{nearpeak}(x) = \frac{0.012}{Q\sqrt{2\pi}} \exp\left(-\frac{\ln(-x+2)^2}{2(\sigma)^2}\right) \tag{2.7.3}$$

$$f_{near}(x) = (x^{0.1} - 0.1) + f_{nearpeak}(x) \tag{2.7.4}$$

$$\sigma = 0.05 \tag{2.7.5}$$

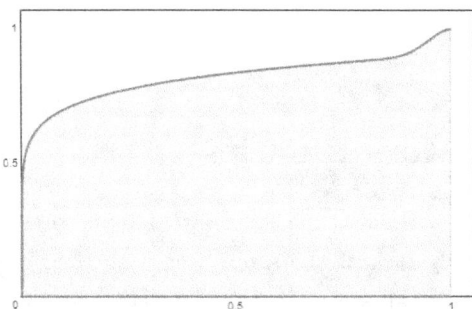

Figure 2.7.2: The PDF function of the secondary stellar mass to the primary stellar mass ratio for tight orbiting binaries

Now, the median mass of the secondary is 0.54 of the primary, the weighted average expected mass of the pairs is then 1.54, shy of the twice of the primary mass.

$$\frac{\int_{0.54}^{1} f_{near}(x)\, dx}{\int_{0}^{1} f_{near}(x)\, dx} = 50\% \tag{2.7.6}$$

Out of tightly orbiting binaries, only 45.156% of which has companions of mass 58.69% or greater relative to the primary.

$$\frac{\int_{0.5869}^{1} f_{near}(x)\, dx}{\int_{0}^{1} f_{near}(x)\, dx} = 0.451556 \tag{2.7.7}$$

and their companions median mass is 80% of the primary which is the benchmark we have set earlier. (The benchmark for companion mass can be any value between 0.1 to 1 of the primary, for a selected companion mass, there is a corresponding percentage range of secondary matching such benchmark. The overall results are similar but we do find that by setting the companion median mass to 80% the result is slightly maximized)

$$\frac{\int_{0.8}^{1} f_{near}(x)\, dx}{\int_{0.5869}^{1} f_{near}(x)\, dx} = 50\% \tag{2.7.8}$$

Based on this assumption, only stars from 0.6643 solar mass to 0.712 solar mass fit our criteria.

Based on the spectral class and the initial mass function, 1.465% of the stars range from 0.6643 to 0.712 solar mass. Out of these stars, 40% of which are binaries or multiples. Out of the binaries and multiples, one needs to find those in tight orbits that allowed a tidal locking free habitable zone. It is difficult to compute the total luminosity of two sources of light when they are separated by a distance. However, we can approximately treat those two sources of light as a single point light source since we are only interested in closely paired binaries. In order for a planet to be in a stable orbit around a binary pair the separation distance between the planet and binary pair has to be at least 3 times or greater (preferentially 5 times or greater) than the distance between the binary pairs. Therefore, if we treat the midpoint between the two closely paired binaries as the source of the emitting light, then the margin of error of total energy received at the stable orbit boundary ranges from 16% to 4.94% and the margin of error decreases as the planet revolves in orbit further away from the pair. Then, the separation distance between the pairs has to be approximately a third of the distance from the binary to the edge of the habitable zone. With pair separation greater than a third of binary's habitable edge distance, all stable orbits lie beyond the outer edge of the habitable zone. With pair separation less than a third of binary habitable edge distance, some stable orbits lie beyond the outer edge of the habitable zone, and one orbit lies on the habitable zone and some lies beyond the inner edge of the habitable zone.

$$j_{habitablebinary} = \sqrt{(0.8 \cdot x)^{4.5} + x^{4.5}} \tag{2.7.9}$$

$$d_{binarypairseparation} = \frac{j_{habitablebinary}}{3} \tag{2.7.10}$$

$$R_{stable} = 3 \cdot d_{binarypairseparation} \tag{2.7.11}$$

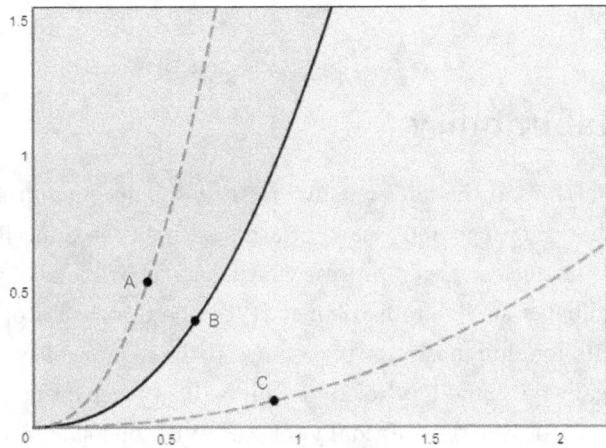

Figure 2.7.3: Case scenario for curve A denotes the pair separation as large as its combined habitable zone radius, as a result, the minimum stable orbit ventures beyond the edge of habitability into the cold side. Case scenario for curve B denotes the pair separation at an exactly one third of the their habitable zone so that the minimum stable orbit of habitable planet falls on the habitable zone. Case scenario for curve C denotes the pair separation at an exact one tenth of the their habitable zone so that the minimum stable orbit ventures beyond the edge of habitability into the hot side but allows the habitable planet to form beyond the minimum stable orbit. The red shaded portion indicates all possible minimum stable orbit placement inside the habitable zone depending on the pair separation. The blue shaded portion indicates all possible minimum stable orbit placement outside the habitable zone.

The habitable zone radius ranges from 0.4657 AU to 0.5443 AU (evaluating on curve B in the graph above) between the binary star and its habitable zone for binary primary mass ranges from $x = 0.6643$ to 0.712 solar mass. We run recursive simulation for binary star's tidal locking time and find that only planet within the habitable zone of binary stars in which the primary with 0.67 ~ 0.712 solar mass achieves spin rate of < 7 days after 4.5 Gyr. This accounts for 1.29% of all stars (6.70×10^9 stars) by computing its total percentage out of the initial mass function:

$$\frac{\int_{.67}^{.712} I_{mf}(x)\, dx}{\int_{0.08}^{\infty} I_{mf}(x)\, dx} = 0.01290 \qquad (2.7.12)$$

The average habitable zone radius for 0.67 to 0.712 solar mass so that the minimum orbit can fall on the habitable zone is $\frac{0.4747+0.5443}{2} = 0.5095$. then the probability of binaries with a separation at or less than $\frac{0.5095}{4} = 0.127375$ AU is 2.27%. (based on statistical data on the probabilistic distribution of binaries of comparable mass)

$$\frac{\int_{\log(0.001)}^{\log(0.130625)} D(x)\, dx}{\int_{-\infty}^{\infty} D(x)\, dx} = 0.022711 \qquad (2.7.13)$$

Out of this total number, 40% of all stars are binaries or multiples and out of those 75% are binaries. Only 2.27% out of which are binaries in an orbit tight enough so that the habitable planet can form in its habitable zone. Out of tightly orbiting binaries, only 44.766% of which has companions of mass 50% or greater relative to the primary to provide enough combined light output to warm the planet. Finally, one takes only those stars formed between 5 Gya and 4 Gya.

$$\frac{\int_{9.199-0.5}^{9.199+0.5} S_{tellar}(x)\, dx}{\int_{0}^{13.799} S_{tellar}(x)\, dx} \times 6.70 \times 10^9 \times (0.40 \times 0.75) \times 0.451556 \times 0.022711 = 987,155 \qquad (2.7.14)$$

This added up to a total of 987,155 extra habitable exoplanets to our list for time period between 5 Gya and 4 Gya regardless of metallicity. With metallicity taking into consideration, there are only 243,867 extra exoplanets to be added to the total.

2.8   Red Dwarves' Habitability

Red dwarves are the most abundant stars in the universe, due to their smaller mass, their absolute luminosity are a small fraction compares to that of orange and yellow dwarves such as the Sun. Red dwarves are known for their long stability once it starts its nuclear fusion process which lasts for trillions of years. As a result, they form a large pool of potential candidates for habitable planets.[121] The drawbacks of the Red dwarf system are also self-evident. Because of its low luminosity, a planet has to be significantly closer to the host star compares to the Sun in order to gain the same level of radiation as it is received on Earth. As a result, all potentially habitable planets around red dwarves are tidally locked. With one side of the planet permanently facing daylight and the other perpetually stares into the darkness, the temperature difference becomes extreme. Prolonged exposure to radiation on the day side brings extreme temperature variations between the day and night side. This is easily verified by seasonal changes on earth, where even a few extra hours of sunlight during the summer month significantly increases the temperature and fewer hours of sunlight during the winter month significantly increases frigid cold storm events. Many have argued that life can be harsh and probably restricted to the dim light zone sandwiched between the hot inferno and the cold dead world, offering very limited adaptive radiation opportunity by the local fauna if any exists at all. Many argue that liquid water may not be sustainable because the water completely evaporates on the day side and condenses into ice on the night side. Some have argued that an atmosphere which is dense enough can distribute the heat more uniformly throughout the planet. Others argued that an atmosphere may be maintained if the red dwarf does not follow a circular orbit around its host star. Eccentric orbit creates tidal heating which in turn generates a magnetic field strong enough to protect the planetary atmosphere from blowing away.[2] Another serious consequence is the loss of its magnetic field. Venus has a similar mass to earth, and its slow rotation and its lack of internal thermal convection, any liquid metallic portion of its core could not be rotating fast enough to generate a measurable global magnetic field. Without a magnetosphere, Venus with an atmosphere comparable to the thickness of earth with a composition of O_2 and N_2 can deplete. A thick atmosphere can reduce depletion loss, as observed on Venus, through its ionosphere.[109] The ionosphere separates the atmosphere from the outer space and the solar wind. This ionized layer excludes the solar magnetic field, giving Venus a distinct magnetic environment. Maintaining such a dense atmosphere seem to give life a chance despite a lack of magnetic field. Even hypothetically a dense O_2 atmosphere can be maintained, it will be extremely flammable, and secondly, life adapted to such dense atmosphere will evolve with higher similarity to aquatic adaptation than terrestrial adaptation. It is known that no opposable thumb and bipedalism is observed in any aquatic species, reducing the chance of the emergence of intelligent tool-using species.

All these assumptions can be valid even if they are not the universal representation of the reality on all terrestrial planets formed around red dwarves. What really separates red dwarves from their more massive cousins are their strong magnetic fields.[1] According to the standard dynamo theory, the magnitude of a magnetic field generated by a heavenly body is proportional to its temperature, its convecting mass, and its rotational rate, and the empirical law can be used to extrapolate the magnetic field strength of different heavenly bodies by its size. The sun, though 10 times greater in mass than Red dwarves, 99 percent of its mass is condensed into such a high density that the heat is transferred through conduction and radiation. The remaining 1% upper layer of the sun, separated from the sun's core by the Tachocline, composed of plasma, is where the convection taking place. Red dwarves with mass smaller than 0.36 solar mass are fully convective. As a result, they produce a magnetic field with strength hundreds of times stronger than that of the Sun. Furthermore, any habitable planets around red dwarves are hundred times closer to their host star than that of the earth. The field strength of magnetic field decreases as the inverse of the cubed of the separation distance from the magnetic field. As a result, the stellar magnetic field strength around such terrestrial planet is a million times stronger than the stellar magnetic field strength observed around the earth, which is around 10^{-9} Tesla. This implies that the stellar magnetic field strength is at 10^{-3} Tesla. If the planet does have its own magnetic field to shield itself from the stellar field to protect its atmosphere like in earth's case, its own magnetic field has to be about three magnitudes stronger, which implies the planet magnetic field strength has to reach 1 Tesla. Studies have shown that large organism such as human cannot tolerate magnetic field strength of 1 Tesla for too long, this has been demonstrated by clinical studies done on patients undergoing MRI scans. This is especially true when one is exposed to a changing field. Organisms can hardly survive in such strong field because all animals move around frequently, as it moves fast, they are subject to moving magnetic fields.

On the other hand, it seems unlikely an Earth-like planet can produce a magnetic field with such strong strength. It is more likely that the planet is vulnerably exposed to the onslaught of the stellar field. Even if the atmosphere is dense enough to be maintained, two major problems arise. First of all, all stars periodically enter active periods. In sun's case, periodic appearance of sun spots and flares which increases the stellar magnetic strength by three orders of magnitude, this is observed on Earth in 1858 during the Carrington Event. Since the field strength can vary from time to time and a solar storm can last from hours to days, a changing field with strength around 1 Tesla inflicts significant damage on organisms even if they assume a sedentary lifestyle. Furthermore, a planet without a magnetic field can no longer divert cosmic rays and radiation particles from reaching the surface of the planet. From earth's polar data, where solar winds strike at the poles, the radiation level reaches 15 msvert, which in a dosage-dependent manner, can render organism sterilized and is lethal to continual exposure for more than three hundred days. Furthermore, in an extremely long stretched imaginative scenario, somehow an extremely radiation resisting organism emerges on such a planet, it may never able to utilize telecommunication technology given the magnetic storm bombardment on a daily basis. In conclusion, every other thing being equal, organisms cannot survive on red dwarves' planets due to the presence of strong magnetic field created by their host star.

To fully appreciate the strength and power of red dwarves' magnetic fields, the detailed calculation is performed. The measured magnetic field strength of a given location is directly proportional to the total mass of the conducting fluid, its temperature, its speed of rotation in radians, and is proportional to the inverse cubed of its distance from the generating dynamo.

$$S = \frac{M \cdot T \cdot \omega}{r^3} \tag{2.8.1}$$

This equation is an approximation and indirect manifestation of the magnetic induction:

$$\frac{\partial \mathbf{B}}{\partial t} = \eta \nabla^2 \mathbf{B} + \nabla \times (\mathbf{u} \times \mathbf{B}) \tag{2.8.2}$$

Whereas $\eta \nabla^2 \mathbf{B}$ describes the magnetic field line placement pattern and flux density. Assuming flux density is proportional to physical density, volume multiplied physical density yields mass and volume multiplies flux density equals total field strength, then the total field strength $\sum \frac{\partial \mathbf{B}}{\partial t}$ is directly proportional to the mass of the planet. For the second term, the ∇ operator simply describes how such magnetic field is rotated. We assume that all planet shares similar pattern of rotation. The term $\mathbf{u}$ describes the velocity of the fluid, which is directly proportional to its internal temperature T and ω, the rotational speed.

We shall start by using the field strength of earth as a reference. [75][67]The Earth, like other planets in the Solar System, as well as the Sun and other stars, all generate magnetic fields through the motion of electrically conducting fluids. The Earth's field originates in its core. This is a region of iron alloys extending to about 3,400 km. It is divided into a solid inner core, with a radius of 1,220 km, and a liquid outer core. The motion of the liquid in the outer core is driven by heat flow from the inner core, which is about 6,000 K, to the core-mantle boundary, which is about 3,800 K. The average temperature is at 4,900 K, and the temperature gradient function for the core is modeled as:

$$T_{earth}(r) = 1.16 \times 10^3 \left(-8.19 \left(6.37 - r \right)^{-1.175} + 5.868 \right) \tag{2.8.3}$$

The average density of outer core is 11.5 $\frac{\text{kg}}{\text{m}^3}$, and the density gradient function is modeled by:

$$D_{earth}(r) = -0.1433 \left(6.37r \right)^{2.251} + 12.25 \tag{2.8.4}$$

The total mass of the outer core amounts to 0.311 Earth mass. The total strength of the magnetic field can be expressed as:

$$\mathbf{B}_{earth} = 1 \times 4\pi \cdot \left(R_{earth} \right)^3 \int_{\frac{1.347}{6.37}}^{\frac{3.518}{6.37}} D_{earth}(r) \, r^2 \times T_{earth}(6.37r) \, dr \tag{2.8.5}$$

The value is multiplied by a factor of 1, indicating a rotational speed of 1,674.4 $\frac{\text{km}}{\text{h}}$. Every other planet's rotation rate in radians will be re-scaled relative to earth's. Finally, the final value is expressed in terms of 1 earth mass. This value is further reduced by 56 folds at the surface of the planet as the strength of magnetic field decreases from the generating dynamo toward the surface based on existing literature. The distance from the generating dynamo to the surface of the planet is 2,970 km. Therefore, the magnetic field strength at the surface of the planet amounts to 26, which is comparable to 25 ~ 65 microtesla (0.25 ~ 0.65 gauss), as it is observed.

$$u_{earth} = \frac{\mathbf{B}_{earth}}{10^3 \mathbf{M}_{earth}} \cdot \frac{1}{56} = 26 \tag{2.8.6}$$

The sun's magnetic field is generated by the fluid convection generated by the convection zone. The convection zone extends from 0.7 solar radii to 0.9992820136 radii. There is a 500 km deep photosphere covering above the convection zone. For approximation purpose, one can consider the convection zone extends from 0.7 solar radii to 1 solar radii. The density of the convection zone ranges from 0.2 $\frac{\text{g}}{\text{cm}^3}$ at the bottom to 0.2 $\frac{\text{g}}{\text{m}^3}$ (about $\frac{1}{6,000}$th the density of air at sea level), a total drop by a million fold. The density profile for radius > 0.7 solar radii is modeled as:

$$D_{solar}(x) = 8.181 \times 10^{-7} r^{-34.785} \tag{2.8.7}$$

The temperature rises from 5,700 K (9,100K) at the surface to 1.5 million K at the base. The temperature profile for radius > 0.7 solar radii is modeled as:

$$T_{solar}(r) = 0.0057 \times 10^6 r^{-15.624} \tag{2.8.8}$$

The total strength of the magnetic field can be expressed as:

$$\mathbf{B}_{sun} = \frac{1}{29.89} \times 4\pi \cdot (R_{sun})^3 \int_{0.7}^{U_{pper}} D_{solar}(r) r^2 \times T_{solar}(r) dr \tag{2.8.9}$$

Which is equivalent to the total mass of convection zone times an average temperature of 1 million K:

$$\mathbf{B}_{sun} = \frac{1}{29.89} \times 4\pi \cdot (R_{sun})^3 \int_{0.7}^{U_{pper}} D_{solar}(r) r^2 dr \times 10^6 \tag{2.8.10}$$

1 million K is the temperature at 0.718 solar radii, the result is justified since half of the convection zone's mass lies between 0.7 and 0.718 radii due to fast dropping density.

$$\frac{4\pi \cdot (R_{sun})^3 \int_{0.7}^{0.718} D_{solar}(r) r^2 dr}{4\pi \cdot (R_{sun})^3 \int_{0.7}^{1} D_{solar}(r) r^2 dr} = \frac{1}{2} \tag{2.8.11}$$

The field strength multiplied by a factor of $\frac{1}{29.89}$ (the rotation speed of sun is 4.293 times faster than earth but translated into radians per second it is only $\frac{1}{30}$ th of earth). The upper reaches of the dynamo can varies depending on the final values one wants to match. The field strength ranges from 1~2 gauss for polar field to 3,000 gauss in sunspots. In either case, the dynamo grows weak and terminates before reaching the surface of the sun. We have already shown, for earth's case, 2,970 km separation from the field decreases the field strength by 56 times. Therefore, we model the field reduction factor for the sun by the following equation with U_{pper} as a variable:

$$F_{upper} = \left(\frac{(1 - U_{pper}) \cdot R_{sun}}{2970} \right)^3 56 \tag{2.8.12}$$

The final field strength in earth mass unit is given by:

$$u_{sun} = \frac{\mathbf{B}_{sun}}{10^3 M_{earth} \times F_{upper}} \tag{2.8.13}$$

The upper reaches of the dynamo is given as:

U_{pper}	Observed	Remainder Mass	Remaining Percent
0.911~0.929	1~2 gauss	0.312~0.159	0.0206%
0.9938	3,000 gauss	0.00366	0.00024%

The polar field strength indicates that the termination boundary of the dynamo lies much deeper inside the sun at the poles than at the sun spot. Since only 0.323 earth mass of material is concentrated beyond 0.911 solar radii due to extremely low density, which is only 0.021% out of a total of 1517 earth mass of convection material $\frac{4\pi \cdot (R_{sun})^3 \int_{0.7}^{1} D_{solar}(r) r^2 dr}{10^3 M_{earth}} = 1517$ between 0.7 to 1 solar radii, the assumption that the dynamo terminates deep inside the sun is justified.

For Jupiter, we find that its convection zone, made of metallic hydrogen, is able to conduct magnetic field sitting above the core and below the liquid and gas hydrogen layers above. No definitive density profile for Jupiter exists, we fine tuned our density profile for the entire planet by satisfying the constraints of a core density of $15 \frac{g}{cm^3}$ and a surface density of $0.01 \frac{g}{cm^3}$ and the integration of the function over radius 0 to 1 must match 317.8 earth mass:

$$D_{jupiter}(r) = \begin{cases} 25 & r < 0.04 \\ 25\,(r+0.955)^{-11.5842r} & r \geq 0.04 \end{cases} \tag{2.8.14}$$

$$\frac{4\pi \cdot (R_{jupiter})^3 \int_0^1 D_{jupiter}(r)\,r^2 dr}{10^3 \mathrm{M_{earth}}} = 317.8 \tag{2.8.15}$$

The average temperature of this region amounts to 23,000 K. The temperature gradient is given as:

$$T_{jupiter}(r) = \begin{cases} 36000 & r < 0.25 \\ 7560r^{-1.12577} & 0.25 \leq r \leq 0.75 \\ 112r^{-18.0786} & 0.75 \leq r \leq 1 \end{cases} \tag{2.8.16}$$

The total strength of the magnetic field can be expressed as:

$$\mathbf{B}_{jupiter} = \frac{24}{9.925} \times 4\pi \cdot (R_{jupiter})^3 \int_{0.178}^{U_{pper}} D_{jupter}(r)\,r^2 \times T_{jupiter}(r)\,dr \tag{2.8.17}$$

The rotational speed of 45,000 $\frac{\mathrm{km}}{\mathrm{h}}$, is 26.88 times faster than earth, translated into $\frac{24}{9.925}$ times the speed of the earth in terms of radians per second. No existing literature discusses the definitive boundary between the core and metallic hydrogen. The lower termination of the dynamo is computed based on the determination that the core must be between 12 to 45 earth mass, or 4%–14% Jupiter mass. We take the mean value of 29 earth mass and 9% Jupiter mass so that the constraints is satisfied as:

$$C_{ore} = 4\pi \cdot (R_{jupiter})^3 \int_0^{0.178} D_{jupter}(r)\,r^2 dr \tag{2.8.18}$$

$$\frac{C_{ore}}{10^3 \cdot M_{jupiter}} = 9\% \tag{2.8.19}$$

$$\frac{C_{ore}}{10^3 \cdot M_{earth}} = 29.36 \tag{2.8.20}$$

Therefore, the core terminates at 0.178 Jupiter radii.

$$F_{upper} = \left(\frac{(1-U_{pper}) \cdot R_{jupiter}}{2970} \right)^3 56 \tag{2.8.21}$$

$$u_{jupiter} = \frac{\mathbf{B}_{jupiter}}{10^3 \mathrm{M_{earth}} \times F_{upper}} \tag{2.8.22}$$

The upper termination radius of the dynamo ranges from 0.632 Jupiter radii given 4.2 gauss at the equator and 0.7506 Jupiter radii given 14 gauss at the poles. Now we can determine metallic hydrogen lies between 0.178 to 0.78 radii with a total of 281.21 earth mass. In either case, just like in the case of the sun, only a negligible amount of convection mass is concentrated beyond the upper termination of the dynamo, so our results are justified.

U_{pper}	Observed	Remainder Mass	Remaining Percent
0.632	4.2 gauss	21.824	7.791%
0.7506	10~14 guass	2.5292	0.897%

For Saturn, we find that its convection zone is also made of metallic hydrogen. No definitive density profile exists, we fine tuned our density profile for the entire planet by satisfying the constraints of a core density of

$13\frac{\text{g}}{\text{cm}^3}$ and a surface density of $0.01\frac{\text{g}}{\text{cm}^3}$ and the integration of the function over radius 0 to 1 must match 95 earth mass:

$$D_{saturn}(r) = \begin{cases} 13 & r < 0.04 \\ 13(r + 0.955)^{-11.5842r} & r \geq 0.04 \end{cases} \tag{2.8.23}$$

$$\frac{4\pi \cdot (R_{saturn})^3 \int_0^1 D_{saturn}(r)\, r^2 dr}{10^3 M_{earth}} = 95.374 \tag{2.8.24}$$

The average temperature of this region amounts to 7,475 K. The temperature gradient is given as:

$$T_{saturn}(r) = \begin{cases} 11700 & r < 0.23 \\ 84r^{-3.35892} & 0.23 \leq r \leq 1 \end{cases} \tag{2.8.25}$$

The total strength of the magnetic field can be expressed as:

$$\mathbf{B}_{saturn} = \frac{24}{10.55} \times 4\pi \cdot (R_{saturn})^3 \int_{0.26}^{0.66} D_{saturn}(r)\, r^2 \times T_{saturn}(r)\, dr \tag{2.8.26}$$

The rotational speed of $35{,}500\ \frac{\text{km}}{\text{h}}$, is 21.2 times faster than earth, or $\frac{24}{10.55}$ times the rotational speed of the earth in terms of radians per second. The lower termination of the dynamo is computed based on the determination that the core must be between 9 to 22 earth mass. We take the max value of 22 earth mass so that the constraints is satisfied as:

$$C_{ore} = 4\pi \cdot (R_{saturn})^3 \int_0^{0.26} D_{saturn}(r)\, r^2 dr \tag{2.8.27}$$

$$\frac{C_{ore}}{10^3 \cdot M_{earth}} = 22 \tag{2.8.28}$$

Therefore, the core terminates at 0.26 Saturn radii.

$$u_{saturn} = \frac{\mathbf{B}_{saturn}}{10^3 M_{earth} \times F_{upper}} \tag{2.8.29}$$

$$F_{upper} = \left(\frac{(1 - U_{pper}) \cdot R_{saturn}}{2970} \right)^3 56 \tag{2.8.30}$$

The upper termination radius of the dynamo is set at 0.66 Saturn radii given 0.2 gauss at the surface. Assuming metallic hydrogen lies between 0.26 to 0.99 radii with a total of 72.515 earth mass. Just like earlier cases, only a negligible amount of convection mass is concentrated beyond the upper termination of the dynamo, so our results are justified.

U_{pper}	Observed	Remainder Mass	Remaining Percent
0.66	0.2 gauss	< 6.71	< 9.25%

The dynamo terminates before the boundary of convection layer based on existing literature for Jupiter and Saturn can be explained in many ways. It is possible that metallic hydrogen's conductivity is lower than its equivalent at a much higher temperature in the plasma state, which is not accounted by our existing model and assumption. It is also possible that the dynamo generating effect does not start at all layers of metallic hydrogen, which is what we assumed for the adjustment. It is more likely to generate from a section of the layer. The adjustment requires further analysis, the validity of the model is not jeopardized because by fine-tuning the parameters within a reasonable range of error of tolerance, the observed and computed values do match.

We then apply our model to that of the red dwarf stars. The model composed the extent of the semi-major axis of planets for stars of different masses. In general, the smaller the star, the less material is required to create one. As a result, the smaller semi-major axis for the stars' hosted planets. The habitable zone model illustrates how does the region where liquid water can form shifts for stars of different mass. Luckily and surprisingly, those two curves coincide fairly well with each other at every possible stellar mass. That is, we can say that if Spectral G class like the sun can host a planet within its habitable zone, so can the rest of the stars regardless of its spectral class.

Then, we model the interplanetary magnetic field strength as observed in the solar system. From the existing literature, it states that the surface of the sun's polar field is around 0.0002 T, but significantly higher at the sunspots and solar prominences. One takes the geometric mean between the polar field and sunspot and chose a value of $\sqrt{1.5 \cdot 3000} = 67$ gauss, or 0.0067 T. This field strength decreases to the inverse cubed from its distance away from the surface. If space were a vacuum, one should observe that the field strength at the earth, theoretically, drops to $1 \cdot 10^{-11}$ Tesla. However, satellite observations show that it is about 100 times greater at around 10^{-9} Tesla.[67] Magneto hydrodynamic (MHD) theory predicts that the motion of a conducting fluid (the interplanetary medium) in a magnetic field, induces electric currents which in turn generates magnetic fields, and in this respect, it behaves like a MHD dynamo. Based on observation, we model our field strength as it decreases away from the source with the following equation:

$$h(d) = \frac{1}{(d+1)^{21}} \qquad (2.8.31)$$

So that when x = 1, representing the field strength at 1 AU, the field strength is only $\frac{1 \cdot 10^{-9}\,\text{T}}{0.0067\,\text{T}} = \frac{1}{6708000}$ of its original value as it is observed on earth.

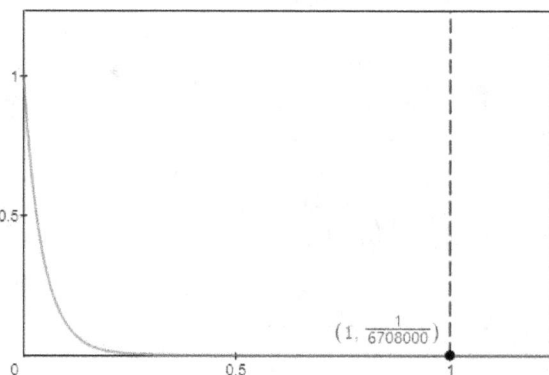

Figure 2.8.1: Magnetic field strength vs distance

Furthermore, the stellar habitable zone with the presence of liquid water shifts ever closer to the star as the stellar mass and luminosity decreases. As a result, the field strength experienced by a habitable planet within the habitable zone of varying stellar mass is:

$$h(x) = \frac{1}{\left(\sqrt{x^{3.5}}+1\right)^{21}} \qquad (2.8.32)$$

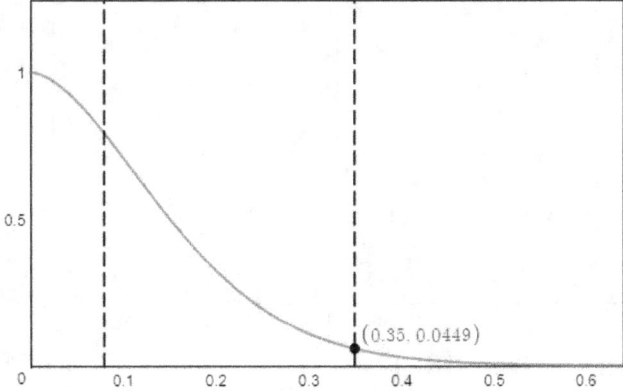

Figure 2.8.2: Field strength experienced by a habitable planet within the habitable zone of varying stellar mass

However, recent studies has introduced a mass-luminosity relationship in which the exponent varies by mass of the star as follows:

$$L(x) = -141.7x^4 + 232.4x^3 - 129.1x^2 + 33.29x + 0.215 \tag{2.8.33}$$

so we can substitute the former equation with:

$$h(x) = \frac{1}{\left(\sqrt{x^{L(x)}} + 1\right)^{21}} \tag{2.8.34}$$

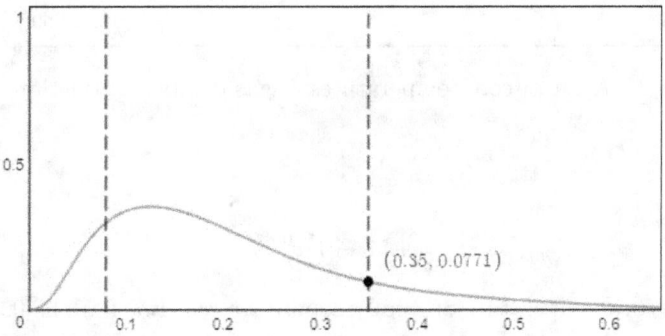

Figure 2.8.3: Field strength experienced by a habitable planet within the habitable zone of varying stellar mass

Lastly, we use our earlier model to predict each celestial sphere's generating magnetic field strength to predict the magnetic field strength of red dwarves. Since red dwarves, those with less than 0.35 solar mass are fully convective. Its entire mass is used to generate a dynamo, at 0.35 solar mass, there is 116,550 earth masses are used to generate the magnetic field inside the star instead of 1512 earth mass as observed for the sun. We use the following equation to estimate the percentage of the star's mass composed of convective layer:

$$y_{scale}(x) = \frac{1}{1 + \exp(20(x - 0.565))} + \frac{1512}{333000} \tag{2.8.35}$$

Anything below 0.35 solar mass is fully convective, and the proportion of convective layer drops to negligible percentage at 0.55 solar mass. At solar mass, it is only $\frac{1512}{333000}$ of the total mass of the star, as it is expected.

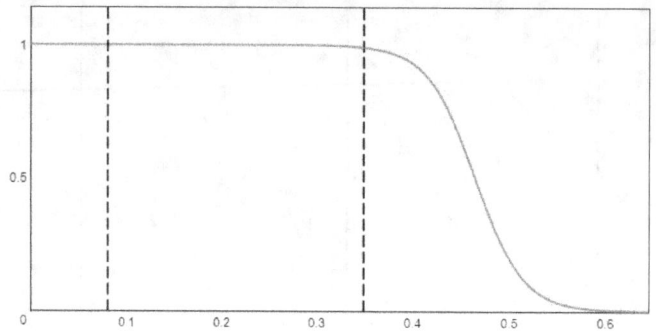

Figure 2.8.4: Convective layer proportion vs stellar mass

The other parameter one has to consider is the internal temperature of the star. Because currently there is a lack of data regarding the internal temperature of other star systems, we have to guesstimate. We find the best curve fit for the core temperature of the star and the surface temperature of the star as a function of its mass. We use the data values for the gas giants Uranus, Neptune, Saturn, Jupiter's core temperature as the lower end and the Sun as the upper end to constraint our curve fiting for the core temperature.

Body Name	Mass Relative to the Sun	Core Temperature in K
Uranus	$\frac{14.536}{333000}$	5,000
Neptune	$\frac{17.147}{333000}$	5,400
Saturn	$\frac{95.159}{333000}$	11,700
Jupiter	$\frac{317.8}{333000}$	35,700
Sun	1	15,700,000

Table 2.8.1: A list of core temperatures of gas giants and the Sun

The best fit curve is:

$$T_{empcore}(x) = 1.57 \cdot 10^7 x^{0.874326} \tag{2.8.36}$$

The curve predicts the internal core temperature of a fully convective red dwarf at the 0.35 solar mass should be comparable to the average temperature of the radiation zone of the sun, at no less than 6.27 million K, and this value decreases as the stellar mass decreases.

Next, we model the surface temperature of stars by different mass:

Body Name	Mass Relative to the Sun	Surface Temperature in K
Uranus	$\frac{14.536}{333000}$	72
Neptune	$\frac{17.147}{333000}$	76
Saturn	$\frac{95.159}{333000}$	134
Jupiter	$\frac{317.8}{333000}$	165
WISE 1828+2650	$\frac{4.5\cdot317}{333000}$	325
0.08 Sun	0.08	2400
0.45 Sun	0.45	3700
0.8 Sun	0.8	5200
1.04 Sun	1.04	6000
1.4 Sun	1.4	7500
2.1 Sun	2.1	10000
16 Sun	1	30,000

Table 2.8.2: A list of surface temperatures of gas giants and stars

$$T_{empsurface}(x) = 6248.39x^{0.565981} \tag{2.8.37}$$

The curve predicts the surface temperature of a fully convective red dwarf at the 0.35 solar mass should be 3449 K, and this value decreases as the stellar mass decreases but at a slower pace than the core temperature. The surface temperature of stars decreases slower than their cores because smaller stars have partial to fully convective internal structures. This is in a sharp contrast to the sun, whereas 99% of sun's nuclear fusion occurred at its non-convective core, and the core temperature is 15.7 million K while the surface temperature is merely 6000 K. A fully convective red dwarf with a smaller internal temperature gradient is more efficient at energy dissipation. A red dwarf at 0.35 solar mass' highest temperature within its convective layer is 6.27 million K and lowest temperature is 3449 K.

Now, we can use both equation to derive the average temperature of the convective layer. Recall that we found for the sun we can treat its temperature of its convection layer as 1 million K while the temperature ranges from 1.5 million K to 6000 K. We found that such value can be achieved by:

$$T_{emp}(x) = T_{empsurface}(x) \left(\frac{T_{empcore}(x)}{T_{empsurface}(x)} \right)^{0.65} \tag{2.8.38}$$

So that $T_{emp}(1) = 10^6$. We model the internal temperature variation of red dwarves as a function of its mass and decreases sublinearly relative to the convective layer's temperature of the sun and represented as a ratio relative to the sun's.

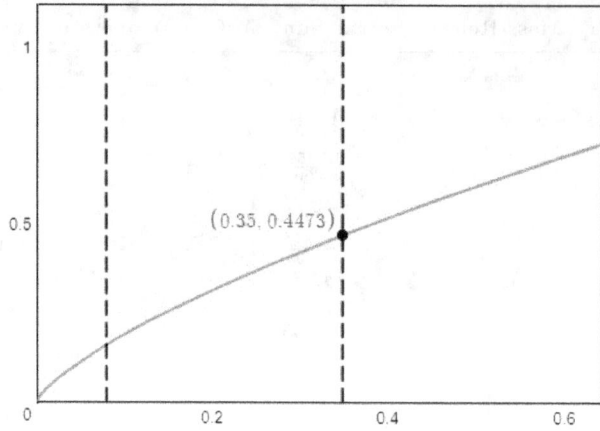

Figure 2.8.5: Convective layer temperature vs stellar mass

Although luminosity decreases by the inverse of 3.5, the temperature drop for red dwarves are much more gentle. One can see that at 0.35 solar mass, the internal temperature of the star's convective layer is predicted at $0.4473 \cdot 10^6 = 447,271$ K.

Since the dynamo terminates before reaching the surface of the star, the strength of the field observed at the stellar surface varies depending on the distance between the termination boundary and stellar surface. The distance varies from 1-0.66=0.34 Saturn radii to 1-0.95=0.05 solar radii. In the case of the sun, only 0.505 earth mass of material lies beyond 0.95 solar radii. In the case of Saturn, nearly 6.17 earth mass of material lies beyond 0.66 Saturn radii. The sun's density ranges from $160\frac{g}{cm^3}$ at the core to $0.2\frac{g}{m^3}$ at the surface. Stars and planet with lower mass has a narrower density profile than the sun. The core density is lower but the surface density is higher than the sun. With shrinking mass and radius, the core density drops slower than the radius shrinkage up to 0.0367 solar mass. Beyond 0.0367 solar mass, a crossover takes place in which the core density drops faster than radius shrinkage. The radius and mass relationship is captured by the equation:

$$R(x) = \left(\frac{4\pi}{3}\right)^{\frac{1}{3}} \left(x \cdot \frac{3}{4\pi}\right)^{\frac{1}{3}} \tag{2.8.39}$$

The relationship between radius and core density is captured by sub-linear equation:

$$R(x)^{0.8212} \tag{2.8.40}$$

Not only stars with lower mass has a narrower range of density distribution than the sun, the range is compressed into a smaller radius, therefore, the density descend gradient from the core to the surface becomes larger. If we assume that dynamo within any star terminates when only 0.505 earth mass of material lies between the dynamo and the surface as observed for the sun, then, one finds that the termination boundary pretty much remain at 0.95 stellar radii despite radius size difference. This is possible since the surface density increases as the star mass and radius decreases as it is observed.

$$T_{ermination}(x) = 0.95x^{0.00521} \tag{2.8.41}$$

However, we know that this model does not fit cases for Jupiter and Saturn with termination boundary around 0.7. The boundary termination based on observation of Saturn, Jupiter, and the sun is given as:

$$T_{observed}(x) = 0.95x^{0.0448266} \tag{2.8.42}$$

There exists a trade off for dynamo generation. For the sun, temperature and heat budget is large enough to fuel a dynamo generation despite low density at its upper radius. For gas giants, there is insufficient heat to generate a dynamo beyond a threshold density. Therefore, a higher internal temperature enables dynamo generation within a medium of low density and low pressure. A low internal temperature enables dynamo generation only within a medium of high density and high pressure.

If one takes our convection layer temperature curve with $T_{ermination}(x)$ and raised to the power of $\frac{1}{18}$:

$$T_{observed}(x) = (T_{ermination}(x) \times T_{emp}(x))^{\frac{1}{18}} \tag{2.8.43}$$

The combined results can match our observation. This shows that temperature does play a role in determining the termination boundary. We will use $T_{observed}(x)$ along with modified version of F_{upper} to determine the field strength reduction required for varying stellar mass:

$$F_{upper}(x) = \left(\frac{(1 - T_{observed}(x)) \cdot R_{sun} \times R(x)}{2970} \right)^3 56 \tag{2.8.44}$$

Finally, we treat the rotation rate for all stars nearly the same or within the range of the error of tolerance. The final equation for magnetic field generation is:

$$\mathbf{B}_{star} = M_{sol} \cdot (x \cdot y_{scale}(x)) \times T_{emp}(x) \cdot \frac{1}{F_{upper}(x)} \times \frac{1}{29.89} \tag{2.8.45}$$

Then, we can almost immediately find that the strength of the magnetic field on a 0.35 solar mass red dwarf is $\frac{0.0919}{0.0067} = 13.72$ times stronger than the sun at its surface.

$$M_{sol} \cdot (0.35 \cdot y_{scale}(0.35)) \times T_{emp}(0.35) \cdot \frac{1}{F_{upper}(0.35)} \times \frac{1}{29.89} = 0.0919\,\text{T} \tag{2.8.46}$$

We combine the above equation with radius of habitability to yield the strength of the stellar magnetic field experienced by red dwarf's planets.

$$S_{magnetic} = \mathbf{B}_{star} \cdot h(x) \tag{2.8.47}$$

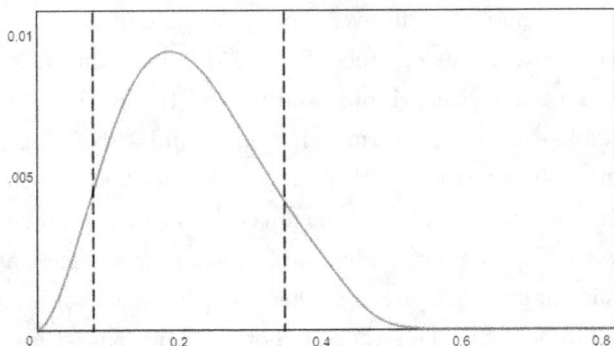

Figure 2.8.6: Magnetic field strength vs stellar mass

The magnetic field strength differs somewhat but non-significantly when the stellar habitable zone applies the newer mass-luminosity relationship:

89

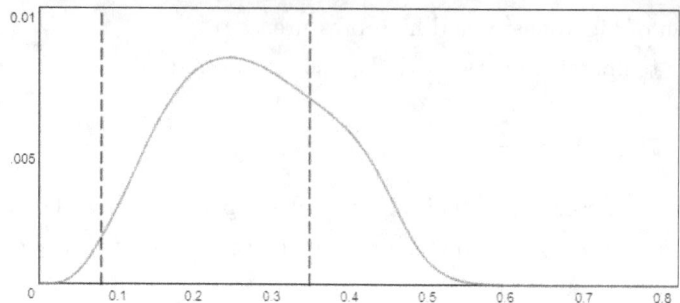

Figure 2.8.7: Magnetic field strength vs stellar mass

It can be shown, then, that the strength of the interplanetary magnetic field increases as the stellar mass decreases as the habitable zone has to move ever closer to the star. At 0.35 solar mass, the field strength is 0.0041~0.0071 Tesla (depending on the mass-luminosity relationship used), and it peaks at 0.0092 Tesla at 0.1824 solar mass, or 0.0083 Tesla at 0.248 solar mass. This is significantly higher than earth's magnetic field strength at $4.5 \cdot 10^{-5}$ Tesla. Since solar magnetic field around the earth is only $4.5 \cdot 10^{-9}$ Tesla (2.04 million times weaker than one observed on a habitable planet around a 0.35 solar mass red dwarf), the earth's magnetic field protects life from solar storm comfortably. However, this is not possible on any red dwarfs. It is impossible for a planet at the size of earth or slightly larger to generate a field not just 0.0083~0.0092 Tesla, but significantly larger by 3 orders of magnitude to shield away from the red dwarf's. As a result, the planet will experience the onslaught of the stellar storm on a daily basis. It is still possible that the atmosphere of the planet maintained as the case of Venus, but certainly, the radiation level can be 3 to 4 times higher than observed on the surface of the earth. It has been shown that space station at low earth orbit, inside the earth's magnetic shield, receives a dosage of 140 millisievert unit of radiation. In the interplanetary space, this value increases to 480 millisieverts. Life could adapt to be more resistant to radiation on such a planet.

However, a compressed magnetic field is not the only problem planets around red dwarf has to deal with. The deadliest is the stellar prominences and stellar flares. it is frequently observed that red dwarfs frequently increases its luminosity in a matter of days and months. In the case of the sun, even a much milder version compares to the red dwarves, creates a major storm every few years. Some notable ones recorded are the Carrington Event (occurred between August 28 to September 2nd, 1859), at the time field strength has increased to 1600 nT, which is 1600 times the average strength observed around the earth. The November 1882, May 1921 geomagnetic storm, March 1989 geomagnetic storm (with minimum Dst of -589 nT, or 589 times than normal.), July 14, 2000 event (with minimum Dst of -301 nT), and October 2003 storm (with a minimum Dst of -383 nT). Since on average, each storm was 2 orders of magnitude above the average strength, the strength of storm experienced on a planet within the red dwarf's habitable zone, even the least affected ones, will be at least 0.83~0.921 Tesla strong. This is not surprising because this result has been confirmed by observation. On April 23, 2014, NASA's Swift satellite detected the strongest, hottest, and longest-lasting sequence of stellar flares ever seen from a nearby red dwarf. The initial blast from this record-setting series of explosions was as much as 10,000 times more powerful than the largest solar flare ever recorded. In other words, a habitable planet will be subject to a field strength of 7,360 Tesla.

Studies have shown that large organism such as human cannot tolerate magnetic field strength of 1 Tesla or greater for more than a few hours before nausea symptoms occur, this has been demonstrated by clinical studies done on patients undergoing MRI scans. [81][59][13][14] This is especially true when one is exposed to a changing field. Organisms can hardly survive in such a strong field because all animals move around frequently, as it moves fast, they are subject to moving magnetic field and can generate electricity. At the same time, the field strength itself also shifts and fluctuates during the storm. Since an average storm each lasts for days at a time, unless organisms stop all activities during these times and hide under extremely thick layers of rocks, it is impossible for them to survive the onslaught of the storm.

2.9   Habitability of Exomoon

Having completed the tally for the number of Earth-like terrestrial planets around their host star and excluded the habitability of red dwarves, now we turn our attention to the number of habitable exomoons.

The existence of moons orbiting around planets within the solar system has been observed. It is, therefore, also hypothesized that such moons must be common in other stellar systems, though no conclusive evidence yet observed.[74]The most interesting or relevant to our assumption are those exo-moons comparable to the size of earth orbiting gas giants within the habitable zones of GFK spectral class stars. (in fact, the exo-moons have to be at least 0.4 earth mass or greater in order to hold enough oxygen concentration, and with mass less than 2 earth mass otherwise it would have enough gravity to accumulate hydrogen in its atmosphere). Studies have shown that earth massed exomoons are indeed possible, many of those can be captured by inward migrating gas giants from further out based on simulation. However, if such exo-moons pre-existed as earth analogs, then they are already counted into our existing habitable exoplanet count. What we are really interested is the chance of co-evolving exomoons with the host planet within its orbit. Since no observations are yet available and research under such topic is rare, we have to resort to an observation made within the solar system. We will make certain assumptions; that is, the total mass of all moons orbiting a host planet is proportional to the hosting planet's mass. We also assume the total mass of all moons as just 1 moon orbiting within the habitable zone of the hosting planet where the radiation is low enough to cause no harm to biological life and the moon is non-tidally locked. We know from our experience within the solar system that only Saturn's moon system approximately fulfills the above assumptions; nevertheless, we shall establish an upper bound on the number of exomoons habitable.

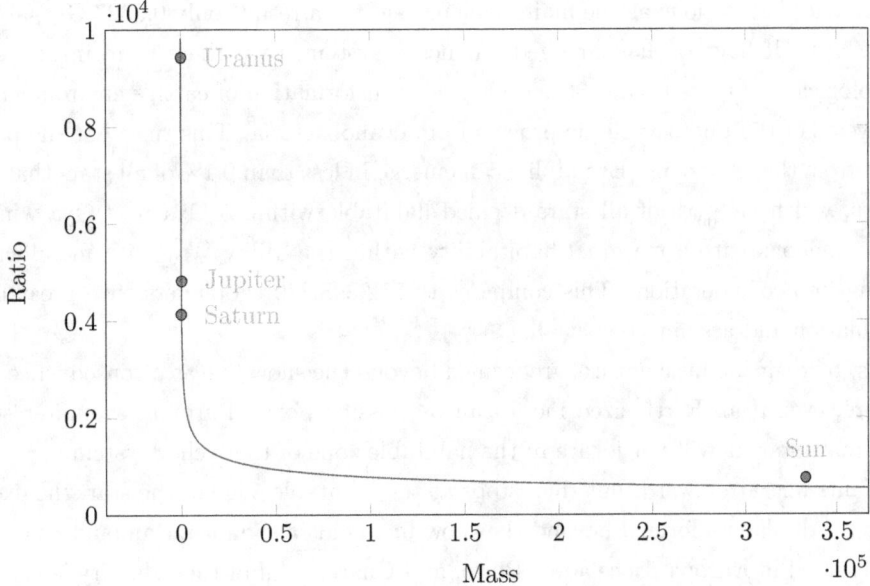

Figure 2.9.1: Primary to satellite mass ratio

The graph above is the ratio of the total mass of the Jovian Gas giants within the solar system to that of the total mass of moons orbiting them respectively. (except Neptune, whose moon Triton exhibits retrograde motion and is suspected to be captured from the Kuiper belt) And the Sun itself in relation to the total mass of the rest of the planets. It is noted that the lighter the planet, the lower the total mass of moons orbiting them. A linear regression is not possible under this model because it would imply that at some point the mass of the moons/planets is greater than that of the planet or the star itself. A power fit is plotted where the equation is obtained. The power fit is further refined by including the mass ratio of exoplanets and its hosting stars. This is called the stellar to planetary mass ratio.[115] The derived empirical law can predict the planetary mass budget or satellite mass budget for any given star mass or planet mass.

To solve for cases where 1 earth mass sized moon is possible, we use equation

$$y = \frac{x}{17,520 x^{-0.2315}} \qquad (2.9.1)$$

The solution indicates that only a planet with 2,793.87 earth masses or 8.78 Jovian masses (2,793.87 earth mass, 0.839% solar mass) or above can produce a satellite with a mass equivalent to earth. (see Chapter 11)

$$y = 0.00839 \cdot M_{sol} \left(17,520 \left(0.00839 \cdot M_{sol} \right)^{-0.2315} \right)^{-1} = 1.0011 \, \mathrm{M_{earth}} \qquad (2.9.2)$$

If we lower the constraints and allow the formation of a satellite with 0.4 earth mass, then we need 1,332 earth masses, or 4.189 Jovian mass to produce a satellite hospitable to life. Since the mass of Jupiter accounted for 70% of all the planetary mass within the solar system, we can assume that for a 4.189 Jovian mass planet to exist, we need at least 5.984 Jovian mass for any stellar system hosting such massive exo-moons. An increase in the total mass of planets also indicates an increase in the mass of the hosting star, as the power law indicates. As a result, we need a star at least 3.857 times the mass of the sun. Even if we take 4.189 Jovian mass as the total mass of the planets (there is only one Jovian planet in the stellar system), the hosting star still has to have a mass 2.888 times the mass of the sun.

Based on the amount of time a star stays on the main sequence:

$$T_{MS} \approx 10^{10} \left[\frac{M}{M_{sol}} \right] \left[\frac{L_{sol}}{L} \right] = 10^{10} \left[\frac{M}{M_{sol}} \right]^{-2.5} \qquad (2.9.3)$$

we know that such star will stay on main sequence 7.055% of the time compares to our sun. That means, that the stars stay on the main sequence for only 0.7055 Gyr. If the earth serves a model, the temporal habitable zone will likely stay stable for half as long as the main sequence age, as a result, only 0.3527 Gyr, about as long as late Devonian up to now. If bacteria has emerged on such a system, then there is an insufficient time to evolve into complex biological creatures because the geological transformation of earth's environment with free oxygen takes 2 billion years of the photosynthetic process from cyanobacteria. This rules out the possibility of an advanced life arising from the exomoons. Even if life can emerge in less than 0.1% of all stars that are greater than 2.888 solar masses, which is $\frac{1}{100}$th of all stars deemed habitable within 5 Gya to 4 Gya window, there are at most 6,800,163 exomoons with a marginal habitability within the Milky Way with metallicity and the temporal window taking into consideration. This compares to 612,398,339 exoplanets with great habitability does not alter our calculation and assumptions at all.

Furthermore, gas giants, by their intrinsic nature, are formed beyond the snow line as a consequence of runaway gas accretion. Therefore, even if an Earth-sized moon can be possibly formed around a Jupiter sized planet circling around a solar mass star, it will not locate in the habitable zone of the stellar system.

Secondly, if such gas giants migrate inward, but then stops at the habitable zone of the star, the exomoon will then be covered in water. All planets formed beyond the snow line holds a significant amount of water in their composition. This is observed in Europa, Enceladus, Pluto, and Charon. All of their density is close to 1 $\frac{g}{cm^3}$, the density of water, implying a significant percentage of its mass is composed of water. Using the generalized empirical law derived based on all stellar and exoplanets data up to date, the upper limit for water budget for the solar system is only $1.1342 \pm^{2.4524}_{0.7750}$ earth mass if we assume that oxygen is always counted toward the composition and accretion of terrestrial planet creation. (10,112 pairs of water molecule per 1 million atoms and if oxygen is counted toward the composition of terrestrial planet formation), 1.1342 earth mass is translated into 1 out of 98.89 of the planetary mass, or 44.5 times in proportion to earth's ocean (which is merely 1 out 4400 of earth's mass) if solar system's water is uniformly distributed. Since the water distribution of the solar system is non-uniform, all outer planets' moons get a greater share of water than the solar system average. As a result, the depth of the ocean must be indeed very high. It is possible to evolve life on such a planet but definitely not intelligent tool using species we are concerned with.

Finally, exomoons formed around gas giants through the accretion process have a semi-major axis well within 5 million km, as a result, all non-captured exomoons tidally lock to their parent gas giants by 0.537 Gyr, in which one side permanently faces the gas giant in a shade of moonlight at the best. The 1.723 factor is the calculated result for the accretion disc size growth for 4.189 Jovian mass planet compares to 1 Jovian mass gas giant.

$$T_{moon2jupter} = \frac{\left(\frac{1}{13750}\right) \cdot \left(1.7239760 \cdot 5 \cdot 10^6 \cdot 10^3\right)^6 \cdot I_{earth} \cdot M_{earth} \cdot 909.0909}{3G \left(1332 \cdot M_{earth}\right)^2 \cdot R_{earth}^3 \cdot 60 \cdot 60 \cdot 24 \cdot 365} \tag{2.9.4}$$

$$= 5.1662962 \cdot 10^8 \text{ years}$$

Therefore, we can confidently predict that all arising extra-terrestrial life originates from habitable terrestrial planets rather than any extra-terrestrial moons.

Very lastly, we can also rule out the rogue planets. Some of the rogue planets may host microbial life, but those lives can not transition to bacteria capable of photosynthesis. Without photosynthesis and its by-product waste oxygen, eukaryotes which based their energy extraction on oxygen and energy-consuming multi-cellular organisms cannot form. Even if it somehow succeeds in so in some unimaginable way, it is impossible for an intelligent, multicellular being to change its mode of living from hunter-gathering, scavenging, to an agricultural one, which depends on the influx of solar energy. It is also predicted that the majority of the rogue planets are ejected early during the formation phase of the stellar system, as a result, cast further doubt on its ability to host complex form of life let alone its maintenance of habitability of complex life after it is ejected from the parent star.

Chapter 3

Number of Earths

3.1 Orbital Eccentricity

The orbital eccentricity is another major selection criteria for the number of potentially habitable planets. For planets with higher orbital eccentricity, the planets can venture beyond its habitable zone so that liquid ocean freeze given a period of time of the year and creating snowball earth. It can also venture further closer to their sun and go through an annual period of unbearable heat, which generates an extreme level of humidity from evaporating ocean at the best and a runaway greenhouse at the worst. Though studies have been done to show that a terrestrial planet with extreme orbital eccentricity can still be habitable under the extraordinary circumstances. If a planet's eccentricity falls within the habitable zone we defined in chapter 2 between 0.840278 AU and 1.0887 AU, we will count them as habitable.

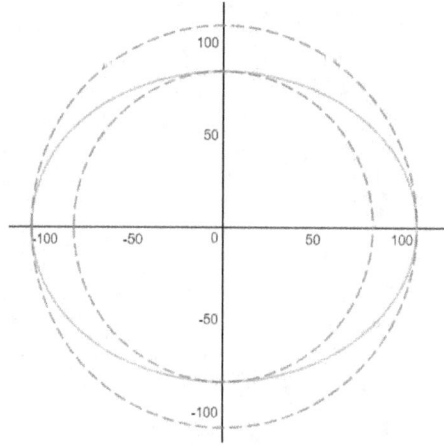

Figure 3.1.1: The inner and outer edge of the habitable zone

As a result, we exclude any planet with an orbital eccentricity greater than 0.6355 from the list of habitable planets.

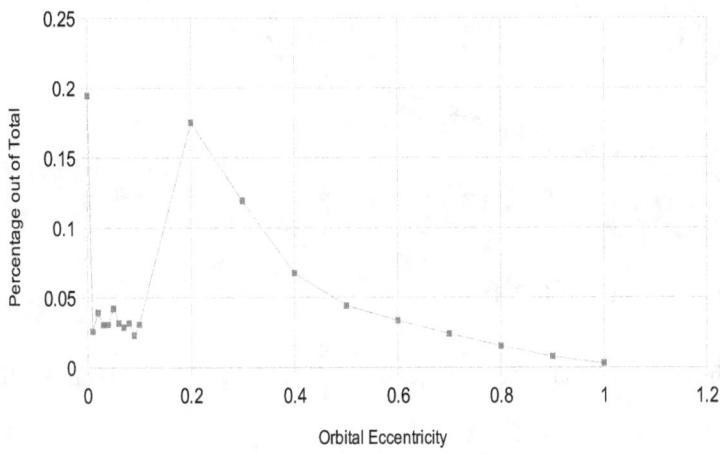

Figure 3.1.2: Percentage of Planets by Orbital Eccentricity

Based on 1,100 exoplanet's data points, *one finds that 95.76% of the exoplanet has eccentricity less than 0.6355.* The data is plotted below:

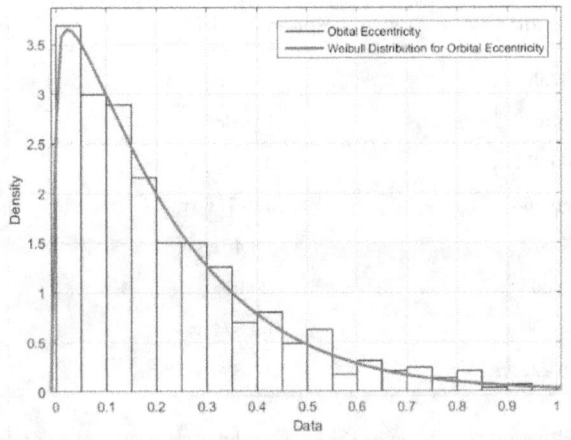

Figure 3.1.3: Weibull distribution for orbital eccentricity

3.2 Orbital Period

Some may also question the orbital period of the earth is unique that it revolves around the sun in 365 days. Planets revolve around the host star in faster or slower orbit may not be stable over the cosmic time scale, or it could be stable but not habitable for certain reason. However, orbital period vs. semi-major axis data derived from Kepler data indicates clearly that earth's orbital period is typical. In fact, plugging into the regression derived from thousands of exoplanets' orbital period yield a result of 356 days for the semi-major axis of 1 AU, which is just 9 days shorter than earth's value. To demonstrate that based on the physical characteristics of earth, the earth is typical as it is evolved from the proto-planetary disk, we have two important properties now observed from thousands of exoplanets. The orbital period vs. its semi-major axis and orbital eccentricity. The graph and its computed best fit power curve predicates that an exoplanet's semi-major axis with a distance of 1 AU has an orbital period of 356 days, closely match earth's value.

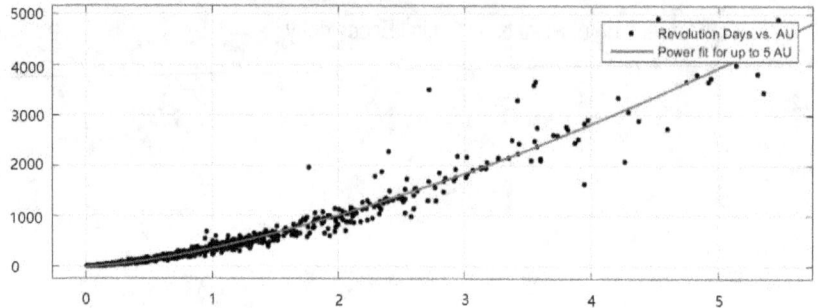

Figure 3.2.1: Orbital period vs. AU

To further instantiate our claim, we find that the orbital speed of all solar system bodies follows the following set of inequality:

[1]

$$v_{orbit} \approx \sqrt{\frac{GM}{r}} \leq v_{orbitalactual} < v_{escape} = \sqrt{\frac{2GM}{r}} \tag{3.2.1}$$

One finds that all solar system bodies' orbital period closely matches the orbit speed, that is, the minimum speed at which the body stays in a circular orbit. This is not a surprise because, after all, all bodies within the solar system has very low eccentricities.

Planet	Orbital Speed Period$\approx$	Escape Speed Period	Actual Period
Mercury	87.967 d	62.202 d	87.969 d
Venus	224.698 d	158.886 d	224.701 d
Earth	365.252 d	258.272 d	365.256 d
Mars	686.963 d	485.756 d	686.971 d
Jupiter	11.873 yr	8.401 yr	11.862 yr
Saturn	29.663 yr	20.975 yr	29.457 yr
Uranus	84.250 yr	59.574 yr	84.02q yr
Neptune	165.222 yr	116.830 yr	164.800 yr

Table 3.2.1: The orbital speed period, escape speed period, and actual period of solar system's planets

3.3   Earth & Moon Separation

3.3.1   Final Separation Distance

The moon is moving slowly away from the earth due to tidal locking. Earth is slowed down and consequently, the moment of inertia decreases. Because earth and the moon system is a closed system where its total energy and momentum is conserved, the decrease of the moment of inertia of earth is transferred to that of the moon, increasing its distance from earth and its angular momentum. The equation is stated as the following:

---

[1]In order for an object to remain in orbit there must be a centrifugal force balancing the gravitational force. It is expressed in the relationship in the orbital equation

$Fg = Fc$ where Fg is the gravitation force and Fc is the centrifugal force.

In other words, $\frac{GM_1 M_2}{r^2} = \frac{M_2 V^2}{r}$. Solving for v gives $v = \sqrt{\frac{GM_1}{r}}$

So, in order to maintain an orbit at a distance of r above the center of the Earth, the object must maintain an orbital speed of v given here.

$$s(x) = \frac{\left(0.3307 \cdot M_{earth} \cdot R_{earth}^2 \cdot \frac{2\pi}{24}\right) + \left(M_{moon}\sqrt{G \cdot M_{earth} \cdot a_{earthmoondist}}\right)^2}{M_{moon}^2 \cdot G \cdot M_{earth}} \tag{3.3.1}$$

This is true because if we rearrange the equation:

$$s \cdot M_{moon}^2 \cdot G \cdot M_{earth} = \left[\left(0.3307 \cdot M_{earth} \cdot R_{earth}^2 \cdot \frac{2\pi}{24}\right) + \left(M_{moon}\sqrt{G \cdot M_{earth} \cdot a_{earthmoondist}}\right)\right]^2 \tag{3.3.2}$$

and taking square root at the same time:

$$M_{moon}\sqrt{G \cdot M_{earth} \cdot s} = \left(0.3307 \cdot M_{earth} \cdot R_{earth}^2 \cdot \frac{2\pi}{24}\right) + \left(M_{moon}\sqrt{G \cdot M_{earth} \cdot a_{earthmoondist}}\right) \tag{3.3.3}$$

The left term is the total angular momentum of the earth and moon system relative to the axis of the center of mass of the earth and moon system when all remainder of earth's moment of inertia is transferred to the orbit of the moon and pushes it into a higher orbit until earth's rate of rotation synchronized with the moon's orbital period. The first term on the right-hand side is the moment inertia of earth at the current rotational speed multiplied by the moment of inertia. 0.3307 is the moment of inertia factor of the earth. Since earth's density is non-uniformly distributed within, the coefficient of the moment of inertia for a sphere (0.4) can not be used. The second term is the angular momentum of the moon relative to the axis of the center of mass of the earth and moon system. The left and right-hand sides must be equal. Hence, we have shown that the equation is valid.

$$s = \frac{\left[\left(0.3307 \cdot M_{earth} \cdot R_{earth}^2 \cdot \frac{2\pi}{24 \cdot 60 \cdot 60}\right) + \left(M_{moon}\sqrt{G \cdot M_{earth} \cdot a_{earthmoondist}}\right)\right]^2}{M_{moon}^2 \cdot G \cdot M_{earth} \cdot 1000} \tag{3.3.4}$$

$$= 556,585.837834 \text{ km}$$

By plugging the equation, the final separation distance between earth and moon is found to be 556,585.84 km, well within the Hill Sphere of the earth. Thus, the moon can be maintained perpetually.

The final separation distance can be modeled for terrestrial planet with different mass range and its corresponding satellite of different mass. Denominator used a factor 1.2204 to rescale earth-lunar pair value to 1. $a = 1$, $M_{earth} = x, G = 1$. Whereas $R_{earth}^2 = \left(\frac{3}{4\pi}x\right)^{\frac{2}{3}}$ because earth's mass can be modeled as:

$$\frac{4}{3}\pi R_{earth}^3 = x \tag{3.3.5}$$

$$R_{earth}^3 = \frac{3}{4\pi}x \tag{3.3.6}$$

$$R_{earth} = \left(\frac{3}{4\pi}x\right)^{\frac{1}{3}} \tag{3.3.7}$$

and $\frac{2}{5} \cdot \frac{3x}{4\pi} \cdot \frac{2\pi}{24} = \frac{\pi}{30}$.

$$Y_{earth1moon} = \frac{\left(\frac{\pi}{30}x \cdot \left(\frac{3}{4\pi}x\right)^{\frac{2}{3}} + 1\sqrt{x}\right)^2}{1^2\,(x) \cdot 1.2204} \tag{3.3.8}$$

$$Y_{earth2moon} = \frac{\left(\frac{\pi}{30}x \cdot \left(\frac{3}{4\pi}x\right)^{\frac{2}{3}} + 2\sqrt{x}\right)^2}{2^2\,(x) \cdot 1.2204} \tag{3.3.9}$$

$$Y_{earth3moon} = \frac{\left(\frac{\pi}{30}x \cdot \left(\frac{3}{4\pi}x\right)^{\frac{2}{3}} + 3\sqrt{x}\right)^2}{3^2\,(x) \cdot 1.2204} \tag{3.3.10}$$

$$Y_{earth4moon} = \frac{\left(\frac{\pi}{30}x \cdot \left(\frac{3}{4\pi}x\right)^{\frac{2}{3}} + 4\sqrt{x}\right)^2}{4^2\,(x) \cdot 1.2204} \tag{3.3.11}$$

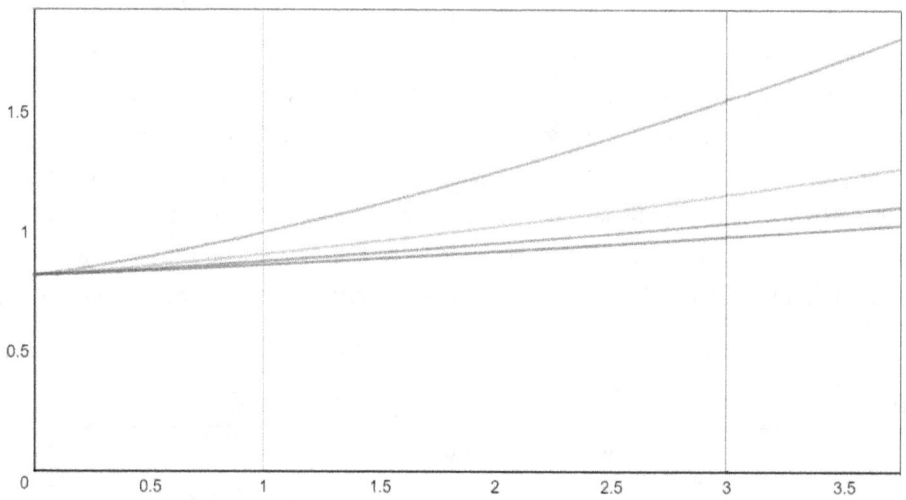

Figure 3.3.1: Terrestrial mass vs final separation

This graph shows the relationship between planet mass and their moon mass and the final separation distance between them when they lock into synchronous orbits. Planets ranges from 1 to 1.6 earth masses are placed within the vertical bars. (typical terrestrial planets arose 5 to 4 Gyr ago does not have metallicity above 0.2, therefore, within each star's habitable zone the mean planetary mass cannot exceed 1.585 earth mass, see "Earth Size"). The green curve represents the locking separation distance for different earth masses with 1 lunar mass satellite, at one earth mass and 1 lunar mass, the separation distance is a unit of 1, represents 556,585.84 km of separation between the earth and moon in a synchronous orbit. One can easily see that as the planet's mass increase, its moment of inertia also increases, and the final separation distance for a moon of the same mass increases. The orange curve represents the locking separation for different earth mass with 2 lunar mass satellite. The blue curve represents the locking separation for different earth mass with 3 lunar mass satellite. The indigo curve represents the locking separation for different earth mass with 4 lunar mass satellite. It can be clearly seen that as the satellite mass increases, the separation distance decreases, though earth with greater mass (higher moment of inertia) still maintains a greater relative separation compares to earth with smaller mass with any given fixed lunar mass satellite. We can also find the final separation distance between 1 earth mass and moons of various sizes.

$$y_{moon} = \frac{\left(\left(0.3307 \cdot 1 \cdot 1^2 \cdot \frac{\pi}{30}\right) + x\sqrt{1 \cdot (1)}\right)^2}{x^2 \cdot 1} \tag{3.3.12}$$

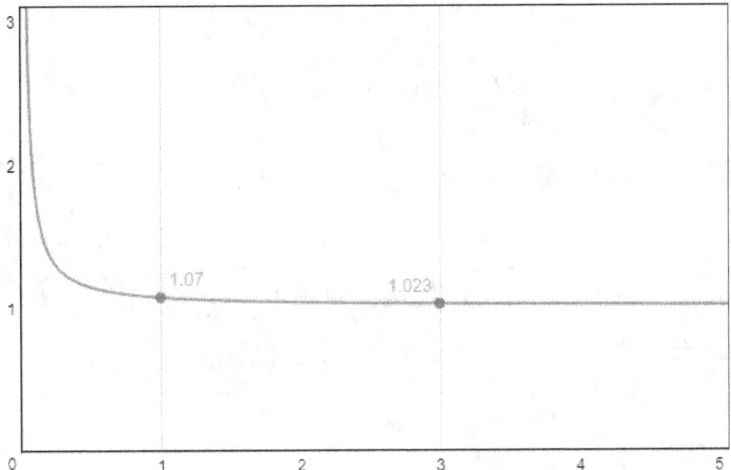

Figure 3.3.2: The final separation distance for tidally locked moons from 0 to 3 lunar mass

For moon's sizes ranging from 0 to 3 lunar mass, the final separation distance for tidal locking decreases sharply from 0 to 0.5 lunar mass and decreases more gently thereafter. The general trend is clear that as the size of the moon increases the separation distance between the bodies decreases.

3.3.2 Earth's Locking Time to the Moon

Furthermore, the time to tidal locking between the two objects is determined to be 47.7 billion years into the future.[27]

$$T_{earthlockingtomoon} = \frac{\left(\frac{1}{13,750}\right)(a_{moon})^6 \cdot I_{earth} \cdot M_{earth} \cdot 909.0909}{3 \cdot G \cdot (M_{moon})^2 \cdot R_{earth}^3 \cdot 60 \cdot 60 \cdot 24 \cdot 365} \tag{3.3.13}$$

$$= 4.7701691912 \cdot 10^{10} \text{ years}$$

At that time, one side of the earth and one side of the moon will constantly face each other and one earth day will take 39.56 days (19.78 days of light and 19.78 days of nights on average), exactly the same as the time to take the moon orbit the earth at that point in time. However, this result is valid only if the semi-major axis remain fixed during the course of earth moon system evolution. The earth-moon system separation is always increasing. As a result, the tidal locking equation is best understood as expected tidal locking time given the current angular velocity and semi-major axis:

$$\frac{dt}{d(\omega, a_{moon})} = \frac{\omega (a_{moon})^6 \cdot I_{earth} \cdot M_{earth} \cdot 909.0909}{3 \cdot G \cdot (M_{moon})^2 \cdot R_{earth}^3} \tag{3.3.14}$$

The derivative of locking time over angular speed and semi-major axis is the tidal locking equation itself, and ω and a_{moon} are also dependent on each other. This is found by substituting $x = \omega$ and $y = a_{earthmoondist}$ Re-arranging our equation:

$$s = \frac{\left[\left(0.3307 \cdot M_{earth} \cdot R_{earth}^2 \cdot \omega\right) + \left(M_{moon}\sqrt{G \cdot M_{earth} \cdot y}\right)\right]^2}{M_{moon}^2 \cdot G \cdot M_{earth}} \tag{3.3.15}$$

$$s \cdot M_{moon}^2 \cdot G \cdot M_{earth} = \left[\left(0.3307 \cdot M_{earth} \cdot R_{earth}^2 \cdot \omega\right) + \left(M_{moon}\sqrt{G \cdot M_{earth} \cdot y}\right)\right]^2 \tag{3.3.16}$$

$$\sqrt{s \cdot M_{moon}^2 \cdot G \cdot M_{earth}} = \left(0.3307 \cdot M_{earth} \cdot R_{earth}^2 \cdot \omega\right) + \left(M_{moon}\sqrt{G \cdot M_{earth} \cdot y}\right) \tag{3.3.17}$$

$$\sqrt{s \cdot M_{moon}^2 \cdot G \cdot M_{earth}} - \left(0.3307 \cdot M_{earth} \cdot R_{earth}^2 \cdot \omega\right) = M_{moon}\sqrt{G \cdot M_{earth} \cdot y} \tag{3.3.18}$$

$$\frac{\sqrt{s \cdot M_{moon}^2 \cdot G \cdot M_{earth}} - \left(0.3307 \cdot M_{earth} \cdot R_{earth}^2 \cdot \omega\right)}{M_{moon}} = \sqrt{G \cdot M_{earth} \cdot y} \tag{3.3.19}$$

$$\left(\frac{\sqrt{s \cdot M_{moon}^2 \cdot G \cdot M_{earth}} - \left(0.3307 \cdot M_{earth} \cdot R_{earth}^2 \cdot \omega\right)}{M_{moon}}\right)^2 = G \cdot M_{earth} \cdot y \tag{3.3.20}$$

$$A\left(\omega\right) = \frac{1}{G \cdot M_{earth}} \left(\frac{\sqrt{s \cdot M_{moon}^2 \cdot G \cdot M_{earth}} - \left(0.3307 \cdot M_{earth} \cdot R_{earth}^2 \cdot \omega\right)}{M_{moon}}\right)^2 \tag{3.3.21}$$

Assuming earth and moon forms synchronization at a distance 556,585.84 km away, this equation predicts at 24 hour spin rate, the earth and moon distance is 384,399 km, 60 earth radius apart. It predicts that at 17.78 hour spin rate, which is recovered from geologic record 2 Gya, earth moon system was 52 earth radius apart, exactly matches other studies. We substitute a_{moon} with $A\left(\omega\right)$ for our tidal locking equation:

$$\frac{dt}{d\left(\omega, A\left(\omega\right)\right)} = \frac{\omega A\left(\omega\right)^6 \cdot I_{earth} \cdot M_{earth} \cdot 909.0909}{3 \cdot G \cdot \left(M_{moon}\right)^2 \cdot R_{earth}^3} \tag{3.3.22}$$

Given a fixed final distance of 556,585.84 km, $A\left(\omega\right)$, and $\frac{dt}{d\left(\omega, A\left(\omega\right)\right)}$, we now define the recursive function that plots the rotational spin vs time as pairs of points:

$$S\left(0\right) = \left(t_1 = 0, \quad \omega = 6\right)$$

$$S\left(1\right) = \left(t_1 = t_1 + \frac{dt}{d\left(\omega, A\left(\omega\right)\right)} \cdot \frac{1}{d}, \quad \omega\left(1 + d\right)\right)$$

$$S\left(2\right) = \left(t_1 = t_1 + \frac{dt}{d\left(\omega\left(1 + d\right), A\left(\omega\left(1 + d\right)\right)\right)} \cdot \frac{1}{d}, \quad \omega\left(1 + d\right)\left(1 + d\right)\right)$$

$$\ldots$$

$$S\left(n\right) = \left(t_1 = t_1 + \frac{dt}{d\left(\omega\left(1 + d\right)^{n-1}, A\left(\omega\left(1 + d\right)^{n-1}\right)\right)} \cdot \frac{1}{d}, \quad \omega\left(1 + d\right)^n\right)$$

$$S\left(n\right) = \left(\sum_{n=1}^{n} \frac{dt}{d\left(\omega\left(1 + d\right)^{n-1}, A\left(\omega\left(1 + d\right)^{n-1}\right)\right)} \cdot \frac{1}{d}, \quad \omega\left(1 + d\right)^n\right)$$

Whereas each step such as Step $S\left(1\right)$ is defined as the time required to decrease the angular spin per hr of earth from ω to $\omega\left(1 + d\right)$. $\frac{dt}{d\left(\omega, A\left(\omega\right)\right)} \cdot \frac{1}{d}$ is the locking time required for an initial spin from ω to $\omega\left(1 + d\right)$. $\frac{dt}{d\left(\omega, A\left(\omega\right)\right)}$ is the total locking time for required for initial spin of ω. d can be made arbitrarily small in simulation to increase precision. The time required to decrease the angular spin varies by each period depending on the current values of ω and $A\left(\omega\right)$ as the parameters for the tidal locking equation.

We assume that an initial spin rate of 6 hr based on previous studies on initial earth rotation, a separation distance of 9 earth radii (which is chosen, accordingly to equation $s\left(x\right)$, leads to a final distance of $556,585$ km with an initial spin rate of 6 hr), and a start time of 0. The curve has to fit our current observation so that it crosses 28 hr spin rate at 4.5 Gyr (It is not 24 hr at 4.5 Gyr because studies have shown that earth had a period of fixed day length which was rather atypical and breaks away from the steady state 1 Gya [48]) and 18 hr spin rate at 2.5 Gyr. We added a factor of 2.4 to $\frac{dt}{d(\omega, A(\omega))}$ to best fit observation. The curve is plotted below:

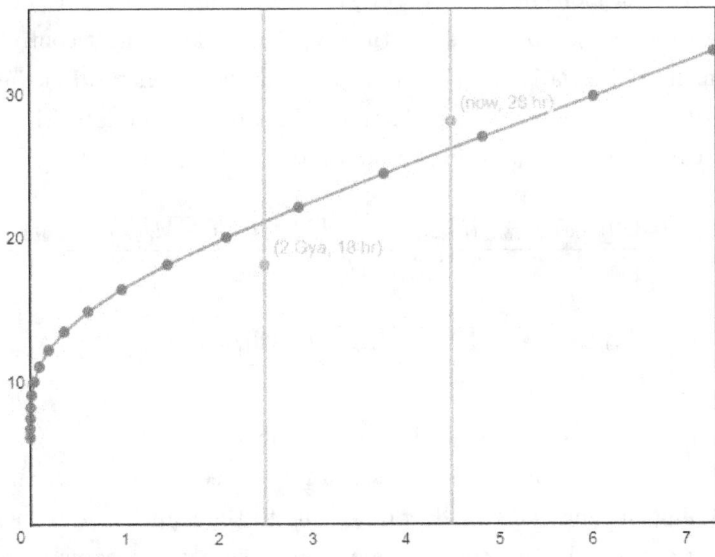

Figure 3.3.3: Plot between 0 and 5 Gyr

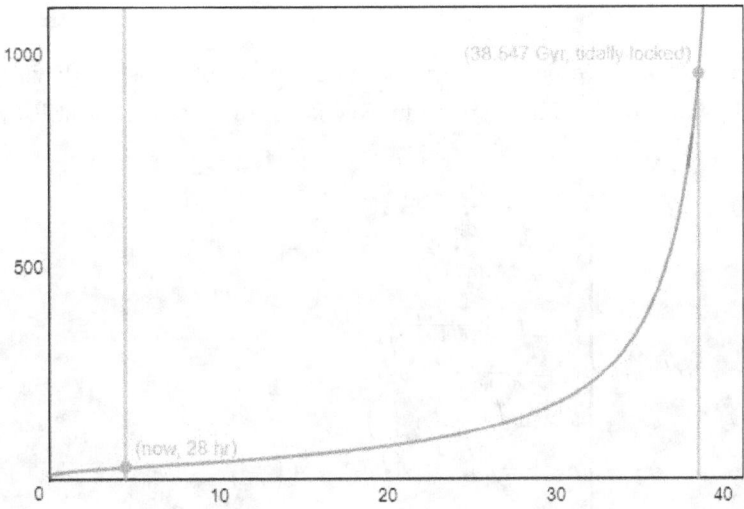

Figure 3.3.4: Plot between 0 and 40 Gyr

It is shown that earth will synchronize with moon's orbital revolution in 38.547 Gyr. Initially, the locking time were very short, earth quickly transfers its own moment of inertia to the moon and contributes a greater separation between them. As the separation increases, much longer tidal locking time is required to decrease the spin by the same factor compares to earlier times, contributing to a phase shift from the rapid spin slow down to a period of slower spin rate slow down. Toward the end of the evolution, the spin rate has slowed down significantly enough and separation distance approaches the final separation distance of 556,585.84 km. Locking time has shortened again and time required to decrease the spin by the same factor decreases toward 0, contributing to a rapid increase in slow down.

Sun also plays a role for tidal locking. In the simplest model, let us first assume that solar tidal force and lunar tidal forces are simply additive. By adding sun's tidal dissipation into consideration, sun's tidal force slows down the spin rate of earth independently of the moon. The share of sun's contribution is understood as the follows. We still chose the time required to slow down the spin from ω to $\omega(1+d)$ by the moon as $t_1 = \frac{dt}{d(\omega, a_{dist})} \cdot \frac{1}{d}$. We also compute the tidal locking time to the sun given the spin of ω as $t_2 = \frac{dt}{d(\omega, a_{earth})}$ (where the mass of moon is replaced by the mass of sun). The ratio of $\frac{t_1}{t_2}$ then gives the solar contribution to the spin slow down. Initially, solar's contribution to the slow down is miniscule but increases over time, and the added sum total contributes to the total slow down as $\omega = \omega \left(1 + d + \frac{t_1}{t_2}\right)$. The slow down by the sun does not contributes toward the transfer of moment of inertia of earth to the moon. As a result, the final expected reachable distance between earth and moon becomes less than 556,585.84 km, and it continues to decrease per step as increasingly more moment of inertia is removed from earth by the sun instead of the moon. Therefore, one also needs to update the synchronization distance per step, which in turn, updates the amount of distance moon has traveled away from earth between two different spin rates.

$$s\left(\omega, a_{dist}\right) = \frac{\left(0.3307 \cdot M_{earth} \cdot R_{earth}^2 \cdot \omega\right) + \left(M_{moon}\sqrt{G \cdot M_{earth} \cdot a_{dist}}\right)^2}{M_{moon}^2 \cdot G \cdot M_{earth}} \tag{3.3.23}$$

$$A\left(\omega, a_{dist}\right) = \frac{1}{G \cdot M_{earth}} \left(\frac{\sqrt{a_{dist} \cdot M_{moon}^2 \cdot G \cdot M_{earth}} - \left(0.3307 \cdot M_{earth} \cdot R_{earth}^2 \cdot \omega\right)}{M_{moon}}\right)^2 \tag{3.3.24}$$

Whereas a_{dist} is the current separation distance of earth from the moon.

We revise the previous recursive function. There is now 2 spin rates to record. ω_2 is actually conceptually the same as ω as before, book keeping the angular spin slow down due to the transfer of moment to the moon. ω is now the composite spin slow down including both the moon and the sun. ω_2 is used to measure the true distance moon has traveled farther away from earth.

$$a_{dist} = a_{dist} + A\left(\omega_2\right) - A\left(\frac{\omega_2}{1+d}\right) \tag{3.3.25}$$

If one were to use ω, the traveled distance becomes overestimated, parts of the slow down is contributed by the sun and does not raise the lunar orbit. The increase in slow down by the moon is still by d as before and expressed as $\omega_2 = \omega_2\left(1+d\right)$.

$$S\left(n\right) = \begin{cases} s = s\left(\omega, a_{dist}\right) \\ \omega_2 = \omega_2\left(1+d\right) \\ t_1 = t_1 + \frac{dt}{d\left(\omega, a_{dist}\right)} \cdot \frac{1}{d} \\ t_2 = \frac{dt}{d\left(\omega, a_{earth}\right)} \\ \omega = \omega\left(1 + d + \frac{t_1}{t_2}\right) \\ a_{dist} = a_{dist} + A\left(\omega_2\right) - A\left(\frac{\omega_2}{1+d}\right) \\ \left(t_1, \omega\right) \end{cases} \tag{3.3.26}$$

We plot the pairs of (t_1, ω) per each step, it shows that tidal locking time becomes much shorter at 18.76 Gyr.

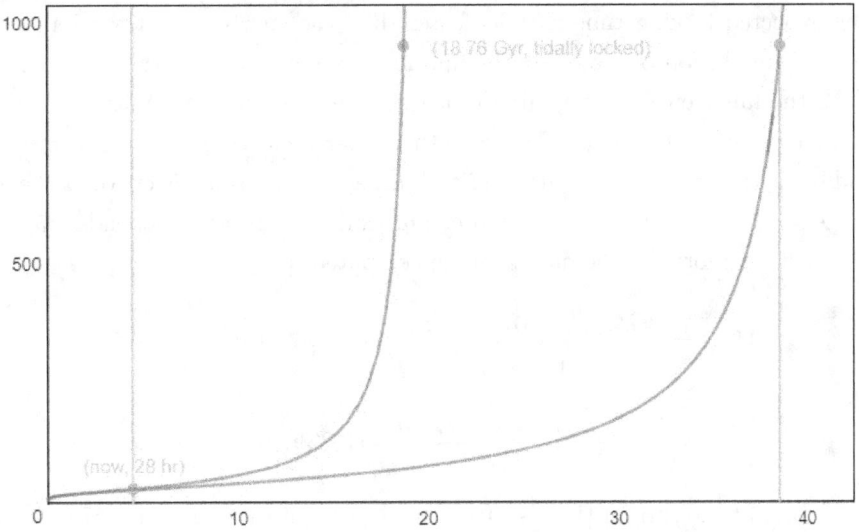

Figure 3.3.5: Plot between 0 and 40 Gyr

We plot the pairs of $(t_1, s(\omega, a_{dist}))$ and (t_1, a_{dist}) per each step:

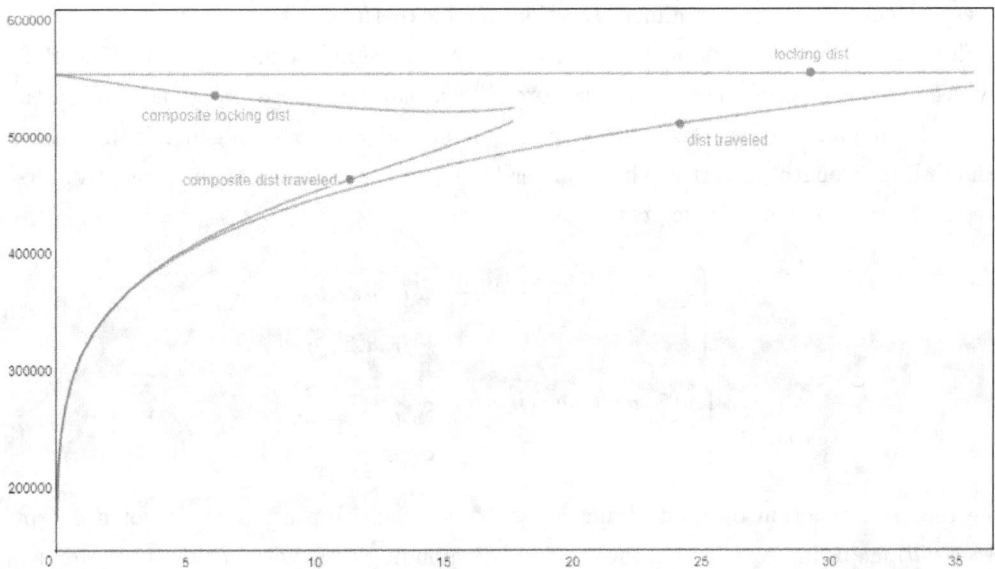

Figure 3.3.6: Locking distance and total distance traveled: combined solar & lunar effect vs. lunar effect alone

The expected locking distance shortens over the course of orbital evolution. At the beginning of orbital evolution, the significant amount of remaining distance needed to be covered by lunar outward migration and solar tidal effect can bring significant shortening of the expected final travel distance, despite its insignificant effect relative to the lunar tidal force. At the late times, there is very little remaining distance untraveled by outward migration, even stronger solar tidal force relative to the waning lunar tidal force does not bring any more significant shortening of the expected travel distance. Therefore, the composite locking distance approaches a constant value.

Throughout and especially evident toward the end of orbital evolution, more distance is covered in the solar-lunar case than the lunar alone case for the same time period. This may feel counter-intuitive. The remaining distance to be covered is shrinking, so does the amount of outward migration distance achieved per same magnitude of spin slow down compares to lunar tidal force alone. The primary cause is the quick shortening of tidal locking time given a fixed spin and semi-major axis per step. Earth moon distance is always increasing, but the slow down in spin eventually overtakes the gradual increase in distance, rapidly shortening the tidal locking time. When the expected locking time shrinks faster than the shortening of the remaining distance to be covered, the outerward migration of moon speeds up. Furthermore, if the total travel time were to be stretched and set equal to the lunar effect alone curve timewise, the moon still out-migrates greater distance for every corresponding time period in the lunar alone case than solar-lunar combined case.

Having defined solar and lunar tidal forces as additive on each other, we can set this curve as the lower bound. In reality, solar tidal forces and lunar tidal forces reinforce and cancel each other depends on configurations. The relative strength of sun's tidal force to the moon can be computed using:

$$M_{pull} = \frac{2GM_{earth}\left(M_{moon}\right)R_{earth}}{\left(a_{moon}\right)^3} = 6.57 \times 10^{18} \tag{3.3.27}$$

$$S_{pull} = \frac{2GM_{earth}\left(M_{sun}\right)R_{earth}}{\left(a_{earth}\right)^3} = 3.02 \times 10^{18} \tag{3.3.28}$$

It is shown that sun's tidal force at the current time is 3.02×10^{18} N, compares to moon's 6.57×10^{18} N. However, earth experienced much stronger lunar tidal force as the moon was significantly closer to earth. It also exists a point in time in the future, when earth and moon distance were 1.29 times greater in separation, sun's tidal force will exceed that of the moon's. The sun's tidal effect is completely additive to the moon's at both full and new moons, and completely subtractive to the moon's when it is at 90 degrees angle with each other, and everything in between. Under the current regimen, the tidal force experienced by earth from the moon ranges between 6.57×10^{18} N (at full and new moons) to 3.55×10^{18} N, averaging at 5.06×10^{18} N. Therefore, solar tidal force actually decreased the strength of lunar tidal force. The solar tidal force experienced by earth ranges between 3.02×10^{18} N (at full and new moons) to 0 N, averaging at 1.51×10^{18} N. In general, the weaker force, on average, exert half of its strength on earth. The stronger force exert an average of the strength of itself and itself minus the weaker force on earth. It is expressed as:

$$m_{oonratio} = \begin{cases} \frac{1}{2}\frac{(M_{pull}+(M_{pull}-S_{pull}))}{M_{pull}} & M_{pull} > S_{pull} \\ \frac{1}{2} & M_{pull} \leq S_{pull} \end{cases} \tag{3.3.29}$$

$$s_{unratio} = \begin{cases} \frac{1}{2}\frac{(S_{pull}+(S_{pull}-M_{pull}))}{S_{pull}} & S_{pull} > M_{pull} \\ \frac{1}{2} & S_{pull} \leq M_{pull} \end{cases} \tag{3.3.30}$$

and we update the recursive function by multiplying $m_{oonratio}$ for each step of recursion for d except for t_1 calculation, where $(1+d)$ factor increase in spin rate to its corresponding time-wise invariant requirement. The solar tidal contribution for slow down $\frac{t_1}{t_2}$ is multiplied by the solar ratio $s_{unratio}$:

$$S(n) = \begin{cases} s = s\left(\omega, a_{dist}\right) \\ \omega_2 = \omega_2\left(1 + d \cdot m_{oonratio}\right) \\ t_1 = t_1 + \frac{dt}{d(\omega, a_{dist})} \cdot \frac{1}{d} \\ t_2 = \frac{dt}{d(\omega, a_{earth})} \\ \omega = \omega\left(1 + d \cdot m_{oonratio} + \frac{t_1}{t_2} \cdot s_{unratio}\right) \\ a_{dist} = a_{dist} + A\left(\omega_2\right) - A\left(\frac{\omega_2}{1 + d \cdot m_{oonratio}}\right) \\ (t_1, \omega) \end{cases} \tag{3.3.31}$$

The plot for pairs of (t_1, ω) is listed below:

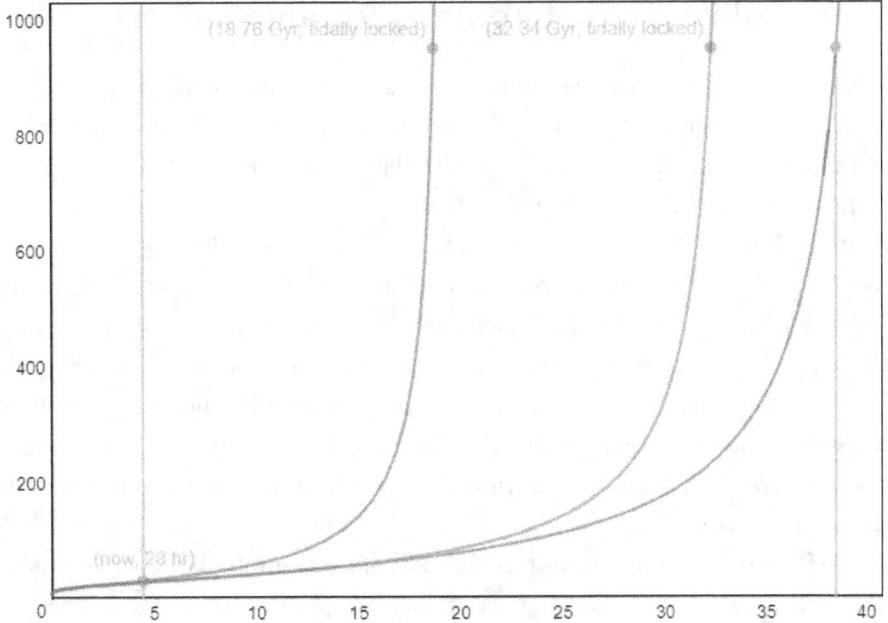

Figure 3.3.7: Plot between 0 and 40 Gyr

Simulation shows that tidal locking occurs at 32.34 Gyr. It has also determined that for most part of the earth moon system's evolution, the pace of tidal dissipation is determined predominantly by the moon. As a result, earlier spin rate of earth remain similar to lunar alone case. Only at the later times (20 Gyr), when earth moon are farther apart, sun's tidal force dominates.

We plot the pairs of $(t_1, s(\omega, a_{dist}))$ and (t_1, a_{dist}) per each step:

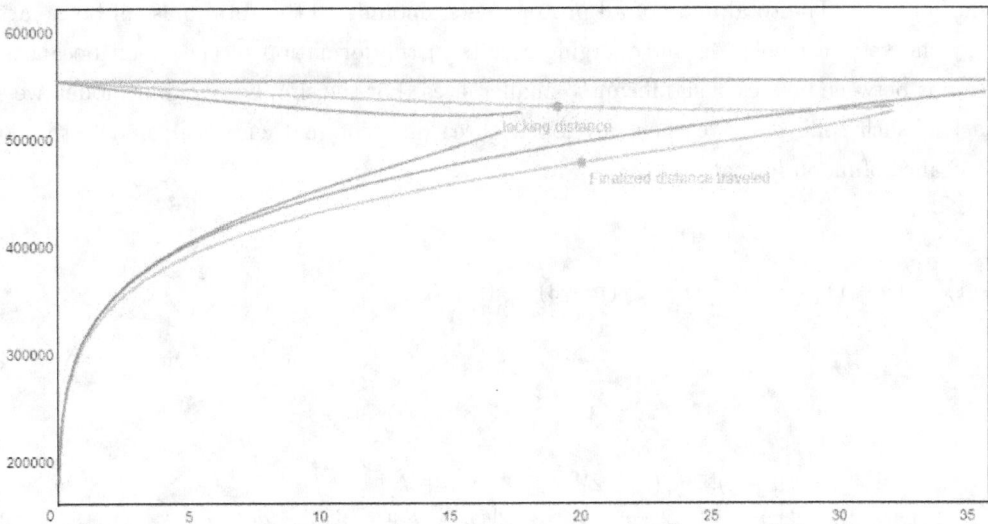

Figure 3.3.8: Locking distance and total distance traveled

The new curve shows that the total distance traveled always falls below the lunar alone case. This is comprehensible since we have shown that solar tidal interaction has on average decreased the lunar tidal force's effect from 6.57×10^{18} N to 5.06×10^{18} N at the current time alone, and it is applicable to all time periods. However, even if solar tidal force is stronger, due to its immense distance, its effect on slowing down the spin is not as effective, determined by the tidal locking time.

3.4 Earth-Moon Collision Probability Explanation

Some argued that the earth and moon system is unique in such a way that binary planetary system which stabilizes the tilt of the planet and reduces climatic shifts and swings which can drastically alter evolutionary trajectories and resets the biodiversity is rare. As a result, the preliminary condition to enable the great diversification of life is rare; and therefore an advanced life with intelligence and manipulative power is also rare. However, the formation of the earth and moon system, following the predominant giant impact hypothesis, is a physical phenomenon, that is subject to the classical laws of physics. Since the classical laws of physics are universally applicable everywhere, the formation of a binary planetary system cannot be ruled as a single, past local event. Most significantly, the Pluto and Charon system, another pair of binary planetary system, located within the solar system, has been simulated and hypothesized to be formed through impact collision.[87][102] Physical simulation concerning material composition, the final angular velocity, and the momentum yields consistent results with the observational data obtained from the Pluto Charon system, confirming the likelihood of a giant impact formation in the past.

Since all terrestrial planets form by a series of collisions and mergers with protoplanets with smaller sizes and masses, simplified mathematical model and simulation can model the probability of the mass ratio of the last major collision between the last remaining protoplanets. It is hypothesized that the origin of the moon is a consequence of a protoplanet Theia smashed into the protoplanet earth. The planet Theia has 10 percent mass of the earth, and 10 percent of which eventually coalesced into two moons and one of the moons pancaked onto the far side of our current moon.[101] Studies and simulation have shown that a single moon and earth system is by far the most stable configuration even if collision creates more than one moon initially. By simulating the final merging process occurred to earth, one can determine the likelihood of terrestrial planet formation with a moon of significant size, and the range of moon mass possible as it is compared to earth.

In order to simulate such results, we assume that earth was formed by the merging of no more than a hundred protoplanets around its orbit. The total mass of all protoplanets amounts to the total mass of both earth and moon system. Only the very last collision and merging results in the formation of the moon because earlier merging and collision is between masses of significantly smaller sizes. For the simplicity of our model, we simply model the very last of such collision. We then developed three different mathematical models to show the likelihood and the chance of moon formation.

$$\text{total} = \frac{1}{2}n\,(n-1) \cdot \frac{1}{2}\,(n-1)\,(n-2) \cdot \frac{1}{2}\,(n-2)\,(n-3) \cdot \frac{1}{2}\,(n-3)\,(n-4) \cdots$$

$$\cdots \frac{1}{2}\,(n-k)\,(n-(k+1)) \cdots 3 \cdot 1 \quad (3.4.1)$$

$$\text{total} = \frac{n!}{2!\,(n-2)!} \cdot \frac{(n-1)!}{2!\,(n-3)!} \cdot \frac{(n-2)!}{2!\,(n-4)!} \cdots \frac{(n-k)!}{2!\,(n-(k+1))!} \cdots \frac{2!}{2!0!} \quad (3.4.2)$$

case 1, case 2, case 3, case 4 ... case k ... case n-3, case n-2: $\quad (3.4.3)$

$$n,\ n-2,\ n-3,\ n-4,\ \cdots n-(k+1) \cdots\ 2,\ 1 \quad (3.4.4)$$

$$\Rightarrow \frac{n\,(n-1)\,(n-2)\,(n-3)\cdots(n-k)\cdots 2 \cdot 1}{\frac{1}{2}n\,(n-1) \cdot \frac{1}{2}\,(n-1)\,(n-2) \cdot \frac{1}{2}\,(n-2)\,(n-3) \cdots \frac{1}{2}\,(n-k)\,(n-(k+1))\cdots 3 \cdot 1} \quad (3.4.5)$$

$$\Rightarrow \frac{1}{\left(\frac{1}{2}\right)^{n-1}(n-1)\,(n-1)\,(n-2)\,(n-3)\cdots(n-k)\cdots 3 \cdot 2} \quad (3.4.6)$$

$$\Rightarrow \frac{1}{\left(\frac{1}{2}\right)^{n-1}(n-1)^2\,(n-2)!} \qquad \Rightarrow \qquad \frac{1}{\left(\frac{1}{2}\right)^{n-1}(n-1)^2 \cdot \frac{(n-2)^n}{e^n}} \quad (3.4.7)$$

Test for convergence:[2]

---

[2]Sterling approximation actually uses formula $n! \approx \sqrt{2\pi n}\left(\frac{n}{e}\right)^n$, but we simplify our steps with $n! \approx \left(\frac{n}{e}\right)^n$, the final result

$$\lim_{n \to \infty} \left| \frac{\frac{1}{\left(\frac{1}{2}\right)^n n^2 \frac{(n-1)^{n+1}}{e^{n+1}}}}{\frac{1}{\left(\frac{1}{2}\right)^{n-1}(n-1)^2 \frac{(n-2)^n}{e^n}}} \right| \tag{3.4.8}$$

$$\Rightarrow \lim_{n \to \infty} \left| \frac{\left(\frac{1}{2}\right)^{n-1}(n-1)^2 \frac{(n-2)^n}{e^n}}{\left(\frac{1}{2}\right)^n n^2 \frac{(n-1)^{n+1}}{e^{n+1}}} \right| \tag{3.4.9}$$

$$\Rightarrow \lim_{n \to \infty} \left| \frac{1 \cdot (n-1)^2 \frac{(n-2)^n}{e^n}}{\left(\frac{1}{2}\right) n^2 \frac{(n-1)^{n+1}}{e^{n+1}}} \right| \tag{3.4.10}$$

$$\Rightarrow \lim_{n \to \infty} \left| \frac{(n-2)^n}{\frac{1}{2} \cdot e^n} \cdot \frac{e^{(n+1)}}{(n-1)^{(n+1)}} \right| \tag{3.4.11}$$

$$\Rightarrow \lim_{n \to \infty} \left| 2 \cdot \frac{e}{n-1} \right| = 0 \tag{3.4.12}$$

In our first model, we simply assume that a hundred protoplanet each with one-hundredth mass of the earth is free to merge with any other in a combinatorial way, from combinatorial derivation, one can see that the protoplanet merges each time with another protoplanet about one-hundredth of the earth mass is very small. In fact, as the number of protoplanets merges toward infinity from the mathematical perspective, the probability that the resulting moon comes from successive rounds of merging between the larger protoplanet with an increasing mass and another protoplanet with a mass of just $\frac{1}{100}$th of the earth approaches 0. This implies that under a majority of the cases, during the merging process, at least one or more times, frequently toward the later stages of merging, merging takes place between two masses of comparable sizes, resulting in Theia and Earth type of collision, resulting in the creation of a moon.

$$\text{total} = \frac{1}{2}n(n-1) \cdot \frac{1}{2}(n-1)(n-2) \cdot \frac{1}{2}(n-2)(n-3) \cdot \frac{1}{2}(n-3)(n-4) \cdots$$
$$\cdots \frac{1}{2}(n-k)(n-(k+1)) \cdots \frac{1}{2} \cdot 3 \cdot 2 \cdot \frac{1}{2} \cdot 2 \cdot 1 \tag{3.4.13}$$

$$\text{total} = \frac{n!}{2!(n-2)!} \cdot \frac{(n-1)!}{2!(n-3)!} \cdot \frac{(n-2)!}{2!(n-4)!} \cdots \frac{(n-k)!}{2!(n-(k+1))!} \cdots \frac{2!}{2!0!} \tag{3.4.14}$$

case 1, case 2, case 3, case 4 ... case k ... case n-3, case n-2: $\tag{3.4.15}$

$n,\ 2,\ 2,\ 2,\ \cdots 2 \cdots 2$ $\tag{3.4.16}$

$$\Rightarrow \frac{n \cdot 2 \cdot 2 \cdot 2 \cdots 2 \cdots 2 \cdot 1}{\frac{1}{2}n(n-1) \cdot \frac{1}{2}(n-1)(n-2) \cdots \frac{1}{2}(n-k)(n-(k+1)) \cdots \frac{1}{2} \cdot 3 \cdot 2 \cdot \frac{1}{2} \cdot 2 \cdot 1} \tag{3.4.17}$$

$$\Rightarrow \frac{2^{n-3}}{\left(\frac{1}{2}\right)^{n-1}[(n-1)(n-2)\cdots(n-k)\cdots 4 \cdot 3 \cdot 2][(n-1)(n-2)\cdots(n-k)\cdots 4 \cdot 3 \cdot 2]} \tag{3.4.18}$$

$$\Rightarrow \frac{2^{n-3}}{\left(\frac{1}{2}\right)^{n-1} \cdot (n-1)! \cdot (n-1)!} \quad \Rightarrow \quad \frac{2^{n-3}}{\left(\frac{1}{2}\right)^{n-1} \cdot \frac{(n-1)^n}{e^n} \cdot \frac{(n-1)^n}{e^n}} \tag{3.4.19}$$

remains the same.

$$\lim_{n \to \infty} \left| \frac{\frac{2^{n-2}}{\left(\frac{1}{2}\right)^n \cdot \frac{n^{n+1}}{e^{n+1}} \cdot \frac{n^{n+1}}{e^{n+1}}}}{\frac{2^{n-3}}{\left(\frac{1}{2}\right)^{n-1} \cdot \frac{(n-1)^n}{e^n} \cdot \frac{(n-1)^n}{e^n}}} \right| \tag{3.4.20}$$

$$\Rightarrow \lim_{n \to \infty} \left| \frac{2^{n-2}}{\left(\frac{1}{2}\right)^n \cdot \frac{n^{(n+1)}}{e^{(n+1)}} \cdot \frac{n^{(n+1)}}{e^{(n+1)}}} \cdot \frac{\left[\left(\frac{1}{2}\right)^{n-1} \cdot \frac{(n-1)^n}{e^n} \cdot \frac{(n-1)^n}{e^n} \right]}{2^{n-3}} \right| \tag{3.4.21}$$

$$\Rightarrow \lim_{n \to \infty} \left| \frac{2}{\left(\frac{1}{2}\right)} \cdot \frac{(n-1)^n}{e^n} \cdot \frac{(n-1)^n}{e^n} \cdot \frac{e^{n+1}}{n^{n+1}} \cdot \frac{e^{n+1}}{n^{n+1}} \right| \tag{3.4.22}$$

$$\Rightarrow \lim_{n \to \infty} \left| 4 \cdot \frac{(e \cdot e)}{n \cdot n} \right| \tag{3.4.23}$$

$$\Rightarrow \lim_{n \to \infty} \left| 4 \cdot \frac{e^2}{n^2} \right| \tag{3.4.24}$$

$$\Rightarrow \lim_{n \to \infty} \left| 4 \cdot \left(\frac{e}{n}\right)^2 \right| = 0 \tag{3.4.25}$$

In our second model, we simply assume that each protoplanet can only merge with its neighbor from the left or right. As a result, we derive several possible cases.

Much like the first model where each protoplanet can merge freely with their neighbors, in the more restricted case that each protoplanet can only merge with their left or right neighbors. Based on combinatorial derivation, the protoplanet merges each time with another protoplanet about one-hundredth of the earth mass is very small, in fact, as the number of protoplanet increases toward infinity, the probability approaches 0. However, careful comparison with the earlier model shows that the n to n matching model without restriction converges to 0 faster, which is not surprising since the restricted case contains fewer choices out of the total number of combinatorial choices given n number of choices, so it takes more rounds to decrease the probability down to zero. In summary, we have demonstrated mathematically that the creation of moon through giant impact between protoplanets with a mass ratio less than 100 to 1 is very common.

$$y_{ntonmatch} = \frac{1 \cdot 100}{(n-1)^2 (n-2)! \left(\frac{1}{2}\right)^{(n-1)}} \tag{3.4.26}$$

$$y_{leftrightmatch} = \frac{2^{(n-3)} \cdot 100}{(n-1)! (n-1)! \left(\frac{1}{2}\right)^{(n-1)}} \tag{3.4.27}$$

---

[3]Sterling approximation actually uses formula $n! \approx \sqrt{2\pi n} \left(\frac{n}{e}\right)^n$, but we simplify our steps with $n! \approx \left(\frac{n}{e}\right)^n$, the final result remains the same.

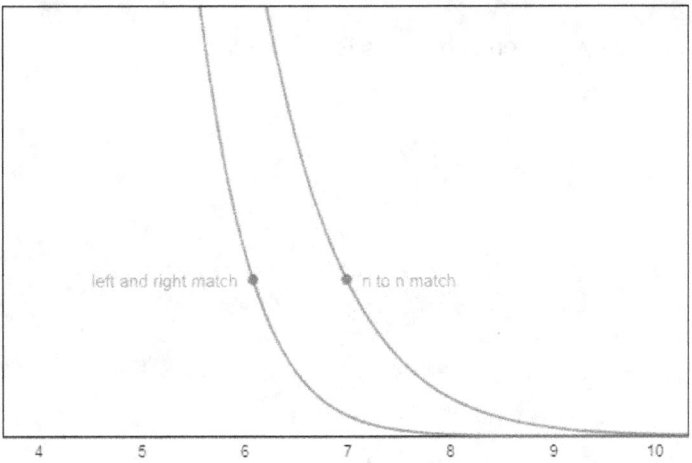

4	5	6	7	8	9	10

left and right match ● ● n to n match

Figure 3.4.1: The probability that the entire oligarchic merging process is non-violent is 0 as the number of mergers increases.

Then, we need to find the mass distribution of the moons created through impact. We first model the protoplanet growth by applying the binomial distribution model, assuming each protoplanet can randomly pair with any other protoplanet. Then the formulas for calculating the mass distribution, our current model uses the case of n = 10, the number of merging protoplanets equals to 10, this number can be increased to a range of values to increase precision and accuracy.

$$f(1) = \Pr(X = 1) = 2\binom{10}{1}0.5^1(1 - 0.5)^{10-1} = 1.953125\% \tag{3.4.28}$$

$$f(2) = \Pr(X = 2) = 2\binom{10}{2}0.5^2(1 - 0.5)^{10-2} = 8.7890625\% \tag{3.4.29}$$

$$f(3) = \Pr(X = 3) = 2\binom{10}{3}0.5^3(1 - 0.5)^{10-3} = 23.4375\% \tag{3.4.30}$$

$$f(4) = \Pr(X = 4) = 2\binom{10}{4}0.5^4(1 - 0.5)^{10-4} = 41.015625\% \tag{3.4.31}$$

$$f(5) = \Pr(X = 5) = 1\binom{10}{5}0.5^5(1 - 0.5)^{10-5} = 24.609375\% \tag{3.4.32}$$

From the results, the probability of 1 lunar mass satellite creation is 1.95%. The probability of 2 lunar mass satellite creation is 8.79%. The probability of 3 lunar mass satellite creation is 23.44%. The probability of 4 lunar mass creation is 41.02%. The probability of 5 lunar mass creation is 24.61%.

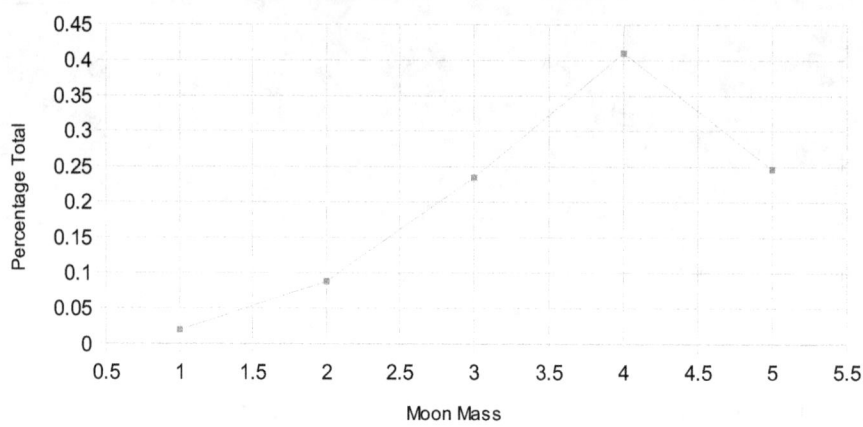

Figure 3.4.2: Percentage of moons by mass binomial distribution method

Next, we model the protoplanet growth by applying the binomial distribution model using restriction, assuming each protoplanet can only pair with its left or right neighbor, our current model uses the case of n = 10, the number of merging protoplanets equals to 10. The formula for calculating the mass distribution for the restricted case is listed below:

$$f(1) = 1 \tag{3.4.33}$$

$$f(2) = 1 \tag{3.4.34}$$

$$f(3) = 3 \tag{3.4.35}$$

$$f(4) = 4\left[f(1)f(3) + f(2)^2\right] = 16 \tag{3.4.36}$$

$$f(5) = 5\left[f(1)f(4) + f(2)f(3)\right] = 95 \tag{3.4.37}$$

$$f(6) = 6\left[f(1)f(5) + f(2)f(4) + f(3)^2\right] = 720 \tag{3.4.38}$$

$$f(7) = 7\left[f(1)f(6) + f(2)f(5) + f(3)f(4)\right] = 6,041 \tag{3.4.39}$$

$$f(8) = 8\left[f(1)f(7) + f(2)f(6) + f(3)f(4) + f(4)^2\right] = 58,416 \tag{3.4.40}$$

$$f(9) = 9\left[f(1)f(8) + f(2)f(7) + f(3)f(6) + f(4)f(5)\right] = 613,233 \tag{3.4.41}$$

$$f(10) = 10\left[f(1)f(9) + f(2)f(8) + f(3)f(7) + f(4)f(6) + f(5)^2\right] = 7,103,170 \tag{3.4.42}$$

$$f(n) = \begin{cases} \text{even} & n\left[f(1)f(n-1) + f(2)f(n-2) + f(3)f(n-3) + \cdots + f\left(\frac{n}{2}\right)^2\right] \\ \text{odd} & n\left[f(1)f(n-1) + f(2)f(n-2) + f(3)f(n-3) + \cdots + f\left(\lfloor\frac{n}{2}\rfloor\right) \cdot f\left(\lceil\frac{n}{2}\rceil\right)\right] \end{cases} \tag{3.4.43}$$

Using the formula, one can count the number of satellites formed with mass from 1 lunar mass to $\frac{n}{2}$ lunar mass. Whereas n is the initial starting number of protoplanets. In our case, n=10. The final results are plotted.

Percentage of Moons by different Mass Restricted Case

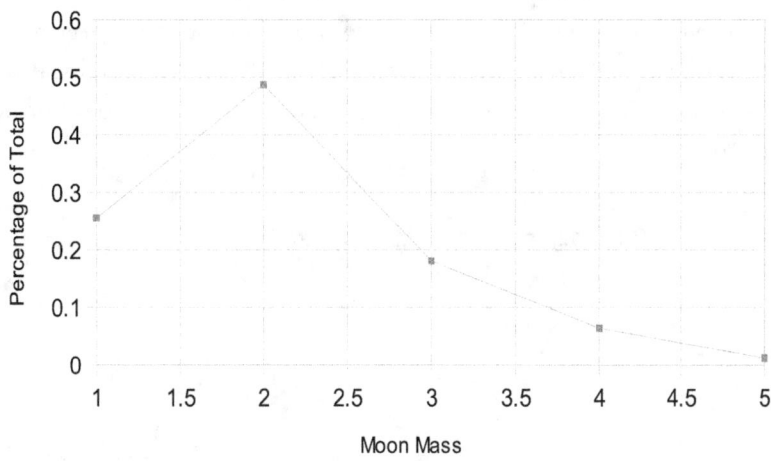

Figure 3.4.3: Percentage of moons by different mass using restricted case

Finally, we shall simulate the growth of protoplanet sizes by gravitational attraction. We start by placing each protoplanet along an orbit with an equal distance between each other. Since each protoplanet is equal in mass and each is separated by an equal distance, no mass moves toward any other. All are in a precarious balance. Then, merging two of the protoplanets results in an imbalance of gravitational force between protoplanets, and the merging process proceeds until all protoplanets merge into one single planet. Although, in reality, it is possible that simultaneous merging of more than two protoplanets is possible, we simplify the simulation by breaking the tie between two simultaneous merging and ordering them in two successive steps. Simulation loops through the steps until all protoplanets merged for protoplanets numbered from 5 up to 100. The results of the simulation are graphed below:

Percentage of Moons by different Mass

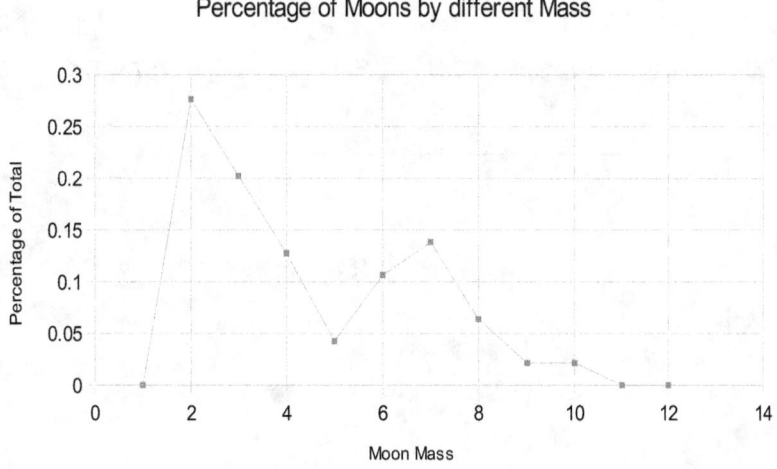

Figure 3.4.4: Percentage of moons of different mass by gravitational simulation

If we summarize the results from our simulation, one can predict the average size of moons created by the terrestrial planet merging process. The final graph shows the composite results of moon creation from the previous three distribution models, whereas the mean lies around 3 lunar mass. It shows that the creation of satellite with 1 lunar mass happens around 9.17 percent of the time. Since it takes another 32.34 billion years for our moon to tidally lock with earth, we found that any moon with a mass greater than 1.3848 lunar mass will tidally lock within the time frame of biological evolution starting with the weighted average of initial spin rate (see section 3.5 "Right Rotational Speed"). In the strictest sense, we exclude those terrestrial planets with large moons and exclude those moons with a mass significantly smaller than ours. *One can conclude that about 23% of the terrestrial planet formed in the universe have moons that bring similar effect to that we observed on earth.* That is, the moon is large enough to stabilize the axis of tilt of the planet so axial wobbling (orbital resonance) observed on Mars is minimized and small enough so that tidal locking does not occur within the time frame of biological evolution.

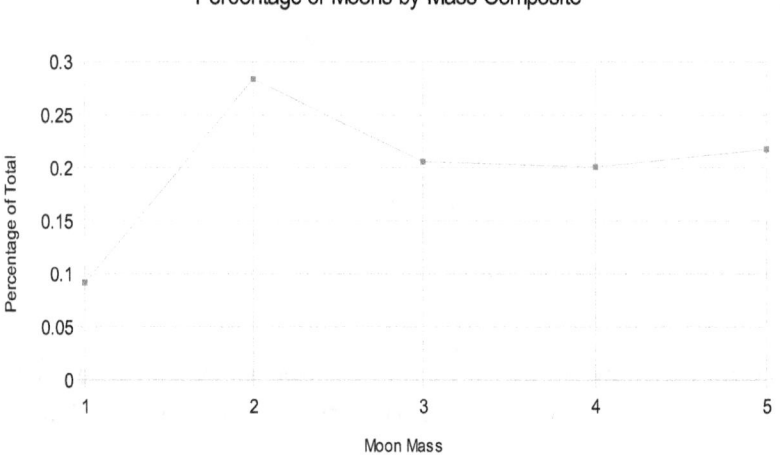

Figure 3.4.5: Percentage of moons of different mass by composite effects

In the final plot, the probability of the creation of moons of various sizes is taken as the average of the three cases.

We use these number points to create the best fit curve for satellites with varying lunar mass creations.

$$S_{ize}(x) = \begin{cases} -0.00961x^4 - 0.004x^3 + 0.13x^2 - 0.024x & 1 \leq x \leq 2 \\ -6.473 \cdot 10^{-4}x^4 + 5.13 \cdot 10^{-4}x^3 + 6.754 \cdot 10^{-2}x^2 - 0.383x + 0.787 & 2 \leq x \leq 5 \end{cases} \tag{3.4.44}$$

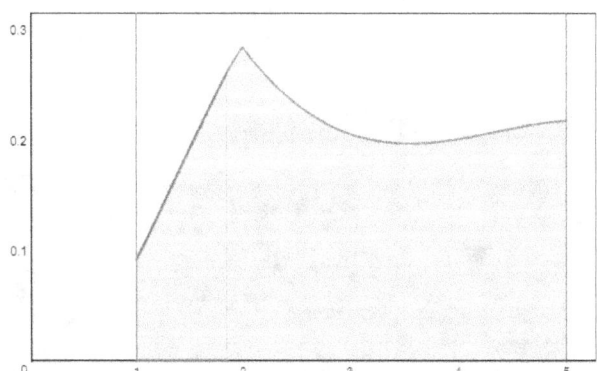

Figure 3.4.6: Best fit curve for satellites with varying lunar mass creations

In reality, the real results may be a weighted average of the three cases or some additional factors unaccounted for. The protoplanets may not all occupy the exact same path, eccentricity, and distance from the sun and may well move at different speeds so that every protoplanet does have a chance to merge with every other. At the same time, each protoplanet does have its own well-defined neighborhood zone, and unbounded interaction with neighboring protoplanet is restricted. The final mass distribution is the compromise between each of these possible scenarios and requires further analysis in the future.

3.5   The Right Rotational Speed

The initial rotational speed of the consolidated planet after several successive mergers with mars sized proto-planets is also an important criterion for habitability. It is important that the initial spin rate of the planet is fast enough to generate earth-like days and nights at the time when intelligent species appears. If a planet starts with a slow rotation, it will quickly tidally lock with its satellite about the earth's moon size, or it could be moonless and retain a slow rotation as Venus. In order to calculate the final spin rate of merged planets, one needs to take into account how much linear momentum, upon collision, is transferred into the angular momentum in either the prograde or the retrograde motion, which in turn, can be either enhanced or partially canceled by the magnitude of existing angular momentum both bodies possess. We assume that upon each collision, both the prograde and the retrograde spin can exist and the magnitude of the linear momentum transferred into the angular momentum is determined by the hitting angle between the two merging mass. During the giant impact stage, the thickness of a protoplanetary disk is far larger than the size of planetary embryos, so collisions are equally likely to come from any direction in a three dimensional space. Therefore, one has to consider the hitting angle in both horizontal and vertical plane.

To simplify the problem and obtain the final result step by step, we first only consider the case whereas collision between protoplanets only occurs at the equator, so variation in striking angle is restricted to the x-y horizontal plane. Since the cross-sectional cut of a sphere is a circle, any arbitrary point tangent to the circle passing through the center of the circle can be defined as the axis of the system. We simply define the axis according to the x-y coordinate plane. Colliding planets are attracted to each other so they approach each other toward their center of mass. Since they eventually merge due to gravitational attraction, we further assume that they approach each other by their attracted escape velocity at the surface of each protoplanet, that is:

$$v_e\left(x\right) = \sqrt{\frac{2x}{\left(x \cdot \frac{3}{4\pi}\right)^{\frac{1}{3}}}} \tag{3.5.1}$$

Whereas x is the given mass of the merging protoplanet, and the final escape velocity is expressed as a fraction of earth's escape velocity:

$$V_{escape}\left(x\right) = \frac{v_e\left(x\right)}{v_e\left(1\right)} \tag{3.5.2}$$

The final linear momentum upon collision between a 0.3 and 0.1 earth mass protoplanet can then be expressed as:

$$p = 0.3V_{escape}\left(0.1\right) - 0.1V_{escape}\left(0.3\right) \tag{3.5.3}$$

According to our defined axis, the colliding angle θ then varies between 0 to 2π. The final linear momentum vector can then be dissected into its x and y components:

$$p_y = p \cdot \sin\theta \tag{3.5.4}$$

$$p_x = p \cdot \cos\theta \tag{3.5.5}$$

The colliding angle θ also determines the moment arm for the linear momentum vector by considering the radius of planet receiving the impact:

113

$$R_y = R \cdot \sin \theta \tag{3.5.6}$$

$$R_x = R \cdot \cos \theta \tag{3.5.7}$$

The moment of momentum is then expressed as:

$$L_1 = R_y \cdot p_x \tag{3.5.8}$$

$$L_2 = R_x \cdot p_y \tag{3.5.9}$$

These 2 moments always turn the sphere in opposite directions (prograde and retrograde), so that the net sum of moment of momentum is simply:

$$L_{total} = R_y \cdot p_x - R_x \cdot p_y \tag{3.5.10}$$

the moment of inertia of the entire system becomes:

$$I = \frac{2}{5} \left(m_1 + m_2 \right) \left(R_{combined} \right)^2 \tag{3.5.11}$$

and the resulting angular velocity of the system is:

$$\omega = \frac{L_{total}}{I} \tag{3.5.12}$$

Simulation shows that under this setup, all cases leads slow final spin rate of earth, which contradicts observation. When the final linear momentum's direction and striking angle coincide, as we illustrated, $R_y \cdot p_x - R_x \cdot p_y \approx 0$ Because there is an equal magnitude of the prograde and the retrograde spin generated in an equal amount and the energy is released in the form of heat and light instead of conserving as the angular momentum. In order to fit observation, the assumption that colliding planets merge into each other toward each other's center of mass is relaxed, allowing linear momentum's direction to be random between 0 to $\frac{\pi}{2}$. Though the colliding angle θ remains the same. We introduce a new angle ϕ, and the final linear momentum vector can then be dissected into its x and y components as:

$$p_y = p \cdot \sin \phi \tag{3.5.13}$$

$$p_x = p \cdot \cos \phi \tag{3.5.14}$$

The simulation then yields much better data, whereas both 6 hours rotational period during the earth-moon formation phase and slower >24 hour spin rate is observed. Therefore, it suggests that the the merging process of protoplanets are more random than initially assumed.

Having finalized rotation in the x-y plane, we now generalize to the x-y-z plane. Collision can occur north or south of the equator. Earth has 24 degrees obliquity. Any collision occurs beyond the equator can be thought as a collision that introduce a new x-y plane with obliquity. Earth can be thought as having a x-y plane with 24 degrees obliquity. Since we already able to derive the rotational spin along the x-y plane of the equator, one can also derive the rotational spin of x-y plane with any arbitrary obliquity. Euler's rotation theorem states that the composition of two rotations is also a rotation. The final spin rate is then given by the composition of the former rotation and the new rotation. The precise computation of merging rotations require rotational matrix and is beyond the scope of this work. However, here we introduce an simple approximate alternative that does compute the rotation composition.

Any rotation can be thought of as a sinusoidal curve.

The new rotational plane can be defined as:

$$y_1 = \alpha_1 \cos(x + \theta_1) \tag{3.5.15}$$

Whereas θ_1 is current colliding angle parallel to the rotational plane. α_1 is the obliquity angle of the rotational plane, which is randomized between 0 and $\frac{\pi}{2}$, (there is no need to generalize to between $-\frac{\pi}{2}$ and $\frac{\pi}{2}$ since θ_1 is randomized between 0 to 2π. An obliquity of $0 \sim \frac{\pi}{2}$ with $\theta_1 = \pi$ is equivalent to an obliquity of $0 \sim -\frac{\pi}{2}$ with $\theta_1 = 0$)

$$y_2 = \alpha_2 \cos(x + \theta_2) \tag{3.5.16}$$

Whereas α_2 is the obliquity of the previous rotational plane, and θ_2 is the last colliding angle parallel to the rotational plane. The final composition of 2 rotational planes can be expressed as:

$$y_{final} = \frac{1}{2}(y_1 + y_2) \tag{3.5.17}$$

However, rarely does both rotation have equal magnitude of angular momentum, therefore, the final rotational plane of the composition of any 2 rotational planes is the weighted results of both, and L_1 and L_2 are the angular momentum of rotation 1 and rotation 2 respectively:

$$y_{final} = \left|\frac{L_1}{L_1 + L_2}\right| y_1 + \left|\frac{L_2}{L_1 + L_2}\right| y_2 \tag{3.5.18}$$

The α_{total} is defined as the max value of y_{final}:

$$\alpha_{total} = \arg\max y_{final} \tag{3.5.19}$$

The angular momentum's component L_{xy} along the x-y plane can be find by:

$$L_{xy} = L_1 \cos\alpha_1 + L_2 \cos\alpha_2 \tag{3.5.20}$$

Knowing the final obliquity angle of the new rotational plane as α_{total}. The angular momentum's component L_z along the z plane can be find by $\tan(\alpha_{total})$ times the angular momentum's component L_{xy} along the x-y plane :

$$L_z = L_{xy}\tan(\alpha_{total}) \tag{3.5.21}$$

and the final angular momentum is defined as:

$$L_{total} = \sqrt{(L_{xy})^2 + (L_z)^2} \tag{3.5.22}$$

and the final angular velocity is defined as:

$$\omega_{total} = \frac{L_{total}}{I} \tag{3.5.23}$$

This approach can yield fairly good approximation for rotation composition. It is almost exact for the composition of two rotations separated by low obliquity. It deviates from reality when two rotations are wider apart from each other. (i.e. one at the equator and another at the pole) Since a sphere is non-euclidean surface, rotation's projection onto a sinusoidal curve in the euclidean flat space results in distortion.

Going to a step further, the previous rotation $y_2 = \alpha_2 \cos(x + \theta_2)$ can be in fact a composite of rotations from previous rounds of merging process. Assuming no new linear momentum arises from the merging process ($y_1 = 0$, $L_1 = 0$), which can occur when 2 equivalent mass collide with each other with the same velocity. The collision between two protoplanets and its final ω requires analysis.

Assuming a perfectly inelastic collision, upon contact, protoplanet 1 contains a rotation with an obliquity angle α_3 and a speed ω_3, which immediately transfer to protoplanet 2. Together as a single mass object, the rotation with obliquity α_3 and at somewhat lower speed $< \omega_3$ now applies to the combined mass of protoplanet 1 and protoplanet 2, regardless of the striking angle between the two. However, if protoplanet 2 also contains a rotation with an obliquity angle α_4 and a speed ω_4, which also immediately transfer to protoplanet 1. Together as a single mass object, the rotation with obliquity α_4 and at somewhat lower speed $< \omega_4$ now applies to the combined mass of protoplanet 1 and protoplanet 2, regardless of the striking angle between the two.

Therefore, contact physics predicts that the composition of two existing rotations of two merging protoplanets behaves similarly as the composition of rotations on a single protoplanet.

protoplanet 1's rotation is defined as:

$$y_3 = \alpha_3 \cos\left(x + \theta_3\right) \tag{3.5.24}$$

protoplanet 2's rotation is defined as:

$$y_4 = \alpha_4 \cos\left(x + \theta_4\right) \tag{3.5.25}$$

and their composite rotation is y_2 that we defined previously:

$$y_2 = \left|\frac{L_3}{L_3 + L_4}\right| y_3 + \left|\frac{L_4}{L_3 + L_4}\right| y_4 \tag{3.5.26}$$

α_2 is defined as the max value of y_2:

$$\alpha_2 = \arg\max y_2 \tag{3.5.27}$$

The angular momentum's component L_{xy} along the x-y plane can be find by:

$$L_{xy} = L_3 \cos\alpha_3 + L_4 \cos\alpha_4 \tag{3.5.28}$$

Knowing the final obliquity of the new rotational plane as α_2. The angular momentum's component L_z along the z plane can be find by $\tan\left(\alpha_2\right)$ times the angular momentum's component L_{xy} along the x-y plane :

$$L_z = L_{xy} \tan\left(\alpha_2\right) \tag{3.5.29}$$

and the final angular momentum is defined as:

$$L_2 = \sqrt{\left(L_{xy}\right)^2 + \left(L_z\right)^2} \tag{3.5.30}$$

y_3 and y_4 in turn can be composition of earlier rotations (as a consequence of both newly introduced and merging of protoplanets), hence we have completed the procedure for computing composition of rotation in x-y-z plane. Next, we need to generalize and compute the final rotational spin rate based on successive collisions of varying masses. In our simple model, we assume that earth is formed by merging with 9 mars sized protoplanets. We will simulate the process through a few rounds of iterations.

Colliding Mass 1	Colliding Mass 2	Final Mass
0.1 M_{earth}	0.1 M_{earth}	0.2 M_{earth}
0.2 M_{earth}	0.1 M_{earth}	0.3 M_{earth}
0.3 M_{earth}	0.1 M_{earth}	0.4 M_{earth}
0.4 M_{earth}	0.2 M_{earth}	0.6 M_{earth}
0.6 M_{earth}	0.3 M_{earth}	0.9 M_{earth}

Table 3.5.1: The merging mass size for each step for the merging process

In the first round, two colliding masses each have only 0.1 earth mass. In the next 3 rounds, each colliding mass 2 will have 0.1 earth mass while the colliding mass 1 accumulates. Starting at round 4, colliding mass 2 increases to 0.2 earth mass and then to 0.3 earth mass, to reflect the weighted average colliding body size during the final merging process based on the previous merging simulation whereas the weighted average mass for 5 lunar mass satellite merging process results in 3 lunar mass (see Section 3.4). At the last step, we could weighted average the results of collision between 0.9 earth mass and 0.1 earth mass (earth's case) up to 0.5 earth mass with 0.5 earth mass. Such weighted average should be similar to just 1 step calculation of 0.6 earth with 0.3 earth mass collision. Whereas each of Colliding Mass 2's final spin rate is also calculated independently before merging with Colliding Mass 1. The merging process is best illustrated as a binary tree diagram below:

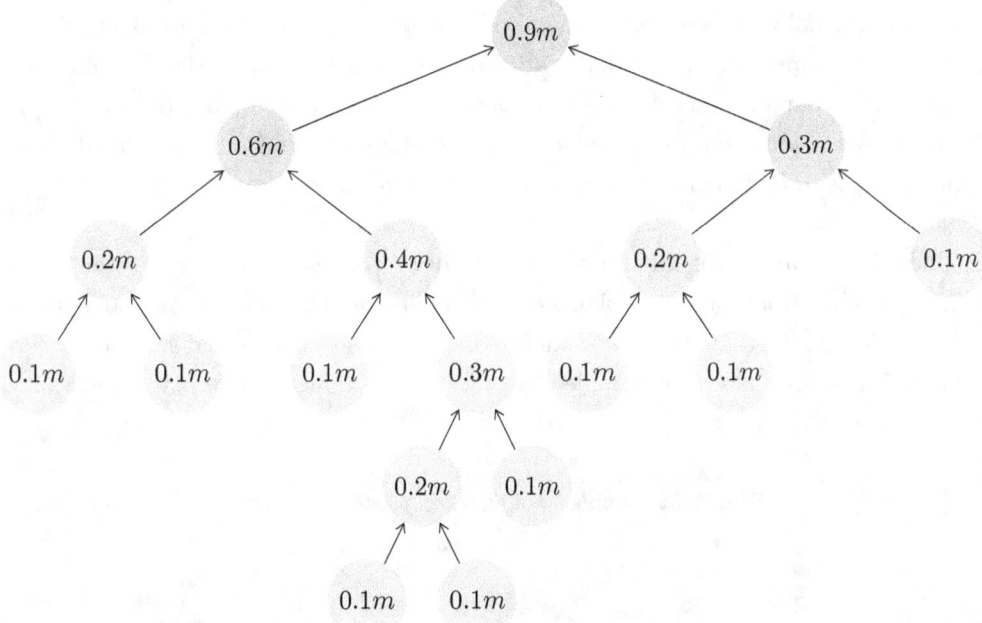

Figure 3.5.1: The merging process demonstrated as a binary tree

We also assume that the initial spin, as a result of the protoplanetary disc, is normally distributed around the mean of a day and then generate the random collision angles.

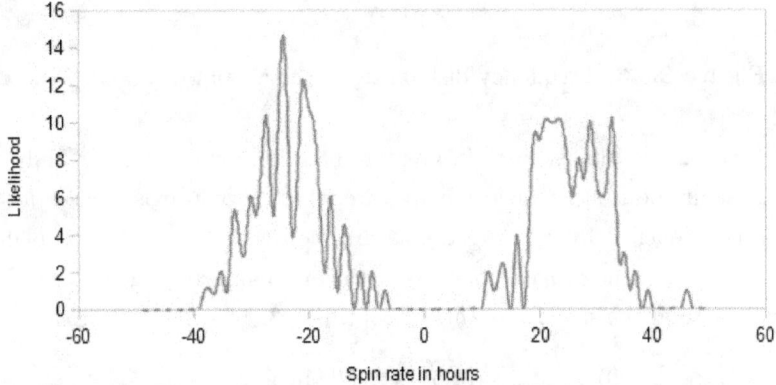

Figure 3.5.2: Frequency distribution of initial spin rate

Although all protoplanet formed within the same accretion disc, we assume that there is some slight variation in rotational plane's obliquity angle ranges between 0 and 10 degrees. We further assumes that the initial rotation hitting angle θ ranges from 0 to 2π. The moment of inertia factor is set to 0.39307 instead of 0.3307 because it is determined that the moon formed 50 Myr after the formation of the earth and it takes 500 Myr for planetary differentiation to take place. Yet, some level of differentiation did take place because moon's density is lower than earth with an exceptionally small iron core while the earth has a larger core than a typical terrestrial planet. We substitute the mass of the protoplanet hitting earth ranges from 0.1, 0.2, to 0.3 earth mass respectively, substitute the earth mass before merger ranges from 0.1, 0.2, 0.3, 0.4, to 0.6 earth mass respectively. We substitute the escape velocity of the earth as the final striking speed before the merger ranges from 0.1, 0.2, 0.3, 0.4, to 0.6 times of earth's escape velocity relative to each other, and the radius of earth ranges from $(0.1)^{\frac{1}{3}}$, $(0.2)^{\frac{1}{3}}$, $(0.3)^{\frac{1}{3}}$, $(0.4)^{\frac{1}{3}}$, and $(0.6)^{\frac{1}{3}}$ times of earth's radii respectively. Upon each merging process, we randomly generate hitting angles θ between 0 to 2π degrees along the x-y plane, randomly generate linear momentum vector direction relative to x-y plane's angle ϕ between 0 and $\frac{\pi}{2}$, and randomly generate the hitting angle α (new x-y rotational plane's obliquity angle) relative to the equator between 0 and $\frac{\pi}{2}$.

At each round, we compute the spin rate ω_1 and the rotational plane y_1 generated by the Colliding Mass 2 by its hitting angles θ and α, the composite weighted average rotational plane y_2 and spin rate ω_2 of the Colliding Mass 2's own initial rotational plane's spin rate and the Colliding Mass 1's own initial rotational plane's spin rate. We find the final composite by weighting the average between y_1 and y_2 and then proceed to the next round of computation.

Simulation actually finds that regardless of the initial spin rate of protoplanets and whether the existence or non-existence of obliquity of the initial rotational plane, even assuming no initial spin rate at all, will converge to similar final spin rate. This implies that the initial spin rate is non-relevant to the final spin rate of terrestrial planet. Almost all of the initial spin rate is determined by oligarchic merging of protoplanets. The final spin rate distribution is plotted below:

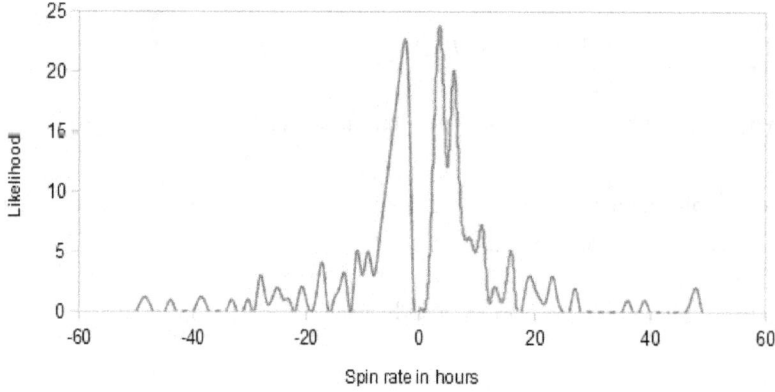

Figure 3.5.3: Frequency distribution of final spin rate

The PDF shows a spike at 3 hours and almost no spin greater than 20 hours. This certainly contradicts our observation. Unless Mars' 24 hour rotation is exceptionally rare, there should exists cases for 24 hour initial rotation. This contradiction is reconciled by relaxing the assumption that the final protoplanets approaches with the escape velocity of each other. Studies have shown that the original formation of the moon requires that protoplanet Theia approaches earth at the speed < 4 km/s,[103] this is 3.998 times slower than our assumption.

$$\frac{(V_{escape}(0.9) + V_{escape}(0.1)) \cdot 11.186 \, \text{km/s}}{4 \, \text{km/s}} = 3.998 \tag{3.5.31}$$

The physical reality once again indicates that movement of protoplanets within the accretion are subject to other interactions are unlikely fully governed by the gravitational attraction alone. Ultimately, we fine-tuned our result by decreasing the approaching speed by a factor of 2, at 7.996 km/s. At this speed of collision, the final generated frequency best represent our observation of the solar system. Whereas earth's initial spin rate of 6 hours is the mode of the distribution and 24 hour initial spin as those observed for Mars is non-rare, and Venusian extreme slow rotation is also possible.

Slowing down of the final spin rate can also be done by assuming there is an non-perfect transfer of linear momentum to angular momentum. This can occur when the coefficient of restitution is between 0 and 1. The details are beyond the scope of this work, but it is shown that the final spin rate can be reduced by reducing the approaching velocity, reducing the momentum transformation efficiency, or both.

Very lastly, we would like to verify our results with energy requirements. It is shown that the initial rotational energy of earth spins at 6 hours is:

$$\omega = \frac{360}{6 \cdot 60 \cdot 60} \cdot \frac{\pi}{180} \tag{3.5.32}$$

$$I = 0.3307 \cdot M_{earth} (6353 \cdot 1000)^2 \tag{3.5.33}$$

$$R = \frac{1}{2} I (w_2)^2 = 3.3726 \times 10^{30} \text{ J} \tag{3.5.34}$$

The total kinetic energy possessed by 0.6 earth and 0.3 earth mass object before impact, reduced by a factor of 2, according to our fine-tuning is:

$$I_{mpact} = \frac{1}{2} \left(\frac{1}{2} (0.3 M_{earth}) \left(\frac{1}{2} V_{escape} (0.6) \right)^2 + \frac{1}{2} (0.6 M_{earth}) \left(\frac{1}{2} V_{escape} (0.3) \right)^2 \right) = 5.168 \times 10^{31} \text{ J} \tag{3.5.35}$$

and we are able to show that:

$$R < I_{mpact} \tag{3.5.36}$$

which is consistent with the hypothesis that majority of the energy is converted into heat and light upon impact and it can be shown that only 4.08% of all impact energy is converted into the initial rotational energy.

$$\frac{R}{I_{mpact}} = 0.04079 \tag{3.5.37}$$

Depending on the final impact velocity, the total amount of energy converted into heat can be calculated. For a 0.6 and 0.3 earth mass impact, the amount of energy converted into heat is:

$$I_{mpact} - R = 5.146 \times 10^{31} \text{ J} \tag{3.5.38}$$

For a 0.9 and 0.1 earth mass impact, the amount of energy converted into heat is:

$$\frac{1}{2} (0.9 M_{earth}) \left(\frac{1}{2} V_{escape} (0.1) \right)^2 + \frac{1}{2} (0.1 M_{earth}) (V_{escape} (0.9))^2 - R = 3.076 \times 10^{31} \text{ J} \tag{3.5.39}$$

In both cases, it indicates that the planet becomes completely molten if one assumes that the planet are composed most of silicon, and the heat of fusion shows that 1.64×10^{31} J of energy is required to molten the entire planet.

$$3.3 \times 10^2 \cdot \frac{(50.21 \cdot 1000)}{(6.02 \cdot 1000)} \cdot 1000 \cdot M_{earth} = 1.64 \times 10^{31} \text{ J} \tag{3.5.40}$$

Whereas heat of fusion of water is 6.02 kJ/mol and heat of fusion of silicon is 50.21 kJ/mol. The energy required to melt 1 g of ice is 3.3×10^2 J.

It is still far below the vaporization point of the planet which is at:

$$3.3 \times 10^2 \cdot \frac{(383 \cdot 1000)}{(6.02 \cdot 1000)} \cdot 1000 \cdot M_{earth} = 1.25 \times 10^{32} \text{ J} \tag{3.5.41}$$

The final plot is given as:

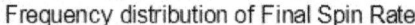

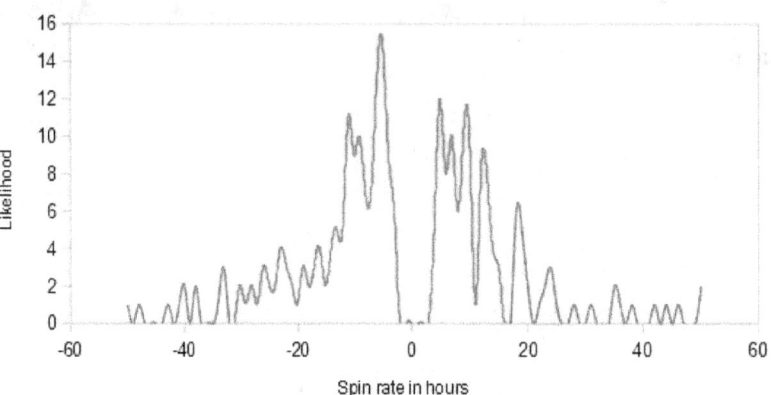

Figure 3.5.4: Frequency distribution of final spin rate adjusted

No spin rate faster than 2 hours 58 minutes in either prograde or retrograde motion has been observed because there is already a limit on the transfer from linear momentum to angular momentum upon collision due to clockwise and counterclockwise torque on rotational spin.

Before we proceed, we run simulation for the creation of 0.42 to 2 earth mass terrestrial planets, covering all possible ranges of habitable terrestrial planet. In the first run, it is simply assumed that the merging process proceeds just as before, except each 0.1 earth mass protoplanet are now ranges from 0.042 to 0.2 earth mass. The results yields statistical equivalent spin rate distributions. In the second run, mass of different planets is determined by the number of mergers. For planet < 1 earth mass, in general, fewer mergers required, and for planet > 1 earth mass, more mergers required. For additional mergers, additional protoplanets with random initial speed and obliquity is created. The mode spin rate of each case is presented below:

Final mass	Spin rate (mode)	Final mass	Spin rate (mode)	Final mass	Spin rate (mode)
0.4 M$_{earth}$	5, 6	1.0 M$_{earth}$	5,6	1.6 M$_{earth}$	7
0.5 M$_{earth}$	11,12	1.1 M$_{earth}$	7	1.7 M$_{earth}$	6,7
0.6 M$_{earth}$	6	1.2 M$_{earth}$	6	1.8 M$_{earth}$	8,9
0.7 M$_{earth}$	5,6,7	1.3 M$_{earth}$	8,9,10	1.9 M$_{earth}$	7
0.8 M$_{earth}$	7,8	1.4 M$_{earth}$	6,7,8,9	2.0 M$_{earth}$	8,9
0.9 M$_{earth}$	6,7	1.5 M$_{earth}$	8,9,10		

Table 3.5.2: The mode of spin rate observed from terrestrial planet formation between 0.4 ~ 2 earth mass

In general, no strong statistical correlation exists between the mass of terrestrial planet creation and the initial spin rate. Therefore, we simply assume that the distribution is one size fits all.
The best fit curve for the data points is found by a lognormal distribution as:

$$R_{otation}(x) = \frac{168}{\sqrt{2\pi}} \exp\left(-\frac{\left(\ln 0.02 x^{1.9}\right)^{1.6}}{2}\right) \tag{3.5.42}$$

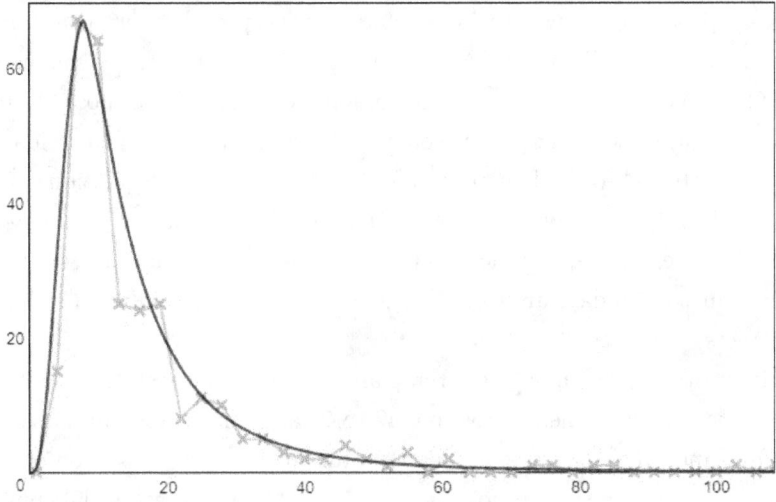

Figure 3.5.5: Best fit for frequency distribution of final spin rate merging both prograde and retrograde spin

To verify the soundness of the result, we also plot the frequency distribution of the final obliquity angle of composite rotational plane. To no surprise, most frequent obliquity angles occurs within 10 to 60 degrees away from the equatorial plane. Earth's 24 degrees tilt is fairly typical among terrestrial planets. It is curious to note that, 90 degrees obliquity of Uranus is extremely rare according to simulation, assuming the terrestrial core of Uranus does not require significantly more rounds of merging than earth does. This seem to substantiate the theory that Uranus' tilt is predominantly a consequence of Jupiter and Saturn's 2:1 resonance and caused it to get torqued onto its side.

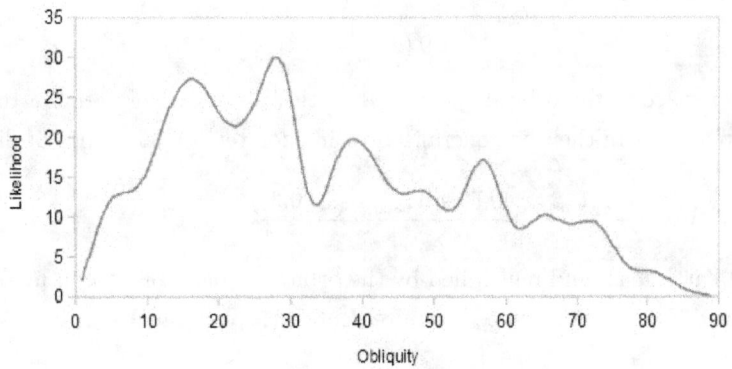

Figure 3.5.6: Frequency distribution of final obliquity

3.6   Non-Locked Moons

We have defined the tidal locking recursive function, which predicts earth moon system will tidally lock to each other in 32.34 Gyr. For such a long timescale, it seems to be irrelevant to our discussion on the emergence and development of intelligent life; however, the final collision and merging of protoplanets, as we have shown, can be different sizes and mass ratios. For the merging and forming planets with similar mass sizes, the mass of the moon can be significantly larger than earth's moon. In such cases, tidal locking happens much sooner, less than the time it takes (4.5 Gyr) for the history of life on earth. This does have some if not a serious challenge for the evolution of life on such a planet. If the process of tidal locking starts after the emergence of life on such a planet, life can gradually adapt to such slow rotation, including a sleep cycle every other month and might function similar to hibernation on earth. The most problematic cases are those massive moons that are quickly tidally locked their parent planet before the emergence of life, long, cold nights can possibly freeze ocean and long, hot days can literally boil water away. These are some of the consequences of such tidally locked planet with its moon. As a result, planets locked to their moons within 4.5 Gyr of its formation are not counted as candidates for potentially intelligent life inhabiting planets. Here we show planets with different sized moons, their locking time, and their final separation distance from their host planet. We shall set stringent selection, any planets' rotation slower than 7 earth days after 4.5 Gyr following its formation will not be counted toward our final list of habitable planets.

We also have derived the initial rotational spin of terrestrial planet and shown the statistical distribution of spin rates universally applies to all terrestrial planets between 0.42 to 2 earth mass. The initial separation distance between the satellite and the planet, in earth case, is 9 earth radii. It is derived based on the final separation distance between the earth and moon in a synchronous orbit of 550,000 km and an initial spin rate of 6 hours. No other information is available and it is assumed that such initial separation is the typical Goldilocks region for the formation of satellite. Furthermore, we show that for any terrestrial planet with varying mass, the initial separation between the satellite and the planet is always 9 planet radii away.

The final collision velocity of two colliding protoplanet is expressed as:

$$S = V_{escape}(0.7) + V_{escape}(0.3) \tag{3.6.1}$$

Whereas 0.7 and 0.3 are the typical colliding protoplanet mass at the last round of merging to form 1 earth size planet, but it can be substitute for any protoplanet mass. Upon impact, both protoplanets shatter. Some of the pieces possessed the colliding velocity, without hitting any other object are ejected into space. We assumed that an initial separation distance of 9 radii and the orbital velocity at such distance is expressed as:

$$v_o(x) = \sqrt{\frac{x}{9\left(x \cdot \frac{3}{4\pi}\right)^{\frac{1}{3}}}} \tag{3.6.2}$$

That is, for pieces ejected into space at the orbital speed of 9 earth radii will likely remain to form its natural satellite. The initial impact velocity can then be generalized to include terrestrial planet of different mass as:

$$V_o = \frac{(V_{escape}(x \cdot 0.6) + V_{escape}(x \cdot 0.3))}{S} \cdot v_o(1) \tag{3.6.3}$$

Which is further divided by 1 earth mass and multiplied by the orbital velocity of 1 earth mass at 9 earth radii. This finalized curve is expressed as the orbital speed requirement for 9 planetary radii given the varying mass of planet, and one find that:

$$v_o(x) = V_o \tag{3.6.4}$$

That is, given the terrestrial planets of varying mass, the rate of change of the initial impact velocity equals the rate of change of the orbital velocity, and we conclude that, if earth's natural satellite forms around 9 earth radii away, then any natural satellite forms around planet of any mass also forms at its own 9 planetary radii away. The formation of planet with smaller mass has smaller impact velocity but is able to match 9 planetary radii's orbital velocity due to smaller gravitational attraction. The formation of planet with larger mass has larger impact velocity but only able to match 9 planetary radii's orbital velocity due to stronger gravitational attraction.

Defining the mass limit of tidally-locked moons, we divide into 4 different categories of stellar mass (4 categories is chosen for convenience) We take the weighted average mass of each category, and use this mass to compute the expected habitable zone radius. So that M_{sun} becomes xM_{sun} and a_{earth} becomes $\sqrt{x^{4.5}}a_{earth}$.

Furthermore, the proportion of each stellar mass category over the range of habitable star is computed from the initial mass function:

$$p_2 = \frac{\int_{0.712}^{0.8} I_{mf}(x)\,dx}{\int_{0.712}^{1.0} I_{mf}(x)\,dx} \quad p_3 = \frac{\int_{0.8}^{0.9} I_{mf}(x)\,dx}{\int_{0.712}^{1.0} I_{mf}(x)\,dx} \quad p_4 = \frac{\int_{0.9}^{1.0} I_{mf}(x)\,dx}{\int_{0.712}^{1.0} I_{mf}(x)\,dx} \tag{3.6.5}$$

Next, we divide the rotational spin distribution into different bins, taking its average rotational spin, which is defined by:

$$A_{vg}(s_1, s_2) = \frac{\int_{s_1}^{s_2} x R_{otation}(x)\,dx}{\int_{s_1}^{s_2} R_{otation}(x)\,dx} \tag{3.6.6}$$

and the lunar mass limit within each bin is computed. The maximum attainable lunar mass within each stellar mass category and rotational spin is found if the rotational spin of the planet 4.5 Gyr after its formation slows to 7 days.

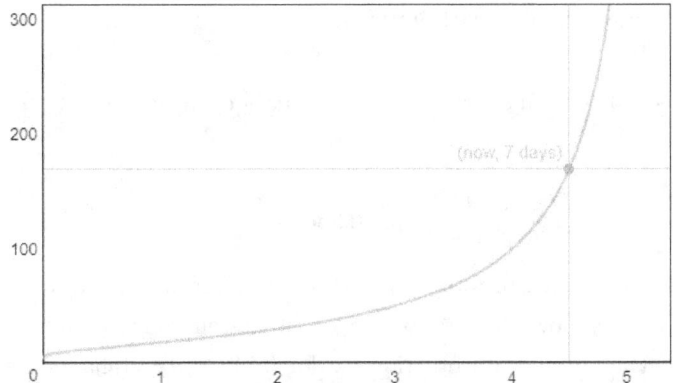

Figure 3.6.1: 4.5 Gyr after satellite and planet formation, the spin of the planet can not exceed 7 days per rotation

Furthermore, the proportion of each rotational spin bin over the possible habitable rotational spin ranges is computed and the percentage for each bin is defined as:

$$h_{r1} = \frac{\int_1^2 R_{otation}(x)\,dx}{\int_1^\infty R_{otation}(x)\,dx} \quad h_{r2} = \frac{\int_2^4 R_{otation}(x)\,dx}{\int_1^\infty R_{otation}(x)\,dx} \tag{3.6.7}$$

$$h_{r3} = \frac{\int_4^6 R_{otation}(x)\,dx}{\int_1^\infty R_{otation}(x)\,dx} \quad h_{r4} = \frac{\int_6^8 R_{otation}(x)\,dx}{\int_1^\infty R_{otation}(x)\,dx} \tag{3.6.8}$$

Because the initial separation between the satellite and the parent planet is consistently 9 planetary radii away based on assumption, our simulations show that tidal locking time is identical across all planetary mass given identical satellite to parent mass ratio. This conclusion saves us from the extra step of computation taking planetary mass range into consideration for each category of stellar mass and rotational spin bin.

Rotational spin	Avg spin	Percentage $h_{r1}, h_{r2} h_3, h_{r4}$	0.712 ~ 0.8	0.8 ~ 0.9	0.9 ~ 1.0
1~2 hr	1.6534 hr	0.33739%	< 3.895	< 3.985	< 3.9825
2~4 hr	3.2310 hr	3.87478%	< 1.965	< 2.159	< 2.172
4~6 hr	5.0919 hr	9.42800%	< 1.117	< 1.410	< 1.432
6~8 hr	7.0280 hr	13.2741%	N/A	< 1.032	< 1.062

Table 3.6.1: The lunar mass limit for each category of stellar mass and rotational spin

It is non-surprising that more rapid spin results in higher allowable limit for lunar mass before tidal locking within 4.5 Gyr. Based on the results, one can quickly conclude that any initial rotational spin slower than 8 hr results in tidal locking of parent planet to the satellite. This means that only 26.914% of initial rotational speed yields habitable planet:

$$\frac{\int_0^8 R_{otation}(x)\,dx}{\int_1^\infty R_{otation}(x)\,dx} = 0.2691434 \tag{3.6.9}$$

Naturally, one can be more precise by breaking the list of number of habitable planets (re-do calculation based on both count number of stars by IMF and Lineweaver method) into its respective ranges and compute the final number of stars fulfilling the initial spin rate requirement:

Mass range	Total	Percentage	Habitable
0.712 ~ 0.8	176,870,839	13.640%	24,125,532
0.8 ~ 0.9	208,353,546	26.914%	56,077,006
0.9 ~ 1.0	227,173,953	26.914%	61,142,396
Total	**612,398,339**		**141,344,934**

Table 3.6.2: The number of stars with initial spin fast enough to avoid tidal locking

This shows that, the more precise calculation shows that *only 23.08% of initial rotational speed yields habitable planet:*

$$\frac{141,344,934}{612,398,339} = 0.230806 \tag{3.6.10}$$

We have just demonstrated the importance of initial spin rates in determining the number of stars deemed habitable, we now proceed to modify how the proportion of each stellar mass category over the range of habitable star is computed from the initial mass function, so for 0.712 ~ 0.8 stellar mass, only 1~6 hr initial spin out of total range of 1~8 hr fits our assumption:

$$p_2 = \frac{\left(\frac{h_{r3}+h_{r2}+h_{r1}}{h_{r3}+h_{r2}+h_{r1}+h_{r4}}\right)\int_{0.712}^{0.8} I_{mf}(x)\,dx}{\left(\frac{h_{r3}+h_{r2}+h_{r1}}{h_{r3}+h_{r2}+h_{r1}+h_{r4}}\right)\int_{0.712}^{.8} I_{mf}(x)\,dx + \int_{0.8}^{1.0} I_{mf}(x)\,dx} \tag{3.6.11}$$

$$p_3 = \frac{\int_{0.8}^{0.9} I_{mf}(x)\,dx}{\left(\frac{h_{r3}+h_{r2}+h_{r1}}{h_{r3}+h_{r2}+h_{r1}+h_{r4}}\right)\int_{0.712}^{.8} I_{mf}(x)\,dx + \int_{0.8}^{1.0} I_{mf}(x)\,dx} \tag{3.6.12}$$

$$p_4 = \frac{\int_{0.9}^{1.0} I_{mf}(x)\,dx}{\left(\frac{h_{r3}+h_{r2}+h_{r1}}{h_{r3}+h_{r2}+h_{r1}+h_{r4}}\right)\int_{0.712}^{.8} I_{mf}(x)\,dx + \int_{0.8}^{1.0} I_{mf}(x)\,dx} \tag{3.6.13}$$

Based on simulated results, the average lunar limit for each stellar mass range is found by weighting the average for each rotational spin bins:

$$m_2 = \frac{1.117 \cdot h_{r3} + 1.965 \cdot h_{r2} + 3.895 \cdot h_{r1}}{h_{r4} + h_{r3} + h_{r2} + h_{r1}} \tag{3.6.14}$$

$$m_3 = \frac{1.032 \cdot h_{r4} + 1.410 \cdot h_{r3} + 2.159 \cdot h_{r2} + 3.985 \cdot h_{r1}}{h_{r4} + h_{r3} + h_{r2} + h_{r1}} \tag{3.6.15}$$

$$m_4 = \frac{1.062 \cdot h_{r4} + 1.432 \cdot h_{r3} + 2.172 \cdot h_{r2} + 3.9825 \cdot h_{r1}}{h_{r4} + h_{r3} + h_{r2} + h_{r1}} \tag{3.6.16}$$

We then find the overall lunar mass limit by weighting the average for all stellar ranges:

$$m_4 \cdot p_4 + m_3 \cdot p_3 + m_2 \cdot p_2 = 1.3848 \tag{3.6.17}$$

That is, on average, within an initial planetary rotational spin between 0 to 8 hr, *satellite with mass ratio < 1.3848 times of lunar to earth mass ratio are free from tidal locking effects within 4.5 Gyr. This means that 5.9456% of all satellites' possible lunar mass configurations are tidally-lock free.*

$$\frac{\int_1^{1.34666} S_{ize}(x)\,dx}{\int_1^5 S_{ize}(x)\,dx} = 0.059456 \tag{3.6.18}$$

Whereas $S_{ize}(x)$ is the probability distribution of the creation of moons of various sizes defined earlier.

The combined probability of appropriate rotational speed and appropriate lunar mass yields a chance of 1.3723%:

$$23.0806\% \cdot 6.482\% = 1.3723\% \tag{3.6.19}$$

Alternatively, the combined probability can be computed more precisely as the follows. The combined probability of having the appropriate rotational speed and appropriate lunar mass within each stellar mass range is found by weighting the average for each rotational spin bin's combined probability of bin's proportion within all rotational spin distribution and lunar mass limit's proportion within all moon size distribution:

$$T = \int_1^5 S_{ize}(x)\,dx \tag{3.6.20}$$

$$s_1 = \frac{\int_1^{1.263} S_{ize}(x)\,dx}{T} \cdot h_{r3} + \frac{\int_1^{2.070} S_{ize}(x)\,dx}{T} \cdot h_{r2} + \frac{\int_1^{3.941} S_{ize}(x)\,dx}{T} \cdot h_{r1} \tag{3.6.21}$$

$$s_2 = \frac{\int_1^{1.117} S_{ize}(x)\,dx}{T} \cdot h_{r3} + \frac{\int_1^{1.965} S_{ize}(x)\,dx}{T} \cdot h_{r2} + \frac{\int_1^{3.895} S_{ize}(x)\,dx}{T} \cdot h_{r1} \tag{3.6.22}$$

$$s_3 = \frac{\int_1^{1.032} S_{ize}(x)\,dx}{T} \cdot h_{r4} + \frac{\int_1^{1.41} S_{ize}(x)\,dx}{T} \cdot h_{r3} + \frac{\int_1^{2.159} S_{ize}(x)\,dx}{T} \cdot h_{r2} +$$
$$\frac{\int_1^{3.985} S_{ize}(x)\,dx}{T} \cdot h_{r1} \tag{3.6.23}$$

$$s_4 = \frac{\int_1^{1.062} S_{ize}(x)\,dx}{T} \cdot h_{r4} + \frac{\int_1^{1.432} S_{ize}(x)\,dx}{T} \cdot h_{r3} + \frac{\int_1^{2.172} S_{ize}(x)\,dx}{T} \cdot h_{r2}$$
$$\frac{\int_1^{3.9825} S_{ize}(x)\,dx}{T} \cdot h_{r1} \tag{3.6.24}$$

We then find the overall probability of appropriate rotational speed and appropriate lunar mass to avoid tidal locking for the planet by weighting the average for all stellar ranges:

$$s_4 \cdot p_4 + s_3 \cdot p_3 + s_2 \cdot p_2 = 1.8966\% \tag{3.6.25}$$

That is, on average, *1.8966% of all possible combinations of satellite mass and rotational speed are free from tidal locking effects within 4.5 Gyr.* Since the latter approach is more precise than the former, we adopt the latter result.

Now, recall that stars with mass between 0.695 and 0.712 solar mass can be tidally locking free if the initial spin rates were between 1 and 4 hr (avg of 3.1 hr, 4.2121% of all possible initial spin rate). Stars with mass between 0.663 and 0.695 solar mass can be tidally locking free if the initial spin rates is between 1 and 2 hr (avg of 1.653 hr, 0.3373% of all possible initial spin rate). We now includes these cases by including them. The equation we defined earlier is modified as:

For case 0.695 ~ 1 solar mass:

$$H = h_{r3} + h_{r2} + h_{r1} + h_{r4} \tag{3.6.26}$$

$$p_1 = \frac{\left(\frac{h_{r2}+h_{r1}}{H}\right) \int_{0.695}^{0.7} I_{mf}(x)\,dx}{\left(\frac{h_{r2}+h_{r1}}{H}\right) \int_{0.695}^{.7} I_{mf}(x)\,dx + \left(\frac{h_{r3}+h_{r2}+h_{r1}}{H}\right) \int_{0.7}^{.8} I_{mf}(x)\,dx + \int_{0.8}^{1.0} I_{mf}(x)\,dx} \tag{3.6.27}$$

$$p_2 = \frac{\left(\frac{h_{r3}+h_{r2}+h_{r1}}{H}\right)\int_{0.7}^{0.8} I_{mf}(x)\,dx}{\left(\frac{h_{r2}+h_{r1}}{H}\right)\int_{0.695}^{.7} I_{mf}(x)\,dx + \left(\frac{h_{r3}+h_{r2}+h_{r1}}{H}\right)\int_{0.7}^{.8} I_{mf}(x)\,dx + \int_{0.8}^{1.0} I_{mf}(x)\,dx} \tag{3.6.28}$$

$$p_3 = \frac{\int_{0.8}^{0.9} I_{mf}(x)\,dx}{\left(\frac{h_{r2}+h_{r1}}{H}\right)\int_{0.695}^{.7} I_{mf}(x)\,dx + \left(\frac{h_{r3}+h_{r2}+h_{r1}}{H}\right)\int_{0.7}^{.8} I_{mf}(x)\,dx + \int_{0.8}^{1.0} I_{mf}(x)\,dx} \tag{3.6.29}$$

$$p_4 = \frac{\int_{0.9}^{1} I_{mf}(x)\,dx}{\left(\frac{h_{r2}+h_{r1}}{H}\right)\int_{0.695}^{.7} I_{mf}(x)\,dx + \left(\frac{h_{r3}+h_{r2}+h_{r1}}{H}\right)\int_{0.7}^{.8} I_{mf}(x)\,dx + \int_{0.8}^{1.0} I_{mf}(x)\,dx} \tag{3.6.30}$$

$$s_1 = \frac{\int_1^{1.106} S_{ize}(x)\,dx}{T}\cdot h_{r2} + \frac{\int_1^{3.369} S_{ize}(x)\,dx}{T}\cdot h_{r1} \tag{3.6.31}$$

$$m_1 = \frac{1.106\cdot h_{r2} + 3.369\cdot h_{r1}}{h_{r2}+h_{r1}} \tag{3.6.32}$$

$$s_2 = \frac{\int_1^{1.0625} S_{ize}(x)\,dx}{T}\cdot h_{r3} + \frac{\int_1^{1.925} S_{ize}(x)\,dx}{T}\cdot h_{r2} + \frac{\int_1^{3.874} S_{ize}(x)\,dx}{T}\cdot h_{r1} \tag{3.6.33}$$

$$m_2 = \frac{1.0625\cdot h_{r3} + 1.925\cdot h_{r2} + 3.874\cdot h_{r1}}{h_{r3}+h_{r2}+h_{r1}} \tag{3.6.34}$$

$$... \tag{3.6.35}$$

Rotational spin	Avg spin	Percentage $h_{r1}, h_{r2}h_3, h_{r4}$	0.695 ~ 0.7	0.7 ~ 0.8	0.8 ~ 0.9	0.9 ~ 1.0
1~2 hr	1.6534 hr	0.33739%	< 3.369	< 3.874	< 3.985	< 3.9825
2~4 hr	3.2310 hr	3.87478%	< 1.106	< 1.925	< 2.159	< 2.172
4~6 hr	5.0919 hr	9.42800%	N/A	< 1.0625	< 1.410	< 1.432
6~8 hr	7.0280 hr	13.2741%	N/A	N/A	< 1.032	< 1.062

Table 3.6.3: The lunar mass limit for each category of stellar mass and rotational spin

For case 0.663 ~ 1 solar mass:

$$p_1 = \frac{\left(\frac{h_{r1}}{H}\right)\int_{0.663}^{0.7} I_{mf}(x)\,dx}{\left(\frac{h_{r1}}{H}\right)\int_{0.663}^{.7} I_{mf}(x)\,dx + \left(\frac{h_{r3}+h_{r2}+h_{r1}}{H}\right)\int_{0.7}^{.8} I_{mf}(x)\,dx + \int_{0.8}^{1.0} I_{mf}(x)\,dx} \tag{3.6.36}$$

$$p_2 = \frac{\left(\frac{h_{r3}+h_{r2}+h_{r1}}{H}\right)\int_{0.7}^{0.8} I_{mf}(x)\,dx}{\left(\frac{h_{r1}}{H}\right)\int_{0.663}^{.7} I_{mf}(x)\,dx + \left(\frac{h_{r3}+h_{r2}+h_{r1}}{H}\right)\int_{0.7}^{.8} I_{mf}(x)\,dx + \int_{0.8}^{1.0} I_{mf}(x)\,dx} \tag{3.6.37}$$

$$p_3 = \frac{\int_{0.8}^{0.9} I_{mf}(x)\,dx}{\left(\frac{h_{r1}}{H}\right)\int_{0.663}^{.7} I_{mf}(x)\,dx + \left(\frac{h_{r3}+h_{r2}+h_{r1}}{H}\right)\int_{0.7}^{.8} I_{mf}(x)\,dx + \int_{0.8}^{1.0} I_{mf}(x)\,dx} \tag{3.6.38}$$

$$p_4 = \frac{\int_{0.9}^{1} I_{mf}(x)\,dx}{\left(\frac{h_{r1}}{H}\right)\int_{0.663}^{.7} I_{mf}(x)\,dx + \left(\frac{h_{r3}+h_{r2}+h_{r1}}{H}\right)\int_{0.7}^{.8} I_{mf}(x)\,dx + \int_{0.8}^{1.0} I_{mf}(x)\,dx} \tag{3.6.39}$$

$$s_1 = \frac{\int_1^{2.984} S_{ize}(x)\,dx}{T}\cdot h_{r1} \tag{3.6.40}$$

$$m_1 = \frac{2.984\cdot h_{r1}}{h_{r1}} \tag{3.6.41}$$

$$... \tag{3.6.42}$$

Rotational spin	Avg spin	Percentage $h_{r1}, h_{r2} h_3, h_{r4}$	0.663 ~ 0.7	0.7 ~ 0.8	0.8 ~ 0.9	0.9 ~ 1.0
1~2 hr	1.6534 hr	0.33739%	< 2.984	< 3.874	< 3.985	< 3.9825
2~4 hr	3.2310 hr	3.87478%	N/A	< 1.925	< 2.159	< 2.172
4~6 hr	5.0919 hr	9.42800%	N/A	< 1.0625	< 1.410	< 1.432
6~8 hr	7.0280 hr	13.2741%	N/A	N/A	< 1.032	< 1.062

Table 3.6.4: The lunar mass limit for each category of stellar mass and rotational spin

Mass range	Total	Percentage	Habitable	Chance	Non-locked	Obliquity	Final
0.712 ~ 0.8	176,870,839	13.640%	24,125,532				
0.8 ~ 0.9	208,353,546	26.914%	56,077,006				
0.9 ~ 1.0	227,173,953	26.914%	61,142,396				
Total	**612,398,339**		141,344,934	1.897%	**2,680,758**	34.593%	**927,360**
0.695 ~ 0.7	9,899,214	4.212%	416,972				
0.7 ~ 0.8	200,683,741	13.640%	27,373,658				
0.8 ~ 0.9	208,353,546	26.914%	56,077,006				
0.9 ~ 1.0	227,173,953	26.914%	61,142,396				
Total	**646,110,454**		145,010,033	1.852%	**2,685,049**	34.268%	**920,104**
0.663 ~ 0.7	73,166,229	0.337%	246,856				
0.7 ~ 0.8	200,683,741	13.640%	27,373,658				
0.8 ~ 0.9	208,353,546	26.914%	56,077,006				
0.9 ~ 1.0	227,173,953	26.914%	61,142,396				
Total	**709,377,469**		144,839,917	1.853%	**2,684,569**	34.380%	**922,953**

Table 3.6.5: The number of stars with initial spin fast enough to avoid tidal locking

It is show that by including additional stars did not yield higher number of habitable planet. By introducing more stars at the lower range of habitability, the weighted importance placed upon star range of 0.8~0.9 and 0.9~1.0 is shifted to the stars with mass 0.695~0.8 and 0.663~0.8 with lower chance of habitability. Therefore, despite higher gain in candidate pool, the habitability chance actually decreased. The overall number of habitable stars remain within 1% difference. By including the obliquity criterion we about to discuss, the other cases indicates even a lower overall number of habitable planets. These results lead to conclusion. First, even by including additional star ranges, the final number of habitable planet does not increase significantly from a base total of 140 million. Secondly, determining the maximum out of the cases is beyond the resolution of our method and requiring repeating it with greater number of bins and smaller interval. Therefore, we conclude that our initial assessment of habitability within 0.712~1 solar mass remain valid.

3.7 Moon's Orbital Obliquity Evolution

The merging of protoplanets and the creation of satellites are common, but not all moons stay with their planet. If the merging occurs without a direct impact, as we have shown earlier that non-direct impact is the norm, then moon formation is inevitable in each case. The evolutionary trajectory of the moon depends both on its orbital obliquity (the angle between planetary spin axis and its orbit normal) to earth and its mass as well as stellar tidal forces. According to simulation and model run by Keiko and Ida, it is shown that five possible fates await moons formed around their host planet[76]. In the first case A1(which includes moons range from 0.01 Earth mass to 0.05 Earth mass, equivalent to less than 1 lunar mass up to 4 lunar mass with varying degrees of orbital obliquity), the moon gradually gains angular momentum and separation from its host planet and decreases the host planet's moment of inertia until both bodies obtain a synchronous rotation and orbits around each other. This is the well-known case we have observed between the Earth and the moon.

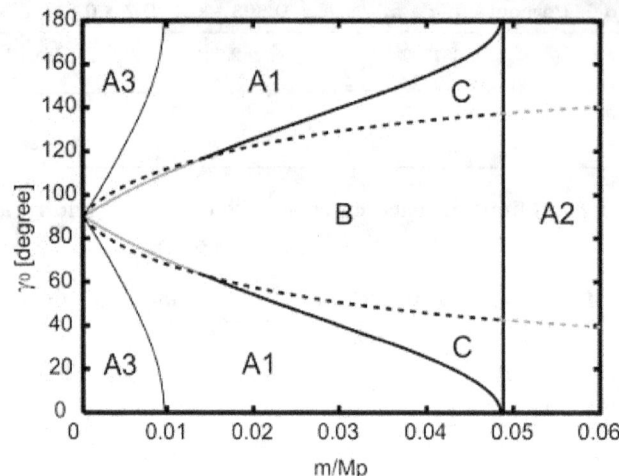

Figure 3.7.1: Lunar mass & orbital obliquity and their ultimate evolution trajectory

In case A2, the satellite with more than 4 lunar mass achieves synchronization before the occurrence of the precession transition. In this case, the obliquity angle are almost conserved as the initial values. In other words, tidally locked to their home planet.

In case A3, the satellite has less than a lunar mass. The stellar tidal torques dominate over the satellite torques. In this case, Ω becomes smaller than n before the obliquity becomes zero, then the satellite begins to decay toward planet very slowly. The subsequent reduction of the planetary spin leads to a synchronous state with planetary mean motion.

In case B which includes scenarios with no less than 40 degrees of initial orbital obliquity, the moon gradually loses angular momentum and turns back onto the host planet. Case B is important because it shows that a significant proportion terrestrial planets evolve to become moonless. It provides justification for observation of Venus and Mars.

In case C, the satellite follows the same evolutionary trajectory as B, but it is locked in a synchronous state before falling onto the planet at a distance of 5~10 earth radii. The timescale of this occurrence happens at (10^6 years), because the satellite orbit turns back at a relatively small radius A_{crit}, resulting in moon and earth experiencing 3.396 earth days per day and no tidal heating contributing to plate tectonics. Therefore, we can only treat case C as marginal habitable.

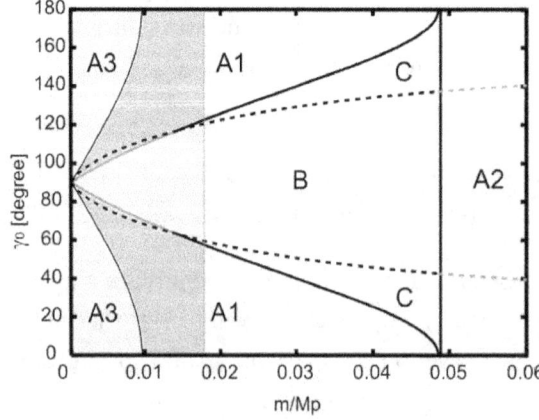

Figure 3.7.2: Lunar mass & orbital obliquity which evolves toward earth-moon relationship

Assuming that there is equally likely chance to generate orbital obliquity ranges from 0 to 180 degrees, which may require farther investigation in the future, we can compute the chance that the moons are stabilized around its host planet. From earlier calculations, we have shown that a satellite with a lunar mass greater than 1.41096 will form a synchronous orbit with their host planet within 4.6 Gyr. Therefore, we compute region A1 and exclude region B and C up to 1.41096 lunar mass by plotting the closest approximate curve to the graph projecting onto Cartesian coordinates: (1.41096 lunar mass is translated into $100 \cdot 0.012300 \cdot 1.38477 = 1.70326$)

$$U = 0.00111408x^4 - 0.0164932x^3 + 0.0939118x^2 - 0.569984x + 2.19873 \tag{3.7.1}$$

$$L = -7.60976x^4 + 11.5748x^3 - 5.71041x^2 - 0.6795x + 2.18666 \tag{3.7.2}$$

$$\frac{\int_0^{1.70326} U\,dx - \int_0^{0.967} L\,dx}{2.2 \cdot 2.2632} = 0.34593 \tag{3.7.3}$$

As a result, *around 34.593% cases, the moon is stabilized around its host planet with various initial obliquity and mass*, this shows that a planet with a stable moon is relatively common though not a universal characteristics of all terrestrial planet. Within the solar system, the earth is the only terrestrial planet hosting a moon of a massive size. On the other hand, dwarf planets such as Pluto and Charon, Eris, Makemake all have moons of significant mass relative to their host planet. In the case of Makemake, the collision occurred relatively recently. This can be implied from the non tidal-locking orbits of their moons and its own fast spin rate.

3.8   Earth like Planet Size Requirement

The stellar to planetary mass ratio indicates that the mass of terrestrial planets likely follows a lognormal, or skewed normal distribution where terrestrial planets ranges from 0.1 to 10 earth masses are possible within the stellar habitable zone. [115] Then the question is, what is the lowest and highest possible limit for a terrestrial planet to be habitable. Lopez and Fortney worked off of data from Kepler and modeled the radii of planets. They determined that planets with radii of less than 1.5 Earth radii will become super-Earths, and planets with radii of greater than 2 Earth radii will become mini-Neptunes. That suggests a radius limit of 2 Earth radii, though most terrestrial planets will probably be under 1.5 Earth radii. The study has been confirmed since there is a lack of exoplanets found between 1.5 earth radii to 2 earth radii. 1.5 earth radii can be translated into 3.375 earth mass assuming similar density. However, a planet does not need to be much larger to start to retain hydrogen gas. According to one study, planets with 1.3 earth mass likely to start capture hydrogen atoms as the planet's escape velocity catching up with the atom's escape velocity at 285 K. Though hydrogen is not poisonous. It is flammable and explosive with oxygen. It is hard to imagine a super earth with a mixed hydrogen and oxygen atmosphere will not burn in flames with the slightest spark of lightning. Another group focused on planets losing their hydrogen envelopes, the gaseous layers of hydrogen that accrete during the early parts of their lives. Their calculations indicate that planets of less than one Earth Mass would accumulate envelopes of masses between 2.5×10^{16} and 1.5×20^{23} kg. The latter is about one-tenth of Earth's mass. Planets with masses between 2 Earth Mass and 5 Earth Mass could accumulate a peak envelope mass between 7.5×10^{20} and 1.5×10^{28} kg, which is substantially more massive than Earth's. The group calculated that planets with masses less than 1 earth mass would lose their envelopes within 100 Myr. They found that planets with masses greater than 2 earth mass retains their envelopes, and so become mini-Neptunes. [45] We take the conservative estimate of 2 Earth mass, which is used as the upper limit of the habitability of terrestrial planets. On the other hand, the lower bound for habitability is cut off at 0.43 earth mass. [71] This is done by a study based on the temperature within the habitable zone and the expected gas loss composing oxygen, carbon dioxide, and nitrogen over the course of evolutionary timescale and have shown that planet with 0.43 earth mass or above can retain an atmosphere. Taking the lower and the upper bound into considerations and using the distribution samples generated based on Kepler's exoplanet data for the exoplanet mass for one solar mass star. [4]

---

[4](See Special Chapter: Stellar to Planetary Mass ratios)

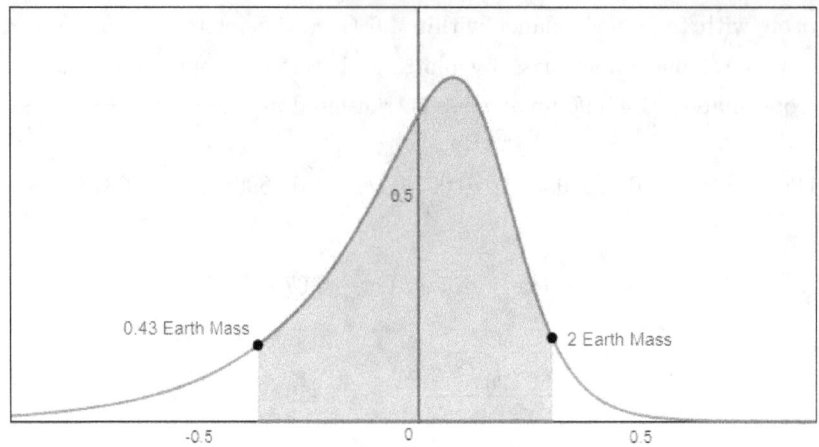

Figure 3.8.1: PDF of terrestrial planets between 0.43 and 2 earth mass

$$f_{earth1}(x) = \frac{4.5}{1(-x+1.4)\cdot\sigma\sqrt{2\pi}}\exp\left(-\frac{\ln(-x+1.9)^{2.5}}{2\sigma^2}\right) \tag{3.8.1}$$

$$\sigma = 0.2$$

$$f_{earth2}(x) = \frac{-1.8}{1+\exp(-16(x-0.16))}+1.8 \tag{3.8.2}$$

$$P(x) = f_{earth1}(x)^{0.4}\cdot f_{earth2}(x) \tag{3.8.3}$$

One can compute the probability of planets falling within this range. *The final probability is obtained to be 85.83%.*

$$\frac{\int_{-0.366}^{0.302}P(x)\,dx}{\int_{-\infty}^{\infty}P(x)\,dx} = 0.858251598868 \tag{3.8.4}$$

3.9 The Chance of Getting Watered

Studies show that terrestrial planet formed linearly as the metallicity increases but then drops sharply as the rate of hot Jupiters also increases sharply as the metallicity rises. The combined effect brings the peak of terrestrial planet creation at the metallicity index of 0.2. Metallicity not only affects the likelihood of terrestrial planet formation. More importantly, depending on the metallicity, the number of hot Jupiter attempts increases between metallicity of -0.4 at 0 percent to 100 percent at metallicity of 0.4. For metallicity 0.4 or greater, an overwhelming majority of the systems hosts hot Jupiters. Therefore, the number of failed hot Jupiter increases with increasing metallicity. This is important because our own Jupiter is also a failed hot Jupiter. The Grand Tack theory posits that Jupiter originated around 3.5 AU, just beyond the snowline of the solar system at its early day of formation (at 2.7 AU). As its protoplanetary embryo gained mass and started a runaway hydrogen accretion, it slowly migrated inward toward the sun due to strong gravitational forces. The migration came to a halt when Saturn formed and began resonate in a 2:3 orbital synchronization with Jupiter. Jupiter ventured as far as 1.5 AU from the sun before being pulled eventually to its current orbit at 5 AU. The theory is proposed to explain the low mass observed for Mars, the void of any planets in the asteroid belt, and the presence of water on earth. The theory is one of many possible fine detailed explanation of how hot Jupiter fails its migration. The true nature and complexity of the possibilities are currently not available. However, the lack of knowledge does not prevent us from arriving at our conclusion. On a system where no hot Jupiters ever arises, the inner terrestrial planets are likely to remain dry given by the understanding of the solar system formation. On a system with hot Jupiters, the inner terrestrial planets are destroyed because every protoplanet is either perturbed by the gravitational effect of the gas giant and ejected, captured, and simply absorbed. Only in cases where hot Jupiters with their migration attempts can possibly disturb the orbits of the inner planets and can bring a bombardment of a significant amount of water. This gave us a clue regarding the likelihood of terrestrial planet covered by water. The chart of the percentage of hot Jupiters, failed hot Jupiters, dry terrestrial planet, and wet terrestrial planet vs. metallicity is plotted below. We made the assumption that the hot Jupiter attempts curve is simply the hot Jupiter formation curve left shifted by 0.2 metallicity. That is, hot Jupiter attempts started at the metallicity of -0.4 but remain unsuccessful until metallicity reaches -0.2.

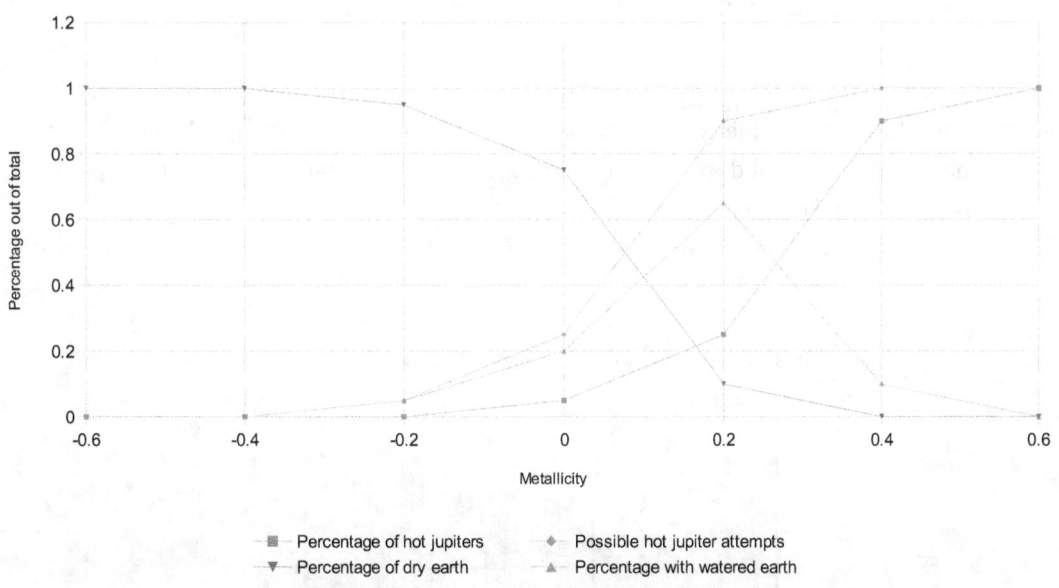

Figure 3.9.1: Watered, dry earth probability vs metallicity

Since the rate of failed hot Jupiter increases with metallicity and peaks at the metallicity of 0.2, we are able to compute the probability of a terrestrial planet gets watered over all possible ranges of metallicity which permits the creation of terrestrial planets in the first place. The probability of wet earth creation is simply the probability of hot Jupiter attempts minus the probability of successful hot Jupiters (the probability of destroying earth defined in Chp 2), and the probability of hot Jupiter attempts is simply the the probability of successful hot Jupiters left shifted by 0.2:

$$P_{JA} = P_{DE}(x + 0.2) \tag{3.9.1}$$

$$P_{WET} = P_{JA} - P_{DE}(x) \tag{3.9.2}$$

Lastly, one needs to multiply by the probability of earth formation given an increase in metallicity:

$$P_{PE}(x) = \begin{cases} (0.625x + 0.625)^{1.68} & -1 \leq x \leq 0.6 \\ 1 & 0.6 \leq x \leq 1 \end{cases} \tag{3.9.3}$$

$$f_{wetearth}(x) = P_{PE} \cdot P_{WET} \tag{3.9.4}$$

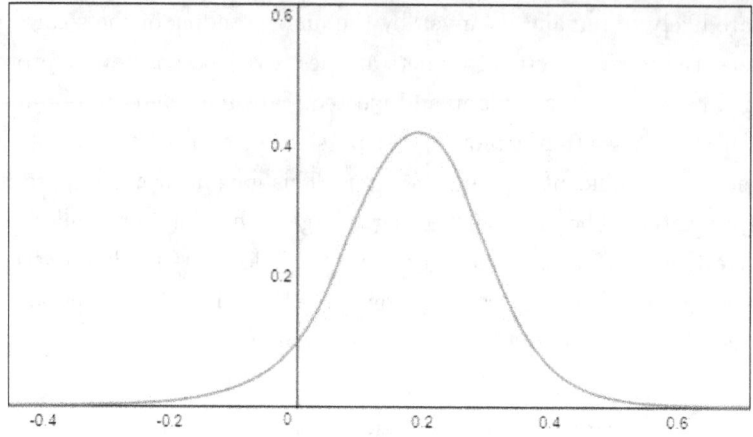

Figure 3.9.2: Metallicity PDF for wet earths

The average metallicity of terrestrial planets changes over the course of cosmic history. For our current investigation, we are only interested in the metallicity distribution from 5 Gyr ago to 4 Gyr ago. Metallicity of stars at any given age is normally distributed, and we can use existing observational data to compute the proportion of stars that will give rise to terrestrial planets.

The average metallicity of the galaxy can be obtained from this graph, assuming the metallicity is normally distributed[17] :

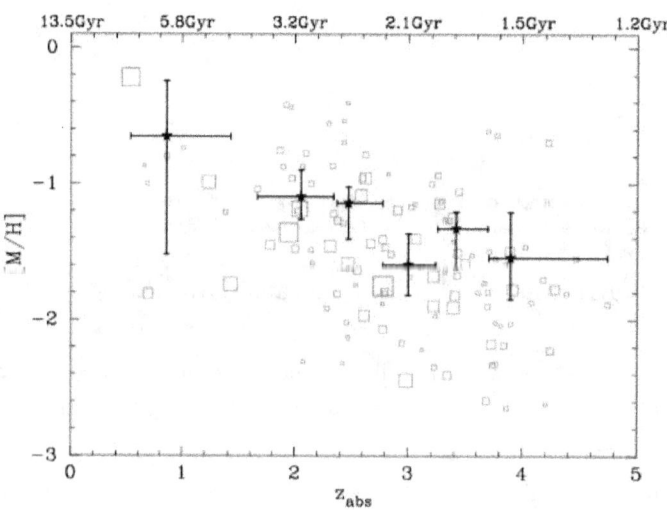

Figure 3.9.3: Average metallicity profile of cosmic historical past

and to keep our calculation consistent, we use the same normal distribution utilized by Lineweaver for the computation of number of earths with an added variable t so that the metallicity is dependent on time:

$$f_{metallicity}\left(x,t\right) = \frac{1}{\sigma\sqrt{2\pi}}\exp\left(-\frac{\left(x+0.3-t\right)^2}{2\sigma^2}\right)$$

(3.9.5)

$$\sigma = 0.3$$

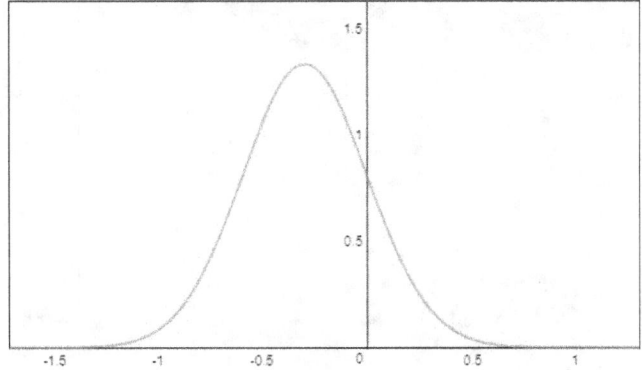

Figure 3.9.4: Observed metallicity PDF 4.6 Gya

Now, as we have obtained the probability distribution of stars by metallicity, we combine this distribution to our existing distribution for the percentage of failed hot Jupiters over a range of metallicity and the final percentage of wet terrestrial planets can be computed:

$$\left(\frac{1}{2\cdot 0.066}\right)\int_{-0.066}^{0.066}\frac{\int_{-1}^{1}f_{metallicity}\left(x,t\right)f_{wetearth}\left(x\right)dx}{\int_{-1}^{1}f_{metallicity}\left(x,t\right)dx}dt \tag{3.9.6}$$

$$= 0.0481139021498$$

Whereas the integration with values between -0.067 to 0.067 is the change of the mean metallicity of the galaxy between 5 Gya and 4 Gya. If one simply assumes that the metallicity does not change, then, the equation simplifies to:

$$\frac{\int_{-1}^{1}f_{metallicity}\left(x,0\right)f_{wetearth}\left(x\right)dx}{\int_{-1}^{1}f_{metallicity}\left(x,0\right)dx} \tag{3.9.7}$$

Based on this result, we need an additional round of computation. Lineweaver's original counting[38] for the number of terrestrial planets used metallicity as the selection criterion. Since the metallicity selection range is more lenient for the terrestrial planets than the selection range for the wet terrestrial planets, the final percentage of the terrestrial planets is higher. We used Lineweaver's distribution with metallicity as a selection criterion for terrestrial planets (Chp 2) and its formation chance:

$$\left(\frac{1}{2\cdot 0.066}\right)\int_{-0.066}^{0.066}\frac{\int_{-1}^{1}f_{metallicity}\left(x,t\right)P_{HE}\left(x\right)dx}{\int_{-1}^{1}f_{metallicity}\left(x,t\right)dx}dt \tag{3.9.8}$$

$$= 0.249119765346$$

with the plotted result:

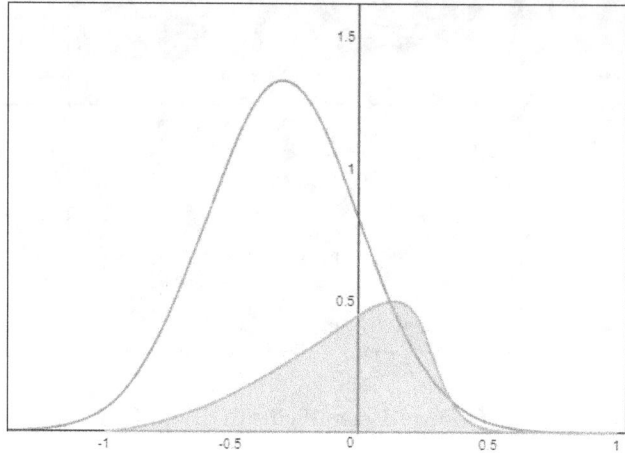

Figure 3.9.5: Metallicity PDF at 4.6 Gya for terrestrial planets overlaying metallicity PDF for dry earth selection

and a final 25% of all stars distribution falls within the Lineweaver's selection criteria.

$$\frac{0.0481139021498}{0.249119765346} = 0.193135627287 \tag{3.9.9}$$

Finally, *the prediction indicates that 19.31% of all terrestrial planets created from 5 Gyr to 4 Gyr ago is covered by ocean.*

3.10 Total Water Budget of Earth

We also need to find the average water budget on the surface of terrestrial Earth-like planets. Is earth's ocean depth typical of all Earth-like planets or is it an anomaly? We have shown earlier the importance of water to foster and create a sustainable environment for the emergence of life. For the rise of intelligent species in particular, the ocean should cover a significant amount of planetary surface to provide a relatively stabilizing climate, yet the total mass of the oceans has to strike a delicate balance enabling continental plates elevated above the sea. If a planet is covered in tens of thousands of meters depth of oceans, intelligent, land-based life manipulating tools and fire will be utterly impossible.

To answer this question, we need to deduce the water budget from several lines of reasoning. [5] First of all, from our previous discussions concerning the average mass budget leftover availability for planet formation for stars of different masses, we know that given one solar mass, the average mass budget available for planet creation is 400 earth masses. Given the metallicity of the sun, we know that 72 percent of the solar nebulae composed hydrogen, and 1.2 percent composed oxygen. Since water molecule composed of one oxygen and two hydrogen molecules, and oxygen reacts with almost every element available, we shall assume that oxygen during the formation of the solar system is readily bonded with some other element, and in particular abundantly with hydrogen. Moreover, helium is a noble gas, and not readily bond with oxygen, so the remaining elements readily bond with oxygen are carbon, iron, sulfur..etc. By finding the percentage of oxygen, as a limiting quantity and the fraction of oxygen that bonds only with hydrogen to form water, we found that the upper limit for the solar system's water budget using the empirical law is 6.107 earth mass (Taking the average of all of the oxygen used in the creation of water and a significant portion used in the construction of terrestrial planets). It is derived based on the solar system's planetary budget empirical law and the water budget is only 1.897 earth mass if we assume that oxygen is always counted toward the composition and accretion of terrestrial planet creation. The upper limit for the solar system's water budget using the generalized empirical law is $3.65^{+7.890}_{-2.495}$ earth mass

---

[5](See Special Chapter: Stellar to Planetary Mass ratios)

derived based on all stellar and exoplanets data up to date, and the water budget is only $1.1342\pm^{2.4524}_{0.7750}$ earth mass if we assume that oxygen is always counted toward the composition and accretion of terrestrial planet creation.

Secondly, we need to settle the issue of the origin of earth's water. Some argue that earth's water was readily available during the formation phase of the earth and is rapidly rose to the surface of the planet as a consequence of planetary differentiation. They further argued that the isotope ratio of earth's ocean differs from meteorite samples, consequently, earth's water cannot be delivered from the outer space. A drawback of indigenous water formation theory is that if water was present during the initial phase of planetary formation, then the total water budget of the earth today will be roughly three percent of earth mass, which is 150 times the total water budget we have, including the underground water reservoirs. Furthermore, all dwarf planets, moons of outer planets beyond the snowline have significant water content higher than the average 3 percent of their body mass. This is easily reflected from the density of Jupiter's moon Europa, Ganymede, Saturn's moon Titan ($1.8798\frac{g}{cm^3}$), Pluto and its moon Charon, which are all close to the density of water. All of this has shown that the distribution of water in the solar system is non-uniform. During the initial phase of solar system formation, the temperature of gas and debris of inner planets exceeded the boiling point of water, as a consequence, a significant amount of water molecules have gained enough energy and momentum and moved beyond the snowline, and rendered the inner terrestrial planets dry.[5] If inner planets were initially dry, then the majority of the water must have been delivered to earth from beyond the snow line. Substantial evidence shows that asteroid from the inner asteroid belt is dry, while dwarf planet Ceres from the outer asteroid belt is icy. Geological evidence has shown that water was present on earth before the late heavy bombardment at 3.8 Gyr ago, then the only other origin of earth's water must come from the asteroid belt. Indeed, the isotopic ratio of earth's ocean closely resembles those of the water found from asteroids of the asteroid belt. The mechanism for this delivery is explained by the migration of Jupiter into the inner solar system and its later migration outward by the pull of Saturn, which was also migrating inward. This is not atypical, in fact, it has been found that around 5% of planetary systems with solar metallicity contains hot Jupiters, and the formation of failed hot Jupiters are more likely, just like in the solar system. As Jupiter migrated toward the sun, it perturbed the protoplanets, asteroids within the asteroid belt and they either gained speed and are ejected from the solar system or lost speed and start to fall into the sun. As the debris falls toward the sun, it intercepts and crosses earth's orbit around the sun. Though a majority of cases, the debris crosses and without gravitationally attracted by the earth, on closer approaches, with a distance at or shorter than the effective Hill radius, and especially shorter than the Roche limit, the debris hit earth, thus delivering water to the surface. Computing the circumference of earth's orbit and weighted effective distance that asteroid can be captured and hit earth, we obtain the final total water budget of the earth.

From this line of reasoning, we can calculate the total amount of water budget the earth can obtain. The total water budget of the solar system beyond the snowline 2.7 AU at the formation of the solar system is 1.1342~1.897 earth masses. We adopt the Nice planetary formation model so that we assume the total water budget is dispersed between 2.7 AU to 26 AU unit, beyond the orbit of Uranus at 20 AU and stretch into the Kuiper belt. Then, we assume that a migrating gas giant could arise from any arbitrary distance away from their star. In the solar system's case, Jupiter started to migrate inward from 3.5 AU and Saturn from 6 AU. We assume that icy comets and asteroids can only be captured and impact earth when they approached 1.5 million km or closer to earth. Twice the Hill sphere distance over the circumference of earth's revolution path around the sun is the fraction of water can be captured by the earth. We find the total deliverable water budget to the planet earth by assuming a gas giant started its migration from the Kuiper belt, then the entire water budget of the solar system can potentially be diverted toward the sun. In the most extreme case, at most 20 times the mass of current ocean will be available to earth.

However, it is estimated an additional 1.5 to eleven times the amount of water in the oceans is contained in the Earth's interior [29] and some have hypothesized that the water in the mantle is part of a "whole-Earth water cycle." [21] The water in the mantle is dissolved in various minerals near the transition zone between

Earth's upper and lower mantle. Direct evidence of the water was found in 2014 based on tests on a sample of ringwoodite. Liquid water is not present within the ringwoodite, rather the components of water (hydrogen and oxygen) are held within as hydroxide ions.

$$w = \frac{26^2}{26^2} \cdot \frac{(1.1342 + 1.897)}{2} \cdot \frac{(2 \cdot 1,500,000 \text{ km})}{2\pi \cdot 149,597,870 \text{ km}} = 0.00483727271819 \quad \text{M}_{\text{earth}} \tag{3.10.1}$$

$$\frac{w}{\left(\frac{1}{4,400}\right) \cdot 2} = 10.64199998 \quad \text{O}_{\text{cean}} \tag{3.10.2}$$

We assumed that oxygen is always counted toward the composition of terrestrial planet creation. As a result, 1.1342 earth mass is the mean total water budget of the solar system based on generalized empirical stellar to planetary mass ratio, and 1.897 earth mass is the total water budget of the solar system based on empirical stellar to planetary mass ratio derived from the solar system only, and we take the average of the two to arrive at the total water budget of the solar system.

We will assume that the total water budget on earth currently is twice of the mass of the ocean, then we arrived a range from 0 to 10.64 times of water deliverable to earth. Since most inward migration of gas giants likely originated near or closer to the snow line because a higher concentration of ice material blown from less than 2.7 AU concentrated just beyond the snow line, it is likely that the total mass of water deliverable to earth is skewed to the left. We used a lognormal distribution to simulate the distribution of water deliverable to earth.

$$g_{ocean}(x) = \frac{5}{.95 x \sigma \sqrt{2\pi}} \exp\left(-\frac{\left(\ln 0.587 x^{0.8}\right)^2}{2\sigma^2}\right) \tag{3.10.3}$$

$$\sigma = 0.63$$

Now based on estimates of continental plates and their different surface area percentage relative to the oceanic plates', the right amount of water which will enable the emergence of a shallow sea which smooth the transition from subduction zone to dry land ranges from 0.0596 earth ocean mass to 2.3 earth ocean mass. So we can integrate this region and find the probability of such ocean budget formation.

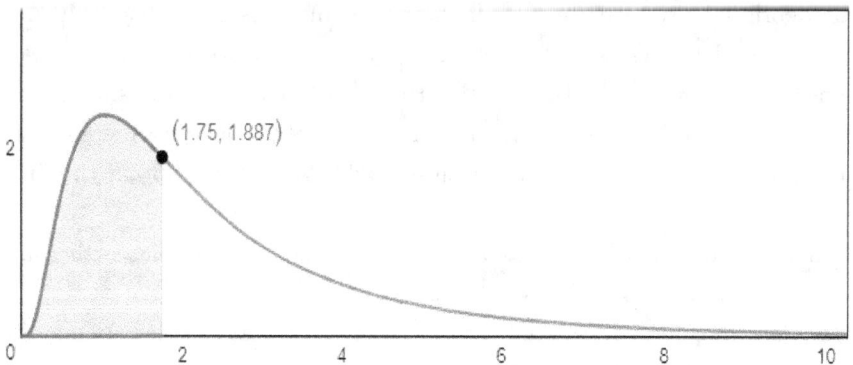

Figure 3.10.1: PDF for various final ocean size on a terrestrial planet

$$\frac{\int_{0.0596}^{2.3} g_{ocean}(x)\, dx}{\int_0^\infty g_{ocean}(x)\, dx} = 0.593152487335 \tag{3.10.4}$$

It is now shown earth at *59.32% chance gets water delivered in the right proportion* that can potentially enable the emergence of dry land and the possibility of land-based life forms.

3.11 Right Ocean and Land Mix

Is earth's water and land ratio typical of any earth-like planet? In order to answer this question adequately, conclusions can only be settled when several different fields and their effects on earth's geologic process is thoroughly analyzed. First of all, the dry part of earth above the sea level, or the continents, are actually cratons made of lighter composition mostly of granite, that semi-floats over the ocean cratons with higher density. Continental plates are thicker in which its upper edges rises above the oceanic plates, and its lower edges sink deeper than the oceanic plates. During the early formation period of earth, planetary differentiation ensured that water, which has much lower density than the crust, covered the entire surface of the earth. Studies have shown that earth's surface temperature was nearly the same compared to that of today, despite the sun with only 75 percent of the luminosity compares to today. In whats being labeled as the Faint Young Sun paradox, the earth was supposed to be frozen as an ice ball. Some have argued that the earth' temperature was much higher as a consequence of higher methane and carbon dioxide level, but equally important, the low albedo of an early ocean planet ensured an absolute higher energy absorption rate. The emergence of continents was not evident until the Archean epoch. During the Hadean phase of earth's development, the leftover heat from radioactivity was three times higher than that of today. The rate of new crust creation along volcanic faults and the rate of existing crust destruction were too quickly for any cratons accretion to take place. By 2.5 Gyr ago, the rate of internal heat has cooled enough enabling the accretion of volcanic arcs (probably similar to the Hawaii islands chains). If the rate of accretion was faster than the rate of destruction, then continental cratons began to form and increases in size until the creation rate significantly slows down as the mantle continues to cool. This trend is clearly observed in earth's geologic history in which 30% of land mass first appeared in the Archean era and 50% in the Proterozoic era and 20% in the Paleozoic. In earth's cases, over 40 % of the surface area is covered by continental cratons, orogenic belts, and platforms. However, this percentage can easily be greater or less depending on the initial endowment of the radioactive leftover of the molecular cloud forming the planet. [86][10][100][99] The primary sources of radioactivity observed on earth come from Uranium 235, Thorium 238, and Potassium 40. On some other planets, each of the radioactive material endowment could be higher or lower than we found on earth which in turn generates different mantle cooling curves and eventually contributing to different continental plate formation sizes. Furthermore, the moon (with a separation distance of 40,000 km when it first formed) was significantly closer to earth in its early days, and must have significantly contributed to tidal heating of the early earth and enabled the accelerated emergence of the growth of the continental plates. Moreover, we have shown that though the moon formation around terrestrial planets is common, the final mass of the moon varies, which again contributes to differential growth and development of the continental plates sizes. In summary, one can conclude that the percentage of continental plates covering any planet which owns one moon and is initially covered by an entire ocean can range from a few percentage of the surface area to completely covering the surface.

We can then formulate a mathematical model to delineate the ratio of drylands to ocean surface for different proportion of continent size. We reinstated the equation describing the continental plates[93][10] in the form of polynomial functions. The initial drop curvature to the right of y-axis in height represents mountain and high plateau, the horizontal leveling portion represents open plains or platforms, and the final drop before hitting the x-axis represents the continental shelf and continental plates' cliff. (i.e. Mariana trenches observed on earth) We formulate a list of curves to mimic the continental plates size from covering a few percentage of the planetary surface to that of the entire surface.

$$y_{smallplate} = -\left(\frac{x-2}{9}\right)^{29} + \frac{4}{x+0.4} + 6.596 \qquad (3.11.1)$$

$$y_{mediumplate} = -\left(\frac{x-8.717}{9}\right)^{11} + \frac{13}{x+0.4} + 6.596 \qquad (3.11.2)$$

$$y_{bigplate} = -\left(\frac{x-13.717}{9}\right)^{5} + \frac{24}{x+0.4} + 6.596 \qquad (3.11.3)$$

We also need to reinstate the oceanic plate curve. For the simplicity of the model, we adopt the linear equation that slopes gradually from the dividing trenches toward the boundary between the continental and oceanic plates.

$$y_{oceanplate} = 0.3\,(x - 11.621) \qquad (3.11.4)$$

The plot results are represented below:

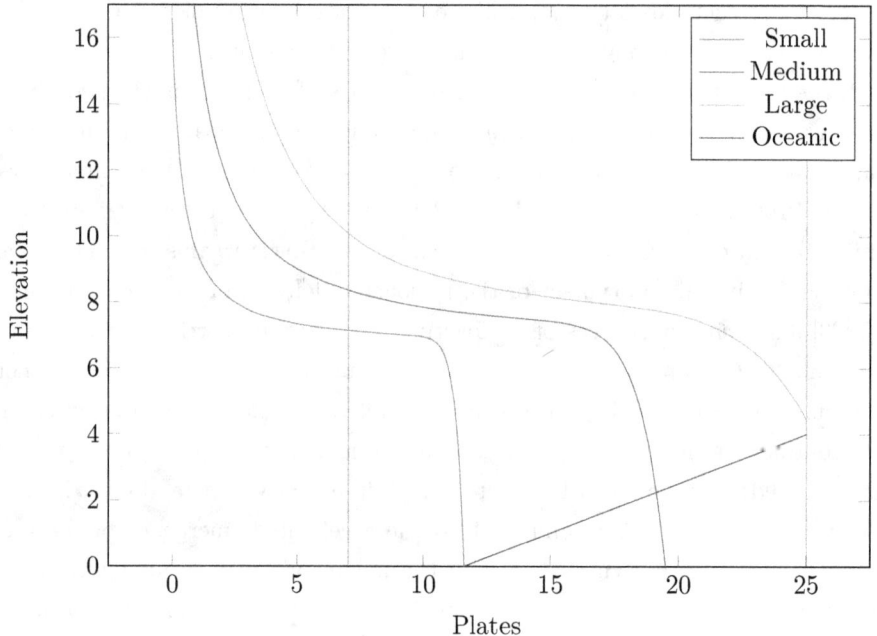

Figure 3.11.1: Planet with small, medium, and large sized continental plates

The vertical line x = 7 represents the proportion of the land and the ocean, where 7 units to the left of vertical line represent 29% of the land surface area and 18 units to the right represent 71% of water surface area. We will integrate and find the area enclosed by the ocean, which represents the total mass of water at the surface of the earth.

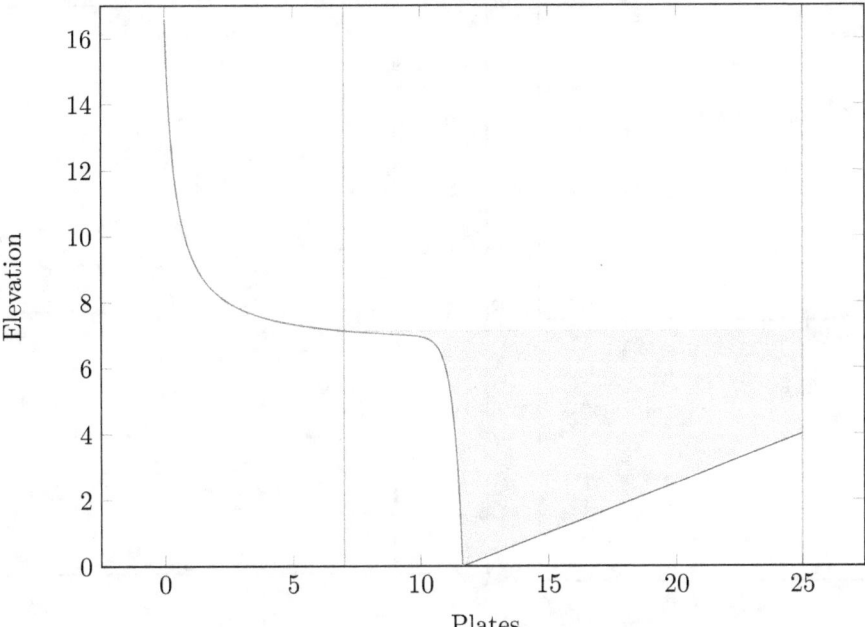

Figure 3.11.2: Earth's case by the model

Using integration, one obtains the final value of 71.4161.

$$(25 - 7.25) \cdot 7.119 - \int_{11.621}^{25} y_{oceanplate}\,dx - \int_{7.25}^{11.621} y_{smallplate}\,dx = 71.4161 \qquad (3.11.5)$$

We will use this value to compute the shoreline of different configurations of plates and oceans. In order to proceed, we need to find the equation which defines the water level for continental plates at different depth and height. The general idea of the equality is expressed as:

The minimal rectangular bounding box of the ocean - the portion occupied by the continental plate - the portion occupied by the oceanic plate = the ocean size

Whereas h_0 is the x coordinate of the intersection between the oceanic plate and the continental plate, x is the x coordinate of the shoreline ranges between 0 to 25 and is the value we are solving for and the equation is:

$$[f_{plates}(x) - f_{plates}(h_0)](25 - x) - \left[\int_{x}^{h_0} f_{plates}(x)\,dx - f_{plates}(h_0)(h_0 - x)\right]$$

$$- \left[\int_{h_0}^{25} f_{ocean}(x)\,dx - f_{ocean}(h_0)(25 - h_0)\right] = 71.141 \quad (3.11.6)$$

$$f_{plates}(h_0) = f_{ocean}(h_0) \qquad (3.11.7)$$

The equation simplifies to:

$$[f_{plates}(x) - f_{plates}(25)](25 - x) \quad - \quad \left[\int_{x}^{25} f_{plates}(x)\,dx - f_{plates}(25)(25 - x)\right] \quad = \quad 71.141 \quad (3.11.8)$$

When ocean completely floats over the continental plates in the most extreme scenarios.

Using this equation, one can then derive the water level and the shoreline and the proportion of ocean and dry land surface ratio.

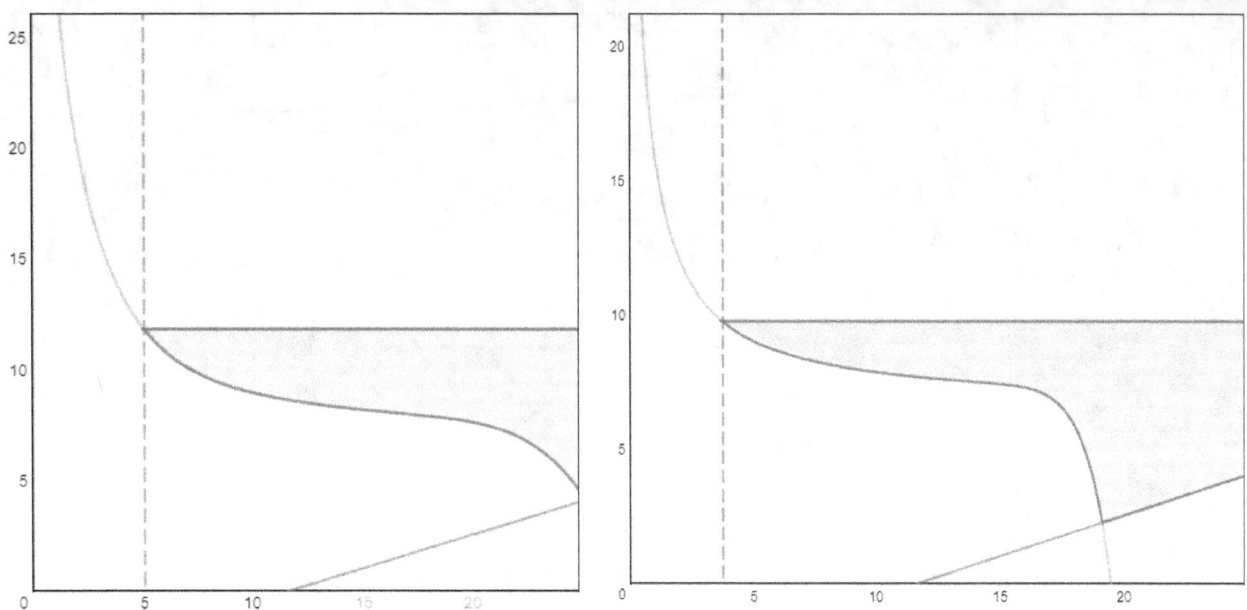

Figure 3.11.3: Water level for 2 different possible continental configurations

We devise a set of equations, mimicking the share of continental crust in proportion to the earth's total surface area from 100%, 77.97%, 53.72%, 49.43%, 46.48%, 44.42%, 38.34%, 33.05%, to 18.08% respectively. Then we find the surface to ocean ratio assuming 1 earth ocean mass.

Plate equations for different dryland coverage

$$y_{plate5}(x) = -\left(\frac{x+8}{9}\right)^{59} + \frac{0.25}{x+0.4} + 6.596 \qquad y_{plate49}(x) = -\left(\frac{x-2.5}{9}\right)^{21} + \frac{5}{x+0.4} + 6.596$$

$$y_{plate10}(x) = -\left(\frac{x+7}{9}\right)^{49} + \frac{0.5}{x+0.4} + 6.596 \qquad y_{plate54}(x) = -\left(\frac{x-3.5}{9}\right)^{19} + \frac{6}{x+0.4} + 6.596$$

$$y_{plate18}(x) = -\left(\frac{x+5}{9}\right)^{39} + \frac{1}{x+0.4} + 6.596 \qquad y_{plate60}(x) = -\left(\frac{x-5}{9}\right)^{19} + \frac{8}{x+0.4} + 6.596$$

$$y_{plate26}(x) = -\left(\frac{x+3.15}{9}\right)^{36.5} + \frac{1.65}{x+0.4} + 6.596 \qquad y_{plate68}(x) = -\left(\frac{x-6.8585}{9}\right)^{15} + \frac{10.5}{x+0.4} + 6.596$$

$$y_{plate34}(x) = -\left(\frac{x+1.3}{9}\right)^{34} + \frac{2.3}{x+0.4} + 6.596 \qquad y_{plate77}(x) = -\left(\frac{x-8.717}{9}\right)^{11} + \frac{13}{x+0.4} + 6.596$$

$$y_{plate38}(x) = -\left(\frac{x+0}{9}\right)^{32} + \frac{3.3}{x+0.4} + 6.596 \qquad y_{plate87}(x) = -\left(\frac{x-10.717}{9}\right)^{7} + \frac{16.5}{x+0.4} + 6.596$$

$$y_{plate41}(x) = -\left(\frac{x-0.75}{9}\right)^{31} + \frac{3.5}{x+0.4} + 6.596 \qquad y_{plate97}(x) = -\left(\frac{x-12.717}{9}\right)^{5} + \frac{20}{x+0.4} + 6.596$$

$$y_{plate46}(x) = -\left(\frac{x-2}{9}\right)^{25} + \frac{4.4}{x+0.4} + 6.596 \qquad y_{plate100}(x) = -\left(\frac{x-13.717}{9}\right)^{5} + \frac{24}{x+0.4} + 6.596$$

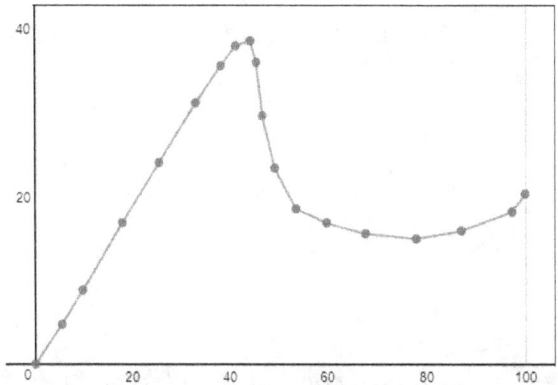

Figure 3.11.4: Continental plate percentage vs dry land percentage as earth's ocean size

Although the graph shows that in all possible cases, dry land is exposed to a significant degree and peaks at 44% for earth's ocean size. Only continental plates with sizes covering from 53.72% to 44.42% of the surface area of the earth are able to accommodate platforms and flat plains that offers feasible agriculture and shallow ocean with a submarine continental shelf which makes a biological transition from marine species to terrestrial ones possible given the endowed water budget on earth. Continental plates covering less than 44.42% of earth surface area has exposed, sub-aerial continental shelf, high cliffs shore render transition from fish to amphibian species impossible. Continental plates covering more than 53.72% have high sea levels due to lower sea depth covering above the continental platform. The shoreline is much further inland, and the elevation rises sharply from the shore. In the best possible case, an intelligent tool-using species, fruit trees, and grass plant can evolve under such a configuration, but it is impossible to develop full-blown agricultural civilization because only very narrow strips of land along the shoreline has low enough elevation with moderate climate enabling cultivation.

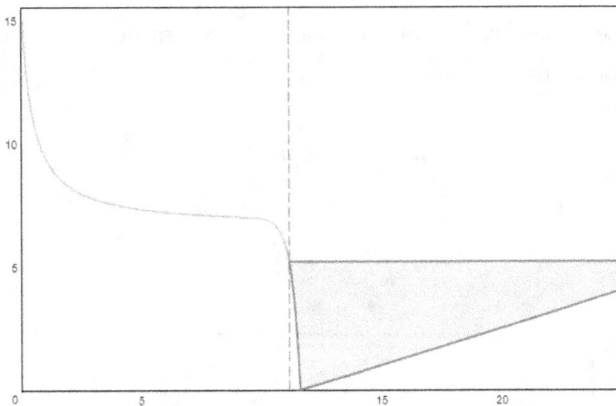

Figure 3.11.5: Planet with shoreline below the continental shelf

As a result, there are 9.3% out of all possible continental covering scenarios can provide the playground for intelligent species to realize its full potential given earth like water budget. Furthermore, parameters tweaking shows that the current continental configuration is susceptible to the total amount of water budget. Although ocean retreats during an ice age, at the peak of ice age 20,000 yrs BP, sea level was 120 meters lower, this only reduces the total mass of the ocean by 3%, unable to expose the continental shelf. Exposing the continental shelf requires a mass reduction by 5% or more. On the other hand, inundating continental shelf platform requires a mass increase by 5% or more. Since the total water budget available to earth enabling the exposure of dry lands ranges from 0 to 11 earth ocean worthy of water and is distributed around a mean of 1, then the chance of having earth-like continental covering ratios and a similar level of ocean mass is merely 0.3596% assuming ocean to land surface coverage ratio stays roughly the same from 23% to 33%.

$$\frac{\int_{0.95}^{1.05} g_{ocean}\left(x\right) dx}{\int_{0}^{\infty} g_{ocean}\left(x\right) dx} = 0.003596 \tag{3.11.9}$$

We then generalize and applies the total probability for all possible cases. Any planet with a continental mass covering the surface provides the necessary condition for the emergence of intelligent life. For each possible continental distribution there lies a narrow range of water budget (a weighted average of from 5% reduction to 5% increase from the baseline).

Continental Proportion	Ocean to Land Ratio	Allowable Water Range relative to Earth's
100.00%	20.29%	0.060~0.270
77.97%	14.92%	0.400~0.580
53.72%	18.50%	0.820~0.960
49.43%	23.36%	0.910~1.042
46.48%	28.00%	0.963~1.083
44.42%	38.56%	1.000~1.100
38.34%	35.62%	1.168~1.282
33.05%	31.13%	1.300~1.350
18.08%	16.83%	1.700~1.750

Table 3.11.1: Ocean to land ratio with required total water budget

Based on the increase and decrease of sea level from the base line, we can devise the upper and lower bound curve for different continental proportions.

$$y_{upper} = 1.68 \cdot \frac{1}{2}x \qquad\qquad (3.11.10)$$

$$y_{lower} = 0.8 \cdot \frac{1}{2}x \qquad\qquad (3.11.11)$$

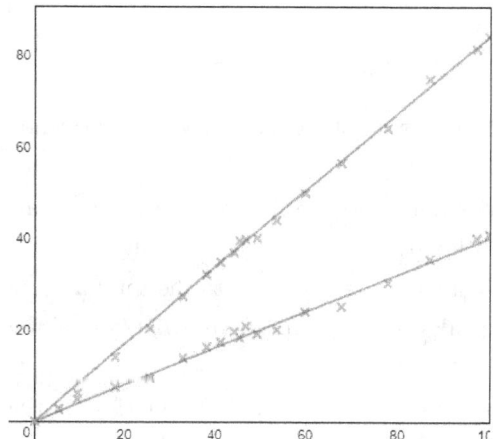

Figure 3.11.6: Curve fitting for the upper and lower bound

Integrating with our earlier curve, we have:

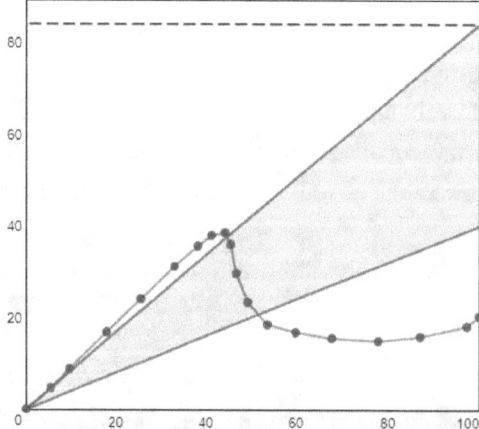

Figure 3.11.7: Continental plate percentage vs dry land percentage as earth's ocean size fitted within the upper and lower permissible bounds enabling the emergence of civilization.

The shaded region represents all possible land to sea ratio for any continent plate coverage for dryland coverage.

One can see that our curve rises above the upper bound at 38.56% land coverage and falls below the lower bound at 18.5% dryland coverage. The portion of the curve falls inside the shaded region is the permissible dryland proportion which enables the emergence of industrial civilization at 1 earth ocean mass across many continent plate coverage ranges.

Furthermore, the proportion of dryland exposure for different percentage of continental plate varies depending on the total mass of the ocean. The portion of the curve falls inside the shaded region, the permissible dryland proportion which enables the emergence of industrial civilization, varies accordingly. As a result, any arbitrary continental configurations, depending on the proportion of ocean endowment, can allow the emergence of civilizations. The 9% out of total continental configurations constraint on the emergence of civilizations given only earth sized ocean budget is relaxed by assuming the ocean size can be variable. We will illustrate later how glaciation can also narrow down the number of potential candidates based on continental configuration. In essence, there is more potential selection criteria that can exclude certain range of continental crusts from considerations.

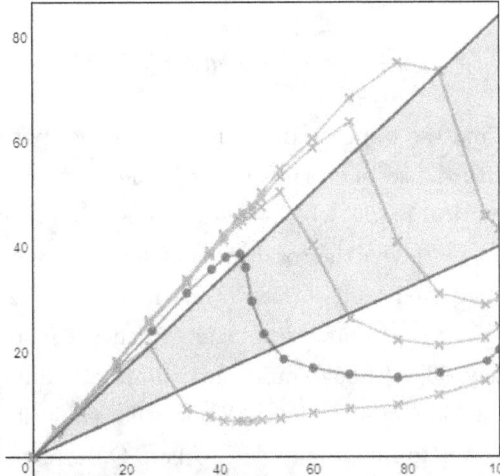

Figure 3.11.8: The proportion of dryland exposure by varying mass of ocean. From the bottom to the top: 1.5, 1, 0.75, 0.5, 0.25 earth ocean mass respectively.

Therefore, the total probability is 5.23%. There are two ways to arrive at this value. One approach is simply taking the weighted average of all permissible ranges for each possible continental configurations, we plot the result:

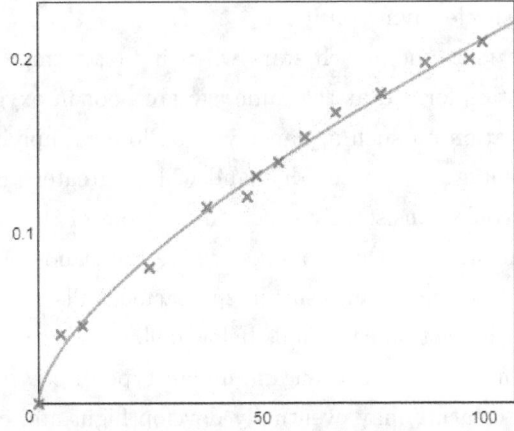

Figure 3.11.9: The permissible ocean range decreases as the continental plate coverage shrinks from 100 to 0 percent

and the best fit is:

$$R_{ange}(x) = 0.0968777x^{0.666504} + 0.0258301 \qquad (3.11.12)$$

and the weighted average is:

$$\frac{1}{100} \int_0^{100} R_{ange}(x)\, dx = 0.1277 \qquad (3.11.13)$$

and we then take the weighted range and divided by the total water budget range that are capable of giving the emergence of intelligent species:

$$\frac{0.1277}{2.3 - 0.0596} = 0.057 \qquad (3.11.14)$$

Another approach is finding the probability that each land's permissible range falling within the PDF of the total water budget of earth which will allow the transition to an intelligent life:

$$\frac{1}{100} \int_0^{100} \frac{\int_{\frac{High(y)}{10}}^{\frac{Low(y)}{10}} g_{ocean}(x)\, dx}{\int_{0.0596}^{2.3} g_{ocean}(x)\, dx} dy = 0.05234 \qquad (3.11.15)$$

The high and low conversion is used to find the upper and lower water budget permissible for each continental plate coverage. We stayed our results with the second one, since it is the more precise definition.

In conclusion, from the previous mathematical model which shows all possible range of cases, dry land occupies at most 40% of the surface area of the planet given the total mass of surface ocean is similar to that on earth. Although nearly all planets do have an exposed land surface if ocean budget < 2.3 earth's, their continental geology will be significantly different. Based on known geologic evidence and research, the first continents formed were much flatter than those today.[93] Although orogenic mountain building process also occurred, the highest mountains are probably around 3,000 meters in height above the sea level or even lower, since most of the continental plates first emerged have yet to merge into each other. On a planet dominated by continental plates land masses frequently bump into each other and creates magnificent mountain building regions. As a result, on a planet dominated by continental plates given earth sized ocean, as soon as one goes inland from the shoreline toward the continents interiors, the sea level rises sharply. With a sharp rise of the continental plates, tropical trees cannot thrive as the temperature drops quickly from the shoreline even if the trees grow near the equator. If we assume any typical intelligent species have to emerge from an environment that is relatively flat, and to cultivate agriculture before its transition into a technological species, then these (island planets and continental planets with steep rises) planets may not be suitable candidates for the emergence of technological civilization despite possibly intelligent species living on it.

Finally, the solar system is one of the more metal-rich stars when it first formed 4.5 Gyr ago, most of the planets revolving around their parent stars formed at the same age are poor in oxygen compares to earth. As a result, the upper limit of water formation on such a planet will be lower, implying ocean with a lower sea level. Under such a scenario, the percentage of exposed dry land will be greater and can exceed 40%. A few interesting facts follow. The shoreline could consistently touch the bottom of the continental shelf. The view of the planet can potentially be spectacular near the shoreline, where thousands of meters of cliff drops from the land to the sea. Rivers discharging into the ocean result in spectacular falls. This type of geology, just like we have shown in cases where the water budget on earth falls below 0.95 earth ocean mass, implies that almost no aquatic species on such a planet can evolve toward an amphibian type of creature. If one really stretches one's imagination, a flying fish type of creature may eventually develop flight and colonize the land, but such probability is astronomically low compares to transition from fish to amphibian where shorelines, lakes, and rivers naturally extend into the ocean. Furthermore, some of the greatest biological diversity is observed within the continental shelf, where the depth of the ocean is no more than a few hundred meters. Within this layer of the ocean, a complete ecosystem comprising food chains and symbiotic relationships develop and co-exist between the top layers of the water (photosynthesis) up to the bottom floor of the sea. Without the existence of continental shelf sea, it is hard to imagine the appearance of many multicellular life forms such as corals, crabs, lobsters, and fish which either directly or indirectly consumes sunlight as well as requiring anchoring on the ground, therefore, it is hard to imagine complex multicellular life (such as flying fish) to evolve at all. One can even stretch one's imagination even further, assume such planet had overcome the insurmountable barriers of high rise of the continents and conquered the land in an astronomically small chance. It is still hard to imagine such species to engage in inter-continental trade and undergoing through an Age of Exploration, which is one of the necessary recipes for the ushering into a technological industrial civilization. An intelligent species on such a planet will know the true meaning of the edge of the world not available in our dictionary.

3.12   Plate Tectonics

The debates have been on whether all terrestrial planets undergo plate tectonics. In fact, no other terrestrial planets undergo plate tectonics as observed on earth. Mars, Europa, Io, Enceladus drives internal heat from the core to the surface by through pipe volcano. Venus undergoes entire planetary resurface. This observation prompts many to propose the initiation of plate tectonics is probably unique to earth, and the number of possible planets that gave rise to the intelligent tool using species is small. However, studies done on the possibility of tectonic activity on super earth indicates that the presence of water on the terrestrial planets, essentially acting as a lubricant, enables tectonic subduction on plate boundaries.[66] Therefore, plate tectonics should be universal on terrestrial planets within the habitable zone with the presence of considerable depth of ocean and plate tectonics is not a selection criteria for filtering the number of potentially habitable planets.

Although plate tectonics may be universal, one has to further investigate the level of intensity of the tectonic movement. If the geologic activity is intense with volcanism and earthquake, it is not conducive gives to the emergence of intelligent life. The model for tectonic movements, though very intricate and complex, can be simplified into a toy model based on two assumptions. First, the rate of tectonic plate creation is directly proportional to the heat release per unit area of the planet. That is, the greater the heat flux, the more active the plate tectonics. The formation of new crusts is a consequence by the convective magma inside the planet as a form of heat dissipation from the planet's original formation in the form of potential energy and radioactive elemental decay such as uranium and thorium. It is, then, no surprise that the young earth billions of years ago had more active geologic activities. Secondly, the rate of tectonic subduction is directly proportional to the gravity acted upon the plates. The oceanic mafic plate emerges from the site of its creation and gradually over the course of millions of years consolidated in density and increased in weight and sloped toward the subduction zone. The subduction zone, such as the Mariana trench, are some of the deepest places on earth, pulls the plate into the mantle upon its own weight, thereby completing the recycling of the oceanic plates. It is no surprise that planets with a higher mass also have a greater surface gravity and a greater surface heat flux. A third factor involving the thickness of the crust is sometimes also considered, but for the simplicity of our argument (a lack of data to correlate crust thickness based on the mass, composition, or the formation condition of the planet. Mars has a thicker crust with 10% earth mass, and Venus has comparable crust thickness to earth with comparable density, yet all three planets have comparable composition), we shall assume that the crust thickness is equivalent in all terrestrial planets of different masses. Finally, the presence of water as a lubricating agent is probably essential for carrying persistent tectonic activities.[66] The question becomes, given an increase in mass of a terrestrial planet, how will the speed of tectonic movement change and by how much. If we assume that all terrestrial planets have a similar density to earth,[34] then the following graph can be used to predict the surface gravity and surface area on terrestrial planets of other masses.

$$r_0 = \left(\frac{3}{4}x\right)^{\frac{1}{3}} \tag{3.12.1}$$

$$y_{plate} = \frac{1}{0.3029}\frac{\frac{4}{3}\pi r_0^3}{4\pi r_0^2} \tag{3.12.2}$$

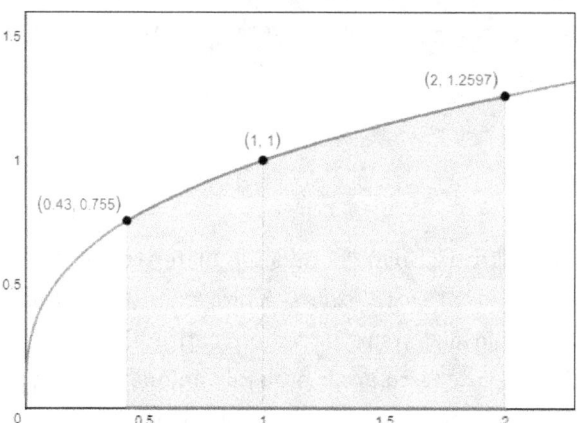

Figure 3.12.1: Geological intensity vs. terrestrial planet mass

The relationship of gravity and heat flux increase is plotted. One can quickly see, from the graph, that tectonics activity does not significantly increase as the mass of the planet increases. Based on the stellar to planetary mass ratio, the maximum attainable super earth size is about 2 earth mass. The surface gravity grows relatively slow compares to the increase in mass, that is, a terrestrial planet at the 2 earth mass will have a surface gravity at only 1.26 times that of the earth. Furthermore, the surface area will also be 1.26 times that of the earth, as a consequence, one can see that radiative convection on this planet will also create new crusts at 1.26 times the speed observed on earth while engulfing old crusts also at 1.26 times the speed on earth. As a result, for a terrestrial planet, put on a limit of no more than 2 earth masses, the plate tectonic movement cannot exceed more than 1.26 times the speed on earth. At this mass range, the speed of tectonic activity is moderately higher than earth. This does justify for the slight differential speed of evolution on different planets (we talked about how species can evolve quickly based on molecular biology but is held in check by the pace of geologic changes)[6] but not sufficient enough to serve as a filter for limiting the number of habitable planets conducive to intelligent life.

---

[6]See Chapter 4 Section 4.5

Chapter 4

Evolution

4.1   Water vs. Other Solvents

Some argue that water as the only solvent for life is probably too limiting, and by including other types of hydrocarbons such as ammonia and methane into consideration is also important. Upon closer examination, ammonia's molecules are composed of nitrogen and hydrogen. Even if nitrogen rarely interacts with other types of atoms, only 960 atoms out of every 1 million atoms in the Milky Way composed of nitrogen. Therefore, the maximum upper bound of ammonia creation is 960 pairs of ammonia molecules out of 1 million atoms. This directly pales in number with water with 10,112 pairs of molecules out of 1 million atoms at its upper bound. Ammonia makes up only 9.49% of the water budget in the galaxy at the most.

Furthermore, ammonia's melting and boiling points are between 195 K and 240 K. Chemical reactions generally proceed more slowly at a lower temperature. Therefore, ammonia-based life, if it exists, might metabolize more slowly and evolve more slowly than life on Earth. Ammonia is also flammable in oxygen, and could not exist sustainably in an environment suitable for aerobic metabolism. Ammonia could be a liquid at Earth-like temperatures, but at much higher pressures; for example, at 60 atm, ammonia melts at 196 K and boils at 371 K. However, the higher atmospheric pressure will guarantee more stabilizing climate and minimize the chance of fluctuating weather patterns. We shown in our discussion regarding the emergence of intelligent species in many ways, one kind of adaptation of intelligent life is its quick responses to ice age uncertainties.[1]

Methane, on the other hand, composed of hydrogen and carbon atoms. The maximum upper bound of methane creation is 4,472 pairs of molecules per 1 million atoms (carbon ready to bind with any other atoms freely). However, it has an even lower melting point at 90 K and boiling point at 112 K, and evolution will proceed even slower than ammonia-based life. With even higher atmospheric pressure, methane may be available at room temperature but as we have shown earlier thick atmosphere minimized the chance of ice ages (in addition to the fact methane is a greenhouse gas). Moreover, it is extremely flammable and may form explosive mixtures with air. It is violently reactive with oxidizers, halogen, and some halogen-containing compounds.

All other hydrogen chalcogenides such as hydrosulfuric acid (H_2S), hydroselenic acid (H_2Se), hydrotelluric acid (H_2Te), and hydropolonic acid (H_2Po) suffer the same handicaps listed earlier, and they are far rarer because sulfur, selenium, tellurium, and polonium are all rarer than nitrogen and carbon. At the same time, their boiling and melting points are all lower than water. Hydropolonic acid is the closest in terms of staying liquid at room temperature at 1 Atm, but it is the rarest of all and very unstable chemically and tends to decompose into elemental polonium and hydrogen; like all polonium compounds, it is highly radioactive.

---

[1]See Chapter 5

Property	H$_2$O	H$_2$S	H$_2$Se	H$_2$Te	H$_2$Po
Melting point	0.0	-85.6	-65.7	-51	-35.3
Boiling point	100.0	-60.3	-41.3	-4	36.1

Table 4.1.1: Melting and boiling points of the list of hydrogen chalcogenides

In conclusion, if one wants to find all forms of life in all environment, then one should also include ammonia, methane, and other hydrocarbons into their targeted list, otherwise, targeting and selecting water as our filter criteria for finding the number of habitable planets hosting intelligent tool using species is sufficient.

4.2   Biocomplexity Explanation

The evolution of biological complexity is one important outcome of the process of evolution. Evolution has produced some remarkably complex organisms, and the assumption that life evolves toward greater complexity is one of the pillar assumptions in calculating the background evolutionary rate. However, it is well known that natural selection does not dictate the direction in any kind of way. Species are equally likely to evolve toward greater complexity with less offspring or evolve toward lower complexity and multiply faster and produce more offsprings if both opportunities are equally available. Then we confront the dilemma, why do we still see an evolution toward complexity as we observed.

Based on the mathematical model, two types of scenarios are possible to enable evolving toward greater complexity despite the non-directional evolution of life. If evolution possessed an active trend toward complexity (orthogenesis), as was widely believed in the 19th century,[84] then we would expect to see an increase over time in the most common value (the mode) of complexity among organisms.[42] Computer models show that the generation of complex organisms is an inescapable feature of evolution.[72][40] This is sometimes referred to as evolutionary self-organization. Self-organization is the spontaneous internal organization of a system. This process is accompanied by an increase in systemic complexity, resulting in an emergent property that is distinctly different from any of the constituent parts.

However, the idea of increasing production of complexity in evolution can also be explained through a passive process.[42] Assuming unbiased random changes of complexity and the existence of a minimum complexity leads to an increase over time of the average complexity of the biosphere.[82] This involves an increase in variance, but the mode does not change. The trend towards the creation of some organisms with higher complexity over time exists, but it involves increasingly small percentages of living things.

In this hypothesis, any appearance of evolution acting with an intrinsic direction towards increasingly complex organisms is a result of people concentrating on the small number of large, complex organisms that inhabit the right-hand tail of the complexity distribution and ignoring simpler and much more common organisms. This passive model predicts that the majority of species are microscopic prokaryotes, which is supported by an estimates of 10^6 to 10^9 extant prokaryotes compared to the diversity estimates of 10^6 to $3 \cdot 10^6$ for eukaryotes.[19][58] Consequently, in this view, microscopic life dominates the Earth, and large organisms only appear more diverse due to sampling bias.

Nevertheless, a passive process can still over time lead to more complex organisms as a consequence of existing biological niches being occupied so only species with increasing novelty in addition to existing faculty can survive, and novelty and faculty lead to ever increasingly sophisticated responses between the Red Queen's predator and prey mechanism.

In order to illustrate the passive growth of biological complexity, we shall resort to a mathematical model. We shall assumed that there are 8 traits (Bipedalism, binocular vision, social, omnivorous, large brain, language, opposable thumbs, and living on land. These traits will appear repeatedly later to illustrate the chance of human emergence, but it can be a list of any trait not necessarily related to human). We further assume, in the beginning of evolution, that each of these traits forms a basic type of species suited to a particular niche. From this basic assumption, we can extrapolate:

If speciation occur by existing species entering another habitat or existing species' habitat has transitioned into a different one (as it is called a convergent evolution in biological evolution).

Then, the new species possessed the existing trait and a new acquired trait suited to the new habitat.

If speciation is indifferent to the ordering of traits acquisition (as it is called a convergent evolution in biological evolution though convergent evolution can simply means a convergence in a particular organ only), then, speciation can be represented as the combination out of the 8 traits listed.

If enough time has passed so that nature altered the environment long enough, alternatively, existing species has successfully entered all possible niches and retained each of its previous acquired traits, then, one can plot the biocomplexity attained by the biosphere as the function of combination of 8. The number of combination can be possibly generated for the acquisition of an number of traits ($1 \leq x \leq 8$) is weighted by the chance of such acquisition occurring in the first place. We also assumed, and later demonstrated by each of these trait is largely independent of one another. Therefore, the chance of such combination occur in the first place is represented by the average chance of observing any particular trait raised to the power of the number of acquired traits.

$$C\left(n,k\right) = \frac{n!}{k!\left(n-k\right)!} \tag{4.2.1}$$

$$S_{comb}\left(T_{rait}, x\right) = \begin{cases} C\left(T_{rait}, 2\left(x - \frac{1}{2}\right)\right) \cdot p^{2\left(x^{1.2} - \frac{1}{2}\right)} & x \leq 1 \\ C\left(T_{rait}, x\right) \cdot p^{x} & x > 1 \end{cases} \tag{4.2.2}$$

$$p = 0.255 \qquad T_{rait} = 8 \tag{4.2.3}$$

We modified values < 1 with a step function because we assumed that a species has to initially possess with at least 1 trait. We assumed that the average chance of observing any particular trait within the biosphere is 25.5%

and the graph is plotted:

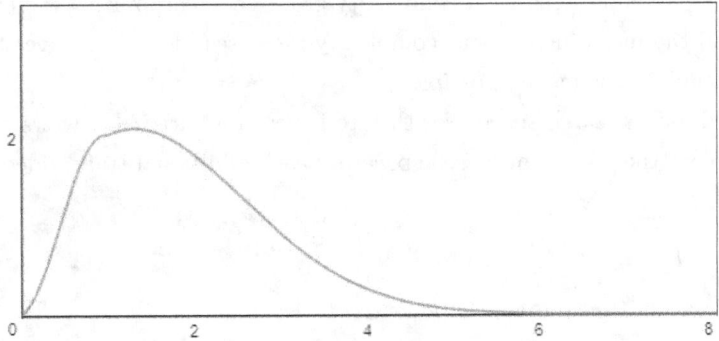

Figure 4.2.1: Total biocomplexity attainable at trait chance of 25.5%

The graph is right skewed for all values of p and arbitrarily large number of traits, showing that the weighted chance of attaining all 8 traits is minimal and the max is reached at 1.31 traits.

As the average chance of trait occurrence increases toward $p = 1$, the graph's right skewness is minimized:

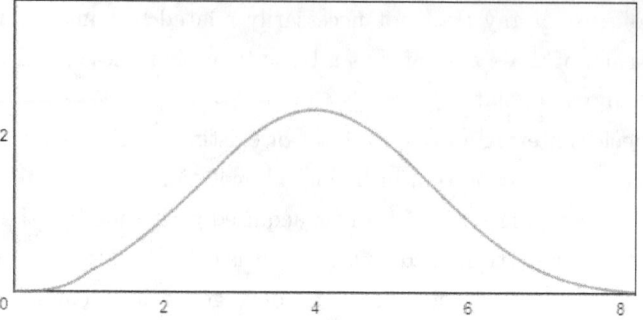

Figure 4.2.2: Total biocomplexity attainable at trait chance of 100%

We then plot the complexity curve for $S_{comb}(16, x)$, $S_{comb}(20, x)$, $S_{comb}(23, x)$, $S_{comb}(26, x)$ against $S_{comb}(8, x)$, assuming overtime the combination of basic traits yields into new basic traits for the species survival in a newly opened niche.

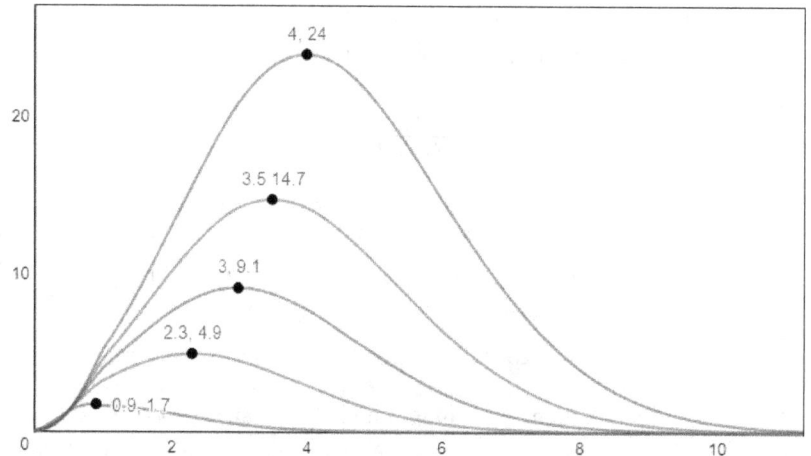

Figure 4.2.3: Total biocomplexity attainable for 8, 16, 20, 23, and 26 traits respectively

One can see that as more traits are added, the total complexity grows by incorporating the existing complexity as its subset, and the max and the mean increases as complexity increases. Thus, we have shown that as time passes, a passive growth of complexity is indeed possible.

Then, $S_{comb}(T_{rait}, x)$ is tested for its search space growth rate for each additional new traits added. We used the following formula to calculate the growth of search space per each additional trait added:

$$\int_0^{n+1} S_{comb}(n+1, \; x) \, dn - \int_0^n S_{comb}(n, \; x) \, dn \tag{4.2.4}$$

We computed traits from 3 up to 25, and the best exponential fit is:

$$0.122063 \, (1.158)^x \tag{4.2.5}$$

Although we have shown that the search space grows exponentially, it seems to be bounded by the exponential curve observed in nature. If one were to consider each of these traits represent a major change in earth's biological history (Chapter 6), we have shown that the emergence of each major trait takes roughly 40 Myr since Cambrian explosion. Therefore, our fit predicts that the biological search space grows $(1.158)^{\frac{1}{0.4}} = 1.44$ times per 100 Myr.

However, we have shown in Chapter 7, that the search space growth observed on earth likely ranges $x > 2.3$. Therefore, it seems that search space created by combination alone is not growing fast enough to match observation.

In reality, the ordering of traits acquisition plays a role at speciation. The order of acquisition leads to different species is another manifestation of divergent evolution. If each different ordering leads into a distinct speciation event, then, speciation can be represented as the permutation out of the 8 traits listed.

$$P(n,k) = \frac{n!}{(n-k)!} \tag{4.2.6}$$

$$S_{permu}(T_{rait}, x) = \begin{cases} x \cdot P\left(T_{rait}, 2\left(x - \frac{1}{2}\right)\right) p^{2\left(x^{1.7} - \frac{1}{2}\right)} & x \le 1 \\ P(T_{rait}, x) \cdot p^x & x > 1 \end{cases} \tag{4.2.7}$$

$$p = 0.255 \qquad T_{rait} = 8 \tag{4.2.8}$$

and the graph is plotted:

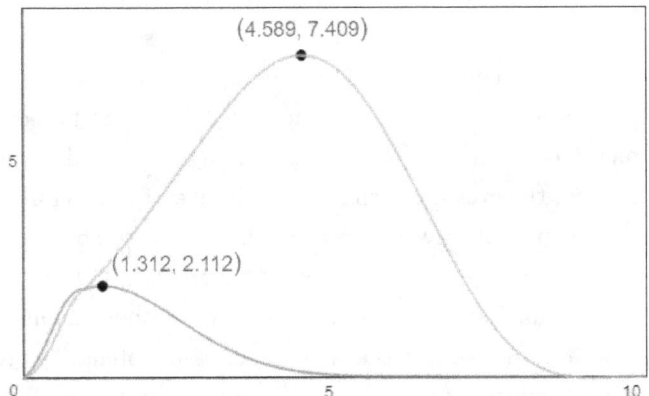

Figure 4.2.4: Total biocomplexity attainable with ordering vs non-ordering at trait chance of 25.5%

In comparison to biocomplexity attainable by combination alone, the search space by permutation is much larger. The mean and the max also shifts further to the right and curve is right skewed when $p < 0.222$ and left skewed when $p > 0.222$ for 8 traits and the criteria for fitting a right skewed curve requires the value of p to be even lower for higher number of traits.

If the total search space is dominated by permutation, the total search space will eventually become left skewed even if the composite curve started as right skewed. A right skewed curve fits well with the current understanding of evolution. That is, evolution is non-goal driven and nature's gradual transcendence toward evermore complex creature is a consequence of an increase in variance, but the mode does not change. The trend towards the creation of some organisms with higher complexity over time exists, but it involves increasingly small percentages of living things. On the other hand, a total search space dominated by permutation implies that evolution is goal driven toward organisms with a greater complexity even if it not specifically goal driven toward intelligent species. This would predict the order of species abundance in the reverse order as it is observed on earth. The greatest number of species are the vertebrate lineage followed by invertebrates, eukaryotes, and finally prokaryotes. It would predict that species with the greatest number of traits acquired in any order is the mode of the distribution.

In order to solve this problem, we utilize the concept of EROEI to evolution. That is, a trait is only maintained if the net energy it helps to bring or save to the organism is strictly positive or at least breaking even. For those addition that brings an energy loss to the organism will be de-evolved (such as the dinosaur like tails for birds and tails for man) Furthermore, the addition of each new trait inevitably lead to a higher cost of maintenance (such as elephant's trunk and peacock's tail). Therefore, more complex species requires more energy intake to survive. However, any fixed habitat's total biomass does not increase in correlation to a species increase in biocomplexity in a short period of time. As a result, less energy per capita is available to a species formerly with fewer traits. As a result, the population will remain lower than its former stage with fewer traits. Most importantly, it is well documented in nature that many species of simpler organisms such as butterfly that share the basic traits of wings, antenna, and straw mouth parts, can co-exist in similar habitat by niche differentiation. That is, certain butterflies only attract and extract nectar and pollens from flowers of certain shapes and colors. This differentiation is only possible when each species' energy requirement per capita is low. This differentiation is remarkably demonstrated even in interspecies co-existence with small energy requirement per capita and small body mass. Although the hawk moth and humming bird descended from completely different lineages more than 500 Myr apart, both derive their energy from flower nectar from a shared selected groups of flowers. If each species energy requirements were higher, competition ensues despite niche differentiation since the extraction of pollen and nectar from a particular colored flower can not satisfy the energy need of the species. This implies that the number of species and biodiversity decreases as the complexity of the species increases. This inverse relationship can be captured from the following plots, which was originally used to illustrate the mass and species abundance relationship. That is, larger animals requires more energy intake for its survival has the least level of abundance in any given environment. Species with more complicated traits sometimes is so successful that it leads to a positive feedback to further enhance its traits. This is observed in human with increasingly larger brain, elephants with longer trunks, and giraffe with longer necks. In each case, there is an ever increasing net intake of energy for the benefited species. The net energy increase can be manifested as an increasing body size as observed in giraffe and elephant, or an increasing population size as observed in humans, or both. Since the habitat's total biomass remain fixed, the ever increasing energy intake of the complex organism eventually reaches the energy limit ceiling by driving out any competitors with these combined traits or blocking any potential future competitor from ever arising in the first place. This trend is indicated by the emergence of Homo sapiens and the extinction of megafauna species on every major continent human migrated to and the extinction of closely related species such as the Neanderthals. In either case, potentially greater search space for more complex organisms is undermined by energy constraints. Finally, organisms with multiple traits evolves by successive geologic environmental changes, which takes geologic time to experiment. Despite potentially greater search space, nature does not have enough time to exhaust the entire search space, and the search space of fewer traits are thoroughly experimented by nature earlier. In conclusion, the overall search space for greater number of traits is reduced by both energy and temporal constraints. Therefore, in general, there is not enough time and energy in nature for more complex species to fully exhaust its potential search space even though it is exponentially larger than simpler organism's search space.

In order to cope with this understanding, we add an extra factor w to downplay the exponential increase in potential search space as the number of traits selected out of a total pool of fixed number of traits increases. Furthermore, this factor is further reduced as the number of total traits increases.

$$S_{permuadjusted}\left(T_{rait}, x, w\right) = S_{permu}\left(T_{rait}, x\right) \cdot w^x \qquad (4.2.9)$$

We used $w = 0.95^x$ and the curves remain right skewed at least up to 170 total number of traits tested.
Then, $S_{permuadjusted}\left(T_{rait}, x, 0.95^x\right)$ is tested for its search space growth rate for each additional new traits added. We used the following formula to calculate the growth of search space per each additional trait added:

$$\int_0^{n+1} S_{permuadjusted}\left(n+1, x, 0.95^x\right) dn - \int_0^n S_{permuadjusted}\left(n, x, 0.95^x\right) dn \qquad (4.2.10)$$

We computed traits from 3 up to 25, and the best exponential fit is:

$$0.017632\left(1.56382\right)^x \qquad (4.2.11)$$

The search space grows exponentially and the fit predicts that the biological search space grows $(1.56382)^{\frac{1}{0.4}} = 3.058$ times per 100 Myr. This result matches the search space growth observed on earth likely ranges $x > 2.3$. Therefore, it seems that search space created by permutation alone does growing fast enough to match observation.

In a realistic scenario, the complexity of the biosphere is unlikely to be composed of exclusively convergent or divergent evolution. We have observed both in nature.

We then formulate an equation by combining the search space for permutation and combination based on their percentage share of the search space.

$$S_{total}\left(T_{rait}, x\right) = f \cdot S_{permuadjusted}\left(T_{rait}, x, 0.95^{x}\right) + (1 - f) \cdot S_{comb}\left(T_{rait}, x\right) \tag{4.2.12}$$

We used a factor f to modulate the proportion of permutation and combination for the search space dominance. For $f = 1$, the complexity search space becomes the complexity search space for permutation. For $f = 0$, the complexity search space becomes the complexity search space for combination. For values between $0 \leq f \leq 1$, the total range of possibilities is represented by the shaded region below:

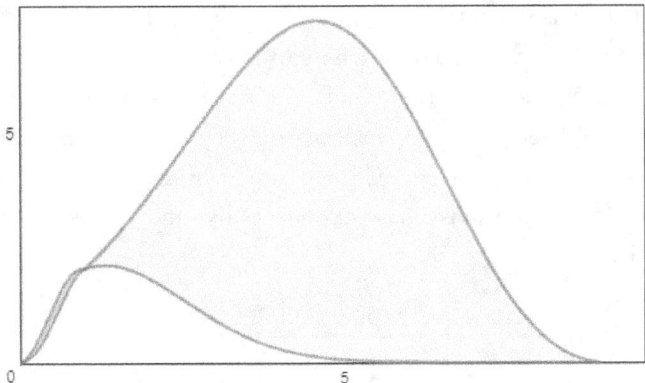

Figure 4.2.5: The shaded region represents the total range of possibilities for total biocomplexity at trait chance of 25.5%

and the total search space is:

$$\int_{0}^{8} S_{permu}\left(8, x\right) dx - \int_{0}^{8} S_{comb}\left(8, x\right) dx \tag{4.2.13}$$

which can be generalized into n traits as:

$$\int_{0}^{n} S_{permu}\left(n, x\right) dn - \int_{0}^{n} S_{comb}\left(n, x\right) dn \tag{4.2.14}$$

One can then fine tune the value of f for $S_{total}\left(T_{rait}, x\right)$ to best match the exact real observation in nature for the search space exponential growth per 100 Myr. A caveat must be raised. The search space growth in nature actually also depend on the branchiation factor and the rate of speciation. The tree of life suggests that species undergoes speciation by branching into two or more new species during speciation. As new river divided a formerly dryland, a species can diverge into two species. Later, river bank on one side further evolved into two habitats, woodland and grassland. The river bank on the other side further divided in half by orogenic mountain creation. This suggests at least 4 species can potentially emerge from initially just 1 species. This suggests that the intrinsic speciation rate is at least 2 per 100 Myr. However, this rate can be much lower. First of all, extinction rate can be high so that not all branches survives. Secondly, the weighted average speciation event per lineage can occur much slower so that the speciation rate falls below 2 per 100 Myr despite a branchiation factor higher than 2 at each speciation event. As a result, the potential search space can grow much faster than the speciation rate. Alternatively, the speciation rate can grow much faster than the search space. Therefore, the actually biological complexity growth is bounded by the minimum of the two:

$$C_{omplexity} = \min\{S_{peciation}, S_{total}\} \tag{4.2.15}$$

Since both speciation and search space grows exponentially fast, we have shown that biocomplexity grows exponentially fast over time.

In conclusion, biocomplexity search space (**BCS**) can be represented as a distribution of trait occurrence frequency given by a mixture of combination and permutation of traits. More complex organisms are represented as those which possessed more traits. The mode of the distribution is dominated by simpler organisms possessed fewer number of traits given a range of possible number of traits. Organisms possessed large number of traits are rare.

For each successive round, the size of the biocomplexity search space increases. The size increase is attributed to an increase in a larger number of traits to create a larger distribution with combination and permutation. Consequently, the new distribution has a greater distribution width and a larger deviation. With additional traits, the mode of the distribution shifts to organisms with higher number of traits, so the mode peak shifts to the right. As a result, as long as the number of traits represented by organisms increases per round, mode peak shifts right, the distribution spreads out further.

Moreover, maintaining an exponential growth in **BER** (the gap between mode peaks), the number of traits must not only increases but increases exponentially per round. We have shown that combination with increasing traits alone can not exceed a BCS of 2.783, and it is achieved only by partial permutation and combination combined. But in order to fit exponentially growing BCS and exponentially growing BER (distance between mode peaks), the number of traits per round must grow exponentially, and can be expressed as (for the t-th round the number of traits grow by d^t):

$$(n + d^t, x) = \frac{(n + d^t)!}{x!\,((n + d^t) - x)!}\,(p)^x \tag{4.2.16}$$

4.3 Probability on the Emergence of Prokaryotes from Amino Acids

Life emerged quickly as the condition of the earth becomes favorable. Just like many other significant milestones achieved later such as the Great Oxygenation Event (post the emergence of continental plates and shelf seas), the appearance of eukaryotes (post the Great Oxygenation event), and the emergence of complex multicellular organisms (post high oxygen build up in the atmosphere,). Life is very opportunistic and taking advantage of new niches. The earliest evidence of life occurred just 0.2 Gyr after the formation of the ocean, pointing toward and confirming the belief that life is easy to generate.[116]

But just how hard or how big a jump is it from generating an organic molecule to that of the first cell is the key question. We need to quantify the difficulty of abiogenesis. In order to quantify this jump, we count the number of atoms in each successive stage of evolution. We count the first amino acid (10 atoms), the first prokaryote ($9 \cdot 10^{10}$ atoms), the first eukaryote ($1 \cdot 10^{14}$ atoms), the start of multicellular life ($1 \cdot 10^{14}$ atoms), and the first multicellular fish ($7 \cdot 10^{27}$ atoms). We specify their emergence at 4.364 Gya, 3.95 Gya, 2.15 Gya, 0.85 Gya, and 0.45 Gya respectively. Counting the number of atoms is a simple and elegant way to capture the complexity of the organism obtained at each stage. Based on Galileo's squared cubed law, organisms not only follow the constraints and selection through natural selection but also subject to the law of physics. Organisms experience a totally different world as their size grows when certain forces dominant over some others, such as the strong capillary action at the microscopic level and gravity at the macroscopic level. A fish is not merely macroscopic-sized eukaryotic cell. As a result, organisms cannot simply just grow in size using their existing surviving strategy. Instead, it has to increase and alter their own information storage and protein creation in order to create new intercommunication protocol and cooperation to grow in size. It is exactly one observed from the transition of prokaryotes to eukaryotes (generally now believed to be the merging of archeon and bacteria) and the subsequent multicellular life forms which are only based on the innovation achieved at eukaryote level. From 3.95 Gya to 2.15 Gya, the number of atoms increased by a factor of 1.48 per 100 Myr. This is a relatively stable period of growth,

$$T_{prokaryote2eukaryote} = 9 \times 10^{10}\,(1.48)^{18}$$

154

$$= 1.0446080939 \times 10^{14} \text{ atoms}$$

followed by a period of stasis from 2.15 Gya to 0.85 Gya. From 0.85 Gya to 0.45 Gya, the number of atoms increased by a factor of 1,200 per 100 myr.

$$T_{1eukaryote2fish} = 1 \times 10^{14} \left(1,200\right)^{4.5} \tag{4.3.1}$$

$$= 7.1831611091 \times 10^{27} \text{ atoms}$$

This is what many generally termed the Cambrian explosion. It shows that evolution, under the right conditions (possibly adequate free oxygen and nitrogen) can accelerate fast. If one assumes that life at the earliest stage also followed a similar track of growth due to favorable conditions, then life could have evolved from simple amino acids to that of the prokaryotes in 0.424 Gyr starting at 4.264Gya.

$$T_{amioacid2life} = 10^1 \left(1,200\right)^{3.235} \tag{4.3.2}$$

$$= 9.1443182195 \times 10^{10} \text{ atoms}$$

If the increase factor is decreased to 223, or 18.6% the speed observed during the Cambrian explosion, then, we can push the start of evolution from simple amino acids to 4.364 Gya when earth's ocean just formed.

$$T_{amioacid2life} = 10^1 \left(223\right)^{4.24} \tag{4.3.3}$$

$$= 9.0534189785 \times 10^{10} \text{ atoms}$$

To further strengthen our argument, one can go a step closer. The viroid, supposedly the smallest pathogen known with a single-stranded RNA without a protein coat, represents the most plausible RNAs capable of performing crucial steps in abiogenesis, the evolution of life from inanimate matter. Many believed that viroid represents the living fossils of a class of species evolved during the RNA world of life evolutionary history which predates the current DNA world. Since viroids are capable of replication, it is then subject to natural selection. *Avocado sunblotch viroid* and *Coconut cadanf-casanf viroid*, two of the smallest of the viroids, consist only 246 nucleotides. Assuming each nucleotide contains 35 atoms, then the smallest structure subject to natural selection contains just 8,610 atoms. Though evolution itself is directionless, the passive growth in complexity is nevertheless inevitable, and it will take only 228 Myr to evolve toward the complexity of prokaryotes if the rate of growth comparable to that of Cambrian explosion.

$$T_{viroid2life} = 8,610^1 \left(1,200\right)^{2.28} \tag{4.3.4}$$

$$= 9.0268534635 \times 10^{10} \text{ atoms}$$

In an RNA world or viroid world, different sets of RNA strands would have had different replication outputs, which would have increased or decreased their frequency in the population, i.e. natural selection. As the fittest sets of RNA molecules expanded their numbers, novel catalytic properties added by mutation, which benefited their persistence and expansion, could accumulate in the population. Such an autocatalytic set of ribozymes, capable of self-replication in about an hour, has been identified. It was produced by molecular competition (in vitro evolution) of candidate enzyme mixtures.

It is possible that such a quick transition occurred because the available free energy in the early ocean limits the size of species. Nevertheless, the free energy is abundant enough to enable the growth from simple viroids to the prokaryotes.

We still have to show the probability of the aggregation from simple amino acids with 10 atoms to that of the smallest viroid with 8,610 atoms. Since this stage of evolution, is the earliest, and possibly does not or at the best only partially replicate its own data, the rule of natural selection is not applicable. Nucleotides, the basic unit of viroids, are the fundamental molecules that combine in series to form RNA. They consist of a nitrogenous base attached to a sugar-phosphate backbone. RNA is made of long stretches of specific nucleotides arranged so that their sequence of bases carries information.

The RNA world hypothesis holds that in the primordial soup (or sandwich), there existed free-floating nucleotides. These nucleotides regularly formed bonds with one another, which often broke because the change in energy was so low. However, certain sequences of base pairs have catalytic properties that lower the energy of their chain being created, enabling them to stay together for longer periods of time. As each chain grew longer, it attracted more matching nucleotides faster, causing chains to now form faster than they were breaking down. Using this hypothesis, we can derive the probability of abiogenesis. Assuming the most dominant bonding occurred between the pairing of two followed by the pairing of three and then the pairing of four..etc, then, the number of steps leading to the simplest viroid from amino acids requires $\log_n 861$ steps ($\frac{8610 \text{ atoms}}{10 \text{ atoms}} = 861$ nucleotides), where n is the pairing of two, three, or more chains of the molecular nucleotides. Pairing of fewer nucleotides leads to greater number of steps to the smallest viroid, but pairing between fewer nucleotides also comes with greater frequency. We further assumed that nucleotides can compose or groups of nucleotides can compose any number of atoms greater than 10. However, the occurrence frequency of pairing lowers in a geometric way. So that the pairing between two nucleotides is 50%, pairing between three nucleotides is 25%.

Pairing number	Number of steps to viroid	Occurrence Frequency
2	9.74987	0.500
3	6.15148	0.222
4	4.87493	0.125
5	4.19904	0.040
6	3.77176	0.027

Table 4.3.1: The chance of pairing between nucleotides

The occurrence frequency distribution is described as:

$$\frac{2}{x^2} \tag{4.3.5}$$

and its integration up to the pairing between infinite number of nucleotides is 100%:

$$2 \int_2^\infty \frac{1}{x^2} dx = 1 \tag{4.3.6}$$

Then, in the simplest model, one can compute that:

$$2 \cdot \int_2^\infty \frac{1}{x^2} \frac{\log 861}{\log x} dx \tag{4.3.7}$$

$$= 5.18115739293$$

It shows that the expected number of steps leading to the simplest viroid requires, on average, 5.18 steps. If the chance of each step of successful bonding between the pairing of two, three, four up to n pairs is 50 percent, the total sum chance of all bonding leading to the smallest viroid is $\left(\frac{1}{2}\right)^{5.18} = 2.756\%$. We can call this the lower bound estimate because if one further assumes that considerable time between each step of aggregations, then, each step may have 100 percent chance leading to a longer chain. This not unreasonable since pairing between 2 nucleotides up to 7 nucleotides, just to illustrate, requires on average 5 steps of aggregation to reach the complexity threshold of viroids. Assuming 196 Myr time frame from the simplest amino acid to the viroids, it gives each step 39.38 Myr time to consolidate their bonds, that is, the building and creation of ever more complex molecules can be spaced out in a long time frame to guarantee its success. The primordial condition on earth is very different from that of today. Atmosphere composed of CO_2 and methane with intense atmospheric pressure, high temperature, high rate of volcanism and the recycling of crusts, and the ubiquity of hydrothermal vents. If life emerged around hydrothermal vents, then the energy intensity of early earth allowed a high frequency of experiments at every local level which would take billions of years to produce the first viroid at today's rate. Therefore, the emergence of life can be an inevitable consequence of any early earth's chemical experiments and can only occur on a young, geologically active planet. In a sense, it is not that life is easy to produce, rather it depends on the frequency of nature's experiment.

If one takes the upper bound of total sum chance of all bonding leading to the smallest viroid is 100%, then *the geometric mean value of life emergence is 16.6%.*

$$\sqrt{2.756 \cdot 100} = 16.60\% \tag{4.3.8}$$

Nevertheless, the true rate of life emergence can still lower than 16.6 percent based on various factors beyond the scope of this paper. We shall denote the additional probability on the emergence of life **as** an factor x in our final calculation to show the lower and upper bound of our model. The calculation does show, however, that life is not extremely implausible to start with.

4.4 Probability on the Emergence of Eukaryotes, Sex, and Multi-cellularity

From our mathematical model on the finding the average speed of evolutionary change[2], we find that biological system can form new function or species relatively quickly and such changes and pace is largely driven by geologic changes. This can be generalized to simpler and earlier evolutionary times in the earth's past. *The emergence of Eukaryotes, for example, tightly followed the onset of the Great Oxygenation Event.* The Great Oxygenation Event, in turn, is driven by the emergence of continental plates for the first time in earth's geologic history as finally the earth has cooled enough. The process of Earth's increase in atmospheric oxygen content is theorized to have started with the continent-continent collision of huge land masses forming supercontinents, and therefore possibly the creation of the first supercontinent mountain ranges. These super mountains would have eroded, and the mass amounts of nutrients, including iron and phosphorus, would have washed into the oceans, just as we see happening today. The oceans would then be rich in nutrients essential to photosynthetic organisms,[93] which would then be able to respire mass amounts of oxygen. All eukaryotic cells use mitochondrion to process oxygen to obtain energy. Consensus agrees that the first proto-eukaryotic cell formed as archaea and prokaryotes merged into each other and to perform one specific function for the benefit of the whole. Merging with the addition of chloroplast also occurred in plant cell lineage at around the same time. This shows that symbiotic merger is a common occurrence. Mitochondrion bacteria utilizing oxygen must have formed only possible after the onset of free oxygen in the ocean. Soon as Mitochondrion bacteria formed and as it propagates through the earth's ocean, merging process logically follows. *The onset of Cambrian explosion, again, is a consequence of a significant rise in oxygen level.*[35] By Cambrian, cyanobacteria for the first time have produced enough oxygen as its waste product not only filled the ocean's oxygen sink as well as the land's. As a result, the formation of ozone layer prevented the incoming of ultra-violet radiation from reaching the surface of the earth, and enough free oxygen available in the ocean and on the land reaching levels similar to that of today. The overabundance of oxygen provides enough fuel for the flourishing diversification of the Cambrian fauna.

---

[2](Chapter 4, Section 4.5 "Speed of multicellular evolution")

Another major biological breakthrough, the emergence of sexual reproduction, is rather peculiar, it is first observed between the emergence of the eukaryotic cell (2.1 Gyr ago) and Cambrian explosion (0.58 Gyr ago) around 1.5 Gyr ago, in a period called the boring billion. When the supercontinent Rodinia and Columbia were maintained, and the climatic condition is generally stable, and no major biochemical and geologic changes occurred on earth. The viral origin of sexual reproduction posits that the cell nucleus of eukaryotic life forms evolved from a large DNA virus, (possibly a pox-like virus such as the lysogenic virus is a likely ancestor because of its fundamental similarities with eukaryotic nuclei. These include a double-stranded DNA genome, a linear chromosome with short telomeric repeats, a complex membrane-bound capsid, the ability to produce capped mRNA, and the ability to export the capped mRNA across the viral membrane into the cytoplasm. The presence of a lysogenic pox-like virus ancestor explains the development of meiotic division, an essential component of sexual reproduction.) in a form of endosymbiosis within a methanogenic archaeon. The virus later evolved into the eukaryotic nucleus by acquiring genes from the host genome and eventually usurping its role. Since it is estimated that viruses kill approximately 20% of marine micro-organism' biomass daily and that there are 10 to 15 times as many viruses in the oceans as there are bacteria and archaea, they infect and destroy bacteria in aquatic microbial communities. They are one of the most important mechanisms of carbon recycling and nutrient cycling in marine environments. Then, *the chance of evolution of sexual reproduction is high.* The meiotic division arose because of the evolutionary pressures placed on the virus as a result of its inability to enter into the lytic cycle. This selective pressure resulted in the development of processes allowing the viruses to spread horizontally throughout the population. The outcome of this selection was cell-to-cell fusion. (This is distinct from the conjugation methods used by bacterial plasmids under evolutionary pressure, with important consequences.)[91] The possibility of this kind of fusion is supported by the presence of fusion proteins in the envelopes of the pox-viruses that allow them to fuse with host membranes. These proteins could have been transferred to the cell membrane during viral reproduction, enabling cell-to-cell fusion between the virus-host and an uninfected cell. The theory proposes meiosis originated from the fusion between two cells infected with related but different viruses which recognized each other as uninfected. After the fusion of the two cells, incompatibilities between the two viruses result in a meiotic-like cell division.[92]

If the viral origin of sex is valid, then nature is constantly experimenting sexual reproduction as an alternative to binary fission. Therefore, we have found a cogent and reasonable mechanism for the emergence of sex. However, it is still likely that the evolution of sexual reproduction, though emerged, only reached its full potential and glories as a successful survival strategy as a consequence of environmental resource pressure from the competition of prokaryotes. That is, the maintenance of such mechanism and its success requires investigation of organism's living environment. The fossil evidence of Stromatolites indicates that prokaryotes reached its greatest extent around 12 Gyr before its sharp decline. Sexual reproduction has been observed, indeed, related to environmental stress. Animals such as Hydra are capable of both sexual and asexual reproduction, depending on the environmental conditions. When resources and food are readily available, hydra reproduces by asexual reproduction in the form of binary fission. When food and environmental condition is harsh, it produces sperms and egg cell which falls to the bottom of the sea floor and gave to the birth of new hydra once the environmental condition becomes suitable again. Such mechanism may explain the maintenance of sexual reproduction. After the great oxygenation event, earth contains enough free oxygen making the evolution of eukaryotes possible but not enough to fuel its explosive growth. As a result, prokaryotes continue to flourish. As prokaryotes continue to flourish and compete for nutrients with eukaryotes, energy-hungry eukaryotes face a crisis. Eukaryotes then well adopt the strategy of sexual reproduction, which carried several advantages over asexual reproduction.

First, by sacrificing itself and disperses its own genetic material into the surrounding, it is able to preserve itself from intense resource competition with very little energy consumption and able to wait until the environmental condition becomes more suitable for its re-emergence. Secondly, sexual reproduction enables the faster emergence of new traits and genotypes, allows eukaryotes to diversify and enter new niches. In what termed as the Hill-Roberson Effect, the benefit of sexual reproduction becomes self-evident. In a population of finite size which is subject to natural selection, random linkage disequilibrium will occur. These can be caused by genetic drift or by mutation, and they will tend to slow down the process of evolution by natural selection.[9] This is most easily seen by considering the case of disequilibrium caused by mutation:

Consider a population of individuals whose genome has only two genes, a and b. If an advantageous mutant (A) of gene a arises in a given individual, that individual's genes will through natural selection become more frequent in the population over time. However, if a separate advantageous mutant (B) of gene b arises before A has gone to fixation, and happens to arise in an individual who does not carry A, then individuals carrying B and individuals carrying A will compete. If recombination is present, then individuals carrying both A and B (of genotype AB) will eventually arise. Provided there are no adverse epistatic effects of carrying both, individuals of genotype AB will have a greater selective advantage than aB or Ab individuals, and AB will hence go to fixation. However, if there is no recombination, AB individuals can only occur if the latter mutation (B) happens to occur in an Ab individual. The chance of this happening depends on the frequency of new mutations, and on the size of the population, but is in general unlikely unless A is already fixed, or nearly fixed. Hence one should expect the time between the A mutation arising and the population becoming fixed for AB to be much longer in the absence of recombination. Hence recombination allows evolution to progress faster.[9] If these assumptions hold, it implies that the appearance of sexual reproduction, though partially attributed to environmental stress related to resource competition with existing prokaryotes, is an inevitable consequence of the evolution of more complex singled cell organism, especially as a logical consequence of viral origin of the eukaryotic cell nucleus. (Nature was trying repeatedly to create sexual recombination at the cellular level) The evolution of sexual reproduction, then, unlike the appearance of more complex eukaryotes and the appearance of complex multicellular life, does not require significant geologic and biochemical environmental changes as sexual reproduction itself does not consume more energy than the survival requirements of eukaryotes. Though with more stressful environmental conditions, it can appear faster in geological history as a successful survival strategy. It does not have to wait for 0.8 Gyr since the appearance of the first eukaryotes. Nevertheless, eukaryotic cells can resort to both asexual and sexual reproduction, and larger multicellular eukaryotes exclusively reproduces through sexual means as a means of cost control and adaptation to environmental uncertainties given the long lifespan of multicellular species and the amount of resources needed to maintain the body.

By elucidating the causes of the timing of the appearance of each major evolutionary changes, Several interesting predictions can be made. First of all, the appearance of eukaryotes is a consequence of cyanobacteria filling up the ocean's oxygen sinks. The filling of earth's oxygen sink, in turn, is contributed by the emergence of continental platforms from the ocean and the creation of the shelf sea.[93] As a result, every other condition being equal, a planet with lower sea levels that comes with a smaller ocean surface area (Ocean depth is largely non-important since only the surface oxygen sinks needed to be filled, studies have shown deep ocean remain anoxic even today.[57] Moreover, nearly all cyanobacteria thrive near the surface of the ocean to convert sunlight) can lead to the appearance of eukaryotes much earlier than that we found on earth. Secondly, the diversification of multicellular eukaryotic organisms is a consequence of significant oxygen buildup which filled not only the ocean's oxygen sinks but as well as the land's. It is directly related to the coverage surface area of ocean and land, instead of its depth. Therefore, a planet with greater portions of land masses (less surface covered by oceans) will take much longer time than what we observed on earth to evolve from eukaryotic cell to multicellular organisms. The reverse is also true, where a planet with higher sea level or entirely covered by the ocean is likely to take even greater delay than observed on earth to evolve oxygen-utilizing eukaryotes but quickly transitioned toward multicellular life forms. To be concise, *the proportion of ocean to land coverage of a planet determines the timing gap between the appearance of prokaryotes and eukaryotes and the timing gap between the appearance of eukaryotes and multicellular life forms.* Thirdly, the size of the planet is a non-determining factor in which organisms evolve assuming the bacteria colony size on all habitable planets is comparable within a magnitude of difference. On smaller planets where the ocean surface area to land area ratio are similar to earth, fewer bacteria using sunlight as their energy source and produce oxygen as its waste to fill up smaller oxygen sinks, leading to a similar interval between the timing of each major evolutionary change. For super earths, there are more thriving bacteria producing more oxygen, but there are also more oxygen sinks in both ocean and exposed land proportional to its surface area to fill, leading to a similar interval between the timing of each major changes.

Therefore, the time requirements between prokaryotes to eukaryotes, and from eukaryotes to multicellular life may differ, the final timing on the emergence of multicellular life should be identical, after the entire planet surface's oxygen sinks are saturated. Since we are computing the number of earth formed between 5 Gya and 4 Gya, the timing of the emergence of multicellular life forms should follow the chronological order of the terrestrial planet formation date. Since earth was formed at 4.5 Gya, *earth's current biocomplexity should be placed as the median of all habitable terrestrial planets within this bracket range, at exactly 50%.* This filter criterion, along with other factors, will be used as a filter to select the current number of habitable planets in Chapter 8.

4.5 Speed of Multicellular Evolution

With the number of bases known in the human genome (3 billion base pairs) and the mutation rate of eukaryotes, one can calculate the theoretical upper bound on how fast human can evolve. In general, the mutation rate in unicellular eukaryotes and bacteria is roughly 0.003 mutations per genome per cell generation.[36][46][119]This means that a human genome accumulates around 64 new mutations per generation because each full generation involves a number of cell divisions to generate gametes.[36] Human mitochondrial DNA has been estimated to have mutation rates of $3\times$ or 2.7×10^{-5} per base per 20-year generation [52] these rates are considered to be significantly higher than the rates of human genomic mutation at 2.5×10^{-8} per base per generation.[39] Using data available from whole genome sequencing, the human genome mutation rate is similarly estimated to be 1.1×10^{-8} per site per generation.[108]

1.1×10^{-8} per site per generation $\cdot$ 3 billion base pairs = 33 base pairs. This implies that each child on average differs from their parents by 33 base pairs. If human lineages were limited to very few individuals for a very long period of time and no mutation repeats at the same base, $\frac{3\text{ billion base pairs}}{33\text{ base pairs}}=$ 90,909,090 generations later (1.818 billion years). Mutations would have turned every base pair once. However, Homo genus population from millions of years ago numbered 10,000 to 100,000 at least. With these many individuals per generation, mutations would have mutated every base within the population once only 181,818 years to 18,181 years! If bottleneck existed and only the most adaptable human survived, then, the emergence of Homo sapiens from ape-like creature can happen very quickly. Much quicker than the fossil record suggested.

To make the argument even more convincing, the key genes for the development of the thumb, brain muscle, language development are now identified as few as just 100 different genes. Furthermore, many genes altered in Homo Sapiens compares to chimpanzees are master switch genes, that controls other switch genes, this implies that by mutating certain key bases, an escalating number of bases are affected.

This calculation confirms the punctuated equilibrium model of evolution, where new species appeared suddenly within rock strata. This is also confirmed by adaptation of black and white moth in Britain during the industrial revolution, the domestication history of agricultural plants such as maize and animals such as pet goldfish, cats, and dogs. Then, it is certain that animal species can easily alter its morphology and form within a very short timescale but why is it not observed in nature? Does nature set a speed limit on how fast species should evolve or is it an interplay between nature and species themselves? *This is an important issue to address and resolve since by elucidating the mechanism and come forth with explanation with observed stasis of animal form in nature and its theoretical maximum limit of rapid evolution potential, we truly confirm the forces and factors shaping the evolutionary rate of speciation and the background evolutionary rate.* We use such measure to compute our years ahead against this background evolutionary rate and closest living arising extraterrestrial industrial civilization.

Stabilizing selection is occurring at many different levels. At the physical level, the square cubed law and gravity applies to all biological species.[50] An animal can only be of a certain size while roaming on land and for arboreal species such as primate living on trees must be even smaller, and birds which have to adapt to an aerial lifestyle have to be smaller still. As a result, nature's physical law places constraints on biological creatures which limit their evolutionary experiments into any random direction.

Primarily, ecological constraints set the limit on how fast evolution changes. Since breaking and rejoining continents and mountain creation occurs slowly over geologic timescale, allopatric, peripatric, and parapatric speciations, three of the four primary drivers of speciation condition can only occur over geologic timescale as well. As a result, individuals with unique mutations which are even beneficial if it underwent allopatric, peripatric, and parapatric speciations are not able to persist in a given population due to genetic recombination, genetic drift. Secondly, the climate changes gradually over geologic timescale, except those during the ice ages. As a result, a species suited well to a given climate and feeds on specific food will continue to survive and thrive in such climate, which perpetuates in geologic timescale. Any deviations do not confer any immediate benefit for such species, and such deviation is then not selected for by natural selection.

There also exists for the case that sexual selection favors those of its own species that conform to the norm. By choosing a devious individual, a partner risk itself in making the wrong decision at the cost of its own gene. This is demonstrated in Homo sapiens as well in which both sexes prefer to mate with the healthy, handsome, pretty and intelligent individuals of the opposite sex, thereby maintaining the uniformity of the gene pool. Therefore, devious members of the species do not even have an equal opportunity at interbreeding, further contributing their removal from a population's gene pool.

Secondly, a majority of the mutations are harmful, that is, newborn species tend to develop into different types of congenital symptoms that either succumbs to such problem prior to birth or die shortly after birth. As a result, many paths of nature's experiment lead to dead ends. Semi-harmful mutations also exist. Newborns with extra teeth, two-toed feet, and six-toe feet have been reported.

Thirdly, with the absence of bottlenecks and isolated habitat scenarios, a large population of a given species breeds randomly with each other and remain constant in number from generation to generation. The Hardy–Weinberg principle states that within sufficiently large populations, the allele frequencies remain constant from one generation to the next unless the equilibrium is disturbed by migration(gene flow), genetic mutations, natural selection, mate choice, genetic drift, and meiotic drive. Mathematically it can be shown that the multinomial expansions for n allele frequencies at time iteration $t+1=$ time iteration t when $t>0$. Below is the mathematical proof for 2 alleles frequencies with binomial expansion.

$$f_t(A) = f_t(AA) + \tfrac{1}{2}f_t(Aa) \tag{4.5.1}$$

$$f_t(a) = f_t(aa) + \tfrac{1}{2}f_t(Aa) \tag{4.5.2}$$

$$f_1(AA) = p^2 = f_0(A)^2 \tag{4.5.3}$$

$$f_1(Aa) = pq + qp = 2pq = 2f_0(A)f_0(a) \tag{4.5.4}$$

$$f_1(aa) = q^2 = f_0(a)^2 \tag{4.5.5}$$

$$f_1(A) = f_1(AA) + \tfrac{1}{2}f_1(Aa) = p^2 + pq = p(p+q) = p = f_0(A) \tag{4.5.6}$$

$$f_1(a) = f_1(aa) + \tfrac{1}{2}f_1(Aa) = q^2 + pq = q(p+q) = q = f_0(a) \tag{4.5.7}$$

$$[f_{t+1}(AA), f_{t+1}(Aa), f_{t+1}(aa)] =$$

$$f_t(AA)f_t(AA)\,[1,0,0] + 2f_t(AA)f_t(Aa)\,[\tfrac{1}{2},\tfrac{1}{2},0] + 2f_t(AA)f_t(aa)\,[0,1,0]$$

$$f_t(Aa)f_t(Aa)\,[\tfrac{1}{4},\tfrac{1}{2},\tfrac{1}{4}] + 2f_t(Aa)f_t(aa)\,[0,\tfrac{1}{2},\tfrac{1}{2}] + f_t(aa)f_t(aa)\,[0,0,1] \tag{4.5.8}$$

$$= \left[\left(f_t(AA) + \tfrac{1}{2}f_t(Aa)\right)^2, 2\left(f_t(AA) + \tfrac{1}{2}f_t(Aa)\right)\left(f_t(aa) + \tfrac{1}{2}f_t(Aa)\right), \left(f_t(aa) + \tfrac{1}{2}f_t(Aa)\right)^2\right]$$

$$= \left[f_t(\mathrm{A})^2, 2f_t(\mathrm{A})f_t(\mathrm{a}), f_t(\mathrm{a})^2 \right] \qquad (4.5.9)$$

Through genetic recombination, new arising mutation is quickly diluted. The law of large numbers is applicable, and only the mean behavior and morphology is maintained. By applying the Hardy-Weinberg Principle, one can also conclude that the genotypes within a given population stabilize within two generations of interbreeding and mixing.

Furthermore, genetic drift will quickly eliminate new arising mutations. When the allele frequency is very small, drift overpowers selection in large populations. For example, while disadvantageous mutations are usually eliminated quickly in large populations, new advantageous mutations are almost as vulnerable to lose through genetic drift as are neutral mutations. Not until the allele frequency for the advantageous mutation reaches a certain threshold will genetic drift have no effect.[79] The above statement can be stated in the following mathematical equation:

$$T_{fixed} = \ln \left(\frac{-10 \cdot N_e \left(1 - p \right) \ln \left(1 - p \right)}{p} \right) \qquad (4.5.10)$$

where T is the number of generations, N_e is the effective population size, and p is the initial frequency for the given allele. The result is the number of generations expected to pass before fixation occurs for a given allele in a population with given size (N_e) and allele frequency (p). If one plots the graph, one can quickly see that as the new arising allele has a very low frequency is equivalent of stating that the dominant allele has a very high frequency, which leads to the removal of the low-frequency alleles in a very short period of time.

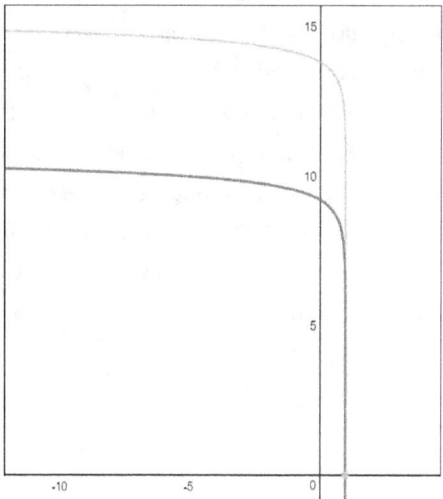

Figure 4.5.1: Allele frequency between 0 and 1 and expected fixations in generations for n=10 and n=1000

Therefore, given the large sampling size, beneficial mutations occur much more frequent in a large population but quickly diluted and drifted and selected away. Beneficial mutations occur much less likely in a small population, but it is likely to remain and fixed once it emerges. In the face of a bottleneck or isolation, beneficial mutations arising from a large population are selected by natural selection in the remaining surviving / small founding population, which generally diluted away in a pre-crisis, large population, to become dominant and fixed in a given population thereafter.

We will illustrate this fact by the calculation taking punctuated equilibrium into account. Homo sapiens and Chimpanzee differs by 40 million base pairs. 35 million single nucleotide changes and 5 million insertion/deletion events are recorded. Therefore, one needs to flip $4 \cdot 10^7$ changes in order to turn an ape into a man. Each generation mutate by at most 33 random sites. We further assume that genetic drift plays minimal role so that new beneficial and harmful mutations are equally likely to be contributed to the gene pool. This implies that, on average, each individual per generation flip 0.44 sites that are related to the evolution toward Homo sapiens.

$$\frac{4 \cdot 10^7}{3 \cdot 10^9} \cdot 33 = 0.44 \tag{4.5.11}$$

The rest 32.56 mutations have little to do with evolution toward Homo sapiens and so be labeled as neutral or harmful mutations.

We also assume that on average, 100,000 individuals thrive at any moment. Therefore, the total number of sites tried per generation evolving toward Homo sapiens is $100,000 \cdot 0.44 = 44,000$. Then the number of generation required to turn an ape into a man is 909 generations.

$$\frac{\left(4 \cdot 10^7\right)}{0.44 \left(100,000\right)} = 909.0909 \tag{4.5.12}$$

Assuming a generation time of 20 years in the Hominid lineage, the total number of years required to introduce 40 million base pairs of change is merely:

$$\frac{4 \cdot 10^7}{1 \cdot 0.44 \cdot 100000} \cdot 20 \tag{4.5.13}$$

$$= 18,181.8 \text{ Years}$$

Furthermore, the number of generations required to introduce all 40 million beneficial base modifications alone guaranteed the introduction of new non-beneficial mutations at some beneficial site within the entire group, due to high non-beneficial to beneficial mutation ratio (33 to 0.44). In 909 generations, the cumulative chance of beneficial gains mutated is then 50 percent and only 25,279,788 base altered in reality.

$$B_{ase} = 100000 \cdot 0.44x \tag{4.5.14}$$

$$\frac{1}{4 \cdot 10^7} \int_0^{909} B_{ase} \cdot \frac{(1) \cdot 10^5 \cdot 33}{3 \cdot 10^9} dx = 0.4999 \tag{4.5.15}$$

Simulation shows that in 4,100 generations, the number of beneficial gains can eventually converges to 40 million base pairs despite the introduction of new non-beneficial mutations at some beneficial sites. This conclusion can be simulated with the following recursive step function:

At $S_{tep}(0)$, we have:

$$S_{tep}(0) = \begin{cases} t_{otal} = 0 \\ M_{utation} = 0 \end{cases} \tag{4.5.16}$$

So that initially neither any beneficial mutation has been introduced nor any neutral or harmful mutations has altered newly introduced beneficial mutations. For $S_{tep}(1)$ to $S_{tep}(n)$, the number of total beneficial mutations introduced are cumulatively increasing, but such increase leads to a higher chance of the introduction of neutral or harmful alteration on these beneficial gains. Therefore, the corrupted beneficial gains are deducted from the overall gains per each step.

$$S_{tep}(n) = \begin{cases} t_{otal} = t_{otal} + 10^5 \cdot 0.44 \\ M_{utation} = t_{otal} \cdot \left(\frac{10^5 \cdot 32.56}{3 \cdot 10^9}\right) \\ t_{otal} = t_{otal} - M_{utation} \end{cases} \tag{4.5.17}$$

This implies that it takes 98,000 years to introduce 40 million base pairs.

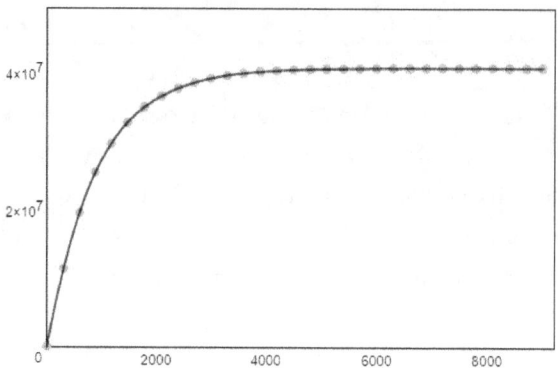

Figure 4.5.2: Number of generations to introduce 40 million base pairs

However, new allele introduced into a large population takes a long time before its fixation due to genetic drift. Individuals possessed all 40 million base pair changes will be exceedingly rare within the population through sexual reproduction based on the Hardy-Weinberg principle. For a population of 100,000, one expect fixation to occur in 10^6 generations. With 20 years per generation, it implies that at least 20 million years of accumulated molecular changes by mutation rates required to turn Chimpanzee into Homo sapiens without the presence of directed selection. If one were to assume that 20 million years required after each generation, then it will take $4900 \cdot 2 \cdot 10^7 = 98$ Gyr, 21 times the age of earth. It certainly contradicts our observation because the divergence took place within 6 million years.

This implies that directional natural selection played role in the evolution of Homo sapiens by expediting the process. If the average beneficial mutation is 0.44 per individual per generation and the number of beneficial mutations obtained per individual per generation is probabilistically distributed, then only those survived a directional population bottleneck event can reproduce. We assume that the probabilistic distribution is:

$$g_{distr}(x) = \frac{3}{.95x^{1.85}\sigma\sqrt{2\pi}} \exp\left(-\frac{\left(\ln 0.456x^{0.9}\right)^2}{2\sigma^2}\right) \tag{4.5.18}$$

$$\sigma = 1.26$$

and only top 1% of the original 100,000 population survives. Then, it implies that

$$\frac{\int_{11.5}^{\infty} g_{distr}(x)\,dx}{\int_{0}^{\infty} g_{distr}(x)\,dx} \approx 1\% \tag{4.5.19}$$

The remaining top 1% population each on average had at least 11.5 beneficial mutations and a weighted average of 20.96 beneficial mutations toward the directional selection of Homo sapiens.

$$\frac{\int_{11.5}^{\infty} x g_{distr}(x)\,dx}{\int_{11.5}^{\infty} g_{distr}(x)\,dx} = 20.96 \tag{4.5.20}$$

Now, assuming the population recovery takes place by increasing each generation by 6%, then the total number of years expected for the surviving population to regain its loss after the bottleneck takes:

$$T_{double} = \frac{\log(100)}{\log(1.06)} \cdot 20 \tag{4.5.21}$$

$$= 1,580.66 \text{ Years}$$

So, a recovery completes every 1,580 years. Moreover, T_{fixed} takes place in 10,000 years for a population of 1,000. In the fastest possible scenario, the very next bottleneck, directional selection event occurs immediately after its newly introduced genes becomes fixed within the population and the population recovered from its initial loss. We further assumed that the bottleneck lasted as long as the fixation time and the population remained small (such as during a harsh glacial period, a super volcanic eruption, an extended drought, habitat isolation due to sea level changes). Any mutations that altered the beneficial gain will be removed by directional selectional pressure during the bottleneck. Beneficial gains still proceeds during the bottleneck period and we assumed the rate is at 0.44 per individual per generation. Since 10,000 years takes 500 generations, $500 \cdot 0.44 = 220$ beneficial mutations are gained per individual in all generations. Then:

$$T_{total} = T_{fixed} + T_{double} \tag{4.5.22}$$

Since $T_{fixed} > T_{double}$, T_{fixed} dominates. During the recovery phase of 1,580 years, directional selectional pressure is removed and mutations are allowed to alter previous beneficial gain. It takes only 79 generations to reach a population of 100,000. The average population size during this period was only 21.5% of the normal level. Therefore, the number of beneficial bases altered are negligible.

$$\frac{1}{4 \cdot 10^7} \int_0^{79} B_{ase} \cdot \frac{(0.215) \cdot 10^5 \cdot 33}{3 \cdot 10^9} dx = 0.00081 \tag{4.5.23}$$

It remains a problem, if the onset of the next bottleneck event is delayed long enough on the order of 10,000 years. Then, new mutations are likely to overwhelm the beneficial gain of the previous round. This is where stabilizing selection comes to aid. During a period of geologic stasis, a stable environment ensure the stability of the gene pool, minimizing alteration on beneficial gains. Furthermore, genetic drift eliminates arising mutations quickly in large populations. Genetic recombination quickly diluted new arising mutation. Therefore, evolution toward human is composed of cycles of directional selection and stabilizing selection. That is, stabilizing selection acts on the population during the recovery phase and in between the bottleneck events.

Then, the fastest time possible to transition from Chimpanzee to Homo sapiens is:

$$\frac{4 \cdot 10^7}{(20.96 + 220) \cdot 1,000} \cdot 11,580 \tag{4.5.24}$$

$$= 1,922,310 \text{ Years} \tag{4.5.25}$$

Since 1.92 Myr<6 Myr, the calculation match reality. This also indicates that each bottleneck events do not immediately follow one and another. They are likely spaced out by 24,600 years to match 6 Myr. This speed is still not the theoretically fastest achievable. We assumed that the recovered population followed each directional bottleneck selection event still generates a probabilistic distribution centered on the mean of 0.44 beneficial mutations per individual per generation. However, it is more likely that the recovered population generates a probabilistic distribution with a higher number of beneficial mutations across all ranges, as they shifted further toward the Homo sapiens' prototype and mutations occur at non-random patterns and tend to occur at selected concentrated sites such as master switch genes. Furthermore, a higher average number of beneficial mutations can also be justified, at least partially, by sexual selection. Sexual selection can be directed by both sexes. An alpha male with more human like traits able to exploit more resources leading to greater bargaining power with females and greater reproductive success. Then, the gene pool shifts toward both a higher number of beneficial mutations and a quicker convergence of its gene's fixation within a given population than random mating. If no alpha male is present at directing faster gene fixation, females can actively choose a selected group of males with enhanced features and fitness to mate with and refuse to mate with the inferior, typical, and average male. A caveat must be raised. This does not imply that sexual selection dominates directional selection. We simply assumed that directional selection, during a bottleneck event, has lifted the constraints placed upon by stabilizing selection so that directed evolutionary change are permitted to take place. The directed evolutionary change can be achieved by either sexual selection (faster), natural selection (slower), or both. When the directed evolutionary change has reached new constraint ceilings and environment reached new equilibrium as the bottleneck event ended, both sexual and directional selection will be replaced again by stabilizing selection. Moreover, we discussed how the total bio-complexity increases overtime exponentially and how stabilizing selection can not be "stabilizing" after all. (See Chp 8). The mode of species sharing certain traits such as brain size, in the absence of directional selection, can still grow exponentially larger if more diverse environment lifted the constraint ceiling. With increased abundance of animal and plant diversity, a species can now extend its food choices and habitat ranges. Any previous deviations do not confer any immediate benefit for such species now becomes beneficial. In the case of human, thanks to the exponentially increasing biodiversity, intelligent manipulation of environment offered strategic advantage not realizable in the more distant past. In the most extreme case imaginable, the mean number of beneficial mutations followed each recovery becomes $21, 21^2, 21^3... 21^n$. As a result, the top 1% population of each round of selection contributes $21, 21^2, 21^3... 21^n$ beneficial genes individually, whereas 1,000 individuals survived each round and $T_{fixed} = 10,000$ years. During the bottleneck, each individual per generation contributes $21, 21^2, 21^3... 21^n$ beneficial mutations, and there are 500 generations.

$$(21 + 21 \cdot 500) + \left(21^2 + 21^2 \cdot 500\right) + ... + \left(21^n + 21^n \cdot 500\right) = 501 \cdot 21 \cdot \frac{21^n - 1}{21 - 1} \qquad (4.5.26)$$

$$501 \cdot 20.96 \left(\frac{20.96^{1.4279} - 1}{20.96 - 1}\right) \cdot 1,000 \text{ individuals} \approx 4 \cdot 10^7 \qquad (4.5.27)$$

$$1.4279 \cdot 11,580 = 16,535 \text{ Years} \qquad (4.5.28)$$

Then, in the most extreme case, it takes only 16,535 years to transition from Chimpanzee to Homo sapiens.

In reality, the speed occur in nature falls within these ranges. Hence, we have mathematically shown that the evolution of Homo sapiens is directed by a series of directional punctuated bottleneck events with periods of stasis.

At this point, we can conclude that most species stabilized by slow changes in earth's climate and continent's configuration, during which time, stabilizing selection is favored over the directional and disruptive selection. Moreover, genetic recombination and genetic drift in a large, stable population will converge the gene pool toward uniformity. From this, we can conclude that the earth's geologic changes played the most critical role in speciation, while genetic drift and recombination played a secondary role by reinforcing genomic uniformity while speciation opportunity does not arise.

Then, the question remains. How does a gradual change in geology sometimes lead to a sudden change in climate or habitat so that an existing species can no longer maintain its status quo and is mandated by disruptive, directional selection? This can be demonstrated by the separation of two continents, while South America and Africa start to split apart and the geologic process continued for tens of millions of years, certain sections of land bridges continue to connect to the two continents. The break up of two continents making migration across the two continents increasingly difficult but still possible. However, once the last landmass bridge is severed, land-based population exchange and gene flow stop completely, which is an one time, sudden disruptive change relative to the past. The reverse is true as well, the terror bird, a predatory bird lived on the island continent of South America, was suddenly forced to compete with northern invaders as the Isthmus of Panama was formed between North and South America. Although South America is drifting north toward North America for millions of years, the habitat of terror bird stabilized for millions of years with no apex predators to compete. Only when the two land masses connected physically by land, a sudden change in its habitat occurred.

When two bodies of water, once freely flow from one side to another, is increasingly blocked by land formation, the aquatic species flow is still possible throughout the plate creation process until the exchange of two bodies of water completely stopped. Then, a sudden change in species habitat takes place. A species formerly able to move freely between two bodies of water and access food resources are now constrained to just one. It may also well adapted to the water temperature at a given range are now forced to adapt to a different one because the water flow to equalize the temperature of bodies of water is no longer possible.

The reverse is also true when two bodies of water are separated by land bridges. Either side is well adapted to its local fauna and predators. As the land bridge stretched thinner and thinner by plate tectonics, two bodies of water are continued to be separated and maintained its status quo. When the land bridge separates the bodies of water disappeared, two bodies of water flow freely, causing temperature change, flow directional change, local fauna change, and exchange of animal species all occur within a very short timescale.

The formation of mountain ranges such as those in Tibetan plateau starts slowly. As mountain creation gradually increases in height, so do bird species adapt by flying higher to get across the mountain ranges. Despite the emergence of a physical barrier, gene flow continued on both sides of the range, so a species' gene pool maintained its uniformity. However, as the barrier continues to rise in height, there is a point reached when the cost of crossing over the mountain range searching for food outweigh the benefit of food access, then the gene flow stops, and speciation occurs at both sides of the mountain.

The reverse again is true. An ancient mountain range is gradually lowered due to weathering and erosion. While species of animals for millions of years are kept to each side of the range, were suddenly able to cross over, leading to drastic habitat change.

A lake is gradually evaporated away in depth nevertheless is able to sustain an aquatic ecosystem until it is finally completely dry, leading to the extinction and destruction of the entire system. The reverse is also true. A freshwater lake is gradually gaining size and its habitat continue to thrive until it joins with the nearest ocean. Suddenly, saline water exchanges with freshwater, bringing invading species and change the entire ecosystem.

The gradual movement of continents and land formations also contribute to sudden climate change which perpetuated for millions of years of stability. South America joined Antarctica before the opening of the Drake Passage. As long as a land bridge existed between them, the oceanic flow circulates both continents and is able to transform frigid polar flow into a warm tropical one and warms Antarctica as it returns. However, once the land bridge disappeared, frigid cold ocean circulates Antarctica, drastically alter the landscape of Antarctica, and turning it into an icy world.

Lastly, sympatric speciation requires a little more discussion, the first three types of speciation are all associated with geological change, yet sympatric speciation concerns that a single species diverged into two by selecting different survival strategies. It seems that sympatric speciation has little to do with geologic changes, but it is. Let consider a thought experiment. It is assumed that a species existed for a long time with increasing numbers and intraspecies competition develops two different feeding strategies on different types of fruits. However, if both types of fruit trees co-existed for a long time, it is likely that this species could have exploited two different niches long time ago instead of now. This contradicts with our assumption. Then if the fruit tree is just introduced now, it must be a consequence of three other types of speciation (from some other isolated environment), and it is just spreading its habitat into the fruit-eating species territory. The fruit tree can not just arise within the species territory because we have shown that stabilizing selection does not favor speciation in a static environment. Therefore, fruit-eating species speciation into two different subspecies by adapting different feeding strategy is an indirect consequence of allopatric, peripatric, and parapatric speciations, which in turn is a consequence of geologic change.

By now, it is clear that geologic movement continues, but its gradual, incremental quantitative change does not bring significant climate and habitat alteration until a critical threshold is reached, whether it is the joining of two separated landmass or the separation of the two. Thereafter, the incremental quantitative change led to a qualitative leap in the environment. Species then have to quickly adapt to avoid extinction. This is confirmed by the fossil record as sudden appearance and disappearance of genus and is the essence of punctuated equilibrium. As a result, *there is no such thing as the average Evolutionary rate of species. The final value of a computed evolutionary rate is actually the combined rate of geological change leading to drastic environmental alteration and the intrinsic rate of speciation in biological creatures.* This is conceptually similar to the final speed at which a box is moving across a rough surface, which is broken down into the input force minus resisting friction. Nevertheless, a value can be obtained regarding this evolutionary rate, which is a factor between 1.23 to 4 per

100 million years. But one must understand that intrinsically biological creature is able to evolve to this rate at much shorter timescale than 100 million years. The rate of geological change leading to drastic environmental alteration is the brake on biological evolution, without drastic change and fluctuation in the environment, natural selection favors static, non-changing morphology and behavior. From this, we can again confirm the importance of ice age in contributing the rise of Homo sapiens, a period of chaotic, ever changing climate and weather patterns.

4.6 BCS, BER, and Evolutionary Speed

4.6.1 BCS

Having elucidated the mechanism setting the nature's pace for evolution, we resort to calculating the **Biocomplexity Search Space** (**BCS**), and the **Background Evolutionary Rate** (**BER**) which are the critical key concepts to abstract evolution into its simplest mathematical form. From mathematical standpoint, as we have already demonstrated and is addressed in detail in Chapter 7 and 8, that the total biodiversity at any given time can be expressed as a lognormal distribution (or tailed, skewed distribution) where species with more acquired traits that differentiates from the rest are generally rarely occur in nature, due to greater amount of time for nature to successfully perform such combination/permutation, it is costly to maintain all traits acquired, or both. The integrated area under the distribution curve is the **Biocomplexity Search Space** (**BCS**) at the current time. The area represents all possible solutions and their frequency distribution by nature's experiment in any given time.

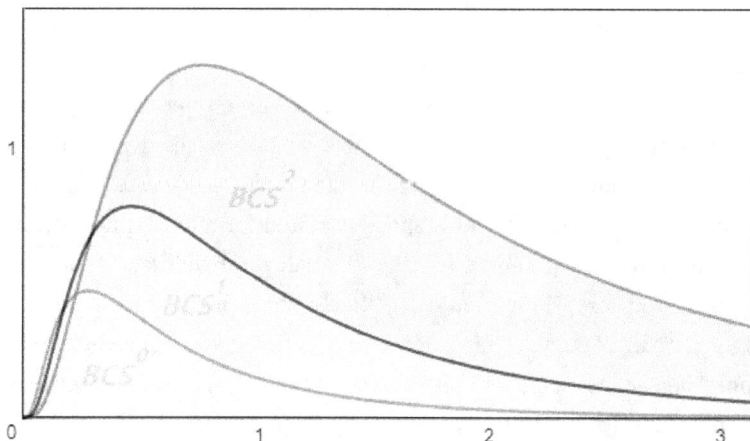

Figure 4.6.1: BCS expressed mathematically for current period followed by 100 Myr and 200 Myr later

Furthermore, it is observed that the total number of species, especially the terrestrial ones, are increasing over time and (must be increasing since there is no terrestrial species 400 Mya and plenty observed today) . Therefore, **BCS** must be increasing over time.

We attempt to compute **BCS** using paleo-biological data. We use the apparent marine fossil diversity during the Phanerozoic to compute **BCS** increase per 100 Myr. The best approximate curve is given by:

$$T\left(x\right) = -0.000026x^4 - 0.02667x^3 + 0.41x^2 - 1.916x + 3.678 \tag{4.6.1}$$

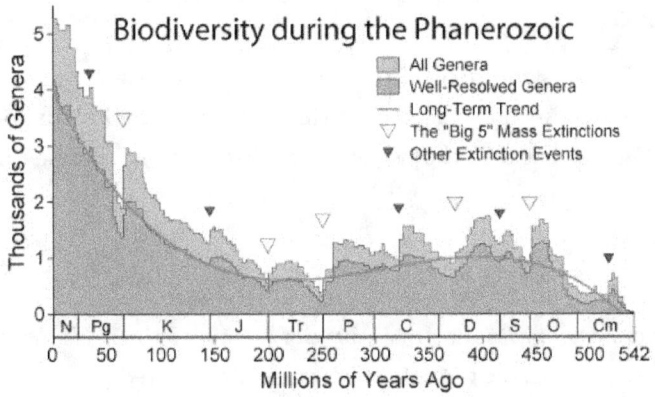

Figure 4.6.2: Historical trend of biocomplexity change

We have shown earlier that an increase in biodiversity is applicable only during the breaking-up phase of supercontinent cycle. Therefore, we compare the most recent 100 Myr to that of the averages of biodiversity achieved between 100 Mya and 200 Mya as well as those achieved between 200 Mya and 300 Mya (those of the supercontinent phase of supercontinent cycle).

$$\frac{\int_0^{1.65} T(x)\,dx}{\frac{1}{2}\left(\int_{1.65}^{2 \cdot 1.65} T(x)\,dx + \int_{2 \cdot 1.65}^{3 \cdot 1.65} T(x)\,dx\right)} = 2.453998 \tag{4.6.2}$$

Alternatively, one can compares the total number of genera achieved by birds (2,172), mammals (1,229), and reptiles (912) during the current Cenozoic to that of the number of genera of dinosaurs (500+ discovered so far and a total of 3,400 predicted, peak dinosaur excavation is expected by the mid century) during the Mesozoic. The number of genera is compared instead of number species since not all species are likely conserved as fossil specimens. The total number of current genera is divided by 66 Myr to represent number of genera attainable within 100 Myr (we exclude those extinct genera of birds, mammals, and reptiles during the earlier phase of Cenozoic, if one were to include them, the estimate should be higher):

$$M_0 = \frac{(2172 + 1229 + 912)}{0.66} = 6534.848 \tag{4.6.3}$$

The total number of estimated dinosaur genera during the entire Mesozoic is divided by the time span of Mesozoic era at 167.23 Myr to derive the average number of genera per 100 Myr.

$$D_0 = \frac{3400}{(2.3323 - 0.66)} = 2033.128 \tag{4.6.4}$$

So that **BCS** increases by 3.214 for terrestrial species exploiting inland regions:

$$\frac{M_0}{D_0} = 3.21418 \tag{4.6.5}$$

Later, we actually shown that the emergence of angiosperm actually contributed a factor of 28 increase per 100 Myr. (see 6.8 "Probability of Angiosperm")

$$\frac{369,000 + 12,421}{12,421 + 1,191} = 28.021 \tag{4.6.6}$$

However, we believed that it is atypical for species increase by this magnitude within every 100 Myr, and we used this atypicality to assert the rarity and early arrival of civilization on earth with agricultural revolution.

4.6.2   Evolutionary Speed

When **BCS** increases, the mode of the distribution shifts toward species possessed more traits, thus, the mode changes. The difference in the modes between two successive time periods is the **Evolutionary Speed**. Mathematically, it is defined as the difference between the peaks of two distributions representing the biocomplexity of earth 100 Myr apart.

$$\text{Speed } t_1 = P_{eak}(t+1) - P_{eak}(t) \tag{4.6.7}$$

The **Evolutionary Speed** is a *dimensioned* scalar quantity per a unit of time. The speed can be expressed numerically by 2 dimensional units. First, as the traits (mode's value of x) shifted per a unit of time and secondly, as the amount of overlap between two successive distributions per a unit of time, signifying the percentage of species made it the next round of evolution x years apart. These 2 measures are not interchangeable and describes 2 different aspects of evolutionary speed. Ideally, both measures are available.

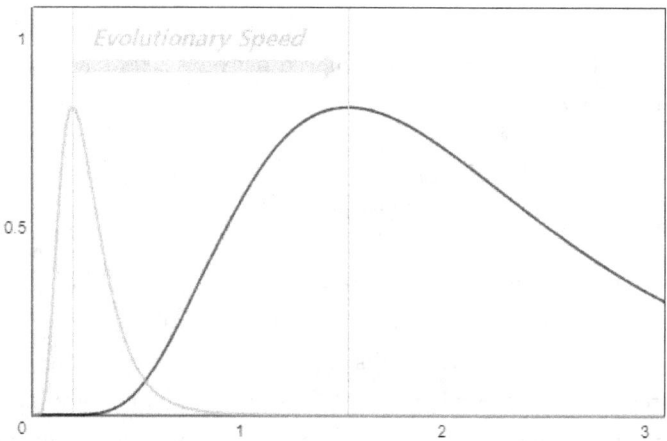

Figure 4.6.3: Evolutionary speed demonstration

4.6.3   BER

When **BCS** increases, the mode of the distribution shifts toward species possessed more traits, thus, the mode changes. The rate of change in evolutionary speed, or equally the rate of change of the difference in the modes between two successive time periods is the **Background Evolutionary Rate (BER)** of change of Evolutionary Speed. Mathematically, it is defined as the ratio of the evolutionary speed from period 1 to the evolutionary speed from period 2, whereas the duration of those two time periods are equal.

$$\textbf{BER} = \frac{\textbf{Speed } t_1}{\textbf{Speed } t_0} = \frac{P_{eak}(t+1) - P_{eak}(t)}{P_{eak}(t) - P_{eak}(t-1)} \tag{4.6.8}$$

It is also the rate of shift of the mode's position, so **BER** can also be expressed as the **Background Evolutionary Rate** of change of the mode:

$$\textbf{BER} = \exp\left(\frac{\ln\left(\frac{(P_{eak}(t))}{P_{eak}(t-1)} \right)}{t - (t-1)} \right) \tag{4.6.9}$$

BER is a *dimensionless* scalar quantity per a unit of time that can be used to transform either evolutionary speed or the mode's position. **BER** ≥ 1. That is, **BER** has to be strictly equal to or greater than 1, but it can be greater, equal, or smaller than **BCS**, depending on the placement configuration of successive distributions.

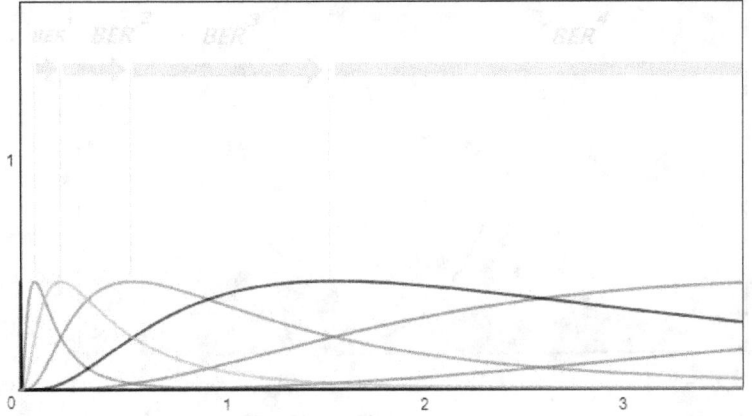

Figure 4.6.4: Cases when **BER** = **BCS**

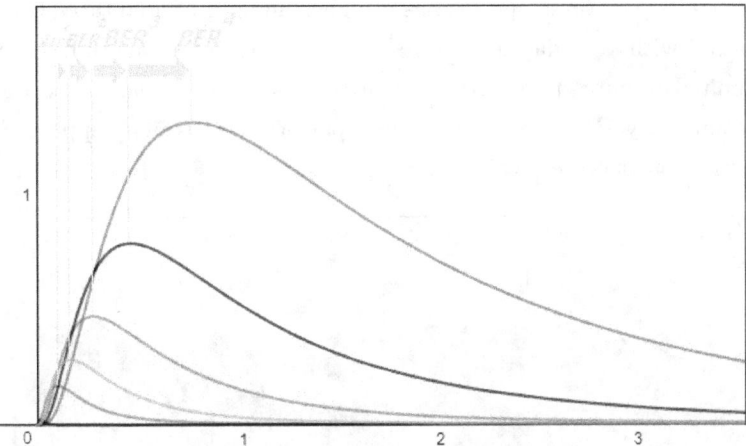

Figure 4.6.5: Cases when **BER** < **BCS**

Nevertheless, regardless of the placement of **BCS**, an increase in **BER** is always proportional to an increase in **BCS**. That is, for the same placement of **BCS**, **BCS** increase with higher exponential rate of growth also provides a **BER** with a higher exponential rate.

$$BER \propto BCS \qquad (4.6.10)$$

4.6.4 Altering the Speed of Evolution

Increase in **BCS** contributes to an increase in **BER**, which in turns increases the speed of evolution, However, it may not be the only factor in contributing to the speed of evolution. Assume that **BCS** stays constant between 2 time periods, but for the 2nd time period, it shifts horizontally by a limited number of traits, so that the mode now centers on the species possessed more traits. In this case, both **BCS**=1 and **BER**=1 but the mode still increases. The shift and skipping number of traits is called the **Selection factor** by altering the speed of evolution regardless of the value of **BCS** and **BER**.

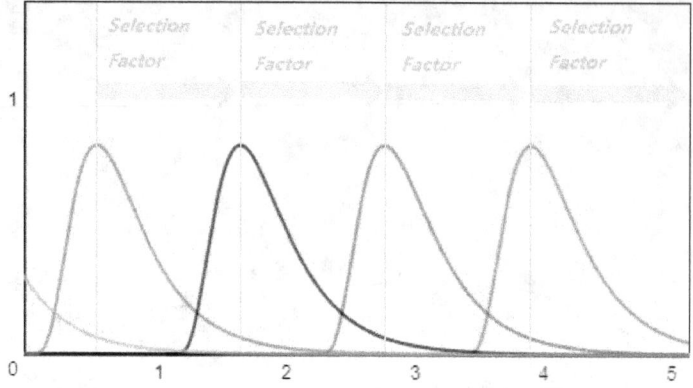

Figure 4.6.6: Cases when **BCS** =1, **BER** = 1, and **BER** = **BCS** with traits skipping per 100 Myr

It is plausible when one considers catastrophic extinction event that periodically wiped out genera and families of species with less adaptable traits, recovered species has a higher number of adaptable traits. Alternatively, it can just simply be that species with less adaptable traits are out-competed by more flexible species over the course of 100 Myr given a limited resource pool with only the top performers make it to the next round. In reality, both increase in biocomplexity **BCS**, **BCS** placement pattern, and traits skipping **Selection factor** are likely to play a role in setting the speed of evolution.

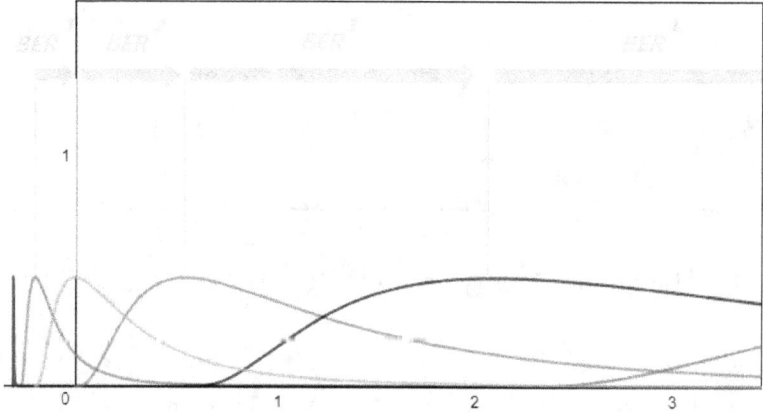

Figure 4.6.7: Cases when **BCS** >1, **BER** > 1, and **BER** = **BCS** with traits skipping per 100 Myr

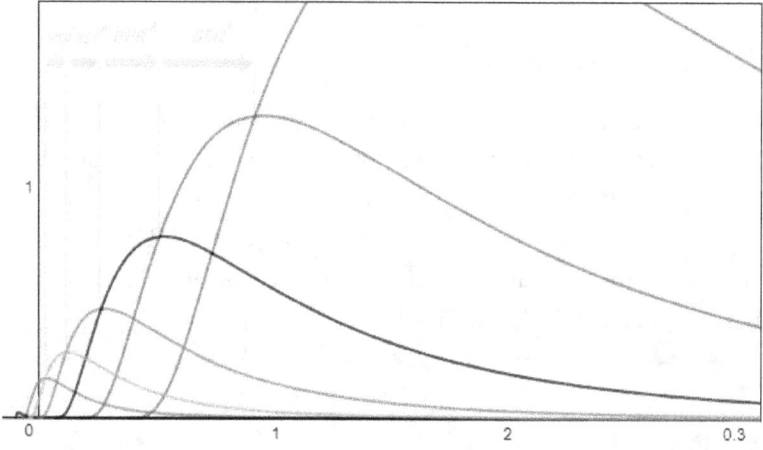

Figure 4.6.8: Cases when **BCS** >1, **BER** > 1, and **BER** < **BCS** with traits skipping per 100 Myr

4.6.5   Measuring BCS and BER

The background evolutionary rate prior to multicellularity is highly predictable. The onset of the great oxygenation event starts roughly 2 billion years after the appearance of photosynthesis, this time should remain largely fixed since smaller ocean surface is compensated by smaller ocean oxygen sinks near the surface of ocean. The onset of multicellularity starts roughly 1.5 billion years after the great oxygenation event, but the total time spent can increase or decrease depending on the exoplanet's land surface area to sea surface area ratio in comparison to that of the earth. However, we are more concerned about the background evolutionary rate at the multicellular stage of evolution. Alexei and Gordon have shown a method of calculation using DNA complexity, and they showed that genome complexity grows by 1.23 every 100 Myr.

However, genome complexity may not strongly associate with functional acceleration of species adaptation to the environment such as flying higher or running faster. In Chapter 7, we determined the background evolutionary rate by comparing the Encephalization quotient of mammals to that of reptiles using the following general formula:

$$\textbf{BER} = \exp\left(\frac{\ln\left(\frac{(P_{eak}(0))}{P_{eak}(2.25)}\right)}{2.25}\right) \tag{4.6.11}$$

Whereas the typical mammal of today had 10 times greater EQ than the last common ancestor of reptiles and mammals 225 Mya. This 10 fold difference is translated into a rate that is equivalent to 2.783 fold increase per 100 Myr. We made the assumption that the **selection factor** played a minimal role in the speed of evolution for the sake of simplicity in the demonstration of the model in Chapter 8.

Later, we did shown a more complicated case where the **selection factor** played a role in the speed of evolution. If **selection factor** played a role in the speed of evolutionary change, one has to determine what percentage of genera from the last 100 Myr survived into the next 100 Myr compared to the percentage made to the next round assuming minimal involvement by the selection factor. (The overlapping area size between two adjacent PDF) This will determine how much **selection factor** has played a role in determining the speed of evolution. In general, one can use four critical pieces of information to predict the characteristics of evolution.

1. One has to count the number of genera from each epoch 100 Myr apart. By constructing a complete census, one is able to determine the rate of **BCS** increase per 100 Myr precisely.

2. One can determine the **BER** by selecting a trait, in our case, Encephalization quotient. One then needs to find a way to effectively measure the characteristics of the trait. For example, by measuring the length of the snout, horn, wingspan, or by gauging the capacity of the skull. Alternatively, one can measure the performance of the trait by recording the speed of running or flying, measuring the height of flight or jump. Then, one can determine the rate of the characteristics or performance change of the trait over a unit of time period (in our case, 100 Myr). This rate of change constitutes the **BER**.

3. One can also determine the **BER**, or cross-validate **BER** derived from the previous method by piecing together the distribution curve from the past epoch so that the mode and the deviation (width) can be derived from the complete fossil records. The width/variance of the previous distributions determine the placement pattern by altering variable k (see Chapter 8 "Generalized Model"). The smaller the difference between the width of the previous distributions, the greater the k value, and distribution more stacked on top of each other as more conservative Darwinian evolution and **BER** approaches the value of 1. Finally, we later shown that $\textbf{BER} = (\textbf{BCS})^{\frac{1}{k}}$, so **BER** can be derived based on knowing **BCS** and k.

4. One has to determine what percentage of genera from the last 100 Myr survived into the next 100 Myr. This will determine how much **selection factor** has played a role in determining the speed of evolution.

For the sake of further simplicity, *the value of **BCS** and **BER** in the model is both set to 2.783.* It is set to 2.783 also because it is approximately the value of e, providing great ease at numerical manipulation as later equations rely on natural logarithm. Later, they are altered to different values under generalized cases.

The exponential increase in both **BCS** and **BER** is not a contradiction to our earlier discussion on the pace of evolution constrained by geology. We illustrated in Chapter 7, the increase in biodiversity does follow a positive feedback loop as greater biodiversity, greater biocomplexity search space, provides a greater ease at speciation and more niches opening during major geologic changes. Species will continue to emerge due to major geologic changes. The rate of major geologic changes will remain nearly constant throughout the history of multicellular life, but the number of newly emerged species and opened niches per each major geologic change increases exponentially. [3]

4.7   Continent Cycle

Although the average rate of background evolutionary rate is now known, the evolutionary rate leading to greater diversity is non-uniform throughout the geologic history. This can be seen from the biodiversity plot. It can be shown that biodiversity started to emerge in Cambrian and took a dive by Permian and continue to grow exponentially thereafter. The growth is especially fast since the Jurassic. The graph correlates well with the configurations of earth's continents' positions. The breaking up of Pannotia supercontinent certainly aided the start of multicellularity besides an increase in atmospheric oxygen, and as the continents merged to form Pangea supercontinent 300 million years ago at the start of Permian, the diversity not only stabilized but also dropped. The Permian-Triassic extinction, the deadliest one in earth's geologic history, caused the extinction of 90% of animal species. Pangea started to break up 170 million years ago during the mid-Jurassic. Thereafter, the diversity increased exponentially.

This slowing and speeding observed in geologic record cannot be associated with the rise and fall of oxygen level in the atmosphere. Although oxygen level is closely associated with the emergence of super-sized insects and enabled the appearance of multicellularity and speeded up biological evolution significantly, there is little correlation between biodiversity and atmospheric oxygen level. Cambrian, Ordovician, Silurian, and Devonian had oxygen level 63%, 68%, 70%, and 75% of modern level respectively, yet the biodiversity was increasing throughout this period. During the following Carboniferous and Permian era, the oxygen content was 163% and 115% of modern level respectively, yet biodiversity stabilized and even dropped.

Carbon dioxide level cannot correlate with biodiversity as well. Throughout all earth's history, fast and slow periods of biodiversity growth occurred during both high and low concentration of CO_2. During some high times of carbon dioxide level such as Cambrian (16 times modern level), Ordovician (15 times modern level), and Cretaceous (6 times modern level), biodiversity was increasing. During other high times of CO_2 levels such as Silurian (16 times modern level), Devonian (8 times modern level), Jurassic (7 times modern level), and Triassic (6 times modern level), biodiversity was stagnant or decreasing. At some low times of carbon dioxide level such as Permian (3 times modern level) and Carboniferous (3 times modern level), biodiversity stabilized. At other low times of CO_2 such as the Paleogene (2 times modern level), however, biodiversity increased exponentially.

Therefore, we can rule out the atmospheric composition played any significant role in the diversification and the rate of evolutionary change. If all terrestrial life-friendly planets go through similar continents-supercontinents cycle, then, we should expect that the background evolutionary rate on any particular planet follows a sinusoidal curve where the evolutionary rate and the diversification occur faster during the separation of continents, slows down, and even drops when continents merge. (the background evolutionary rate, the computed average of all habitable planets, however, will exhibit a smooth exponential curve since some planets at a given period go through evolutionary stasis while others are evolving rapidly). Nonetheless, biodiversity should follow the general increasing trend because new species of plants and animals establish in previously uninhabitable regions, altering the biochemistry and environment and rendering them habitable. This has repeatedly happened in

---

[3](more on this please follow chapter 7 on section " YAABER for Evolution of Homo Sapiens").

earth's history. For example, the cyanobacteria's metabolic process has transformed the earth's atmosphere by providing free oxygen. The establishment of land plants enabled the habitability by land animals later on. Moreover, new species opens new niches on existing habitable environment. For example, the evolution of fruit trees enabled the evolution of arboreal species in the primate family.

The loss of diversity can be expressed mathematically based on the configuration of continents. In the simplest model, island continents are surrounded by bodies of oceans, which brings precipitation to their shorelines. Marine climate further extends inland, however, the central part of the supercontinent remains dry and arid, receiving little precipitation with extreme temperature swings due to continental climate, reducing its chance to host a diversity of biological life. This simplified assumption generally applies to the climate currently observed on all continents across earth except the equatorial, tropical regions of the Amazon rain forest and central African jungle. A more complicated version of the concept is presented in here:

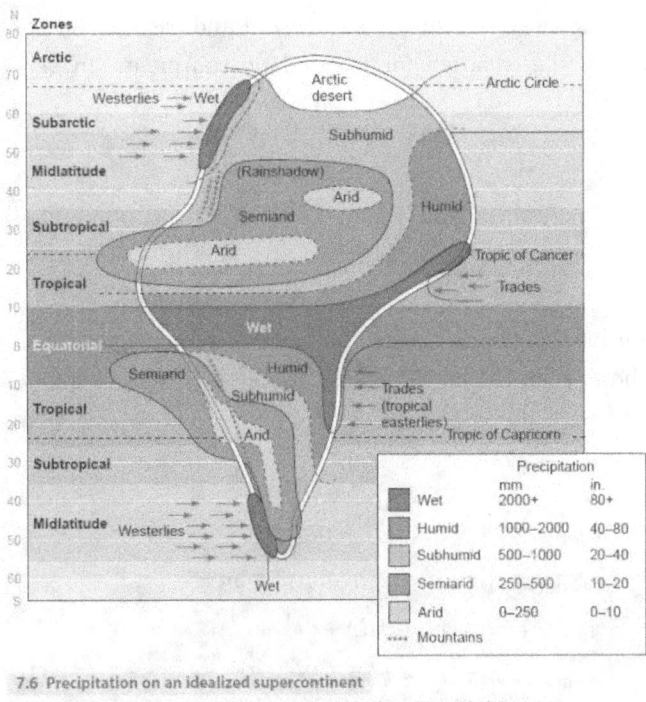

7.6 Precipitation on an idealized supercontinent
A schematic diagram of annual precipitation over an idealized continent and adjoining seas.

Figure 4.7.1: Precipitation on an idealized supercontinent

Additionally, a perimeter surrounding each island continent with an extended continental shelf of shallow seas provides a great biodiversity for marine life. It has been measured that marine life biodiversity decreases with the depth, due to decreasing sunlight penetration disabling photosynthesis and decreasing temperature. If each island continent can be abstracted into the shape of a circle, then, one can make the following deductions.

Assuming each island continent size is small enough that is completely covered by marine climate to guarantee its biodiversity, and its radius is r. Then, the total size of m island continents composes a total region providing biodiversity:

$$i_{island} = m\pi r^2 \tag{4.7.1}$$

If m island continents merged into one supercontinent; then, the radius of the supercontinent is:

$$\pi R_{super}^2 = m\pi r^2 \tag{4.7.2}$$

$$R_{super}^2 = mr^2 \tag{4.7.3}$$

$$R_{super} = \sqrt{mr^2} \tag{4.7.4}$$

and assuming that biodiversity only extends a distance of d inland from the shoreline, then the total habitability of the supercontinent is given by the equation:

$$i_{supercontinent} = m\pi r^2 - \pi(\sqrt{mr^2} - d)^2 \tag{4.7.5}$$

$$i_{supercontinent} = m\pi r^2 - \pi(mr^2 - 2\sqrt{mr^2}d + d^2) \tag{4.7.6}$$

$$i_{supercontinent} = (m\pi r^2 - \pi mr^2) + 2\pi\sqrt{mr^2}d - \pi d^2 \tag{4.7.7}$$

$$i_{supercontinent} = 0 + 2\pi\sqrt{mr^2}d - \pi d^2 \tag{4.7.8}$$

$$i_{supercontinent} = \pi\left(2\sqrt{mr^2}d - d^2\right) \tag{4.7.9}$$

If one assumes that biodiversity extends a unit distance of 1 inland and each island continent's radius is also a unit distance, and omitting π, then the equation for total island continents diversity and supercontinents diversity can be simplified into:

$$i_{land} = m \tag{4.7.10}$$

$$y_{land} = 2\sqrt{m} - 1 \tag{4.7.11}$$

Moreover, one should also take continental shelf's marine biodiversity into account. Assuming the continental shelf extends a distance of R offshore, then the total zone of marine biodiversity for m island continents and supercontinent is given by:

$$z_{island} = m\pi\left(r + R\right)^2 - m\pi r^2 \tag{4.7.12}$$

$$z_{island} = m\pi\left(r^2 + 2Rr + R^2\right) - m\pi r^2 \tag{4.7.13}$$

$$z_{island} = m\pi\left(r^2 + 2Rr + R^2 \quad r^2\right) \tag{4.7.14}$$

$$z_{island} = m\pi\left(r^2 - r^2 + 2Rr + R^2\right) \tag{4.7.15}$$

$$z_{island} = m\pi\left(2Rr + R^2\right) \tag{4.7.16}$$

$$z_{supercontinent} = \pi\left(\sqrt{mr^2} + R\right)^2 - \pi\left(\sqrt{mr^2}\right)^2 \tag{4.7.17}$$

$$z_{supercontinent} = \pi\left(mr^2 + 2\sqrt{mr^2}R + R^2\right) - \pi mr^2 \tag{4.7.18}$$

$$z_{supercontinent} = \pi mr^2 + 2\pi\sqrt{mr^2}R + \pi R^2 - \pi mr^2 \tag{4.7.19}$$

$$z_{supercontinent} = \left(\pi mr^2 - \pi mr^2\right) + 2\pi\sqrt{mr^2}R + \pi R^2 \tag{4.7.20}$$

$$z_{supercontinent} = \pi\left(2\sqrt{mr^2}R + R^2\right) \tag{4.7.21}$$

If one assumes that biodiversity extends a unit distance of 1 offshore and each island continent's radius is also a unit distance, and omitting π, then the equation for total island continents marine diversity and supercontinents marine diversity can be simplified into:

$$i_{sea} = 3m \tag{4.7.22}$$

$$y_{sea} = 2\sqrt{m} + 1 \tag{4.7.23}$$

176

combining land diversity with marine diversity, the following mathematical graph can be extrapolated:

$$i_{total} = 4m \tag{4.7.24}$$

$$y_{total} = 4\sqrt{m} \tag{4.7.25}$$

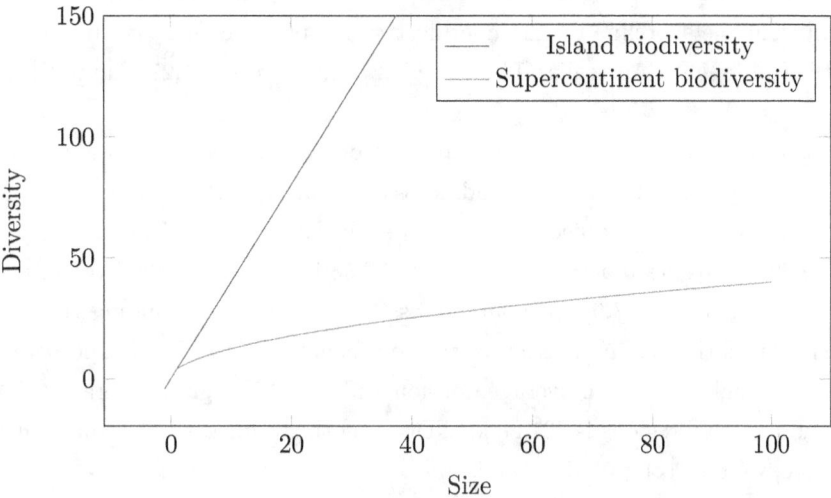

Figure 4.7.2: Biocomplexity of island continents and supercontinents vs land mass size

It shows that under ideal conditions, in all possible number of island continents leading up to supercontinents, island continents has as much as all their land areas nourished by moderate to adequate precipitation, moderated by sea currents to reduce temperature extremes. On the other hand, a supercontinent in size comparable to the total area of many island continents left vast stretches of its inland in arid, dry climate with huge temperature swings. Therefore, island continents provide a greater biodiversity than supercontinents in all ranges. Of course, taking into consideration the functioning of Hadley cell and its explanation for major dry, arid bands of latitudes stretching both the northern and the southern hemispheres, further refinements to the model are possible. Furthermore, instead of cutting off biodiversity at a certain distance inland, the drop in biodiversity is more likely to be gradual. However, the basic assumption that biodiversity correlates positively with separated island continents remains.

Of course, supercontinent such as those of Pangea must have had episodes of increased biodiversity, but such spurts of biological diversification do not alter the general trend of the epoch. After the Carboniferous rainforest collapse, first, local recovery simply filled the previously vacated niche. Since each local fauna recovered on its own and no two local fauna exchanged gene flows, self-imposed barrier existed between each region. As local fauna evolved and resorted to adaptive radiation and increased biological diversity (when new niches opened faster than stabilizing genetic drift), the total diversity increased on the continent. However, as these species started to re-establish themselves among all others on the vast continent, they break the previously self-imposed barrier. Competition ensued and eventually led to stabilized or drop in the total biological biodiversity across the entire continent.

4.8 Supercontinent and Island continent biodiversity in depth-analysis

Having demonstrated our simple model, we now refine the model for more detailed analysis and will later be used to calculate the emergence probability of Homo sapiens. We divide our analysis of planetary biodiversity based on three configurations: island, supercontinent, and supercontinent within the tropical regions.

First of all, we continue to maintain the average radius of each island continent to $r = 1$. Therefore, the total island land biodiversity is $m\pi\left(1\right)^2$.

To compute the supercontinent land biodiversity, one needs to define the distance d, which is how much further inland precipitation and moderate climate maintains biodiversity. We now assumes that a distance of d inland $= 0.574$. We arrived at this value by first treating the total surface area of the planet can hold at most 100 island continent configurations. That is, $100\pi r^2 = 5.1\cdot10^8$ km^2. Since earth's surface was covered by 29% of dryland, there can be at most 29 island continent configurations. Then each island continent covers 5,100,684.93 km^2 and with a radius of 1,274 km from the shoreline to the innermost region of the continent, this is comparable and somewhat smaller than the continent of Australia. Therefore, 1,274 km $= r$ and 1,274 km is 1 unit distance for r.

We then compute the average stretch width of the wet regions on earth. The wet region with considerable precipitations stretches an average of 1,324.76 km inland across East and Southeast Asia, 2,097.025 km inland across the entire Europe (855.44625 km averaged over all 4 sides for Eurasia) 439.5 km inland across northern and eastern Australia (329.63 km averaged across all sides), 1,972.53 km inland across North America (493 km averaged on four sides of the ocean), 1,517.6 km inland across South America excluding the Amazons (758.8 km averaged on 2 sides since it is the tipping portion of the continent), 1,174.18 km inland across Sub-Sahara Africa excluding the tropical jungle (880.6 km averaged on four sides) The weighted average taking the size of each continent into account across all regions is 731.28 km. Dividing this value over the unit r, this is translated into $\frac{731}{1274} = 0.574$. Therefore, we now computed $d = 0.574$.

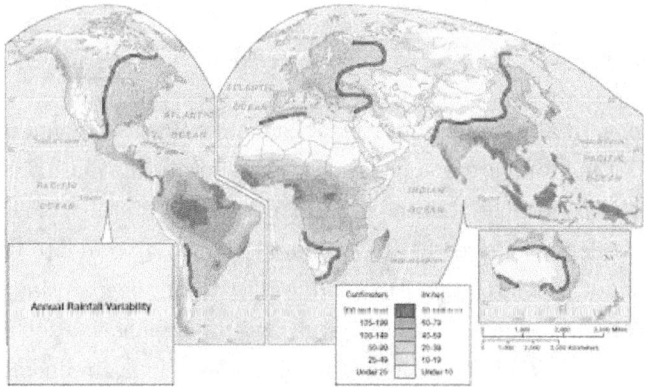

Figure 4.8.1: World arid and wet region demarcation lines

We intentionally excluded the tropical region from our consideration. Those regions such as the central African and Amazon jungle receives significant precipitation even at the innermost corners away from the shoreline. The tropical regions' biodiversity will be treated as islands.

As a result, the supercontinent land diversity is now expressed as:

$$i_{supercontinent} = \pi\left(2\sqrt{mr^2}\cdot 0.574 - 0.574^2\right) \tag{4.8.1}$$

$$i_{supercontinent} = 1.148\sqrt{m} - 0.574^2 \tag{4.8.2}$$

Revisiting the land biodiversity for island continents, we just defined that:

$$i_{island} = m\pi\left(1\right)^2 \tag{4.8.3}$$

This is assuming that an island continent at the size smaller than Australia receives considerable precipitation even at its innermost regions. However, we have just shown that on average, adequate precipitation can only extend up to 57.4% of the measured radius from the shoreline to the innermost region. Therefore, the renewed land biodiversity of island continents is lowered to:

$$i_{island} = m\pi (1)^2 - m\pi (1 - 0.574)^2 \tag{4.8.4}$$

$$i_{island} = 0.818524m \tag{4.8.5}$$

The supercontinent marine diversity depends on the stretch width of the continental shelf. The width of continental shelf R varies depending on the proportion of continental plate submerged under the ocean. We have already shown in chapter 3 that for various continental plate proportion, there exists an upper and lower bound which permits the emergence of industrial civilizations with flat plains. Equivalently, one can derive a range of continental plates with varying percentage submerged under the ocean yet all converge to the same dryland to ocean surface area ratio. We use these equations to find the weighted average continental shelf proportion relative to the exposed dryland surface for all ranges of land to sea ratio from 0 to 84 percent. First, we derive the inverse for the upper and lower bound for different continental configurations.

$$I_{upper}(x) = \frac{x}{0.84} \tag{4.8.6}$$

$$I_{lower}(x) = \begin{cases} 2x & x \le 50 \\ 100 & x > 50 \end{cases} \tag{4.8.7}$$

We used a higher lower bound than the original by multiplying the original by 1.25, so that the lower bound comes:

$$y_{lowerhigh} = \frac{1}{2}x \tag{4.8.8}$$

Although certain land to sea ratio permits the emergence of flat plains, not all continental shelves are shallow. We are only interested in the creation of shallow continental shelves with adequate sunlight penetration. As a result, the requirements for the creation of shallow continental shelf is more stringent than the creation of flat plains on dryland.

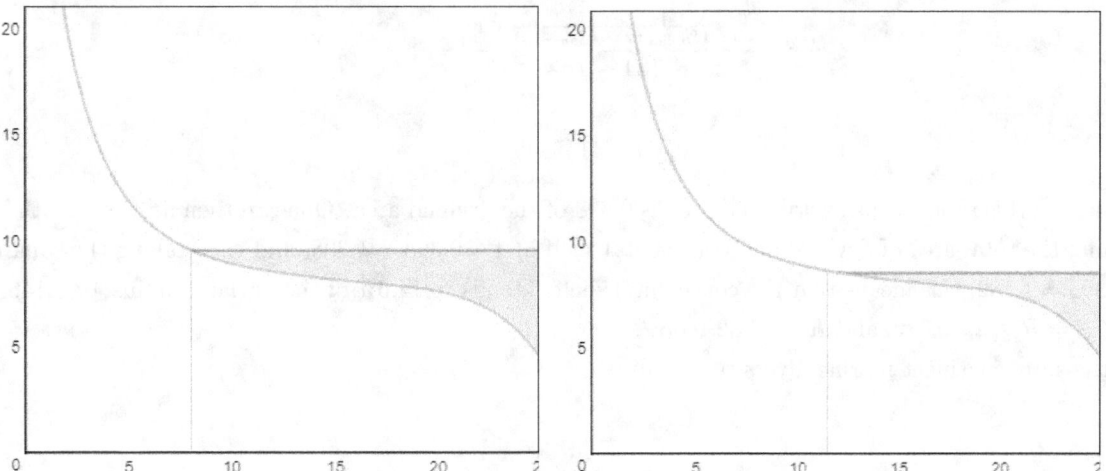

Figure 4.8.2: A example of deep vs shallow continental shelf for the same continental plate coverage but different land to ocean ratio.

Then, the ratio of the continental shelf to the exposed dryland is given by:

$$y_{shelf} = \frac{y_{upper} - x}{y_{upperhigh}} \tag{4.8.9}$$

179

With the inverse of the lower and upper bound formulated, one can find a range of continental configurations converges to a given land to ocean ratio. A specific example is illustrated for the case of dryland coverage at 41.6 percent of the planetary surface:

$$y_{shelf} = \frac{y_{upper} - 41.6}{y_{upperhigh}} \tag{4.8.10}$$

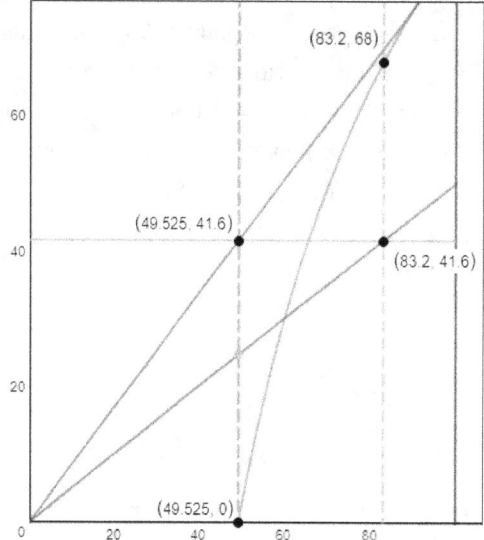

Figure 4.8.3: The shelf proportion grows as 0 percent relative to the dryland area at the upper bound for dryland exposure over the entire continental plate to 68 percent relative to the dryland area at the lower bound for dryland exposure over the entire continental plate for the case of dryland coverage at 41.6 percent of the planetary surface:

There appears to be two stages, for dryland coverage between 0 to 50 percent, the continental shelf proportion relative to the exposed dryland surface of earth is 0.398.

$$\frac{\int_0^{50} \frac{\left(\int_{I_{upper}(x)}^{I_{lower}(x)} y_{shelf} dx \right)}{(I_{lower}(x) - I_{upper}(x))} dx}{(84 - 0)} \tag{4.8.11}$$

$$= 0.398274157445$$

Integrating across this range, continental shelf area is 0.398 of the dryland area. One can then find the width R by first finding the total area of dryland and continental shelf at $1 + 0.398 = 1.398$, and then taking the square root to $\sqrt{1.398} = 1.182$, this means that the continental shelf extends to 14.6% of the dryland radius, or 0.146r. As a result, $R = R_{super} \cdot 0.182$, and $R = 0.182 \cdot \sqrt{mr^2}$.

Therefore, the supercontinent marine diversity is defined as:

$$z_{supercontinent} = \pi \left(2\sqrt{mr^2} R + R^2 \right) \tag{4.8.12}$$

$$z_{supercontinent} = \pi \left(2\sqrt{mr^2} \left(0.182\sqrt{mr^2} \right) + \left(0.182\sqrt{mr^2} \right)^2 \right) \tag{4.8.13}$$

$$z_{supercontinent} = \pi \left(0.292 mr^2 + 0.182^2 mr^2 \right) \tag{4.8.14}$$

$$z_{supercontinent} = 0.364m + 0.146^2 m \tag{4.8.15}$$

$$z_{supercontinent} = 0.398m \tag{4.8.16}$$

Between 50 to 84 percent dryland coverage, the shelf area to dryland area ratio continue to shrink from 0.398 to

0 as increasingly greater dryland coverage is satisfied by ever smaller range of continental plates. With less than 50 percent dryland coverage and at the lower bound (the greatest extend in which a shallow shelf inundating the plate), continental plates with greater coverage can be continually added to the entire range of continental plates satisfying the land to sea ratio. With greater than 50 percent dryland coverage, the lower bound's continental plate coverage remain at 100 percent (the entire surface of earth) and greater dryland gain is satisfied by lower sea level. Eventually as the sea level continues to drop, the upperbound (the smallest extend in which a shallow shelf inundating the plate and renders marine species' evolution onto land feasible and practical) satisfying the designated dryland coverage ratio converges on the lower bound at 100 percent continental plate and 84 percent dryland coverage.

This inverse relationship is captured by the equation:

$$R\left(x\right) = \frac{1}{100}\left(0.00566243x^2 - 1.92724x + 121.982\right) \tag{4.8.17}$$

Applying our method, we have:

$$z_{supercontinent} = m\left(2\left(\sqrt{1+R\left(x\right)}-1\right) + \left(\sqrt{1+R\left(x\right)}-1\right)^2\right) \tag{4.8.18}$$

Then the combined equation across all ranges is:

$$z_{supercontinent} = \begin{cases} 0.398m & 0 \le x \le 50 \\ m\left(2\left(\sqrt{1+R\left(x\right)}-1\right) + \left(\sqrt{1+R\left(x\right)}-1\right)^2\right) & 50 \le x \le 100 \end{cases} \tag{4.8.19}$$

We then compute the average continental shelf width of earth for island continents. Based on our previous derivation, a slightly smaller land mass than Australia occupies a surface area of 1% of the earth, which translates into continental plate coverage between 1.19% to 2%, substituting our equation, we have:

$$\frac{\int_{1.19}^{2} \frac{\left(\int_{I_{upper}(x)}^{I_{lower}(x)} y_{shelf}dx\right)}{\left(I_{lower}(x) - I_{upper}(x)\right)}dx}{\left(2 - 1.19\right)} \tag{4.8.20}$$

$$= 0.398274157445$$

That is, the weighted average of continental shelf proportion relative to the exposed dryland surface occupying $\frac{1}{100}$th surface area of earth is 0.398. This implies that the continental shelf size shrinks slightly slower than the permissible dryland ranges. However, this prediction does not completely match real observation and real observation offers a whole range of values. The continental shelf extending from the shores of New Zealand is roughly 3 folds thicker than the island itself, the proportion falls to 2 for Baja California. The proportional ratio for Iceland stood at exactly 1. Greenland's continental shelf stretches one half as far as its radius toward its dryland interiors. At the most extreme cases, Cuba and Madagascar has almost no continental shelf. This mismatch still does not contradict our observation because we have shown that earth's ocean budget can readjust so that flat plains of island continents to shallow continental shelf can match.

Then, the island biodiversity is defined as:

$$z_{island} = m\pi\left(2Rr + R^2\right) \tag{4.8.21}$$

$$z_{island} = m\pi\left(2(0.182)r + (0.182r)^2\right) \tag{4.8.22}$$

$$z_{island} = 0.398m \tag{4.8.23}$$

One can also notice that island marine biodiversity now matches the supercontinent's marine biodiversity. Their discrepancy only occur when land to total surface ratio higher than 50 percent's lower bound can only be maintained by a continental plate covering 100% surface of the planet. Finally, the combined biodiversity for supercontinent is then:

$$T_{supercontinent} = z_{supercontinent} + 1.148\sqrt{m} - 0.574^2 \qquad (4.8.24)$$

and the combined biodiversity for island continent is then:

$$T_{island} = 1.217m \qquad (4.8.25)$$

The updated supercontinent biodiversity curve approaches the island biodiversity because the continental shelf with size proportional to the size of the continental plate now offers great opportunity for speciation as the size of continents grow. On the other hand, the biodiversity on dryland stays nearly the same as our simplified model. Since the emergence of human depends heavily on the dryland biodiversity, our future calculation determining the emergence of human shall rely on the island and the supercontinent land biodiversity only. This result also suggests that, every other condition being equal, it is far more likely that planets are dominated by greater marine biodiversity and far smaller land biodiversity.

4.9   Continental Movement Speed

Having shown that continental cycles drive the cyclic pace of evolution, we now focus on the continental movement speed across all terrestrial planets. Continental movement speed varies between terrestrial planets of different sizes. The model for tectonic movement, though very intricate and complex, and be simplified into a toy model based on two assumptions. First, the rate of tectonic plate creation is directly proportional to the heat release per unit area of the planet. That is, the higher the heat flux, the more active the plate tectonics. The formation of new crusts is a consequence by the convective magma inside the planet in the form of heat dissipation from the planet's original formation in the form of potential energy and radioactive elemental decay such as uranium and thorium. It is, then, no surprise that young earth billions of years ago had more active geologic activities. Secondly, the rate of tectonic subduction is directly proportional to the gravity acted upon the plates. The oceanic mafic plate emerges from the site of its creation and gradually over the course of millions of years consolidated in density and increased in weight and sloped toward the subduction zone. The subduction zones, such as the Mariana trench, are some of the deepest places on earth, pulls the plate into the mantle upon its own weight, thereby completing the recycling of the oceanic plates. It is no surprise that planets with higher mass also has greater surface gravity and greater surface heat flux. A third factor involving the thickness of the crust is sometimes also considered, but for the simplicity of our argument (a lack of data to correlate crust thickness based on the mass, composition, or the formation condition of the planet. Mars has a thicker crust with 10% earth mass, and Venus has comparable crust thickness to earth with comparable density, yet all three planets have comparable compositions), we shall assume that the crust thickness is equivalent in terrestrial planet of different masses. The question becomes, given an increase in mass of a terrestrial planet, how will the speed of tectonic movement change and by how much. If we assume that terrestrial planets have a similar density to earth, then the following graph can be used to predict the surface gravity and surface area on terrestrial planets of other masses.

$$r_0 = \left(\frac{3}{4}x\right)^{\frac{1}{3}} \qquad (4.9.1)$$

$$y_{plate} = \frac{1}{0.3029} \frac{\frac{4}{3}\pi r_0^3}{4\pi r_0^2} \qquad (4.9.2)$$

182

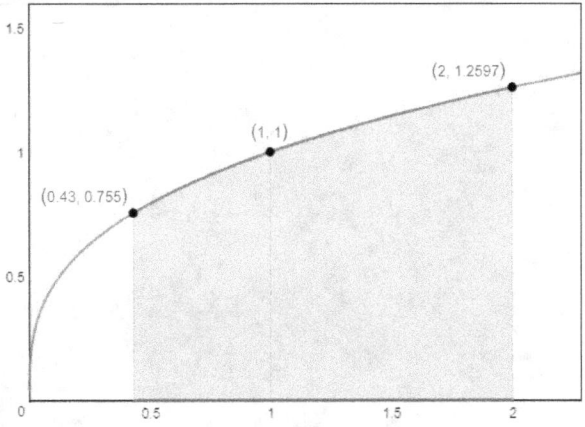

Figure 4.9.1: Geological intensity vs. terrestrial planet mass

The surface gravity grows relatively slow compares to the increase in mass, that is, a terrestrial planet at the mass of 2 times of earth will have a surface gravity at only 1.26 times that of the earth. Furthermore, the surface area will also be 1.26 times that of the earth. As a consequence, one can see that radiative convection on this planet will also create new crusts at 1.26 times the speed observed on earth while engulfing old crusts also at 1.26 times the speed on earth. As a result, for a terrestrial planet, put on a limit of no more than 2 earth mass, the plate tectonic movement cannot exceed more than 1.26 times the speed on earth. On the other hand, plate tectonics movement on planets with mass smaller than earth slows down significantly, with only a fraction observed on earth. We have deduced in the section regarding biodiversity cycle and its correlation with the continent-supercontinent cycle. It is shown the formation of supercontinent is not conducive to the maintenance of diversity relative to the island continent configurations. However, a drop in biodiversity is not only compensated by the breaking up phase of the supercontinent but with further increase in biodiversity. The first niche exploited by biological life on earth is the ocean, then life moved onto land as their next habitat, and then biological species created its own niche habitat in the form of forests and trees, and finally, biological creatures exploited the sky. One needs to stretch on more imagination to imagine what else is possible if the continental supercontinental cycle continues. For example, increased biodiversity and photosynthesis leads to a much denser atmosphere thus creating a new niche with lower density compares to the ocean but much higher density than the air we accustomed to on earth today. Species semi-adapted to both flying and walking will become possible within such a niche. Seahorse like creatures swim through the air near the ground. Anything is possible. In the graph plotted below, we compare the upper limit at which the evolutionary rate can occur on super-earths analogs and the lower limit at which evolutionary rate can occur on mini-earths, the middle line is earth itself, assuming their multicellular life all started at the same time.

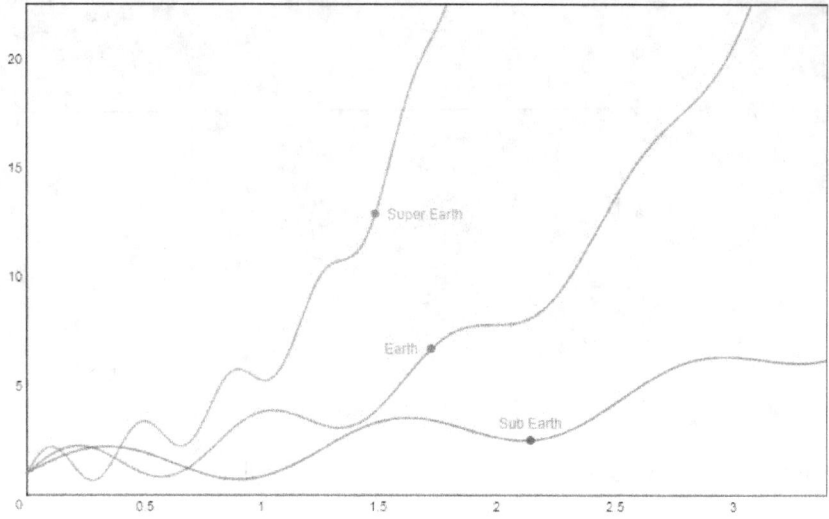

Figure 4.9.2: Super-earth, earth, sub-earth's idealized tectonic rates and their subsequent biocomplexity development

It can be seen that super earth goes through a shorter cycle of continental drifts, as a result, achieves greater biodiversity at a faster rate than earth and faster background evolutionary rate compares to earth. On the other hand, mini-earth goes through a longer cycle of continental drifts, as a result, achieves greater biodiversity at a slower rate than earth and slower background evolutionary rate compares to earth. It is likely that life is sustained on such planet, will progress very slowly, since the movement of plates is the ultimate determining factor in biological evolution. It is also to be noted that, at times, especially toward the beginning of emergence, a mini earth's high point in biodiversity can eclipse that of the other super earth's low point in biodiversity, even though ultimately super earth led to a faster rate of evolution. In summary, by analyzing the different speed of evolutionary rate as a consequence of plate drift rate, *we achieved a physical explanation for the abstract mathematical concept that background evolutionary rate varies. Some planets achieve emergence of intelligence faster than others, and the sum total of all background evolutionary rate on all planets contributes to the* **cosmic background evolutionary rate**, centered on the mean and represented as a probabilistic distribution with measurable deviation.

Chapter 5

Glaciation and Super Continent Cycles

5.1   Ice Age as an Accelerator and Its Causes

Many have argued that the emergence of Homo sapiens is tightly intertwined with the current Ice Age. The quick, fluctuating climate and weather pattern made animal adaptations by means of natural selection difficult. At the onset of the current inter-glacial 10,000 BP, many major animal groups particularly the woolly mammoth became extinct. Though human hunters have played some role in accelerating its demise, its gigantic size and surviving strategy do not adapt well to a much humid, temperate epoch especially the interglacial summers. Throughout the last few ice ages, animals survived a glacial period can become extinct by the next interglacial, and the lucky ones survived the interglacial but ill-prepared for the harsh glacial period becomes extinct by the next one. Homo genus, on the other hand, adapts environment by tool usage, fire control, coordinated teamwork, and culture transmission of experience to the next generation. As a result, Homo genus not only survived each of the glacial and interglacial periods, and actually prospered.[3][28] Without the current ice age, the human ancestor likely remains on the trees in African tropical forests, and no evolutionary pressure forces them to roam the ground to search for food. They will either become extinct as a species or eventually adapt to walking on the ground tens of millions of years into the future. Since earth's continent continues to move, in millions of years further into the future, Africa plate will further shift north and join the Eurasian plate, causing fauna and environmental changes in current East and Subsahara Africa. Therefore, in the absence of fast tectonic movement to drastically alter the living environment within a short geological period, an Ice Age acts as an accelerating contributing factor to the rise of Homo Sapiens.

In summary, ice age creates greater speciation opportunity with fluctuating climate but such diversity increases and wanes quickly with the climate, and the macro-trend of evolution in the long term should remain largely unchanged. Unless a species such as Homo sapiens which emerged from the uncertainty but survives beyond changes, most species does not adapt well. It is almost as if day trader on a stock market, buys and sells frequently, supposedly gained significant income, instead made as much gains as its losses, so overall his wealth remain the same. Nevertheless, even if species appeared and disappeared within short geologic time frame and no net change in final biodiversity, the total number of speciation events per the time period increases, manifested as t_{Win} increases. There remains the possibilities that an ice age can still contributes to a net gain in biodiversity increase. If increase in biodiversity occurred, it can be a consequence of increasing the overall variance of PDF representing the biodiversity by introducing more advanced species with greater number of traits and shifting the mode. It can also be a consequence of keeping the mode and variance but increase the height of the PDF by increasing the evolutionary window size. It can also increase the biodiversity by some combination of both previous approaches.

We illustrate the first approach mathematically as: Whereas $P(0, x)$ (see Chapter 8) denotes the PDF representing the biocomplexity at the current time, and $t_{Win} \int_7^\infty P_{df}(0, x)\, dx$ denotes the emergence chance of homo sapiens. An acceleration on evolutionary speed shifts the variance and the mode to higher values, corresponding to a time into the future. As a result, an ice age accelerate evolution by increasing the value of time t which in turn increase the variance and the mode. By taking the limit:

$$\lim_{t \to -\infty} t_{Win} \int_{7}^{\infty} P_{df}(t, x)\, dx = \infty \tag{5.1.1}$$

It is shown that as t increases toward the future, the chance giving rise to intelligent, tool-using species increases toward infinity.

We illustrate the second approach mathematically as: Whereas $t_{Attempt}$ (see Chapter 6 and Chapter 7) is the number of speciation attempt of the entire Cenozoic era which enables the total emergence of 2.4 million species of birds, mammals, reptiles. $\frac{P(7,7)}{G_2(7)^{-1} \times C(7,7)} \cdot R_p(7) A_P(7) \left(\frac{\left(\frac{1}{7}\right)^7}{p^7} \right) \left(\frac{28}{7} \right)$ (see Chapter 7) is the minimum number of species to experiment by nature before the emergence of intelligence.

$$\frac{\frac{P(7,7)}{G_2(7)^{-1} \times C(7,7)} \cdot R_p(7) A_P(7) \left(\frac{\left(\frac{1}{7}\right)^7}{p^7} \right) \left(\frac{28}{7} \right)}{\left(\frac{1}{1.5} \right) t_{Attempt}} = 4 \tag{5.1.2}$$

It has been determined that it takes 4 Cenozoic time at the current Background evolutionary rate (see Chapter 6 and Chapter 7) to guarantee the emergence of Homo sapiens. A glaciation acts to accelerate nature's experiment by unpredictable climate and catastrophic ecological changes faster than geologic rates. This is equivalent to increase $t_{Attempt}$.

By taking the limit:

$$\lim_{t_{Attempt} \to \infty} \frac{\frac{P(7,7)}{G_2(7)^{-1} \times C(7,7)} \cdot R_p(7) A_P(7) \left(\frac{\left(\frac{1}{7}\right)^7}{p^7} \right) \left(\frac{28}{7} \right)}{\left(\frac{1}{1.5} \right) t_{Attempt}} = 0 \tag{5.1.3}$$

It is shown that as $t_{Attempt}$ increases, the time required for giving rise to intelligent, tool-using species decreases toward 0.

If Ice Age is critical to give rise to the intelligent species, we should resort to calculate the probability of any given geologic time period in which earth falls under one. The formation of earth's ice age is an interplay of solar radiation output, atmospheric composition (methane, carbon dioxide, and oxygen concentration), and the earth plates positions. The formation of Ice age occurred a few times during the earth's geologic past, some of which is attributed primarily to the changing atmospheric composition such as the Huronian, Cryogenian, and Karoo Ice Age.

Huronian glaciation extended from 2.4 billion years ago to 2.1 billion years ago caused by Cyanobacteria's evolution of photosynthesis. Their photosynthesis produced oxygen as a waste product expelled into the air. At first, most of this oxygen was absorbed through the oxidization of surface iron and the decomposition of life forms. However, as the population of the cyanobacteria continued to grow, these oxygen-sinks became saturated.[98] This led to a mass extinction of most life forms, which were anaerobic, as oxygen was toxic to them. As oxygen filled the mostly methane atmosphere, and methane bonded with oxygen to form carbon dioxide and water, a different, thinner atmosphere emerged, and Earth began to lose heat. From our calculation on the continuously habitable zone, the earth was beyond the outer edge of the habitable zone with current atmospheric conditions and composition until 1.3 Gya. Without the presence of methane as a strong greenhouse gas, earth plummeted into an ice age until solar output eventually matched the loss of methane as a greenhouse gas.

The Karoo Ice Age was caused by the evolution of land plants from the earlier Devonian period which led to significantly higher oxygen content, and the global carbon dioxide went below the 300 parts per million level. However, it can be argued that without a significant presence of land mass near the south pole at the time, this ice age with biological origin can not perpetuate for long.

If we assumed that oxygen content and atmosphere density should reach levels comparable to that of earth on any other extraterrestrial planets enabling the emergence of intelligent species, then the ice age caused by biological process preparing for such prerequisite condition should be excluded from our investigation. The earth, after each of such preparatory changes, readjusted itself toward an ice-free world.

We can also assume that the sun's increasing radiation is slow compares to the geologic process, tectonic movement by at least a magnitude, then we can reasonably conclude that the movement of cratons and plates is the most significant contributing factor to the onset of an ice age.

Therefore, atmospheric changes and solar output can trigger and start an ice age, but in order to perpetuate one with a length comparable to a geologic period and an intensity comparable to our current glaciation, the configurations of plates are the necessary though insufficient condition.

The Quaternary glaciation is the most well-understood due to its recency. From the plate tectonics and ocean current theory, the long-term temperature drop is related to the position of the continents relative to the poles. This relation can control the circulation of the oceans and the atmosphere, affecting how ocean currents carry heat to high latitudes. Throughout most of the geologic time, the North Pole appears to have been in a broad, open ocean that allowed major ocean currents to move unabated. Equatorial waters flowed into the polar regions, warming them with water from the more temperate latitudes. This unrestricted circulation produced mild, uniform climates that persisted throughout most of the geologic time.

The formation of Antarctica ice sheet is a consequence of the formation of Drake passage that separates South America from Antarctica started 43 million years ago. The separation created the Antarctic Circumpolar Current that completely circles the continent. This current does not exchange with the warmer currents closer to the equator. Prior to the separation, currents alongside the southern continent flowed toward the equator, circled the entire South America continent before reaching Antarctica again with warmer currents. Over the course of millions of years, the cold Antarctic Circumpolar Current changed the continent's climate and cooled it significantly. At the same time, North America continent, Greenland, and Eurasian continent moved north with the north pole in a small, nearly landlocked region of the Arctic Ocean. A nearly landlocked ocean again can not exchange its colder currents with that of the warmer currents nearer the equator, resulting in the polar ice cap. Currents can also warm up regions. The Gulf Stream of Mexico which flows from the equator toward European continent helped Europe to be significantly warmer than the rest of the world at similar latitudes. Therefore, the role of currents in shaping climate is critical, and its direction is predicted by the position of continents. If one traces further back one can also find that the position of continents also played a significant role in earlier glaciation events. During the Karoo Ice Age from 360 million to 260 million years ago, Pangea supercontinent covers the entire south pole. Unlike Antarctica today, the shoreline to the south pole is significantly farther away. Even in the absence of Antarctic Circumpolar Current at the time, warm currents and its effect on the climate is limited since its impact can only reach so much inland. Although the north pole is wide open, making current flow from tropics to the pole possible, North America and Angaran region of the supercontinent stretches well into 60 degrees in latitude north of the equator, making a complete exchange of ocean currents between the two hemispheres difficult.

Andean-Saharan glaciation occurred earlier during the Ordovician epoch showed a remarkably similar continent layout to that **of** the Permian epoch, a vast supercontinent with a significant landmass covered the south pole. From these encouraging observation, one can create a simplified model for the formation of Ice Age.

1. That is, when significant continental mass located at the poles

2. One craton located at the pole and separates from the rest by the ocean

3. A polar ocean is landlocked by the continents

It can be said if any one of the three conditions is fulfilled, then ice accumulation occurs on earth, a more severe form of ice age occurs when 2 or all 3 conditions are met. The Quaternary Ice Age is satisfied by condition 2 and 3.

5.2 Expected Ice Age Interval

Since earth shifts through continental cycles, the probability of ice age occurrence and its accelerated effects on evolution are essential in our discussion on the probability of the emergence of Homo sapiens if any planet already evolved flower plants and animals with capabilities similar to Reptiles, Birds, and Mammals. The onset of an early ice age on a planet with these prerequisite conditions will generate intelligent, tool-using species faster. Computing the probability of earth entering into any severe ice age as the one we observed right now, where both poles covered in a significant depths of snow, is required. From our earlier discussion, we know that earth can enter into an ice age if there is two Antarctica sized plates sits on both poles. Glaciation initiates if there is significant continental mass covering the poles, as observed during Ordovician ice age and the Karoo ice age. Glaciation also initiates when one pole is covered by one Antarctica sized plate and another pole encircled by two or three Antarctica sized plates (our case observed today).

After listing the conditions for the creation of an ice age, we need to convert the complex everchanging continental movement of earth crusts into a much simpler problem we can tract and at the same time provide a reasonably good estimation. The configurations can be simplified into a toy mathematical model of the following configurations. First, dividing the surface of the earth into 34 blocks of equal sizes, where each block represents a plate. The number of subaerial blocks totaled 10, corresponding to 29 percent of the surface of earth emerged above water. Each block sized 3,872 km per side, with a total surface area of 14,992,384 km^2, this roughly comparable to the size of Antarctica. This is to show that over the geologic time period, any continent is free to move into any one of these 34 blocks of regions. Out of these 34 blocks of regions, only 1 block is reserved for the north pole and another one for the south pole, so the chances of moving into any one of these blocks is a matter of random combination. Furthermore, the nearest blocks to each of the pole, three of them circles the north pole, and another three circles the south pole. If a pole is unoccupied, yet all 3 adjacent blocks are occupied by the continents, then a current earth's north pole type of scenario emerges. On the other hand, if the south pole is occupied by a continent and all 3 adjacent blocks are unoccupied; then, current earth's Antarctica scenario emerges. If the pole is occupied, yet only 1 or 2 of the adjacent blocks is also occupied by a landmass; then, a south America attached to Antarctica scenario emerges and no ice accumulation at this pole. If all 3 adjacent blocks along with the pole block are occupied by landmass, then Pangea type of supercontinent configuration during the Karoo glaciation and the Andean-Saharan glaciation scenario emerges.

Moreover, Europe, Australia, South America, and Antarctica is represented as one block. Africa is represented by two blocks. North America is represented by only one block while its total surface area corresponds to 1.5 block sizes. However, it is not critical for our analysis because any land mass south of Canadian border can be discarded because we are only interested in a continental configuration which circled the pole and its discarded mass is added as an extra to Australia. Finally, Asia is represented by 3 blocks.

Then, one can see that the configuration of the blocks can be represented as a string of digits, where the leftmost digit represents the first block of the north pole. The second, third, fourth blocks represent the plates that surround the pole plate with a surface area size of Antarctica extending to 40 degrees south from the north pole, or 7,744 km south of north pole. The rest of digits represents different blocks comprising the mid-latitude Pacific Ocean and Atlantic Ocean. The last block represents Antarctica sits on top of South pole. Once we formulated the rule, we number our blocks from 1 to 34 and arrange them into 34th to the 1st digit of a binary number representation.

Each region can be labeled as either 1 or 0. Each subaerial block can be represented as a digit of 1 (1 being occupied by a continent), and each submarine block can be represented as a digit of 0 (0 as being filled by the ocean). Total possible cases of 0's and 1's for such an arrangement can be represented by enumerating the binary number of 34 digits ranging in value from 1 to 2^{34}-1. We then go through by hand or a computer program to pick and count the results which simulate ice age scenarios. *We take this total and divide by the total number of scenarios possible we end up with the probability of an ice age emergence.*

The enumerating table will be too large to exhibit. [1]Fortunately, no exhaustively listing of possibility is needed.

---

[1]however, a simpler case of total 10 blocks, or a binary number composed of 10 digits ranging from 1 to 2^{10}-1 is listed at the

One can find the probability by picking the combinatorial results by giving different combinatorial configurations. For the combination of 10 chose out of 34, there are 131,128,140 total possibilities. This is the total number of possible land over ocean configurations.

$$\Pr(X = 10) = \binom{34}{10} = 131,128,140 \tag{5.2.1}$$

With both poles covered by Antarctica sized continents, there are 10,518,300 possibilities (choose 8 out of 32).

$$\Pr(X = 8) = \binom{32}{8} = 10,518,300 \tag{5.2.2}$$

With one pole covered by Antarctica sized plate and the other surrounded by large plates as it is observed today on earth, there are 475,020 possibilities (choose 6 out of 29). This case is symmetric in regards to north or south pole, so it is doubled to 950,040 possibilities. The choose base is 29 instead of 30 because 1 block of sea is surrounded by 3 blocks of land.

$$\Pr(X = 6) = 2\binom{29}{6} = 950,040 \tag{5.2.3}$$

The Andean-Saharan ice age observed during the Ordovician and Karoo ice age observed during the Carboniferous have significant landmass buildup around the south pole, this suggests that large landmass centered around one pole also promotes glaciation, adding an additional 593,775· 2 possibilities due to symmetry.

$$\Pr(X = 6) = 2\binom{30}{6} = 1,187,550 \tag{5.2.4}$$

The total probability giving rise to glaciation can then be computed. $\frac{12,655,890}{131,128,140} = 9.65\%$, or 1 out of 10.361. If we take the average speed of plate movement around 9 $\frac{cm}{year}$, and each block has a side length of 3,872 km, then, $10.9267\cdot3,872 \text{ km}\cdot1,000\frac{m}{km}\cdot\frac{100}{1}\frac{cm}{m}\cdot\frac{1}{9}\frac{year}{cm} = 445,753,244$ years.

It implies that it takes about 10 different random combinations of earth plate to generate one glaciation phase, and each combination lasts 43 million years. If one compares this result with what we have observed from the paleogeological record, we found the timespan between Andean-Saharan ice age and Quaternary ice age closely match the computation. The Karoo ice age, then, is interesting, because it occurred between these two episodes. Two explanations can be made regarding this case.

First of all, the earth's continental movement was not completely random, and prior to Permian, a majority of the land masses were consistently located south of the equator. Consequently, two episodes of glaciations closely followed one and another by merely 60 million years apart.

Secondly, the ice age was caused not primarily by the land mass configuration alone, as it is observed during the Quaternary. The evolution of land plants with the onset of the Devonian Period began a long-term increase in planetary oxygen levels. Giant tree ferns, growing to 20 m high, were secondarily dominant to the large arborescent lycopods (30–40 m tall) of the Carboniferous coal forests that flourished in equatorial swamps stretching from Appalachia to Poland, and later on the flanks of the Urals. Oxygen levels reached up to 35%,[96] and global carbon dioxide got below the 300 parts per million level,[47] which in today is associated with glacial periods. This reduction in the greenhouse effect was coupled with lignin and cellulose (as tree trunks and other vegetation debris) accumulating and being buried in the great Carboniferous Coal Measures. The reduction of carbon dioxide levels in the atmosphere would be enough to begin the process of changing polar climates, leading to cooler summers which could not melt the previous winter's snow accumulations. The growth in snowfields to 6 m deep would create sufficient pressure to convert the lower levels to ice.

Thirdly, the Karoo ice age had more moderate climatic effects than the current Quaternary ice age. With only one pole covered in ice, continents near the equators enjoyed milder climatic fluctuation than those observed

end of the chapter.

today, as it is evident from the graph below.

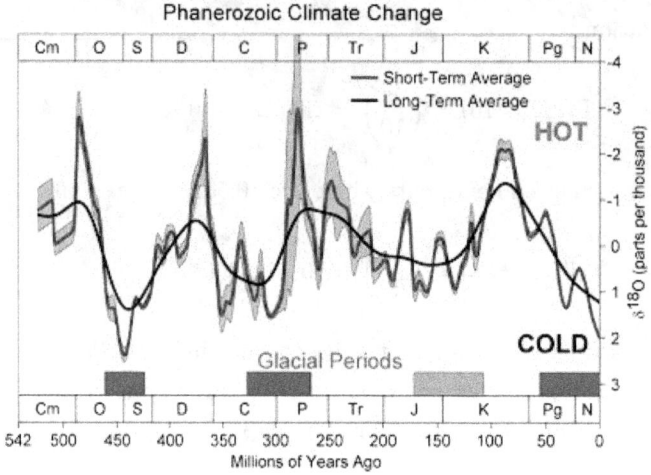

Figure 5.2.1: Earth's climate in the past 542 Myr

Nevertheless, we shall compute the probability of entering into an ice age at our current epoch by assuming that the last ice age ends 260 Mya. Since it would take another 472.35 Myr to guarantee the onset of the next glaciation which sets a limit to 212.35 Myr into the future, *the chance of entering another glaciation event at the current epoch is 55%.*[2]

Generalization can be made based on this model. Our model indicates the probability leading to glaciation with 29% land coverage and 71% ocean coverage. Applying the same approach, one can find the probabilities distribution of different land and ocean coverage.

$$P(n, k) = \frac{n!}{k!(n-k)!} \tag{5.2.5}$$

$$T_0 = P(34, x) \tag{5.2.6}$$

$$T_1 = P((34-2), (x-2)) \tag{5.2.7}$$

$$T_2 = 2P((34-5), (x-4)) \tag{5.2.8}$$

$$T_3 = 2P((34-4), (x-4)) \tag{5.2.9}$$

$$G(x) = \frac{3,872 \cdot 10^3 \cdot 10^2}{9 \cdot 10^8} \left(\frac{T_1 + T_2 + T_3}{T_0}\right)^{-1} \tag{5.2.10}$$

---

[2]This is a special case from a whole range of permissible values, the final generalized results show to be 70%, please check section 5.4

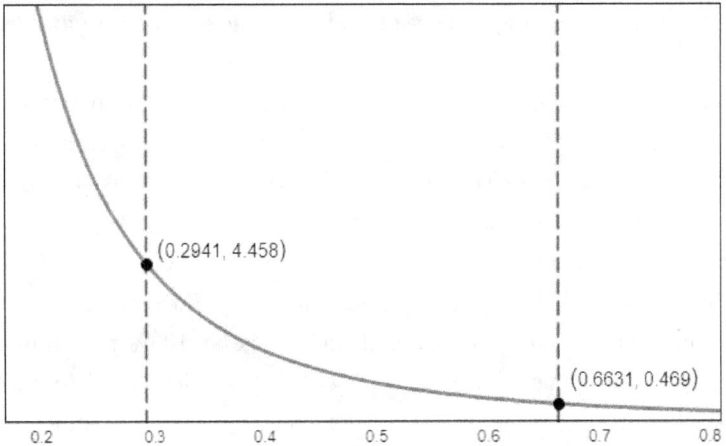

Figure 5.2.2: Dryland percentage coverage and the expected timing of the next glaciation event for 100% dryland coverage to 0% dryland coverage

In general, smaller land coverage results in a smaller probability of glaciation and more land coverage results in a higher probability of glaciation. On a hypothetical planet with scattered island continents, glaciation occurs at much longer intervals, and the accelerated evolution of Homo sapiens cannot occur. On the other hand, on a hypothetical desert planet, where land covers more than 66.3 percent of the planet surface, glaciation dominates the planet at every epoch, with 100% probability in any given time. Extreme, extended period of glaciation impedes the evolution and development of complex multicellular life, as it is evidenced by the Cryogenian ice age prior to Cambrian explosion. Atmospheric oxygen is already reaching a significant proportion at the start of the glaciation, but multicellular evolution was kept in check by advancing snowball earth.

We investigated the distribution of the percentage of ocean coverage given different total surface areas of lightweight granite continental plate in Chapter 3. It is shown that earth analogs (with both land and sea coverage on their surface in a delicate proportion to allow a transition from ocean-based multicellular life to a terrestrial one) includes a wide range (0 to 84 percent) of dryland to the total planetary surface ratio.

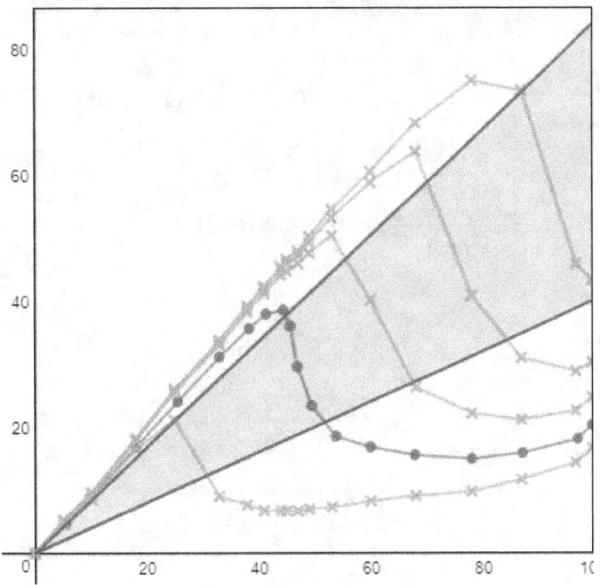

Figure 5.2.3: The proportion of dryland exposure by varying mass of ocean. From the bottom to the top: 1.5, 1, 0.75, 0.5, 0.25 earth ocean mass respectively. As the ocean mass decreases, the permissible range of exposed dryland with flat plains occupying the total planetary surface area increases.

Depending on the ocean mass budget, the possible range of dryland to the total planetary surface varies. We will eventually compute the weighted average chance of emergence across all dryland coverage ranges. Currently, we shall focus on deriving a special case assuming the ocean budget is one earth ocean and use it as a reference to serve our purpose.

Since we derived the relationship between continental plate percentage to oceanic plates and the total ocean coverage in Chapter 3 and 4, we now have to derive the range of ocean to land coverage most likely results in an ice age at the current epoch just like we observed on earth. We take 12.92 blocks out of 34 blocks, or 38% as the upper limit because the computed results show that terrestrial land mass can at most extend to 38% of land coverage in all possible configurations for an ocean of exactly earth's size.

The lower limit is obtained by assuming if the current epoch has a certain chance of an ice age occurring (between 0 and 1) and by pushing on the expected onset of an ice age at 100% probability further into the future to the mean expected appearance time of intelligent, tool-using species on all terrestrial planets. It is computed to be 334.4 Myr into the future (see Chapter 7). Then, the appearance of an ice age occurring at an interval greater than the mean appearance time on all terrestrial planets do not contribute at all, at accelerating the emergence of intelligent, tool-using species. We found that 8 out 34, or dry land covering less than 23.53% of the planetary surface satisfies the aforementioned condition. We then reached the conclusion that dry land coverage ranges from 23.53% to 38% can broadly be labeled as ice age capable configurations along the y-axis. This is translated into a range of continental plate coverage within the lower and upper bound of sea levels. We compute the area of this range as $A_{selectedrange}$, and we compute the total range of all continental plates within the lower and upper bound of sea levels as A_{total}. *This amounts to 26.49% of all possible configurations.*

$$A_{total} = \int_0^{100} y_{upper} dx - \int_0^{100} y_{lower} dx \tag{5.2.11}$$

$$A_{selectedrange} = \int_{S_{upper}(23.53)}^{S_{lowerwrap}(38)} y_{upper2} dx - \int_{S_{upper}(23.53)}^{S_{lowerwrap}(38)} y_{lower2} dx \tag{5.2.12}$$

$$S_{upper}(x) = \frac{x}{0.84} \tag{5.2.13}$$

$$S_{lower}(x) = 2.5x \tag{5.2.14}$$

$$S_{lowerwrap}(x) = \begin{cases} S_{lower}(x) & 0 < S_{lower}(x) < 100 \\ 100 & S_{lower}(x) \geq 100 \end{cases} \tag{5.2.15}$$

$$\frac{A_{selectedrange}}{A_{total}} = 0.2649818731 \tag{5.2.16}$$

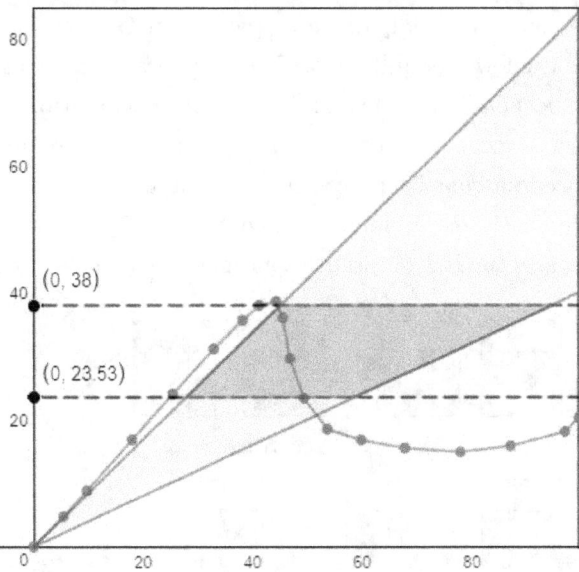

Figure 5.2.4: 23.53% to 38% dryland coverage translated into a range of continental plate coverage within the lower and upper bound of sea levels.

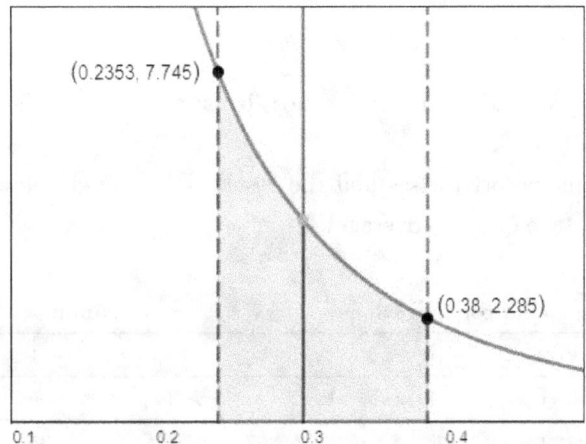

Figure 5.2.5: The expected timing of next glaciation event for the lower (23.53% dryland coverage) and upper bound (38% dryland coverage) coverage enabling glaciation. At the lower bound, every 228.5 Myr guarantees an ice age. At the upper bound, every 774.5 Myr guarantees an ice age, and currently 260 Myr passed since the last ice age, guaranteeing the appearance of another ice age in at most in 514.5 Myr.

5.2.1 Proportion of island to supercontinent - lower bound biodiversity calculation

Of course, the rate of emergence is non-uniform within ice age capable range from 23.53% to 38% of dry landmass coverage. The rate of emergence depends on the proportion of island continents to supercontinents at any given time. That is, more island continent configurations leads to greater biodiversity. We used a toy model representing the earth's surface by composing 9 blocks fitted in a 3 by 3 grids and assumed that land coverage ranges from 0 to 100% by running the combinations of placing 0, 1, 2, and up to 9 blocks of dryland over 9 blocks of sea. We define an island continent as an individual land block that does not touch any other land block at the top, bottom, left, and right side. Two land blocks can touch each other diagonally but it is not counted as a connected landmass because each block is well surrounded by the ocean. We define a supercontinent as three land blocks connected with each other by the top, bottom, left, or right side. Three blocks of drylands translated into 33% of land coverage, roughly equivalent to the proportion of earth's total land coverage to the surface area. Any land configurations composing more than three blocks eventually leading to a desert planet scenario is treated as a megacontinent. On the other hand, land configurations composing two blocks are labeled as mini-supercontinent, comparable to the size of Gondwana observed on earth. This island to supercontinent ratio is also the lower bound for biodiversity because we consider that supercontinent lying at the tropical region does not benefit from an overall increase in precipitation at its interiors. In reality, heavy precipitation is observed in the Amazons and central African jungles. Therefore, supercontinents placed within the tropics can be served conceptually as island continents from an evolutionary perspective despite its vast size due to its habitability more closely resembling island continents. This is where island continent from an evolutionary perspective differs conceptually from island continent from a strict geologic perspective.

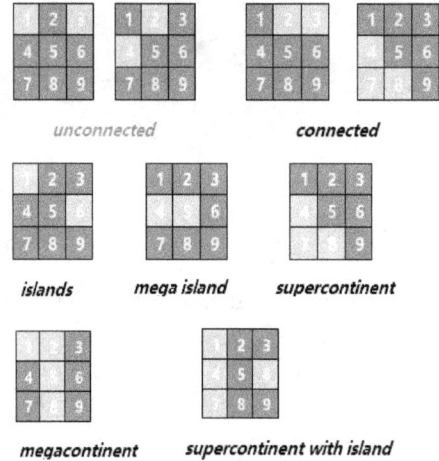

Figure 5.2.6: The graphically illustrated possible scenarios from the model

We first enumerated all possible combinatorial cases and the result is presented below (Island continent and supercontinent breakdown by percentage of land coverage):

Cases	Sea	land configurations	Number	Subtotal
no land	9	0	1	**1**
$\frac{1}{9}$ land	8	1	9	**9**
$\frac{2}{9}$ land	7	1+1	24	
		2	12	**36**
$\frac{3}{9}$ land	6	1+1+1	22	
		2+1	40	
		3	22	**84**
$\frac{4}{9}$ land	5	1+1+1+1	6	
		2+2	12	
		3+1	44	
		2+1+1	28	
		5+4	36	**126**
$\frac{5}{9}$ land	4	1+1+1+1+1	1	
		4+1	36	
		3+2	16	
		3+1+1	16	
		2+2+1	8	
		5	49	**126**
$\frac{6}{9}$ land	3	5+1	20	
		4+2	8	
		3+3	4	
		4+1+1	4	
		6	48	**84**
$\frac{7}{9}$ land	2	6+1	4	
		7	32	**36**

194

Cases	Sea	land configurations		Number	Subtotal
$\frac{8}{9}$ land	1	8		9	9
all land	0	9		1	1

Under each case, we compute the probability of island continent formation and the supercontinent formation. Each island continent is awarded with 1 point and then multiplied by its appearance frequency within each case. A supercontinent with a size of 2 blocks is awarded with $\frac{2}{3}$ point, a supercontinent with a size of 3 blocks is awarded with 1 point, and supercontinent with a size of 4 blocks is awarded with $\frac{4}{3}$ point. Some configurations contain a mixture of island continents and supercontinents. The probability is then computed for both island continents and supercontinents multiplied by their appearance frequency under each configuration and is added to the total probability of hosting an island continent and supercontinent under each case respectively.

The finalized curve fitting for both island continent formation probability and supercontinent formation probability is listed in the graph below:

$$y_{super} = \begin{cases} 0 & 0 \leq x \leq 0.11469 \\ -9.052x^5 + 25.126x^3 - 18.8x^2 + 6.593x - 0.546 & 0.11469 \leq x \leq 1 \end{cases} \tag{5.2.17}$$

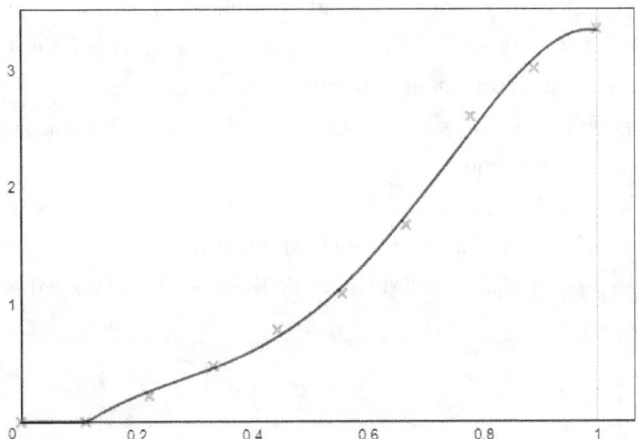

Figure 5.2.7: Supercontinent formation probability

$$g_0(x) = \frac{1}{39.4}x^{-\left(\frac{\ln 80(x)}{\ln 3.7}\right)} \tag{5.2.18}$$

$$S_{down} = 0.5\left(\tanh\left(-7\left(x - 0.7\right)\right) + 1\right) \tag{5.2.19}$$

$$y_{island} = g_0(x) \cdot S_{down} \tag{5.2.20}$$

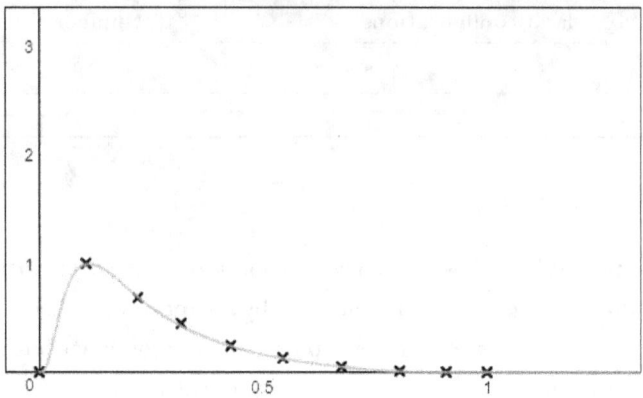

Figure 5.2.8: Island continent formation probability

Alternatively, it can be fit nearly as accurately but with a simpler function as:

$$y_{island} = \left(\exp\left(-x^{0.358865} + x^{0.318023} - 0.02988\right)\right)^{170.896} x \qquad (5.2.21)$$

Both graph plotted along the same axis shows that the unscaled probability is much in favor for supercontinent formation with dryland coverage greater than 33.02%. It also shows that island continent formation rate reaches its zenith at 11.08% of dryland coverage over the surface area. The crossover point occurs at 33.02% in which any less coverage by dryland results in more chance in island continent configuration and more land coverage results in more chance of supercontinent formation. In earth's case, continent's configurations are in slight favor of island continent configurations over supercontinent configurations at a chance of $\frac{0.484}{0.484+0.324} = 59.9\%$ This is confirmed through geologic history where earth goes through periodic continent and supercontinent cycles in which each period lasts as long as the other one.

We then re-run the toy model with 4 and 4 grids, and achieved similar results as our 3 by 3 grids. In earth's case, continent's configurations are still in a slight favor over island continent configurations over supercontinent configurations at a chance of $\frac{0.484}{0.484+0.324} = 55.36\%$, slightly lower than the previous value.

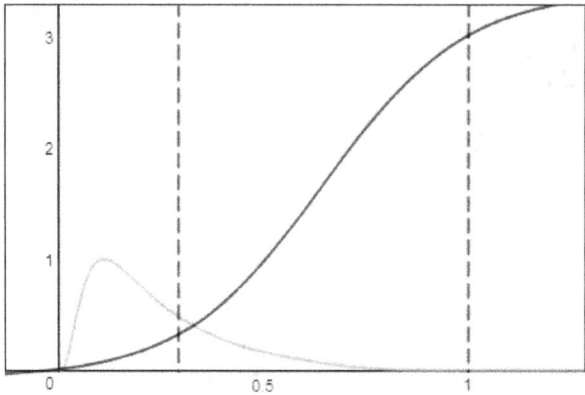

Figure 5.2.9: Island continent and supercontinent formation probability combined

5.2.2   Upper bound biodiversity calculation

We demonstrated that it is far more likely generating supercontinent than islands as the dryland surface area increases. We now focus on computing the upper bound on biodiversity by conceptually treating continents within the tropics as islands. First, one needs to consider the size of tropical region. Any plates configurations falls under this region will be treated as island configurations. We want to know how much out of the total number of 9 blocks does the tropical region represents. The tropical rain forests lies between 12.5 degrees away from the equator for both hemispheres.

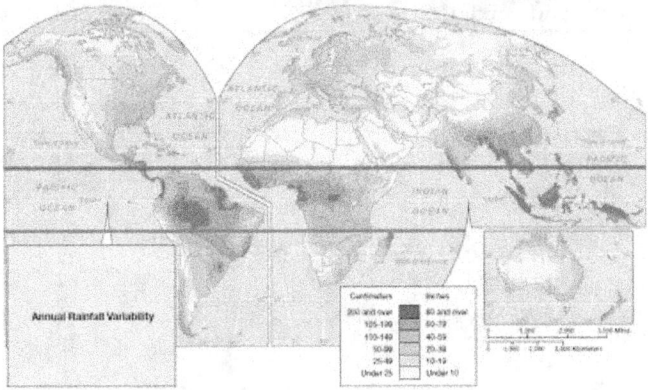

Figure 5.2.10: Bands of tropics

The height relative to the equator for one hemisphere is defined as:

$$h = 1 - \frac{12.5}{90} \tag{5.2.22}$$

and the surface area of hemisphere is 2π, so the sectional area of tropics is defined as:

$$A_{tropcs} = 2\pi h \tag{5.2.23}$$

and the proportion of surface area occupied by tropics for both hemispheres is then:

$$2 \cdot \left(1 - \frac{A_{tropcs}}{A_{half}}\right) = 0.2777 \tag{5.2.24}$$

This translates into 2.5 block regions out of total 9 blocks. We can rounded it up to 3 blocks. Then, the three blocks at the middle tier can be roughly labeled as the tropical region. A new rule for supercontinent classification is formulated as the follows. If any blocks of two or three occupies the central row within a grid of 9 blocks. Then each of these blocks is labeled as an island continent. Therefore, a supercontinent composed of 3 blocks lies within the central row is listed as the sum of three island continents. The labeling for the top and the bottom row remains the same as before. As a result, no supercontinent greater than 3 blocks is possible. The table below lists island continent and supercontinent breakdown by percentage of land coverage by conceptually treating supercontinents lying inside tropical regions as islands.

Cases	Sea	land configurations	Number	Subtotal
no land	9	0	1	1
$\frac{1}{9}$ land	8	1	9	9
$\frac{2}{9}$ land	7	2+0	10	
		1+1	23	
		(1+1)+0	3	36
$\frac{3}{9}$ land	6	2+1	32	
		3+0	13	
		1+1+1	20	
		(1+1+1)+0	1	
		(1+1)+1	18	84
$\frac{4}{9}$ land	5	2+2	9	
		3+1	28	
		2+1+1	23	
		4+0	16	
		1+1+1+1	5	
		(1+1+1)+1	6	
		(1+1)+2	18	
		(1+1)+1+1	21	126
$\frac{5}{9}$ land	4	4+1	16	
		3+2	8	
		3+1+1	8	
		2+2+1	6	
		1+1+1+1+1	1	
		5+0	12	
		(1+1+1)+2	4	

197

Cases	Sea	land configurations	Number	Subtotal
		(1+1+1)+1+1	11	
		(1+1)+2+1	36	
		(1+1)+3	6	
		(1+1)+1+1+1	18	**126**
$\frac{6}{9}$ land	3	5+1	4	
		4+2	4	
		3+3	1	
		4+1+1	2	
		6+0	8	
		(1+1+1)+2+1	12	
		(1+1+1)+3	2	
		(1+1+1)+1+1+1	6	
		(1+1)+3+1	18	
		(1+1)+2+2	12	
		(1+1)+2+1+1	12	
		(1+1)+1+1+1+1	3	**84**
$\frac{7}{9}$ land	2	(1+1+1)+3+1	6	
		(1+1+1)+2+2	4	
		(1+1+1)+2+1+1	4	
		(1+1+1)+1+1+1+1	1	
		(1+1+1)+2+3	12	
		(1+1)+3+1+1	6	
		7+0	3	**36**
$\frac{8}{9}$ land	1	(1+1+1)+2+3	4	**9**
		(1+1+1)+3+1+1	2	
		(1+1)+3+3	3	
all land	0	(1+1+1)+3+3	1	**1**

Based on the new labeling, we recompute the upperbound probability of island continent formation and the supercontinent formation weighted by their appearance frequency. The finalized curve fitting for both island continent formation probability and supercontinent formation probability is listed in the graph below:

$$y_{superupper} = \begin{cases} 0 & 0 \leq x \leq 0.1171 \\ 1.94x^3 - 4.37x^2 + 3.43x - 0.345 & 0.1171 < x \leq 1 \end{cases} \qquad (5.2.25)$$

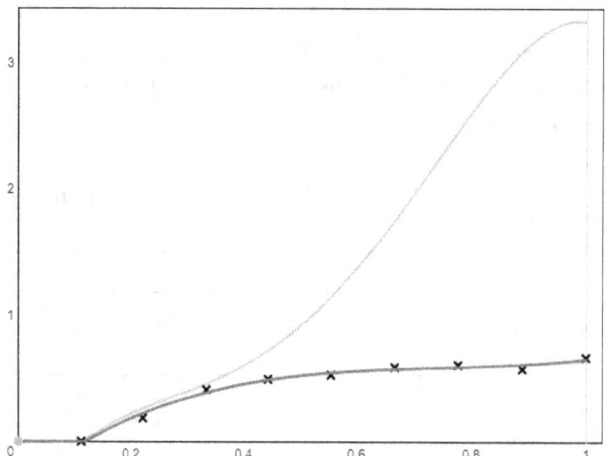

Figure 5.2.11: Supercontinent upperbound vs. lowerbound

$$y_{islandupper} = \begin{cases} g_0\left(x\right) & 0 \leq x \leq 0.18804 \\ 5x^4 - 14.79x^3 + 15.45x^2 - 7x + 1.67 & 0.18804 < x \leq 1 \end{cases} \qquad (5.2.26)$$

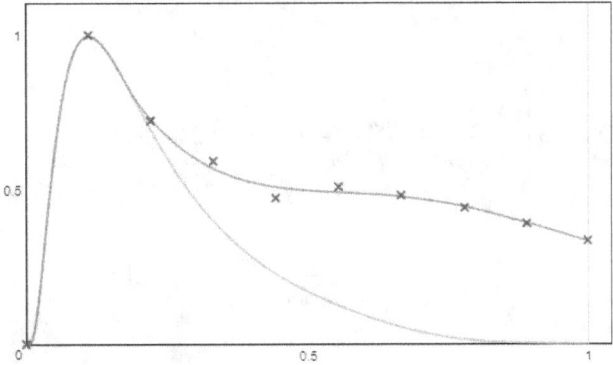

Figure 5.2.12: Island continent upperbound vs. lowerbound

For the purpose of biodiversity calculation, we take the average of the upper and lower bound. This step is justified since tropical region on earth indeed can be conceptually treated as island continent. At the same time, geologic islands such as Indonesian islands still receives more precipitation than both central African jungle and the Amazons. So the final equation is listed below:

$$T_{super} = \frac{1}{2}\left(y_{super} + y_{superupper}\right) \tag{5.2.27}$$

$$T_{island} = \frac{1}{2}\left(y_{islandupper} + y_{island}\right) \tag{5.2.28}$$

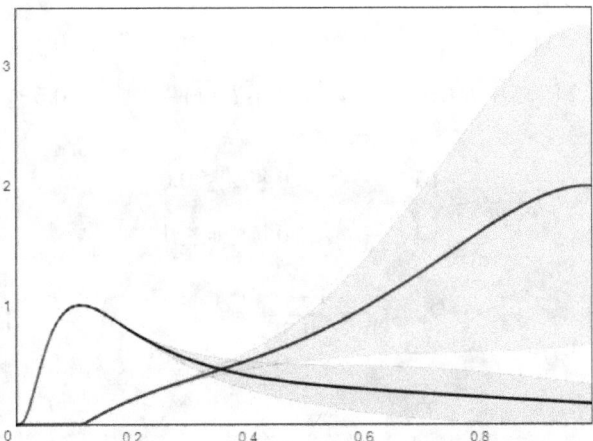

Figure 5.2.13: Island continent and supercontinent emergence probability ranges and their mean value

We then apply our earlier equations accounting for the biodiversity for island and supercontinent configuration. Wheres one takes the island continent formation chance divided by the composite chance of both super and island continent and multiplied by island continent biodiversity curve $34x$ and combines with the supercontinent formation chance divided by the composite chance and multiplied by supercontinent biodiversity curve $\sqrt{34x}$. When one considers only land biodiversity:

$$y_{land}(x) = T_{tectonics} \cdot D_{perim} \cdot D_{inland} \left(\left(\frac{y_{island}}{y_{super} + y_{island}} \right) \cdot 0.818524 \cdot (34x) + \right.$$

$$\left. + \left(\frac{y_{super}}{y_{super} + y_{island}} \right) \left(1.148\sqrt{34x} - 0.574^2 \right) \right) \quad (5.2.29)$$

When one considers both land and marine biodiversity:

$$y_{marine}(x) = T_{tectonics} \cdot D_{perim} \cdot D_{sea} \left(\left(\frac{y_{island}}{y_{super} + y_{island}} \right) \cdot (0.398 \cdot 34x) + \right.$$

$$\left. + \left(\frac{y_{super}}{y_{super} + y_{island}} \right) (y_{marineland}(x)) \right) \quad (5.2.30)$$

$$R(x) = \frac{1}{100} \left(0.00566243 \, (100x)^2 - 1.92724 \, (100x) + 121.982 \right) \quad (5.2.31)$$

$$y_{marineland}(x) = \begin{cases} 0.398 \cdot (34x) & 0 \leq x \leq 0.5 \\ 34x \left(2 \left(\sqrt{1 + R(x)} - 1 \right) + \left(\sqrt{1 + R(x)} - 1 \right)^2 \right) & 0.5 \leq x \leq 1 \end{cases} \quad (5.2.32)$$

$$D_{perim} = \begin{cases} 1 & 0 \leq x \leq 0.5 \\ \left(\frac{(1-x)}{0.5} \right)^{\frac{1}{2}} & 0.5 \leq x \leq 1 \end{cases} \quad (5.2.33)$$

$$D_{inland} = \frac{1-x}{0.5} \quad (5.2.34)$$

$$D_{sea} = \frac{1-x}{0.5} \quad (5.2.35)$$

$$T_{tectonics} = \left((1 - (x+1))^7 + 1 \right)^{50} \quad (5.2.36)$$

Supercontinent biodiversity is further multiplied by D_{perim} and D_{inland}. We derived earlier that the total biodiversity of supercontinent can be expressed as:

$$\pi \left(2\sqrt{mr^2}R + R^2 \right) + \pi \left(2\sqrt{mr^2}d - d^2 \right) \quad (5.2.37)$$

Whereas the continental shelf extends a distance of R offshore and non-extreme habitat extends a distance of d inland, and assuming R=d, then we have:

$$\pi \left(2\sqrt{mr^2}R \right) + \pi \left(2\sqrt{mr^2}d \right) \quad (5.2.38)$$

$$\Rightarrow \pi \left(2\sqrt{mr^2}(R+d) \right) \quad (5.2.39)$$

$$\Rightarrow r\sqrt{m}d \quad (5.2.40)$$

One notices that the total biodiversity of supercontinent is directly proportional to both the radius and the distance d. This implies that that the biodiversity is also proportional to the perimeter of the supercontinent. The perimeter of the supercontinent increases until its coverage exceeds 50% of planetary surface. With coverage beyond 50%, the continent shore line decreases as the surface area of ocean decreases. This relationship is captured by D_{perim}. As a result, as the dryland coverage approaches 100%, the biodiversity drops to 0. Furthermore, as the size of the ocean coverage decreases, milder tropical storms and rains are less likely to bring precipitation to vast stretches of dryland. Therefore, the non-extreme habitat range also decreases. We find that D_{inland} is a good approximation to the drop of ocean's effect on land. The equation is obtained at one extreme by assuming at an ocean coverage at 50% of the planetary surface guarantees a diverse biocomplexity on land by ensuring the maintenance of the water cycle. Any greater surface coverage results in a linear increase in precipitation to exposed lands due to linear increase in total area of absorption of sunlight by the ocean surface. On the other extreme, the Mediterranean Sea covering 2.5 million km^2, or about $\frac{1}{200}$th the surface area of earth, evaporated within a thousand years during the Messinian salinity crisis 6 Mya. We assumed that the curve touches the point (0.995, 0.01), that is, there is about 1% chance that a Mediterranean Sea sized ocean coverage is able to maintain the water cycle because the Mediterranean Sea was not completely dry during the period and Mediterranean Sea is shallower than a typical ocean. Finally, D_{sea} is a factor added to marine biodiversity and grows linearly as the total surface area of ocean coverage increases and enabling greater photosynthesis opportunities regardless its proximity to shelves. Greater energy converting capacity eventually serves as an upper bound to the total attainable marine biodiversity since all levels of consumers feed on the energy produced by primary producers. One can also see that $D_{sea} = D_{inland}$.

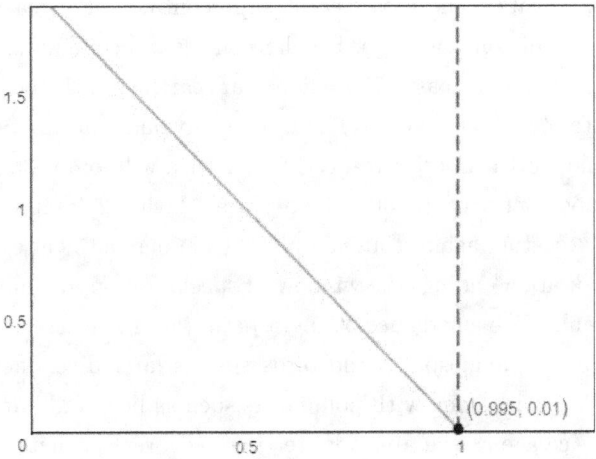

Figure 5.2.14: D_{inland} curve

Finally, the overall diversity is multiplied by the factor $T_{tectonics}$ because we assume that a significant presence of water, acting as a lubricant, is critical in ensuring mechanism of plate tectonics, which in turn is essential in generating biodiversity in the first place. We simply assumed that no plate tectonics is possible when the entire planet is covered by 75.6% of dryland or more.

One could see that biodiversity reaches its highest level at 29.41% of dry land coverage for landbased biodiversity, matching earth's 29% dryland coverage. Any greater land mass coverage results in a higher chance of forming supercontinents with greater climate extremes less suitable for land life. Any smaller land mass decreases the chance of supercontinent formation, but smaller land mass provides smaller niche space for biodiversity. Biodiversity reaches its highest level at 37.86% for marine based biodiversity. We have shown earlier that greater dryland coverage between 0 to 50 percent can be supported by greater range of continental plate configurations with varying ocean mass budgets. Greater range provides greater total shelf area for marine biodiversity. Finally, the marine curve sits below the land curve. This does not imply that marine biodiversity is lower since we simply assumed that the same biodiversity and complexity density per unit area. Marine biodiversity can easily outstrip the land based one if its biodiversity density is higher.

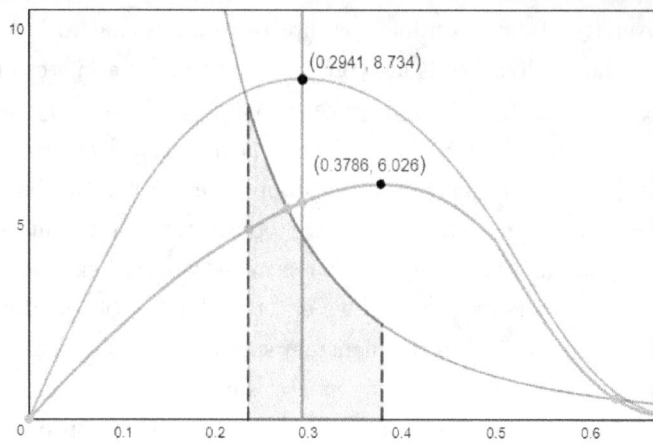

Figure 5.2.15: Total biodiversity curves between 0 to 60 percent dryland coverage

5.3   Supercontinent Cycle and Ice Age Cycle

We have discussed earlier (Chapter 4) that the formation of supercontinent stabilizes the rate of speciation, that is making newer species more difficult to emerge. In earth's case, a cycle of the formation of the supercontinent and breaking up into smaller ones repeats. The most recent supercontinent Pangea was formed 300 million years ago and started to break up 175 million years ago. Secular continental configuration is transitioning from fragmentation toward reunification, and for the past 175 million years earth's continents were scattered as island continents, but such scattering pattern is about to end. The Indian subcontinent and South America continent have already joined with Eurasia and North America respectively. Africa will join with Europe 50 million years into the future, forming the first supercontinent in 200 million years. Within 250 million years, Pangea Ultima will form and join all the continents together again; thus completing a supercontinent cycle in 550 million years. With the scattered continent configuration ending, the window of speciation opportunity closes, making future emerging species' life prospect difficult. The emergence of angiosperm (later fruit trees) and birds synchronized in time with the breaking up of Pangea, angiosperm and birds species later diversified and developed into a symbiotic relationship during the Cenozoic, along with pollinators such as bees and butterflies. Thanks to such diversification of fruit trees and an ice age at the appropriate time, arboreal primate species is able to evolve opposable thumbs, binocular vision, and partial bipedalism. Since the chance of glaciation onset at the current epoch is 1 in 2, and the onset of glaciation could have delayed as long as 235 million years into the future (in a more likely scenario), a hypothetical bipedal ape (non tool-use, non fire-control *Australopithecus Afarensis* type) evolved due to ice age then walks on the ground. It will not only find itself with fewer resources available with decreased biodiversity on a super-continent, but will also face predators not only confined to Africa but the Eurasia and Indian subcontinent. Under such a scenario, the bipedal ape can be out competed by predators and gone extinct.

If glaciation occurred earlier during the breaking up of Pangea (though glaciation more likely to occur now than earlier, so this is the less likely scenario), at a time with lower overall biodiversity, then glaciation's accelerating effects on evolution will not be as helpful as at later times. It can be manifested in terms of mathematics. We shall get ahead of ourselves by introducing our PDF distribution representing biodiversity from Chapter 7 and 8. If we simply assumed that ice age increases the speciation events without altering the total number of unique traits all species possessed, so t_{Win} increased by a factor F in proportion to existing t_{Win} , that is, the total number of species ever appeared during a given time period increased is directly proportional to the existing number of species before ice age. Assuming that two time periods 100 Myr apart, the first time period's biodiversity before ice age is $t_{Win1} = 1$ and the second time period representing biodiversity 100 Myr later before an ice age is $t_{Win2} = B_{cs}$. That is, it is 2.783 times greater than t_{Win1}. After the ice age, the absolute increase in biodiversity of second time period exceeds the first by $F \times B_{cs} - B_{cs} - F + 1$:

$$F \times t_{Win1} - t_{Win1} < F \times t_{Win2} - t_{Win2} \qquad (5.3.1)$$

$$F - 1 < F \times B_{cs} - B_{cs} \qquad (5.3.2)$$

$$0 < F \times B_{cs} - B_{cs} - F + 1 \qquad (5.3.3)$$

It can also be graphically illustrated as:

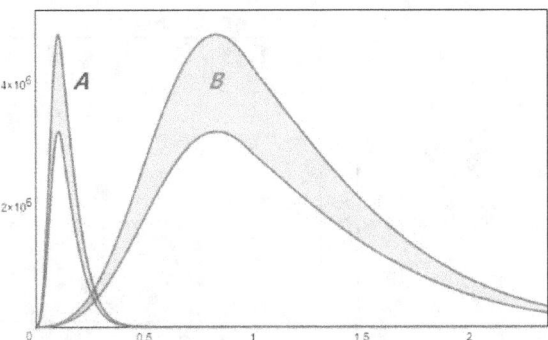

Figure 5.3.1: After an ice age, the total biodiversity increase is greater for a later time period because shaded area B > shaded area A

Of course, not all species survives through ice age. If F is expressed as variable, we have an inequality:

$$F(B_{cs} - 1) + 1 - B_{cs} > 0 \qquad F > 1 \qquad (5.3.4)$$

It is shown that as long as ice age brings a non-zero positive increase to the total biodiversity, a time period with greater initial biodiversity achieves greater increase in biodiversity after the ice age.

Therefore, even if an earlier onset of glaciation helps to usher in crop plant species, grass plants earlier than 30 Mya, such glaciation will not able to accelerate the evolution of hominid lineage since it is yet to emerge. As a result, diversification of grass plants occurred but human ancestors appeared after the opportunity window ended and still lived on the trees. By the time a bipedal ape (non tool-use, non fire-control *Australopithecus Afarensis* type) evolved 50 million years into the future due to a cooling climatic trend as Africa continent shifts north, it will again face predators not only confined to Africa but the Eurasia, Americas, and Indian subcontinent. More importantly, the rejoining of the continents eventually reduces the living space of all species as the interiors of the merging supercontinent subject to more temperature fluctuation extremes. As a result, human will face significant challenges under such scenario.

Therefore, we have three constraints on the temporal placement of the supercontinent cycle on the existing glaciation cycle, and we define a complete cycle as a breaking up phase of a supercontinent followed by a rejoining phase.

1. The previous glaciation must occur at or before the onset of the current breaking up phase of the cycle, minimizing possible disturbances on the biodiversity increase during the island continent phase.

2. It is assumed that in the last 200 Myr on any earth like planet, it requires, on average, 170 Myr to generate the biodiversity as we observed on earth today.

3. The following glaciation must occur at or before the biodiversity reaches its maximum and began returning toward a supercontinent configuration.

With the constraints listed above, the possible temporal placement of supercontinent cycle ranges from 0 Myr to 221 Myr after the termination of the previous glaciation. It can be no later than 221 Myr because we assumed that it takes 179 Myr to generate the biodiversity of the Cenozoic at the current epoch. The following glaciation is guaranteed to occur 400 Myr after the previous one and lasts another 50 Myr, completing an ice age cycle in 450 Myr. Therefore, 400 Myr − 179 Myr = 221 Myr is the latest possible placement. Then, one can see that only $\frac{221 \text{ Myr}}{400 \text{ Myr}} = 51.1\%$ of the possible glaciation cycle can potentially foster the emergence of an intelligent creature. This implies that the emergence of intelligence can only occur at the later phase of any glaciation cycle when the chance of the next glaciation reaches from 40% to 100%. *Therefore, on average, the chance of glaciation contributing to the emergence of human is 70%.* This is a revision from our prediction from section 5.3. This is more generalized and more accurate than our earlier calculation for earth's unique temporal placement of supercontinent cycle over the glaciation cycle (89 Myr following the Karoo ice age).

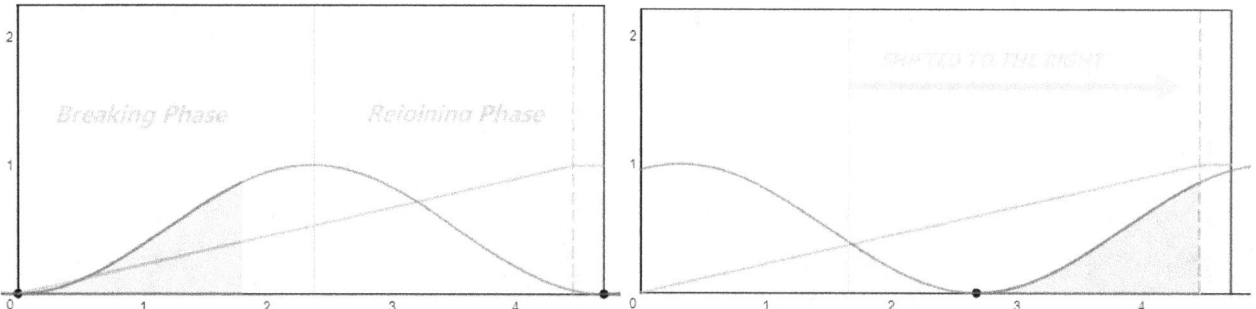

Figure 5.3.2: The permissible placement range for the supercontinent cycle: from immediately after the last glaciation up to 221 Myr after the previous glaciation

Within a breaking up phase of the supercontinent cycle, the probability of giving the emergence of intelligence is non-uniform. Three critical factors, the rate of the biodiversity increase, the biocomplexity transformation, and the increasing chance of the next onset of ice age determine the final probabilistic outcome on the chance of emergence. We define the biodiversity increase as the derivative of the supercontinent cycle. As the breaking up phase initiates, the rate of biocomplexity increases. The rate of increase decreases to 0 as the breaking up phase terminates, and the maximum is reached at the midpoint.

$$w_{continent} = |\sin A (t + p_0)^{2.0}| \tag{5.3.5}$$

$$y_{diversitychange} = \frac{d}{dt} w_{continent} \tag{5.3.6}$$

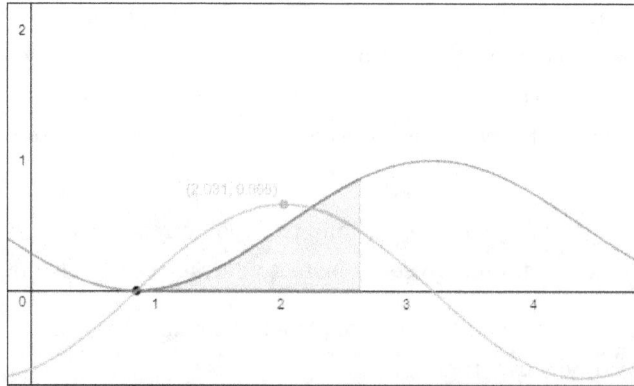

Figure 5.3.3: According to this simplified model, the rate of biodiversity change turns positive at 179 Mya in the mid Jurassic when Pangea started to break apart and peaked at 203.1 Myr after the Karoo ice age, or 60.9 Mya at the early Cenozoic, and the rate of biodiversity change turns negative again in 57.2 Myr when Africa joined with Europe and Asian joined with North America.

The biocomplexity transformation factor, which is discussed in Chapter 6 and Chapter 7, basically states that the total biodiversity BCS increases over time at the rate of 2.783 per 100 Myr. That is, the number of species within all genera and new genera increases by 2.783 folds per 100 Myr. Finally, one applies the increasing chance of the onset of the next glaciation. The composite curve pushes the maximum likelihood of the emergence to 160 Myr after the start of the breaking up phase.

$$y_{emergeman} = \left(\frac{d}{dt} w_{continent} \right) \cdot 2.783^t \cdot y_{ice} \tag{5.3.7}$$

$$y_{ice} = 0.223564t \tag{5.3.8}$$

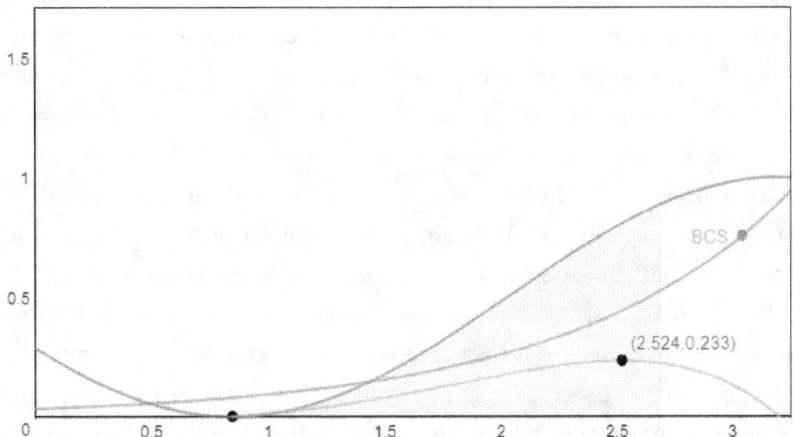

Figure 5.3.4: The rescaled biocomplexity transformation factor and glaciation occurrence probability curve applied to the simplified model of the rate of biodiversity change shows that the peak of the likelihood of the emergence of man is 11.6 Mya, at the mid Miocene.

This is fairly close to our current time. Since the constraint requires that the next glaciation to occur 170 Myr after the diversification of species, preparing for the emergence of intelligence, up to the time when the breaking up phase ends, then we derived the probability on the appropriate timing for the onset of ice age acting as an accelerator at this period±15 Myr to be 22.09%.

$$\frac{\int_{2.49}^{2.79} y_{emergeman} dt}{\int_{0.85}^{3.211} y_{emergeman} dt} = 0.220946083607 \tag{5.3.9}$$

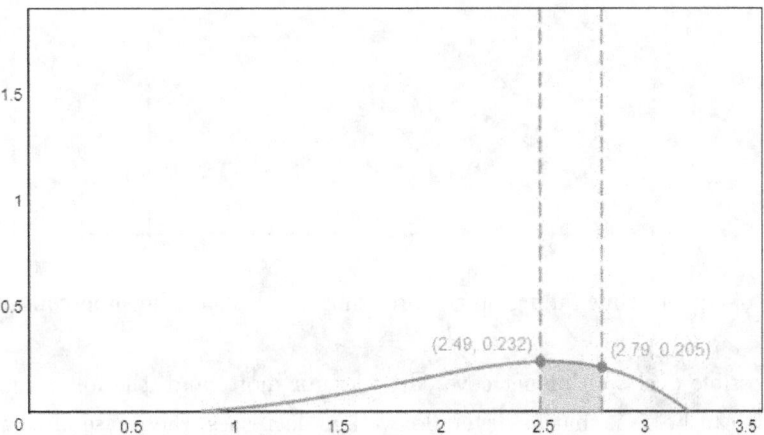

Figure 5.3.5: The emergence of intelligence toward the late times of the breaking up phase is favored

This shows that, regardless of the temporal placement of supercontinent cycle relative to the glaciation cycle within the permissible ranges, the emergence of intelligence toward the late times of the breaking up phase is favored. Nevertheless, there is a nearly 80% chance of glaciation happening earlier during the breaking up phase of the cycle if it were to occur at all (but its chance of occurrence is lower than the average chance of glaciation within the permissible range of 70%), disrupting the chance of the emergence of human.

One yet to address the fact that the highest probability of emergence predicted to occur earlier at mid Miocene. This implies that all earth like planet undergoing a breaking up phase which initiated 170 Mya should emerge earlier than us. This seemingly contradiction is resolved if one considers the timing of the emergence of Homo sapiens. If Homo sapiens emerged 15 to 1 Mya, at a time when Africa was still separated by the sea from Asia, the Bering strait were wider, and the Isthmus of Panama still did not yet exist, they will have much more difficulty in migration and colonization of the surface of the planet. One can then speculate, on many planets that are undergoing similar transformation as earth but with earlier emergence, the intelligent species, though fully emerged, is stucked on their own continent, still separated by wide bodies of ocean, and have to wait for another 15 Myr to 1 Myr before they rejoin and cross over. As a result, they arose early but their domination of the planetary surface proceed at the same time as ourselves. Furthermore, the factors enabling human domination is actually bidirectional. Majority of domesticated crops and animals originated outside of human's native Africa. Ancestors of dogs, horses and camels first evolved in North America and crossed into Asia and later Africa 8 Mya. Cats and chickens evolved in Asia. Only sheep, goat, and buffalo were found in both Asia and Africa. All major crop plants originated from North America except potato evolved in South America. In fact, the spread of grassland may well be a consequence of fauna migration out of North America. Without the rejoining of the continents, grassland may not even evolve in Africa by 2 Mya. This is intriguing because it shows that it is not a coincide that we find ourselves dominating the planet at a particular time as it is now. Whereas the island continent configuration is just starting to transition toward a supercontinent configuration and all continents are just barely connected to each other.

Having determined the permissible secular placement of continent cycle relative to the glaciation cycle and the probability of the occurrence of life within the breaking up phase of the cycle. We can now think the glaciation cycle and the supercontinent cycle as one interwoven cycle. The current time, represented as a point on the interwoven cycle, can occur at the breaking up phase or the rejoining phase of the continent cycle. For the simplicity of the model, we assume that each phase lasts half as long of the entire cycle, though, we have shown earlier in Section 5.3, that 29% of dryland coverage results in 59% breaking up configuration and 41% supercontinent configurations. As a result, the chance that any time period on all earthlike planet falls on a breaking up phase is simply 50%.

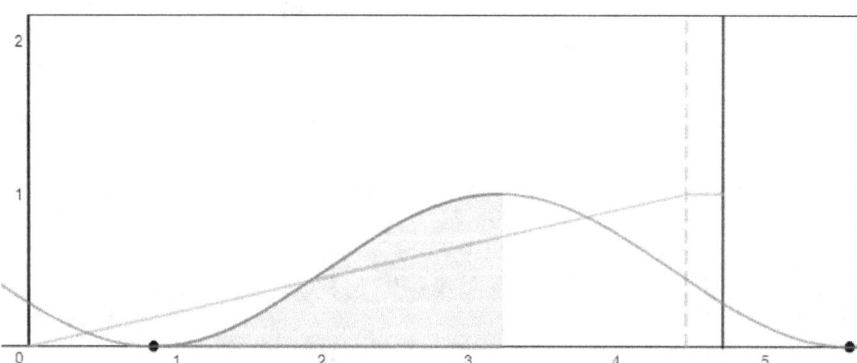

Figure 5.3.6: The chance of current time falling on the breaking up phase of the supercontinent cycle is 50%

In order to illustrate the possible cycles on planet covered by less or more land, the following sinusoidal wave and their graph is shown. It can be seen that as dryland coverage decreases, the phase of island configuration dominates over the supercontinent phase, and vice versa. Of course, in reality, the cycles are not necessarily perfectly sinusoidal, but the macro trend stays. Moreover, it is also to note the probability of glaciation increases slower throughout the glaciation cycle in the case of less dryland coverage, and the probability of glaciation approaches 1 throughout all periods in the case of more dryland coverage. Different continent cycle length are discussed in detail in the consecutive section.

$$y_{nearlyoceanplanet} = |\sin A \, (t + p_0)^{0.48}| \tag{5.3.10}$$

$$y_{earthtypical} = |\sin A \, (t + p_0)^{2.0}| \tag{5.3.11}$$

$$y_{nearlydesertplanet} = |\sin A \, (t + p_0)^{121}| \tag{5.3.12}$$

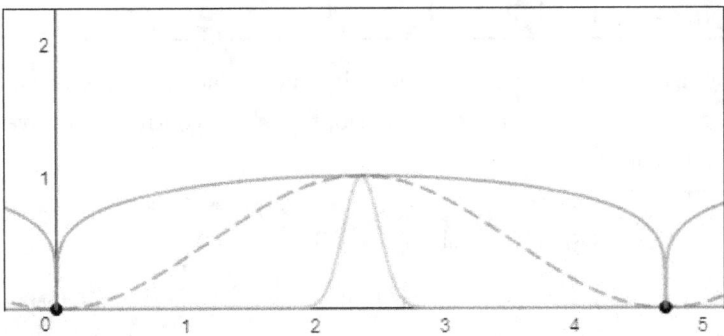

Figure 5.3.7: Nearly ocean planet, earth typical (dashed), and nearly desert planet continent cycle

As a result, one can confirm that glaciation at the right time and its subsequent consequence on habitat change otherwise taking geologic time scale serves as an accelerator rather than a detractor. If one assumes that the emergence of hominid lineage 15 Mya is the start of the opportunity window for the onset of an ice age acts as an accelerator, then the opportunity window ends with Africa colliding into Eurasia and the closing the Mediterranean ocean in 15 million years.

$$P_{permissibleRange} \cdot P_{endofbreakingphase} \cdot P_{chanceinbreakingphase} \tag{5.3.13}$$
$$0.511 \cdot 0.212 \cdot 0.50 = 0.054166$$

Then, only 5.4166% of an expected ice age interval of 470 Myr duration will ice age results in an accelerated evolutionary pace.

5.4 Weighted Emergence Rate across All Dryland Ranges: Intro

Now, glaciation events and the supercontinent cycle (see Section 5.4) interweaving and creating complicated effect on the final probability on the likelihood of human emergence of any given continent coverage size to ocean coverage size. The timing of the onset of glaciation is critical to the emergence of intelligent species. Based on the constraint criteria listed under Section 5.4, the final chance on the appropriate ice age timing for a particular type of land to ocean ratio to host the emergence of intelligent life depend on the permissible range of supercontinent cycle placement within its glaciation cycle, the weighted average chance of glaciation within this placement range, the weighted number of emerging civilization within each glaciation cycle, and the cycle length relative to earth's case. As we already indicated, a greater land coverage results a shorter duration between episodes of glaciation and a disproportionally longer timespan on a supercontinent configuration within the continent cycle. A smaller land coverage results a longer duration between episodes of glaciation and a disproportionally longer timespan on a island configuration within the continent cycle. We run the simulation

207

for land coverage over 21.992%, 24.862%, 26.836%, 28.291%, 29.41% (earth's case), 30.69%, 32.99%, and 38% and their chance on the appropriate ice age timing accelerating the emergence of intelligent species. The cyclic function is modeled as the follows:

Whereas A is the amplitude and parametrically determined to be 0.66516888705.

Plate equations for different dryland coverage

$$C_{15}(t) = |\sin A (t + p_0)^{121}|$$
$$C_{20}(t) = |\sin A (t + p_0)^{50.12}|$$
$$C_{25}(t) = |\sin A (t + p_0)^{27.4}|$$
$$C_{50}(t) = |\sin A (t + p_0)^{6.0}|$$
$$C_{75}(t) = |\sin A (t + p_0)^{3.12}|$$
$$C_{90}(t) = |\sin A (t + p_0)^{2.4}|$$
$$C_{100}(t) = |\sin A (t + p_0)^{2.0}|$$
$$C_{110}(t) = |\sin A (t + p_0)^{1.84}|$$

$$C_{120}(t) = |\sin A (t + p_0)^{1.68}|$$
$$C_{125}(t) = |\sin A (t + p_0)^{1.6}|$$
$$C_{150}(t) = |\sin A (t + p_0)^{1.28}|$$
$$C_{200}(t) = |\sin A (t + p_0)^{0.96}|$$
$$C_{300}(t) = |\sin A (t + p_0)^{0.64}|$$
$$C_{400}(t) = |\sin A (t + p_0)^{0.48}|$$
$$C_{1000}(t) = |\sin A (t + p_0)^{0.48}|$$

These equations are derived by using the equation describe a dryland surface coverage of 11% and a glaciation cycle that lasts 10 times as long as earth's. Under such scenario, the island continents always remain separated and never join to form supercontinents.

$$C_{1000}(t) = |\sin A (t + p_0)^{0.48}| \tag{5.4.1}$$

As a result, the result of the integration of this equation is used as 100% chance of forming isolated island continents, and every other cycles with a lower chance of forming isolated continents is compared with it. Every other case is identical to the base case except the value of its exponent. We determined the exponent c of each possible case by using the chance of forming island continent across different dryland surface coverage we derived earlier as the constraint.

Then, one arrived at a particular value for c so that the integration within 1 earth's glaciation cycle (472.3 Myr) of a particular dryland surface coverage over the integration of dryland surface coverage subject to 100% forming isolated continents within 1 earth's glaciation cycle has to equal to the chance of forming island continent for that particular dryland surface coverage.

$$y_{breaking}(x) = \frac{y_{island}(x)}{y_{super}(x) + y_{island}(x)} \tag{5.4.2}$$

$$\frac{\int_0^{4.723} C_x(t)\,dt}{\int_0^{4.723} C_{1000}(t)\,dt} = \frac{\int_0^{4.723} |\sin A (t + p_0)^c|\,dt}{\int_0^{4.723} C_{1000}(t)\,dt} = y_{breaking}(x) \tag{5.4.3}$$

Next, we sum up the area between 0 to 1.79 times of earth's case for glaciation cycle lasts 472.3 Myr long. This area is the minimum time requirement for biodiversity we have established enabling the emergence of human given every other favorable conditions are met. This correspond to the split of the continent Pangea from early Jurassic up to today. For smaller dryland coverage but greater chance forming island continents supporting greater biodiversity, less time is needed. The reverse is also true. We compute the minimum time requirement t_{MIN} for each possible case using the following equation:

$$\frac{\int_0^{t_{MIN}} C_x(t)\,dt}{\int_0^{1.79} C_{100}(t)\,dt} = 1 \tag{5.4.4}$$

And we determined the values as follows:

Glaciation cycle length	Min time (100 Myr)	Glaciation cycle length	Min time (100 Myr)
0.15 Earth	2.3615	1.20 Earth	1.702
0.20 Earth	2.3615	1.25 Earth	1.677
0.25 Earth	2.3615	1.50 Earth	1.562
0.50 Earth	2.259	2.00 Earth	1.416
0.75 Earth	2.001	3.00 Earth	1.227
0.90 Earth	1.8795	4.00 Earth	1.1096
1.00 Earth	1.79	10.00 Earth	0.858
1.10 Earth	1.7481		

The best fit is expressed as:

$$t_{MIN}(x) = \begin{cases} 2.3615 & x \leq 44.628 \\ 8.47452x^{-0.336399} & x > 44.628 \end{cases} \qquad (5.4.5)$$

Notices that for significant dryland coverage, $t_{MIN} = 2.3615$. This is the maximum time attainable since each continent cycle lasts only 472.3 Myr long. This is exactly the half of the cycle that falls under the island configuration phase. Those t_{MIN} actually does not satisfy the time requirements of the emergence of intelligence. Later, we will add a partial chance factor for these cases. Their chance to give the emergence to intelligence is then somewhat minimized within 1 cycle.

5.5   The Permissible Range Factor

First, we need to find the permissible range factor for each dryland coverage case which narrows the total number of habitable earth across all times. We will use it later to fine-tune the parameters of our distribution function. The total number of habitable earth giving birth to intelligence *across all times* is not conceptually equivalent to finding the number of earth giving emergence to human like intelligence *at the current time* or within a *selected temporal range*. A habitable planet potentially giving emergence to intelligence may have the appropriate supercontinent placement within a glaciation cycle, but at the current time or within the temporal window it can be experiencing a glaciation or running the course of a supercontinent phase or at the early stage of breaking up phase. At one extreme, it is also possible that the planet is dominated by island continent configurations and has consistently undergoing emergence event opportunities but island continents rarely rejoining with each other, so that a species exchange and full blown agriculture potential can not be realized until much later. At the other extreme, for a semi-desert planet dominated by supercontinents, the permissible placement range becomes so narrow between frequent glaciation cycles that there is not enough time to guarantee for the emergence of 1 civilization at the current time. Nevertheless, as long as the planet contains the correct placement of continent cycle over glaciation cycle, such planet can eventually give birth to civilizations despite its suboptimal conditions per each emergence event and its delay of emergence. However, there exists a set of placements in which the placements are forever in mismatch between the continent and glaciation cycles. No matter how long one will wait, it will never experience a chance of intelligence emergence. Then, the total number of habitable earth giving birth to intelligence across all times is defined by the probability of the permissible range of placement of the continent cycle within the glaciation cycle.

The continent cycle duration remains constant for earth sized planet due to similar rate of tectonic movement speed (Chapter 3). Additionally, mini earth and super earth's tectonic movement strength does not differ significantly from earth. Furthermore, the weighted average of all habitable planets' tectonic movement speed closely match earth's case. Although supercontinent cycle duration remains constant, the proportion of island configurations to the supercontinent configurations varies depending on the dryland coverage ratio. Most importantly, greater dryland ratio contributes to shorter duration between each episodes of glaciation. As a result, the glaciation cycle duration is dependent on the dryland coverage size. A supercontinent cycle with constant duration but variable proportion of breaking up phase vs joining phase of continents placed within a glaciation cycle with a variable duration creates an inverse relationship between dryland coverage vs. its chance within a permissible range.

The chance of any placement of continent cycles over glaciation cycles can give rise to intelligence as one of the permissible range can be computed as the follows:

Assume each cycle is initially placed at $p_0 = 0$, so that:

$$C_x(t) = \sin A (t + p_0)^c \qquad (5.5.1)$$

209

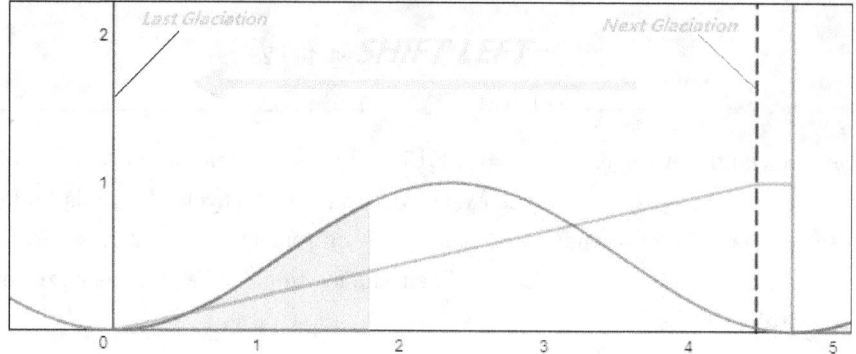

Figure 5.5.1: The starting requirement

As one can see, one emergence event is initiated immediately after the end of the last glaciation event, and one then slides left by assigning values for $p_0 > 0$ (simulating all possible placement cases of continent cycles over glaciation cycles). As the continent cycle placements shift immediately to the left, the placements disrupt an emergence event until the next emergence slides within the glaciation cycle as:

$$C_x(t) = \sin A (t + C_{End})^c \tag{5.5.2}$$

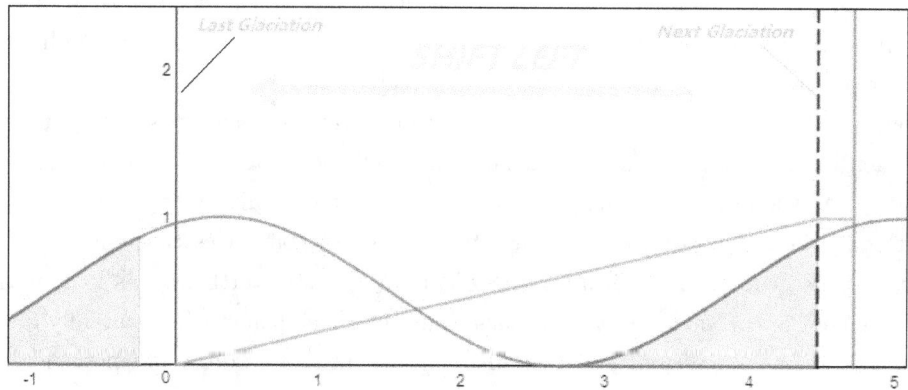

Figure 5.5.2: The ending requirement

So 1 minus the time span under C_{End} expressed in units of 100 Myr over the duration of one earth's continent cycle length of 472.3 Myr is the chance of possible placements of continent cycles over the glaciation cycle.

$$P_{permissible} = 1 - \frac{C_{End}}{4.723} \tag{5.5.3}$$

Furthermore, for significant dryland coverage, the emergence of island continents takes place only briefly during the entire continent cycle. As a result, the starting position can not be assumed at $p_0 = 0$, as we just stated earlier. Instead, we assumed that as long as 99.9% of all shaded area, representing the total time requirement for biodiversity we have established enabling the emergence of human given every other favorable conditions are met, then it gives the emergence of intelligence. Therefore, the initial starting position of the equation becomes:

$$C_x(t) = \sin A (t + C_{Start})^c \tag{5.5.4}$$

and C_{Start} must satisfies the requirement of:

$$100 \left(1 - \frac{\int_0^{t_{MIN} - C_{Start}} C_x(t)\, dt}{\int_{0 - C_{Start}}^{t_{MIN} - C_{Start}} C_x(t)\, dt} \right) = 0.1\% \tag{5.5.5}$$

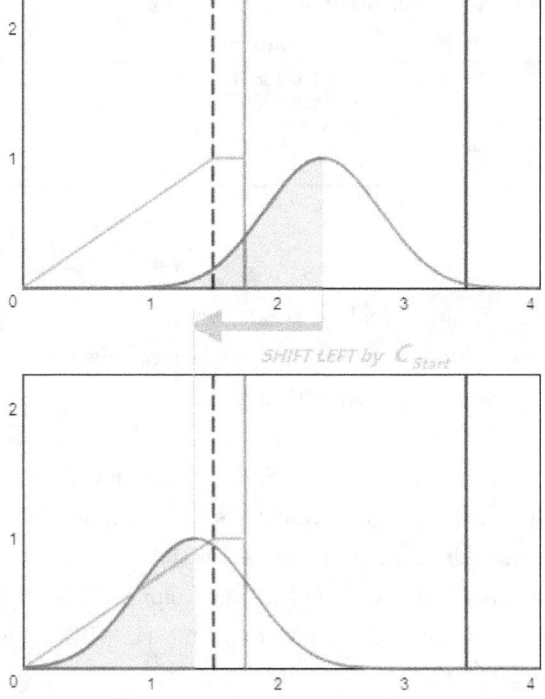

Figure 5.5.3: An illustration of the difference between the placement for $p_0 = 0$ and $p_0 = C_{Start}$ for $C_{37}(t)$

So the finalized chance of placements of continent cycle over glaciation cycle is:

$$P_{permissible} = 1 - \frac{C_{End} - C_{Start}}{4.723} \quad (5.5.6)$$

The computed results is listed in the table below (for permissible chance over 1 with more than 1 emergence chance per cycle is capped at 1, C_{End} becomes negative when even at the initial starting position the next continent cycle is well within the current glaciation cycle for cycle length 1.5 times earth's cases, so one has to shift right to meet the condition that the second emergence event placed just at the start of the next glaciation cycle):

Whereas the best fit for C_{Start} is:

$$C_{Start} = 2.92349\,(0.0526399)^x + 0.0214772 \quad (5.5.7)$$

and the best fit for the dryland proportion to glaciation cycle length is given by:

$$d_{Ratio} = \exp\left(-\frac{1}{2.2778}\ln\left(\frac{x + 7.50603}{236144}\right)\right) \quad (5.5.8)$$

Land	Glaciation cycle length	Permissible range chance	C_{End}	C_{Start}
10.98%	10.00·Earth	1	-3.62	0.00271
16.27%	4.00·Earth	1	-3.36	0.0103
18.45%	3.00·Earth	1	-3.25	0.0179
21.99%	2.00·Earth	1	-3.05	0.0403
24.86%	1.50·Earth	1	-0.55	0.0716
25.595%	1.40·Earth	1	0	0.0842
25.83%	1.37·Earth	0.9932246454	0.12	0.088
26.84%	1.25·Earth	0.8644505611	0.75	0.1098
27.29%	1.20·Earth	0.8136777472	1	0.12
28.29%	1.10·Earth	0.7059072623	1.53	0.141
29.41%	1.00·Earth	0.602583104	2.04	0.163
30.69%	0.90·Earth	0.4960829981	2.6	0.22
32.99%	0.75·Earth	0.3421554097	3.43	0.323
35.5%	0.62·Earth	0.2140588609	4.17	0.458
38.47%	0.50·Earth	0.1105229727	4.87	0.669
40.64%	0.43·Earth	0.0565318653	5.3	0.844

Land	Glaciation cycle length	Permissible range chance	C_{End}	C_{Start}
42.86%	0.37·Earth	0.0332415837	5.58	1.014
48.93%	0.25·Earth	0.0059919543	6.15	1.4553
52.55%	0.20·Earth	0.0046580563	6.38	1.679
57.41%	0.15·Earth	0.0019267415	6.63	1.9161
64.59%	0.10·Earth	0.0032606394	6.86	2.1524

However, the above is the simplified version of the reality. In actuality, things gets more complicated. For the glaciation cycle's length that is non-whole number divisor of the supercontinent cycle length such as 0.37·Earth or 0.43·Earth, even if its initial starting cycle was placed within the permissible range, its successive cycle will gradually shift beyond the permissible range and eventually placed outside the permissible range as there is a non-whole number factor mismatch in length between the glaciation cycle and the supercontinent cycle. Nevertheless, depending on the exact mismatching length, glaciation cycle will guarantee to synch with supercontinent cycle again given an arbitrary number of additional cycles in which the misalignment becomes aligned again. Some mismatching requires fewer cycle repeats while others requires more. This implies, one needs to modify the definition of permissible range. Even if the cycle is originally placed within the permissible range, it is non-guarantee that it will remain within the permissible range at every successive cycle. On the other hand, even if an initial placement of continent cycle superimposed on a glaciation cycle was outside of the permissible range, it can still be within the permissible range following a number of successive repeating cycles. In order to fully cover all possible cases, one needs to compute all possible initial placement positions. Moreover, different starting placement positions between the continent and glaciation cycle within the permissible range produce differential chance of hosting emergence chance, albeit the differences remain small based on simulation runs.

Making the process even more complicated, for the glaciation cycles' length that is a whole number divisor of the supercontinent cycle length, such as 1·Earth or 0.5·Earth, for initial placements position within the permissible range guarantees an emergence per every cycle. However, every initial placement position outside the permissible range guarantees no emergence at all per every cycle. We run simulation up to 2,500 mini-steps between glaciation cycle length from 10% to 140% of earth's case, and we find that there is a constant 26.5% of all dryland ratios falls under a whole number divisor of the supercontinent cycle length scenarios. Out of these, 90% of them are found between glaciation cycle length from 10% to 40% of earth's case. This means that out of this dryland ratios, for initial placement outside the permissible range, there exists a set of placements in which the placements are forever in mismatch between the continent and glaciation cycles. No matter how long one will wait, it will never experience a chance of intelligence emergence. This fits well with our original definition of the permissible range.

There is no simple function to describe this relationship and we wrote a program to simulate all possible scenarios by utilizing the properties we defined earlier and run hundreds of different dryland coverage ratios. For each ratio, we run through every possible initial placement of continent cycle over glaciation cycle at the steps of 100 within a supercontinent cycle length. We then find the number of successful emergence cases per 100 cycles per each dryland coverage ratios.

The result is plotted:

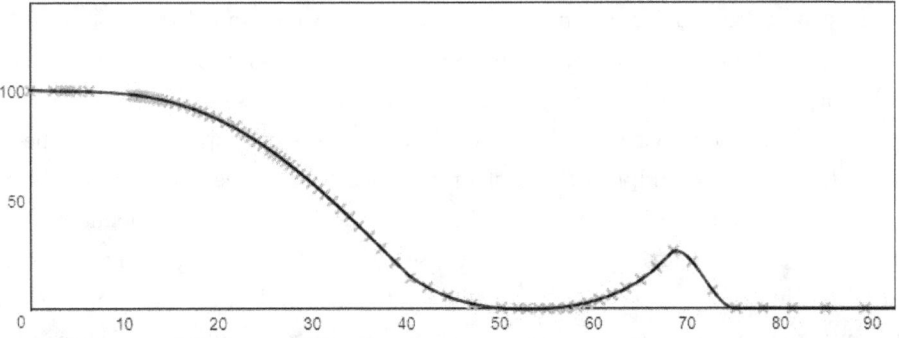

Figure 5.5.4: The permissible range of continent cycle over glaciation for varying dryland coverage

The best fit for the data is formulated as:

$$S_{tep1}(x) = \frac{-100}{1 + \exp\left(-\frac{1}{6.111}(x - 31.8566)\right)} + 100 \tag{5.5.9}$$

$$S_{tep2}(x) = 1.89737 \cdot 10^{-4} x^4 - 0.0402998 x^3 + 3.27566 x^2 - 120.608 x + 1693.1 \tag{5.5.10}$$

$$S_{tep3}(x) = \frac{-26.1}{1 + \exp\left(-\frac{1}{0.93212}(x - 72.1134)\right)} + 26.1 \tag{5.5.11}$$

$$y_{permissible}(x) = \begin{cases} S_{tep1}(x) & 0 \leq x \leq 37.459 \\ S_{tep2}(x) & 37.459 \leq x \leq 68.462 \\ S_{tep3}(x) & 68.462 \leq x \leq 100 \end{cases} \tag{5.5.12}$$

One can notice that less dryland coverage offers a higher chance of emergence. This is true only because longer glaciation cycle permits more than 1 emergence event following a breaking up phase despite the universal presence of non-permissible placement ranges for any initial starting placement. For example, at 16.27% dryland coverage, the glaciation cycle lasts 4 times longer than earth's case. There are then 4 emergence events possible within the glaciation cycle; however, certain range of placement permits only 3 emergence events per glaciation cycle. Nevertheless, the minimum requirement of at least 1 emergence cycle per glaciation cycle is satisfied. There is also a slight peak at 70% dryland coverage. This is due to a range of short glaciation cycle length that is a whole or partial whole number divisor of the supercontinent cycle length, so that at least a portion of all possible placements guaranteeing emergence for every cycle. This peak subsides as the dryland coverage ratio increases because the glaciation cycle becomes impossibly short and no emergence cycle's duration is short enough to fit into any of them.

In conclusion, we started using permissible range expecting to completely exclude a percentage of habitable planet with wrong placement of continent cycle over glaciation cycle prohibiting the emergence to intelligence. In actuality, only 26.5% falls within our original definition and expectation. For the rest 73.5%, the definition of permissible range is non-applicable. Nevertheless, it is shown that per 1 continent cycle, only a fraction of these placement results in an emergence condition. Since we computed earth among the first batch of habitable planets formed between 5 Gya and 4 Gya, and any planet underwent at most 1 supercontinent cycle ahead or behind earth's case, we can modify our requirement for the total number of habitable earth giving birth to intelligence *across all times* to that of the total number of habitable earth giving birth to intelligence *within 1 continent cycle of 472.3 Myr*. This time duration is considerably shorter than all times into the future, but significantly longer than any current time or short selected temporal range. If one were to use the strict definition of permissible range only, the final number of habitable earth will be significantly higher but this does not reflect the reality within the next 500 Myr. One can wait for a billion years to observe the appropriate timing and emergence events within glaciation cycles, but we are only interested in calculating the chance of intelligence emergence at the most recent 500 Myr.

Nevertheless, a caveat must be raised. Later in our distribution model, we assumed that the total biodiversity achieved similar level of development and size (i.e. 3.6 million species generated per most recent 100 Myr) regardless of BCS and BER at the current time. Along with the current assumption that only 500 Myr of multi-cellular window is possible, models with low BCS and BER (suggests that up to 500 Mya the biodiversity stays nearly the same as that of today) implies the first appearance of multi-cellular life will be much more drastic and dramatic than Cambrian explosion. However, the pace of evolution will become much more stagnant and lackluster thereafter. This interpretation is simply a consequence of our assumptions and mathematical inference. If we assumed that initial emergence of multi-cellular life must be less dramatic or even significantly less than one observed during the Cambrian explosion and typical multi-cellular life history lasts much longer than 500 Myr, then, the permissible range should then be fixed to the total number of habitable earth giving birth to intelligence *across all times*, and all habitable terrestrial planets within the permissible range to be 73.5%, as we already mentioned.

With the caveat in mind, we continue to compute the chance inside the permissible range within 1 continent cycle of 472.3 Myr. Now, we use our best fit for permissible range and the following (defined in Section 5.2) to calculate the selected dryland coverage range translated into a percentage of continental plate configurations range out of all possible continental plate configurations creates favorable emergence conditions:

$$A_{total} = \int_0^{100} y_{upper}\,dx - \int_0^{100} y_{lower}\,dx \qquad (5.5.13)$$

We define the U_{pper} and L_{ower} with 2 percent increments between 11.469 to 84 percent dryland coverage. Values < 11.469 percent has 0 chance in observing any supercontinent configurations enabling species and diversity exchange, and for values > 84 percent there is insufficient ocean to guarantee a transition from aquatic species to a terrestrial one.

If we define each step of U_{pper} and L_{ower} with infinitesimal small increments, we could define the chance as:

$$\int_{11.469}^{84} \frac{S_{lowerwrap}(x) - S_{upper}(x)}{A_{total}} \cdot y_{permissible}(x)\,dx \qquad (5.5.14)$$

and our integration results:

$$= 29.0260894545 \qquad (5.5.15)$$

Therefore, all habitable terrestrial planets within the permissible range to be 29.03%.

5.6 Weighted Emergence Rate across All Dryland Ranges: Detailed Analysis

5.6.1 Chance within a breaking up phase

We just have shown how the emergence conditions within glaciation cycle varies within each dryland coverage. We now show the probability that current time on all earth like planets falls into a breaking up phase. Although the supercontinent cycle remain constant in duration, the chance of forming an island continent configuration increases as dryland to ocean ratio drops. The chance of entering an island continent phase is captured by our earlier equation $\frac{y_{island}}{y_{super}+y_{island}}$. We use this equation to estimate the chance within a breaking up phase at the current time.

$$y_{breaking}(x) = \frac{y_{island}(x)}{y_{super}(x) + y_{island}(x)} \qquad (5.6.1)$$

5.6.2 Chance of rejoining

We have also shown that in order for an intelligent species to realize its full potential, its timing of appearance has to coincide with the rejoining of the continents. The great biodiversity created by the speciation opportunity during the breaking phase has to be exchanged between island continents during a rejoining phase so that the potential of a full blown agriculture can be realized. This chance of exchange is maximized when the dryland coverage is high. This relationship is the direct inverse relationship to the chance within a breaking up phase and is equivalent to the chance within a supercontinent phase:

$$y_{join}(x) = 1 - \frac{y_{island}(x)}{y_{super}(x) + y_{island}(x)} = \frac{y_{super}(x)}{y_{super}(x) + y_{island}(x)} \tag{5.6.2}$$

5.6.3 Glaciation chance

The chance at the current time that a glaciation event initiates is the weighted average chance of the opportunity window between the start of a permissible $+ t_{MIN}$ and the end of a permissible range just before the onset of glaciation. It is show that this chance increases toward 100% as the dryland coverage increases. At the other extremes, this chance decreases toward 50% as the dryland coverage decreases to 0%. The onset of glaciation at the appropriate times serves as an accelerator to the evolution of intelligence.

The results is listed in the table below:

Land	Glaciation cycle length	Glaciation chance
10.98%	10.00·Earth	0.5091027033
16.27%	4.00·Earth	0.5294844974
18.45%	3.00·Earth	0.5434334363
21.99%	2.00·Earth	0.5747988256
24.86%	1.50·Earth	0.6090350428
25.60%	1.40·Earth	0.6199349282
25.83%	1.37·Earth	0.6228998908
26.84%	1.25·Earth	0.6385982755
27.29%	1.20·Earth	0.6460056113
28.29%	1.10·Earth	0.6624876145
29.41%	1.00·Earth	0.6818689917
30.69%	0.90·Earth	0.7074012048
32.99%	0.75·Earth	0.7548409143
35.50%	0.62·Earth	0.8100893864
38.47%	0.50·Earth	0.8765095903
40.64%	0.43·Earth	0.9236645722
42.86%	0.37·Earth	0.9499135231
48.93%	0.25·Earth	0.9864356702
52.55%	0.20·Earth	0.9912899511
57.41%	0.15·Earth	0.9857672592
64.59%	0.10·Earth	0.9703103914

The best fit for the data is formulated as a logistic function:

$$y_{glaciation}(x) = \frac{47.729}{1 + e^{-\frac{1}{5.156}(x-33)}} + 52.116 \tag{5.6.3}$$

The relationship is plotted:

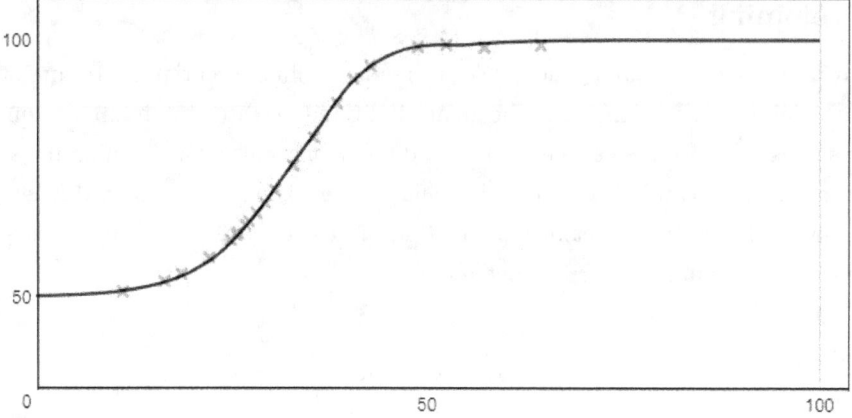

Figure 5.6.1: Glaciation chance

5.6.4 Partial chance of emergence

Very lastly, dryland coverage $\geq 33\%$ (glaciation cycle duration ≤ 0.75 Earth's) poses another problem to the emergence of intelligence. As the glaciation cycle shortens, the permissible range for the breaking up phase placement shrinks accordingly. The permissible range out of the shrinking glaciation cycle becomes so tight that there is not enough time lapsed to guarantee the biodiversity observed in earth's case (We assumed in Section 5.3 that it takes at least 170 Myr of breaking up phase to guarantee the emergence of human) before the breaking phase is interrupted by the next onset glaciation event at 100% chance. If the breaking phase is interrupted by the next glaciation, we compute the total amount of biodiversity already achieved over the amount of biodiversity desired to derive the probability of human emergence despite its interruption. That is, the chance shall be strictly less than 100% but strictly greater than 0.

It is computed based on the equation for selected x ≤ 0.62 Earth cycle length:

$$\frac{\int_0^{2.362} C_x\left(t\right)dt}{\int_0^{1.79} C_{100}\left(t\right)dt} \tag{5.6.4}$$

The results is listed in the table below:

Land	Glaciation cycle length	Partial chance of emergence
38.47%	0.50·Earth	0.9208027303
40.64%	0.43·Earth	0.8195776149
42.86%	0.37·Earth	0.7154689933
48.93%	0.25·Earth	0.4603543657
52.55%	0.20·Earth	0.3417925089
57.41%	0.15·Earth	0.2206196671
64.59%	0.10·Earth	0.1029939417
77.90%	0.05·Earth	0.0150814501

The best fit for the data is formulated as:

$$y_{partial} = \begin{cases} 100 & x \leq 36.855 \\ 294613x^{-2.15567} - 23.7087 & x \geq 36.855 \end{cases} \tag{5.6.5}$$

The relationship is plotted:

216

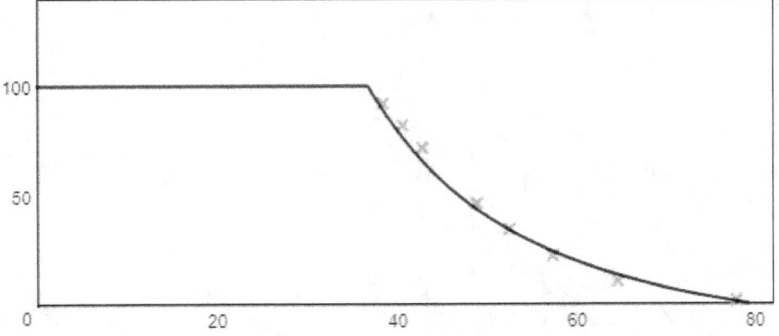

Figure 5.6.2: Partial chance of emergence

We combine all these factors into a single equation with normalized raw emergence chance relative to earth's case as (assuming earth's island configuration chance and joining chance combined at 25% is rescaled as 100%):

$$P_{iceage}(x) = y_{permissible}(x) \cdot \frac{y_{breaking}(x)}{y_{breaking}(0.2941)} \cdot \frac{y_{join}(x)}{y_{join}(0.2941)} \cdot y_{glaciation} \cdot y_{partial} \qquad (5.6.6)$$

and the result is plotted:

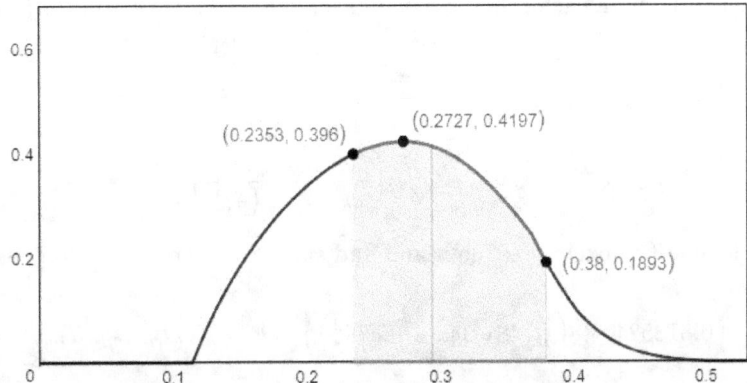

Figure 5.6.3: The *overall* raw emergence chance curve between 15 to 50 percent dryland coverage

The combined equation illustrates the landscape of the likelihood of emergence given different dryland surface coverage within 1 full earth's glaciation cycle duration of 472.3 Myr. One can draw several conclusions based on the data. First of all, the percentage of permissible range of supercontinent cycle placement within its glaciation cycle grows as the duration of the hiatus between episodes of glaciation grows. As a result, the emergence chance per glaciation cycle increase as the glaciation cycle lengths, as the length of continent cycle becomes shorter than the length of the glaciation cycle. On other hand, greater land mass coverage with shorter, more frequent ice age cycle results lower percentage of permissible ranges which only occur at the late times of the glaciation cycle. A mismatch on the length of continent and glaciation cycle can adversely effect events per cycle such that there always exists some continent cycle places itself completely on the non-permissible range when others are placed within the permissible range. This is especially true for greater land coverage than earth such as 38% land coverage case in which only 18.93% emergence chance relative to earth's case is possible per every continent cycle. As a result, the *overall chance* of emergence indicates that lower dryland surface coverage dominates over greater land coverage as $\int_0^{0.2941} P_{iceage}(x)\,dx > \int_{0.2941}^1 P_{iceage}(x)\,dx$

However, out of those continent cycles *within the permissible range* of the glaciation cycle, the situation is reversed.

$$P_{iceage}(x) = \frac{y_{breaking}(x)}{y_{breaking}(0.2941)} \cdot \frac{y_{join}(x)}{y_{join}(0.2941)} \cdot y_{glaciation} \cdot y_{partial} \qquad (5.6.7)$$

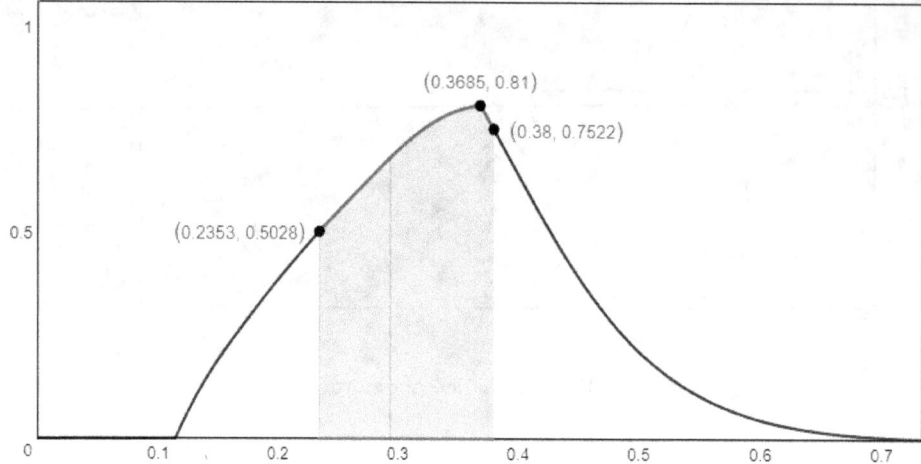

Figure 5.6.4: The raw emergence chance curve *within the permissible range* between 15 to 50 percent dryland coverage

Though the overall chance of emergence is lower for greater dryland coverage, those dryland coverage falls within the permissible range actually offers a greater chance of emergence due to higher chance of glaciation and a higher chance of continents rejoining with each other. So one have: $\int_0^{0.2941} P_{iceage}(x)\,dx < \int_{0.2941}^1 P_{iceage}(x)\,dx$ The composite effect on biodiversity and evolutionary rate for different dryland coverage based on the overall raw chance is:

$$E_{mergence}(x) = P_{iceage}(x) \cdot \left(\frac{y_{diversity}(x)}{y_{diversity}\left(\frac{10}{34}\right)}\right)^2 \tag{5.6.8}$$

In which the diversity curve is modified excluding the island and supercontinent formation chance:

$$y_{diversity}(x) = \left(0.818524x + \left(1.148\sqrt{34x} - 0.574^2\right)\right) \cdot D_{perim} \cdot D_{inland} \cdot T_{tectonics} \tag{5.6.9}$$

This is possible because $P_{iceage}(x)$ includes factor $\frac{y_{breaking}(x)}{y_{breaking}(0.2941)} \cdot \frac{y_{join}(\tau)}{y_{join}(0.2941)}$, which means that for all island continents a successful civilization emergence requires a chance of continents rejoining to exchange fauna. Therefore, we have:

$$(0.818524x)\frac{y_{breaking}(x)}{y_{breaking}(0.2941)} \cdot \frac{y_{join}(x)}{y_{join}(0.2941)} \cdot D_{perim} \cdot D_{inland} \cdot T_{tectonics} \tag{5.6.10}$$

On the other hand, $\frac{y_{join}(x)}{y_{join}(0.2941)} \cdot \frac{y_{breaking}(x)}{y_{breaking}(0.2941)}$ means that for all super continents a successful civilization emergence also requires a chance of peripheral island continents rejoining to exchange fauna. Therefore, we have:

$$\left(1.148\sqrt{34x} - 0.574^2\right)\frac{y_{join}(x)}{y_{join}(0.2941)} \cdot \frac{y_{breaking}(x)}{y_{breaking}(0.2941)} \cdot D_{perim} \cdot D_{inland} \cdot T_{tectonics} \tag{5.6.11}$$

Since both shared the same factors, the sum of both becomes:

$$\left(0.818524x + \left(1.148\sqrt{34x} - 0.574^2\right)\right)\frac{y_{breaking}(x)}{y_{breaking}(0.2941)} \cdot \frac{y_{join}(x)}{y_{join}(0.2941)} \cdot D_{perim} \cdot D_{inland} \cdot T_{tectonics} \tag{5.6.12}$$

Since $\frac{y_{breaking}(x)}{y_{breaking}(0.2941)} \cdot \frac{y_{join}(x)}{y_{join}(0.2941)}$ along with $y_{glaciation} \cdot y_{partial}$ forms $P_{iceage}(x)$, we can remove this factor from the diversity curve when $y_{diversity}(x)$ multiplied with $P_{iceage}(x)$.

A particular dryland coverage offers a certain chance of island and supercontinent formation within one continent cycle. The cumulative chance offered in the equation $y_{breaking}(x)$ states the total chance within 1 cyclic period, but it does not state the chance of island continent formation at any particular time within the cycle. Given

218

earth's case, there is 50% cumulative chance of island continent formation. In reality, every period of the cycle there is some portion of all lands falls under the island configuration and others under supercontinent configuration. The proportion ranges from 0% to 100%. From mathematical perspective that is quantitatively equal, one can assume that 50% of the time during the cycle, the earth is completely in a supercontinent configuration and switches over into an island configuration for the rest of the period immediately thereafter. With this simplified assumption, we can just use the island biodiversity curve to calculate the total amount of biodiversity generated under the island configuration, and supercontinent biodiversity curve to calculate the total amount of biodiversity under the supercontinent configuration. Coupled with $P_{iceage}(x)$, it leads to a high emergence chance for dryland coverage larger than earth and a lower chance for dryland coverage smaller than earth. Despite shorter opportunity time window given by t_{MIN} for the emergence of intelligence for larger land coverage, the total summed island continental area is larger during the opportunity window, providing more niche space for flourishing biodiversity and compensating its overall shorter time. One could even argue for a a_{MIN}. That is, assuming earth's dryland coverage area under island configuration is the minimum requirement for the emergence of intelligence as well as t_{MIN} given other conditions are met. Then, every other land ratio's t_{MIN} and a_{MIN} requirements can be computed by fixing earth's case as the invariant.

with the plotted graph:

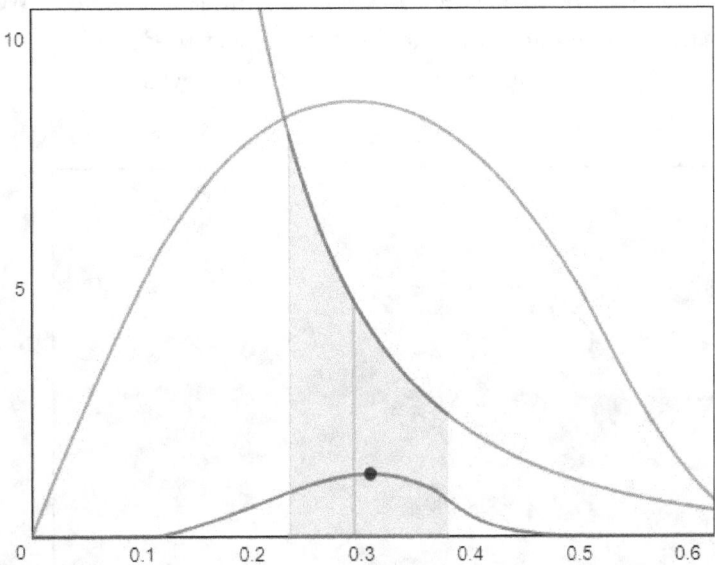

Figure 5.6.5: The composite effect curve on biodiversity between 0 to 60 percent dryland coverage. A maxima of emergence is reached at 30.88% dryland coverage level; however, emergence chance does not drop to 0 beyond our earlier assumption that the birthing of civilization can only occur between 23% and 38% dryland coverage.

The biodiversity is compared with the baseline of 29% dryland coverage as earth's case and raised to the 2nd power. The reason for doubling the ratio (or some power) is that an increasing biodiversity has a non-linear effect in creating further biodiversity and accelerating evolutionary change. Starting at the Quaternary glaciation, the Hominid lineage retained opposable thumb, gained bipedalism, omnivorous diets, enlarged brain, and language communication. The lineage also lost almost all of its hair. Out of these attributes, at least the retainment of the opposable thumb, the evolution of omnivorous diets and the enlarged brain is strengthened by the increasing biodiversity of the Cenozoic era.[88] Opposable thumb in a non-arboreal habitat seem useless. However, freed hands can gather a diverse range of resources such as nectar, fruits, nuts, and grass plant roots only available since the Cenozoic and directly influenced the evolution of omnivorous diets. Feeding on omnivorous diet expedite the evolution of a larger head. With a larger cranial capacity capable of abstract and creative thinking, Hominid is able to take the advantage of its increasingly biological diverse environment by experimenting with different combination of resources. In turn, the continual success of the conscious thinking leads to further evolution of dexterous hands and a greater tolerance to a diverse range of food in a positive feedback loop. Furthermore, limited resources will prevent an arising civilization to fulfill its full potential. The Aztecs and Incas were transitioning into an agricultural society. However, a lack of domesticable games such as horses and oxes with considerable muscle strength significantly hamper the rate of their progress. With less abundant resources for manipulation and exploitation, the traits of intelligent, tool-making species are less selected for by evolution since its full-blown potential cannot be realized.

Based on the final composite results, the overall chance on the appropriate ice age timing accelerating the emergence of intelligent species is approximately the same across 23 to 29 percent of dryland coverage and drops considerably beyond 40 percent. There is a slight peak around 31% land mass coverage, but in general a widening mismatch on the length of continent and glaciation cycle for dryland coverage greater than earth's case, essentially smaller duration between glaciation episodes, can increasingly adversely effect emerging chance per cycle. Smaller surface land coverage resulting in a greater chance in an island configuration thus is favored and can offset the decline in emergence chance per continent cycle. However, greater permissible placement ranges offered by smaller land coverages decreases the weighted chance of glaciation and the chance of island continents rejoining with each other for exchanging biodiversity. Both conditions adversely effect the chance of overall emergence.

One major conclusion can be drawn is that emergence chance does not drop to 0 beyond our earlier assumption that the birthing of civilization can only occur between 23% and 38% dryland coverage. We will evaluate the weighted chance of emergence across all dryland coverage cases in the next section.

The composite effect on biodiversity and evolutionary rate for different dryland coverage based on the raw chance within the permissible range is:

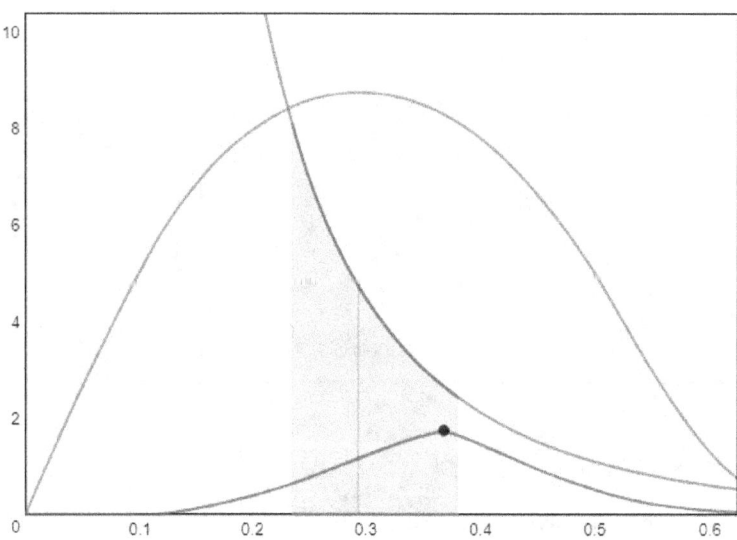

Figure 5.6.6: The composite effect curve on biodiversity between 0 to 60 percent dryland coverage for those within the permissible range

Despite overall composite emergence is low for land surface coverage higher than earth, for those do falls within the permissible range, the chance of emergence is actually higher than earth's case due to higher glaciation chance, greater chance of land rejoining, and most importantly, more niche space for flourishing biodiversity. That is, if such a cycle is selected, an intelligent species, under the stress of fluctuating ice age climate, has a greater manipulative possibilities on nature given the greater biodiversity (due to greater summed island continental area), which serves as the basis of a greater chance of its survival. There are a greater number of domesticable species offering larger combinatorial search space of biological ingredients.

On a last note, earth's mass is just the typical average among a range of values of habitable terrestrial planets. We have defined in Chapter 3 that sub-earth to super-earth's ranges from 0.43 to 2 earth mass, and the surface area ranges from 0.57 to 1.5874.

$$S_{surface}(x) = \frac{4\pi \left(r_0(x)\right)^2}{4\pi \left(r_0(1)\right)^2} \qquad (5.6.13)$$

We define the lower and upper bound on the emergence curve based on these ranges, and the final shaded region is the total possible emergence ranges across all dryland ratios on all earth like planets.

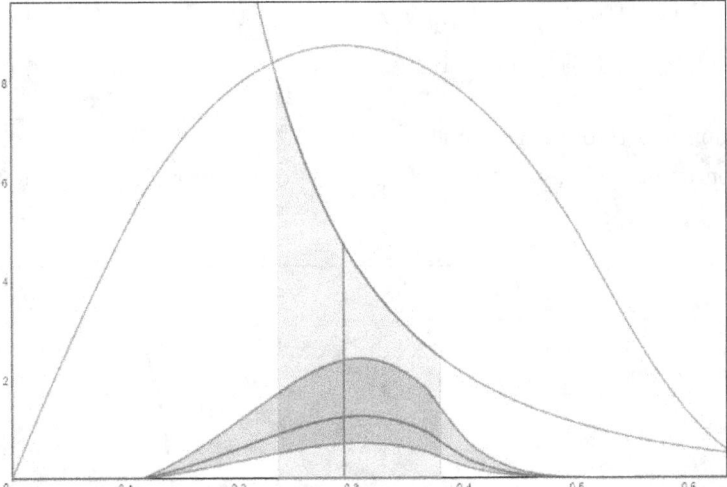

Figure 5.6.7: The deeply shaded region represents the total possible emergence ranges across all dryland ratios on all earth like planets

This chapter emphasized the importance of glaciation on the emergence of intelligence and have shown the intricate relationship between glaciation and continent cycles. By now, one should rationalize the importance of an ice age and its non-trivial chance across all land ranges. If earth did not enter an ice age, the retainment of the opposable thumb, the evolution of bipedalism, omnivorous diets, enlarged brain, and language communication can still happen as the total biodiversity increases. However, it will occur at a much slower pace. The early bipedal Hominid lineages such as *Australopithecus Africanus* was exclusively vegetarian, based on their dental fossil records. They may nevertheless use hands to exploit more non-meat food resources, but they were not evolving toward an omnivorous diet. The unpredictable change of climate during a glaciation can cause a drastic change of fauna in their habitat within a generation's time, and the adaptation of an omnivorous diet can be a choice between life and death. Hominid can only obtain their energy from meat in certain seasons at one extreme and only plants in certain seasons at the other extreme and anything in between. An adoption of omnivorous diet, again, is no guarantee of a quick transition toward a larger brain. If the environment is stable, natural selection places less emphasize on the emergence of a larger brain since planning, memory, abstract, creative thinking is less useful under the regime of a stable climate. As a result, a omnivorous diet will lead to a larger brain but at a much slower pace. This fact is observed in many bird species. Many omnivorous birds such as crow has EQ between human and other species. In the case of flying birds, their living range is significantly larger than land-based mammals. With the initiation of ice age, birds can choose new habitat by crossing mountain ranges, rivers, seas, and open oceans and settles into regions with more stable climate. Therefore, birds experience less selectional pressure toward the evolution of a large brain with the ability to gather information and to plan for the future. Hominids, on the other hand, have no choice in the face of chaotic climate to evolve larger brain to survive since they are confined by geologic barriers of mountains, seas, rivers, and deserts.

5.7 Generalized Emergence Curve across All Temporal Periods

The glaciation curve can also be generalized to include more than one continent cycle. The equation to include n number of cyclic iteration is stated as:

$$P_{iceage}(x, n) = (1 - (1 - y_{permissible}(x))^n) \cdot (1 - (1 - y_{breaking}(x))^n) \cdot$$
$$(1 - (1 - y_{join}(x))^n) \cdot (1 - (1 - y_{glaciation})^n) \cdot (1 - (1 - y_{partial})^n) \tag{5.7.1}$$

When n = 1, the equation simplifies to our earlier one.

With a greater number of continent cycles, eventually all possible cycles provides the potential for emergence. Plot is shown for iterations for n = 1, 2, 10 and 100:

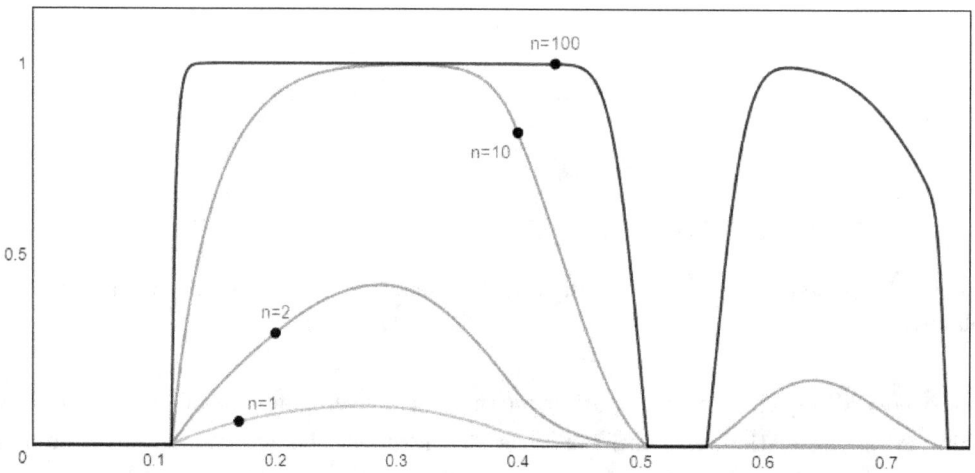

Figure 5.7.1: Generalized glaciation curve for n = 1, 2, 10 and 100

With generalized glaciation curve, one can generalize the emergence curve across time. Whereas t is the number of years relative to us in its initial formation time ($t > 0$ if ahead of us in its initial formation time and $t < 0$ if behind us in its initial formation time),

$$E_{mergence}(x, t) = P_{iceage}\left(x, 1 + \frac{t}{4.723}\right) \cdot \left(\frac{y_{diversity}(x)}{y_{diversity}\left(\frac{10}{34}\right)}\right)^2 \cdot A_{allearth} \tag{5.7.2}$$

To generalize the emergence curve across habitable planet of all size ranges, we find the weighted average size of all earth like planet is at 1.0568 times larger than earth (check Chapter 3):

$$\frac{\int_{-0.366}^{0.024} P(x)\, dx}{\int_{-0.366}^{0.302} P(x)\, dx} \approx 0.5 \tag{5.7.3}$$

Then, the weighted surface area size of all habitable earth is 1.037 times of earth. This is fairly close to earth's surface area, and in our calculation we will simply assume that $A_{allearth} = 1$.

For every 100 Myr, the cumulative BCS increases by 2.783, emergence curve based on BCS alone is expressed as (using the current biocomplexity as its reference):

$$E_{mergence}(x, t) = P_{iceage}(x, 1) \cdot \left(\frac{y_{diversity}(x)}{y_{diversity}\left(\frac{10}{34}\right)}\right)^2 \cdot (B_{cs})^t \tag{5.7.4}$$

by adding an additional BCS factor dependent on time t whereas for the current time t = 0, and t < 0 for the past and t > 0 for the future.

We then finds that the complexity growth based on geologic factor alone is always lower than the total composite biocomplexity achievable. This also shows that biocomplexity search space growth is contributed by both geologic and biologic factors.

$$P_{iceage}\left(x, 1 + \frac{t}{4.723}\right) \cdot \left(\frac{y_{diversity}(x)}{y_{diversity}\left(\frac{10}{34}\right)}\right)^2 \leq P_{iceage}(x, 1) \cdot \left(\frac{y_{diversity}(x)}{y_{diversity}\left(\frac{10}{34}\right)}\right)^2 \cdot (B_{cs})^t \qquad (5.7.5)$$

so we adopt the emergence curve based on BCS.

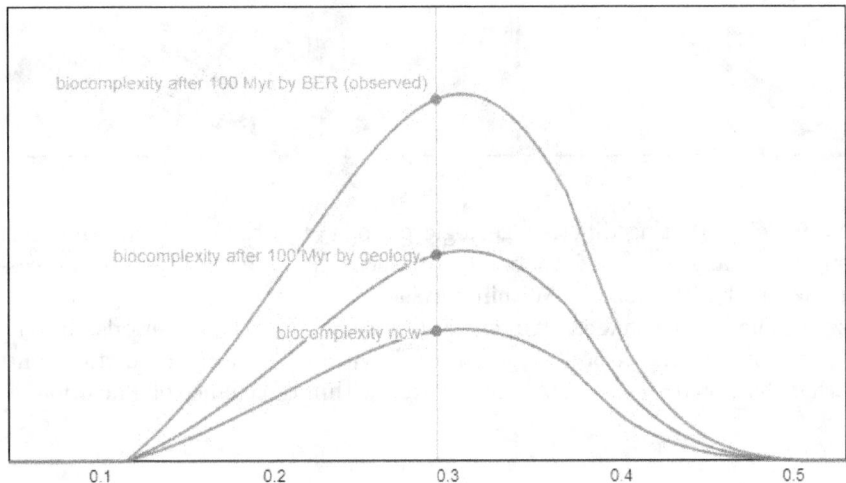

Figure 5.7.2: Emergence curve based on BCS > Emergence curve based on Geological factors only

For planets ahead of us in evolution, they are experiencing a higher biocomplexity at the current time, so that every 100 Myr the curve's height increased 2.783 fold along every point.

Whereas the cumulative complexity at our current time is defined as:

$$E_{mergence}(x, 0) \qquad (5.7.6)$$

$E_{mergence}(x, 0)=$ total biodiversity achievable within 1 cycle since we assumed that earth have just completed 1 full continent cycle of 500 Myr. In the following plot, we illustrate that earth has just successfully completed 1 emergence cycle within the last 500 Myr. Exactly 500 Myr ahead of the youngest planets formed between 5 Gya and 4 Gya.

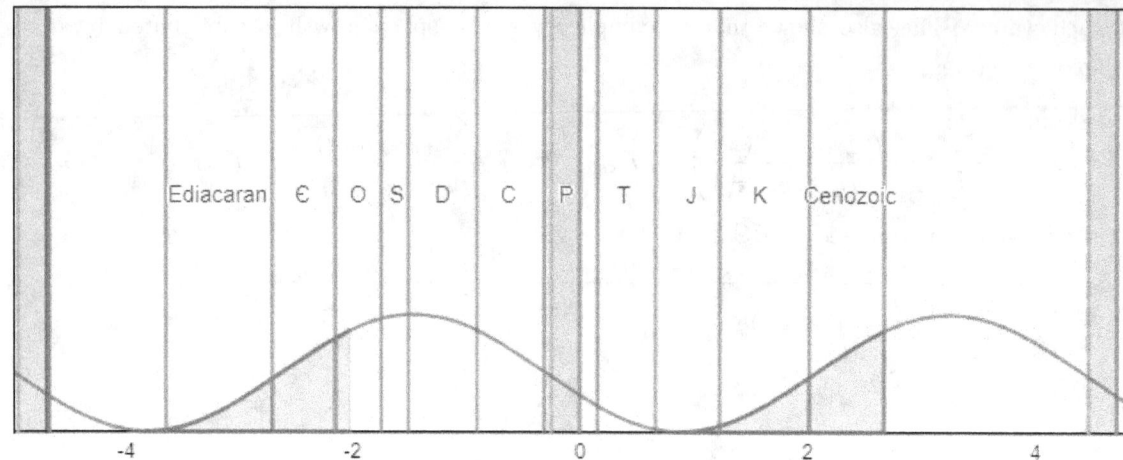

Figure 5.7.3: Simplified model projecting different geologic periods onto the continent and glaciation cycle. Notice that the supercontinent Pangea started to break up since mid Jurassic and the process started to accelerate during the Cretaceous and Cenozoic. Within the last 500 Myr, there is only 1 emergence chance during the current Cenozoic. The orange shaded strip signifies idealized glaciation periods. In reality, the pre-Cambrian glaciation lasted much longer. Karoo ice age falls well within the model prediction, but our current glaciation comes much earlier than predicted at 100% though well within the chance of glaciation defined earlier (between 0 and 100%).

The cumulative emergence complexity 100 Myr, 200 Myr, 300 Myr, and up to 500 Myr ahead of earth is defined as $E_{mergence}(x,1)$, $E_{mergence}(x,2)$, $E_{mergence}(x,3)$, $E_{mergence}(x,4)$, and $E_{mergence}(x,5)$. Since we assumed that earliest habitable earth originated 500 Myr ahead of earth so the maximum value for $t = 5$.

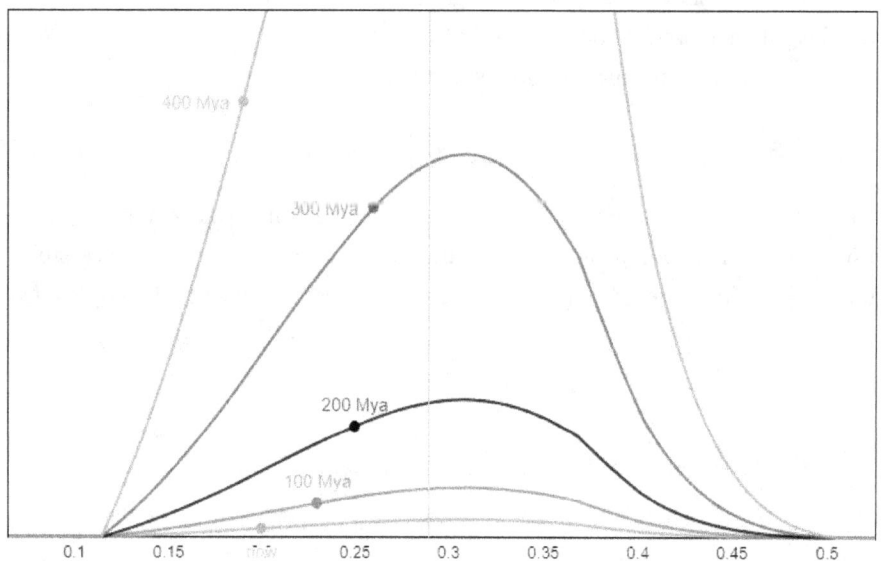

Figure 5.7.4: From bottom to top: $E_{mergence}(x,0)$, $E_{mergence}(x,1)$, $E_{mergence}(x,2)$, $E_{mergence}(x,3)$, $E_{mergence}(x,4)$, and $E_{mergence}(x,5)$ respectively.

Using the generalized cumulative emergence curves across time and planet size ranges, one can compute how earth's current attainable complexity after 1 full 473 Myr continent cycle is compared with those planet emerged at the same time as earth up to 500 Mya and up to 500 Myr later.

Recall we can use the following equation:

$$\frac{\int_{I_{ntercept1}}^{I_{ntercept2}} \frac{S_{lowerwrap}(x)-S_{upper}(x)}{A} \cdot y_{permissible}(x)\,dx}{\int_{11.469}^{84} \frac{S_{lowerwrap}(x)-S_{upper}(x)}{A} \cdot y_{permissible}(x)\,dx} \tag{5.7.7}$$

$I_{ntercept1}$ and $I_{ntercept2}$ are the lower and upper intercepts of any particular emergence curve at a well-defined time t crossing the current earth's attainable complexity level defined as a horizontal line:

$$E_{earthCurrent} = P_{iceage}\left(0.2941, 1\right) \cdot \left(\frac{y_{diversity}\left(0.2941\right)}{y_{diversity}\left(\frac{10}{34}\right)}\right)^2 \cdot \left(B_{cs}\right)^0 \tag{5.7.8}$$

Whereas the denominator $\int_{11.469}^{84} \frac{S_{lowerwrap}(x) - S_{upper}(x)}{A} \cdot y_{permissible}\left(x\right) dx$ is the percentage of all possible continental configurations within the permissible range defined earlier, so the final percentage is expressed as a ratio within the permissible range.

On a side note, for continent configuration at the current epoch that contributes to a lower biodiversity compares to earth, its evolutionary lag in years can be expressed as:

$$t_{years} = \frac{\ln\left(\frac{E_{mergence}(x,0)}{E_{mergence}(0.2941,0)}\right)}{\ln\left(B_{cs}\right)} \tag{5.7.9}$$

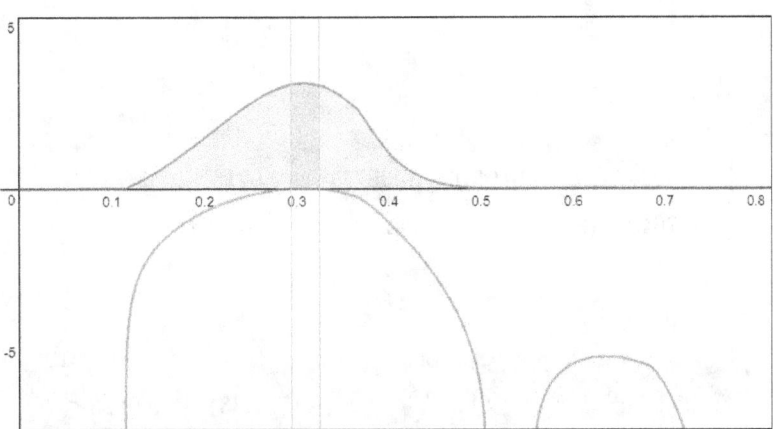

Figure 5.7.5: The green curves indicates that the vast majority of dryland to ocean ratio within 1 supercontinent cycle translates into biocomplexity of earth hundreds of million years into the past

That is, despite these planets emerged at the same time as earth, due to their overall lower biodiversity level constrained by their geology, they are currently at the evolutionary stage comparable to earth that occurred t_{years} ago.

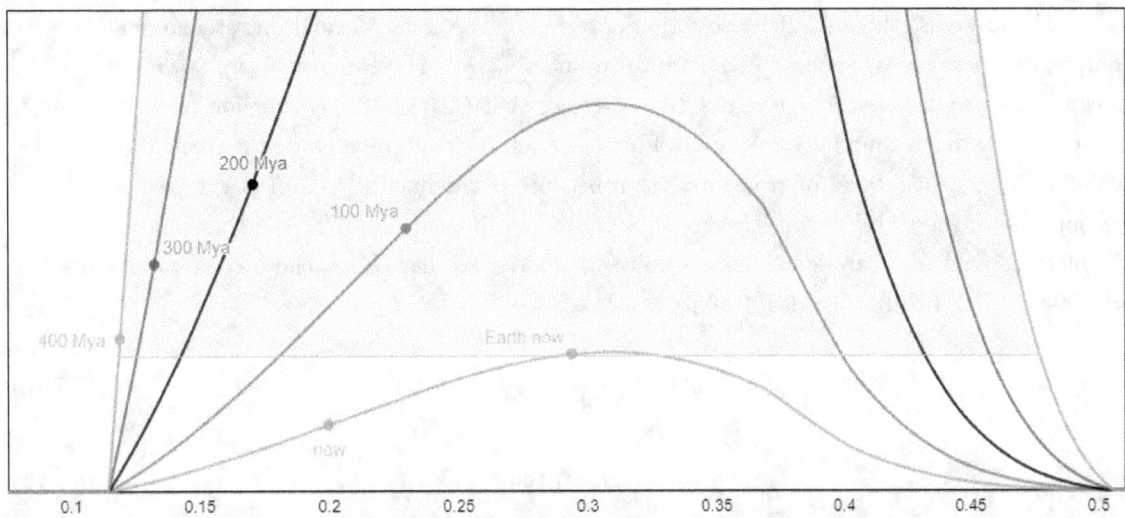

Figure 5.7.6: The shaded portion indicates across different time periods different dryland to surface ratio achieves biocomplexity above earth's current level.

The finalized results is plotted:

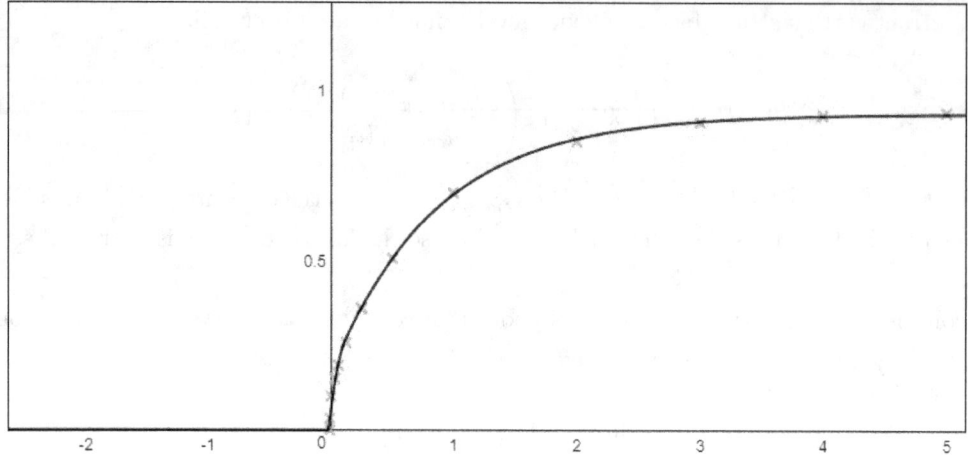

Figure 5.7.7: Complexity curve as a function of time, whereas the future < 0 and the past > 0

and the best fit is:

$$T_{op}(t) = \begin{cases} 0 & t \leq -0.0213 \\ -0.322476\,(0.00018146)^t + 0.38734 & -0.0213 \leq t \leq 0.1024 \\ -0.764934\,(0.312476)^t + 0.932897 & 0.1024 \leq t \end{cases} \tag{5.7.10}$$

Alternatively as:

$$T_{op}(t) = \begin{cases} 0 & t \leq -0.0213 \\ t\,(891.974t^{0.01645} - 2598.6t^{0.00581878} + 1707.33) & -0.0213 \leq t \end{cases} \tag{5.7.11}$$

We use this result, along with our earlier assumption on the timing of multicellular emergence is dependent on the initial formation time of the planet, to show that our current complexity is the top 40.06% of the entire candidate pool between 5 Gya and 4 Gya.

$$\frac{1}{10}\int_{-5}^{5} T_{op}(t)\,dt = 0.4006 \tag{5.7.12}$$

More importantly, we have shown that terrestrial planets formed at later times are favored since it is more likely to be watered. So we use the metallicity selection criterion on the stellar formation rate curve along with earth's biocomplexity curve across time and sizes, and we find that *our current complexity is the top 35.75% of the entire candidate pool between 5 Gya and 4 Gya*. Notices that $C_0(t)$ is the conversion factor to convert the metallicity change through time into units of 100 Myr so that T_{op} can be computed correctly. (i.e. -0.066 = -5, 0=0, 0.066 = 5). T_{op} also takes on a negative sign because our defined direction of the past and future is exactly the opposite of the metallicity selection. This can be easily changed. $C_1(t)$ is the conversion factor to convert the metallicity change through time into units of 100 Myr so that S_{tellar} can be computed correctly. (i.e. -0.066=9.199-0.5, 0=9.199, 0.066 = 9.199+0.5).

$$C_0(t) = \frac{5}{0.066}t \tag{5.7.13}$$

$$C_1(t) = \frac{5}{0.66}t + 9.199 \tag{5.7.14}$$

$$\frac{\int_{-0.066}^{0.066}\int_{-1}^{1} f_{metallicity}(x,t) \cdot f_{wetearth}(x)\,dx \cdot T_{op}(C_0(-t)) \cdot S_{tellar}(C_1(t))\,dt}{\int_{-0.066}^{0.066}\int_{-1}^{1} f_{metallicity}(x,t) \cdot f_{wetearth}(x)\,dx \cdot S_{tellar}(C_1(t))\,dt} = 0.3575 \tag{5.7.15}$$

Very lastly, we generalize earth's position not only across all time periods but also for different values of BCS.

Different BCS is used later in Chapter 8 to set the constraints and limit of the model based on observation. The results for BCS of 1, 1.125, 1.25, 1.5, 2, 2.783, 4, 10, and 40 is computed, and the plot is presented as:

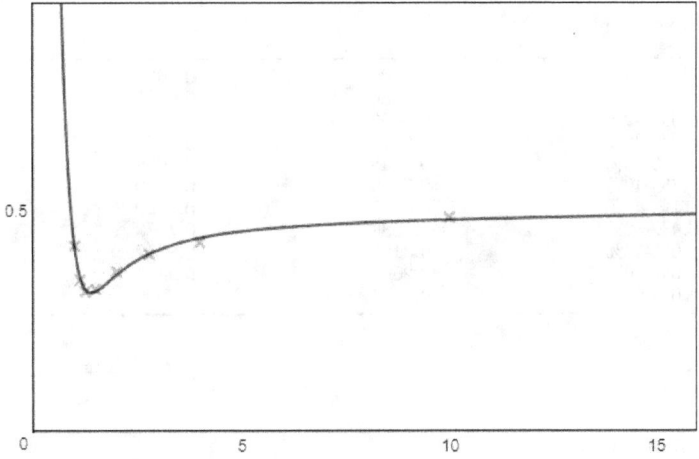

Figure 5.7.8: Earth's position over cumulative time period for different values of BCS

and the best fit is:

$$C\left(B_{cs}\right) = \ln\left(\frac{0.00225\left(B_{cs}\right)^{4} + 1.595\left(B_{cs}\right)^{3} + 0.06582\left(B_{cs}\right)^{2} - 1.5367\left(B_{cs}\right) + 1.38454}{B_{cs}^{3}}\right) \tag{5.7.16}$$

5.8 Chance of Human Emergence Recalibration

Earlier in section 5.3 we have derived a special case of the emergence of human sapiens assuming that the permissible range of human emergence falls between 23 and 38 percent dryland coverage and assumed that emergence can occur exclusively within this coverage range. Later, in section 5.4 we have defined how super-continent cycle's placement over the glaciation cycle placement reduces the chance on emergence. In section 5.5, we have derived the general expression for the weighted emergence rate across all dryland ranges. So now we can compute the weighted average emergence rate of human by taking into considerations of continental plate configurations, dryland to sea ratio, supercontinent cycle, and glaciation cycle across all dryland coverage ranges, not just between 23 and 38 percent. We will evaluate the soundness of our initial assumption by blindingly assuming that emergence can only occur between 23 and 38 percent of dryland coverage.

We use the following equation to calculate the emergence chance within the selected dryland coverage range:

$$E_{ratio} = \frac{\int_{Lower}^{U_{pper}} E_{mergence}dx}{U_{pper} - L_{ower}} \tag{5.8.1}$$

We use the following equations (defined in Section 5.2) to calculate the selected dryland coverage range translated into a percentage of continental plate configurations range out of all possible continental plate configurations:

$$A_{total} = \int_{0}^{100} y_{upper}dx - \int_{0}^{100} y_{lower}dx \tag{5.8.2}$$

$$A_{selectedrange} = \int_{Lower}^{U_{pper}} \left(S_{lowerwrap}\left(x\right) - S_{upper}\left(x\right)\right)dx \tag{5.8.3}$$

$$D_{ratio} = \frac{A_{selectedrange}}{A_{total}} \tag{5.8.4}$$

The computed results is listed in the table below:

Selection range	Lower to Upper dryland coverage	Emergence chance within selected range E_{ratio}	Dryland coverage range to continental plate ranges D_{ratio}	Weighted chance $E_{ratio} \cdot D_{ratio}$
0.00%	29.4%-29.4%	1.0000	0.0000	0.0000
3.40%	27.7%-31%	0.9780	0.0595	0.0582
5.88%	26.5%-32%	0.9663	0.1029	0.0994
14.5%	23.5%-38%	0.8822	0.2654	0.2341
17.36%	21%-38%	0.8448	0.3039	0.2567
22.0%	18%-40%	0.7549	0.3850	0.2906
37.53%	11%-48%	0.5025	0.6190	0.3110
46%	6.4%-52%	0.4080	0.7178	0.2929
60%	0%-60%	0.3103	0.8442	0.2619
70%	0%-70%	0.2665	0.9470	0.2524
84%	0%-84%	0.2221	1.0000	0.2221

The best fit for the data is formulated as:

$$W_{ratio}(x) = -2.44 \cdot 10^{-6}x^4 + 6.573 \cdot 10^{-4}x^3 - 0.06396x^2 + 2.46x - 1.14 \tag{5.8.5}$$

and the result is plotted:

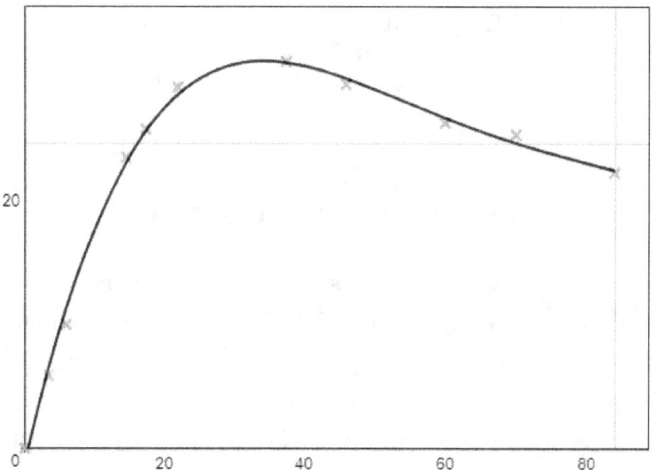

Figure 5.8.1: The weighted emergence chance for different range of dryland coverage

Earlier in the simplified model, we set that the lower and upper limit for habitability extending from 23.5% to 38% of dryland coverage based on the assumption of an ocean with 1 earth ocean mass. Based on this assumption, the chance of emergence of human is $W_{ratio}(14.5) = 22.99\%$. This initial assessment, in fact, does not deviate much from the weighted total average:

$$\frac{\int_0^{84} W_{ratio}(x)\, dx}{84} = 24.877\% \tag{5.8.6}$$

This weighted average line is drawn on the graph in comparison to the curve. In general, as the gap between the lower and the upper limit narrows, the total biodiversity is under represented and the chance of emergence drops to 0. On the other hand, as the gap between the lower and the upper limit widens, the chance of emergence does not grow much since 90% of biodiversity is represented between 10% to 50% dryland coverage.

Our conclusion to our assessment is as the follows:

If one were to consider all dryland coverage ranges, then one can no longer maintain the identical level of emergence as it is observed between 23% to 38% coverage ranges. In fact, the weighted emergence rate becomes only 22.21% vs. 88.22% for dryland coverage between 23% and 38%. If one were to increase the chance of emergence, one has to reduce the range of dryland coverages. A maxima is reached for dryland coverage between 11% to 48%. In this case, $\frac{\int_{0.1}^{0.48} E_{mergence} dx}{\int_{0}^{0.84} E_{mergence} dx} = 99.6\%$ of all emergence chance is captured within this range of dryland coverage, yet the weighted emergence chance lowered to only 50.25% relative to earth's max attainable set as 100% at dryland coverage of 30.88%.

The weighted total average of 24.877% can be interpreted conceptually as taking considerations of all dryland coverage ranges, the weighted emergence chance is at 24.877% relative to current max attainable chance at 30.88% dryland coverage. Alternatively, it can be interpreted as 100% emergence chance is clumped into 24.877% of all possible dryland coverages. As a result, our initial assessment by assuming that emergence can only occur exclusively within 23% to 38% ranges, which translated into 26.54% of all possible dryland coverages, is a pretty good guess. *However, we shall now adopt with confidence a better chance at 24.877%.* Nevertheless, we have shown that every 100 Myr, BCS helps to increase the overall biocomplexity on any planet by 2.783 times. Therefore, dryland coverage selection criterion is largely confined for planets initially formed within 100 Myr apart from each other across space, the criterion is largely overridden by the selection criterion by exponential growth of biocomplexity through time as defined by the T_{op} function in the earlier section. As a result, this selection criterion is already embedded into the earlier selection by T_{op} function. Its only other role is re-adjusting the earliest window for counting habitable earth, which is discussed in 8.1.

Chapter 6

Homo Sapiens Emergence Probability

6.1 Why Human Did not Appear Earlier

Ice age and its fluctuating climate pattern act as tremendous accelerators on the emergence and the diversification new species. In order to quantify the magnitude of acceleration, we can resort to the annual cranial capacity growth rate of Australopithecus Afarensis at 405 cc to that of Neanderthal at 1600 cc, and one obtains 0.0000376248 percent. This compares to the rate of growth of EQ of early mammals at 1 to that of Chimpanzees at 2.35, at an annual evolutionary growth rate of 0.0000014046, or 26.786 times faster than the background evolutionary rate. To put in perspective, this increase in growth rate is comparable to human's rate of progress during post-Industrial Revolution compares to that of hunter-gatherers.

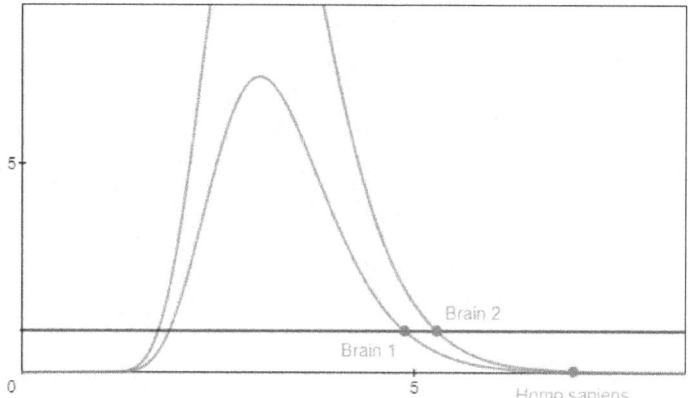

Figure 6.1.1: Assuming biocomplexity doubled during the ice age but keeping the mode fixed, then the number of species with previous cranial capacity at Brain 1 doubled, and the number of species with cranial capacity Brain 2 reaches parity with previous Brain 1. Homo sapiens brain size is at 7. If ice age at its best doubles biocomplexity and but only shifts brain size $5 < x < 7$, and the horizontal displacement between point Brain 2 and Brain 1 falls short aforementioned speed shifted horizontally, then other factors such as the new distribution predisposes rescinding unevenly at species shared greater number of traits, or sexual selection may accelerate brain size toward the Homo sapiens value at 7.

In earlier ice ages, clearly illustrated from the formation of supercontinent Pangea and the Karoo ice, which drastically alter the humid climate from earlier epoch and earth entered a period of dry, cold climate. These ice ages also gave rise to novel adaptations but no adaptations of intelligent, tool-using species. Upon the drastic climatic change, amphibians which requires adaptation to moisture and close proximity to bodies of water to procreate the young evolved the mechanism of nurturing the young within hard-shelled eggs and gave rise to the reptiles. Moreover, the emergence of seed gave rise to Gymnosperms, which protects the plant seed from drying out, providing additional protection. Earliest species adapted to arboreal locomotion such as late

Permian synapsid *Suminia getmanovi* and tree climbing dinosaurs such as *Deinonychus* exists during the early Cretaceous. However, neither species eventually transitioned to upright walking species with flexible hands. Besides the fact that no ice age with fluctuating climate pattern occurred since Permian before the Quaternary ice age, which is unable to turn arboreal species to a ground walking one, as in human's case. A very important and often overlooked fact is that the evolution of intelligence is not particularly beneficial for an organism adapting to its environment. When the total biological diversity is low in the earlier epochs of earth's history, organism's strategy is to maximize their body size, their running speed, their visual acuity. If any species adapts for flexibility, it may well be flexibility specialized in a particular way. For example, both chameleon and octopus can alter their skin color to fit its environment quickly, which hides the species from both predators and preys. But only as biological diversity bloomed during the Cenozoic era, with the abundance of fruit-bearing trees and wild berries to provide energy, furry animals with skin hide capable of providing warmth, and beehive to provide honey, can the evolution of intelligence benefits outweigh its costs. An intelligent species is able to combine different species' material based on its strength and characteristics to accomplish yet unseen impossible tasks, such benefit grows exponentially over time as the manipulative power and potential search space becomes ever greater. From this perspective, the evolution of intelligence, even though a passive evolutionary process not seeking any goal, will be inevitable. On a further note, the evolution of intelligent creatures in both ocean and land are equally likely, as indicated by the EQ of dolphins and killer whales. However, only the cohort of terrestrial species have the chance to develop an industrial civilization, provided with a cataclysmic event such as the ice age (Ice age has a moderate effect on ocean temperature not as drastic climate shifts as those on land).

6.2 Why intelligent species can not emerge from Arthropods

Arthropods were the first group of species to colonize terrestrial space. During the Carboniferous period, arthropod were the dominant terrestrial species and grown to enormous sizes. However, insects never regained its status after the emergence of vertebrates on land. In order to evaluate the inevitability of the dominance of vertebrates, one has to resort calculation on allometry. The bone strength, the bone maintenance cost, and breathing efficiency of arthropods and vertebrates of varying sizes are compared.

The vertebrate's bones strength is expressed as in proportion to their body and limb's cross sectional area. The weight of their body is distributed and supported by their bones. The heavier the species, the thicker their bones. It is expressed in the simplest form as the cross sectional area of their body and the cross sectional area of their bones:

$$S_{vertebrate} \propto \pi r^2 \tag{6.2.1}$$

For arthropods, its exoskeletons support their weight. The exoskeleton has a negligible cross sectional area compares to their body cross sectional area. Therefore, it is expressed in the simplest form as the perimeter of the cross section of their body:

$$S_{anthropod} \propto 2\pi r \tag{6.2.2}$$

The cost of vertebrate bone maintenance is expressed as in proportion to the volume of the bone mass where $h = 2r$:

$$C_{vertebrate} \propto \pi r^2 h \propto \frac{4}{3}\pi r^3 \tag{6.2.3}$$

The cost of arthropod bone maintenance is expressed as the total surface area of its body composing all exoskeletons:

$$C_{anthropod} \propto 4\pi r^2 \tag{6.2.4}$$

The breathing efficiency of vertebrates is expressed as the total volume of lungs occupied its torso, we simply assume that 20% of a typical vertebrate specie' torsal volume is reserved for its lungs so that breathing efficiency stays constant as the body size increases. This is expressed as:

$$R_{vertebrate} \propto \frac{\frac{1}{5} \cdot \frac{4}{3}\pi r^3}{\frac{4}{3}\pi r^3} = \frac{1}{5} \tag{6.2.5}$$

The breathing efficiency of arthropods is expressed as the surface area to total body size volume. Since only small openings on the thorax and the abdomen were used for gas exchange via diffusion and pumping, we set 5% of a typical arthropod's exoskeleton's surface area is used for breathing.

$$R_{anthropod} \propto \frac{1}{20} \cdot \frac{4\pi r^2}{\frac{4}{3}\pi r^3} = \frac{3}{20} \cdot \frac{1}{r} \tag{6.2.6}$$

Finally, the combined factors is expressed as:

$$V_{rtebrate} \propto \frac{S_{vertebrate}}{C_{vertebrate}} \cdot R_{vertebrate} \tag{6.2.7}$$

$$A_{anthropod} \propto \frac{S_{anthropod}}{C_{anthropod}} \cdot R_{anthropod} \tag{6.2.8}$$

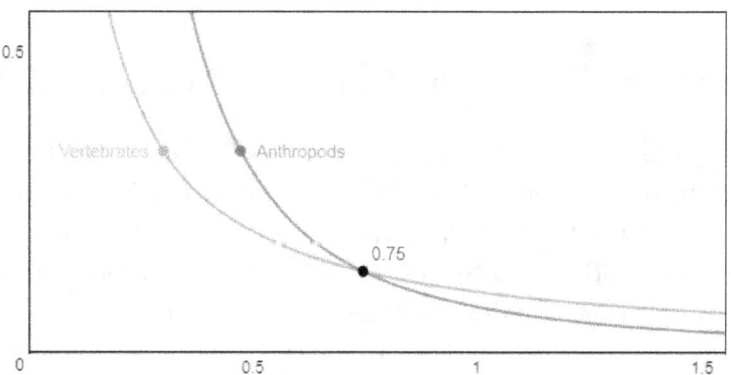

Figure 6.2.1: Allometric analysis of vertebrates vs. arthropods

Though it is shown that smaller body size is favored by both vertebrates and arthropods alike due to increasing bone maintenance cost, vertebrates holds a comparative advantage over arthropods over larger size. The graph shows that crossover comes at a body size radius of 0.75 m. That is, in our very rough allometric analysis, we are able to demonstrate that arthropod body plan is best suited for smaller body size and vertebrates for larger size. In the absence of vertebrates during Carboniferous, arthropods can still grow to larger sizes and exploit ecological niches. However, as vertebrates claim the land, larger versions of them were out-competed but they are still well-adapted at the smaller niches. By exploiting small niches, they share the majority of the biodiversity due to their smaller sizes and low energy requirements compares larger vertebrates. One may then ask, vertebrates surely out-competed arthropods at larger sizes, but if vertebrates does not exist at all, will arthropods eventually to become intelligent, tool-using species? Unfortunately, this is still highly unlikely. Arthropods do not have a closed circulatory system as vertebrates does. Although oxygen can be delivered to tissues through book lungs and skins, nutrients still have to be delivered. They achieves such delivery through their own body movements. Without a strong heart to pump nutrient and creates internal blood pressure to work against the pull of gravity, nutrients delivery for a raised head would be costly. This is exactly what one observes in nature. Out of all living species of insects, only the order of Mantodea walk in a semi-bipedal posture. This only accounts to $\frac{2,400}{5,500,000} = 0.044\%$ of all insects. This is in direct contrasts to vertebrates. Out of our cohorts of terrestrial vertebrates, $\frac{10,089}{25,483} = 39.59\%$ are bipedal. Without gaining a bipedal posture, it is impossible to evolve into an intelligent species. It is only possible, when a lineage of arthropods evolves closed circulatory system. Even then, as we have shown allometrically, its overall flexibility will trail behind vertebrate equivalents with comparable large size.

Of course, evolution has only experimented with species growing exoskeletons and internal skeletons on earth, the two extremes of a whole spectrum of possibilities. It is possible whole classes of species with both exoskeletons and internal skeletons, or with partial exoskeletons and partial internal skeletons can evolve. All different combinations are possible. However, given earth like environment and condition, vertebrates still likely to dominate given its high strength and low maintenance cost. Since the range of possibilities are endless on different planets, it will be no surprise that given the conditions on some of all exoplanets where natural selection favors some portion of the entire spectrum over others.

6.3 The Probability of the Hominid Lineage

In order to compute the probability of giving rise to Hominid lineage, we can not simply pick our denominator as the current total number of living species of birds, mammals, and reptiles. Neither can we use computational molecular biology, we have shown already that species can adapt and evolve quickly to changing environment (Chapter 4). A suitable sequence of quick environmental change can accelerate the evolution of Homo sapiens in timescale much faster than even those we have observed. If earth's environment is the brake, the pacemaker, and the cookie cutter of evolution, then the success and the pervasiveness of each trait leading to Homo sapiens adapted to each niche is the key in understanding our likelihood of emergence. What we really need to evaluate is the probability of each particular trait that is critical for the emergence of an intelligent tool-using species. Since each trait is largely independent of other traits, then the probability of Homo sapiens can be defined as the product of the probabilities of each trait. Caveats must be thrown, however, because the probability of each trait inevitably depends on current cohorts of species which adapted to the current climate, the probability of animals possessing these traits can fluctuate throughout different epochs of earth's history. (Even if one were to use the current data samples, the megafaunal mass extinctions of 10,000 BP may have distorted our calculation since extinctions across all major land masses are correlated with the arrival of human.) Each of these traits has been observed in at least one species since Mesozoic, but no species possessed all of them at once until the emergence of man. This observation indicates that each trait is a local evolutionary maximum where life converged to quickly and arose early. Certain traits, such as opposable thumbs and bipedal locomotion, can be inversely correlated. That is, advancing forest gives more opportunity to thriving arboreal residents but squeezes the living space of land-based bipeds on grasslands. (This is equivalent to total biodiversity remain fixed and the proportion of land-based bipeds and arboreal residents varies) If it holds, then, the total probability giving rise to Homo sapiens does not deviate in orders of magnitudes throughout all time periods because certain characteristics become more common at the expense of rendering other traits rarer. However, in the long run, more species are appearing from geologic data records, and if more species are appearing within every niche, then the total probability should remain constant if every other condition stays the same. (This is equivalent to total biodiversity increased by BCS over 100 Myr period, yet the proportion of land-based bipeds and arboreal residents remain fixed) Ultimately, to adequately sample the probability of each trait throughout the entire Mesozoic and Cenozoic, its mean value, and its relationship with climate change, and the general trend over time is critical. The most tantalizing problem though is that, even by sampling data across all time period, our picture may remain incomplete in regards to the actual probability giving rise to intelligent, tool-using species. It is possible that in the real course of evolution, not all possible paths and combinations were adopted by nature. In fact, nature only tried the combination of a particular subset of opposable thumb (excluding those of panda and tree frog and favoring the primate family). It tried a particular type of bipedal locomotion (excluding bipedal locomotion observed in birds, hopping animals such as the kangaroo), a particular type of language communication using larynx vocalization, and a particular type of brain structure (mammalian brain with neocortex). These types of traits may be easiest to evolve first or easiest to evolve given the available opening ecological niche. It is also possible for nature to adopt all possible combinations but it requires significantly more time than we observed and requires a length of period beyond the habitability of the planet. So in a sense, every trait ultimately becomes dominant within all groups, and currently observed percentage is its transitional chance from its initiation and its final adaptation by all surviving species after rounds of competitions. All of these are problems for future paleontologists and researchers and is beyond the scope of this paper.

6.3.1   Binocular Vision

The first trait we need to compute for its probability is the binocular vision. Binocular vision creates depth perception and is found in Primates for fruit searching and arboreal locomotion. Carnivora and birds of prey used depth perception for prey capture.

Mammal	Species	Bird	Species
Primate	450	Accipitriformes	261
Carnivora	286	Cathartidae	7
Bats	1,240	Strigiformes	200
–	–	Coraciimorphae	6
–	–	Cariamiformes	1
–	–	Falconiformes	75
Sum Total			**2,526**

Table 6.3.1: Species breakdown by binocular vision

There are 450 extent Primate species (including Homo sapiens), 286 extant species of Carnivora, and 1,240 species of bats. For the birds of prey, there are 261 species of Accipitriformes (1 species of Sagittariidae, 4 species of Pandionidae, 256 species of Accipitridae), 7 species of Cathartidae (New World vultures), 200 species of Strigiformes (owls), 6 species of Coraciimorphae, 1 species of Cariamiformes, and 75 species of Falconiformes (one species of Cariamidae, 63 species of Falconidae, 11 species of Polyborinae). We have an added total of 2,526 species with binocular vision.

6.3.2   Large Cranial Capacity

Next, we compute the total number of species possessing a large cranial capacity in a ratio relative to their body mass, or what one calls as high EQ. For certain animals where EQ cannot be obtained, those that passed the mirror test, which supposedly tests self-awareness, is used as a criterion for inclusion. There are 56 extant species of dolphins (with 3 species of the humpback whale, fin whale, and sperm whale). 120 species of Corvidae (including crows, raven, and magpies). 167 species of tegu lizards. 79 species of monitor lizards. 600 species of anolis lizards, 200 species of owls, 41 species of falcons, and 7 species of Hominidae, totaling 1,270 species.

Mammal	Species	Bird	Species	Reptile	Species
Dolphin	56	Corvidae	120	Tegu lizard	167
Hominidae	7	Owl	200	Monitor lizard	79
–	–	Falcons	41	Anolis lizard	600
Sum Total					**1,270**

Table 6.3.2: Species breakdown by cranial capacity

6.3.3   Language

Next, we compute the number of species possessing language communication skills. Communication can take place within water such as those generated by dolphins and whales, and in the air such as birds and bats. Species with simple alarm calls are not included in the list. The species included are those that evolved relatively more complex call systems that beyond a mere reflex, that is, it is able to synthesize new sounds based on different combination of existing patterns and symbols. The diverse array of sound symbol manipulation may reflect an advanced overall neural developments such as human, or merely an advanced functioning of a particular organ such as the the tongue of greyparrots' used in mating and social signaling. There are 1 species of Hominidae, 51 species of dolphin, 44 species of whales, 4,000 species of birds, 1,240 species of bats, totaling 5,336 species.

Mammal	Species	Bird	Species
Hominidae	1	Song birds	4,000
Dolphin	51	–	–
Whales	44	–	–
Bats	1,240	–	–
Sum Total			5,336

Table 6.3.3: Species breakdown by language communication

6.3.4 Bipedal

Next, we compute the number of species capable of bipedal locomotion. There are 65 species of macropods, 22 species of kangaroo rats and mice, 2 species of springhares, 4 species of hopping mice, 8 species of pangolins, 1 Hominidae, and all species of birds (10,000 species), totaling 10,089 species.

Mammal	Species	Bird	Species
Macropods	65	Birds	10,000
Kangaroo rats and mice	22	–	–
Springhares	2	–	–
Hopping mice	4	–	–
Pangolins	8	–	–
Hominidae	1	–	–
Sum Total			10,089

Table 6.3.4: Species breakdown by bipedal locomotion

6.3.5 Thumbs

Next, we compute the number of species possessing opposable thumbs. All such species have adapted to arboreal locomotion. There are 145 species of old world monkeys, 7 species of Hominidae, 18 species of gibbons, 2 species of giant pandas, 6 species of pencil-tailed tree mice, 4 species of Vandeleuria, 9 species of hopping mice, 28 species of Phalangeridae, 1 species of koala, 103 species of opossums, totaling 353 species.

Mammal	Species	Mammal	Species
Old world monkeys	145	Pencil-tailed tree mouse	6
Hominidae	7	Vandeleuria	4
Gibbons	18	Hopping mouse	9
Giant panda	2	Phalangeridae	28
Opossums	103	Koalas	1
Sum Total			353

Table 6.3.5: Species breakdown by opposable thumbs

6.3.6 Social

Next, we compute the number of species that developed social organizations. The social organization has to be complex enough to extend beyond the immediate family members. The ability to organize and cooperate for the common good is a necessary step before the adaptation to a much more complex organization as those created by Homo sapiens. There are 3 species of phodopus, 1,240 species of bats, 21 species of cockatoos, 42 species of Callitrichidae, 18 species of tamarins, 22 species of marmosets, 45 species of corvus, 53 species of dophins, 4 species of elephants, 1 species of starling, 103 species of gerbils, 1 species of guinea pigs, 7 species of hominidae, 7 species of horse, 4 species of hyenas, 1 species of killer whales, 63 species of rabbits, 1 species of lion, 1 species of meerkat,1 species of orange-fronted parakeet, 3 species of paracheirodons, 152 species of Tetra, 21 species of penguins, 10 species of Psittacidae, 1 species of sea otter, 64 species of rats, 1 species of wolves, and 1 species of Zebra finch. The number of social species totaled 1,901 species. It is 3,586 species if one considers all primates and 1,000 species of migratory birds as social animals.

Mammal	Species	Mammal	Species	Bird	Species
Phodopus	3	Gerbil	103	Cockatoos	21
Bats	1240	Guinea pigs	1	Corvidae	120
Primate	445	Hominidae	1	Starling	1
dwarf mongoose	1	Horse	1	Orange-fronted parakeet	1
naked mole rat	1	Hyena	4	Penguin	21
Dolphin	53	Killer whales	1	Psittacidae	148
Elephants	4	Rabbits	1	Zebra finch	1
Meerkats	1	Lion	1	Stork	19
Sea otter	1	Rats	64	Crane	15
Wolf	1	Mole rats	1	Migratory birds	1,000
Elephant seals	2	Zebra	3	Common pheasant	1
Red deer	1	Wildebeests	2	Greater rhea	1
African buffalo	1	Sheep	1	-	-
Bison	1	Goat	1	-	-
Lion	1	-	-	-	-
Sum Total					**3,586**

Table 6.3.6: Species breakdown by prosocial characteristics

6.3.7 Omnivorous Feeding

Finally, we compute the number of species possessing omnivorous feeding behaviors.

First, we count the number of omnivorous mammals. Among mammals, there are 1 species of pig, 11 species of badgers, 8 species of bears, 4 species of coati, 20 species of civets, 17 species of hedgehogs, 103 species of opossums, 12 species of skunks, 6 species of sloths, 285 species of squirrels, 1 species of raccoon, 25 species of chipmunks, 30 species of mice, 64 species of rats, 7 species of Hominidae, 385 species of tree shrews, and 43 species of Erinaceidae, 103 species of Gerbil, totaling 1,125 species. The majority of mammals are either herbivores or insectivores, the proportion of omnivores are lower than birds.

Species	Number	Species	Number	Species	Number
Pig	1	Opossum	103	Mouse	30
Badger	11	Skunk	12	Rats	64
Bear	8	Sloth	6	Hominidae	7
Coati	4	Squirrels	285	Tree shrews	385
Civet	20	Raccoon	1	Erinaceidae	43
Hedgehog	17	Chipmunk	25	Gerbil	103
Sum Total					**1,125**

Table 6.3.7: Omnivorous mammal species

Secondly, we count the number of omnivorous birds. We do not have an accurate description of all species of birds with an omnivorous diet, but we do have the catalog of species of birds that have more specialized dieting habits. Therefore, we will work our way backward. Among carnivores, there are 60 species of eagles, 200 species of owls, 31 species of shrikes. Among Crustacivores, there are 1 species of crab plover and 212 species of rails. 16 species of detritivores. Among folivores, there are 1 species of hoatzin and 6 species of mousebirds. Among frugivores, there are 26 species of turacos, 240 species of tanagers, 42 species of birds-of-paradise. Among granivores, there are 146 species of geese and 25 species of grouses, 142 species of estrildid finches. Among herbivores, there are 8 species of whistling ducks, 1 species of ostrich, and 1 species of mute swan. Among insectivores, there are 177 species of cuckoos, 83 species of swallows, 150 species of thrushes, 25 species of drongos, and 240 species of woodpeckers. Among Nectarivores, there are 1,039 species of hummingbirds, 132 species of sunbirds, and 58 species of lorikeets. Among piscivores, there are 4 species of darters, 5 species of loons, 8 species of pelicans, 20 species of penguins, and 19 species of storks. Among sanguinivorous, there are 2 species of oxpeckers and 1 species of sharp-beaked ground finch. Among Saprovores, there are 16 species of vultures and 37 species of crows. The number of non-omnivorous bird species totaled 3,174. Since there are 10,000 species of birds, the number of omnivorous bird species totaled 6,826, proportionally significantly higher than mammals.

Carnivores		Crustacivores		Detritivores	
Eagles	60	Crab plover	1	–	16
Owls	200	Rails	212	–	–
Shrike	31	–	–	–	–
Folivores		**Frugivores**		**Granivores**	
Hoatzin	1	Turacos	26	Geese	146
Mousebirds	6	Tanager	240	Grouse	25
–	–	Birds-of-paradise	42	Estrildid finches	142
Herbivores		**Insectivores**		**Nectarivores**	
Whistling ducks	8	Cuckoo	177	Hummingbirds	1,039
Ostrich	1	Swallows	83	Sunbirds	132
Mute swan	1	Thrush	150	Lorikeets	58
–	–	Drongos	25	–	–
–	–	Woodpecker	240	–	–
Piscivores		**Sanguinivorous**		**Saprovores**	
Darter	4	Oxpecker	2	Vultures	16
Loon	5	Ground finches	1	Crow	37
Pelican	8	–	–	–	–
Penguin	20	–	–	–	–
Stork	19	–	–	–	–
Sum Total					**3,174**

Table 6.3.8: Non-omnivorous birds

Thirdly, we count the number of omnivorous reptiles. Turtles (327 species) are predominantly omnivorous. They are exclusively carnivorous before reaching adulthood and herbivorous once reaching adulthood. Tortoises (155 species), or land-based turtles, are herbivorous. 98 percent of lizards (the rest are herbivores), snakes, and worm lizards are carnivorous, totaling 9,600 species. All species of Crocodilia (25 species) are carnivorous.

Omnivorous	Species	Carnivores	Species	Herbivores	Species
Turtles	327	Snakes	–	2% of Lizards	30
–	–	98% of Lizards	–	–	–
–	–	Worm lizards	–	–	–
Sum Total					**10,108**

Table 6.3.9: Reptile species breakdown by feeding behaviors

Among 10,108 species of reptiles, only 327 species are omnivorous. Reptiles are predominantly carnivorous. The total number of omnivorous species then numbered 8,175 species.

Before we proceed, we do need to verify that the traits are independent from each other. In order to confirm, one needs to take a combinatorial approach and find the product of the probability of two traits out of all listed and denote the probability as $P_{predicted}$. One then needs to manually examine the species that indeed share both traits and divided by the total number of species we counted, which is 25,483 and denote the probability as P_{actual}. One could take the combinatorial up to the product of no more than three traits because the product of thumb, brain, and binocular vision yields the predicted number of species among 25,483 is only 1.67. If one were continue to multiply with additional traits, it will take a larger cohorts of all extent species to verify the prediction which is not unavailable to us. If $P_{predicted} = P_{actual}$, it implies that the two traits are exactly independent from each other as mathematics would predict. If $P_{predicted} > P_{actual}$, it implies in reality these two traits are more unrelated than mathematics would predict. if $P_{predicted} < P_{actual}$, it implies that two traits are dependent on each other, so they are not independent variables. We define relatedness by $R = \frac{P_{actual}}{P_{predicted}}$ and ranked them in the order of the most related to the least. Finally, we want to find the total product of all R, as:

$$T = \prod_{n=0}^{m} R_n \tag{6.3.1}$$

Whereas:

$$T \neq \frac{P_{actual}(Trait_0, Trait_1, Trait_2...Trait_n)}{\prod_{n=0}^{m} Trait_n} \tag{6.3.2}$$

T does not equal to the number of observed species over the total product of the probabilities of all traits. It can be illustrated from a simple example by assuming one wants to define T as the total product of all R for cross examination of the traits of omnivorous, binocular, bipedal, and social.

$$R_0 = \frac{P_{actual}\left(O_m, b_{inocular}\right)}{O_m \cdot b_{inocular}} \tag{6.3.3}$$

$$R_1 = \frac{P_{actual}\left(O_m, b_{ipedal}\right)}{O_m \cdot h_{ipedul}} \tag{6.3.4}$$

$$R_2 = \frac{P_{actual}\left(O_m, S_{ocial}\right)}{O_m \cdot S_{ocial}} \tag{6.3.5}$$

$$T = \prod_{n=0}^{m} R_n = \frac{P_{actual}\left(O_m, S_{ocial}\right) \cdot P_{actual}\left(O_m, b_{ipedal}\right) \cdot P_{actual}\left(O_m \cdot b_{inocular}\right)}{O_m^3 \cdot S_{ocial} \cdot b_{ipedal} \cdot b_{inocular}} \tag{6.3.6}$$

$$T \neq \frac{P_{actual}\left(O_m \cdot S_{ocial} \cdot b_{ipedal} \cdot b_{inocular}\right)}{O_m \cdot S_{ocial} \cdot b_{ipedal} \cdot b_{inocular}} \tag{6.3.7}$$

Trait 1	Trait 2	$P_{predicted}$	P_{actual}	R
thumb	binocular	0.00137	0.00679	4.94412
social	binocular	0.01395	0.06636	4.75719
social	thumb	0.00195	0.00667	3.42227
social	language	0.02844	0.09453	3.32407
omnivorous	bipedal	0.12861	0.26790	2.08308
language	binocular	0.02003	0.04140	2.06663
brain	binocular	0.00472	0.00973	2.06087
bipedal	language	0.08001	0.15701	1.96229
omnivorous	language	0.06565	0.10719	1.63269
social	brain	0.00670	0.00712	1.06242
omnivorous	thumb	0.00450	0.00432	0.95927
social	bipedal	0.05571	0.05298	0.95088
omnivorous	social	0.04571	0.03487	0.76279
bipedal	brain	0.01886	0.01421	0.75317
language	brain	0.00963	0.00685	0.71125
bipedal	binocular	0.03924	0.02162	0.55096
thumb	brain	0.00066	0.00027	0.41625
omnivorous	brain	0.01548	0.00498	0.32204
thumb	language	0.00280	0.00075	0.26633
omnivorous	binocular	0.03220	0.00263	0.08165
thumb	bipedal	0.00548	0.00039	0.07155

Table 6.3.10: Cross examination

The ranking indicates that opposable thumbs, binocular vision, social, and language are not independent from each other. They are more likely to find on the same species occupying the arboreal niche. Social & language are highly related because the majority of the song bird species are social. Omnivores & bipedalism and language & bipedalism are related because the majority of the bipedal species are song birds feeding on an omnivorous diet. Language & binocular vision are related because mini bats using echolocation and binocular vision to capture its tiny insect prey comparable to its own body size. Large cranial capacity & binocular vision is related because predatory bird species feeding on meat also requires considerable flexible intelligence to catch its prey comparable to its own body size. High protein intake also enables predatory birds to gain a larger brain, completing a positive feedback loop. On the other end of the spectrum, one finds that bipedalism & brain are more unrelated than prediction since most bipedal bird does not have a large brain. Language & brain is not as related because many mini-bats species uses echolocation for survival but no large brain is needed to capture insects which are proportionally small compare to its own body size. Bipedalism & binocular vision is unrelated mainly because most birds species are non-predatory on other bird species. Additionally, all bat species with binocular vision are non-bipedal. Thumb & brain is unrelated because the majority of the intelligent species such as dolphins, corvidae, owls do not possess thumb and does originate from arboreal habitats. Omnivorous diet & brain are not related since only high protein intake with meat guarantees more energy can be invested to the development of the brain. Human is utterly an exception in this case. Some speculate that human's initial enlargement of the brain is due to extraction of bone marrow by using tool. This is a rather peculiar route at achieving a larger brain rather than the typical path of first becoming a carnivore. Thumb & language are not related because many species with a highly developed auditory capacity such as song birds and mini-bats do not possess opposable thumbs. Language seems to be evolved in the settings of dim light environment (echolocation in nocturnal bats and dolphin) and social interactions (song birds, raven, and human) in species experiencing lesser predatory pressures. Most arboreal primates are diurnal and social but quiet to avoid predation in open day light. The benefit of language communication must outweigh the cost of broadcasting one's location. This can be achieved in species subject to fewer predations, or species formed a very strong defense system against predators. Omnivorous diets & binocular vision are unrelated because binocular vision is essential and critical on the survival of predatory carnivorous species. Binocular vision is evolved in the arboreal habitat, and primates using binocular vision to thrive on the trees and feeds on a frugivorous diet. Therefore, frugivorous diet and carnivorous diet are the peaks in a landscape of binocular vision utilization and omnivorous diet sits in the deep valley. There is no evolutionary pressure for omnivores to acquire depth perception for capturing its prey. Ultimately, thumb and bipedal are highly unrelated because all bird species does not possess a thumb and all mammalian species possessing opposable thumb thrives in an arboreal habitat.

Finally, we find the total product of all traits cross-examined:

$$T = \prod_{n=0}^{m} R_n = 0.07 < 1 \tag{6.3.8}$$

and we find the total product is less than 1, this implies that though some of the traits cross-examined are dependent on each other, so they are not independent variables. Other traits such as thumb & bipedalism is so rare in nature that these two traits are more unrelated than mathematics would predict. Therefore, the overall result concludes that all traits possessed by human is largely independently related from each other.

We then apply the traits cross examination for 3 traits combined (we only listed some of the most related and the least related):

Trait 1	Trait 2	Trait 3	$P_{predicted}$	P_{actual}	R
binocular	thumb	social	0.0002	171	34.7279
binocular	language	social	0.0028	1055	14.6860
binocular	cranial	bipedal	0.0019	242	5.0794
cranial	language	social	0.0014	175	5.0543
binocular	cranial	thumb	0.0001	7	4.1993
language	bipedal	omnivorous	0.0260	2731	4.1239
language	social	omnivorous	0.0092	890	3.7794
cranial	thumb	social	0.0001	7	2.9580
cranial	social	omnivorous	0.0022	121	2.1804
bipedal	social	omnivorous	0.0181	923	2.0003
binocular	bipedal	thumb	0.0005	1	0.0722
binocular	social	omnivorous	0.0045	8	0.0693
bipedal	thumb	social	0.0008	1	0.0508
language	thumb	omnivorous	0.0009	1	0.0432
binocular	cranial	language	0.0010	1	0.0411
language	bipedal	thumb	0.0011	1	0.0354
bipedal	thumb	omnivorous	0.0018	1	0.0220
binocular	bipedal	social	0.0055	1	0.0071
binocular	language	omnivorous	0.0065	1	0.0060
binocular	language	bipedal	0.0079	1	0.0049
binocular	bipedal	omnivorous	0.0127	1	0.0031

Table 6.3.11: Cross examination

The results shows both more correlated features and more independent features at both extremes. Binocular & thumb & social traits defines the primates. Binocular & language & social defines bats. Binocular & cranial & bipedal defines predatory birds. Binocular & cranial & thumb defines the Homininid. At the other extremes, binocular & bipedal & omnivorous is rare in any species because predatory birds always eat meat, omnivorous bird not binocular, chimps are not bipedal. Binocular & language & bipedal is rare because predatory birds do not communicate, song birds do not possess binocular vision, hominid are not bipedal except human. Binocular & language & omnivorous are rare because song birds do not possess binocular vision. Binocular & bipedal & social are rare because predatory birds not social. Bipedal & thumb & omnivorous rare because birds did not evolve from the trees and possess no thumb. Language & bipedal & thumb are rare because birds do not possess thumbs and primates generally are quiet and non-bipedal. Binocular & cranial & language is rare because most Homininid produces no speech, predatory birds possess limited language capability, bats has small brain, dolphin has no binocular vision. Language & thumb & omnivorous is rare because birds possessed only wings and most primates are non-omnivorous. Bipedal & thumb & social are rare because birds only possess wings, and primates except human are non-bipedal.

The overall result concludes that all traits possessed by human is largely independently related from each other.

$$T = \prod_{n=0}^{m} R_n = 3.5791 \cdot 10^{-14} < 1 \tag{6.3.9}$$

There is 25,483 total number of extant species of birds, mammals, and reptiles (5,450 species of mammals, 9,925 species of birds, and 10,108 species of reptiles). So the total probability is computed as the follows:

$$p = \frac{1}{\prod_{n=0}^{m} Trait_n} \tag{6.3.10}$$

$$p = \left(\frac{25,483}{2,526}\right)\left(\frac{25,483}{1,214}\right)\left(\frac{25,483}{353}\right)\left(\frac{25,483}{10,089}\right)\left(\frac{25,483}{5,150}\right)\left(\frac{25,483}{8,278}\right)\left(\frac{25,483}{3,586}\right) \tag{6.3.11}$$

$$p = \frac{1}{4,179,613.11129} \tag{6.3.12}$$

$$= 2.3925659466 \times 10^{-7}$$

Once we computed the total probabilities of all traits that made Homo sapiens unique as a tool using intelligent species, we find that nature needs to experiment on average 4,179,613 speciation attempts to create a species similar to the Hominid family per any particular permutation paths. However, there are 7!=5040 possible permutation paths for 7 out of 7 traits, Therefore, if every permutation path (ordering of trait acquired does not matter) can lead into a human equivalent species, we need only 4179613/5040= 829 speciation attempts to create a species similar to the Hominid family.[1] However, careful examination reveals that the ordering of the evolved characteristics of a human-like creature is also important. Homo sapiens are lucky enough that we took one of the shortest paths of trait acquisition. *Although in theory it is possible to start a transition toward Homo sapiens from any of the 7 listed traits, we will show in later section, some are very strong dead ends and others less so. Some traits have to be lost first in order to gain others before it is regained. This leads to extra steps that takes longer time in evolution.* In fact, at least two traits a large, complex brain and bipedal locomotion cannot be the initial conditions of the shortest paths of trait acquisition for a human-like creature. At the same time, certain traits such as the evolution of opposable thumb have to occur before the emergence of large brain and bipedalism. Other traits, such as language, sociality, and binocular vision, within the more relaxed assumptions, can happen either before or after the evolution of opposable thumb. Bipedal locomotion as observed in kangaroo and birds offers great advantages that these species will not sacrifice their existing beneficial feature to trade for a lesser one just for the sake of evolving opposable thumbs. From a mathematical perspective, in order for bipeds to evolve toward a tree climbing one, it has to give up a huge local optimum choice, climbing a high cost hill before getting into another local optimum, which is somewhat an inferior choice than the one it started. In case of the large brain, or high EQ, all organism evolved high EQ requires some carnivorous diet. A species with high EQ adapt to arboreal lifestyle have to forego its carnivorous or omnivorous diet to become almost exclusively herbivorous. Such lifestyle eventually led to a reduction of EQ by natural selection, a large brain relative to body mass can no longer be maintained because the energy intake has been lowered. In some species such as the genus Homo and Elephant, the switch can be even more absurd. For

---

[1]The expected emergence of human will not be further delayed even if other non-human traits such as wing, feather, snout, and horn are taking into consideration during the course of evolution. The chance of other traits combined leading to other species can be either higher or lower than human's case. Therefore, its emergence can be earlier or later than human arrival. As a matter of fact, the very time required before our arrival is the time spent on the evolution of other species possessing other traits that have higher chance of emergence than ourselves.

species with such a large brain and body mass, trees branches, in general, do not have the strength to support. As a result, out of the seven traits listed, two of them can to be excluded from the list as non-starting conditions for the shortest paths. Since ordering matters, the rest of possible choices with anyone as the initial starting traits that lead to homo sapiens can be expressed as a simple permutation n!. 5! equals 120 possible paths leading to Homo sapiens. However, 7! equals 5,040 possible paths is also important. It shows that only $\frac{1}{42}$ out of all paths leading to Homo sapiens (not starting with big brain and bipedalism) as the shortest paths. If we consider at least some species with large EQ with a small light body can somehow adjust such as anolis lizards, we can also consider excluding bipedal trait only, that leaves us 6! which equals 720 possible paths leading to Homo sapiens, showing $\frac{1}{7}$ out of all paths leading to Homo sapiens. We treat those two cases as the upper and the lower bound determined that on average there is $\frac{1}{14}$ chance leading to Homo sapiens out of a total path of 5,040.

We also show that in the most stringent case and more likely case, human emergence chance for the shortest paths can drop to as low as $\frac{1}{1008}$. This can happen if one assumes that 6 out of 7 traits has to be occur in sequence. The species has to be in an arboreal habitat so that it first evolves opposable thumbs. Before it leaves the habitat, it has to gain binocular vision. In theory it could gain binocular vision later as a carnivorous species, but acquiring carnivorous diet and then evolves omnivorous diet takes an extra step. After it left the trees, it has to gain bipedal locomotion. After which, a land based omnivorous diet is required to fuel the growth of brain. Only once a large brain appeared it can gain advanced language capability. One may argue that language as exhibited by birds and bats does not require extremely large brain. However, language as a trait is primarily exhibited by non-ground based species facing fewer predators. Incessant, loud communication noise easily expose one's location to predators. Only when a terrestrial creature starts to face fewer predators the trait becomes advantageous. The remaining trait, social, can be placed anywhere except it has to precede the development of brain and language. Therefore, 5 out of 7 total possible paths can be deemed the shortest. This is $\frac{5}{7}\left(\frac{1}{6}\cdot\frac{1}{5}\cdot\frac{1}{4}\cdot\frac{1}{3}\cdot\frac{1}{2}\right)=\frac{5}{7!}=\frac{1}{1008}$ chance. In a more relaxed case, even if one further assume that an omnivorous diet and bipedal locomotion's place are interchangeable, This is still $\frac{1}{504}$ chance.

Path 1	Path 2	Path 3	Path 4	Path 5
social*	thumb	thumb	thumb	thumb
thumb	social*	binocular	binocular	binocular
binocular	binocular	social*	bipedal	bipedal
bipedal	bipedal	bipedal	social*	omnivorous
omnivorous	omnivorous	omnivorous	omnivorous	social*
cranial	cranial	cranial	cranial	cranial
language	language	language	language	language

Table 6.3.12: Shortest paths

It shows that we need between 417961 and 835922 speciation attempts to create a homo sapiens since the start of the Cenozoic.

$$4179613\left(\frac{504}{5040}\right) < A_{ttempts} < 4179613\left(\frac{1008}{5040}\right) \tag{6.3.13}$$

Now we compute the total number of species of birds, reptiles, and mammals arose since the Cenozoic. At this stage, we simply took the number of fossil species along the genus Homini, which numbered 13 species and one extent living species and use it as the filter factor for every 2.85 million years. That is, on average one of out of every 13 species survived to the current day for every 2.85 million years. Then, the total number of species (birds, mammals, reptiles) ever lived since Cenozoic is 7,555,486 (25,096 species $\times \frac{66\text{ myr}}{2.85\text{ myr}} \times 13$) species which serves as the upper bound. This is assuming ice age played a decisive role in speciation. We apply the same methodology to the entire family of Hominidae lineage, which contains 7 extant species and 69 extinct ones. It can be inferred that on average one out of every 10.857 species survived for every 14 million years. This means that the total number of species ever lived since Cenozoic is 1,284,534 (25,096 species $\times \frac{66 \text{ myr}}{14 \text{ myr}}$ $\times 10.857$), which serves as the lower bound. This is before the onset of ice age. (Assuming a linear increase so that earlier Cenozoic follow the same pace of speciation, which in reality, can be lower due to positive feedback loop, even without considering ice ages). The author is inclined toward the upper bound because, we have shown earlier, that glaciation helps to accelerate climate fluctuations and indirectly accelerate speciation. Nevertheless, we conclude, on average, then, there should be around 2,345,199 ever lived species generated our selected cohorts since the Cenozoic based on a weighted geometric mean taking both the upper and lower bound into consideration.

Furthermore, we show in Chapter 7, the final number of species ever realized < the number of speciation attempted, which is especially true for species possessed higher number of traits. As a result, during the Cenozoic, there should be > 2,345,199 speciation attempts. Which shows that human should have arrived earlier or more human equivalent species other than hominid should have been observed at the current time:

$$417,961 < 835,922 < 2,345,199 \tag{6.3.14}$$

The logic behind human could have arrived earlier as the follows. The number of final speciation is bounded by speciation attempts, we later show that in human's case, there are 7 traits with 8 trait acquisition attempts needed, so in total of 7^8 attempts. At the start of Cenozoic, there were only 7 species each possessed an unique trait which eventually enabled human emergence (The initial number of species can be > 7 so that fewer number of acquisition attempts are needed and we assume each attempt interval between attempt round is longer). Then, at the next round, each species within its habitat had thoroughly acquired an additional trait from the rest of possible trait choices either though voluntary migration to new habitat or passive adaptation to geologic changes. (i.e. a species possessed trait 1, at the next time period evolved into 7 species, 11, 12, 13, 14, 15, 16, 17) As a result, at the end of each round, there is an exponentially greater number of speciation attempts. So that we have:

Round Number	Cumulative Attempts Registered	Date
1st Round	7	66 Mya
2nd Round	7^2	57.75 Mya
3rd Round	7^3	49.50 Mya
4th Round	7^4	41.25 Mya
5th Round	7^5	33.00 Mya
6th Round	7^6	24.75 Mya
7th Round	7^7	16.50 Mya
8th Round	7^8	0 Mya

Each round of speciation attempts are spaced out in geologic time frame by evenly dividing the Cenozoic time period in blocks of 8.25 Myr. Since we need between 417961 and 835922 speciation attempts, so homo sapiens equivalent is achieved in

$$6.65 = \frac{\ln(417961)}{\ln 7} < t < \frac{\ln(835922)}{\ln 7} = 7 \tag{6.3.15}$$

between the 6.65th round (19.38 Mya) and the 7th round (16.5 Mya). One may question the validity of achieving human equivalent in less than 7 rounds since it takes 7 acquisition attempts to gain 7 traits. However, that is under most ideal situation. It is possible some particular species under particular habitat had more drastic pace of change that outpace the average performance of the rest and other habitat may stagnant longer than typically assumed. Though there is a minimum bound time required to achieve human equivalent. That is, even if we assume every permutation path leads to human equivalent with 829 speciation attempts, it will take $\frac{\ln(829)}{\ln 7} \geq 3.45$ rounds at 45.75 Mya.

On the other hand, if we assume that no outpacing is possible, then all human equivalent is achieved in the 7th and 8th round acquisition attempts, and we should observe more human equivalents since:

$$13.792 = \frac{7^8}{417961} > \frac{7^8}{835922} = 6.896 \tag{6.3.16}$$

This problem is resolved by further reducing the chance for the emergence of homo sapiens by modifying the chance of certain traits within all cohorts of interests, It is justified, since by assuming the start of simulation occurred at 65 Mya, it is reasonable to assume that certain traits were less dominant at the time.

Trait name	Currently Observed chance	Revised chance for 65 Mya
binocular vision	$\frac{2526}{25483}$	$\frac{2526-450}{25483}$
large cranial capacity	$\frac{1214}{25483}$	$\frac{1214 \times 0.2}{25483}$
opposable thumbs	$\frac{353}{25483}$	$\frac{353 \times 0.4}{25483}$
bipedal	$\frac{10089}{25483}$	$\frac{10089 \times 0.732}{25483}$
language	$\frac{5150}{25483}$	$\frac{5150 \times 0.87}{25483}$
omnivorous	$\frac{8278}{25483}$	$\frac{8278 \times 0.917}{25483}$
social	$\frac{3586}{25483}$	$\frac{3586}{25483}$

As a result, it shows that we need between 10,885,601 and 21,771,202 speciation attempts to create a homo sapiens since the start of the Cenozoic, and now it takes 21598 speciation attempts if ordering is irrelevant.

$$108856012 \left(\frac{504}{5040} \right) < A_{attempts} < 108856012 \left(\frac{1008}{5040} \right) \tag{6.3.17}$$

Then, we have:

$$2,345,199 < 10,885,601 < 21,771,202 \tag{6.3.18}$$

if one uses speciation under ice age as a reference, we have:

$$7,555,486 < 10,885,601 < 21,771,202 \tag{6.3.19}$$

Now we able to show that, in earth's case, ice age gives homo sapiens a head start, although this number still fall below the probability of giving rise to Homo sapiens at the current time. In fact, it predicts the rise of Homo sapiens type of intelligent species guaranteed 93~187 million years into the Cenozoic based on the current BER. (assuming break up phase of island continent continues and ignoring supercontinent's disruptive nature on biocomplexity in the next 50 Myr onward)

If ordering and steps of successive trait gaining are not important, then human equivalent species should emerge very quickly within 34 Myr $\left(\frac{\ln(21598)}{\ln 7} - 1 \right) \times 8.25 \text{Myr}$, so it should have already emerged 32 Mya. Now, we added the ordering, then, we would expect, it takes 300~600 million years ($\frac{504 \times 21598}{2345199} \times 65 \text{Myr}$ and $\frac{1008 \times 21598}{2345199} \times 65 \text{Myr}$) to guarantee the evolution of the next Homo Sapiens at the current **BCS** (Biological Complexity Search Space), assuming **BER**=1 and k=∞ and evolution speed=0. For k< ∞ and **BER** > 1, when the evolutionary speed for the mode of species is non-zero, the timing to guarantee the evolution of the next Homo Sapiens is < 300~600 million years. (see Chapter 8 "Generalized Model") The final emergence window can be found based on various assumptions. If we assume that the factor of 7 trait multiplication based on existing species can no longer be maintained for beyond the 8th attempt so i.e. for the 9th acquisition attempt, we have < 7^9 and the factor gradually diminishes to unity within every 100 Myr, then we can ideally assume that every 100 Myr thereafter **BCS** grows by 2.783, as we originally assumed as the weighted multiplication factor over all rounds of attempt acquisitions within every 100 Myr, then we have:

$$\frac{504 \times 21598}{2345199 \cdot (B_{cs})^{1.5}} = 1 \qquad (6.3.20)$$

$$\frac{1008 \times 21598}{2345199 \cdot (B_{cs})^{2.176}} = 1 \qquad (6.3.21)$$

That is, with 1.5 and 2.176 additional rounds of 100 Myr, it takes 215~367.6 Myr to guarantee the evolution of Homo Sapiens at the current **BCS,** assuming BER=1. In reality, BER > 1 for the next few 100 Myr. Therefore, the arrival time < 215~367.6 Myr. Instead, if we assume that the factor of 7 trait multiplication based on existing species can be maintained for beyond the 8th attempt so i.e. for the 9th acquisition attempt, we have 7^9, then we can ideally assume that every 8.25 Myr thereafter total attempts grows by 7, then we have:

$$\frac{504 \times 21598}{2345199 \cdot 7^{0.7889}} = 1 \qquad (6.3.22)$$

$$\frac{1008 \times 21598}{2345199 \cdot 7^{1.145}} = 1 \qquad (6.3.23)$$

That is, with 0.7889 and 1.145 rounds of acquisition attempts, it takes 71.5~74.4 Myr to guarantee the evolution of Homo Sapiens. We use the same approach for expected humanity arrival if the ordering of trait acquisition matters with speciation size accounting for ice age as:

$$\frac{504 \times 21598}{7555486 \cdot (B_{cs})^{0.356}} = 1 \qquad (6.3.24)$$

$$\frac{1008 \times 21598}{7555486 \cdot (B_{cs})^{1}} = 1 \qquad (6.3.25)$$

That is, with 0.356 and 1 round of acquisition attempts, it takes 101.6~166 Myr to guarantee the evolution of Homo Sapiens, and if the factor of 7 trait multiplication based on existing species can be maintained for beyond the 8th attempt:

$$\frac{504 \times 21598}{7555486 \cdot 7^{0.1877}} = 1 \qquad (6.3.26)$$

$$\frac{1008 \times 21598}{7555486 \cdot 7^{0.5439}} = 1 \qquad (6.3.27)$$

That is, with 0.1877 and 0.5439 round of acquisition attempts, it takes 67.54~70.49 Myr to guarantee the evolution of Homo Sapiens. Then, one can arrive at the following conclusion regarding the probability of Homo sapiens' emergence.

List of cases	Years required at current BCS and BER=1 and k=∞ and evolution speed=0	Years required in reality	Chance Factor
Ordering non-important	34.06 Myr	N/A[2]	1
Ordering important	301.7 Myr ~ 603.41 Myr	< 215~367.6 Myr	×504~1008
Ordering important with Ice age	93.65 Myr ~ 187.29 Myr	< 101.6~166 Myr	×156~312
Ordering important with the shortest path	34.06 Myr	N/A	1
Ordering important with the shortest path with Ice age	29.1 Myr	N/A	×0.31

Table 6.3.13: List of possible ordering cases

If the ordering of traits for the emergence of Homo sapiens is not important, one would expect the emergence in the first 34.06 million years into the Cenozoic.

Since the ordering is important, it would take 300~600 million years at the current **BCS** but fixed **BER** = 1 to guarantee the emergence of an intelligent, tool-using species with speciation attempts. Moreover, if k < ∞ and **BER** > 1, the speed of emergence will be faster in the future as both **BCS** increases and **BER** increases (the mode of new species appearing in a 100 million year period should have shared greater number of traits and lead to a shorter time of emergence by nature's trial and error. That is, some traits from the previous round become universal among the species in the new round after a major extinction, and previously the emergence of human requires the permutation of 7 traits becomes the permutation of < 7 traits, resulting in faster human equivalent emergence and ultimately greater chance of emergence) It would take, in reality, only < 215~367.6 Myr into the future (also see Chapter 8 Section 8.9 "Complexity Transformation") for the emergence.

When the ordering is important and nature took the shortest path without ever had a single trial and error and beating on dead ends of wrong order of trait acquisitions, one would expect the emergence in the first 34 million years into the Cenozoic, but it will be a probabilistically very small chance $< \frac{10}{5040}$.[3]

Finally, if ordering is important and the planet enters an ice age, and it luckily takes the shortest path, then, one would expect the arrival of intelligent species within $34\text{Myr} + \frac{\ln\left(\frac{2345199}{7555486}\right)}{\ln 7} \times 8.25\text{Myr} = 29$ million years. In the last case scenario, given its known unlikelihood prior probability $< \frac{10}{5040}$, the probability of the emergence of Homo sapiens is increased by more than 504~1008 folds, and possibly as much as 1500~3000 folds compares to the case whereas ordering is important but takes the longest path.

Homo sapiens, even in the slowest possible scenario, inevitably rise within the next 600 million years at the current **BCS** and **BER**=1. However, an arrival in the first 65 million years of Cenozoic is somewhat early and on the slightly lesser possible side, not even counting the probability of the appearance of grass plant as as a pre-condition for ultimate civilization emergence.

---

[2]It is not possible to consider accelerated effect of BCS/BER, since BCS/BER is defined based on 100 Myr period and the expected appearance time < 100 Myr. The species increase within 100 Myr period can be very rapid initially after a post major extinction recovery followed by slower increase due to increasingly restrictions on permutation paths generating new species and ultimately vertical scaling on BCS only if acquisition attempts > fixed number of traits.

[3]This refers to the chance taken by any permissible permutation out of all permutations of 7, assuming every permutation path takes an average expected arrival time since gaining the next trait is a probablistic event. There exists an even more extreme scenario, in which every speciation attempt results in the right order and right trait acquired, so emergence occurs within 7 speciation attempts, but such chance is 1/108856012.

6.4 The Probability of Alternative Intelligence

We will now cross-examine our results with that of other species. Assuming humans are gone, determining the timing for the emergence of next intelligent, tool-using species.

In order to calculate the probability of the rise of the intelligent, tool-using species, we do need to list the major features of Homo sapiens that distinguishes us from the rest of other species. We have stated earlier, that our species is differentiated from the rest by large cranial capacity, manifested as having high EQ, opposable thumbs, bipedal locomotion, binocular vision, omnivorous diet, language communication, and social organization. We have discussed earlier that each of the listed traits are independently evolved. That is, the opposable thumb does not increase the chance of evolving toward a high EQ or an omnivorous diet. Each trait can stand alone as an independent variable. We have calculated the number of species currently thriving possessed each of these traits divided by the total number of species of mammals, reptiles, and birds. The computed probability for each trait possessed by the cohorts under consideration gives us a general overview how successful a trait (whether evolved only once or repeatedly by convergent evolution) is ensuring the survival of the species in question at the current time.

This probability is time biased; that is, we can only compute the probability for the current geologic period. Because fossil records are incomplete, it is hard if not impossible to compute the average probability for each trait totaled under each epoch. We do need to take some faith in that data is unbiased though natural selection at different epoch may favor one type of traits or behavior more over the other. All major traits and behavior have been explored and established by the Mesozoic such as bipedalism, flight, increasingly large brain; therefore, the probability computed may not truly reflect the usefulness of the trait across all times. But the margin of error should be within the error of tolerance and validate and strengthening our argument. There is a general trend, though, that all traits becomes more dominant/common within all groups over time.

Secondly, this probability is location biased. This probability is observed and only observed on earth, the only habitable planet we are currently able to investigate. Aside from temporal and spatial limitation of our data, the total probability of the emergence on Homo sapiens can be computed by multiplication of the probability of each independent variable.

What does the multiplication mean in this case? To state simply, the multiplication implies the chance that a species have well adapted into $environment_1$ with its possessed $trait_1$ with given probability P_1 has at some later time either voluntarily or involuntarily changed into $habitat_2$ and evolved $trait_2$ ($behavior_2$). Since $habitat_2$ and $trait_2$ adapted to $environment_2$ can be known based on existing species, and we can label it with probability P_2. Then, the total chance that this species possessed $trait_1$ and $trait_2$ then is simply $P_1 \times P_2$. If the species possessed n traits, then the total probability for the emergence of that species is $P_1 \times P_2 \times P_3...P_n$. A caveat to this problem is that one can not over-interpret the mathematical formula to real evolutionary settings. As a matter of fact, $P_1 \times P_2 \times P_3...P_n! = P_n \times P_{n-1} \times P_{n-2}...P_1$.

To understand their non-equivalence, *that is the ordering in the multiplication is important*, as we stated earlier. Let us use real examples to illustrate the asymmetry.

Human evolution toward intelligent, tool-using creature went under the following sequence:

First, the earliest primate adapted arboreal lifestyle and evolved partially opposable thumb. Then it evolved binocular vision, social organization, bipedal locomotion, omnivorous diet, enlarged cranial capacity, and finally language communication.

For eagles, the sequence would be bipedal locomotion, binocular vision, enlarged cranial capacity. For certain songbirds, it would be bipedal locomotion, omnivorous diet, enlarged cranial capacity, social organization, and language communication.

In order to evolve the additional trait of an opposable thumb, birds have to first de-evolve into a quadrupedal terrestrial species. However, birds have no chance to claim the ground casually given the number of fast running predators. It is only to occur if a mass extinction kills all land-based predators. If it succeeds, then it has to become first smaller in size, then climbs back on trees, and then descend from the trees.

The greatest challenge is that there is no short route to achieve the next major trait leading to intelligent, tool use species, instead of seemingly taking just one additional step, it has to take many more steps before it can fully gain a given trait. One may object that there is a possible short route by bird evolving opposable claws on its wings. However, early ancestors of birds all had claws on wings. As soon as flight ability and specialized beak fully evolved, they are able to survive by adopting these traits, and claws become unnecessary. It is possible that if angiosperm based fruit tree evolved earlier, there is a chance that some of the bird species may maintain their claws by gliding from tree to tree and using their claws to extract fruit. This shows that the timing of the appearance of new ecological niche is critical to the emergence of an associated trait. If a niche does not exist, even potentially very beneficial traits are removed by selection.

For a different case, one can consider that of dolphin, which evolved fins and tails adapting to the ocean, then carnivorous diet, and then social organization, enlarged cranial capacity, and finally language communication. However, in order to gain traits such as bipedal locomotion, it has to first return to dry land.

Vertebrates evolved onto land seem happened only once because existing land predators quickly kill transitional forms. However, a transition back to water is easier. It is because ancestors of dolphin could use their legs in the shallow water and swim and retract back to land when it is necessary. Therefore, their legs served a dual purpose until it is completely transitioned toward the fins. The reverse, however, is difficult, fins are adopted in the aquatic environment but almost helpless once on land. As a result, their fins, an existing trait cannot be used in a different setting, making a transition difficult.

The only scenario in which a dolphin reclaim on land if a major extinction event occurs and all land predators and herbivores no longer able to compete with aquatic competitors.

Once it regained its hold on the land, it has to re-evolve quadrupedal locomotion. By living on land, it has to adopt different vocalization range because their voice generated underwater is difficult to duplicate in the medium of air. Because it no longer able to chase its food source, it has to re-adapt into a herbivorous diet, reducing their caloric consumption and their cranial capacity. Furthermore, in order to gain opposable thumbs, it has to reduce its size and climb on trees. As a result, by gaining an additional trait required to become intelligent tool user, it has to not only go through many more meandering steps, and significantly decreasing its chance of becoming one, it has also to lose many of the traits it gained before.

If species such as birds and dolphins' YAABER (read more about in chapter 7) is plotted against the rest of animal cohorts, these species can be said to have YAABER millions even tens of millions of years ahead of the average value. However, they could not keep ahead forever because it takes many more steps to gain the additional features to become intelligent tool-using species. As a result, they either become stagnant in their position while the rest of the species catch up in millions of years or they re-adapt into a new niche and loses existing traits associated with intelligent tool-using species and their YAABER retract.

If we summarize the major breakthroughs since the Cambrian explosion, we have the following major evolutionary events leading to intelligent, tool-using species:

Event Name	Epoch
Multicellularity	Pre-Cambrian
Evolution of Vertebrate*	Cambrian
Plants moved on land*	Ordovician
Tetrapod moved on land*	Devonian
Evolution of Gymnosperms*	Carboniferous
Evolution of amniote egg-bearing tetrapod*	Carboniferous
Evolution of Mammals*	Triassic~Jurassic
Evolution of angiosperms*	Cretaceous
Evolution of birds*	Cretaceous
Evolution of opposable thumbs (primate)*	Paleocene
Evolution of binocular eyes (primate)	Paleocene
Evolution of bipedalism (ape)	Neogene
Omnivorous diet (Homo)	Neogene
Enlarged brain	Neogene
Language communications	Neogene

Table 6.4.2: Major evolutionary innovation and their first emergence

After the evolution of primate in the mammalian lineage, the evolution rate started to race ahead of the background rate. Therefore, We consider 9 asterisked cases as the frequency of major evolutionary change by the background evolutionary rate for a period spanning from 542 million years ago to 3.2 million years ago. There is a period of supercontinent Pangea which lasted 170 million years with no increase in diversity. Discounting this time period, so on average, 45.42 million years a major change occurs either by the tetrapod lineage themselves or plant lineage opens new biological niche, notice that such timing correlates well with the average time tectonic movement transitioned from an existing configuration to a new one from our previous derivations.

Song Birds: (Number of years expected to become intelligent, tool-using species)

Event Name	Years
A major extinction event	108.4 Myr (average mass extinction gap observed)-66 Myr = 42.4 Myr
Regain foothold on land	45.42 Myr
Regained quadrupedalism	45.42 Myr
Climb on Tree (opposable thumb)	45.42 Myr
Binocular vision, (Bipedalism if glaciation happened)	45.42 Myr (100% at the initiation of the next glaciation)
Expected time required	**212.3 Myr**
The reign of Pangea Ultima Supercontinent	130 Myr (supercontinent not conducive to evolutionary complexity and diversity)
Bipedalism	45.42 Myr
Max time required	**399.5 Myr**

Table 6.4.4: A hypothetical evolutionary trajectory for song bird gaining transcendence

Dolphin: (Number of years expected to become intelligent, tool-using species)

Event Name	Years
A major extinction event	108.4 Myr (average mass extinction gap observed)-66 Myr = 42.4 Myr
Regain foothold on land	45.42 Myr
Adopt different vocalization range	45.42 Myr
Reduce body size	45.42 Myr
Climb on Tree (opposable thumb), (Binocular vision and Bipedalism if glaciation happened)	45.42 Myr (100% at the initiation of the next glaciation)
Expected time required	**212.3 Myr**
The reign of Pangea Ultima Supercontinent	130 Myr (supercontinent not conducive to evolutionary complexity and diversity)
Binocular vision	45.42 Myr
Bipedalism	45.42 Myr
Max time required	**444.92 Myr**

Table 6.4.5: A hypothetical evolutionary trajectory for dolphin gaining transcendence

A careful reader may point out extra time is required because one needs to wait for the emergence of crop plants, and especially grass plant family. However, over the course of another 212.3 Myr, it is expected that grass plants have been evolved as the biodiversity grows among all genera, and it is assumed that once it is evolved, its form persisted and the transition from hunter-gatherer to agricultural societies becomes possible. It is also taken for granted that the metallicity of the home planet is high enough so that at least project PACER type of nuclear fusion is economically feasible to sustain the expanding industrial civilization.

In retrospect, Homo sapiens and earth itself took one of the shortest paths possible (by first hanging on trees) to achieve an industrial civilization, and it is likely the typical path of any early intelligent extraterrestrial intelligence's path to attain transcendence. Homo sapiens is fortunate because many traits evolved have already been used in earlier niches and served dual purposes during transitional periods and none of the critical traits gained earlier enabling an intelligent tool user have been lost. (other than none essential traits such as tail, hair growth). *This luck may also partially be attributed to the meteorite impact at Yucatan 66 Mya.* Without the extinction of dinosaurs, the chance of tree climbing frugivores diminishes. Although lizards and chameleons are arboreal, they retained their reptilian feeding behavior of predominantly carnivorous diet even today. Since no major species adapted the arboreal niche, they can only feed on insects or tiny creatures and unable to utilize fruits as an energy source. As a result, their own sizes decrease to lower the energy requirements based on their energy intake. Birds, on the other hand, feed on fruits but they fasten themselves using claws, and no development of opposable thumb is necessary. Reptilian arboreal frugivores may eventually emerge but possibly much later than the emergence of primates. We will discuss more on catastrophic extinction rates and its effect on evolution in Chapter 8 under the cases of conservative, classic, and progressive evolutionary scenarios.

6.5 Probability of the Emergence of Homo Sapiens within the Genus Homo

We have defined the probability of the emergence of bipedalism, opposable thumb, binocular vision, large cranial capacity, and complex communication into a single intelligent, tool-using species.

Basically, we have defined the probability giving rise to any species within the genus of Homo. However, not all members of the genus are created equal. All earlier ancestral species possessed all traits described as human except complex language communication.

Earlier we have shown that the number of species possessing language communication skills totaled 5,336 species, which is a 5,336 out of 25,483 chance. Then, the probability indicates that the chance Homo sapiens emerges from the hominid lineage is at 20.939%, or 1 out of 4.7756 chance.

Next, we cross-examine this result with real data. We simply took the number of fossil species within the genus Homini, which numbered 13 species and one extent living species and use it as the filter factor in the last 2.85 million years. That is, on average one out of every 13 species survived to the current day.

Species Name	Existence	Species Name	Existence
H. habilis	2.8 Mya	H. rhodesiensis	0.4 Mya ~ 0.12 Mya
H. naledi	2 Mya	H. helmei	0.259 Mya
H. ergaster	1.9 Mya ~ 1.3 Mya	H. neanderthalensis	0.25 Mya ~ 0.028 Mya
H. rudolfensis	1.9 Mya	H. sapiens sapiens	0.195 Mya ~ now
H. gautengensis	1.9 Mya ~ 0.6 Mya	H. tsaichangensis	0.19 Mya ~ 0.01 Mya
H. erectus	1.9 Mya ~ 0.07 Mya	H. sapiens idaltu	0.16 Mya
H. antecessor	1.2 Mya ~ 0.8 Mya	H. floresiensis	0.094 Mya ~ 0.013 Mya
H. heidelbergensis	0.6 Mya ~ 0.3 Mya	Cro-magnon	0.05 Mya
H. paleojavanicus	0.5 Mya	Denisovans	0.041 Mya
H. cepranensis	0.5 Mya ~ 0.35 Mya		

Table 6.5.1: List of discovered species within the genus Homo

This is 1 out of 13 chance, or 7.69% of the emergence of truly intelligent, tool-using species that ultimately transitioned into an industrial civilization. This probability is lower than purely derived based on the chance of evolving additional complex language as an adaptation of communication. This suggests that it is harder to evolve into Homo sapiens even with language as an additional trait taking into account. *This suggests that a factor of $\frac{13}{4.7756}$, or 1 out of 2.722 (36.73%) should be applied further to the emergence of truly intelligent, tool-using species that ultimately transitioned into an industrial civilization.* This shows that the rise of Homo Sapiens is not inevitable even if bipedalism, opposable thumb, binocular vision, and large cranial capacity is evolved within the lineage in the presence of an ice age at the current time (though guaranteed at future times with greater emergence chance $\frac{d}{dt}C_{dt}(t)$). This is at least partially justified because even our closest cousin, Neanderthals did not exhibit complex ritualistic behavior, extensive artworks, and very likely being displaced by the migration of Homo sapiens. It can also be assumed, if the earth's evolutionary history rewind and unfold again from the start of Cenozoic even with the onset of an ice age, there is a chance that Homo sapiens will not emerge.

6.6   Probability of Fruit Trees

Of course, Primates is not the first family of species to embark on this shortest route out of many possible trait combinations to attain transcendence, by first adapting to the arboreal niche. Many had attempted but failed. The earliest documented from the fossil records traced back to Permian. This adaptation was not successful because the species can not fully adapt to an arboreal lifestyle. Tree species of the late Paleozoic are dominated by gymnosperms, hard to chew and woody. Even if it did adapt to such niche, the energy content obtainable from such source is very low. This can be reflected from the energy content such as lettuce, spinach in contrast to fruit such as pear, almond, apricot, apple, and banana. Creatures living on such low energy content diet cannot evolve enlarged cranial capacity and its body size likely remained small.

The next attempt came in Cretaceous of the late Mesozoic. *Deinonychus*, a species of bipedal dinosaur's claws has been investigated, and its strength was not significant enough to cause fatal harm to prey, but its likely adapted to arboreal climbing. Although it is likely that it lived on trees before reaching adulthood, it had an exclusive niche on the ground as an adult. This showed that just before the emergence of angiosperm, gymnosperms do not offer significant ecological niche to any potential explorers. *Deinonychus* went extinct 74 million years ago, before the KT boundary.

Therefore, the emergence of Primates is strongly depended on the diversification of fruit trees, and if fruit tree evolved independently from the rise of primates, then we need to multiply the chance of the emergence of fruit trees into the probability giving rise to Homo sapiens.

Fruit tree's speciation, continuation, and dispersion are almost entirely independently evolved from the emergence of Primates. Primates, at most, played a marginal role in the dispersion of fruit tree.

First, fruit tree pollination is regulated by both biological and physical factors. Physical factors such as wind and gravity played an important role in flower pollination. More importantly, biological vectors such as insects (bees, fruit flies, butterflies, ants, and beetles) developed a symbiotic relationship with the plants. In fact, plant response to ant adaptation has evolved hundreds of times independently, indicating a strong correlating, non-independent relationship. The dispersion of fruit tree species is done by both physical and biological vectors. Fruit seed can be dispersed by wind, river flow, ocean currents, and gravity. Some seeds are even dispersed by exploding mechanism. Biological vectors are dominated by birds, mammals, and insects. Birds frequently eat fruits and scattered seeds through their digestive tract. Hairy mammals carried sticky seeds along with them on their fur. Insects carried seeds as a form of food for storage in their nests such as ants, giving a chance for the seeds to germinate. Primates disperse seeds by both digestions and sticking to their fur. However, as we observed, even with the absence of primates, fruit trees will continue to diversify.

Given a complete list of all fruit trees, it accounted for a total of 667 species, out of total 295,383 species of angiosperms, one can see that 1 in 442.853 chance gives rise to fruit trees. This number is significantly lower than the portion of cohorts of birds, mammals, and reptiles surviving on the arboreal niche. Even by the most conservative estimate based on the number of primates, 256 species out of the total of 25,616 cohorts of mammals, reptiles, and birds lived on trees, 1 in 100 species adapted the arboreal niche. This implies that arboreal habitat is exceptionally nourishing despite their rare occurrence as species in the angiosperm family. However, we will not include fruit tree as a filter criterion for the emergence of Homo sapiens. First, tree accounted for 25% of all plant species' diversity, indicating its commonality. Second, angiosperm is differentiated from gymnosperm by enclosing seeds into fruit bodies; therefore, fruit-bearing is a universal trait among all angiosperms.

6.7   Probability of Crop Plants

In Chapter 6, we discuss how and why the emergence of crop plants play a crucial role in the transition from hunter-gatherer to a feudal society, which enabled the development of city-states and the continuation of civilizations. The passage and the accumulation of knowledge and technology eventually ushered in the industrial revolution. Crop plants allowed the harvest of solar energy at an unprecedented scale and resulted in a population explosion and the division of labor. It is, therefore, essential to compute the lower and upper bound on the probability of the emergence of crop plants essential for human lives.

To establish an upper bound of all plant species that are able to feed a very large population base, we count the number of species within all the family groups that contains all the crop species that gave us the agricultural revolution.

10,035 of them belongs to the family of Poaceae, also called true grasses, are a large and nearly ubiquitous family of monocotyledonous flowering plants. With more than 10,035 domesticated and wild species, the Poaceae are the fifth-largest plant family. This family includes rice, wheat, barley, oats, rye, sorghum, millet, and maize, providing more than half of all calories eaten by humans.[7][65] Of all crops, 70% are grasses, and are members of this family.[55]

The Fabaceae family contains 19,500 species, also known as the bean family, contains soybean, pea, alfalfa, and peanut. Both peanut and soybean have a higher level of energy content than rice and wheat, enabling these crops to nourish complex agricultural societies with significant population base.

Solanaceae family contains 2,460 species, some of the most important species within this family that contribute to the rise of complex agricultural civilization are the potato and the eggplant. Pepper and tomato are also members of this family, but it is not in our interest of research since tomato and pepper do not contain enough energy content to aid a large population base.

Polygonaceae family contains 1,200 species and includes buckwheat, a high energy content seed.

The total number of species, therefore, is 31,995 for all three families. We assume since all member species within such family group is more genetically closely related to each other, these species all have a significant chance of evolving into crops for the benefit of the agricultural society than the other comparing groups.

Since all living species of flowering plants contain 295,383 species; therefore, *1 out of every 8.898 plant species are potentially domestic-able and give rise to agricultural revolution.* It shows that the rise of Homo sapiens is much rarer than the rise of domestic-able plants. Indeed, there are several species such as maize, rice, potato, and beans that are able to independently sustain an agricultural society. If one of such species does not exist, one or more alternative can be used as a substitute. However, Homo sapiens cannot be substituted by any other species such as *Homo Neanderthals*. This shows that the requirements for domesticable plant species are simply able to store a significant amount of energy (very relaxed), the requirements for an environment-altering and self-altering species is much more stringent (very rigorous). Unlike accounting for the probability of the rise of Homo sapiens, we can not just multiply the probability of the emergence of crop species with extinction rates of angiosperms (which is again roughly 100 within 10 million years for 1% survival rate of any species within a 10 million years temporal window). It is because a very suitable plant species could arise before the emergence of human, and it could lead to agricultural revolution because it is passively selected and breed-ed by earlier arising intelligent species. We can only assume that given our current temporal window, the number of crop species is an average, typical of all temporal period since Cenozoic era.

We may be still interested in calculating the lower bound of the probability of crop producing species. In order to calculate such lower bound, we have to sum up all species for each type of major crops on earth.

Oats contains 17 wild species and 5 cultivate ones as indicated from the Avena genus. There are 19 species within the rice genus Oryza. There are 9 species within the rye genus Secale. There are 28 species within sorghum's genus Sorghum. 4 species of Zizania or wild rice. 15 species of buckwheat under the genus of Fagopyrum. 23 species of Wheat. 38 species within the genus Hordeum which contains barley 6 species. Within the genus Zea which comprises maize. 27 species within the genus Glycine which contains soybean. 3 species within the genus Pisum which includes pea. 80 species within the genus Arachis which includes peanuts. 2,000 species within the genus Solanum which contains potato and eggplant. The total number of species is; therefore, 2,274 species. Since all living species of flowering plants include 295,383 species; therefore, *1 out of every 129.89577 plant species are potentially domestic-able and give rise to agricultural revolution at the lower bound.*

However, much like the way we treated the fruit trees, we will not use the probability of crop plant as a filter criterion for the emergence of Homo sapiens. We simply assumed that regardless of the lower and upper bound, as long as angiosperm biodiversity reaches the level currently observed on earth, then the existence of crop plant is assured.

6.8 Probability of Angiosperm

After one examined the probability of crop plants and fruit trees, one needs to take a closer examination of angiosperms, the class of flowering plants. We have exempt the probability of both as considerations on the emergence of civilization, but we can not exclude the emergence chance of angiosperms. The emergence and diversification of angiosperm seem to be a natural, logical consequence of the evolutionary change, serving as an exemplary case for exponentially increasing biological species diversity. Upon closer examination, the increase may be ahead of the average rate of growth, and this may have set ourselves apart from the rest of the habitable planets. Given for example, the earlier dominant class of seed plants the gymnosperms includes only 1,080 species. One can not argue that gymnosperm is out competed by angiosperms because 80% of all temperate and high latitude forests are composed of gymnosperms. Even earlier plants such as ferns, the very first vascular plants, contains 10,560 species. The total number of plant species, before the emergence of angiosperms, are roughly in proportion to the number of vertebrate animal species. With the emergence of angiosperms, the number of plant species dramatically outpace the number of vertebrates. The vast increase in plant diversity enables the emergence of primates and agricultural revolution. Using our exponentially increasing evolutionary transformation factor of 2.783^4, one should expect an increase of 37,884 species of angiosperms at most. It is derived by taking into account the total number of vascular land plant species in all clades excluding existing number of angiosperms and multiplied by the transformation factor. We assumed that the number of non-angiosperm species stayed constant since 100 Mya and all increase in species number is attributed to the emergence of angiosperm. There are 766 species of club mosses, 15 species of horsetails, 10,560 species of ferns, and 1,080 species of gymnosperms, with non-angiosperm species totaled 12,421. We can then construct an exponential function to reflect the vascular plant species biodiversity growth curve:

$$P_{lant} = (12421) \cdot 2.783^{(t+1)} \tag{6.8.1}$$

Angiosperms diverged from gymnosperms between 245 and 202 Mya (so averaged divergence time of 223.5 Mya), we assumed that at the time there was only 1 stem angiosperm species in existence and 223.5 Myr of evolution results in the number of angiosperm species of 369,000. Then, the trajectory of angiosperm species growth can be modeled by a simple exponential function as:

$$A_{ngiosperm} = 1 \cdot (B_{ase})^{(t+2.235)} \tag{6.8.2}$$

$$B_{ase} = \exp\left(\frac{\ln 369,000}{2.235}\right) = 309.6275 \tag{6.8.3}$$

With the presence of both curves, angiosperm overtook expected vascular plant species biodiversity growth curve 50 Mya in the early Cenozoic.

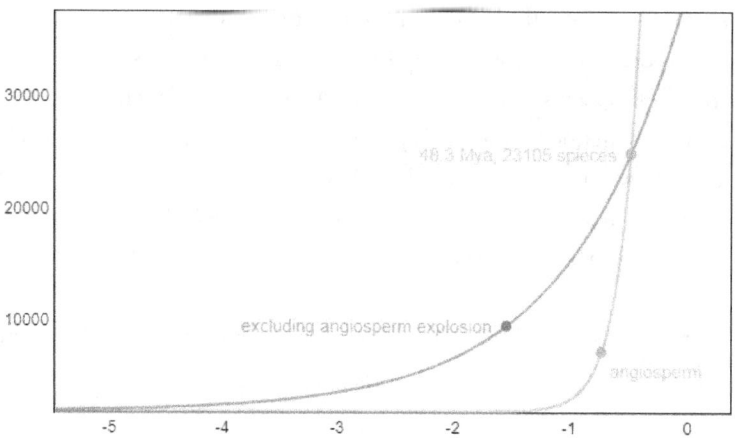

Figure 6.8.1: Angiosperm diversity over took the general vascular plant diversity trend $(12421 + 1191) \cdot 2.783^{(t+1)}$ 48.3 Mya

The model predicts that 100 Mya only 1,191 species of angiosperm are in existence and we add this number to the number of non-angiosperm vascular plants 100 Mya , and the total number of vascular plants predicted at the current time is adjusted and multiplied by 2.783:

$$(12,421 + 1,191) \times 2.783 = 37,884 \tag{6.8.4}$$

---

[4]See Chapter 8 Section 8.5 "Complexity Transformation"

If one takes the predicted value of 37,884, comparing to the number of angiosperm species of 369,000 and non-angiosperm species of today, the total vascular plant diversity on earth is then 10.068 times greater than the expected value.

$$\frac{369,000 + 12,421}{37,884} = 10.068 \tag{6.8.5}$$

Assuming an exponential acceleration on growth in planet diversity by a factor of 2.783 per 100 Myr, earth is 225 Myr ahead of the average rate of expected evolution diversity.

$$\frac{\ln(10.068)}{\ln(2.783)} = 2.2562 \tag{6.8.6}$$

If one simply assumed that any planet with 500 Myr of multicellular evolution should have produced an expected 37,884 number of vascular plant species, and we assumed that the chance of producing higher number of species can be a simple inverse linear relationship. Then the chance of this happening on earth is $(\frac{1}{2.783})^{2.256}$, or 9.93%. Additionally, if one were to give it a margin of tolerance, assuming 100 Mya, the total number species of non-angiosperm existed but was displaced by the emergence of angiosperm and was driven to extinction can range from 0 to the total number of angiosperm one observed currently, we can set the chance by taking the geometric mean between 9.93% and 100% and rounding to the nearest integer fraction:

$$\sqrt{\frac{1}{2.783^{2.256}}} \approx \frac{1}{4} \tag{6.8.7}$$

That is, the chance on the emergence of great abundance and diversity of angiosperm observed on earth occur with a chance of 25%, which enabled the emergence of fruit trees, creating arboreal habitat for primates, and the emergence of crop plants, enabling transition to civilization.

Chapter 7

The Distribution Model

7.1 Mathematical Model for Human Evolution

To capture all possible scenarios and represent them abstractly we need the tool of mathematics. Knowing that life can potentially be abundant on all habitable exoplanets, and yet highest attainable life form similar to human composed of different attributes each stands for independently evolves through local evolutionary forces. Then, we can expect a Gaussian/log normal distribution of all extra-terrestrial life forms in the Milky Way and beyond. Normal distribution should be used because of its most general form, under the conditions (which include finite variance), states that averages of random variables independently drawn from independent distributions converge in distribution to the normal, that is, become normally distributed when the number of random variables is sufficiently large. Physical quantities that are expected to be the sum of many independent processes (such as measurement errors) often have distributions that are nearly normal.[16].

Binocular vision, bipedal locomotion, opposable thumbs and grabbing fingers, little to no tails, omnivorous diet, land-dwelling, and big brain are each independent attributes observed across many different genera and species of animals on earth. The only partially correlated attributes in human are the big brain and sophisticated manipulation of language communication. To further demonstrate that binocular vision is not a byproduct of a big brain, we found mice, which resembles the earliest ancestor of mammals before adaptive radiation 65 million years ago, had partial depth perception. Carnivorous cats, a different genus of mammal, also have binocular vision for catching prey yet much smaller Encephalization quotient compares to human. Birds such as owls and eagles evolved through the Cenozoic era. Both have binocular visions. On the other hand, dolphins with highest Encephalization quotient other than Homo Sapiens, do not have binocular vision. Bipedal locomotion does not directly correlate with a big brain. Ostrich, and extinct Dodo bird and bipedal dinosaurs most have EQ<1, *Troodon* from the late Cretaceous may be an exception compares to its contemporary cohorts; however, its EQ is still less than 1. *Australopithecus Afarensis* of the Hominid lineage had highly developed bipedal locomotion but with a small cranial capacity of 350 cc. Opposable thumb has evolved on many tree-dwelling animals ranging from tree shrews, monkeys, to amphibian tree frogs, and reptilian Chameleon. Most of these animals have Encephalization quotient comparable to 1 or even lower. Therefore, the gripping power of fingers and claws contributes little to the expansion of brain, and vice versa. Animals with little to no tails may first appear a significant achievement of hominid lineage, but a closer examination reveals that early ancestors of amphibian frog-like creatures already shed their tails after fully metamorphosed into an adult in the Carboniferous epoch, some 325 Mya. Later, some species of sea turtles become tailless during the Mesozoic, and some mammals species also evolved to become tailless. Primates lineage certainly re-evolved long and thick tails to balance on trees, in a sense regressed from the average norm of the evolutionary prototype of the mammalian ancestor. We also found raven and magpie which has a feathered tail, score high on self-cognition and measured Encephalization quotient; Dolphin lives in the ocean and has no legs but a tail used for aquatic propulsion. Therefore, brain size has no strong correlation with tail size. Brainy animals could have a long, short, or no tail. Human eats both vegetables, fruits, and meat, yet many species of insects, birds, mammals have shown to exhibit similar behavior. Dolphin and cetacean have some of the highest brain sizes among all living animals but they live in the ocean, and they lack bipedal locomotion, grabbing fingers, and are carnivorous.

So each of these attributes is evenly likely distributed among different species of animals, then, we expect human to be the rightmost outlier in the normally distributed data set because we possess all these attributes. Some have argued that we are not evolving toward higher intelligence and one can well devise a normal distribution dataset with criteria, so that elephant with long nose becomes the rightmost outlier.[37] The argument is valid that datasets can be rearranged to show the differential importance of each attribute or particular species of animal's possessed characteristics enables it to be plotted as the rightmost outlier in the distribution. However, what unique about Homo sapiens is that our set of biological attributes enables us to change and adapt at a rate much faster than natural selection. So that over time, our position shifts further to the right and becoming ever more and increasingly outlying compares to the mode/mean value. That is, our position on the distribution changes while other animals held stationary in sub-geologic time scale (at time scale too short to observe significant biological evolutionary changes x $< 10^7$ yrs). It is true that evolution does not dictate a predetermined path to industrial civilization. However, with increasing biodiversity on earth, which is evident from the graph below, that as certain animals adapted to certain niches become saturated, a new differentiated species must develop new attributes or more exaggerated existing features to occupy new niches and avoid competition. As biodiversity increases along with gradual geological change periodically leading to drastic change, the chance and propensity for nature evolving both brainy, long feathered, great wingspan, long-nosed, or some combinations with these attributes increases. In summary, an extra-terrestrial civilization's host planet must have a great diversity of animals species and genera sufficiently guarantee the rise of an organism sharing all functional equivalent attributes of Homo sapiens.

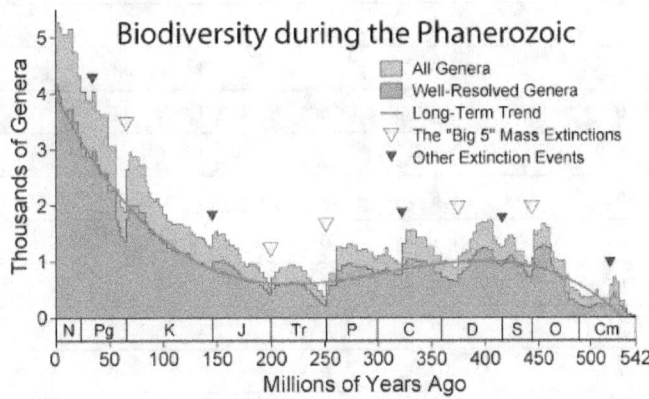

Figure 7.1.1: Historical trend of biocomplexity change

It is important to emphasize, that once an organism achieves fully anatomical or functional equivalent of Homo sapiens, they will alter their environment in degrees according to the level of mastery of the sophistication of technology, so that natural selection has less an effect on them than other animals. Yet as long as they remain biologically unaltered, they will subject to biological constraints on food, resources, temperature swings, aging, sickness, and death. As a result, a log-normal distribution or a heavy-tail distribution will fit better in model and forecasting. Furthermore, for simplifying our analysis, we can divide our data sets by temporal epochs of evolutionary development. For example, *evolving from common mammalian ancestor up to anatomical and functional Homo sapiens can be captured and modeled by approximately Gaussian Normal Distribution. Evolving from hunter-gatherers up to post-singularity civilization follows lognormal or a heavy tail distribution. A post-singularity civilization follows a uniform probability distribution.* More will be discussed in detail regarding analysis simplification in later sections. For the simplicity of our argument, we will use normal and log normal distribution in our discussion for now.

7.2 Background Rate

7.2.1 Sample Data

After we have selected our mathematical prototype model, how do we calculate the mean/mode, the deviation of such Gaussian normal/lognormal distribution given limited data available on astrobiology? For now, fortunately and unfortunately, the best data we can gather is from the earth itself and paleontology. We choose an average mammal with encephalization quotient=1 as the average/mode attained by evolutionary sample from all habitable exoplanets. Then, we have to use bio-informatics, genomic, epigenetic, and functional complexity to calculates how much more accelerated biological progression occurred in the Hominid lineage since its divergence from the main lineage following the K-T extinction event 65 million years ago. The following table lists some common animals by their encephalization quotient and other essential characteristics possessed by Homo sapiens. Where each animal obtains a score under each category, and their summed final score is listed in the very right column, arranging their final scores from high to low, we have some expected and surprise results:

Animal	Brain	Bipedal	Binocular Vision	Language	Landbased	Thumb	Omnivorous	No tail	Total
Homo sapiens sapiens	1.00	1	1	1	1	1	1	1	**8.00**
Chimpanzee	0.52	0.2	1	0.3	1	0.5	1	1	**5.52**
Crow	0.51	0.5	0	0.3	1	0.3	1	0.75	**4.36**
Pig	0.14	0	0.7	0.1	1	0	1	0.9	**3.84**
African grey parrot	0.49	0.5	0	0.6	1	0.3	0.1	0.8	**3.79**
Gorilla	0.14	0	1	0.1	1	0.4	0	1	**3.64**
Rhesus macaque	0.21	0	1	0.1	1	0.5	0.3	0.3	**3.41**
Dog (husky)	0.18	0	1	0.2	1	0	0	0.8	**3.18**
Cat	0.18	0	1	0.1	1	0.1	0	0.7	**3.08**
Mouse	0.07	0.1	0.6	0.1	1	0.2	1	0	**3.07**
Baboon	0.18	0	0	0.15	1	0.5	1	0.2	**3.03**
Wolverine	0.21	0	1	0	1	0	0	0.8	**3.01**
Rat	0.05	0	0.6	0.1	1	0.2	1	0	**2.95**
Lion	0.09	0	1	0	1	0	0	0.85	**2.94**
African elephant	0.49	0	0	0.3	1	0	0	0.85	**2.64**
Hummingbird	0.20	0.3	0	0	1	0	0	0.75	**2.25**
Rabbit	0.08	0.1	0	0	1	0	0	0.95	**2.13**
Tegu lizard	0.49	0	0.25	0	1	0.3	0	0	**2.04**
Monitor lizard	0.49	0	0.25	0	1	0.3	0	0	**2.04**
Anole	0.49	0	0.5	0	1	0.3	0	0	**2.04**
Giraffe	0.19	0	0	0	1	0	0	0.85	**2.04**
Horse	0.14	0	0	0	1	0	0	0.9	**2.04**
Cattle	0.09	0	0	0.1	1	0	0	0.85	**2.04**
Zebra	0.12	0	0	0	1	0	0	0.9	**2.02**
Nile crocodile	0.04	0	0.35	0	0.8	0.1	0	0	**1.65**
Saltwater crocodile	0.04	0	0.35	0	0.8	0.1	0	0	**1.65**
Hippopotamus	0.05	0	0	0	0.65	0	0	0.9	**1.60**
Sulcata tortoise	0.09	0	0.2	0	0.2	0	0	0.9	**1.39**
Giant octopus	0.15	0	0	0	0	0.2	0	1	**1.35**
Bottlenose dolphin	0.73	0	0	0.6	0	0	0	0	**1.33**
Killer whale	0.49	0	0	0.4	0	0	0	0	**0.89**

Animal	Brain	Bipedal	Binocular Vision	Language	Landbased	Thumb	Omnivorous	No tail	Total
Manta ray	0.20	0	0	0	0	0	0	0.5	**0.70**
Walrus	0.20	0	0	0	0.5	0	0	0	**0.70**
Giant cuttlefish	0.18	0	0.2	0	0	0	0	0	**0.38**
Sperm whale	0.07	0	0	0.2	0	0	0	0	**0.27**
Elephant fish	0.17	0	0	0	0	0	0	0	**0.17**

leveraged results considering all essential biological attributes contributed to the rise of Homo sapiens versus other animals, we found human score 8 on top and followed by Chimpanzee, which is somewhat expected, and then followed by crow. Most surprisingly, dolphin scored extremely low on the ranking despite their big brain because dolphin pretty much failed on every other category essential for the emergence of functional equivalent of the human species. Each listed attribute is essential for a capable biological species to adapt eventually to an industrial civilization.

Needless to say, a big brain is a requirement for comprehension of the environment, abstract concepts, new idea construction, and communication and complex ideas comprehension through language.

Binocular vision enables depth perception. Without depth perception, it is very difficult to develop geometric theories, advanced mathematics and creating tools that fit one part into another. Our brain is evolved and fine-tuned with binocular vision so that we have an innate understanding of geometry, shapes just like bats marvelously able to interprets rebounding high-frequency sound for obstacle detection.

An omnivorous diet is essential for the development of industrial civilization. First of all, carnivorous and omnivorous animals tend, on average, have greater cranial capacity because they are able to obtain more proteins from their intake, especially when they had similar biological attributes and lives in similar habitats. This is illustrated in omnivorous crow which has EQ score of 4.5 compares to vegetarian African Grey Parrot at 3.75, and omnivorous Chimpanzee at 4.8 compares to vegetarian Gorilla at 2.1. Dolphins and killer whales both are carnivorous and have sufficient protein to support their large brains. Omnivores feed on meat. Animals feeding on meat requires greater flexibility, agility, planning, and canniness to catch its prey. Those are the essential quality selected by natural evolution is also essential for the successful development of a civilization. Most importantly, omnivores are also adapt well to vegetables and starch. In order to transition from a hunter-gatherer society to an industrial society, an intermediate agricultural society phase requires each member of the species consume a significant amount of vegetation such as rice, wheat, and rye. A species can only digest meat can not significantly expand their population beyond scattering hunter-gatherer bands; therefore, trap on a stable local maximum level of energy extraction from the locality and unable to form into a flourishing industrial civilization.

Bipedal locomotion is essential because highly advanced technological society (ladder, building, tunnels, bridges, airplane, and auto) requires a biological hand to construct. Walking on hind legs freed the arms to perform that tasks. However, opposable thumb is not a consequence of bipedalism. Tree shrews have grabbing power without standing upright. Neither do ostrich and birds with bipedal locomotion developed opposable thumbs. It is the independent development of bipedal locomotion in combination with opposable thumb brings significant advantage to human. Human with a dexterous hand is able to manipulate and create objects, but to carry and move tools over long distances, requires bipedal locomotion which freed the forelimb for carrying. Later toolset and contemporary artifacts/edifices of Homo sapiens require creations which made up many parts originating from great distances from each other. It is utterly unthinkable that human is able to achieve greater technological improvements if bipedal locomotion is not evolved and forelimb is not freed to carry these parts across great distances, so tools creation can only be confined locally. Human can surely hold tool parts in their mouth to carry over great distance, but human jaw muscles are adapted for an omnivorous diet. If human had been exclusively carnivorous, greater jaw muscle would able to hold greater goods over large distances without bipedal locomotion and developed into an intelligent species almost identical to human except walking on four legs. However, we have just concluded that a carnivorous species cannot successfully transform its mode of living from a hunter-gatherer to an agricultural society. As a result, no transformation into industrial civilization is possible. On the other hand, Omnivorous but non-bipedal tree shrews with dexterous hands can develop tools using their hands, but its living range will be limited to the trees. Its forelimbs and hind legs are not well adapted to walk over great distances, so no complex tool making (requires materials from far away) is possible for this species. Most importantly, by limiting its own living range on the trees, it will never transition from hunting and gathering lifestyle to that of an agricultural mode of living (agriculture crops requires flat land for cultivation not on the trees. One could argue these animals can cultivate their host trees so that it becomes its own living habitat as well as crop producing warehouse. So a sort of horticultural revolution is possible. This reasoning is flawed in 2 ways. First, trees take a significant growth cycle because they are perennial. Artificial breeding and selection will take extremely long time to see significant improvement in food production, and the costs outweigh the effort to start such a transition. Secondly, a tree, no matter how finely tuned, will not produce as much food compares to staple crops, so energy return versus energy invested will always be less than a human agricultural society. A significant amount of energy is invested by the tree in its own maintenance of its trunk, bark, branches, and roots. As a result, the population supported by such a horticultural revolution will still hold lower population density in a given region than one started by human agricultural revolution. Horticulture society is also unable to undergo crop rotation which increases food intake diversity. If a tree with weaker trunk is selected by the species to breed in exchange for greater energy return in the forms of fruit production, then the species is on a suicide journey because its own survival is dependent on the sturdiness of the tree trunk itself to escape from land predators) even if it achieves characteristics just like human such as omnivorous diet (have the potential to expand its population density significantly), big brain, language, opposable thumb, land dwelling (the potential to use fire), binocular vision except not developing bipedal locomotion on flat land surfaces. It is noted that monkeys have greater grabbing power on their hands than even human. Paradoxically, stronger grabbing power trades with lesser precision control in tool making. Therefore, bipedal locomotion, once freed monkeys opposable thumb from branch grabbing, refined it for sophisticated tool making. So hand evolved ever more manipulative of objects as a self-reinforcing positive feedback loop. A luxury neither enjoyed by quadrupedal nor by arboreal species. Therefore, one can argue that there is a positive correlation between refined opposable thumbs at human level precision and bipedal locomotion but opposable thumb as an independently evolved feature must already present at the time when bipedal locomotion is evolved. Indeed, human lineage developed quickly after the emergence of convincing bipedal locomotion found in fossil remains of *Australopithecus Afarensis*. As a result, Gaussian normal, lognormal distribution should be right skewed, and at least sub log-normal distributed even just consider the data set from average mammalian sample to the emergence of anatomical Homo sapiens. For the simplicity of our argument and calculation, we shall treat the data as Gaussian normally distributed for now.

Furthermore, an animal has to be a land dweller. The use and control of fire enabled human to first transition from stone to bronze tools, and then to iron tools. With iron molding technology, human eventually constructed steel furnace and ushered in the industrial revolution. If human evolved in the ocean, no matter how smart we become, we would never be able to contain fire (fire manipulation is not possible underwater by all means) inside a furnace and transition from biological muscle power to steam power. Curiously enough as a thought experiment, it is possible that a hypothetical smart aquatic species with dexterous hands can utilize the steam vents from the ocean floor by casting a stone furnace around it and do useful work. However, such device cannot store an energy source and can not be transferred from one location to another. So their device resembles a localized medieval waterclock rather than a steam engine even though they capture energy from steam emitted by vents so be called a steam-powered engine. However, one should not be confused by the language verbiage tricks from its intrinsic property.

The opposable thumb is, of course, essential for the development and continued progress of human civilization. We have already mentioned that bipedal locomotion truly freed human hands for other tasks. Human hands create tools, and bipedal locomotion helps human to carry these tools over great distances. Palm with one opposable thumb may not be the only functional equivalent to a dexterous biological appendage enabling technological civilization. A hand with two or more opposable thumbs, or some other anatomically bendable structure is possible. It is likely such landscape of possibilities can also be quanta-sized by mathematics, but it is beyond the scope of this paper.

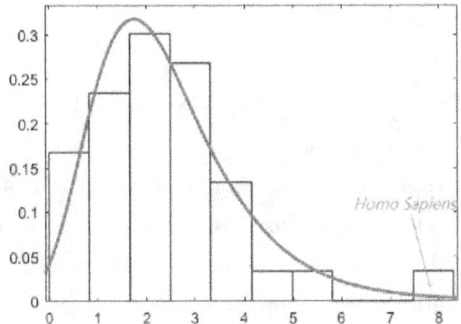

Figure 7.2.1: Lognormal distribution of species possessing different number of Homo sapiens' traits

The total attributes plotted for probabilistic distribution shows skewed normal/lognormal with human as an extreme outlier.

7.2.2   Permutation attempts and Generalized Biocomplexity

Recall previously we have demonstrated the growth of biological complexity through the manipulation of combination and permutation on the number of traits that constitutes a Homo sapiens. We expand and generalize the previous discussion by introducing the concept of permutation attempts. We will demonstrate how permutation attempts, permutation, and combination acting together mathematically constructs the lognormal distribution of species possessing different number of Homo sapiens' traits. A permutation attempt is defined as a permutation with repeated elements. Assume we have 7 traits initially with all species possessed only one of the 7 traits within each respective specific niche and each of the trait is equally divided into all species so that the chance a species possessed a trait is $\frac{1}{7}$ and the chance of gaining additional trait is also $\frac{1}{7}$, and after a given time period (a period of hundreds of millions of years), each initial species were given a chance to acquire new traits up to 7 attempts due to geologic changes or its migration. A species or its successor species may acquire new traits each time or remain unchanged, if we label each trait as an integer number, then the number sequence 1122344 and the number sequence 1112334 are permutation wise equivalent since they all reduces to the distinct permutation sequence of 123, so each permutation with repeats is treated as a permutation with a lower number of traits. That is, after a period of evolution, the finally evolved species possessed 3 traits with trait label in an increasing order. By simply grouping these species by traits acquired in an order, one could not distinguish the minute detail of trait acquisition. One species may stay without acquiring the next trait longer than the other. Since at each step there are 7 choices with equal chance, then, we will have a total of 7^7 possible choices. In general, given number of traits n, and t number of times new trait acquisition is allowed, whereas $t \leq n$, we have n^t possible choices. Out of these choices, the total number of permutation is given as:

$$\sum_{m=1}^{t} P(n, m) = P(n, 1) + P(n, 2) + ... + P(n, t) \tag{7.2.1}$$

Whereas $P(n, 1)$ represents permutation of 1 trait, and $P(n, 2)$ represents permutation of 2 traits, up to $P(n, t)$ represents permutation of t traits. Likewise the total number of combination is given as:

$$\sum_{m=1}^{t} C(n, m) = C(n, 1) + C(n, 2) + ... + C(n, t) \tag{7.2.2}$$

One can see that:

$$n^t > \sum_{m=1}^{t} P(n,m) > \sum_{m=1}^{t} C(n,m) \qquad (7.2.3)$$

The total possible choices greatly out-number the total distinct permutations and total distinct combinations. In reality, we have:

$$\sum_{m=1}^{t} C(n,m) \subset \sum_{m=1}^{t} P(n,m) \subset n^t \qquad (7.2.4)$$

That is, all extra repeats derives from the rudimentary number of combinations and permutations for a given number of traits, and distinct combination is a subset of distinct permutations, which, in turn, is a subset of all permutation with repeats, which is the total sum of all combinations, permutations, and partial permutations with repeats. In fact, the number of permutation with repeated attempts for any given distinct permutation can be determined using the following steps by first assuming the simplest case that $t = n$:

For species eventually possessed only 1 trait out of t speciation attempts, it is trivially to show that there are only n possible choices, since there are 8 slots with only 1 choice for each slot 1^7, and there is $P(n,1) = n$ permutations, so we have

$$1^7 \times P(n,1) = n \qquad (7.2.5)$$

For species eventually possessed 2 traits out of t speciation attempts, we start by counting the number of permutation with repeated element for each distinct permutation of 2 traits, we starts placing the distinct non-repeated sequence at the start of all slots, and for the remaining slots, each slot 2 trait attempts are allowed, and we shift the distinct 2 traits in permutation order one slot to the right, for the remaining slots, each slot 2 trait attempts are allowed, and we repeat the process until the last digit of the distinct permutation occupies the last slot

Slot 3+4+5+6	Slot 4+5+6	Slot 5+6	Slot 6	no slot
$13(1111)$	$113(111)$	$1113(11)$	$11113(1)$	111113

so we have in total:

$$\sum_{m=0}^{n-2} 2^m = 2^{t-2} + 2^{t-3} + 2^{t-4} + ... + 2^1 + 2^0 \qquad (7.2.6)$$

For slots to the left of the distinct permutation of 2 traits, those slots are not counted to avoid over-counting. It can be shown by simple counter examples that considering those slots ending up generating attempts already covered under previous scenario. The total number of 2 traits out of t speciation attempts is:

$$\sum_{m=0}^{n-2} 2^m \times P(n,2) \qquad (7.2.7)$$

For species eventually possessed 3 traits, the search space explodes and the number of repeats recursively depends on the number of repeats for each distinct permutation of 4 traits, and this pattern of recursion continues until counting for the repeats for distinct permutation of t traits. In fact, previous repeats with 1 and 2 traits can also be derived based on known number of permutation with repeats for 3 traits. Therefore, we work our way backwards.

For species eventually possessed t traits, each ordered permutation fills all t slots, so there is 0 remaining slot for n possible choices. We have $n^0 = 1$ for each permutation path, and the total number of n traits out of t speciation attempts is:

$$n^0 \times P(n, t) = \frac{n!}{(n-t)!} \tag{7.2.8}$$

For species eventually possessed $t - 1$ traits, with computer simulation for verification, as a demonstration for 3 traits out of 4 slots,

Slot 4	Slot 3	Slot 2
$123(1)$	$12(1)3$	$1(1)23$
$123(2)$	$12(2)3$	
$123(3)$		

we start placing the distinct non-repeated distinct permutation at the start of all slots with the last slot with 3 attempts, and then shifts the last digit of the permutation to the last slot, making slot 3 with 2 attempts, and then shifts the last two digits of the permutation to the last slots, making slot 2 with 1 attempt, the reason that last 2 shifts can not allow for all 3 attempts due to over-counting from other possible distinct permutations allowing repeats.

As a result, counting the number of permutation with repeated elements for each distinct permutation of $t - 1$ traits, we have:

$$\sum_{m=1}^{t-1} m = (t-1) + (t-2) + (t-3) + \ldots + 2 + 1 \tag{7.2.9}$$

The total number of $t - 1$ traits out of t speciation attempts is:

$$\sum_{m=1}^{t-1} m \times P(n, t-1) \tag{7.2.10}$$

For species eventually possessed $t - 2$ traits, based on computer simulation for verification, as a demonstration for 4 traits out of 5 slots, we start placing the distinct non-repeated distinct permutation at the start of all slots with the last 2 slots each with 3 attempts, and shifts the last digit of the permutation one slot to the right, and then the last digit 2 slots to the right, and then the entire permutation 1 slot to the right, and then on top of it shifting the last digit of permutation one slot right, and finally, shifting the entire permutation 2 slots right.

Slot 4+5	Slot 3+5	Slot 3+4	Slot 1+5	Slot 1+5	Slot 1+2
$123(11)$	$12(1)3(1)$	$12(11)3$	$(1)123(1)$	$(1)12(1)3$	$(11)123$
$123(12)$	$12(1)3(2)$	$12(12)3$	$(1)123(2)$	$(1)12(2)3$	
$123(13)$	$12(1)3(3)$	$12(21)3$	$(1)123(3)$		
$123(21)$	$12(2)3(1)$	$12(22)3$			
$123(22)$	$12(2)3(2)$				
$123(23)$	$12(2)3(3)$				
$123(31)$					
$123(32)$					
$123(33)$					

Again, the reason that all shifts other than the first can not allow for all 3 attempts as 3^2 choices due to over-counting from other possible distinct permutations allowing repeats. So we have the following setup:

3 × 3	2 × 3	2 × 2	1 × 3	1 × 2	1 × 1
11	11	11	11	11	11
12	12	12	12	12	
13	13	21	13		
21	21	22			
22	22				
23	23				
31					
32					
33					

As a result, counting the number of permutation with repeated elements for each distinct permutation of $t-2$ traits, we have:

$$[(3 \times 3) + (2 \times 3) + (1 \times 3)] + [(2 \times 2) + (1 \times 2)] + [(1 \times 1)]$$
$$= 3 \times (3 + 2 + 1) + 2 \times (2 + 1) + 1 \times (1)$$

$$\sum_{a_{ttempts}=1}^{t-2} a_{ttempts} \times \sum_{m=1}^{a_{ttempts}} m = (t-2) \times \sum_{m=1}^{t-2} m + (t-3) \times \sum_{m=1}^{t-3} m + ... + 1 \times \sum_{m=1}^{1} m \qquad (7.2.11)$$

The total number of $t-2$ traits out of t speciation attempts is:

$$\left(\sum_{a_{ttempts}=1}^{t-2} a_{ttempts} \times \sum_{m=1}^{a_{ttempts}} m \right) \times P(n, t-2) \qquad (7.2.12)$$

For species eventually possessed $t-3$ traits, we start to see the pattern, if we treat the number of permutation with repeated elements for each distinct permutation of $t-2$ traits as multiplication of 2 slots with ever decreasing count for each possibilities to avoid over-counting, then the number of permutation with repeated elements for each distinct permutation of $t-3$ traits is multiplication of 2 slots with ever decreasing count for each possibilities to avoid over-counting. As a result, counting the number of permutation with repeated elements for each distinct permutation of $t-3$ traits, we have:

$$\sum_{a_2=1}^{t-3} a_2 \times \left(\sum_{a_1=1}^{a_2} a_1 \sum_{m=1}^{a_1} m \right) =$$
$$(t-3) \times \left(\sum_{a_1=1}^{t-3} a_1 \sum_{m=1}^{a_1} m \right) + (t-4) \times \left(\sum_{a_1=1}^{t-4} a_1 \sum_{m=1}^{a_1} m \right) + ... + 1 \times \left(\sum_{a_1=1}^{1} a_1 \sum_{m=1}^{a_1} m \right)$$

The total number of $t-3$ traits out of t speciation attempts is:

$$\left[\sum_{a_2=1}^{t-3} a_2 \times \left(\sum_{a_1=1}^{a_2} a_1 \sum_{m=1}^{a_1} m \right) \right] \times P(n, t-3) \qquad (7.2.13)$$

In general, for species eventually possessed $t-j$ traits, counting the number of permutation with repeated elements for each distinct permutation of $t-j$ traits, we have:

$$\underbrace{\sum_{a_j=1}^{t-j} a_{j-1} \left[...a_3 \left[\sum_{a_2=1}^{a_3} a_2 \left(\sum_{a_1=1}^{a_2} a_1 \sum_{m=1}^{a_1} m \right) \right] \right]}_{\times j \text{ times}} \tag{7.2.14}$$

and the total number of $t - j$ traits out of t speciation attempts is:

$$\left\{ \underbrace{\sum_{a_{j-1}=1}^{t-j} a_{j-1} \left[...a_3 \left[\sum_{a_2=1}^{a_3} a_2 \left(\sum_{a_1=1}^{a_2} a_1 \sum_{m=1}^{a_1} m \right) \right] \right]}_{\times j \text{ times}} \right\} \times P(n, t-j) \tag{7.2.15}$$

Whereas the number of nesting summation is determined by the value of j. Notice that when $j = t - 1$, so that $t - j = 1$, that is the number of 1 trait out of all trait attempts, we have:

$$\underbrace{\sum_{a_j=1}^{1} a_j \left[...a_3 \left[\sum_{a_2=1}^{a_3} a_2 \left(\sum_{a_1=1}^{a_2} a_1 \sum_{m=1}^{a_1} m \right) \right] \right]}_{\times t-1 \text{ times}} =$$

$$1 \times \underbrace{[...1 \times [1 \times (1 \times 1)]]}_{\times t-2 \text{ times}} = 1$$

So one can possibly extrapolate step by step, in an inverse order, the total number of $t - 1$, $t - 2...3$, 2, and 1 traits out of t speciation attempts.

For our toy model, we list the number of permutation attempts with acquired 1, 2, 3, 4, 5, 6, and 7 traits given 7 initial trait choices and 7 trait acquisition attempts as:

Trait number	Repeated permutation attempts registered	Permutation $P(7,x)$	Combination $C(7,x)$	Total number 7^7/attempts ratio	Attempts/permutation ratio	Permutation/combination ratio
1 trait	7	$P(7,1) = 7$	$C(7,1) = 7$	7^6	1	1
2 traits	2,646	$P(7,2) = 42$	$C(7,2) = 21$	311.24	63	2
3 traits	63,210	$P(7,3) = 210$	$C(7,3) = 35$	13.03	301	6
4 traits	294,000	$P(7,4) = 840$	$C(7,4) = 35$	2.80	350	24
5 traits	352,800	$P(7,5) = 2520$	$C(7,5) = 21$	2.33	140	120
6 traits	105,840	$P(7,6) = 5040$	$C(7,6) = 7$	7.78	21	720
7 traits	5,040	$P(7,7) = 5040$	$C(7,7) = 1$	163.40	1	5040

Although the total number / repeated permutation attempts can only be evaluated at an integer number of traits as:

$$R_p(x) = \frac{t_{Attempt}}{\left\{ \underbrace{\sum_{a_j=1}^{t-x} a_j \left(... \left(a_1 \sum_{m=1}^{a_1} m \right) \right)}_{\times t-x \text{ times}} \right\} \times P(n, x)} \tag{7.2.16}$$

it does not prevent us from finding a best fit continuous function that connects all 7 trait numbers (for different each distinct pairing of trait number n and attempts number t there is a corresponding best fit function that needs to be derived), we assume that there exists such a function as:

$$R_p(x) \approx \exp\left(a(x)^{bx} + c(x+d)^{x^f}\right) \tag{7.2.17}$$

Exponential is used since regression fitting is easier based on the log of the ratio of the total number / repeated permutation attempts. The total number of species (BCS, or alternatively called the PDF of biocomplexity if normalized) possessed any number of traits can then be derived as:

$$S_{total}(T_{rait}, x, t_{Attempt}) = \frac{t_{Attempt}}{R_p(x)} \tag{7.2.18}$$

Likewise, the best fit continuous function that connects all counted numbers for distinct permutation based on each finally acquired trait number can be expressed as:

$$P(T_{rait} = 7, x) = P(x) \approx \frac{7^7}{R_p(x) \times A_P(x)} \tag{7.2.19}$$

$$A_P(x) \approx \exp\left(a(x+b)^c + d(x+f)^g\right) \tag{7.2.20}$$

Whereas $A_P(x)$ is the best fit continuous function that connects all attempts to permutation ratio, which in exact formula should be expressed as:

$$A_P(x) = \frac{\left\{\sum_{a_j=1}^{t-x} a_j\left(\ldots\left(a_1\sum_{m=1}^{a_1}m\right)\right)\right\} \times P(n,x)}{P(n,x)} = \underbrace{\sum_{a_j=1}^{t-x} a_j\left(\ldots\left(a_1\sum_{m=1}^{a_1}m\right)\right)}_{\times t-x \text{ times}} \tag{7.2.21}$$

that is, we do not directly express the continuous function $P(T_{rait}, x)$ due to smoothly phasing in speciation based predominately on distinct permutation and combination and phasing out speciation based predominately on permutation allowing repeats, as the number acquired traits increase, which is discussed as follows:

We have demonstrated before, due to the passive nature of biocomplexity, the likelihood under normalized model, or the absolute values at the right hand tail for species with increasing number of traits acquired must be reduced to reflect that not all permutation with repeats, or permutation path can lead to speciation. This can be done by introducing a new function $G(x)$ which initially started at 0 and increases exponentially:

$$G(x) = \ln\left(\left(x - (T_{rait} - 1)\right)^{\frac{(x-(T_{rait}-1))^{ax}}{d}}\right) + v \tag{7.2.22}$$

which helps to gradually phase out permutation with repeats and phase in distinct permutation and ultimately distinct combination. It has an initial intercept value of v, which helps to smoothly joins two step function without discontinuity. Its value can be determined by solving for the v in the equality:

$$\frac{t_{Attempt}}{v A_P(3) R_p(3)} = \frac{t_{Attempt}}{R_p(3)}$$
$$v = \frac{R_p(3)}{A_P(3) R_p(3)} = \frac{1}{A_P(3)} \tag{7.2.23}$$

$$S_{total}(T_{rait}, x, t_{Attempt}) = \begin{cases} \frac{t_{Attempt}}{R_p(x)} & T_{rait} \leq 3 \\ \frac{t_{Attempt}}{G(x) \times R_p(x) \times A_P(x)} & T_{rait} > 3 \end{cases} \tag{7.2.24}$$

In this example given, all permutation with repeated attempts with up to 3 acquired traits all have successfully evolved into species. For permutation with repeated attempts with higher number of acquired traits, increasing smaller proportion of repeated permutation attempts realized into species.

The second conditional can be further simplified as follows, by substituting $R_p(x)$ and $A_P(x)$ with their exact formula:

$$\frac{t_{Attempt}}{G(x) \times R_p(x) \times A_P(x)}$$

$$= \frac{t_{Attempt}}{G(x) \times \frac{t_{Attempt}}{\left\{\sum_{a_j=1}^{t-x} a_j \left(\dots (a_1 \sum_{m=1}^{a_1} m)\right)\right\} \times P(n,x)} \times \sum_{a_j=1}^{t-x} a_j \left(\dots (a_1 \sum_{m=1}^{a_1} m)\right)}$$

$$= \frac{P(n,x)}{G(x)}$$

so that the equation simplifies to:

$$S_{total}(T_{rait}, x, t_{Attempt}) = \begin{cases} \frac{t_{Attempt}}{R_p(x)} & T_{rait} \leq 3 \\ \frac{P(T_{rait}, x)}{G(x)} & T_{rait} > 3 \end{cases} \tag{7.2.25}$$

The initial intercept value of v has to be re-calibrated, which helps to smoothly joins two step function without discontinuity. Its value can be determined by solving for the v in the equality:

$$\frac{P(T_{rait}, 3)}{v} = \frac{t_{Attempt}}{R_p(3)}$$

$$v = \frac{P(T_{rait}, 3) R_p(3)}{t_{Attempt}} \tag{7.2.26}$$

In general, the remaining proportion can be determined by parametric manipulation on the coefficients of a or d of function $G(x)$. The best fit and exact value can be derived based on underlying data, which is discussed in the next section.

7.2.3 Best Fit

Recall in Chapter 4 we have defined biocomplexity in terms of permutation and combination of essential traits. Now we have just generalized the biocomplexity concept by introducing the permutation with repeats. We can now apply them on our dataset. We define our biocomplexity given 7 selected traits. Whereas the geometric mean of each trait is computed based on our earlier work on the emergence of human with 7 traits as:

$$p = \left(\prod_{i=1}^{7} p_i\right)^{\frac{1}{7}} \approx \frac{1}{8.8282} \tag{7.2.27}$$

For each statistical bin representing the number of traits possessed by a particular type of species, we obtain the percentage of the number of species received score within the designated bin of our dataset, designated as $r_{Observed}$, compared with the percentage of species predicted by $S_{total}(T_{rait}, x, t_{Attempt})$. Whereas bin is the designated bin ranging from 1 to 7 out of total 7 traits. The steps of bins are separated by one apart, but this number can be lowered for increased data resolution and precision.

$$\ln\left(\frac{r_{Observed}}{\left(\frac{\int_{bin}^{bin+1} S_{total}(T_{rait}, x, t_{Attempt})dx}{\int_0^8 S_{total}(T_{rait}, x, t_{Attempt})dx}\right)}\right) \tag{7.2.28}$$

There is many way to manipulate function $G(x)$, here we simply manipulate d, We first blindly assumes that $d = 2000$, and we find that bin < 2 and Bin > 5 is over-represented. There are several possible implications. It may imply that more species receiving a score between 2 and 5 should have been included but was not included in the original data set. Most species received a score between 2 and 5 are terrestrial based vertebrates especially mammals, suggesting more of them should be included. Bin > 5 is over-represented, is also due to our data selection bias, since our data set is too small to give emergence to chimpanzee and human but we nevertheless included them. It may also imply, that one should not treat each trait as equally likely to be gained in the environment. The existence of some traits such as bipedal and omnivorous diet are over-represented in typical mammals and birds, so that attempts that should have been attributed to the formation of species possessing higher number of traits given higher rounds of trait acquisitions instead contributed toward species sharing lower number of traits. Therefore, over-representation of species with lower number of shared traits is possible with our model by assuming initial 8 trait within all species has non-equal chance of existence. We discussed more in detail on this later.

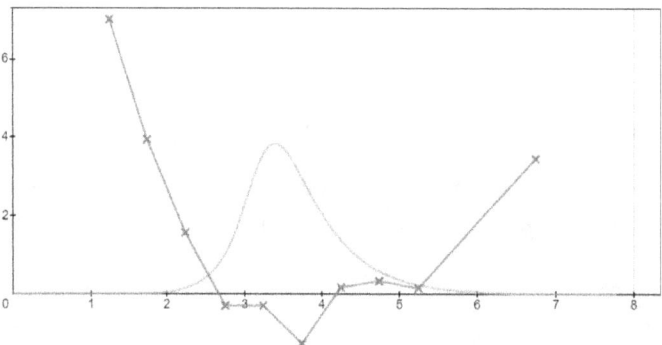

Figure 7.2.2: Assuming order of traits acquisition is partially relevant and $d = 2000$

Of course, our data set for the illustrative purpose is very incomplete. Assuming the data set is complete, one can minimize the error rate between the data set and $S_{total}\left(T_{rait}, x, t_{Attempt}\right)$ by fine tuning the parameter $d = 1920$ so that such distribution best represents the dataset by using the threshold test:

$$\sum_{bin=0}^{7} \left| \ln \left(\frac{r_{Observed}}{\left(\frac{\int_{bin}^{bin+1} S_{total}(T_{rait}, x, t_{Attempt}) dx}{\int_{0}^{8} S_{total}(T_{rait}, x, t_{Attempt}) dx} \right)} \right) \right| = 13.1712 \tag{7.2.29}$$

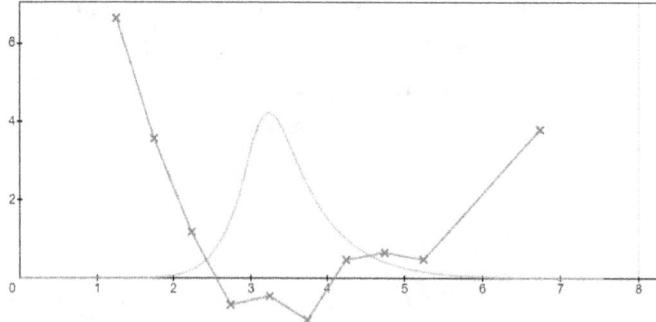

Figure 7.2.3: Assuming order of traits acquisition is partially relevant and $d = 1920$

This implies that biocomplexity is best represented by permutation with repeats, a subset of distinct permutation. If one were to extrapolate this trend well into higher number traits, we find that based on the data fitting for number of species at the lower number of trait attainment, it predicts nearly 100 species sharing all 7 traits given 7 trait acquisition attempts through evolutionary time.

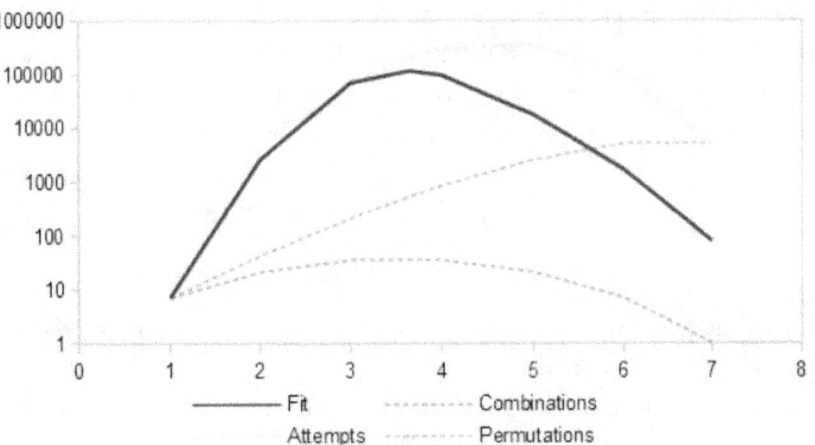

Figure 7.2.4: On a logscale, the best fit extrapolates 100 species with 7 traits, while based on combination alone (the lowest dashed curve) there should be just 1 species with 7 traits, the middle dashed curve is based on distinct permutation only, and the highest dashed curve is based on all possible permutation with repeats.

In reality, the conclusion implies that combination of traits alone is a necessary but insufficient condition defining species. There must exists other species in which the order of the traits attainment does not follow the unique evolutionary path leading to human but nevertheless eventually possessed all 7 traits uniquely identified with human. We illustrate this by showing a hypothetical species that evolves to become an intelligent, tool-using species in an alternative evolutionary setting.

That is, a particular descendants of dinosaurs (possibly *troodon*), either escaped the asteroid impact or earth avoided the impact altogether, could have achieved its transcendence through the steps of language, semi-bipedalism, binocular vision, social, opposable thumb, fully bipedalism, omnivorous diet, and large cranial capacity.

Figure 7.2.5: A hypothetical species that evolves to become an intelligent, tool-using species possessed all traits associated with human such as bipedalism, binocular vision, opposable thumbs, language, social, large cranial capacity, and omnivorous diet yet it is not human. This is to show that the order of traits acquisition (permutation) also determines the total number of intelligent species emergence. Notices that the species possess an ornamental feathered crown, much like human hair, due to sexual selection.

On the other hand, if one is confident that only =1 species sharing all 7 traits given 7 trait acquisition attempts through evolutionary time has been achieved or possible given 7 most traits uniquely identified with human. Then a more drastic, steeper decline on the number of possible speciation given increasing traits shared is required. One approach is adding the third conditional to the existing step wise function of $S_{total}\left(T_{rait}=7,x,t_{Attempt}\right)$. That is, when $G\left(x\right)=1$ and $x_0=5.645$, at the point where $S_{total}\left(T_{rait}=7,x,t_{Attempt}\right)=\frac{t_{Attempt}}{R_p(x)\times A_P(x)}=P\left(T_{rait},x\right)$, we introduce a new phasing function $G_2\left(x\right)$ as:

$$G_2\left(x\right)=\begin{cases}\ln\left(\left(x-\left(x_0-1\right)\right)^{\frac{\left(x-\left(x_0-1\right)\right)^x}{100}}\right)+v_2 & G_2\left(x\right)<1 \\ 1 & G_2\left(x\right)\geq 1\end{cases}$$

(7.2.30)

This phasing function is capped at 1 because we assumed that at least one speciation is possible based on all possible permutation paths leading to a distinct combination.

It has an initial intercept value of v_2, which helps to smoothly joins three step function without discontinuity. Its value can be determined by solving for the v_2 in the equality:

$$\frac{t_{Attempt}}{v_2 A_P\left(x_0\right)R_p\left(x_0\right)P_C\left(x_0\right)}=\frac{t_{Attempt}}{A_P\left(x_0\right)R_p\left(x_0\right)}$$
$$v_2=\frac{A_P\left(x_0\right)R_p\left(x_0\right)}{A_P\left(x_0\right)R_p\left(x_0\right)P_C\left(x_0\right)}=\frac{1}{P_C\left(x_0\right)}$$

(7.2.31)

The third conditional is then added:

$$S_{total}\left(T_{rait}=7,x,t_{Attempt}\right)=\begin{cases}\frac{t_{Attempt}}{R_p(x)} & T_{rait}\leq 3 \\ \frac{t_{Attempt}}{G(x)\times R_p(x)\times A_P(x)} & 3<T_{rait}\leq 5.645 \\ \frac{t_{Attempt}}{G_2(x)R_p(x)A_P(x)P_C(x)} & 5.645<T_{rait}\end{cases}$$

(7.2.32)

$$P_C\left(x\right)=\frac{P\left(T_{rait},x\right)}{C\left(T_{rait},x\right)}$$

(7.2.33)

Whereas $P_C\left(x\right)$ is the intrinsic permutation to combination ratio. The $C\left(T_{rait},x\right)$ is the raw combination of choosing x out of given number of traits.

Again, the equation can be simplified as:

$$S_{total}\left(T_{rait},x,t_{Attempt}\right)=\begin{cases}\frac{t_{Attempt}}{R_p(x)} & T_{rait}\leq 3 \\ \frac{P(T_{rait},x)}{G(x)} & 3<T_{rait}\leq 5.645 \\ \frac{C(T_{rait},x)}{G_2(x)} & 5.645<T_{rait}\end{cases}$$

(7.2.34)

The third condition is simplified through:

$$\frac{P\left(T_{rait},x\right)}{G\left(x\right)G_2\left(x\right)P_C\left(x\right)}=\frac{P\left(T_{rait},x\right)}{1\times G_2\left(x\right)\frac{P(T_{rait},x)}{C(T_{rait},x)}}=\frac{C\left(T_{rait},x\right)}{G_2\left(x\right)}$$

Whereas the intercept v_2 of $G_2\left(x\right)$ is modified as:

$$\frac{P\left(T_{rait},x_0\right)}{G\left(x_0\right)v_2 P_C\left(x_0\right)}=\frac{P\left(T_{rait},x_0\right)}{G\left(x_0\right)}$$

(7.2.35)

$$\frac{P\left(T_{rait},x_0\right)}{G\left(x_0\right)v_2 P_C\left(x_0\right)}=\frac{P\left(T_{rait},x_0\right)}{G\left(x_0\right)}$$
$$v_2=\frac{1}{P_C\left(x_0\right)}$$

(7.2.36)

The final result is shown below:

270

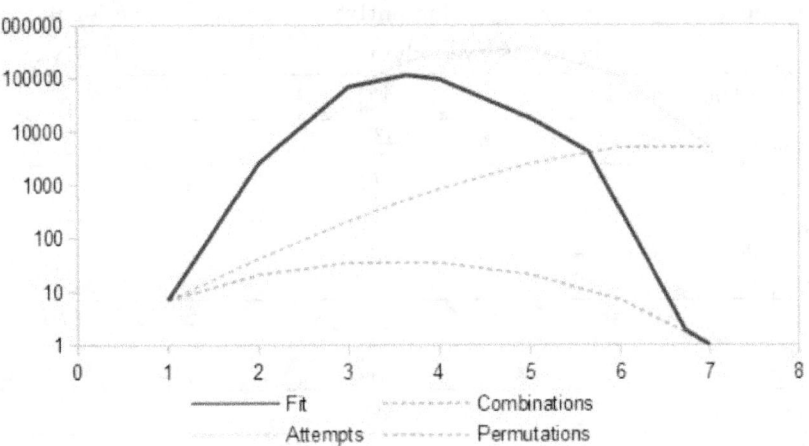

Figure 7.2.6: On a logscale, the best fit now descends faster and merges with the combination curve (the lowest dashed curve) at 6.724 traits and extrapolates 1 species possessed 7 traits, the middle dashed curve is based on distinct permutation only, and the highest dashed curve is based on all possible permutation with repeats.

On the other hand, if one is confident that only <1 species sharing all 7 traits given 7 trait acquisition attempts through evolutionary time has been achieved, which is possible since we have shown that given 7 traits most uniquely identified with human, their geometric mean emergence chance per each trait is $p < \frac{1}{7}$. It is therefore, even by exhaustively performing all distinct 7^7 attempts, a less than unity chance for the emergence of each permutation path of acquiring 7 traits.

$$\left(\frac{7^7}{p^7}\right) < 1 \tag{7.2.37}$$

Furthermore, not only $p < \frac{1}{7}$, our observation of the chance of our cohort of species of mammals, reptiles and birds possessing each particular traits of social, thumb, binocular vision, bipedal, omnivorous, large cranial capacity, and language is non uniformly $< \frac{1}{7}$. That is:

$$p_1 \neq p_2 \neq p_3 \neq p_4 \neq p_5 \neq p_6 \neq p_7 \tag{7.2.38}$$

Certain traits such as bipedal is almost universally observed in birds, and omnivorous diet is common in all clades, and some other traits such as opposable thumb is a much less frequent adaptation primarily dominated by arboreal species. As a result, species possessed lower number of traits is over represented and those with higher number of traits become under represented, thus the PDF of species profile becomes left shifted. In order for such left shifted PDF of species profile remain fixed in size and the relative emergence of each permutation path of acquiring traits fixed as $\left(\frac{7^7}{p^7}\right)$ in our example, we have to introduce the following assumption:

1. The total sum $p_1 + p_2 + \dots + p_{Trait} = \sum_{i=1}^{Trait} p_i = 1$.

2. The geometric mean of all traits $\left(\prod_{i=1}^{Trait} p_i\right)^{1/Trait} < \frac{1}{Trait}$, yet $\frac{\arg\max p_i}{\arg\min p_i} < \epsilon$

For assumption 1, at the start of rounds of trait acquisition, all traits are assumed to be independent of each other, by multiplicative property, only when the total probability is at unity, the final multiplicative results remain at unity. When the total probability < unity, the final multiplicative results < unity. When the total probability > unity, the final multiplicative results > unity. When the total probability $\neq$ unity, rescaling/normalization is feasible but the relative emergence of each permutation path of acquiring traits is no longer maintained. In order to fulfill this criteria, we have to reduce the chance of binocular vision, large cranial capacity, opposable thumbs, bipedal, language, and omnivorous diet by a certain percentage so that $\sum_{i=1}^{Trait} p_i = 1$.

Trait name	Currently Observed chance	Predicted chance for 65 Mya
binocular vision	$\frac{2526}{25483}$	$\frac{2526-450}{25483}$
large cranial capacity	$\frac{1214}{25483}$	$\frac{1214\times0.2}{25483}$
opposable thumbs	$\frac{353}{25483}$	$\frac{353\times0.4}{25483}$
bipedal	$\frac{10089}{25483}$	$\frac{10089\times0.732}{25483}$
language	$\frac{5150}{25483}$	$\frac{5150\times0.87}{25483}$
omnivorous	$\frac{8278}{25483}$	$\frac{8278\times0.917}{25483}$
social	$\frac{3586}{25483}$	$\frac{3586}{25483}$
Total sum	1.22418	1

Although these traits are largely independent of each other, some species already possessed more than one of each these traits, by assuming the start of simulation occurred at 65 Mya, it is reasonable to assume that certain traits were less dominant at the time.

For assumption 2, it is an alternative way to state that no 2 traits should share the same probability so that

$$p_1 \neq p_2 \neq \ldots \neq p_{Trait} \tag{7.2.39}$$

As long as discrepancies exist between traits and the sum of all traits equals unity, then the gain in chance in some traits becomes the loss in chance in others and results in a geometric mean $< \frac{1}{Trait}$. Geometric mean is strictly less than arithmetic mean for arbitrary number of traits when there is difference of ε between them. A full proof is not given but for base case with 2 factors is easily demonstrated as:

$$(a + \varepsilon)(a - \varepsilon) = a^2 - 2a\varepsilon - \varepsilon^2 \tag{7.2.40}$$

Since $a > 0$ and $\varepsilon > 0$, we have:

$$(a + 0)(a - 0) = a^2 > a^2 - 2a\varepsilon - \varepsilon^2 = (a + \varepsilon)(a - \varepsilon) \tag{7.2.41}$$

$$\left(a^2\right)^{1/2} > \left(a^2 - 2a\varepsilon - \varepsilon^2\right)^{1/2} \tag{7.2.42}$$

and:

$$(a + \varepsilon) + (a - \varepsilon) = 2a = 2a = (a + 0) + (a - 0) \tag{7.2.43}$$

$$\frac{1}{2} \times 2a = \frac{1}{2} \times 2a \tag{7.2.44}$$

At the same time, $\frac{\arg\max p_i}{\arg\min p_i} < \epsilon$ states that the ratio between the trait with the largest value and the trait with the smallest should be be bounded by ϵ. When one trait becomes almost universally dominant among all species under analysis it fails our assumption for prediction analysis. For example, vertebrate with thermoregulation, egg laying/womb bearing had become universal trait among all species by 65 Mya. In general, as species evolves through time, more traits becoming universal among them through bottleneck events in which only species possessed more traits survived, creating founders effect.

Due to differential chance per each trait, the process for computing the weighted proportion of species shared x traits is generalized as:

$$P_r(x) = \sum_{k=1}^{C(T_{rait},x)} \frac{P(T_{rait},x)}{C(T_{rait},x)} \times S_{um}(x, \{x\}_k) \tag{7.2.45}$$

$$S_{um}(x, \{x\}) = \underbrace{\sum_{j_n=1}^{j_n \in \{x\}} \cdots \sum_{j_2=1}^{j_2 \in \{x\}} \sum_{j_1=1}^{j_1 \in \{x\}}}_{\times t-x \text{ nested } \Sigma} \left(\prod_{i=1}^{i \in \{x\}} p_i \right) \times \underbrace{(p_{j1}p_{j2}...p_{jn})}_{\times t-x \text{ times}} \times \left(\frac{A_P(x)}{x^{t-x}} \right) \tag{7.2.46}$$

$$A_P(x) = \underbrace{\sum_{a_j=1}^{x} a_j \left(... \left(a_1 \sum_{m=1}^{a_1} m \right) \right)}_{\times t-x \text{ times}} \tag{7.2.47}$$

The number of possible permutation is multiplied with $S_{um}(x)$, $S_{um}(x)$ is the sum of combined chance of each *permutation with repeats* derived from the *same permutation*. For example, for 3 out 4 traits, the permutation with repeats given the same permutation 134 is:

134(1)	(1)134
134(3)	(3)134
134(4)	(4)134

$\prod_{i=1}^{i \in \{x\}} p_i$ is the combined chance of given selected x traits. The set $\{x\}_k$ is defined as any collection of x number of traits equivalent to one of the possible combinations of $C(T_{rait}, x)$. If set $\{A\} = \{1, 2...n\}$ is defined as the total set of all possible number of traits, then $\{x\} \subset \{A\}$. Nested loops are required to enumerate all possible pairing between all possible choices within the set $\{x\}$. The number of loops required depends on the number of total attempts t excluding the number of attempts needed for acquiring x distinct traits, as a result, $t - x$ times. Since each loop requires the summation of x choices as the number of traits within the set $\{x\}$, the total number of uniquely defined combined chance for permutation with repeats is x^{t-x}. A given permutation can be broken apart and placed at several starting positions within all possible permutation attempts with repeats for the same permutation, so $A_P(x) \geq x^{t-x}$, whereas each starting position gives to x^{t-x} number of uniquely defined combined chance for permutation with repeats. Recall, whereas $A_P(x)$ is the ratio of permutation with repeats over the number of permutations, or equivalently, the number of permutation with repeats per each unique permutation. Given $\frac{A_P(x)}{x^{t-x}}$, it gives us the factor, or multiple for each uniquely defined combined chance for permutation with repeats.

Since the final result of multiplication does not depend on the ordering, $S_{um}(x)$ is identical among all permutation paths derived from the same combination, hence it is multiplied by $\frac{P(T_{rait},x)}{C(T_{rait},x)}$, the number of permutation per each unique combination, finally, the sum of combined chance of all *permutation with repeats* derived from the *same combination* is summed up with sum of combined chance of all permutation with repeats from *every possible combination* of x traits.

If $S_{um}(x)$ is divided by $A_P(x)$, (the number of permutation with repeats per each unique permutation), one arrives at the weighted chance of each permutation path with repeats:

$$S_{um}\left(x, \{x\}_{k \in \{1,...,C(T_{rait},x)\}}\right) \times \frac{1}{A_P(x)} \tag{7.2.48}$$

and $P_r(x)$ can be modified as:

$$P_r(x) = \sum_{k=1}^{C(T_{rait},x)} \frac{A_P(x) P(T_{rait},x)}{C(T_{rait},x)} \times S_{um}(x, \{x\}_k) \times \frac{1}{A_P(x)} \tag{7.2.49}$$

Whereas $\frac{A_P(x)P(T_{rait},x)}{C(T_{rait},x)}$ is the the number of permutation with repeats per each unique combination.
The total number of species sharing $x$trait is then easily derived by:

$$P_r(x) \times (T_{rait})^{T_{rait}} \tag{7.2.50}$$

Simulation based on our assumption yields the following left shifted distribution:

Trait number	Repeated permutation attempts registered	Permutation $P(7,x)$	Combination $C(7,x)$	Total number 7^7/attempts ratio	Attempts/permutation ratio	Permutation/combination ratio
1 trait	277	277	277	2,980.23	1	1
2 traits	27,237	432	216	30.27	63	2
3 traits	211,999	704	117	3.89	301	6
4 traits	378,656	1,082	45	2.18	350	24
5 traits	184,030	1,315	10.95	4.48	140	120
6 traits	21,933	1,044	1.45	37.59	21	720
7 traits	367	367	0.07	2,248.28	1	5040

Notice that the original attempts to permutation ratio and permutation to combination ratio remains unchanged, but the total number of permutation and combination changes positively with the number of permutation with repeats.

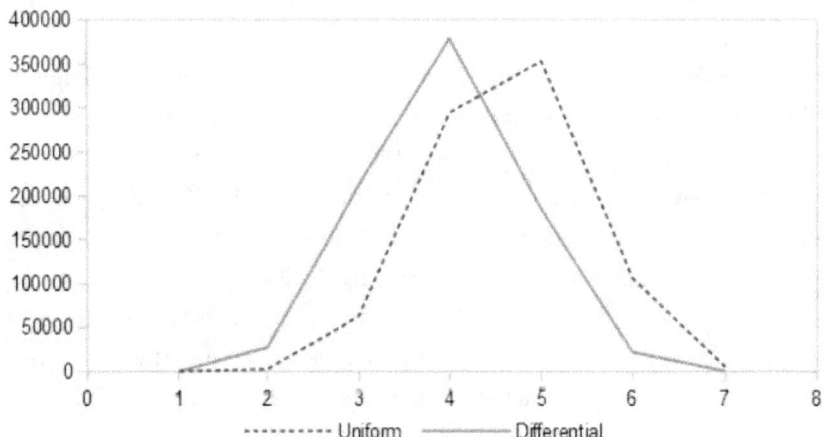

Figure 7.2.7: With differential chance for each trait within all species across all times and a geometric mean $< \frac{1}{7}$, species with more traits are now under-represented and with fewer traits are over-represented.

By using this new distribution, we perform the next fit function for $R_p(x)$ and $A_P(x)$ as:

$$R_p(x) \approx \exp\left(ax^b + c(x+d)^{f^x}\right) \tag{7.2.51}$$

Which is slightly simpler than the earlier fit and revised permutation function since it is no longer $P(T_{rait}, x)$:

$$P'(x) \approx \exp\left(ax^b + c(x+d)\right) \tag{7.2.52}$$

and revised combination function since it is no longer $C(T_{rait}, x)$:

$$C'(x) \approx ax^b + c(x+d) \tag{7.2.53}$$

In order to best fit existing data, $G(x)$'s $d = 8480$, $v_1 = \frac{p'(2.334)R_p(2.334)}{7^7}$, $G_2(x)$'s $d = 70$, $v_2 = \frac{C'(5.786)}{P'(5.786)}$, so a new best fit is derived as:

$$S_{total}\left(T_{rait} = 7, x, t_{Attempt}\right) = \begin{cases} \frac{t_{Attempt}}{R_p(x)} & T_{rait} \leq 2.334 \\ \frac{P'(x)}{G(x)} & 2.334 < T_{rait} \leq 5.785 \\ \frac{C'(x)}{G_2(x)} & 5.785 < T_{rait} \end{cases} \qquad (7.2.54)$$

Notice that the joining point between the first two sections have shifted to $x = 2.334$, since the entire distribution has been left shifted, and the evaluation at $x = 2.334$ matches the value $x = 3$ formerly. Other possible merging points are $x = 2$ and $x = 3$.

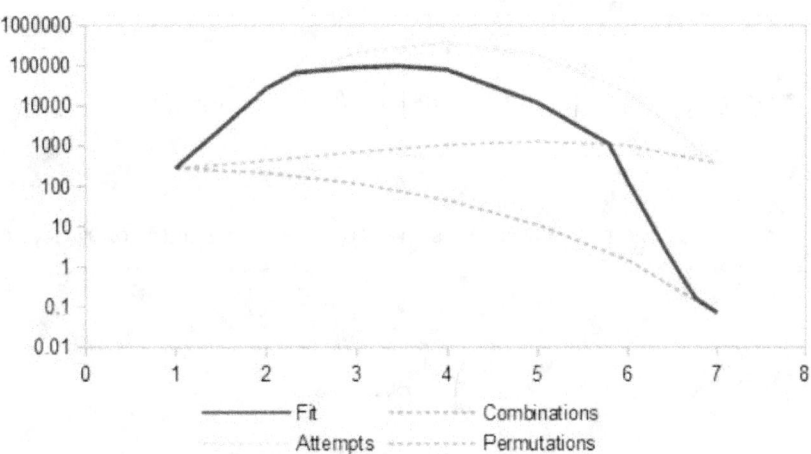

Figure 7.2.8: On a logscale, the best fit now descends faster and merges with the combination curve (the lowest dashed curve) at 6.769 traits and extrapolates 7% of 1 species possessed 7 traits, the middle dashed curve is based on distinct permutation only, and the highest dashed curve is based on all possible permutation with repeats. It is possible to extrapolates less than a unity of species possessed 7 traits due to limited number of attempts to guarantee an emergence. Nevertheless, the assumption that at least one speciation is possible based on all possible permutation paths leading to a distinct combination continues to hold.

However, the total number of species realized with this fit function falls below our original assessment of cumulatively 1 million to 4 million species evolved during the Cenozoic since:

$$\int_1^7 S_{total}\left(T_{rait} = 7, x, t_{Attempt}\right) dx = 214,521 < 1 \times 10^6 \qquad (7.2.55)$$

Therefore, data fitting alone can only determine the relative proportion/frequency of species possessed all ranges of traits but not the absolute number of species. We can find a better match with the absolute number of species evolved by assuming given $n = 7$ traits with $t = 8$ speciation attempts. The computation procedure is similar to the case when $n = t$, except that this time the attempts to permutation ratio shifts one bracket so that trait 7: $\sum\limits_{m=1}^{7} m \times P(n,n)$

trait 6: $\sum\limits_{a_1=1}^{6} a_1 \times \left(\sum\limits_{m=1}^{a_1} m \right) \times P(n,n-1)$

trait 5: $\sum\limits_{a_2=1}^{5} a_2 \times \left(\sum\limits_{a_1=1}^{a_2} a_1 \sum\limits_{m=1}^{a_1} m \right) \times P(n,n-2)$

...

trait j: $\underbrace{\sum\limits_{a_{n-j}=1}^{j} a_{n-j} \left(... \left(a_1 \sum\limits_{m=1}^{a_1} m \right) \right)}_{\times n-j+1 \text{ times}} \times P(n,n-(n-j))$

In general, as $t > n$, the difference between $t - n$ determines the attempts to permutation ratio, i.e. if $t = n + 3$, we have:

trait 7: $\sum\limits_{a_2=1}^{n} a_2 \times \left(\sum\limits_{a_1=1}^{a_2} a_1 \sum\limits_{m=1}^{a_1} m \right) \times P(n,n)$

trait 6: $\sum\limits_{a_3=1}^{n-1} a_3 \times \left(\sum\limits_{a_2=1}^{a_3} a_2 \sum\limits_{a_1=1}^{a_2} a_1 \sum\limits_{m=1}^{a_1} m \right) \times P(n,n-1)$

trait 5: $\sum\limits_{a_4=1}^{n-2} a_4 \times \left(\sum\limits_{a_3=1}^{a_4} a_3 \sum\limits_{a_2=1}^{a_3} a_2 \sum\limits_{a_1=1}^{a_2} a_1 \sum\limits_{m=1}^{a_1} m \right) \times P(n,n-2)$

...

trait j: $\underbrace{\sum\limits_{a_{n-j+2}=1}^{j} a_{n-j+2} \left(... \left(a_1 \sum\limits_{m=1}^{a_1} m \right) \right)}_{\times n-j+3 \text{ times}} \times P(n,n-(n-j))$

Likewise, as $t < n$, the difference between $n - t$ determine the attempts to permutation ratio in reverse order, i.e. if $t = n - 3$, we have:

trait 4: $1 \times P(n,t)$

trait 3: $\sum\limits_{m=1}^{t-1} m \times P(n,t-1)$

trait 2: $\sum\limits_{a_1=1}^{t-2} a_1 \times \left(\sum\limits_{m=1}^{a_1} m \right) \times P(n,t-2)$

trait 1: $\sum\limits_{a_2=1}^{t-3} a_2 \times \left(\sum\limits_{a_1=1}^{a_2} a_1 \sum\limits_{m=1}^{a_1} m \right) \times P(n,t-3)$

...

trait j: $\underbrace{\sum\limits_{a_{t-j-1}=1}^{j} a_{t-j-1} \left(... \left(a_1 \sum\limits_{m=1}^{a_1} m \right) \right)}_{\times t-j \text{ times}} \times P(n,t-(t-j))$

We list the number of permutation attempts with acquired 1, 2, 3, 4, 5, 6, and 7 traits given 7 initial trait choices and 8 trait acquisition attempts with differential chance of 7 traits as:

Trait number	Repeated permutation attempts (assuming $p = 1/7$	Repeated permutation attempts registered	Permutation $P'(x)$	Combination	Total number 7^8/attempts ratio	Attempts/permutation ratio	Permutation/combination ratio
1 trait	7	650	650	650.40	8,918.11	1	1
2 traits	5,334	127,057	1,000	500.22	45.65	127	2
3 traits	202,860	1,382,803	1,431	238.58	4.19	966	6
4 traits	1,428,840	2,883,806	1,695	70.64	2.01	1701	24
5 traits	2,646,000	1,317,178	1,254	10.45	4.40	1050	120
6 traits	1,340,640	87,759	330	0.46	66.09	266	720
7 traits	141,120	1,068	38	0.0076	5,428.70	28	5040

Again, we perform the next fit function for $R_p(x)$ and $A_P(x)$ as:

$$R_p(x) \approx \exp\left(ax^b + cx^{dx} + f\right) \tag{7.2.56}$$

and revised permutation function since it is no longer $P(T_{rait}, x)$:

$$P'(x) \approx \exp\left(ax^b + c(x+d)\right) \tag{7.2.57}$$

and revised combination function since it is no longer $C(T_{rait}, x)$:

$$C'(x) \approx ax^b + c(x+d) \tag{7.2.58}$$

In order to best fit existing data, $G(x)$'s $d = 40890$, $v_1 = \frac{p'(3)R_p(3)}{7^8}$, $G_2(x)$'s $d = 140$, $v_2 = \frac{C'(6.64)}{P'(6.64)}$, so we have:

$$S_{total}(T_{rait} = 7, x, t_{Attempt}) = \begin{cases} \frac{t_{Attempt}}{R_p(x)} & T_{rait} \leq 3 \\ \frac{P'(x)}{G(x)} & 3 < T_{rait} \leq 6.64 \\ \frac{C'(x)}{G_2(x)} & 6.64 < T_{rait} \end{cases} \tag{7.2.59}$$

The total number of species realized with this fit function falls within our original assessment of cumulatively 1 million to 4 million species evolved during the Cenozoic since:

$$1 \times 10^6 < \frac{1}{1.4}\int_1^7 S_{total}(T_{rait} = 7, x, t_{Attempt})\, dx = 2,224,351 < 4 \times 10^6 \tag{7.2.60}$$

A factor of $\frac{1}{1.4}$ is introduced to lower the species count from 3 million to 2.2 million. (Actually, a direct rescaling can only be an approximation, a true rescaling requires a new fit for $R'_p(x)$ since the increase in magnitude for species sharing each number of trait gradually increases.

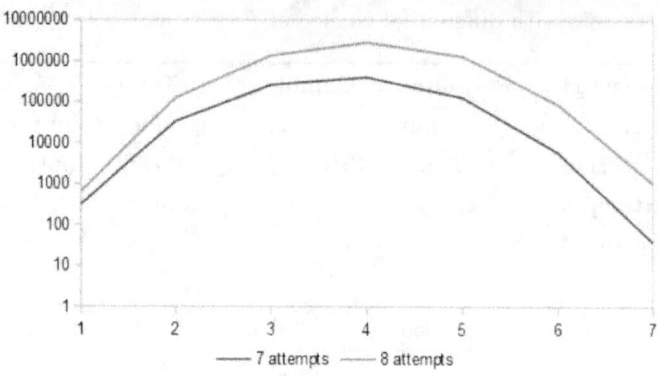

Figure 7.2.9: The increase in magnitude for species sharing each number of trait is non-fixed

As a result, one needs to compute 7^7 as the lower bound and 7^8 as the upper bound. Since there are 7 folds increase between the two, to derive the number of species shared a given number of traits, one have to see the following formula to derive the base for an exponential function:

$$b = \exp\left(\frac{\ln\left(\frac{R_p(x)_{upper}}{R_p(x)_{lower}}\right)}{6}\right) \tag{7.2.61}$$

and in between 2 attempts size increment can be found as:

$$R'_p(x) = \left\{ R_p(x)(b)^t \quad 1 \le t \le 6 \right. \tag{7.2.62}$$

[1] It indicates that the process of performing 7^8 evolutionary speciation attempt is still ongoing and yet to complete. If enough attempts$> 7^8$ have been made and all possible unique permutation with repeats can be observed, then there should be 141,120 species possessed all 7 traits associated with human when permutation allowing repeats. However, since our time window is limited. Only $S_{total}(7, x, 7^8)$=1068 species without any data fit as additional restrictions for 7^8 speciation attempts. (check the last row from the table above), that is, there should be 1008 species possessed all 7 traits associated with human when permutation allowing repeats. Moreover, we capped the attempt ceiling by $\frac{1}{1.4}$, so only $\frac{1}{1.4} \times 1068 = 762$ species remain. Finally, recall that only 5~10 out of 5040 unique permutation paths lead to a possibly human sapiens equivalent given 7 speciation attempt, as a result, the chance is reduced to $\frac{1}{2}\left(\frac{762}{1008} + \frac{762}{504}\right) = 1$. Furthermore, by the 8th attempt, $7 + 6 + 5 + 4 + 3 + 2 + 1 = 28$ times more possible choices per each unique permutation pattern becomes possible (see the last of row of the table). We make an additional selection criterion. That is, permutation with repeats have to follow the exact same order as the previous 7 attempts, i.e. 12(2)34567 and 1234(4)567 are acceptable choices but 12(5)34567 and 1234(7)567 are not. Since there are 7 possible insertion places for 7 digit attempts (1 additional insertion shares same results with other 7), $28/7 = 4$, an additional factor $\frac{1}{4}$ is required for recognizing the number of possible speciation as $\frac{1}{4} \times \frac{1}{2}\left(\frac{762}{1008} + \frac{762}{504}\right) = \frac{1}{4}$. Since $S_{total}(7, x, 7^8)$=1068 species with data fit alone (2nd conditional) predicts 9 species with 7 traits, this clearly is larger than real prediction at $\frac{1}{4}$. Therefore, we add the third conditional. By adding the third conditional, which still is above the chance for a single speciation per only unique combination, as it is expected.

The final result is shown below:

---

[1] Capping is possible by assuming we selectively choose the ones out of all possible permutations with repeats out of 7^8. At the start of 7^8 simulation, all previous 7 digit cases under 7^7 are extended with one additional digit which repeat the 7th digit. Then, the last digit can switch to the rest of 6 traits.

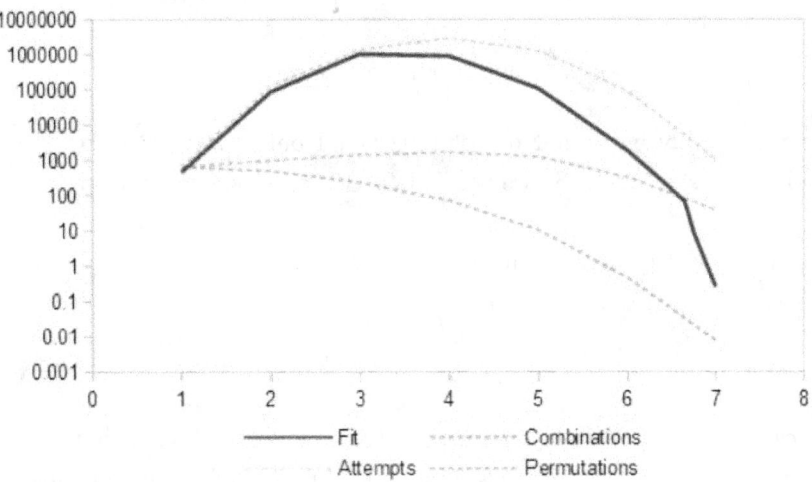

Figure 7.2.10: On a logscale, the best fit now descends faster and crosses the permutation curve (the middle dashed curve) at 6.64 traits and extrapolates 1/4 species possessed 7 traits, the lowest dashed curve is based on unique combination only, and the highest dashed curve is based on all possible permutation with repeats.

Therefore, we deemed that 7 traits with 8 trait acquisition attempts describes the biocomplexity since the Cenozoic more accurate than with 7 trait acquisition attempts. In general, if the final number of traits acquired is predetermined but final integration value after data fitting falls short of the expected range of species, one can always increases the number of trait acquisition attempts one step at a time, if after such an adjustment is made, the new integration value becomes larger than the expected value, one can easily fix to lower values by capping on $t_{Attempt}$ so that $t_{Attempt} < n^t$, $t_{Attempt} = n^t$ corresponds to complete attempt numbers given n traits with t trait acquisition attempts, but it is possible that such process is yet to complete given limited time span so we observed $t_{Attempt} < n^t$ instead.

7.2.4   Multinominal Distribution In-Depth Analysis

The multinominal distribution is explained as the follows: At the third section of the 3 step function and $P(T_{rait} = 7, x) \geq S_{total}(T_{rait} = 7, x, t_{Attempt}) \geq C(T_{rait} = 7, x)$:

$$S_{total}(T_{rait} = 7, x, t_{Attempt}) = \frac{t_{Attempt}}{G_2(x) R_p(x) A_P(x) P_C(x)} \tag{7.2.63}$$

and for those higher number of traits attempts, since not all permutations leads to new speciations (otherwise there will be $P(7, x)$ species for organisms share x traits) and only a subset of all permutations leads to new species. For species possessing a few number of traits, both of its distinct permutation and distinct combination space as a sub-unit of permutation attempts with repeating traits can be exhaustively searched quickly within geologic time frame and more than one species shared same number of traits appear. It is also permissible that the ordering of limited number of trait acquisition repeats several times. However, for increasing number of traits acquired, additional repeats of existing traits becomes difficult if not impossible as certain traits addition creates evolutionary local maximum and the cost of transition becomes to high, furthermore, with EROEI applied to evolution, that a trait is only maintained if the net energy it helps to bring or save to the organism is strictly positive or at least breaking even. With increasing limitation, the search space becomes large for the creation of species with additional number of traits, in the case of our best fit for 7 traits attainment, it would require a generation up to $G_2(7) R_p(7) A_P(7) P_C(7) = 14,431,197$ speciation attempts, with $\left(\frac{1}{p}\right)^7 = 108,233,978$ species attempts required to guarantee a speciation per any permutation path, but there are in total $P(7, 7) = 7! = 5040$ permutation paths, so we have:

$$\frac{\left(\frac{1}{p}\right)^7}{P\left(7,7\right)} = 21474 \text{ attempts per 7 trait permutations} \tag{7.2.64}$$

Furthermore, we have shown that only between 1 out of 1008 and 1 out of 504 paths, or $\frac{1}{504} \sim \frac{1}{1008}$ chance that any permutation path leads to speciation. So we need 14,431,197 species attempts per each species with 7 traits.

We can also think it as because species emergence chance is:

$$S_{total}\left(7,7,7^8\right) = \frac{\frac{1}{1.5}\left(7^7\right)}{\frac{1}{7.5} \times \left(\frac{1}{p}\right)^7} \times \frac{28}{4} \approx \frac{1}{4} \tag{7.2.65}$$

The final 8th digit adds 28 cases but only those repeats with the same order as the unique permutation counts, so there is 7 additional species per each unique permutation. and if just 1 combination possible:

$$S_{lower} = \frac{\frac{1}{1.5}\left(7^7\right)}{1 \times \left(\frac{1}{p}\right)^7} \times \frac{28}{4} \approx \frac{1}{28} \tag{7.2.66}$$

and $\frac{S_{total}\left(7,7,7^8\right)}{S_{lower}} = 7.5$, so the upperbound $S_{total}\left(7,7,7^8\right)$ creates 7.5 species by permutation compares to just 1 species by 1 combination for 7 traits only. Because there are $P\left(7,7\right) = 7! = 5,040$ permutation paths possible, the chance of speciation is $\frac{5}{5040} \sim \frac{10}{5040} \approx \frac{1}{1008} \sim \frac{1}{504}$.

One can express $G_2\left(x\right)P_C\left(x\right)$ as:

$$G_2\left(x\right)P_C\left(x\right) = \frac{G_2\left(x\right)P\left(T_{rait},x\right)}{C\left(T_{rait},x\right)} = \frac{P\left(T_{rait},x\right)}{G_2\left(x\right)^{-1}C\left(T_{rait},x\right)} \tag{7.2.67}$$

Conceptually, $G_2\left(x\right)$ is the ratio representing the best data fit with a proportion $G_2\left(x\right)$ of out all possible permutation $P\left(T_{rait},x\right)$, which is then divided by the total possible number of unique combination only, which is $C\left(T_{rait},x\right)$. One can think $G_2\left(x\right)^{-1}$ as the combination factor, since our best data fit is now expressed as a multiple of the possible number of unique combination (for a given number of traits) only.

$$G_2\left(x\right)^{-1} = \text{Combination Factor} \tag{7.2.68}$$

Then, one can derive the convergence factor by taking the total number of permutations over the total number of re-scaled combination as:

$$\frac{P\left(T_{rait},x\right)}{G_2\left(x\right)^{-1} \times C\left(T_{rait},x\right)} = \frac{\text{total permutation of } x \text{ out of } n \text{ traits}}{\text{Combination Factor} \times \text{combination } x \text{ of } n \text{ traits}} \tag{7.2.69}$$

$$= \text{Convergence Factor} \tag{7.2.70}$$

This can be demonstrated by a toy example, assuming there are 4 traits, and we are interested in the investigation of combination with 3 traits out of 4 traits, and assuming $A_P\left(x\right)$ selects distinct permutation from permutation allowing repeats based on their initial slot positions, (i.e., a distinct permutation 123 is selected from permutation allowing repeats sequences 123___, instead of selection from sequences ___123___) and we have 4 different combinations and 24 different permutations with 6 permutations per each combination:

		(1,2,3)			
(1,2,3)**	(1,3,2)	(2,1,3)*	(2,3,1)	(3,1,2)*	(3,2,1)
		(1,2,4)			
(1,2,4)**	(1,4,2)	(2,1,4)*	(2,4,1)	(4,1,2)*	(4,2,1)
		(1,3,4)			
(1,3,4)**	(1,4,3)	(3,1,4)*	(3,4,1)	(4,1,3)*	(4,3,1)
		(2,3,4)			
(2,3,4)**	(2,4,3)	(3,2,4)*	(3,4,2)	(4,2,3)*	(4,3,2)

The ratio $\frac{P(T_{rait,x})}{C(T_{rait,x})}$ is simply $\frac{24}{4}$. That is, the only successful speciation per each combination are the ones marked with double asterisks. The convergence factor becomes 6. Since we are taking into account the combination factor (combination + partial permutation), the convergence factor becomes 2 if we assumed that one out of every permutation that starts with an unique number (those permutation marked with one or more asterisks) results in speciation. Then, there are 3 successful speciation events per combination, so Combination factor = 3, and $\frac{P(T_{rait,x})}{\text{Combination Factor} \times C(T_{rait,x})} = \frac{24}{3 \times 4} = 2$. That is, it takes 2 permutation paths to guarantee a successful speciation per any unique combination for choosing 3 out 4 traits.

The minimum number of species attempts needed per any unique combination for a successful speciation is then:

$$\text{Convergence Factor} \times R_p(x) A_P(x) = \text{Convergence Factor} \times \text{attempts per perm path} \tag{7.2.71}$$

$$= \text{Convergence Factor} \times \frac{\text{attempts}}{\text{permutation}} \tag{7.2.72}$$

$$= \text{Attempts needed per combination} \tag{7.2.73}$$

Knowing that it takes $R_p(x) A_P(x)$ attempts of speciation before a species with x traits will appear assuming initially every permutation results in a successful speciation and knowing that only 1 out of every n (the Convergence Factor) permutations results a successful speciation, the number of speciation attempts needed before a successful speciation sharing x traits is Convergence Factor $\times R_p(x) A_P(x)$. If we assume that the geometric mean of all possible $P(4,3)$ is $\frac{1}{4}$, then $R_p(x) = \frac{4^4}{144}$, and $A_P(x) = \frac{144}{P(4,3)}$, it is $2 \times \frac{4^4}{P(4,3)} = 21\frac{1}{3}$ speciation attempts. Otherwise if the geometric mean of all possible $P(4,3)$ is p whereas $p < \frac{1}{4}$, then $R_p(x) = \frac{4^4}{144 \times \frac{p^4}{(\frac{1}{4})^4}}$, and $A_P(x) = \frac{144}{P(4,3)}$, it is $2 \times \frac{4^4}{P(4,3)} \times \left(\frac{1}{4}\right)^4 \frac{1}{p^4}$ speciation attempts. It is also likely that the combined chance of particular combination of 3 traits such as 1233 results differs from another particular combination of 3 traits such as 2344. Then $R_p(x) = \frac{4^4}{\left[\frac{144}{C(4,3)} \times \frac{p_1^4}{(\frac{1}{4})^4}\right] + \left[\frac{144}{C(4,3)} \times \frac{p_2^4}{(\frac{1}{4})^4}\right] + \left[\frac{144}{C(4,3)} \times \frac{p_3^4}{(\frac{1}{4})^4}\right] + \left[\frac{144}{C(4,3)} \times \frac{p_4^4}{(\frac{1}{4})^4}\right]}$,

that is, each p_i represent the geometric mean of the combined chance of a particular combination. (Recall that $p_i = S_{um}\left(3, \{3\}_{i \in \{1,...,C(4,3)\}}\right) \times \frac{1}{A_P(3)}$) $A_P(x) = \frac{144}{P(4,3)}$, it is $2 \times \frac{4^4 C(4,3)}{P(4,3)} \times \left(\frac{1}{4}\right)^4 \left(\frac{1}{p_1^4 + p_2^4 + p_3^4 + p_4^4}\right)$ speciation attempts. Factor $\frac{144}{C(4,3)}$ is the number of permutation with repeats for each possible combination. In general, finding $R_p(x)$ for a given number of trait when each combination holds differential combined chance can be derived as:

$$R_p(x) = \frac{\text{Trait}^{\text{Trait}} C(\text{Trait}, x)}{P(\text{Trait}, x)} \times \left(\frac{1}{\text{Trait}}\right)^{\text{Trait}} \times \frac{1}{\sum_{i=1}^{C(\text{Trait}, x)} p_i^{\text{Trait}}} \tag{7.2.74}$$

Whereas now $R_p(x)$ becomes the weighted arithmetic mean of each combination of x out of total number of traits available.

Given an evolutionary time window, say 100 Myr with an expected number of speciation attempts within this period, one can derive the total number of species that can be generated per combination by different permutation paths.

$$\frac{\text{Evolutionary Window}}{\text{Attempts needed per combination}} = \text{Species per Combination} \tag{7.2.75}$$

The final integrated interpretation of the equation is described as:

$$S_{total} = \frac{\text{attempt window size}}{\frac{\text{total permutation of } x \text{ out of } n \text{ traits}}{\text{Combination Factor} \times \text{combination } x \text{ of } n \text{ traits}}} \times \text{attempts per permutation} \tag{7.2.76}$$

Since we have shown earlier that only 2,345,199 species of mammals, birds, and reptiles have ever existed since the Cenozoic, which corresponds to $t_{Attempt} = 7^8$, the chance of emergence for acquiring all 7 traits associated with Homo sapiens becomes less than unity, assuming there is a fixed geologic time frame and evolutionary time window within one's investigation.

$$7.5 \times \left(\frac{\left(\frac{1}{1.5} \right) 7^7}{\left(\frac{1}{p} \right)^7} \right) \left(\frac{28}{4} \right) = \frac{1}{4} \tag{7.2.77}$$

Recall that the final number of possible speciation 7.5 is derived from (assuming $R_p(7) = \frac{7^7}{P(7,7)}$):

$$\frac{7^7}{\frac{P(7,7)}{G_2(7)^{-1} \times C(7,7)} R_p(7) A_P(7)} = 7.5 \tag{7.2.78}$$

If the time window does not exist, that is, we assume there is an infinitely long waiting time, then, every 4 Cenozoic time at the current BCS and mode peak's value is guaranteed to host the emergence of 1 human equivalent species:

We first modify $R_p(7)$ by introducing the weighted chance factor $\frac{1}{\frac{p^7}{\left(\frac{1}{7} \right)^7}} = \frac{\left(\frac{1}{7} \right)^7}{p^7}$, which equivalent to $\left(\frac{7^7}{\left(\frac{1}{p} \right)^7} \right)^{-1}$:

$$R_p(7) = \frac{t_{Attempt} = 7^8}{\left(\frac{7}{28} \right) \left(\sum_{m=1}^{7} m \right) P(7,7) \left(\frac{p^7}{\left(\frac{1}{7} \right)^7} \right)} \tag{7.2.79}$$

The extra factor $\left(\frac{7}{28} \right)$ for numerator is due to the fact that 7^8 attempts increases each of the 5~10 possible speciation with 7 traits under 7 attempts by $\left(\sum_{m=1}^{7} m \right) = 28$ times, since $R_p(7)$ is now defined based on 7^8, not on 7^7 any more. Out of these, one out of every 4 path is selected as feasible (i.e. repeats along the same permutation order is allowed such as 1(22)34567 or 1234(55)67), so 7 fold attempt increase also increase possible speciation by an equal amount, so the number of speciation attempts needed remains the same at 14,384,551 attempts. then we have:

$$\frac{\frac{P(7,7)}{G_2(7)^{-1} \times C(7,7)} \cdot \left(R_p(7) \frac{1}{\frac{p^7}{\left(\frac{1}{7} \right)^7}} \cdot \frac{1}{\frac{7}{28}} \right) A_P(7)}{\left(\frac{1}{1.5} \right) t_{Attempt} = 7^8}$$

$$= \frac{\frac{P(7,7)}{G_2(7)^{-1} \times C(7,7)} \cdot R_p(7) A_P(7) \left(\frac{\left(\frac{1}{7} \right)^7}{p^7} \right) \left(\frac{28}{7} \right)}{\left(\frac{1}{1.5} \right) t_{Attempt} = 7^8} = 4$$

Alternatively, this implies an emergence chance of only $\frac{1}{4}$ per 65 Myr for $t_{Attempt} = 7^8$, the number of speciation attempts per 65 Myr.

In total 7.5 species can emerge taking at least 30 Cenozoic time at the current BCS and mode peak's value:

$$N_{species} = \frac{S_{total}(7, 7, t_{Attempt})}{C(7,7)} = 7.5 \tag{7.2.80}$$

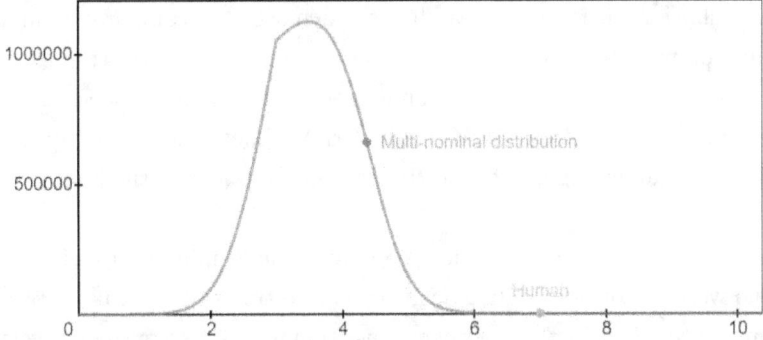

Figure 7.2.11: Time restricted cumulative multi-nominal distribution

To summarize, for species that shared a few number of traits, very few number of speciation attempts are needed for a successful speciation and every speciations occurred within a fraction of the total evolutionary time window. Regardless of repeated generation and speciation along the same permutation path multiple times, the saturation and exhaustion of the search space until it reaches the ceiling of permutation allowing repeats for a given number of shared traits is guaranteed.

For species that shared a large number of traits, vast number of speciation attempts are needed for a successful speciation, the final observed speciation events within the observational time window will becomes a fraction of the total speciation events realizable, and the fraction becomes increasing small as more traits are considered and the time window shrinks. Because our window size is large, the final curve is almost unaffected by the window size and only when the number of trait $x > 5.889$, $\frac{p^{5.889}}{7^8} > 1$ the required speciation attempts exceeds the time window and the emergence chance is further reduced.

7.2.5 Transforming Multinomial Distribution to Lognormal Distribution

This completes the presentation for distribution expressed in terms of combination, permutation and permutation allowing repeats. Moreover, we are more concerned with the cumulative distribution over the course of 100 Myr. Recall that the cumulative distribution evaluated at $x = 7$ is $\frac{1}{4}$ per 65 Myr of evolutionary time window instead of $\frac{1}{14431197}$ for distribution representation for any species at the moment.

$$S_{total}\left(7, 7, 1\right) = \frac{1}{\frac{1}{7.5} \times \left(\frac{1}{p}\right)^7} \approx \frac{1}{14431197} \tag{7.2.81}$$

$$S_{total}\left(7, 7, 7^8\right) = \frac{\frac{1}{1.5}\left(7^7\right)}{\frac{1}{7.5} \times \left(\frac{1}{p}\right)^7} \times \frac{28}{4} \approx \frac{1}{4} \tag{7.2.82}$$

In order to standardize with our later equations, the chance of human emergence has to be converted into units of 100 Myr. Ideally, one should increase the attempt size $> 7^8$ to accommodate greater number of species. However, 2 problems appear. First, based on geological history, a major extinction event occurs on average every 65 Myr; therefore, one can not simply increase attempt size $> 7^8$. Rather, as we have later shown in Chapter 8, it is likely only those shared more traits are favored after a major extinction, and a new round of simulation is initiated based on an additional set of newly evolved traits.

Secondly, if we assume that each simulation cycle can last as long as 100 Myr, and human emergence represented by number of traits is fixed, and we do not have particular preference for the start date of the simulation. A 100 Myr simulation cycle could have happened 100 Mya and it either falls short, reaches, or exceeds the current emergence chance achieved with a cycle initiated 65 Mya. Under idealistic scenario, despite longer simulation duration and greater search space, the simulation started at earlier times with typical species possessed more primitive traits, so it is expected that the emergence chance of human is comparable to simulation initiated 65 Mya. Likewise, simulation initiates later than 65 Mya implies initial species shared more advanced traits, but the simulation duration is shorter with smaller search space achieved, so comparable human emergence is achieved. Furthermore, simulation with 100 Myr duration initiates later than 100 Mya but extended beyond the current time into the future is guaranteed to give rise to human equivalent species, however its time prediction occur in the future. [2]Therefore, conversion from 65 Myr into 100 Myr depends on its initiation date and can only be at 100 Mya, so that the human emergence chance for the past 100 Myr is the human emergence chance for the past 65 Myr.

As a side note, this is the emergence chance assuming ice age plays little to minimal role in accelerating human emergence. First of all, if one were to consider only glaciation at late times of a breaking phase played a role, and assuming all planet formed later than earth experienced less than one supercontinent cycle, then no planet formed later than earth should be included at all. (though later planets contribute little to the cumulative galaxy civilization emergence chance if they are included, due to left shifted mode despite the presence of smaller but non trivial BCS) Those formed earlier than earth but within one supercontinent cycle, have decreased biodiversity during its joining phase, and should use its highest attained biodiversity from the past during the cycle as when accelerated ice age initiates. Those formed much earlier than earth and experienced more than one supercontinent cycle should be treated separately as biodiversity is high enough to guarantee civilization emergence regardless of ice age. However, this analysis requires joining several cases.

Secondly, ideally our later successive distributions according to cases above requires some distributions stagnant at joining phase in terms of BER and BCS. This renders our model over-complicated and computationally intensive for our platform. It can be expressed as:

$$C_{df}(t) = \int_{t_1}^{t_2} P_{planet}(-s) \left(t_{Win}(B_{cs})^{K(s)} \int_{18}^{\infty} P_{df}(-K(s), x) \, dx \right) ds \qquad (7.2.83)$$

Whereas $K(s)$ can be any type of sinusoidal function with increasing peaks. We instead assume BCS and BER was exponentially increasing during the entire cycle.

Therefore, we resort to minimal glaciation effect on emergence. We show that a minimal glaciation effect on emergence does not necessarily mean a distortion on the final emergence. Recall in Chapter 5 and 6 we repeatedly emphasized that ice age increases human emergence chance at the most recent 100 Myr toward 1. We have shown that if we assumed ice age played a role, then one should take the upper bound on speciation per 65 Myr at 7,555,486, with an attempt window size $7^8 < t_1 < 7^9$ Whereas all of the increase occurred in the last 2 Mya. Then, $S_{total}(7, 7, t_1) \geq 1$. However, recall that the chance of a constructive ice age that brings accelerating evolution at the late times of a breaking phase is simply 21.2%, and average glaciation chance is 68%. The combined chance actually decreased to $\approx \frac{1}{7}$. On the other hand, human emergence without taking the accelerated ice age and with an average glaciation chance of 50% is $\approx \frac{1}{4}$, which is just the original assumption. This is possible since glaciation occurs at earlier times of the current round of simulation with lower established number of species at the time guarantees a negligible increase in final speciation size. On the other hand, glaciation at the later rejoining phase can only at best compensate a loss of biodiversity at the time. Check Section 5.3. Therefore, taking accelerated ice age raises human emergence chance at the expense of fewer number of planet candidates, the overall effect is comparable to not taking it into consideration.

In general, one can imagine the cumulative distribution representing all species as a dynamically growing distribution function of the moment $S_{total}(T_{rait}, x, 1)$ in which the likelihood of organisms shared any number of traits grow in 100 Myr so that the cumulative distribution function $\propto S_{total}(T_{rait}, x, 1)$. In earth's case, the cumulative distribution is just $S_{total}(T_{rait}, x, 1)$ increased by the factor of $t_{Attempt}$:

$$C_{umulative}(T_{rait}, x) = S_{total}(T_{rait}, x, t_{Attempt}) = S_{total}(T_{rait}, x, 1) \times t_{Attempt} \qquad (7.2.84)$$

The distribution based on combination and permutation is *then transformed/approximated into a lognormal*

---

[2]In fact, we later show that 100 Myr simulation cycle initiated at the current time and extended into the future is equivalent to a cycle initiated at 100 Mya and extended up to now except its time parameter t, and can be expressed simply as $P_{df}(t, x)$ whereas $t < 0$ to signify a time period into the future, and when $t = 0$ for $P_{df}(0, x)$ stands for a 100 Myr cycle with 100 Mya initiation up to now.

distribution which we defined in Chapter 8.

$$P_{df}(0, x) \propto C_{umulative}(T_{rait}, x) \tag{7.2.85}$$

One could use the current multi-nominal distribution to compute the detection chance of expanding civilizations. *Lognormal is used because it is one of the most familiar heavy-tailed distribution to the most with better manipulative flexibility. Furthermore, multi-nominal distribution terminates at the maximum number of traits associated with Homo sapiens, while lognormal distribution extends to ∞ based on existing trend extrapolation and can evaluate on non-discrete trait values.* The lognormal distribution is proportional to the distribution based on composite combination and permutation but not exactly equivalent. Ideally, it translates the value of $C_{umulative}(T_{rait}, 7)$ into the value of lognormal distribution $P_{df}(0, 7) \cdot t_{Win}$ for $x=7$ as well as any other values other than $x=7$. The lognormal's distribution mode's x and y value should also ideally matches with the mode of the former. The lognormal is proportional to the former because $P_{df}(0, x) \cdot t_{Win} \propto C_{umulative}(T_{rait}, x)$ for whatever likelihood value both distributions hold. Both reflect the size of BCS at any given time.

They are nevertheless not exactly equivalent. One could minimize the discrepancy by modifying σ for the distribution, and it is actually what we did later with the lognormal distribution in order to fit the observational constraint of the emergence of civilization at 1 per 3 galaxies with a deviation of 18 so that:

$$t_{Win} \int_{18}^{\infty} P_{df}(0, x)\, dx = \frac{1}{3} \tag{7.2.86}$$

However, under such a scenario, the lognormal likelihood at 7 signifying the EQ of human becomes much larger than its expected value of the composite multinominal distribution:

$$t_{Win} P_{df}(0, 7) > t_{Win} \int_{7}^{\infty} P_{df}(0, x)\, dx > \frac{1}{4} \tag{7.2.87}$$

Therefore, we conclude the lognormal distribution is an approximation of the original distribution at its best by exaggerating extreme values.

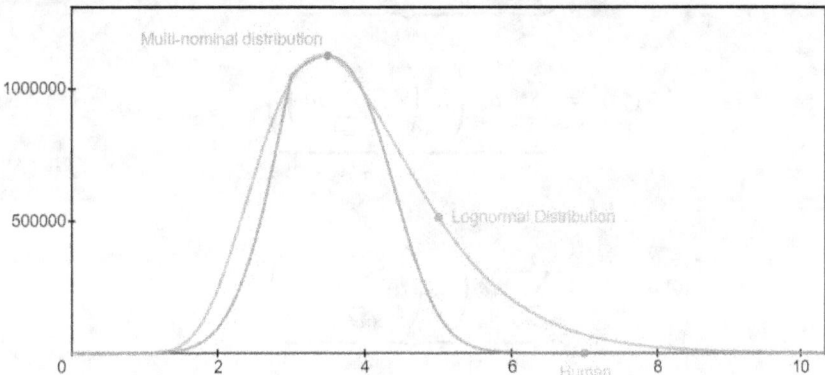

Figure 7.2.12: When $t_{Win} \int_{18}^{\infty} P_{df}(0, x)\, dx = \frac{1}{4}$, lognormal distribution resembles the multinominal distribution $C_{umulative}(7, x)$ but drops slower than the multinominal distribution at std of 7.

The lognormal distribution is later generalized to include a spectrum of placement possibilities through variable k (Check Chapter 8 "Generalized Model"). Likewise, the multinominal distribution can be manipulated similarly. It is achieved by re-scaling proportionally in both horizontal and vertical directions. By increasing $t_{attempt}$ in multiples of $(T_{rait})^x$, one increases the vertical height of the multinominal distribution. By increasing the number of traits for simulation $P(T_{rait}, x)$, one increases the horizontal variance/width of the multinominal distribution. The horizontal and vertical rescaling mutually enhance the final emergence rate (the integral area to the right of the point L_{imit} designated as human) as the height re-scaling ratio on the emergence chance expressed as an area size, which is later expressed in lognormal form as:

$$C_{df}(t) = \exp\left(\overbrace{t\left(\frac{k-1}{k}\right)\ln(B_{cs})}^{\text{increase by } B_{cs}} + \overbrace{\frac{t}{k}\ln(B_{cs})}^{\text{increase by } B_{cs}} + \overbrace{\ln\left(1 + \frac{B_{cs}^{\frac{t}{k}} - 1}{L_{imit} - B_{cs}^{\frac{t}{k}}}\right)}^{\text{displacement ratio}} + \underbrace{\frac{b}{k} + \frac{a}{k^d}}_{\text{Conversion Factor}}\right) \tag{7.2.88}$$

$$\underbrace{}_{\text{Height Ratio}} \qquad \underbrace{}_{\text{Width Ratio}}$$

When $k \to \infty$, we have:

$$C_{df}(t) = \exp\left(t\left(\frac{k-1}{k}\right)\ln(B_{cs})\right) = B_{cs}^t \tag{7.2.89}$$

Whereas $t\left(\frac{k-1}{k}\right)\ln(B_{cs})$ stood for the vertical re-scaling, which is translated conceptually in multinominal distribution as the follows. Recall that given a fixed number of traits n, when the number of attempts $t > n$, the number of permutation with repeats for species shared j traits increases exponentially as

$$\text{trait } j: \underbrace{\sum_{a_{n-j+2}=1}^{j} a_{n-j+2}\left(\cdots\left(a_1 \sum_{m=1}^{a_1} m\right)\right)}_{\times n-j+i \text{ times}} \times P\left(n, n - (n-j)\right) \tag{7.2.90}$$

Recall that function $A_P(j)$ is defined based on trait number = acquisition attempts:

$$A_P(j) = \underbrace{\sum_{a_k=1}^{j} a_k\left(\cdots\left(a_1 \sum_{m=1}^{a_1} m\right)\right)}_{\times n-j \text{ times}} \tag{7.2.91}$$

We generalize this expression with operator $\langle A_P(j)\rangle$, which we defined as follows:

$$\langle A_P(j)\rangle^0 = \underbrace{\sum_{a_k=1}^{j} a_k\left(\cdots\left(a_1 \sum_{m=1}^{a_1} m\right)\right)}_{\times n-i+0 \text{ times}}$$

$$\langle A_P(j)\rangle^1 = \underbrace{\sum_{a_k=1}^{j} a_k\left(\cdots\left(a_1 \sum_{m=1}^{a_1} m\right)\right)}_{\times n-j+1 \text{ times}}$$

$$\cdots$$

$$\langle A_P(j)\rangle^i = \underbrace{\sum_{a_k=1}^{j} a_k\left(\cdots\left(a_1 \sum_{m=1}^{a_1} m\right)\right)}_{\times n-j+i \text{ times}}$$

Whereas i stands as the number of acquisition attempts $>$ the number of fixed traits. As a result, new speciation for species possessed more traits becomes more likely.

Assuming that $L_{imit} = n - 1$ trait mark position. Then, the previous multinominal distribution defined region to the right of L_{imit} is expressed as $\langle A_P(n)\rangle^0 \times P(n,n)$, and the rightmost boundary of the region is defined by n, yet the matching lognormal distribution extends from $[L_{imit}, \infty)$. If one assumes that ideally the lognormal matches in y value with the multi-nominal at L_{imit} and L_{imit} is defined on a whole integer trait number j, then we have the following inequality:

When $L_{imit} = n$ and $n = t$:

$$\langle A_P(n)\rangle^0 P(n,n) \le \langle A_P(n)\rangle^0 \int_n^{\infty} P_{df}(n,x)\, dx \tag{7.2.92}$$

When $L_{imit} < n$ and $n = t$:

$$\langle A_P\left(L_{imit}\right)\rangle^0 P\left(n, L_{imit}\right) + \ldots + \langle A_P\left(n\right)\rangle^0 P\left(n, n\right) \le \langle A_P\left(L_{imit}\right)\rangle^0 \int_{L_{imit}}^{\infty} P_{df}\left(n, x\right) dx$$

...

$$\sum_{j=L_{imit}}^{n} \langle A_P\left(j\right)\rangle^0 \times P\left(n, j\right) \le \langle A_P\left(L_{imit}\right)\rangle^0 \times \int_{L_{imit}}^{\infty} P_{df}\left(n, x\right) dx$$

and when $L_{imit} < n$ and $n < t$:

$$\langle A_P\left(n\right)\rangle^i P\left(n, n\right) \le \langle A_P\left(n\right)\rangle^i \int_{n}^{\infty} P_{df}\left(n, x\right) dx$$

$$\langle A_P\left(L_{imit}\right)\rangle^i P\left(n, L_{imit}\right) + \ldots + \langle A_P\left(n\right)\rangle^i P\left(n, n\right) \le \langle A_P\left(L_{imit}\right)\rangle^i \int_{L_{imit}}^{\infty} P_{df}\left(n, x\right) dx$$

...

$$\sum_{j=L_{imit}}^{n} \langle A_P\left(j\right)\rangle^i \times P\left(n, j\right) \le \langle A_P\left(L_{imit}\right)\rangle^i \times \int_{L_{imit}}^{\infty} P_{df}\left(n, x\right) dx$$

That is, the approximated lognormal always over-estimates the emergence chance, regardless the number of attempts per fixed number of traits and the position of L_{imit}.

We show that under this case, the number of species shared n traits increases exponentially as additional steps of number of acquisition attempts increase. We assume initially species shared n traits contains an attempts to permutation ratio of

$$\sum_{m=1}^{n} m = n + (n-1) + (n-2) \ldots + 2 + 1 = \frac{n(n+1)}{2} \tag{7.2.93}$$

With an additional step of acquisition attempts, it becomes:

$$\sum_{a_1=1}^{n} a_1 \sum_{m=1}^{a_1} m =$$

$$n \times (n + (n-1) + (n-2) \ldots + 2 + 1) = \frac{n^2(n+1)}{2} +$$

$$n - 1 \times ((n-1) + (n-2) \ldots + 2 + 1) = \frac{(n-1)^2 n}{2} +$$

$$n - 2 \times ((n-2) \ldots + 2 + 1) = \frac{(n-2)^2 (n-1)}{2} +$$

...

$$2 \times (2 + 1) = \frac{(2)^2 (3)}{2} +$$

$$1 \times (1)$$

We compare the first term with the sum of our previous ratio as:

$$\frac{n(n+1)}{2} < \frac{n^2(n+1)}{2} = 1 < n \tag{7.2.94}$$

and we find that the first term of the new sum is greater than the sum of our previous ratio by n.

Then we compare the first term with the sum of our previous ratio as:

$$\frac{n(n+1)}{2} < \frac{(n-1)^2 n}{2} \tag{7.2.95}$$

287

In this case, only when $n > 3$, the inequality holds, based on solving $\frac{(n-1)^2 n}{n(n+1)} > 1$, and only $n \to \infty$, we have $1 < n$.

Likewise, for the rest of cases, similar pattern emerges, for

$$\frac{n(n+1)}{2} < \frac{(n-2)^2(n-1)}{2} \tag{7.2.96}$$

only when $n > 5$, the inequality holds, based on solving $\frac{(n-2)^2(n-1)}{n(n+1)} > 1$, and only $n \to \infty$, we have $1 < n$. In general, depend on the number of traits shared by the species under investigation, there are limited number of additional terms for which the inequality holds. For any additional terms in which the inequality does not hold, it still add some minimal but non-zero contribution toward the final factor. Therefore, clearly we have:

$$1 < n + \sum_{}^{i_0} \epsilon_0 \tag{7.2.97}$$

Whereas ϵ_0 is the factor value for each of next terms divided by the sum of the previous ratio. Whereas i_0 is the number of additional terms for a selected bin of species with shared given number of traits.

With an additional step of acquisition attempts, it becomes:

$$\sum_{a_2=1}^{n} a_2 \sum_{a_1=1}^{a_2} a_1 \sum_{m=1}^{a_1} m =$$

$$n \times \left(\frac{n^2(n+1)}{2} + \frac{(n-1)^2 n}{2} + \frac{(n-2)^2(n-1)}{2} + ... + 1 \right) +$$

$$n-1 \times \left(\frac{(n-1)^2 n}{2} + \frac{(n-2)^2(n-1)}{2} + ... + 1 \right) +$$

$$n-2 \times \left(\frac{(n-2)^2(n-1)}{2} + ... + 1 \right) +$$

$$...$$

$$2 \times \left(\frac{(2)^2(3)}{2} + 1 \right) +$$

$$1 \times (1)$$

We compare the first term with the sum of our previous ratio as:

$$\left(\frac{n^2(n+1)}{2} + ... + 1 \right) < n \times \left(\frac{n^2(n+1)}{2} + ... + 1 \right) = 1 < n \tag{7.2.98}$$

Then we compare the 2nd term with the sum of our previous ratio as:

$$\left(\frac{n^2(n+1)}{2} + ... + 1 \right) < n-1 \times \left(\frac{(n-1)^2 n}{2} + ... + 1 \right) \tag{7.2.99}$$

For the 1st term within the 2nd term, if $n \geq 4$, the inequality holds, based on solving $\frac{(n-1)^3 n}{n^2(n+1)} > 1$, for the 2nd term within the 2nd term, if $n > 4$, the inequality holds, based on solving $\frac{(n-1)(n-2)^2(n-1)}{(n-1)^2 n} > 1$, for the 3rd term within the 2nd term if $n > 5$, based on solving $\frac{(n-1)(n-3)^2(n-2)}{(n-2)^2(n-1)} > 1$, in general, one needs to find the solution to the following inequality:

$$\frac{(n-i)(n-i-k_1)^2(n-i-k_1+1)}{(n-k_1)^2(n-k_1+1)} > 1 \tag{7.2.100}$$

This expression can be further generalized to any steps of acquisition attempts whereas $s \geq 4$ as:

288

$$\frac{(n-i)\overbrace{(n-i-k_1)\ldots}^{\times\,(s-4)\text{ times}}\left(n-i\overbrace{-k_1\ldots-k_n}^{\times\,(s-4+1)\text{ times}}\right)^2\left(n-i\overbrace{-k_1\ldots-k_n}^{\times\,(s-4+1)\text{ times}}+1\right)}{\underbrace{(n-k_1)\ldots}_{\times\,(s-4)\text{ times}}\left(n\underbrace{-k_1\ldots-k_n}_{\times\,(s-4+1)\text{ times}}\right)^2}>1 \qquad (7.2.101)$$

We defines $s=1$ as the base case in which the permutation with repeats to permutation ratio is unity, and $s=2$ in which the permutation with repeats to permutation ratio is $\sum_{m=1}^{n} m$. As the number of acquisition attempt increases, increasing number of terms are multiplied, as a result, given a trait acquisition attempt s, we have $s-4$ extra possible terms, it is removes a constant number 4 because ratio always ends with *last two factors* of identical value due to results from $s=3$, as well as numerator requires an extra term $n-i$ for acquisition step s and the counting starts initially at $s=2$. Likewise, the iteration continues until the the last two factors, which has to be further reduced in value by $-k_1\ldots-k_n$ terms, which includes an extra term k_n to distinguish itself from the previous term. Whereas the range of variables $i,-k_1\ldots-k_n$ are:

$$0\le i\le j-1$$
$$0\le k_1\le j-1$$
$$0\le k_2\le j-1$$
$$\ldots$$
$$0\le k_n\le j-1 \qquad (7.2.102)$$

The termination condition is determined by:

$$i+k_1+\ldots+k_n\le j-1 \qquad (7.2.103)$$

Likewise, the denominator is just like the numerator except the term i is omitted since each new attempt sinferior terms are compared against the superior terms of $s-1$.

An illustrative demo is presented for $s=3$, whereas the offset indexes $i,k_1\ldots k_n$ for the terms

$$(n-i)\overbrace{(n-i-k_1)\ldots}^{\times\,(s-4)\text{ times}}\left(n-i\overbrace{-k_1\ldots-k_n}^{\times\,(s-4+1)\text{ times}}\right)^2 \qquad (7.2.104)$$

are extracted from the numerator and offset indexes for the terms $\underbrace{(n-k_1)\ldots}_{\times\,(s-4)\text{ times}}\left(n\underbrace{-k_1\ldots-k_n}_{\times\,(s-4+1)\text{ times}}\right)^2$ are extracted from the denominator. The selected trait number $j=5$

	$k_1=0$
$i=1$	$\frac{1(1)}{(0)}$
$i=2$	$\frac{2(2)}{(0)}$
$i=3$	$\frac{3(3)}{(0)}$
$i=4$	$\frac{4(4)}{(0)}$

for $s=4$:

	$k_1=0$	$k_1=1$	$k_1=2$	$k_1=3$
$i=1$	$\frac{1(11)}{(00)}$	$\frac{1(22)}{(11)}$	$\frac{1(33)}{(22)}$	$\frac{1(44)}{(33)}$
$i=2$	$\frac{2(22)}{(00)}$	$\frac{2(33)}{(11)}$	$\frac{2(44)}{(22)}$	
$i=3$	$\frac{3(33)}{(00)}$	$\frac{3(44)}{(11)}$		
$i=4$	$\frac{4(44)}{(00)}$			

for $s=5$:

		$k_2=0$	$k_2=1$	$k_2=2$	$k_2=3$
$i=1$	$k_1=0$	$\frac{1(111)}{(000)}$	$\frac{1(122)}{(011)}$	$\frac{1(133)}{(022)}$	$\frac{1(144)}{(033)}$
	$k_1=1$	$\frac{1(222)}{(111)}$	$\frac{1(233)}{(122)}$	$\frac{1(244)}{(133)}$	
	$k_1=2$	$\frac{1(333)}{(222)}$	$\frac{1(344)}{(233)}$		
	$k_1=3$	$\frac{1(444)}{(333)}$			
$i=2$	$k_1=0$	$\frac{2(222)}{(000)}$	$\frac{2(233)}{(011)}$	$\frac{2(244)}{(022)}$	
	$k_1=1$	$\frac{2(333)}{(111)}$	$\frac{2(344)}{(122)}$		
	$k_1=2$	$\frac{2(444)}{(222)}$			
$i=3$	$k_1=0$	$\frac{3(333)}{(000)}$	$\frac{3(344)}{(011)}$		
	$k_1=1$	$\frac{3(444)}{(111)}$			
$i=4$	$k_1=0$	$\frac{4(444)}{(000)}$			

Parenthesis is shown to exclude the leading extra term for each $s$the attempt vs previous $s-1$th attempt.

If the sum of all terms within a term in which inequality holds, then the inequality holds, otherwise, any additional terms in which the inequality does not hold, it still add some minimal but non-zero contribution toward the final factor. Therefore, clearly we have:

$$1 < n + \sum^{i_1} \epsilon_1 \tag{7.2.105}$$

Whereas ϵ_1 is the factor value for each additional terms divided by the sum of the previous ratio, which differs from the earlier round. Whereas i_1 is the number of additional terms for a selected bin of species with shared given number of traits. The inequality difference after t additional steps of acquisitions, we have:

$$n^t < \prod_{i=0}^{t} \left(n + \sum^{i_i} \epsilon_i \right) \tag{7.2.106}$$

It is also true that

$$\sum^{i_{i+1}} \epsilon_{i+1} < \sum^{i_i} \epsilon_i \tag{7.2.107}$$

so that

$$\lim_{t \to \infty} \prod_{i=0}^{t} \left(n + \sum^{i_i} \epsilon_i \right) \approx n^t \tag{7.2.108}$$

Hence, we have shown that the number of permutation with repeats grows exponentially with additional steps of acquisitions.

We add additional constraints α so that the ratio of new speciation under new time period t compares to the previous time period remains constant across species possessed any number of traits. (By best data fit we have mentioned previously and conceptually as not all permutation paths leads to speciation). Then, multi-nominal equivalent expression for $C_{df}(t) = B_{cs}^t$ becomes:

$$\frac{\alpha_2 \langle A_P(j) \rangle^1 P(n,j)}{\alpha_1 \langle A_P(j) \rangle^0 P(n,j)} = \frac{\alpha_2 \langle A_P(j) \rangle^1}{\alpha_1 \langle A_P(j) \rangle^0} = B_{cs}^t \tag{7.2.109}$$

and the area ratio is expressed as:

$$A\left(L_{imit}\right) = \frac{\sum_{j_2=L_{imit}}^{n} \alpha_{j_2} \left\langle A_P\left(j_2\right)\right\rangle^{i} P\left(n, j_2\right)}{\sum_{j_1=L_{imit}}^{n} \alpha_{j_1} \left\langle A_P\left(j_1\right)\right\rangle^{0} P\left(n, j_1\right)} = B_{cs}^{t} \tag{7.2.110}$$

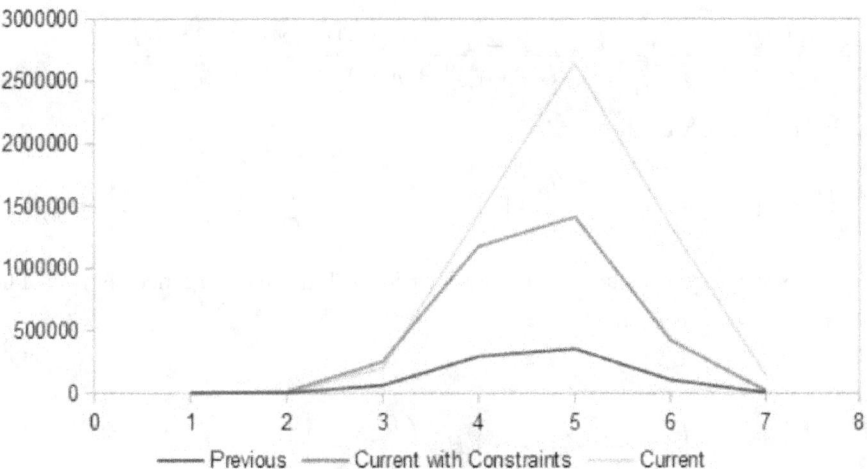

Figure 7.2.13: For $k \to \infty$, by adding the filtering criterion α_2, current distribution's every point along the curve and its final area size becomes a constant multiple of the previous.

When $1 < k < \infty$, every term within the expression of $C_{df}\left(t\right)$ is considered. Its corresponding multi-nominal distribution is one with $n + i \leq t + i$. That is, the current multi-nominal distribution has included i extra number of traits and i extra number of acquisition attempts. As a result, the mode of the distribution become right shifted, the mode itself achieved greater height.

The area ratio expressed in multi-nominal distribution is:

$$A\left(L_{imit}\right) = \frac{\sum_{j_2=L_{imit}}^{n} \alpha_{j_2} \left\langle A_P\left(j_2\right)\right\rangle^{i} P\left(n+i, j_2\right) + \sum_{j_2=n+1}^{n+i} \alpha_{j_2} \left\langle A_P\left(j_2\right)\right\rangle^{0} P\left(n+i, j_2\right)}{\sum_{j_1=L_{imit}}^{n} \alpha_{j_1} \left\langle A_P\left(j_1\right)\right\rangle^{0} P\left(n, j_1\right)} \tag{7.2.111}$$

Therefore, we have:

$$C_{df}\left(t\right) = A\left(L_{imit}\right) \tag{7.2.112}$$

Based on Chapter 8, (also check the next diagram analysis) we know that

$$\exp\left(\ln\left(1 + \frac{B_{cs}^{\frac{t}{k}} - 1}{L_{imit} - B_{cs}^{\frac{t}{k}}}\right)\right) \tag{7.2.113}$$

is the horizontal displacement ratio, so that the horizontal displacement ratio in lognormal is equivalent to the difference in mode between the current and the previous distribution:

$$1 + \frac{B_{cs}^{\frac{t}{k}} - 1}{L_{imit} - B_{cs}^{\frac{t}{k}}} = \frac{\left[\left(L_{imit} - \text{Mode}_2\right) + \left(\text{Mode}_2 - \text{Mode}_1\right)\right]}{\left(L_{imit} - \text{Mode}_2\right)} \tag{7.2.114}$$

We know that

$$\exp\left(\ln\left(1 + \frac{B_{cs}^{\frac{t}{k}} - 1}{L_{imit} - B_{cs}^{\frac{t}{k}}}\right) + \frac{t}{k}\ln\left(B_{cs}\right)\right) \tag{7.2.115}$$

is the horizontal displacement ratio and the horizontal re-scaling by B_{cs}^{t} relative to the previous half variance in lognormal is equivalent to the cross sectional half variance represented by the difference between L_{imit} and the mode of the current distribution plus the the horizontal displacement ratio we have already defined.

$$\left(1 + \frac{B_{cs}^{\frac{t}{k}} - 1}{L_{imit} - B_{cs}^{\frac{t}{k}}}\right) B_{cs}^{\frac{t}{k}} = 1 + \frac{(\text{Mode}_2 - \text{Mode}_1)}{(L_{imit} - \text{Mode}_2)} (B_{cs})^{\frac{t}{k}} \tag{7.2.116}$$

in which the BCS Bio-complexity search space expressed in terms of multi-nominal distribution is:

$$B_{cs} = \frac{\sum_{j_2=1}^{n} \alpha_{j_2} \langle A_P(j_2) \rangle^i P(n+i, j_2) + \sum_{j_2=n+1}^{n+i} \alpha_{j_2} \langle A_P(j_2) \rangle^0 P(n+i, j_2)}{\sum_{j_1=1}^{n} \alpha_{j_1} \langle A_P(j_1) \rangle^0 P(n, j_1)} \tag{7.2.117}$$

Since $1 \leq k \leq \infty$, the width ratio can vary between:

$$1 + \frac{(\text{Mode}_2 - \text{Mode}_1)}{(L_{imit} - \text{Mode}_2)} \leq \left[1 + \frac{(\text{Mode}_2 - \text{Mode}_1)}{(L_{imit} - \text{Mode}_2)}\right] (B_{cs})^{\frac{t}{k}} \leq \left[1 + \frac{(\text{Mode}_2 - \text{Mode}_1)}{(L_{imit} - \text{Mode}_2)}\right] (B_{cs})^t \tag{7.2.118}$$

and the terms $\exp\left(\frac{b}{k} + \frac{a}{k^d}\right)$ converts such width ratio to area ratio for human emergence chance. The area ratio for only horizontal width rescaling ratio is then:

$$\left(1 + \frac{B_{cs}^{\frac{t}{k}} - 1}{L_{imit} - B_{cs}^{\frac{t}{k}}}\right) B_{cs}^{\frac{t}{k}} \left(\frac{b}{k} \times \frac{a}{k^d}\right) \tag{7.2.119}$$

$$= A(L_{imit}) \times \frac{1}{(B_{cs})^{t\left(\frac{k-1}{k}\right)}} \tag{7.2.120}$$

That is, it is the original area ratio expressed in multi-nominal distribution divided by the height rescaling factor $(B_{cs})^{t\left(\frac{k-1}{k}\right)}$.

Finally, the height ratio relationship between the value at L_{imit} and the previous distribution's height can be expressed as:

$$\frac{\arg\max \langle A_P(j) \rangle P(n, j)}{\langle A_P(L_{imit}) \rangle P(n, L_{imit})} \tag{7.2.121}$$

Whereas $\arg\max$ determines the height of the previous distribution with maximum number of species exhibited j number of shared traits.

and the height ratio relationship between the current distribution's height and previous distribution's height can be expressed as:

$$\frac{\arg\max \langle A_P(k) \rangle^i P(n+i, k)}{\arg\max \langle A_P(j) \rangle P(n, j)} \tag{7.2.122}$$

$\langle A_P(k) \rangle^i$ is increased by i because we have assumed that $L_{imit} \leq n$. That is, the defined human emergence position must be located to the left of the maximum number of traits possible of the previous distribution, and we defined that $L_{imit} \geq B_{cs}^{\frac{t}{k}}$. That is, the defined human emergence position must locates to the right of the mode peak of the new distribution, so that the final partial area of the new distribution dictated by L_{imit} is solely proportional to the raised height of the distribution. Therefore, regardless of the value of chosen k, the permutation with repeats to permutation ratio must be $\arg\max \langle A_P(k) \rangle^i$. When both the current and the previous distribution's mode peak happens on the same trait, we have $k = j$, so:

$$\frac{\arg\max \langle A_P(j) \rangle^i P(n+i, j)}{\arg\max \langle A_P(j) \rangle P(n, j)} \tag{7.2.123}$$

In general, multi-nominal distribution can faithfully match cases for $2 \leq k \leq \infty$.

We show that under this case, the number of species shared n traits also increases exponentially as additional number of traits increases. As the number of traits increases, the corresponding number of attempts grows accordingly, therefore, the steps of number of acquisition attempts increase. We already shown that the number of species shared n traits increases exponentially as additional steps of number of acquisition attempts increase.

$$\langle A_P(j) \rangle^i P(n+i,j) \tag{7.2.124}$$

so that the rate of increase for attempts to permutation is $> j^i$.

We need to also show that $\frac{P(n+x,j)}{P(n,j)} > 1$ so that the number of permutation also grows exponentially. Assume that the new distribution shares $n+1$ traits and we are inspecting species shared j traits, so we have:

$$\frac{(n+1)!}{(n+1-j)!} = (n+1) \times \underbrace{n(n-1)\ldots(n+1-j+1)}_{\text{shared terms with previous distribution}} \tag{7.2.125}$$

and the previous distribution contains n traits, so we have:

$$\frac{n!}{(n-j)!} = \underbrace{n(n-1)(n-2)\ldots(n-j+2)}_{\text{shared terms with current distribution}} \times (n-j+1) \tag{7.2.126}$$

The ratio between the two becomes:

$$\frac{n+1}{n-j+1} > 1 \tag{7.2.127}$$

When the new distribution shares $n+2$ traits we have:

$$\frac{(n+2)!}{(n+1-j)!} = (n+2)(n+1) \times \underbrace{n(n-1)\ldots(n+2-j+1)}_{\text{shared terms with previous distribution}} \tag{7.2.128}$$

$$\frac{n!}{(n-j)!} = \underbrace{n(n-1)(n-2)\ldots(n-j+3)}_{\text{shared terms with current distribution}} \times (n-j+2)(n-j+1) \tag{7.2.129}$$

The ratio between the two becomes:

$$\frac{(n+2)(n+1)}{(n-j+2)(n-j+1)} > 1 \tag{7.2.130}$$

Notice that the additional factor

$$1 < \frac{n+2}{n-j+2} < \frac{n+1}{n-j+1} \tag{7.2.131}$$

In general, for the new distribution shares $n+i$ traits, the expression becomes:

$$\frac{(n+i)\ldots(n+1)}{(n-j+i)\ldots(n-j+1)} = \prod_{i=1}^{i \leq j} \frac{n+i}{n-j+i} > 1 \tag{7.2.132}$$

and each additional traits added, an additional factor $\frac{n+i}{n-j+i}$ is multiplied to existing chance, and as

$$\lim_{i \to \infty} \frac{n+i}{n-j+i} \to 1 \tag{7.2.133}$$

Hence every factor contributes > 1 chance and consequently $\prod_{i=1}^{i \leq j} \frac{n+i}{n-j+i} > 1$. In reality, the number of factors can only increases up to j before overlapping terms disappears. Consequently, both numerator and denominator shares the same number of terms designated by the number of traits under consideration, so we have:

$$\prod_{m=1}^{j-1} \frac{n+i-m}{n-m} > 1 \tag{7.2.134}$$

and $i \geq j$ since $n+j \leq n+i$. Notice that for each step increase of i, each term within the numerator increases by 1, so clearly:

$$\prod_{m=1}^{j-1} \frac{n+i+1-m}{n-m} > \prod_{m=1}^{j-1} \frac{n+i-m}{n-m} > 1 \qquad (7.2.135)$$

so combining both cases we have:

$$\begin{cases} \prod_{i=1}^{j} \frac{n+i}{n-j+i} > 1 & i \le j \\ \prod_{m=1}^{j-1} \frac{n+i-m}{n-m} > 1 & i \ge j \end{cases} \qquad (7.2.136)$$

Hence we have shown that the number of permutation also grows exponentially.

Finally, since we know that the increased number of attempts for permutation with repeats with fixed number of traits (which results in the exponential rate of increase for attempts to permutation ratio only) logically correlates to B_{cs}. One may wonder there exists a logical connection between $\exp\left(\frac{b}{k} + \frac{a}{k^d}\right)$ terms and the increase on the number of permutation. However, new distribution with more traits can be broken into 2 parts, horizontal displacement and vertical re-scaling, as we have mentioned earlier.

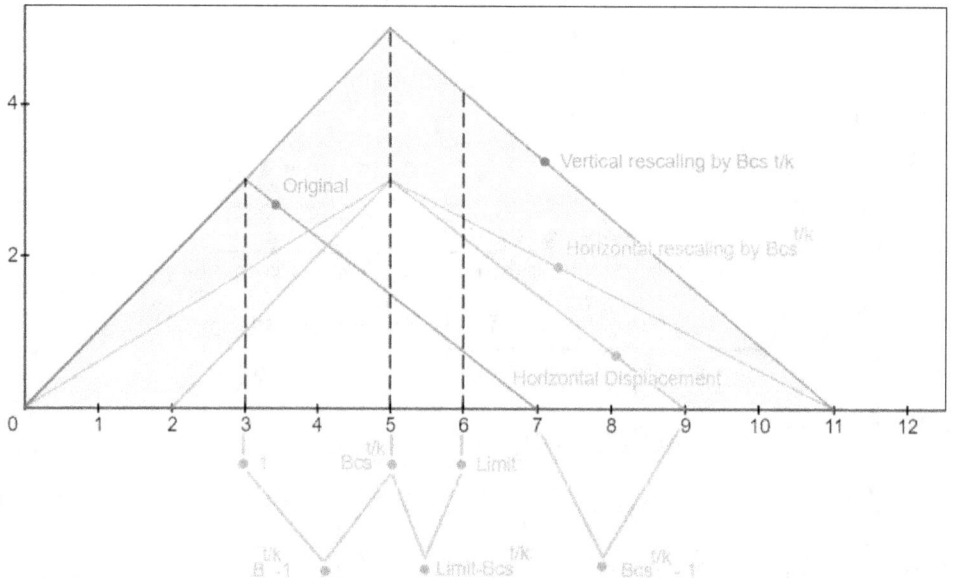

Figure 7.2.14: The geometric breakdown analysis for cases $1 < k < \infty$, The bottom lognormal variables correspond with given number of shared traits for species, establishing their logical connections.

For horizontal displacement, one can imagine that there exists a distribution with the same area size as the previous except every traits number under consideration shifted left horizontally by $\text{mode}_2 - \text{mode}_1$ traits. As a result, trait j becomes trait $j - (\text{mode}_2 - \text{mode}_1)$. Its subsequent attempts to permutation ratio becomes $\langle A_p (j - (\text{mode}_2 - \text{mode}_1))\rangle^0$. Its permutation becomes $P(n, j - (\text{mode}_2 - \text{mode}_1))$. Since $\text{mode}_2 \le j - (\text{mode}_2 - \text{mode}_1) < j$, we have:

$$\langle A_p(j)\rangle^0 < \langle A_p(j - (\text{mode}_2 - \text{mode}_1))\rangle^0 \qquad (7.2.137)$$

Note that after the shift, trait j uses the attempt to permutation ratio represented for species shared $j - (\text{mode}_2 - \text{mode}_1)$ traits, it does not imply that species shared j traits are now sharing $j - (\text{mode}_2 - \text{mode}_1)$ traits only. Its simply a conceptual equivalent in mathematical quantity. We are dividing the number of species sharing j traits into its respective parts for the analysis.

and

$$P(n, j - (\text{mode}_2 - \text{mode}_1)) < P(n, j) \qquad (7.2.138)$$

That is, not only horizontal displacement constitutes both an increase in attempts to permutation ratio but as well as a *decrease* in the number of permutation. Hence, there is no logical connection between $\exp\left(\frac{b}{k} + \frac{a}{k^d}\right)$ terms and the increase on the number of permutation. Finally, on top of horizontal displacement, we have the vertical rescaling, and we know that after vertical rescaling, we have:

$$\langle A_p(j)\rangle^0 < \langle A_p(j)\rangle^i \tag{7.2.139}$$

$$P(n, j) < P(n + i, j) \tag{7.2.140}$$

That is, vertical rescaling results in both increase on attempts to permutation ratio and increase in permutation. Hence, B_{cs} can logically correlates with both increase on attempts to permutation ratio and permutation and there is one to one correspondence between them. In reality, B_{cs} correlate with both indirectly through the additional α factor we have mentioned as:

$$B_{cs} = \alpha \langle A_p(j)\rangle^i P(n + i, j) \tag{7.2.141}$$

when $k = 1$, $C_{df}(t)$ is simplified as:

$$C_{df}(t) = \exp\left(t\ln(B_{cs}) + \ln\left(1 + \frac{B_{cs}^t - 1}{Limit - B_{cs}^t}\right) + b + a\right) \tag{7.2.142}$$

In this case, the height of the new distribution is forced to be at the same height with conversion factor as the previous so that:

$$\arg\max \langle A_P(k)\rangle^i P(n + i, k) \times \frac{1}{\exp\left(t^{\frac{k-1}{k}}(B_{cs})\right)} = \arg\max \langle A_P(j)\rangle P(n, j) \tag{7.2.143}$$

The loss in height is compensated by an expansion in width $\exp\left(t^{\frac{k-1}{k}}(B_{cs}) + t^{\frac{1}{k}}(B_{cs})\right) = \exp\left(t(B_{cs})\right)$ so that the total BCS remains the same. Furthermore, the leftmost expansion edges must fall on species shared only 1 trait. Therefore, the mode of the current distribution must be shifted to position B_{cs}^t instead of $B_{cs}^{\frac{t}{k}}$. No multi-nominal distribution with larger number of traits give rise to a distribution with same height as those with lower number of fixed trait. After rescaling, the original multi-nominal distribution has a greater variance with an expansion of width by B_{cs} and consequently greater shift in mode value, implying even larger number of traits to be considered than originally started. Therefore, for $k = 1$, there is no correspondingly naturally forming multi-nominal distribution without modification according to our definition.

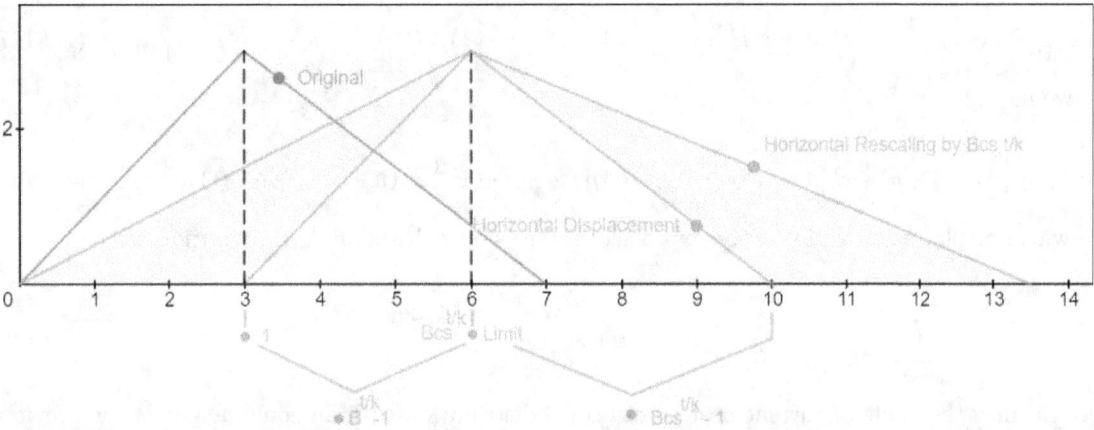

Figure 7.2.15: The geometric breakdown analysis for cases $k = 1$, The bottom lognormal variables correspond with given number of shared traits for species, establishing their logical connections. Notice that in our demo $B_{cs}^{t/k} = Limit$, but in reality it can be $B_{cs}^{t/k} \geq Limit$.

Nevertheless, its correspondence with multi-nominal distribution can be established. One can imagine an original current distribution that represents species with at most $n+i$ traits. Then, with a flattening procedure, additional k traits are introduced, so in total $n+i+k$ traits possible. One then starts to shift permutation with repeats falling under traits $n+i$ into traits $n+i+k$. Shifting stops when traits $n+i+k$'s quota is fulfilled. The quota requirement for each trait can be determined by horizontal displacement and horizontal re-scaling. Then, if traits under $n+i$ still remain, one continues to move them into traits $n+i+k-1$. If trait $n+i$ becomes empty, then one shifts permutation with repeats from $n+i-1$ into $n+i+k-1$. The process is repeated until all permutation with repeats are shifted to the right of the mode. The area ratio expressed in multi-nominal distribution is:

$$A\left(L_{imit}\right) = \frac{\sum_{j_2=L_{imit}}^{n+i+k} Q\left(j_2\right)}{\sum_{j_1=L_{imit}}^{n} \alpha_{j_1} \left\langle A_P\left(j_1\right)\right\rangle^0 P\left(n, j_1\right)} \tag{7.2.144}$$

Whereas $Q\left(j_2\right)$ is the quota function which predetermines the allocated number of permutation with repeats. We define the set $\mathbb{Z}$, which includes all permutation with repeats before the shifts.

$$\mathbb{Z} = \left\{\alpha_1 \left\langle A_P\left(1\right)\right\rangle^i, \alpha_2 \left\langle A_P\left(2\right)\right\rangle^i ... \alpha_{L_{imit}} \left\langle A_P\left(L_{imit}\right)\right\rangle^i, \right. \tag{7.2.145}$$

$$\left. \alpha_n \left\langle A_P\left(n\right)\right\rangle^i, \alpha_{n+1} \left\langle A_P\left(n+1\right)\right\rangle^0 ... \alpha_{n+i} \left\langle A_P\left(n+i\right)\right\rangle^0\right\} \tag{7.2.146}$$

So that permutation with repeats that re-categorized into each of the expanded number of traits j_2 contains elements from traits $j_2 - \delta$ and $j_2 - \delta - 1$, or earlier traits respectively. δ is the offset difference between the trait number the borrowed element originates and the final settling trait number. β_i stands for the proportion of the elements from the original trait set shifted to the current trait set. For example, a previous distribution contains 8 traits and current distribution contains 12 traits, in order for the current distribution conforms to the same mode peak height as the previous, the number of traits are expanded to 16. Trait bracket 15~16 receives elements from former traits bracket 11~12, Trait 14~15 receives 70% of elements from former trait bracket 10~11, Trait bracket 13~14 receives remaining 30% of elements from trait bracket 10~11 and 20% from trait bracket 9~10. Trait bracket 12~13 receives remaining 80% of elements from trait bracket 9~10. Trait bracket 11~12 receives 80% of elements from trait bracket 8~9. Trait bracket 10~11 receives remaining 20% of elements from trait bracket 8~9 and 40% from trait bracket 7~8. Trait bracket 9~10 receives remaining 60% of elements from trait bracket 7~8 and 20% from trait bracket 6~7.

If $j_2 - \delta \leq n$, we have:

$$Q\left(j_2\right) \in \left\{\beta_1 \left\langle A_P\left(j_2 - \delta\right)\right\rangle^i P\left(n+i, j_2 - \delta\right), \beta_2 \left\langle A_P\left(j_2 - \delta - 1\right)\right\rangle^i P\left(n+i, j_2 - \delta - 1\right)...\right\} \tag{7.2.147}$$

If $j_2 - \delta > n$, we have:

$$Q\left(j_2\right) \in \left\{\beta_1 \left\langle A_P\left(j_2 - \delta\right)\right\rangle^0 P\left(n+i, j_2 - \delta\right), \beta_2 \left\langle A_P\left(j_2 - \delta - 1\right)\right\rangle^0 P\left(n+i, j_2 - \delta - 1\right)...\right\} \tag{7.2.148}$$

As before the width displacement ratio can be expressed in terms of multi-nominal distribution as:

$$\left(1 + \frac{B_{cs}^t - 1}{L_{imit} - B_{cs}^t}\right) B_{cs}^t = 1 + \frac{\left(\text{Mode}_2' - \text{Mode}_1\right)}{\left(L_{imit} - \text{Mode}_2\right)} \left(B_{cs}\right)^t \tag{7.2.149}$$

Whereas Mode_2' is now the mode of current post-transformed distribution with an equal height as the previous distribution, and the overall equation is expressed as:

$$\left(1 + \frac{B_{cs}^t - 1}{L_{imit} - B_{cs}^t}\right) B_{cs}^t \left(\frac{b}{k} \times \frac{a}{k^d}\right) \qquad (7.2.150)$$

$$= \left[1 + \frac{\left(\text{Mode}_2' - \text{Mode}_1\right)}{\left(L_{imit} - \text{Mode}_2\right)} (B_{cs})^t\right] \left(\frac{b}{k} \times \frac{a}{k^d}\right) \qquad (7.2.151)$$

$$= A\left(L_{imit}\right) \qquad (7.2.152)$$

Hence, we have shown how multinominal distribution corresponds to the lognormal distribution in all possible k.

Furthermore, neither multinomial nor lognormal distribution describes values > 7 accurately. We later stated that x > 7 on the distribution curve indicates contribution by crop plants and the diversity of angiosperm enabling agricultural and eventually industrial civilization by sustaining large population. It is shown that the distribution curve should be decreasing much gentler than both distributions dictate so that the rate of likelihood decrease should be replaced by a much gentler decreasing function. (Since the explosive biodiversity of angiosperm does not require orders of magnitudes of smaller chance beyond the emergence of human) A better approach would be creating a composite piece-wise distribution in which x < 7, a typical distribution is used, and for x > 7, a slower decreasing slope is adopted.

In fact, we took advantage of this shortcoming by assuming that the multinominal distribution was replaced by a much slower decreasing slope for x > 7 and assuming that the lognormal distribution remain the same. Since we later defined observational constraint as the sum of all extreme values satisfies the emergence requirements for human civilization or even beyond as:

$$C_{ivilization} = t_{Win} \int_{18}^{\infty} P_{df}\left(0, x\right) dx \qquad (7.2.153)$$

Then, the integration of the lognormal for extreme values converges quickly despite having a much higher likelihood at values equivalent to the emergence of human civilization comparing to the multinominal distribution such that:

$$t_{Win} P_{df}\left(0, 18\right) > C_{umulative}\left(7, 18\right) \qquad (7.2.154)$$

The multinominal distribution, despite having a much lower likelihood at values equivalent to the emergence of human civilization, converges much slower so that the integration can reach same value as the lognormal's. We find that by setting the emergence chance of explosive diversity of angiosperm to $\frac{1}{4}$ th of human emergence chance with additional extra 18-7.6=10.4[3] gain in deviation from the mode (so that the slope of the slow dropping function becomes $\frac{\frac{1}{4}-1}{10.4}$), the value of integrations for the composite multinomial and lognormal distribution ($\sigma = 0.4811$) are equal.

$$t_{Win} \int_{18}^{\infty} P_{df}\left(0, x\right) dx = \int_{18}^{\infty} C_{umulative}\left(7, x\right) dx \qquad (7.2.155)$$

The tail of the multinomial distribution is replaced with:

$$f_2\left(x\right) = \frac{1}{4} \times \frac{1}{\left[\sqrt{\frac{1}{\frac{x_1 - x_0}{3}}} \left(x - \left(x_0 - \frac{x_1 - x_0}{3}\right)\right)\right]^2} = \frac{1}{4} \times \frac{1}{\left(\sqrt{\frac{1}{\frac{11}{3}}} \left(x - \frac{10}{3}\right)\right)^2} \qquad (7.2.156)$$

Whereas x_0 represents the deviation achieved by Homo sapiens as a species, and x_1 represents the deviation achieved by Homo sapiens driven civilization before technological singularity. $\frac{1}{x^2}$ is chosen because $\int_1^{\infty} f_2\left(x\right) dx = 1$. The factor $\sqrt{\frac{1}{\frac{x_1 - x_0}{3}}}$ is used for rescaling so that $\frac{\int_{x_1}^{\infty} f_2(x)dx}{\int_{x_0}^{\infty} f_2(x)dx} = \frac{1}{4}$. It is divided by 3 due

---

[3]it is no longer at 7 because we later shown that EQ of human is 7.6

to $\int_{(1+3)}^{\infty} f_2(x)\,dx = \frac{1}{4}$. A horizontal displacement $x_0 - \frac{x_1-x_0}{3}$ is also required so that the integration from $[x_0, \infty)$ is unity as $\int_{\frac{11}{3}+\frac{10}{3}=7}^{10^5} f_2(x)\,dx = 1$ is required. Finally a factor of $\frac{1}{4}$ is added to represent the sum of all values $x > x_0 = \frac{1}{4}$ as the chance of human emergence at the current time. Since the original underlying data spans entire trait 7, the single point value of $\frac{1}{4}$ at $x_0 = 7$ should have been extended into a range from $6.5 \le x_0 \le 7.5$. As well as for the rest of traits under the multinominal distribution and modfiying the entire distribution stepwise.

We also find that the integration of both multinominal and lognormal distribution representing the past (by shifting the mode to the left and with smaller deviation such as $P_{df}(0.11, x)$ and $C_{omposite}(0.11, 1, x)$) or the future (by shifting the mode to the right and with larger deviation) results in exponentially smaller or larger values (based on $A_p(x)$ and $P(T_{rait}, x)$ we have discussed earlier) so that:

$$t_{Win} \int_{18}^{\infty} P_{df}(T_{ime}, x)\,dx \propto \int_{18}^{\infty} C_{umulative}(T_{raits}, x)\,dx = \int_{18}^{\infty} C_{omposite}(t, 1, x)\,dx \qquad (7.2.157)$$

Whereas the $P_{df}(T_{ime}, x)$'s $T_{ime} \ne 0$ represents time other than now and $C_{omposite}(t, 1, x)$'s $t \ne 0$ represents biodiversity of the past or future.

To further demonstrate this exponential rate of change in emergence given by L_{imit} is straightforward based on our previous proofs. For a toy demo, assuming we initially have a distribution with 7 traits and 7 acquisition attempts, and assume that $L_{imit} = 5$, then the number of species shared 5 or more traits is simply expressed as:

$$\langle A_p(5) \rangle^0 P(7,5) + \langle A_p(6) \rangle^0 P(7,6) + \langle A_p(7) \rangle^0 P(7,7) \qquad (7.2.158)$$

Then, the number of species shared 5 or more traits for a new distribution with 8 traits and 8 acquisition attempts expressed in terms of 7 traits and 7 acquisition attempts is:

$$\langle A_p(5) \rangle^1 P(8,5) + \langle A_p(6) \rangle^1 P(8,6) + \langle A_p(7) \rangle^1 P(8,7) + \langle A_p(8) \rangle^0 P(8,8) \qquad (7.2.159)$$

for 9 traits and 9 acquisition attempts:

$$\langle A_p(5) \rangle^2 P(9,5) + \langle A_p(6) \rangle^2 P(9,6) + \langle A_p(7) \rangle^2 P(9,7) + \qquad (7.2.160)$$
$$+ \langle A_p(8) \rangle^0 P(9,8) + \langle A_p(9) \rangle^0 P(9,9) \qquad (7.2.161)$$

for 10 traits and 10 acquisition attempts:

$$\langle A_p(5) \rangle^3 P(10,5) + \langle A_p(6) \rangle^3 P(10,6) + \langle A_p(7) \rangle^3 P(10,7) + \qquad (7.2.162)$$
$$+ \langle A_p(8) \rangle^0 P(10,8) + \langle A_p(9) \rangle^0 P(10,9) + \langle A_p(10) \rangle^0 P(10,10) \qquad (7.2.163)$$

One can see that for each respective trait number, there is an exponential increase on the rate of both attempts to permutation ratio $A_p(j)$ and permutation $P(n, j)$, so that for 8 traits and 8 acquisition attempts the rate of increase becomes:

$$\left(5 + \sum^{i_0} \epsilon_0\right)\left(\frac{7+1}{7-5+1}\right), \left(6 + \sum^{i_0} \epsilon_0\right)\left(\frac{7+1}{7-6+1}\right), \left(7 + \sum^{i_0} \epsilon_0\right)\left(\frac{7+1}{7-7+1}\right) \qquad (7.2.164)$$

for 9 traits and 9 acquisition attempts the rate of increase becomes:

$$\left(5+\sum^{i_0}\epsilon_0\right)\left(5+\sum^{i_1}\epsilon_1\right)\left[\left(\frac{7+1}{7-5+1}\right)\left(\frac{7+2}{7-5+2}\right)\right],\tag{7.2.165}$$

$$\left(6+\sum^{i_0}\epsilon_0\right)\left(6+\sum^{i_1}\epsilon_1\right)\left[\left(\frac{7+1}{7-6+1}\right)\left(\frac{7+2}{7-6+2}\right)\right],\tag{7.2.166}$$

$$\left(7+\sum^{i_0}\epsilon_0\right)\left(7+\sum^{i_1}\epsilon_1\right)\left[\left(\frac{7+1}{7-7+1}\right)\left(\frac{7+2}{7-7+2}\right)\right],\tag{7.2.167}$$

for 10 traits and 10 acquisition attempts the rate of increase becomes:

$$\prod_{i=1}^{3}\left(5+\sum^{i_i}\epsilon_i\right)\left[\prod_{i=1}^{3}\frac{7+i}{7-5+i}\right],\prod_{i=1}^{3}\left(6+\sum^{i_i}\epsilon_i\right)\left[\prod_{i=1}^{3}\frac{7+i}{7-6+i}\right],\prod_{i=1}^{3}\left(7+\sum^{i_i}\epsilon_i\right)\left[\prod_{i=1}^{3}\frac{7+i}{7-7+i}\right]\tag{7.2.168}$$

which can be generalized as:

$$\prod_{i=1}^{m}\left(j+\sum^{i_i}\epsilon_i\right)\left[\prod_{i=1}^{m}\frac{n+i}{n-j+i}\right],\prod_{i=1}^{m}\left((j+1)+\sum^{i_i}\epsilon_i\right)\left[\prod_{i=1}^{m}\frac{n+i}{n-(j+1)+i}\right],...\tag{7.2.169}$$

At each step, for the number of species shared a given number of traits j, it increases by at least a factor of j for attempts to permutation ratio and increases by at least a factor > 1 for permutation. Hence, species sharing a given number of traits increases exponentially with trait and acquisition attempt increase. Furthermore, for the distributions including higher number of traits, additional species sharing a higher number of traits becomes possible so that for 8 traits and 8 acquisition attempts, there exists an extra unmatched number of species as:

$$\langle A_p(8)\rangle^0 P(8,8)\tag{7.2.170}$$

and for 9 traits and 9 acquisition attempts:

$$\langle A_p(8)\rangle^0 P(9,8)+\langle A_p(9)\rangle^0 P(9,9)\tag{7.2.171}$$

and for 10 traits and 10 acquisition attempts:

$$\langle A_p(8)\rangle^0 P(10,8)+\langle A_p(9)\rangle^0 P(10,9)+\langle A_p(10)\rangle^0 P(10,10)\tag{7.2.172}$$

As a result, the final rate of emergence of species sharing j trait for $n+i=m$ traits and $n+i=m$ acquisition attempts is:

$$\prod_{i=1}^{m}\left(j+\sum \epsilon_i\right)^{i_i}\left[\prod_{i=1}^{m}\frac{n+i}{n-j+i}\right]\left(\frac{\langle A_p(j)\rangle^m P(n+i,j)}{(n+i)^{n+i}}\right)+$$

$$\prod_{i=1}^{m}\left((j+1)+\sum \epsilon_i\right)^{i_i}\left[\prod_{i=1}^{m}\frac{n+i}{n-(j+1)+i}\right]\left(\frac{\langle A_p(j+1)\rangle^m P(n+i,j+1)}{(n+i)^{n+i}}\right)+$$

$$...$$

$$\prod_{i=1}^{m}\left(n+\sum \epsilon_i\right)^{i_i}\left[\prod_{i=1}^{m}\frac{n+i}{n-n+i}\right]\left(\frac{\langle A_p(n)\rangle^m P(n+i,n)}{(n+i)^{n+i}}\right)+$$

$$\langle A_p(n+1)\rangle^0 P(n+i,n+1)\left(\frac{\langle A_p(n+1)\rangle^0 P(n+i,n+1)}{(n+i)^{n+i}}\right)+$$

$$...$$

$$\langle A_p(n+i)\rangle^0 P(n+i,n+i)\left(\frac{\langle A_p(n+i)\rangle^0 P(n+i,n+i)}{(n+i)^{n+i}}\right) \tag{7.2.173}$$

For each term, an extra weighted factor is required since the number of species shared given number of traits differs as (whereas the denominator $(n+i)^{n+i}$ is the total number of attempts possible):

$$\left(\frac{\langle A_p(j)\rangle^m P(n+i,j)}{(n+i)^{n+i}}\right) > \left(\frac{\langle A_p(j+1)\rangle^m P(n+i,j+1)}{(n+i)^{n+i}}\right) > \left(\frac{\langle A_p(n)\rangle^m P(n+i,n)}{(n+i)^{n+i}}\right)$$

$$> \left(\frac{\langle A_p(n+1)\rangle^0 P(n+i,n+1)}{(n+i)^{n+i}}\right) > ... > \left(\frac{\langle A_p(n+i)\rangle^0 P(n+i,n+i)}{(n+i)^{n+i}}\right) \tag{7.2.174}$$

If we denote $R(j)$ as the rate of emergence based on $L_{imit}=j$, then we have:

$$\prod_{i=1}^{m}\left(j+\sum \epsilon_i\right)^{i_i}\left[\prod_{i=1}^{m}\frac{n+i}{n-j+i}\right] < R(j) < \langle A_p(n+1)\rangle^0 P(n+i,n+i) \tag{7.2.175}$$

That is, the final emergence ratio must fall between the species shared the lowest number of traits and species shared the highest number of traits.

It is reasonable to assume, based on our simulation, as long as $j \geq \text{mode}_1$, that the following holds,

$$\prod_{i=1}^{m}\left(j+\sum \epsilon_i\right)^{i_i}\left[\prod_{i=1}^{m}\frac{n+i}{n-j+i}\right] < R(j) < \prod_{i=1}^{m}\left((j+1)+\sum \epsilon_i\right)^{i_i}\left[\prod_{i=1}^{m}\frac{n+i}{n-(j+1)+i}\right] \tag{7.2.176}$$

That is, the final emergence ratio must fall between the species shared the lowest number of traits and species shared the 2nd lowest number of traits. These assumptions remain valid since, the L_{imit} position must be located to the right of the peak of the previous distribution and the slope of descent is steep, so that the contribution of successive terms to the weighted average becomes negligible. On the other hand, when $j < \text{mode}_1$,

$$R(j) \to \frac{(n+i)^{n+i}}{n^n} \tag{7.2.177}$$

That is, the final emergence ratio approaches the overall ratio of two distributions since $L_{imit}=j < \text{mode}_1$ implies the majority of the distributions already lies to the right of L_{imit}.

Finally, one may try to establish the correspondence between the extra unmatched terms from distribution with higher number of traits with the horizontal displacement of the lognormal distribution. In reality, the effect of horizontal displacement is uniformly distributed among the matched and unmatched terms of the new distribution. Therefore, the effect of horizontal displacement does correspond with unmatched terms but such correspondence is not exclusive. It is best illustrated from a picture diagram below:

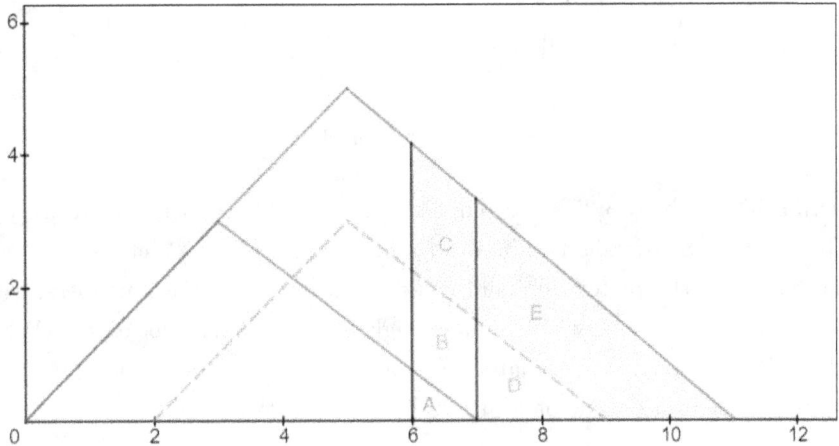

Figure 7.2.16: Area A denotes the number of species with shared number of traits $> L_{imit}$, observed under the previous distribution. Area B denotes the extra number of species emerged with simply a horizontal displacement, Area C denotes, on top of area A+B, the extra number of species emerged given that the new distribution is by a factor of B_{cs}^t higher. Area D denotes the extra number of species sharing unmatched terms with horizontal displacement only. Area E, on top of area D, denotes the extra number of species emerged given that the new distribution is by a factor of B_{cs}^t higher.

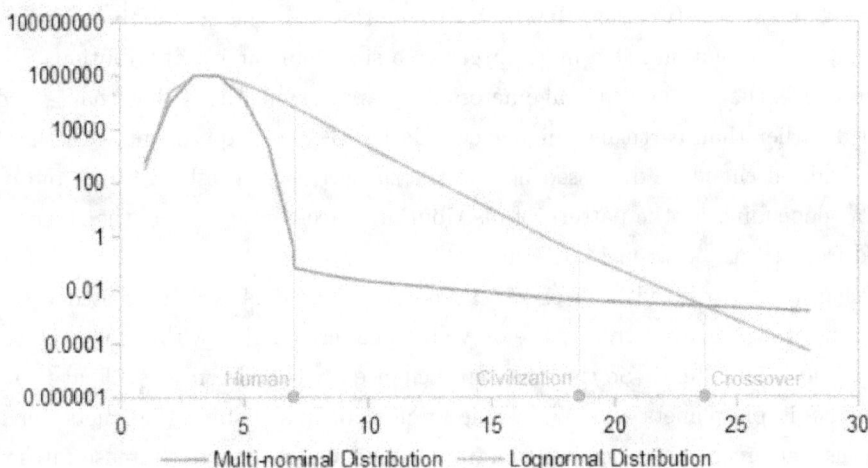

Figure 7.2.17: Both distributions plotted on log scale so that their discrepancies can become obvious. Human denotes the deviation in which human emerged, notice that piece-wise gentler slope afterward denoting the chance of emergence of angiosperm is alot larger than multinomial distribution dictates. At the point civilization is defined, lognormal overestimates the likelihood. At the crossover, lognormal's faster dropping rate eventually overtakes the multinomial distribution.

Despite these discrepancies, lognormal does a fair job at simplifying our model and provides an excellent general assessment of our model. These discrepancies provide room for future research in which more appropriate distribution can be substituted in place of lognormal distribution for better approximating the distribution based on combination and permutation.

Finally, we can generalize our distribution by taking into account all earth like terrestrial planets (with those achieved lower and higher BCS than earth based on their initial formation time and surface size). The final cumulative distribution size can be expressed as the sum of all earth like planets:

$$\sum_{n=0}^{N_{all}} C_{umulative}\left(T_{rait}, x\right) \tag{7.2.178}$$

or alternatively, as the average of all earth like planets:

$$\frac{1}{N_{all}} \sum_{n=0}^{N_{all}} C_{umulative}\left(T_{rait}, x\right) \tag{7.2.179}$$

We have shown that earth achieves the top 35% biocomplexity comparable to other earth like planets formed between 5 Gya and 4 Gya. The BCS size falls within a range of 2.783^{-5} to 2.783^{5} relative to earth's 2.783^{0} as more advanced planets achieved much greater BCS and less evolved ones achieved much less, with an average size of $2.783^{3.185}$ across all habitable planets, which is much higher than BCS size for earth. (We later actually show that the threshold window can be moved to date (45 Mya) much closer to earth formation time to fit the observational requirement of 1 civilization out of nearby 3 galaxies, so that the average BCS size is close to earth's typical size)

Ideally, one needs subtract number of years from the **Deviation** so that the mode's new position plus **Deviation** remain at the current position defining the emergence of hominid. As a result, a smaller **Deviation** signifies a shorter timespan required for the typical organism reaching human equivalent species on some habitable planets within the Milky Way rather than just on earth alone. Shifting the mode of the distribution to the left decreases the emergence and shifting the mode to the right increases the emergence if the civilization's position is fixed on the x axis.

The finalized distribution is proportional to earth's case by keeping the civilization's position fixed on the x axis. There are only two differences between earth's BCS distribution and the cumulative distribution of all planets. First, the cumulative distribution has a much larger area size than earth's distribution. Secondly, the emergence chance for human is the weighted average across all planets, which is higher than $\frac{1}{119}$ due to top 35% of all planets formed earlier than earth had higher cumulative BCS to experiment with the creation of intelligent species. The finalized chance is discussed now in the subsection "Threshold Test" and one obtains between 14.32% and 94%, depending on the pattern of distribution placement. That is, the emergence chance of hominid has increased from $\frac{1}{119}$ to between 14.32% and 94%.

Finally, the emergence chance has to fit with observational constraints. It is later discussed that there is a high confidence (70%) chance that we are currently the only emerging civilization within the nearest 3 galaxies (Milky Way, Andromeda, and Triangulum). So the emergence chance is overwritten to $\frac{1}{3}$. That is, it is at least as rare as 1 out of $3N_{all}$ habitable planets and N_{all} is the number of final habitable planets per galaxy. In order to achieve this status, the emergence of civilization at the current time has to decrease further since the cumulative BCS of all earlier formed planets greatly increased the search space available. We later mentioned that various other factors may are at play to act as the final filters. The further reduction can be attributed to intrinsic factors of the habitable planets (such as the explosive adaptive radiation of angiosperm is atypical for habitable planets, BCS and BER ultimately follows a logarithmic growth pattern) or extrinsic factors (such as the the number of habitable planets per galaxy is actually lower than expected due to such as stellar migration, and earliest habitable planet not much older than earth) On the other hand, keeping up with observational constraints by minimally decreasing the current emergence chance of civilization requires adjusting the earliest permissible time for habitable planet, negating the assumption that the earliest habitable planet formed 500 Myr earlier than earth with 500 Myr head start in evolution, which is discussed in Chp 8 Earliest threshold window. All of these justifications do not invalidate our earlier analysis on emergence. By stating the derivation of emergence chance from a step by step process, one has a complete grasp on the validity of the final conclusion.

7.2.6 Threshold Test

By taking the division of $S_{total}(7, x, t_{Attempt})$ and $C(7, x)$ curve, one finds that given infinitely long time, if ordering matters, our best fit and $p = \left(\frac{1}{108,233,978}\right)^{\frac{1}{7}}$ shows that eventually 7.5 human equivalent species should emerge instead of just 1.

$$N_{species} = \frac{S_{total}(7, 7, t_{Attempt})}{C(7,7)} = 7.5 \tag{7.2.180}$$

Since there are in total 7! possible permutation representing the total search space of traits acquisition by different orders, this indicates a convergence ratio of 672. That is, 1 out of every 672 different ordering lead to a new speciation.

$$C_{onvergeratio} = \frac{7!}{N_{species}} = 672 \tag{7.2.181}$$

Recall we established that it is likely at least 1 out of every 14 chance successfully leads into a human (bipedalism and large brain can not be the starting traits leading to human equivalent) given a fixed geologic time window of 100 Myr. so, one finds that the convergence ratio is over 48 times higher than our observation.

$$\frac{C_{onvergeratio}}{14} = 48 \geq 1 \tag{7.2.182}$$

In general, the convergence ratio should always be more stringent than our observation. There can always be extra factors we are unaccounted for regarding the order leading to speciation so that some traits must be gained prior to others and this restriction has to increase by 48 times. We do know, from computing the chance of emergence of Homo sapiens from the Hominid lineage, the chance of emergence is further reduced by $\frac{1}{2.722}$. Therefore, a remaining filtering factor with a value of 17.634, beyond our current understanding, is explained by the more stringent ordering requirements we have mentioned earlier. Alternatively, it could also imply that, out of remaining possible paths, every 17.634 permutation path or steps of acquisition of traits leads to the same species as convergent evolution. That is, more than 1 permutation paths per each combination may lead to the same speciation, but no more than 1 speciation can result from the same permutation path per each combination attempt. Thus, the mapping from permutation path to speciation per each combination attempt is surjective in nature (reattempt along the same permutation allows new speciation since we assumed that each re-attempt occur separately in space and time, and altered external environment and interaction with other emerging species allows new species to emerge even along the same permutation path) We can illustrate the difference between **Scenario 1**: combination allowing all permutation paths, **Scenario 2**: only a subset of permutation, **Scenario 3**: permutation with convergent evolution, and **Scenario 4**: only 1 permutation by each combination by the following tables:

Type	1st	2nd	3rd	4th	5th	6th	...	22th	23th	24th	Total
Scenario 1	1	2	3	4	5	6		22	23	24	n
Scenario 2	$\frac{1}{3}$	$\frac{2}{3}$	$\frac{3}{3}$	$1\frac{1}{3}$	$1\frac{2}{3}$	$1\frac{3}{3}$	...	$7\frac{1}{3}$	$7\frac{2}{3}$	$7\frac{3}{3}$	$\frac{n}{3}$
Scenario 3	1	1	1	2	2	2		8	8	8	$\left\lceil \frac{n}{3} \right\rceil$
Scenario 4	$\frac{1}{24}$	$\frac{2}{24}$	$\frac{3}{24}$	$\frac{4}{24}$	$\frac{5}{24}$	$\frac{6}{24}$		$\frac{22}{24}$	$\frac{23}{24}$	$\frac{24}{24}$	$\frac{n}{24}$

Table 7.2.2: The number of distinct species observed

Type		1st	2nd	3rd	4th	5th	6th	...	22th	23th	24th
Scenario 1		1	1	1	1	1	1		1	1	1
Scenario 2		$\frac{1}{3}$	$\frac{1}{3}$	$\frac{1}{3}$	$\frac{1}{3}$	$\frac{1}{3}$	$\frac{1}{3}$		$\frac{1}{3}$	$\frac{1}{3}$	$\frac{1}{3}$
Scenario 3	Method 1:	1	0	0	1	0	0	...	1	0	0
	Method 2:	1	1	1	1	1	1		1	1	1
Scenario 4		$\frac{1}{24}$	$\frac{1}{24}$	$\frac{1}{24}$	$\frac{1}{24}$	$\frac{1}{24}$	$\frac{1}{24}$		$\frac{1}{24}$	$\frac{1}{24}$	$\frac{1}{24}$

Table 7.2.3: The speciation chance experienced by any permutation path

303

Type	1st	2nd	3rd	4th	5th	6th	...	22th	23th	24th	Value	
Scenario 1	1	$\frac{2\cdot1}{2}$	$\frac{3\cdot1}{3}$	$\frac{4\cdot1}{4}$	$\frac{5\cdot1}{5}$	$\frac{6\cdot1}{6}$		$\frac{22\cdot1}{22}$	$\frac{23\cdot1}{23}$	$\frac{24\cdot1}{24}$	$=1$	$\frac{n}{n}$
Scenario 2	$\frac{1}{3}$	$\frac{2}{3\cdot2}$	$\frac{3}{3\cdot3}$	$\frac{4}{3\cdot4}$	$\frac{5}{3\cdot5}$	$\frac{6}{3\cdot6}$	...	$\frac{22}{3\cdot22}$	$\frac{23}{3\cdot23}$	$\frac{24}{3\cdot24}$	$=\frac{1}{3}$	$\frac{\frac{n}{3}}{n}$
Scenario 3	$\frac{1}{1}$	$\frac{1}{2}$	$\frac{1}{3}$	$\frac{2}{4}$	$\frac{2}{5}$	$\frac{2}{6}$		$\frac{8}{22}$	$\frac{8}{23}$	$\frac{8}{24}$	$1\sim\frac{1}{3}$	$\frac{\lceil\frac{n}{3}\rceil}{n}$
Scenario 4	$\frac{1}{24}$	$\frac{2}{24\cdot2}$	$\frac{3}{24\cdot3}$	$\frac{4}{24\cdot4}$	$\frac{5}{24\cdot5}$	$\frac{6}{24\cdot6}$		$\frac{22}{24\cdot22}$	$\frac{23}{24\cdot23}$	$\frac{24}{24\cdot24}$	$=\frac{1}{24}$	$\frac{\frac{n}{24}}{n}$

Table 7.2.4: The emergence chance for distinct species per permutation attempt

In the table we assume that a particular combination with 4 traits results in a total of 24 possible permutations and a convergence factor 3.

1. **Scenario 1** : The speciation chance experienced by any permutation path is 100%, each attempt results in an additional distinct speciation observed. Therefore, the speciation chance experienced by any permutation path and the emergence chance for distinct species per attempt is equivalent. The emergence chance for distinct species is generalized as $\frac{n}{n}$ per attempt given n distinct species observed with n attempts made. After 24 attempts, there are $\frac{n}{n} \times 24 =24$ species emerged.

2. **Scenario 2**: Only 1 out every 3 permutations lead to a successful speciation. The speciation chance experienced by any permutation path is $\frac{1}{3}$. Each attempt results only a $\frac{1}{3}$ increase in the chance of additional distinct speciation observed. Again, the speciation chance experienced by any permutation path and the emergence chance for distinct species per attempt is equivalent. The emergence chance for distinct species is generalized as $\frac{\frac{n}{3}}{n}$ per attempt given $\frac{n}{3}$ distinct species observed with n attempts. Only every 3 attempts guarantee a speciation. After 24 attempts, there are $\frac{\frac{n}{3}}{n} \times 24 =8$ species emerged.

3. **Scenario 3**: If every 3 permutations converge into the same species, then, every attempt will result in the same successful distinct speciation. Only after every 3 attempts result in an additional distinct speciation. There are 2 ways to count the speciation chance under this scenario. In the first way, whichever path attempted first blocks the speciation chance of the remaining paths. The first attempted path results in a speciation and the rest 2 are blocked since the niche space is pre-occupied. In this scenario, the speciation chance experienced by any path is equivalent to **Scenario 2** before the first attempt among the 3 permutations. After the 1st attempt, the speciation chance experienced by the remaining paths drop to 0 while the 1st increased to 1. In the alternative way, each path repeatedly contributes to the emergence of the same species (possibly subspecies). In this scenario, the speciation chance experienced by any path is guaranteed and is equivalent to **Scenario 1.** However, only after 3 permutations guarantee an additional distinct species emergence. Therefore, the emergence chance of distinct speciation < the speciation chance per path allowing repetition. For both counting methods, a non-symmetry arises in the speciation chance experienced by any permutation path and the emergence chance for distinct species per attempt. In essence, if one is only concerned with the emergence chance of the first distinct species possessing 4 traits regardless of path, i.e any species with 4 traits, then, the emergence chance for just 1 distinct species observed is the same as Scenario 1, regardless of the ways to count because $\frac{\lceil\frac{1}{3}\rceil}{1} = \frac{1}{1}$. If one is concerned with the maximum emergence chance of distinct species per attempt given $\lceil\frac{n}{3}\rceil$ distinct species observed,

then, the maximum emergence chance is: $\dfrac{\dfrac{n}{C_{onvergeratio}}}{C_{onvergeratio} \times \left\lceil \dfrac{n}{C_{onvergeratio}} \right\rceil - (C_{onvergeratio}-1)} = \dfrac{\lceil\frac{n}{3}\rceil}{3\lceil\frac{n}{3}\rceil-2}$, which

infinitely approaches Scenario 2 as n approaches infinity. $\lim_{n\to\infty} \dfrac{\lceil\frac{n}{3}\rceil}{3\lceil\frac{n}{3}\rceil-2} = \frac{1}{3}$.

4. **Scenario 4**: Only one permutation path can lead to speciation. The speciation chance experienced by any permutation path is $\frac{1}{24}$. Each attempt results only a $\frac{1}{24}$ increase in the chance of additional distinct speciation observed, The emergence chance for distinct species is generalized as $\frac{\frac{n}{24}}{n}$ per attempt given $\frac{n}{24}$

distinct species observed with n attempts. Only all 24 attempts guarantee a speciation. After 24 attempts, there is $\frac{n}{24} \times 24 = 1$ species emerged.

The emergence chance for distinct species per attempt given n attempts across all scenarios is generalized to be:

$$P(n) = \begin{cases} \dfrac{\overline{\frac{n}{C_{onvergeratio}}}}{n} = \dfrac{1}{C_{onvergeratio}} & \text{non repetitive convergence} \\[3ex] \dfrac{\left\lceil \frac{n}{C_{onvergeratio}} \right\rceil}{n} & \text{repetitive convergence} \end{cases} \tag{7.2.183}$$

and for finding the maximum emergence chance per attempt given n attempts under a repetitive convergence:

$$P_{max}(n) = \frac{\left\lceil \dfrac{n}{C_{onvergeratio}} \right\rceil}{C_{onvergeratio} \times \left\lceil \dfrac{n}{C_{onvergeratio}} \right\rceil - (C_{onvergeratio} - 1)} \tag{7.2.184}$$

If convergence ratio per attempt is less stringent than our observational logic dictates, one has to re-examine the data sample, exclude re-attempt over-counts, or re-examine human evolution until the threshold invariant is regained and surjection is conserved.

Having demonstrated that the emergence of the hominid lineage is reduced to $\frac{1}{672}$ out of all 5040 permutation paths at the current time instead of $\frac{1}{14 \cdot 2.722}$ based on our more stringent selection and data fitting for the distribution. Now, we can compute the overall emergence rate by considering those planets formed earlier by up to 500 Myr based on our 5 Gya to 4 Gya time window. Recall that cumulative BCS increases the biocomplexity search space by 2.783 every 100 Myr. For planet formed 500 Myr earlier, the biocomplexity search space reaches $2.783^5 = 167$ times higher than current earth. Since there is an exponentially greater space for nature's experimentation, this implies that the chance of emergence is exponentially increased.

We compute the overall emergence chance across all earlier forming planets by first computing the emergence chance based on the biocomplexity search space formed t years earlier than earth, whereas t is in units of 100 Myr:

$$E_{merge}(t) = \frac{1}{4} \cdot (B_{cs})^t \tag{7.2.185}$$

Then, we compute $E_{merge}(0)$, $E_{merge}(1)$, $E_{merge}(2)$, $E_{merge}(3)$, $E_{merge}(4)$, and $E_{merge}(5)$ respectively to represent the total cumulative search space for planet emerged 100 Myr, 200 Myr, 300 Myr, 400 Myr, and 500 Myr ago earlier than earth:

Then, we have established the pairing data of $(t, E_{merge}(t))$, and the best fit for such data points is:

$$H_{ominid}(t) = E_{merge}(t) \tag{7.2.186}$$

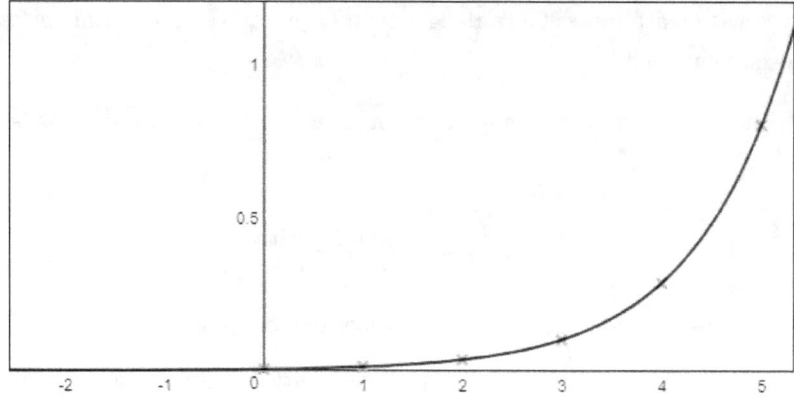

Figure 7.2.18: Hominid lineage emergence chance across time whereas t < 0 represents planets formed later and t > 0 represents planets formed earlier

We have shown earlier that no planet hosts cumulative biocomplexity higher than earth across all dryland coverages formed 2.13 Myr later than earth, then a simplified computation on the emergence chance of hominid lineage equivalent on all planets formed earlier than earth is:

$$\frac{1}{(5 + 0.0213)} \int_{-0.0213}^{5} H_{ominid}\left(t\right) dt = 0.27136 \tag{7.2.187}$$

We can improve on the previous results by taking into consideration that there are fewer suitable habitable candidates earlier in cosmic history by taking metallicity selection into effect, whereas 0.000281 denotes 2.13 Myr into the future:

$$\frac{\int_{-0.066}^{0.000281} \int_{-1}^{1} f_{metallicity}\left(x,t\right) \cdot f_{wetearth}\left(x\right) dx \cdot H_{ominid}\left(C_0\left(-t\right)\right) \cdot S_{tellar}\left(C_1\left(t\right)\right) dt}{\int_{-0.066}^{0.000281} \int_{-1}^{1} f_{metallicity}\left(x,t\right) \cdot f_{wetearth}\left(x\right) dx \cdot S_{tellar}\left(C_1\left(t\right)\right) dt} = 0.25286 \tag{7.2.188}$$

The emergence chance of hominid lineage equivalent on all planets formed between 2.13 Myr behind to 500 Myr ahead of earth with higher cumulative biocomplexity than earth at the current time given different dryland surface coverage is 25.29%.

Unfortunately, one may have to go another step further. The emergence of Hominid lineage is also depended on BER. The previous results are only applicable as a special case when BER=1. If each successive 100 Myr the mode of the search space progressively shifts toward organisms with a higher number of traits, then the emergence chance is higher than simply having an exponentially increasing BCS with a mode centering on simpler organism with a fixed number of traits (static BER of 1). In general, for BER > 1, the emergence chance for earlier formed planet increases much faster than BCS. However, this chance can only be computed based on the finalized lognormal function we shall define. It is not possible to compute this chance based on BCS or BER alone. The total area (the integration) represented by the lognormal distribution is the average BCS value we find for different planets across different times. The information regarding the probability within a selected distribution area range representing the emergence of intelligence is lost due to dimensionality reduction. Given distributions with the same BCS, the likelihood of human comparable organism or more advanced is biased toward a mode closer to human and a distribution with a higher BER. The final realistic chance is computed based on the multi-regressional equation derived from the emergence of the hominid lineage given in Section 8.7.5 "Generalized Model and Emergence".

$$\frac{d}{dt} C_{df}\left(t\right) = \exp\left(\frac{a}{k^d} + t \ln\left(B_{cs}\right) + \frac{b}{k}\right) \tag{7.2.189}$$

With all previous considerations recognized, assuming BER=BCS=2.783, one sets the lognormal distribution $P_{df}\left(0,x\right)$ at the current time evaluated to be $C_{df}\left(0\right) = \frac{1}{119}$ and one computes the area of the integration to the right of deviation value of 18. Then, one establish the pairing data of $(t, C_{df}(-t))$, and the best fit for such data points is:

$$H_{ominid}(t) = \begin{cases} -51.9689\,(137601)^t + 51.981\,(137728)^t & t \leq 0.306 \\ 1 & t > 0.306 \end{cases} \qquad (7.2.190)$$

That is, for any habitable planet formed 30 Myr earlier than earth, the combined contribution of a shifting BER of 2.783 per 100 Myr and an exponential increase of BCS at 2.783 per 100 Myr guaranteed 100% of at least 1 hominid species emergence chance on the planet. For habitable planets formed between 0 and 30 Myr earlier than earth, there are a smaller but non-zero chance of at least 1 hominid species emergence chance on the planet. As for earth's case, it drops to $\frac{1}{183.12}$ as it is expected. The revised curve shows that it rises much faster than the earlier one. The raw emergence chance is given as:

$$\frac{1}{5} \int_0^5 H_{ominid}(t)\,dt = 0.953 \qquad (7.2.191)$$

It shows that nearly all habitable planets with greater than earth complexity give rise to at least 1 hominid lineage species with extra 30 Myr of evolution. We can improve on the previous results by taking into consideration that there are fewer suitable habitable planet candidates earlier in cosmic history by taking metallicity selection into effect, whereas 0.000281 denotes 2.13 Myr behind earth formation time:

$$\frac{\int_{-0.066}^{0.000281} \int_{-1}^{1} f_{metallicity}(x,t) \cdot f_{wetearth}(x)\,dx \cdot H_{ominid}(C_0(-t)) \cdot S_{tellar}(C_1(t))\,dt}{\int_{-0.066}^{0.000281} \int_{-1}^{1} f_{metallicity}(x,t) \cdot f_{wetearth}(x)\,dx \cdot S_{tellar}(C_1(t))\,dt} = 0.94374 \qquad (7.2.192)$$

In general, depending on the placement of successive distributions and varying BER ($1 < x <$ BCS), the chance of hominid lineage emergence varies between those two extreme cases. If the emergence exceeds our observation for at most 1 civilization out of 3 galaxies, one has to adjust the earliest possible window for habitable planet emergence, which is discussed in Chapter 8.

7.2.7 Projection onto 3D Space

We want to stress that this distribution can be projected onto a 3 dimensional space, whereas our current lognormal distribution is a sectional cut among the overall landscape of log normal distribution of species possessing different sets of all animal traits. There are other traits and features such as wing, feather, snout, horn, and neck which are non-critical representation of human characteristics but nevertheless are important features possessed by the examined cohorts. The total sum of all 2 dimensional lognormal distributions giving emergence to all possible species currently exists or not forms this 3 dimensional log normal distribution landscape. The topmost horizontal line denotes the base of the lognormal distribution in the perspective of Homo sapiens. A class of line with slopes can be drawn to represent the base of lognormal distribution of other species, whereas S_{core} is the ranking score they received from the perspective of its similarity index to human, with 8 being human and 0 being sharing no common traits at all with human.

$$y = -\frac{8 - S_{core}}{8} x \qquad (7.2.193)$$

The concentric circles are given by:

$$x^2 + y^2 = S_{core}^2 \qquad (7.2.194)$$

and the position of human, i.e. the similarity index of human to other species must fall on the intersection of concentric circles and slope lines.

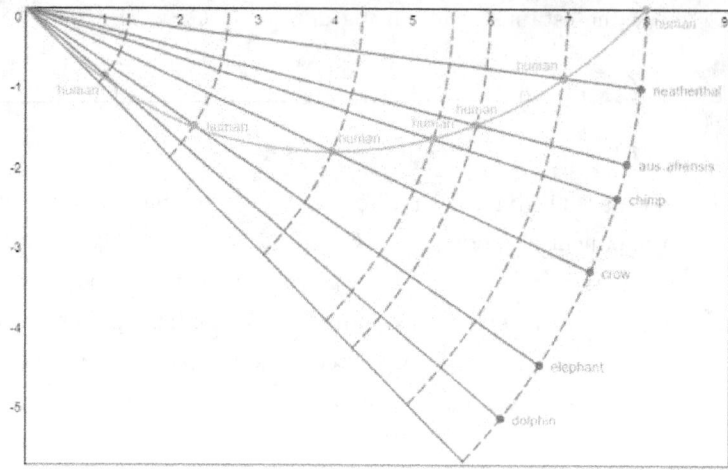

Figure 7.2.19: Human's position in other species perspective based on lognormal distribution slices viewed from top down on a 3D plane

Based on the graph, one can see how Homo sapiens position is ranked by other species. The lower the score ranked by Homo sapiens, the lower the score human received assigned by other species reciprocally. An equal distance curve can be drawn for human as well as every other species, which designates its position from the perspective of other species. The equal distance curve can be fit as the follows:

$$f_{in}(x) = \sqrt{x^2 \left(1 + \frac{(8-x)^2}{64}\right)} \tag{7.2.195}$$

$$T_{ran}(x) = x \left(\frac{f_{in}(x)}{x}\right)^{-1} \tag{7.2.196}$$

One plot the pairs of:

$$\left(T_{ran}(x), \frac{8-x}{8} \cdot T_{ran}(x)\right) \tag{7.2.197}$$

and derives the best fit:

$$E(x) = 0.00114x^4 - 0.0187x^3 + 0.208x^2 - 1.049x - 0.02 \tag{7.2.198}$$

Although the distance is well-defined on such a plane for Homo sapiens relative to other species, the distance between each other species is non-accurate. For example, the distance between elephant and dolphin suggests that dolphin and elephant are closely related despite their great dissimilarity. In order to solve this problem, the 3D plane has to be able to fold and spread as a manifold, This lead to non-euclidean geometry combined with statistics. That is, in order to truly calculate the distance between elephant and dolphin, one should able to expand the distance between these two species by adding additional traits (such as blow hole, dorsal fins, submarine sonar detection unique to dolphin, tusks, ear flaps, and long snouts unique to elephant). Mathematically, it is expressed by adding additional concentric circles beyond the listed 8 traits.

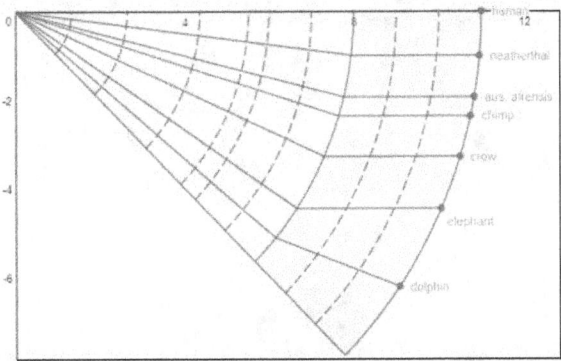

Figure 7.2.20: Increased distance between dolphin and elephant by unfolding 3 hidden concentric circles (shaded parts) representing the traits blow hole, dorsal fins, submarine sonar detection unique to dolphins.

The plot shows that the first 3 additional concentric circles representing blow hole, dorsal fins, submarine sonar detection unique to dolphin not only enlarged the separation distance between dolphin and elephant, but with the rest of the terrestrial species as well.

It is a bit more tricky when one expands 3 additional concentric circles representing tusks, ear flaps, and long snouts unique to elephant. By specifying these 3 traits, the separation distance is enlarged for elephant between both dolphins and the rest of terrestrial species, as illustrated below:

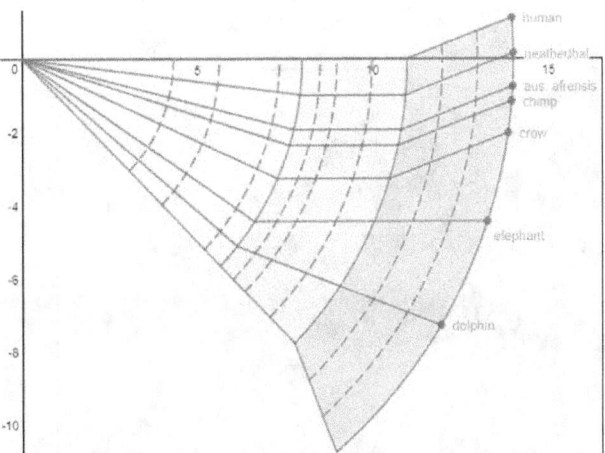

Figure 7.2.21: Increased distance between dolphin and elephant by unfolding 3 additional hidden concentric circles (deeply shaded parts) representing the traits tusks, ear flaps, and long snouts unique to elephant.

However, these three traits are non-applicable to other species other than elephants, so the distance between dolphins and the rest of terrestrial species should be unchanged and running in parallel. The inconsistency is resolved as we mentioned earlier, by treating the 3D plane as an flexible manifold, so that the extra separation created by elephant can be folded up like an origami (doing the reverse operation of spreading) when true distance between dolphins and the rest of the terrestrial species is compared, as indicated below:

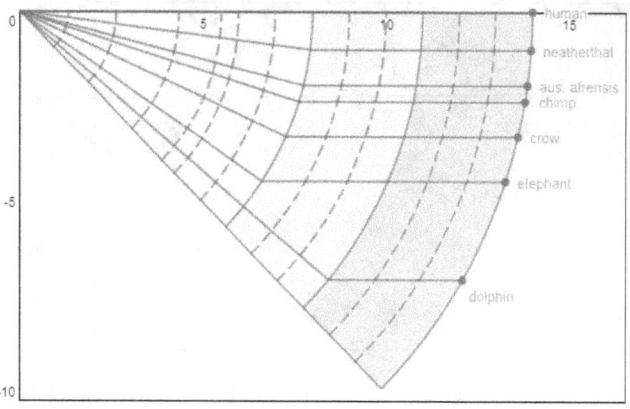

Figure 7.2.22: Folding up, the reverse operation of spreading, on the elephant so that the true distance between dolphin and the rest of the terrestrial species is conserved

This method can then be universally applied to any species.

Finally, the equal distance curve for Homo sapiens position ranked by other species been shifted 3 concentric circles can be shifted outward. We assume that for a terrestrial intelligent species, it is possible for it further go on attaining tusks, ear flaps, and long snouts but not blow holes, dorsal fins, and submarine sonar detection. Therefore, human's potential position ranked by all species is shifted.

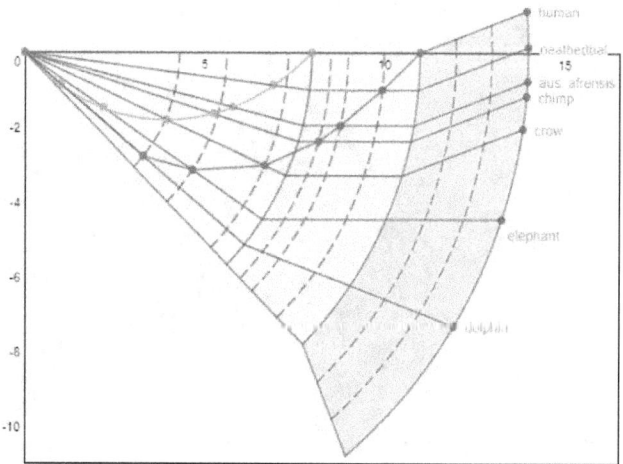

Figure 7.2.23: The new equal distance curve for Homo sapiens position is ranked by other species vs. the old one

7.2.8 Conclusion

By now, we have confirmed that animals on earth follow lognormal distributed attributes curve, then we have to compute human attributes as an outlier, how much more advanced or faster we evolved compared to the average rate of evolution. We need to set a background evolutionary rate. We do not know the mean overall evolutionary rate within our galaxy and beyond, but we can choose different mean rate based on cohorts of animals here on earth. Mammals, birds, and reptiles are chosen because they are most similar to human and more evolved than fish, amphibians, and invertebrates. If we can show that human emergence is rare even if assuming species functional equivalent to an average mammal, birds, and reptiles roamed on all Earth-like planets, then we set an upper bound and complied better with the principle of Mediocrity.

7.3 Counting Deviation and YAABER

Using the number of traits acquired to represent the deviation of homo sapiens is not the only approach, we now introduce a different approach in which each trait has a non-equal share in the contribution toward the final deviation, which can complement current deviation calculation, yet both achieves similar results. Here we introduce the equation for calculating the **deviation** value of Homo sapiens from the mammalian mode. We also introduce a concept called **Years Ahead Against the Background Evolutionary Rate. Or YAABER.** This is an estimation of how many years into the future using the current rate of evolution observed in which the average model organism we compare and contrasts to will diversify and evolve into a species comparable to human today with a probability of 1. We have already shown that the emergence of human at the current epoch on earth, even among the current cohorts of avian, mammalian, and reptilian lineage, is closer to 0 than 1. With increasing bio-diversity, the chance of emerging functionally human equivalent species in any habitable planet will approach 1 in some number of years into the future.

It is important to note that the deviation is calculated by the distance shifted away from the current mode by much higher exponential acceleration of human emergence over the course of a few million years ($t < 50,000,000$ yr) to that of the distance the current mode shifts over the course of 100 Myr by much lower exponential acceleration of the background evolution. Using the mode value obtained at the current time, the relationship between deviation and YAABER is:

$$\textbf{YAABER} = \frac{\text{deviation by Human}}{\text{deviation by BER}} = \frac{(B_{\text{human}})^t \, \textbf{mode} - \textbf{mode}}{(\textbf{BER})^{10^8} \, \textbf{mode} - \textbf{mode}} \tag{7.3.1}$$

since we have assumed that current **mode** value $=1$, the equation simplifies to:

$$\textbf{YAABER} = \frac{(B_{\text{human}})^t - 1}{(\textbf{BER})^{10^8} - 1} \tag{7.3.2}$$

Since the denominator $(\textbf{BER})^{10^8} \, \textbf{mode} - \textbf{mode}$ is expressed as the distance the current mode shifted by exponential acceleration of the background evolution over the course of 100 Myr, one can also think it as the average background evolutionary speed per 100 Myr. Then one have:

$$\textbf{YAABER} = \textbf{YAABES} = \frac{\text{deviation by Human}}{\text{speed by BER}} \tag{7.3.3}$$

Therefore, we can also call **YAABER** as **YAABES**, or **Years Ahead Against the Background Evolutionary Speed.**

To take one more step further, knowing that the average speed of background evolution is expressed as the distance shifted $(\textbf{BER})^t \, \textbf{mode} - \textbf{mode}$ per 100 Myr, the differential speed of background evolution is simply the derivative over the total distance covered:

$$\frac{d}{dx}(\textbf{BER})^t = \ln(\textbf{BER})(\textbf{BER})^t \tag{7.3.4}$$

so YAABER based on speed at the current time and a fixed **deviation** is:

$$\textbf{YAABER} = \frac{\text{Deviation}}{\ln(\textbf{BER})(\textbf{BER})^0} = \frac{\text{Deviation}}{\ln(\textbf{BER})} \tag{7.3.5}$$

Keen readers might quickly point out that Earth has only $5 \cdot 10^9$ years of an effective land dwelling evolutionary window remaining, so the window might be too short to guarantee the emergence of human functional equivalent again on earth or any other planets. I would like to point out and remind them in that somewhat the Sun is the more massive of the GFK spectral class stars, stars with lower mass, not as small as a red dwarf, have significantly longer time span for biological evolution. [4]

---

[4]See Chapter 2

We now introduce the equation for calculating the **weighted deviation**:

$$e_n = e_{n-1} + \sum_{i=0}^{m} f(k_i, e_i) \qquad (7.3.6)$$

$$f(k_i, e_i) = \begin{cases} k_i & k_i > 0 \\ 0 & k = 0 \\ -k_i & -k_i < 0 \end{cases} \qquad (7.3.7)$$

The equation above is a recursive summation, where the term e_n is recursively defined until it reaches e_0, the base case. The term e_n is defined as the mean cohort average we are comparing against with. We currently defined e_{n-1} as the mammalian ancestor of 65 Mya with close similarity to mouse today, along with surviving bird ancestors and reptilian ancestors following the K-T extinction event. Then, e_{n-2} can be defined as the last common ancestor between reptiles and mammals, some 225 Mya. e_{n-3} can be defined as the last common ancestor between amphibians and reptiles. e_{n-4} can be defined as the last common ancestor of amphibians and fish. e_{n-5} can be defined as the last common ancestor between fish and vertebrates. e_{n-6} can be defined as the last common ancestor of vertebrates and invertebrates. e_{n-7} can be defined as the last common ancestor of multicellular eukaryotes and unicellular eukaryotes. e_{n-8} can be defined as the last common ancestor between eukaryotes and prokaryotes. Finally, e_0 can be defined as the last common ancestors for all life forms on earth. However, each term of e does not need to be assigned to a major lineage split in the evolutionary tree of life. It can well be represented by the last common ancestor between subspecies where one of the subspecies led to the human ancestor. For example, *Homo Habilis* can be used as the term e_{n-1}, and *Homo Erectus* as the term e_{n-2}, *Australopithecus Afarensis* as the term e_{n-3}. It is not very practical to use this approach, however, because we will soon see that any evolutionary features take time to evolve and we need greater temporal time span to derive mathematically significant value. The first order approximation of our computed values closely approaches the actual value if each successive ancestors leading up to human were computed in the recursive summation. Higher resolution in temporal aspects leads to more precise computations but takes much significant time, and not all missing links are well-documented.

This recursive summation can be simplified in our discussion to

$$e_n = e_0 + \sum_{i=0}^{m} f(k_i, e_i) \qquad (7.3.8)$$

$$f(k_i, e_i) = \begin{cases} k_i & k_i > 0 \\ 0 & k = 0 \\ -k_i & -k_i < 0 \end{cases} \qquad (7.3.9)$$

Whereas e_0 is the mean cohort average of terrestrial mammals, birds, and reptiles from 65 mya. This is possible because we assume that all habitable planets have biological and functional equivalent creatures to terrestrial mammals, reptiles, and birds roaming on its surfaces to comply better with the principles of Mediocrity. Then, we are only interested in the **weighted deviation** since the Cenozoic era. If ancestors of mammals and reptiles are used as the typical average model organisms on all terrestrial planets relative to Homo Sapiens, then their value for **weighted deviation** will be larger indeed. We do need to pay closer attention to the defined function $f(k_i, e_i)$. In order to appreciate all cases listed and defined for the function, we need to draw a 2 by 3 matrix for each different cases.

K_i	e_0 (trait formerly lacking)	e_1 (trait formerly possessed)
$K_i > 0$	Ancestors lacked the trait, but descent evolved the trait	Ancestors possessed the trait, but descent outperformed the ancestor
$K_i = 0$	Both ancestor and descent lacked the trait	Both ancestors and descent possessed the trait
$K_i < 0$	Ancestor lack the trait, but descent has further lost the trait	Ancestor possessed the trait, but descent has lost the trait

Table 7.3.1: A table lists the deviation value for a list of traits possessed by a particular species compared against the basal mammalian ancestors when that trait is also absent from the ancestor is grouped under column e_0, and a list of traits possessed by a particular species compared against the basal mammalian ancestors when that trait is already present in the ancestor is grouped under column e_1

K_i	e_0	e_1
$K_i > 0$	Homo sapiens' complex language, tool usage	Homo Sapiens' bigger brain, fixed bipedal locomotion, and binocular vision
$K_i = 0$	Homo sapiens' lack of feather, bird-like wings	Homo sapiens' Omnivorous diet
$K_i < 0$	Naked mole rats with little to no vision, bats with poor eyesight	Dolphin's lack of bipedal locomotion and the use of tail

Table 7.3.2: With listed examples of specific traits drawn from Homo sapiens, naked mole rats, bats, and dolphins

The matrix shows different cases of features possessed by Homo sapiens, bats, rats, and dolphins versus the prototypical mammal's features. Traits under e_0 such as complex language and tool usage are completely absent from prototypical mammals, surviving reptiles, and bird species 65 Mya, the complex language and tool usage using opposable thumb exhibited by humans converted into K_i value then will be counted positively to the final **Deviation**. Traits listed under e_1 such as cranial capacity, partial binocular vision, forms of bipedal locomotion is found in birds, and mammals from 65 Mya, but Homo Sapiens has a greater cranial capacity. Therefore, its K_i value is counted positively toward the final **Deviation**. Mammals 65 Mya did not possess bird-like wings and feather, so the trait falls under e_0, and Homo Sapiens do not possess feathers and wings 65 million years later, so no value is counted toward the **Deviation**. On the other hand, Mammals 65 Mya had an omnivorous diet, as it is listed under e_1, Homo Sapiens 65 million years later also had an omnivorous diet. Therefore, no extra value is added toward the **Deviation**. It is important to note, however, primates, from 30 million years ago, had a predominantly, insectivorous diet. As a result, a negative value is added toward the final **Deviation** if we had used primate as an intermediary ancestor in our recursive summation computation in case of computation with a higher resolution. However, this negative value is canceled later in the equation because Homo Sapiens re-evolved omnivorous diet and an equally positive value is added to the final **Deviation**. Therefore, omnivorous diet had a total minimal contribution toward **Deviation**, just as we have computed without taking primates as an intermediary consideration in our recursive summation. Mammals 65 Mya lacks adequate color detection readily found in birds and reptiles. Therefore, color-detection and poor vision for mammals 65 Mya falls under e_0. it is possible, such as naked mole rat living exclusively underground and bats lived predominately inside caves and active during the night, had even poorer vision. As a result, the color vision attributes of naked mole rats and bats had contributed negatively toward the final **Deviation** for each these species. Finally, mammals, birds, reptiles 65 Mya in general walked with legs, so this trait falls under e_1, yet dolphin 65 million years later evolved legs into a tail, therefore, contributed negatively toward the **Deviation** for dolphin.

Having shown the definition of the function $f(k_i, e_i)$, we now proceed to define each computed value of K.

First and foremost, we need to compute the cranial capacity increase of Homo Sapiens relative to the Background Evolutionary rate observed in an average bird, reptile, and mammals since the start of the Cenozoic era. Human has an EQ of 7 while the rest group e_0 had an average of 1.

As we have seen earlier, omnivorous diet contributed minimally toward the final **Deviation** because mammals from 65 Mya had an omnivorous diet (although this was not the universal trait among all species 66 mya), so human had some gains $1 - \frac{8278 \times 0.917}{25483} = 0.702$ on omnivorous diet, on top of the rest group e_0 (including non mammals) which had an average of $\frac{8278 \times 0.917}{25483} = 0.298$.

This is almost equally valid conclusion for Homo sapiens evolved bipedal locomotion because birds 65 Mya had evolved bipedalism and bipedal dinosaurs roamed earth since Triassic lasted up until the K-T extinction event, so human had $1 - \frac{10089 \times 0.732}{25483} = 0.711$ gains on fixed postured bipedalism, on top of the rest group e_0 which had an average of $\frac{10089 \times 0.732}{25483} = 0.289$.

Homo sapiens has opposable thumbs. This attribute contribute some value to the final **Deviation** because primate ancestors from 65 Mya had at least partial gripping power, so human had $1 - \frac{353 \times 0.4}{25483} = 0.9945$ gains on opposable thumb, on top of the rest group e_0 which had an average of $\frac{353 \times 0.4}{25483} = 0.00555$.

Homo sapiens has binocular vision. This attribute contribute some value to the final **Deviation** because primate ancestors from 65 Mya had at least partial binocular vision, so so human had $1 - \frac{2526 - 450}{25483} = 0.9185$ gains on binocular vision, on top of the rest group e_0 which had an average of $\frac{2526 - 450}{25483} = 0.08$.

Homo sapiens is social. This attribute contribute some value to the final **Deviation** because some basal mammals, birds, and reptiles from 65 Mya are social, so so human had $1 - \frac{10089 \times 0.732}{25483} = 0.86$ gains on opposable thumb, on top of the rest group e_0 which had an average of $\frac{3586}{25483} = 0.14$

Homo Sapiens unique language skill does contribute positively to the final **Deviation**, however, human language, as demonstrated by experiments, originated from the Broca's and Wernicke's areas of the brain. The abstract thinking behind complex language and symbol manipulation lie under the frontal cortex. Essentially, language is a by-product of a big brain. We need to exercise extreme cautiousness to avoid double counting values to the final **Deviation**. Human language, viewed from the perspective of the range of vocalization capable by the larynx and vocal chord, seem to suggest more evolved than others, this view is unfortunately undermined by African Grey Parrot such as Alex, which demonstrated stunningly accurate imitation of human sound in different tones, so human had just $1 - \frac{5150 \times 0.87}{25483} = 0.8242$ gains on language, on top of the rest group e_0 which had an average of $\frac{5150 \times 0.87}{25483} = 0.1758$.

Finally, someone may point out that other attributes such as being warm-blooded, having a placenta and having hair may contribute positively to the final **Deviation**. However, warm-blooded, hairiness and bearing young inside one's body is present in mammals 65 Mya. Being warm blooded is even demonstrated in dinosaurs in some degree during the Mesozoic epoch. Therefore, if such traits is ignored in the rest group e_0 and taken for granted, then so does homo sapiens.

The final tally is listed below:

Trait name	Homo Sapiens	The rest group e_0
binocular vision	1	$\frac{2526-450}{25483} = 0.081$
large cranial capacity	7	1
opposable thumbs	1	$\frac{353 \times 0.4}{25483} = 0.0055$
bipedal	1	$\frac{10089 \times 0.732}{25483} = 0.2899$
language	1	$\frac{5150 \times 0.87}{25483} = 0.1758$
omnivorous	1	$\frac{8278 \times 0.917}{25483} = 0.298$
social	1	$\frac{3586}{25483} = 0.14$
Total	**13**	**2**
Total Deviation	**13+2=15**	**2**

The assessment of the qualitative difference of traits other than cranial capacity in comparison to the rest group e_0 is somewhat subjective, in which each trait acquired by human is simply labeled as a gain by an unity. Unlike cranial capacity which can be mathematically well established by given volume size, other traits either are harder to quantity, or in general it does not offer qualitative difference once it is fully acquired. For example, there are no qualitative difference between all omnivorous eaters, they all consume a wide range of resources. The score assessment for traits under the rest group e_0 is simply the weighted average of those possessed the trait divided by the total species under consideration. Future work should focus on providing a better qualitative assessment and ways to quantify on the functional enhancement of the trait acquired. That is, if there exists a range of social interaction among all social groupers, levels of vocalization ability among all language acquirers, qualitative difference between multiple thumbs and a single thumb. The author's intuition is that, unlike the enlargement of brain which can offer unlimited potential, there is limited if any qualitative gain in other traits with enhancement, and explanatory attempt is presented at the end of this section.

In conclusion, on a weighted scale, Homo sapiens' great cranial capacity attribute contribute more than the majority toward the final **Deviation**, (conforms with our intuition). Though other attributes are also essential for the transformation of a biological species into an industrial one, they do not contribute more significantly in our calculation toward the final **Deviation** since these attributes have been evolved in birds, reptiles, and mammals 65 Mya or earlier. Nevertheless, the unique combination is only evolved in the human lineage and their presence along with a large brain created a self-reinforcing positive feedback loop enabling greater cranial capacity. That is, the growth of the cranial capacity as a trait to reach our current size is only possible when other traits help to magnify the advantage and fulfill the potential offered by an advanced brain and makes the a directional selection toward an ever larger brain feasible.

Knowing that the total weighted deviation is 15 for homo sapiens, and weighted deviation is 2 for the rest of species, one can normalize the deviation to 7.5 for homo sapiens and 1 for the rest of the species. The normalization step brings us to greater insight. The 6 traits excluding the cranial capacity had a total contribution of ≈ 1 toward the rest of species, this is based on our original assumption that all 7 traits' total probability must sums up to 1. It implies that if homo sapiens is defined by n traits, then the rest of species must be defined by a weighted average of 1 trait only. We also notice that the cranial capacity of homo sapiens to the rest of species in a ratio of 7 to 1, corresponding to the 6 traits excluding cranial capacity of homo sapiens to the rest of species in a ratio of 6 to 1. Then, we take a major leap of faith and believe that each additional trait defines human increases an unity value toward the final cranial capacity of homo sapiens. Therefore, if human were defined by n traits excluding cranial size, then the corresponding cranial capacity of homo sapiens to the rest of species should be in a ratio of n to 1. [5]As a result, the original EQ of the rest of species of 1 is the consequence of the sum of weighted score of all possible traits. When all 7 traits are present in a species as homo sapiens, the EQ is increased to 7 as in Homo sapiens. Consequently, we have double counted our results. One needs to only count number of traits excluding cranial capacity or count the cranial capacity increase only.

We have shown that not only other traits just contributing toward the formation of a larger brain size, but, in fact, the presence of all additional traits exclusively contributed to the formation of a larger brain. By simply calculating the cranial growth size we also elegantly included the evolutionary pace of other complementary traits against the background average. This also confirms that greater cranial capacity is a less well-adapted feature by evolution, it does not immediately gain a great benefit to the organism otherwise it would have evolved much earlier such as bipedal locomotion or flying. Finally, we are able to address the question we have raised earlier, regarding the existence of a range of quantifiable values for each of the trait with further functional enhancement. With our analysis, it is quite likely that brain size is probably the only trait that will exhibit noticeable range of quantifiable values with further functional enhancement. The cause of such increase is due to continued acquisition of additional traits that further enhances the survival of the organism, which in turn, maintains these traits by increasing the neural coordination within the central nervous system, aka increasing brain size. Even if some traits exhibit a range of quantifiable values with further functional enhancement, the range of quantifiable values with further functional enhancement for the brain will always be larger, since there exists a one to one positive correspondence between the addition of new traits / functional enhancement of existing traits and cranial capacity size.

7.4   Deviation and YAABER for Evolution of Homo Sapiens

If we choose an average mammal with encephalization quotient=1 as the mode of our current population of mammals, birds and reptiles, then, we can determine the background evolutionary rate by comparing the encephalization quotient of average mammals to that of the the last common ancestors of mammals and reptiles. (This assumption is based on our older model, in which the population of mammals, birds and reptiles from 66 mya had an EQ of 1, at the current time, the population have a EQ of 1~2 due to acquisition attempts. We are still keeping this assumption due to 1) it is simplest at the very start to express our later lognormal distribution with a mode value of 1, 2) Adjustment is simple, one needs to set $C_{df}(t)$ to $C_{df}(F(t))$, whereas $F(t) = t - x$. So that the mode of at the current time is centered on a given EQ > 1. The σ does need to be adjusted to 0.38 to accommodate the observation constraint of emergence at most 1 per every 3 galaxies. The consequent wall of semi-invisibility is somewhat steeper, but in general conforms well with previous predictions.)

In general, average mammals' encephalization quotient is a magnitude higher than the reptiles. Since the earliest mammals evolved 225 million years ago in the age of reptiles, we can assume that the median of EQ has grown 10 folds since 225 Mya (Tikitherium) in the age of mammals after the diversification. The definition of mammals is defined as synapsids that possess a dentary-squamosal jaw articulation and occlusion between upper and lower molars with a transverse component to the movement or, equivalently in Kemp's view, the clade originating with the last common ancestor of Sinoconodon and living mammals. Then, we are able to define the **BER**, the background evolutionary rate of the earth.

$$q = 10^{\left(\frac{1}{225,000,000}\right)} \tag{7.4.1}$$

$$q - 1 = 1.0233711656 \times 10^{-8} \tag{7.4.2}$$

$$\mathbf{BER} = q^{100,000,000} = 2.783 \tag{7.4.3}$$

---

[5]If the original EQ of the rest of species is x and $x > 1$, then one should expect the EQ of a species with n traits to be nx, but the ratio remains n to x.

This rate is equivalent to 2.783 times in 100 Myr. This implies that, at the current time, every 100 million years the biological diversity leading to greater EQ increase by 2.783 times. The EQ of the mode has shifted from $\frac{1}{2.783}$ from 100 Mya in Cretaceous to 1 at the current time and becomes 2.783 100 Myr into the future. Therefore, the speed of EQ increase over the course of the last 100 Myr is :

$$1 - \frac{1}{2.783} = 0.64067 \tag{7.4.4}$$

The speed of EQ increase over the course of the next 100 Myr is:

$$2.783 - 1 = 1.783 \tag{7.4.5}$$

Taking the average of the two becomes:

$$\frac{0.64067 + 1.783}{2} = 1.2118 \tag{7.4.6}$$

We can be even more precise by evaluating the differential speed at t=0:

$$\lim_{t \to 0} \frac{d}{dx} \left(\mathbf{2.783}\right)^t = \ln\left(\mathbf{2.783}\right)\left(\mathbf{2.783}\right)^t = 1.0235 \tag{7.4.7}$$

That is, the differential speed increases the EQ by 1.0235 per 100 Myr at the current time.

The background evolutionary rate is also cross-checked with the number of neurons in different species. This is somewhat tricky because one has to pick the right data points to compare with. For example, octopus has 500 million neurons and emerged 323.2 Mya, and lemur contains 254,710,000 neurons but emerged 55 Mya by the earliest. If one compares these two data points, the Background Evolutionary Rate seems to increase negatively. In reality, octopus is probably one of smartest species during the Carboniferous epoch and lies as the farthermost point lying to the right from the median value for the cohorts of all animal species at the time. Lemur, on the other hands, lies much closer to the median value in the Cenozoic.

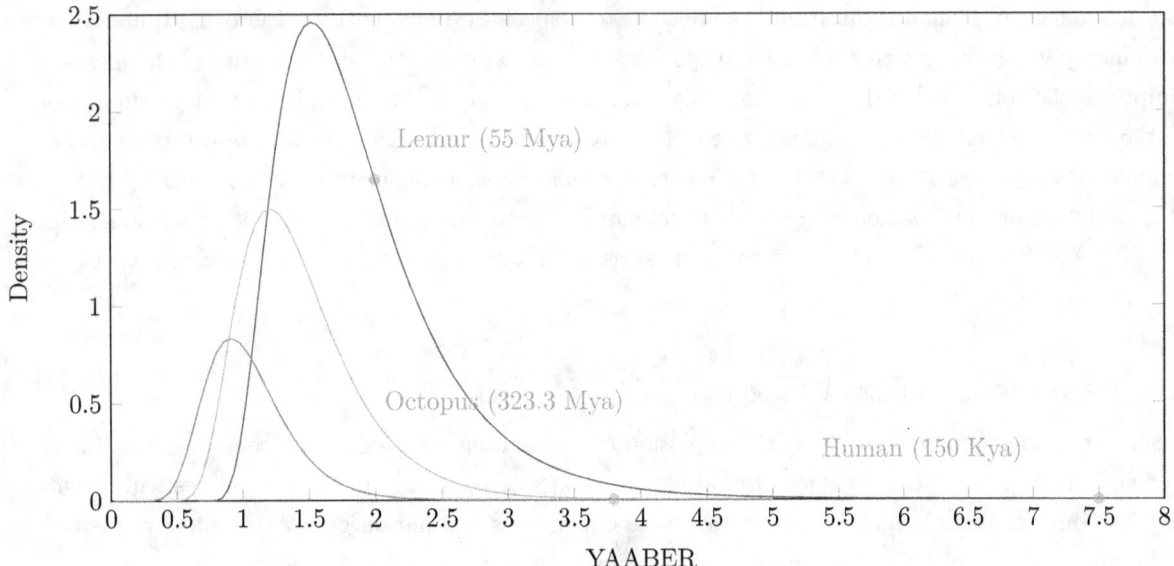

Figure 7.4.1: Hypothetical biocomplexity probability distribution of Cambrian, Carboniferous, and Cenozoic

Moreover, the size of two species across different epochs has to be similar in size. No one would expect two species such as an ant and whale have the same number of neurons even if the EQ of the two species are the same. We eventually selected two data points to compare with. Frog with 16,000,000 neurons which emerged 200 Mya during the Early Jurassic to that of short-tailed shrew and house mouse emerged in the early Cenozoic with 52,000,000 neurons and 71,000,000 neurons respectively. Both frog and mouse has comparable size and yields a BER of 2.3, which is not too different from our earlier calculated result.

$$16,000,000 \, (2.3)^{1.6} = 60,657,501.871 \text{ neurons} \tag{7.4.8}$$

Once we calculated the BER, we can compute the Deviation and YAABER. YAABER for the emergence of human is divided into 2 periods. The first period spans from 55 Mya when the last common ancestor of primates and mammals diverged up to the emergence of Homo sapiens 300 Kya. The first period is dictated by the evolution of larger cranial capacity from an EQ of average of 1 for mammals to 7.6 for humans. The second period spans from 300 Kya up to today with the expansion of human population from a base population to that of 7 billion. The second period can be further divided into hunter gatherer, agricultural, and industrial periods. During the second period, the cranial capacity is fixed at EQ of 7.6 but the change is manifested in increasing energy usage of the entire population.

Measuring **Deviation** or **YAABER** for the emergence of human is obvious. Knowing that currently human holds an EQ of 7.6, then the **Deviation** is simply 7.6. Knowing that background evolutionary rate increases the mode of the EQ of species by 1.0235 times per 100 Myr. Then, the evolution of average species with cranial capacity comparable to human is simply 645 Myr of **YAABER** (645 Myr into the future) in terms of the speed of evolution for the current time:

$$\frac{7.6 - 1}{1.0235} = 6.4484 \tag{7.4.9}$$

Calculation for the second period is trickier. If the deviation is defined based purely on population growth from a population base of 50,000 when Homo sapiens first evolved in East Africa, then at the emergence of civilization at the current time is

$$EQ_1 = \left(\frac{7,500,000,000}{50,000}\right) \cdot 7.6 = 1,140,000 \tag{7.4.10}$$

That is, assuming typical mammal has an EQ of 1, and the rate of evolution increases EQ by 1.0235 times, then the current attainment of human civilization is equivalent to a species evolved with an EQ of 1.14 million. In which the human population growth alone by 150,000 folds is responsible for the vast majority of the increase. This assumption relies on the idea that each doubling from the base population is equivalent to a doubling on the EQ of the evolved species. Since human is one of the most cooperative, organized vertebrate social species that communicate with language, one can treat the entire population as a single entity with increasing EQ. As a result, the evolution of a civilization comparable to human lead by average species alive today is a staggering 1,113 Gyr of YAABER (1,113 Gyr into the future) in terms of the speed of evolution at the current time! We use this value as the upper bound.

$$\frac{1}{1.0235} \cdot \left(\frac{7,500,000,000}{50,000} \cdot 7.6 - 1\right) = 1,113,824 \tag{7.4.11}$$

On the other hand, one can also argue that the explosion of population is an inevitable consequences of agricultural revolution. As long as the presence of crop plants, sooner or later, any intelligent species will reach a population size comparable to human today. In this case, one ignores population growth completely. Instead, one adds the rarity and atypicality of the explosive diversification of angiosperm to the EQ of human. Recall that we actually shown that the emergence of angiosperm actually contributed a factor of 28 increase per 100 Myr. (See 6.8 "Probability of Angiosperms")

$$a_2 = \left(\frac{369,000 + 12,421}{12,421 + 1,191}\right) = 28.021 \tag{7.4.12}$$

The expected number of vascular plant species should be N_{expected} =37,882 without the explosive growth of angiosperm at the current time, and the evolutionary speed increase applied to angiosperm at the current time can be obtained precisely by evaluating the differential speed at t=0:

$$S = \lim_{t \to 0} \frac{d}{dx} 37882 \times (\mathbf{2.783})^t = 37882 \times \ln(\mathbf{2.783})(\mathbf{2.783})^t = 38{,}773.54 \qquad (7.4.13)$$

That is, the differential speed increases the biodiversity of angiosperm by 37,882 species per 100 Myr at the current time.

The evolution reaching the diversity of crop plants observed $N_{\text{observed}} = 369{,}000 + 12{,}421 = 381{,}421$ currently on earth on a typical exoplanet requires 886 Myr of YAABER (886 Myr into the future) in terms of the speed of evolution at the current time.

$$\frac{N_{\text{observed}} - N_{\text{expected}}}{S} = \frac{381421 - 37882.196}{37882} = 8.86 \qquad (7.4.14)$$

Combining the emergence of human with the emergence of angiosperm, one obtains:

$$6.4484 + 8.86 = 15.31 \qquad (7.4.15)$$

1.531 Gyr of YAABER, that is, the evolution of a civilization comparable to human lead by average species alive today utilizing a diverse range of crop plants is 1.531 Gyr of YAABER (1.531 Gyr into the future) in terms of the speed of evolution at the current time. This value is the lower bound, and the corresponding value translated into a species with equivalent EQ, a Deviation of :

$$EQ_2 = 15.31 \times 1.0235 = 15.6683 \qquad (7.4.16)$$

This result can also be derived directly by simply taking the difference between EQ of human and EQ of typical mammal with the ratio of the difference between the biodiversity of angiosperm observed and the biodiversity of vascular plants expected over the biodiversity of vascular plants expected.

$$EQ_2 = (7.6 - 1) + \frac{N_{\text{observed}} - N_{\text{expected}}}{N_{\text{expected}}} = 15.6683 \qquad (7.4.17)$$

The more complex steps are given to show how YAABER and deviation can be expressed interchangeably to express the same deviation.

In reality, both scenarios hold some degrees of truth. Nevertheless, the author is inclined toward the second approach since a lower Deviation and YAABER renders analysis simpler. We will later in Section 8.8 show the soundness of our intuition by constraining the model using observations and finding the lower and upper bound values for the deviation. We will show the intricate relationship between the distribution's deviation, BCS, BER, and extinction rates. We eventually settles on the square root of the geometric mean between the first approach and the second approach as the final deviation and YAABER for the emergence of civilization:

$$\textbf{Deviation} = \sqrt{\sqrt{EQ_1 \times EQ_2}} = 17 \qquad (7.4.18)$$

$$\textbf{YAABER} = \frac{\textbf{Deviation}}{\text{Speed}} = \frac{\textbf{Deviation}}{1.0235} = 16.609 \qquad (7.4.19)$$

[6]

*That is, it takes the evolution of a civilization comparable to human lead by average species alive today utilizing a promised diverse range of crop plants is **1.6609 Gyr** of **YAABER** and a **Deviation** of **17** (1.6609 Gyr into the future with the mode shifted to a species with an EQ equivalent of $1.0235 \times 16.609 = 17$) in terms of the speed of evolution at the current time. 644.84 Myr describes the evolution of homo sapiens and the remaining 1.01606 Gyr measures the progress of hunter gatherer, agricultural, and industrial revolution of Homo sapiens.* We now have a general picture of the deviation and YAABER but how does each period breaks down? We now run detailed analysis for each period.

---

[6]EQ$_1$ is modified by taking into account an expected future human population up to 22.72 billion and an initial hunter gather population of 4,263,300

First, we compute the contribution by the primate lineage leading up to *Australopithecus afarensis*, the first bipedal walking ape. Primates, on average, have an EQ of 2.12 with 55 Myr of evolution compares to typical mammalian average of 1 today. Then, we can compute the Evolutionary rate for the primate lineage and use it to calculate the YAABER.

$$k_{primate} = 2.122^{\left(\frac{1}{55,000,000}\right)} \tag{7.4.20}$$

$$k_{primate} - 1 = 1.368575 \times 10^{-8} \tag{7.4.21}$$

$$T_{prim} = \frac{(k_{primate})^{55,000,000-3,900,000}}{1.0235} \tag{7.4.22}$$

$$T_{prim} - \frac{1}{1.0235} = 0.9885 \times 10^8 \text{ YAABER} \tag{7.4.23}$$

We can see that the YAABER for Primates is 98.85 Myr ahead of the mammalian average, that is, it will take this long for the mammal's average EQ reaches parity with the current primate level using current rate of evolutionary growth. T_{prim} is subtracted by $\frac{1}{1.0235}$ because EQ from 0 to 1 is covered by earlier evolution. An EQ from 0 to 1 took 540 Myr of evolution takes only $\frac{1}{1.0235}$=97.7 Myr based on the current speed.

The evolutionary rate from the emergence of *Australopithecus afarensis* to Homo Sapiens is computed based on the growth of cranial capacity of the species, which is equivalent to EQ, but just with more precision and the YAABER is calculated to be an additional 545.72 Myr ahead of the evolutionary rate. We assumed current Homo sapiens cranial capacity at 1,450 cc and *Australopithecus afarensis* between 380 cc and 430 cc.

$$k_{homo} = \left(\frac{1,450}{\frac{(430+380)}{2}}\right)^{\left(\frac{1}{3,600,000}\right)} \tag{7.4.24}$$

$$k_{homo} - 1 = 0.0000003542866 \tag{7.4.25}$$

One can see that the emergence of Homo sapiens is 25.887 times faster than the evolutionary change of primates based on the cranial size change.

$$\frac{k_{homo} - 1}{k_{primate} - 1} = 25.88725 \tag{7.4.26}$$

The initial mode of the cranial capacity is now assumed to be 2.122, the typical primate average and we subtract the YAABER required for the evolution of up to primate we computed earlier:

$$T_{homo} = \frac{2.122 \, (k_{homo})^{3,900,000-300,000}}{1.0235} - T_{prim} = 5.4572 \times 10^8 \text{ YAABER} \tag{7.4.27}$$

and we are able to confirm that $0.9885 \times 10^8 + 5.4572 \times 10^8 = 6.44 \times 10^8$ YAABER as we have derived earlier.

7.5   YAABER for Hunter Gatherer

Starting from the emergence of homo sapiens, all progress gained shares within the 1.01606 Gyr of YAABER. First and the foremost, the evolutionary rate for the emergence of homo sapiens during the period of 300 kya up to now still played a very minor role in contributing to the YAABER, we take this into account assuming there are possible parallel evolution of hominid lineage such as the Neanderthal. Since it is evolution in parallel to homo sapiens, one should subtract its contribution from 1.01606 Gyr. The derivation for each successive period is computed later and here we simply takes those results for granted.

320

$$A_{homo} = \frac{1}{3.58}(f_{homo} \cdot f_{hunterxhomo} \cdot f_{citystatesxhomo} \cdot f_{middlexhomo} \cdot f_{explorationxhomo}$$
$$\cdot f_{industrialxhomo} \cdot f_{earlyindustrialxhomo} - f_{homo}) \quad (7.5.1)$$

or alternatively can be expressed as:

$$A_{homo} = (k_{homo})^{(3\cdot10^5+5000+2100)} - 1 = 11.49 \text{ Myr} \quad (7.5.2)$$

So that the final total YAABER contributing to all homo sapiens progress is:

$$1.01606 \text{ Gyr} - A_{homo} = 1.00457 \text{ Gyr} \quad (7.5.3)$$

The proportion of hunter gather's contribution to the remaining YAABER can be found if the overall contribution T_{total} to the mode (now assumed to be 7.6) is known. The final re-scaling factor for the current attainment of human civilization's equivalent to a species evolved with an extremely high EQ is derived as:

$$T_{total} = A_{composite} + A_{hunter} + A_{exploration} \quad (7.5.4)$$

So that the final mode becomes 7.6·T_{total}=3,411,708. This value is triple the value we have before because we eventually settled on world population at the end of 21st century instead of now.

Whereas $A_{composite}$ is the re-scaling factor for each period in which the most dominant activities performed by the mainstream human population groups, which constitutes the vast majority of the re-scaling.

$$A_{composite} = f_{citystates} \cdot f_{middle} \cdot f_{exloration} \cdot f_{earlyindustrial} \cdot f_{industrial} \quad (7.5.5)$$

Whereas A_{hunter} is the re-scaling factor according to the rate of increase under the hunter gatherer period extrapolated into the years between 5000 BC and AD 2100, a period dominated by agricultural and industrial revolution. This is a negligible amount. We take this part into account assuming there are human hunter gatherer societies even today in the Amazons.

$$A_{hunter} = f_{hunter}f_{citystatesxhunter} \cdot f_{middlexhunter} \cdot f_{explorationxhunter} \cdot f_{earlyindustrialxhunter}$$
$$\cdot f_{industrialxhunter} - f_{hunter} \quad (7.5.6)$$

Whereas $A_{exploration}$ is the re-scaling factor according to the rate of increase under the late phase of agricultural society extrapolated into years between 1750 to 2100, a period dominated by industrial revolution. This is significantly higher than the previous two, but still significantly smaller than our primary re-scaling. We take this part into account assuming there are plenty of non-industrialized regions during these period.

$$A_{exploration} = f_{hunter} \cdot f_{citystates} \cdot f_{middle}$$
$$\cdot (f_{exploration} \cdot f_{earlyindustrialxexloration} \cdot f_{industrialxexloration} - f_{exploration}) \quad (7.5.7)$$

After Homo sapiens emerged as a new species in Africa by 195,000 years ago, the species spent the rest of 185,000 years as hunter-gatherers. In fact, it is the mode of life just like any of its predecessor species. Because the species eventually transitioned into an agricultural one and the changes occurred during this period was much faster than the epigenetic and functional modularity changes occurred in the hominid lineage. The toolset of the species transitioned from Paleolithic to Mesolithic and eventually into Neolithic stone tools. Around 45,000 BC a profound cultural change occurred where behaviors comparable to modern man has been observed from the archaeological remains. Human started to practice ritualistic burials, painted cave art, sculpted effigies, and started extensive trade networks. The domestication of animals started around 30,000 BC starting with the wolf and eventually led into a full-blown agricultural revolution. Human anatomical structure, however, changed little if at all. As a result, we need to resort to some other parameters to measure the rate of progress compares to the background evolutionary rate, and add this value to the final YAABER. Evolution can be rethought as a form of passive manipulation of matter and energy. Starting from the hunter-gathering phase of human existence, the species actively manipulates matter and energy by using tools. In essence, we need to measure the rate of change in energy manipulation, passive or active. Then, in order to quantify the change occurred during this period of human as hunter-gatherers, the best we can use is to measure the rate of energy acquisition growth during this phase of human evolution and the most direct way to measure it is through measurement on the rate the human population growth. At the emergence of our species, 50,000 individuals lived in East Africa. Mt.Toba eruption had further reduced our numbers to 5,000~20,000 around 70,000 BC. Then, from 70,000 BC to 5,000 BC, human population increased to 5,000,000 around the world just before the start of the Neolithic revolution. Human population is one of the more reliable means to quantify the pace of change during this period. First of all, human, anatomically similar, consumed a similar amount of calories per day extracted from its immediate surroundings, the energy consumption discrepancy between genders or race (if it existed) can be deemed negligible. Secondly, human as hunter-gatherers still subject to the laws of natural selection. If human had exhausted a local supply of food resources and without able to locate newer ones, they would die. The population of a given animal species is fixed in a sub geological time scale as hunter-gatherers. If human has been completely subject under the law of natural selection, we should have observed human population stagnates or fluctuates around a mean value. However, human population during this period was continually increasing. This trend of growth suggests that the total amount of energy extractable by man in its immediate surrounding is increasing and the total energy consumption is also increasing. By utilizing better tools with greater precision and more clear communicating language, humans are more capable of hunting big games previously deemed too dangerous to be accessible, collecting nutrients and energy from nuts too sturdy to crack and consume. Despite a possible hunter-gatherers' version of Malthusian catastrophe awaiting for them, they constantly worked around ecological constraints by exploiting new food niches. The tools sets become a powerful extension of human biological capability, which even at the very best, takes hundreds of thousands of years to evolve. Yet the evolution of tool sets can be accomplished in tens of thousands of years using successive generation of advanced tool sets.

Therefore, measuring the total population growth rate of humans since its emergence is the first order approximation of the actual growth rate occurred during this period if one considers every aspect of human advancement in arts, language, culture, trade, and toolset innovations contributes toward the overall growth of this period. It is an excellent approximation because every human progress eventually can be measured in the advancement of the welfare and well-being of the species, and its total population is a direct manifestation of such transformation.

$$k_{hunter} = \left(\frac{5,000,000}{50,000}\right)^{\left(\frac{1}{300,000}\right)} \tag{7.5.8}$$

$$k_{hunter} - 1 = 0.00001535 \tag{7.5.9}$$

The colonization of the world lead by hunter gatherers is 43.328 times faster than the rate of emergence of Homo sapiens.

$$\frac{k_{hunter} - 1}{k_{homo} - 1} = 43.328 \tag{7.5.10}$$

We can now use the rate of population growth per year to compute f_{hunter}:

$$f_{hunter} = (k_{hunter})^{(307,000-7,000)} = 100 \qquad (7.5.11)$$

$$T_{hunter} = (f_{hunter} - 1) \qquad (7.5.12)$$

We take the ratio of the rescaling of T_{hunter} compares to the overall rescaling by T_{total}:

$$\frac{T_{hunter}}{T_{total}} \cdot 1 \text{ Gyr} = 209,710 \text{ YAABER} \qquad (7.5.13)$$

Of course, we should not neglect the rate of human evolution itself, the passive manipulation of matter and energy encoded into our DNA. This still applies despite the shortness of hunter-gatherer period. Whereas $f_{hunterxhomo}$ is the amount mode shifted by the rate of evolution of human within the hunter gatherer period, and it is rescaled by $\frac{1}{3.58}$ because we now assumed that the initial starting mode position is 7.6, the EQ of human, and the previous starting mode position of average primate becomes $\frac{2.122}{7.6} = \frac{1}{3.58}$ of the current.

$$f_{homo} = (k_{homo})^{(3,900,000-300,000)} \qquad (7.5.14)$$

$$f_{hunterxhomo} = (k_{homo})^{300,000} \qquad (7.5.15)$$

$$A_{homo2} = \frac{1}{3.58}(f_{homo} \cdot f_{hunterxhomo} - f_{homo}) = \frac{1}{3.58} f_{homo}(f_{hunterxhomo} - 1) \qquad (7.5.16)$$

Finally, the amount shifted is the total distance shifted from the 3.9 Mya up to 5 kya minus 3.9 Mya up to 300 kya we have already added to the YAABER.

$$10^8 \times A_{homo2} = 11,213,988 \text{ YAABER} \qquad (7.5.17)$$

It adds an additional of 11.2 Myr compares to the human-directed cultural and technological evolution change of 209,710 yr. Since we have re-scaled and de-emphasized the contribution of population growth to the deviation primarily based on the rarity of angiosperm's diversity explosion, a population increase by a factor of 100 from a base of 50,000 is still far from fulfilling the ultimate potential of population ceiling. Our assumption suggests that, despite rapid population increase, a species with fixed traits and features (evolutionarily stagnant) requires a factor of 529.4 from a base of 50,000 in order to outrun the rate of human evolution itself within a period of 300 Kyr. That is, the re-scaling sets the rate of human evolution equivalent to the speed of increasing net population of homo sapiens by $26,520,000 - 50,000 = 26,470,000$ per 300 kyr.

7.6 YAABER for Feudal Society

No one should underestimate the importance of plant diversity in providing opportunity for species diversification. This importance has been observed in geologic past many times. The formation of Pangea supercontinent ushered in a period of cold, dry climate. This change in climate triggered the evolution of seed plants. By enclosing seed within hard shell, plants are able to populate further inland and to places less hospitable to earlier plant species. Following the speciation of seed plant, reptiles emerged within 20 Myr, also adapted to drier environment and exploited this new ecological niche opened by seed plants. Breaking of the Gondwana supercontinent during the Cretaceous gave the emergence of angiosperm. The emergence of angiosperm created the arboreal niche. The arboreal niche provides the living space for primates and birds. Finally, the advent of grass plants made human civilization possible. The transition from hunter-gatherer to agricultural civilization happened almost simultaneously around the world at the start of the last inter-glacial period. Domestication of wild rice, wheat, rye, and barley enabled energy acquisition to be magnitudes higher than a hunter-gatherer. The shortness of time span required to transition from hunter-gatherers to agricultural society seems to suggest as long as mild climate persists (interglacial) then agriculture revolution seems to be inevitable. However, I would suggest that many intelligent species on many Earth-like planets may never evolve into an agricultural society, consequently, maintained their mode of living throughout its entire existence as hunter-gatherers and never able to transform into an industrial one. Grass plants are unique in their annual growth cycle and little investment in their self-maintenance and a significant portion of investment in their seeds. As a result, high levels of energy density is stored in seed kernels. Its biological adaptation evolved in the recent millions of years, a very recent biological innovation that was possibly co-evolved with herbivores which are the ancestors of goats, cows, sheeps, and horses. Ancestors of humans, however, lived on the trees throughout this time when grass evolved and co-evolved with herbivores on land. *Therefore, the advent of grass plants is an independent evolutionary event from the emergence of intelligent Homo sapiens.* Our ingenuity and dexterity eventually exploited this biological innovation. Therefore, in a sense, we took a free ride from the hard working symbiotic relationship and evolutionary feedback loop between herbivores on land and grass plants persisted millions of years before we walked on the ground. During the latter part of the agricultural revolution, grass also provides the ingredients of papermaking, which made information dissemination much more efficient and cheap and prepared for the transition into an industrial civilization. The following table is a list of seeds and their energy content based in kilo-calories.

Species Name	Energy	Species Name	Energy	Species Name	Energy
Sunflower Seed	2445	Fig	310	Beet	180
Maize	1528	Raspberry	220	Carrot	173
Rice	1528	Apple	218	Onion	166
Broad Beans	1425	Pineapple	209	Cabbage	103
Sorghum	1419	Blackberry	180	Spinach	97
Wheat	1369	Grapefruit	138	Turnip	84
Cassava	670	Lemon	121	Bell pepper	84
Soybean	615	Pumpkin	109	Tomato	74
Yam	494	Eggplant	104	Radish	66
Sweet Potato	360	Cucumber	65	Lettuce	55
Pea	339	–	–	–	–
Potato	322	–	–	–	–
Average	**1042.83**	**Average**	**167.4**	**Average**	**108.2**

Table 7.6.1: Edible plants and their energy content

One can see that rice plants top the list along with sunflower seeds and pine seeds and followed by fruits and then by vegetables such as spinach and lettuce. Although pine seeds have comparable energy content to that of the rice plants, pine trees are perennials and takes a long growth cycle to reach maturity. Then a much smaller amount of energy it captured is stored into its final seed product, a significant portion is invested into its bark, trunk, and leaves maintenance. Therefore, pine trees, from hunter-gatherers' perspective can be too costly to be domesticated. In fact, the cost of domestication is so prohibitively expensive that it will fail to start in the first place. A transition from hunter-gatherer to an agricultural civilization on a planet dominated with pine like trees or lacking a biological equivalent of grass plants on earth will be almost impossible. One may argue, that given enough time, in the scale of tens of millions of years (which will still be a magnitude or two faster than natural selection), pine trees may possibly be successfully domesticated. However, 99% of species went extinct within 10 million years. Therefore, the timescale involved to transform a pine-like tree into a grass plant is impractical. On the other hand, a planet may be dominated with fruits growing on tomato, zucchini like annual plants, a transition to an agricultural civilization is then possible but its population density will be significantly lower than that is attainable on earth. Furthermore, a planet dominated by the sea with scattered island masses will have little carrying capacity even if rice like plants are plenty and their intelligent species transitioned successfully from hunter-gatherers to agriculturalists. Finally, an agricultural society will be much smaller in size on a planet dominated by land with a few seas and a significant supply of grass plants yet a majority of the land are occupied by desert, ice sheets, or high lands, or all of them combined. Of course, we might find the completely the reverse to be true, where abundant land filled with grass plants and technologically capable species. The point of the thought experiment is to appreciate the number of possibilities given the billions of habitable planets confirmed to exist. Given the enormous amount of available data points, each planet, representing a data point of slightly different values from each other if they are ever possibly be sorted from high to low, will form the lognormal curve in our model for extraterrestrial civilizations' advancement index distribution.

A planet dominated by grasslands or highlands with scattered spots capable of agriculture will be particularly interesting. Since highland and grasslands are less suitable for raising the type of plants similar to wheat and rice, pastoral nomads may become the dominant mode of living among its most intelligent creatures. A pastoral society shares some characteristics of the agricultural society in that it is able to raise a higher number of people through animal breeding yet they also roam in order to secure water and strategical resources like hunter-gatherers. The primary disadvantage of such society is its difficulty in fostering scientific and investigative science since it requires a high level of specialization to produce the tools (which requires stationary factories to build and hardly can be moved from place to place) over successive generations with dictated improvements and refinements. A pastoral society primarily concerned with its own subsistence and emphasize self-independence can produce little work specialization beyond a family and tribal complexity. Most interestingly, as observed in the history of earth repeatedly, that the competition between pastoral and agricultural society was intense. The earlier examples are the Huns which looted and attacked the agricultural based Han dynasty, and the Germans and Gothic people attacked the Romans. A later example is the Mongols which attacked the Song dynasty in East Asia and Europe and Arabic Empire in the Western Eurasia continent. Though all agricultural society survived the onslaught and eventually transformed into industrial civilizations, the destruction is nevertheless substantial. Given the sheer size of agricultural land on earth and limited pastoral land and the destructive power possessed by those smaller pockets of pastorals, it is possible that on a planet dominated by grasslands, agricultural society follows cycles of prosperity and bust triggered by the onslaught of the pastorals and can hardly survive at all. Its science follows periods of progress and then regress as information and knowledge are lost due to war and destruction. It is also possible that some pastorals eventually become agriculturalists themselves as they occupied the land once owned by previous agriculturalists. However, as they become the agriculturalists themselves, their fate followed the similar trajectory of their predecessors. Therefore, its civilization can hardly transform into an industrial one and is trapped at the stage of development of pastoral society and agricultural society.

To measure the rate of progress compares to the background evolutionary rate, and add this value to the final YAABER, we resort to both population growth from 10,000 BC with yields per acre. Population data is well-extrapolated and documented for the past 10,000 years. We sample our data from three periods. The first one ranges from 5,000 BC at the start of the full transition from hunter-gatherer to agricultural to 1,000 BC when city-states began merging into empires. We call the first period the period of city-states. Then, from 1,000 BC to AD 1600, it includes the classical, Roman, the middle ages, and the early modern period. Finally, the age of exploration spans from 1600 to 1750. We compute the population growth rate of each period (obtain the annual rate of growth) and then use them to compute the final YAABER.

7.6.1 City states period

$$k_{citystates} = \left(\frac{50,000,000}{5,000,000} \right)^{\left(\frac{1}{4,000} \right)} \tag{7.6.1}$$

$$k_{citystates} - 1 = 0.000576 \tag{7.6.2}$$

$$f_{citystates} = (k_{citystates})^{(7000-3000)} = 10$$
$$f_{citystatesxhunter} = (k_{hunter})^{(7000-3000)}$$

$$A_{hunter3} = f_{hunter} \cdot (f_{citystatesxhunter} - 1) \tag{7.6.3}$$

$$T_{citystates} = (f_{hunter} \cdot f_{citystates} - f_{hunter}) + A_{hunter3} \tag{7.6.4}$$
$$\approx f_{hunter} \cdot f_{citystates} - f_{hunter} \tag{7.6.5}$$

The total amount mode shifted is approximately just the contribution by the progress made under civilization during early antiquity, and the progress contributed by hunter gatherer societies if they still exists is negligible.

$$\frac{T_{citystates}}{T_{total}} \cdot 1 \text{ Gyr} = 1,919,872 \text{ YAABER} \tag{7.6.6}$$

Most importantly, the progress contributed by the evolution in hominid lineages is now negligible.

$$f_{citystatesxhomo} = (k_{homo})^{(7000-3000)} \tag{7.6.7}$$

$$A_{homo3} = \frac{1}{3.58} f_{homo} \cdot f_{hunterxhomo} (f_{citystatesxhomo} - 1) \tag{7.6.8}$$
$$= 157,718 \text{ YAABER} \tag{7.6.9}$$

From this point onward, consciously directed energy change dominates over passive manipulation by natural selection, which requires a geological timescale to add significant value to the YAABER. Because human cultural and technological changes occurred in $x < 10^4$ years, biological contribution to the YAABER still exists but becomes very insignificant.

7.6.2   Middle Ages

$$k_{middle} = \left(\frac{580,000,000}{50,000,000}\right)^{\left(\frac{1}{2,600}\right)} \tag{7.6.10}$$

$$k_{middle} - 1 = 0.000943 \tag{7.6.11}$$

$$f_{middle} = (k_{middle})^{(1000+1600)} = 11.6$$
$$f_{middlexhunter} = (k_{hunter})^{(1000+1600)}$$

$$A_{hunter4} = f_{hunter} \cdot f_{citystatesxhunter} \cdot (f_{middlexhunter} - 1) \tag{7.6.12}$$

$$T_{middle} = (f_{hunter} \cdot f_{citystates} \cdot f_{middle} - f_{hunter} \cdot f_{citystates}) + A_{hunter4} \tag{7.6.13}$$
$$\approx f_{hunter} \cdot f_{citystates} \cdot f_{middle} - f_{hunter} \cdot f_{citystates} \tag{7.6.14}$$

$$\frac{T_{middle}}{T_{total}} \cdot 1 \text{ Gyr} = 22,463,017 \text{ YAABER} \tag{7.6.15}$$

The progress contributed by the evolution in hominid lineages during this period:

$$f_{middlexhomo} = (k_{homo})^{(1000+1600)} \tag{7.6.16}$$

$$A_{homo4} = \frac{1}{3.58} f_{homo} \cdot f_{hunterxhomo} \cdot f_{citystatesxhomo} (f_{middlexhomo} - 1) \tag{7.6.17}$$

$$= 102,636 \text{ YAABER} \tag{7.6.18}$$

7.6.3   Age of Exploration

$$k_{exploration} = \left(\frac{791,000,000}{580,000,000}\right)^{\left(\frac{1}{150}\right)} \tag{7.6.19}$$

$$k_{exploration} - 1 = 0.00207 \tag{7.6.20}$$

$$f_{exploration} = (k_{exploration})^{(1750-1600)} = 1.363 \tag{7.6.21}$$

$$f_{explorationxhunter} = (k_{hunter})^{(1750-1600)} \tag{7.6.22}$$

$$A_{hunter5} = f_{hunter} \cdot f_{citystatesxhunter} \cdot f_{middlexhunter} (f_{explorationxhunter} - 1) \tag{7.6.23}$$

$$T_{middle} = f_{hunter} \cdot f_{citystates} \cdot f_{middle} (f_{exploration} - 1) + A_{homo5} + A_{hunter5} \tag{7.6.24}$$

$$\approx f_{hunter} \cdot f_{citystates} \cdot f_{middle} (f_{exploration} - 1) \tag{7.6.25}$$

$$\frac{T_{middle}}{T_{total}} \cdot 1 \text{ Gyr} = 8,939,712 \text{ YAABER} \tag{7.6.26}$$

The progress contributed by the evolution in hominid lineages during this period:

$$f_{explorationxhomo} = (k_{homo})^{(1750-1600)} \tag{7.6.27}$$

$$A_{homo5} = \frac{1}{3.58} f_{homo} \cdot f_{hunterxhomo} \cdot f_{citystatesxhomo} \cdot f_{middlexhomo} (f_{explorationxhomo} - 1) \tag{7.6.28}$$

$$= 5,924 \text{ YAABER}$$

$$t = \frac{T_{citystates} + T_{middle} + T_{exploration}}{T_{total}} \cdot 1 \text{ Gyr} = 33.322 \text{ Myr YAABER} \tag{7.6.29}$$

We found that the rate of increase is another 33.322 My years faster than the evolutionary background rate of growth. That brings us to another good question. Will agricultural productivity continue to increase without steam power or any machine power in general? To answer this question, we need to know if any growth constraints persist in biological plants themselves. Since all biological products we consume ultimately derives its energy content from that of the sun, then biological conversion efficiency can be measured, and the highest attainable conversion sets the upper bounded constraints on agricultural, domestic selection and breeding. Photosynthesis can be described by the simplified chemical reaction:

$$6H_2O + 6CO_2 + energy \Rightarrow C_6H_{12}O_6 + 6O_2 \qquad (7.6.30)$$

where $C_6H_{12}O_6$ is glucose (which is subsequently transformed into other sugars, cellulose, lignin). The value of the photosynthetic efficiency is dependent on how light energy is defined – it depends on whether we count only the light that is absorbed, and on what kind of light is used. It takes eight photons to utilize one molecule of CO_2. The Gibbs free energy for converting a mole of CO_2 to glucose is 114 kcal, whereas eight moles of photons of wavelength 600 nm contains 381 kcal, giving a nominal efficiency of 30%.[78] However, photosynthesis can occur with light up to wavelength 720 nm so long as there is also light at wavelengths below 680 nm to keep Photosystem II operating. Using longer wavelengths means less light energy is needed for the same number of photons and therefore for the same amount of photosynthesis. For actual sunlight, where only 45% of the light is in the photosynthetically active wavelength range, the theoretical maximum efficiency of solar energy conversion is approximately 11%. In actuality, however, plants do not absorb all incoming sunlight (due to reflection, respiration requirements of photosynthesis and the need for optimal solar radiation levels) and do not convert all harvested energy into biomass, which results in an overall photosynthetic efficiency of 3 to 6% of the total solar radiation.[4] If photosynthesis is inefficient, excess light energy must be dissipated to avoid damaging the photosynthetic apparatus. Energy can be dissipated as heat (non-photochemical quenching), or emitted as chlorophyll fluorescence.

Quoted values sunlight-to-biomass efficiency:

Plant	Efficiency
Plants, typical	0.1%[6], 0.2~2%[8]
Typical crop plant	1~2%[6]
Sugarcane	7~8% peak[6]

Table 7.6.2: Photosynthetic efficiency

The following is a breakdown of the energetics of the photosynthesis process from Photosynthesis by Hall and Rao:[73]

Starting with the solar spectrum falling on a leaf, 47% lost due to photons outside the 400–700 nm active range (chlorophyll utilizes photons between 400 and 700 nm, extracting the energy of one 700 nm photon from each one) 30% of the in-band photons are lost due to incomplete absorption or photons hitting components other than chloroplasts. 24% of the absorbed photon energy is lost due to degrading short wavelength photons to the 700 nm energy level. 68% of the utilized energy is lost in conversion into d-glucose. 35–45% of the glucose is consumed by the leaf in the processes of dark and photorespiration.

Stated another way: 100% sunlight split into 47% non-bioavailable photons as waste, leaving 53% (in the 400–700 nm range). Then, 30% of remaining photons are lost due to incomplete absorption, leaving 37% as the absorbed photon energy. Out of which, 24% is lost due to wavelength-mismatch degradation to 700 nm energy, leaving 28.2% of sunlight energy collected by chlorophyll. Out of this collected sunlight, 32% efficient conversion of ATP and NADPH to d-glucose, leaving 9% of sunlight collected as sugar. 35 to 40% out of the sugar is recycled/consumed by the leaf in dark and photo-respiration, leaving 5.4% net leaf efficiency. Finally, many plants lose much of the remaining energy on growing roots. Most crop plants store 0.25% to 0.5% of the sunlight in the product (corn kernels, potato starch, etc.). Sugar cane is exceptional in several ways, yielding peak storage efficiency of 8%.

According to the cyanobacteria studies, the total photosynthetic productivity of earth is between 1,500 and 2,250 TW, or from 47,300 to 71,000 exajoules per year. Using this source's figure of 178,000 TW of solar energy hitting the Earth's surface,[30] the total photosynthetic efficiency of the planet is 0.84% to 1.26%.

Based on these studies, a typical plant yields a photosynthetic efficiency of 1% and typical crop plant yields at 1.5% and maximum upper bound at 6~8%. Therefore, we can apply the yield per acre growth rate observed from the past two millennia to see at which year into the future had industrial civilization not occurred the agricultural revolution would reach its ultimate potential. Yields per acre data have been accurately preserved for the past three millenniums, especially in China. During each dynasty, one of the most important tasks of the imperial court of China is to take the census and measure the average yield per acre on different types of crops. Using this data,[122] we can back-extrapolate the yield per acre at 10,000 BC and its rate of energy acquisition to that of the evolutionary background. From the table below, we have collected and sampled the yield per acre record from the Eastern Zhou dynasty (771 BC - 256 BC) to Qing dynasty (AD 1644 - AD 1911).

Dynasty	Year	Wheat Yield	Rice Yield
East Zhou	476 BC	0.2	–
Han	8	0.6	0.4
Wei & Jin	300	0.6	0.6
East Jin	400	–	0.9
North & South Dy	500	1.2	–
Tang	800	0.6	0.9
Song	1127	0.6	1.2
Yuan	1300	1.2	2.4
Ming & Qing	1644	1.2	2.4

Table 7.6.3: Historical yields per acre

Regression on wheat yield:

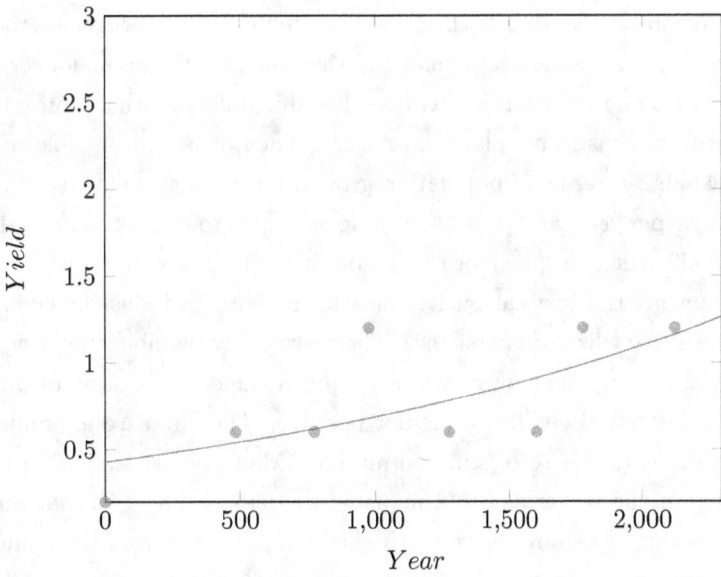

Figure 7.6.1: Regression on wheat yield over 2000 years

$$y = 0.429(1.00048)^x \qquad (7.6.31)$$

with an annual growth rate of 0.048%.

Regression on rice yield:

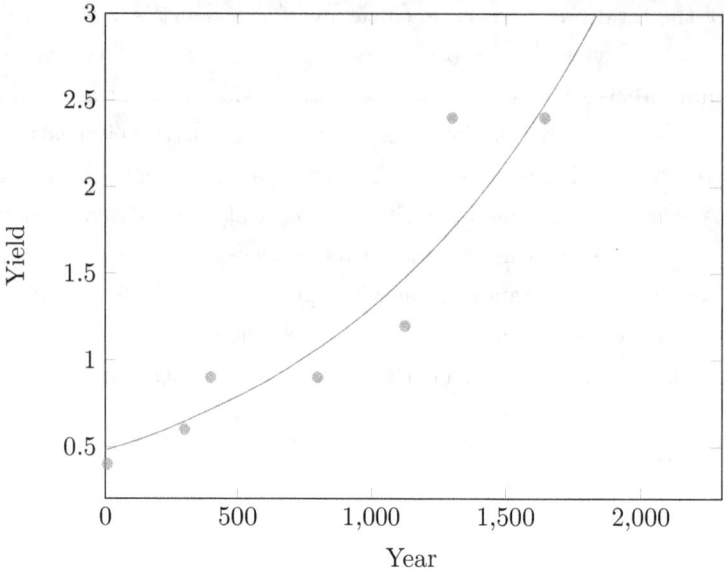

Figure 7.6.2: Regression on rice yield over 2000 years

$$y = 0.480857(1.00102)^x \qquad (7.6.32)$$

with an annual growth rate of 0.102%.

Simply using these rates, one finds the final value to be 1,509 years (rice) and 2,732 years (wheat) into the future to reach 7% efficiency. That is, if industrial revolution had not occurred and human and livestock based muscle power dominated agricultural revolution continues; then, we would expect, on average, another 2,120 years of a steady rise of world total energy output harvested by man and then the growth can no longer be maintained. In the long run, if industrial civilization never occurred, artificial breeding may eventually raise the photosynthetic efficiency to 30% based on chemical constraints placed in nature. Then, it is still possible to extend a sense of continued progress in human affairs in terms of population growth for another 4,000 years given the standard agricultural economic growth rate per year, and possibly as long as 40,000 to 400,000 years if domestic selections to yield higher photosynthetic efficient crop plant proved to be difficult. Thereafter, the human society would stagnate at this level of development. But how can such society never develops industrial civilization if hundreds of thousands of years of agricultural civilization persisted? There are many possible scenarios barred them from further development. First of all, the planet may lack a significant amount of uranium in its crust so that transitioning into a sustainable industrial civilization is not possible. The Faint Young Sun paradox proposes that earth was warm even in the early phase of sun's formation. Many pointed out that greater amount of radioactive material was present at the earlier days of the planet and released heat to compensate the dimming sun in its early days, and we continue to enjoy active earth with internal heat from the abundance of uranium in our earth' crust. Uranium presence may be a necessary but insufficient condition for life formation. This assumption is undermined, however, it is now generally agreed that tidal heating contributes toward a third of the internal heat budget of the earth. That is, uranium contributes some but not all of the internal heat. On a super earth, with even greater tidal heating in its core, even greater amount of heat can be released and keep the planet warm throughout all periods so that the planet remains warm despite its deficiency in uranium. Furthermore, some terrestrial planets, even born at the same time as the sun, likely to be more metal-poor in uranium. In both of these scenarios, intelligent life may use fossil fuels to develop an industrial civilization but found themselves unable to maintain it because the cost of EROEI for uranium extraction is prohibitively expensive or simply non-existent. As we have shown earlier it is also possible that the planet is by dominated by highlands or oceans. It is also possible, though much fewer cases throughout the cosmos, fossil fuels were not well-preserved on a planet even though uranium is abundant in its crusts. Without the cheap, accessible fossil fuel to kickstart an industrial revolution, intelligent life on such a planet stuck at the level of development of agricultural society. One should not underestimate the level of complexity and energy budget available to such a stagnant civilization. A fully matured agricultural society which utilizes all possible land mass to capture solar energy at the maximum photosynthetic efficiency can yield total energy output comparable to our industrial civilization. Let's do some thought experiment, given the total available arable land on earth today, and assume the theoretical maximum photosynthetic efficiency is reached, and we arrived at 31 TJ of energy for such a civilization! However, we will show that even though they are able to achieve such level of energy budget, their society still deemed as non-progressing because all these energy is diverted into the direct consumption by livestock, biomass, and the extremely well-fed species itself. (can still be under-nourished if high population is required to maintain such large energy output) However, no high-density energy driven technology such as airplanes, computers are possible, and no computer and Moore's law to observe. This brings us to one of the most critical assumptions in our overreaching hypothesis. *That is, almost all intelligent species evolved on all existing habitable planets are stagnant in its development, that is, it does not shift its position relative to the mean cosmic evolutionary rate of the static, non-moving lognormal distribution model bounded by geologic time frame, ($x < 10^7$) years.* This is very counter-intuitive to our common sense. Since the rise of Homo sapiens, we learned nothing but change and progress strived and achieved by man. These assumptions, then, implies that human progress is transitory, and ephemeral, and likely either to end soon or not sustainable.

Then, to solve the assumptions that human progress is transitory and ephemeral, and likely either to end soon or not sustainable we need to introduce the uniform distribution model in an AI led technological civilization. For a civilization which successfully transitioned into a sustainable industrial one powered by nuclear fusion and entered a cosmic expansionary phase, a different scenario awaits and at the same time, does not violate the assumption stated above. The resolution of the dilemma regarding an ever-progressing and expanding cosmic civilization in our static, non-moving lognormal distributed model is discussed below.

7.7 YAABER for Industrial Society

Human started the transition from biological muscle dominated society into a fossil fuel based industrial machine civilization in the 18th century. Thanks to James Watt's steam engine, fuel driven devices made progress much faster than agricultural revolution. It is estimated that one gallon of oil is equivalent to 2,000 manual labor hours.[107] With such level of cheap, abundant energy available, an intelligent species is able to exploit uranium in its crust, if available, and make it affordable to transition into a sustainable industrial civilization driven by nuclear fusion. As we have stated in our opening pages, once a civilization developed fusion and gained adequate knowledge on the development of fusion spaceship, it is just a matter of time before it expands and uses the energy of the sun and other stars. Then, its position relative to the cosmic evolutionary rate continually shift, and YAABER will continue to increase in our static, non-moving normal/lognormal distribution model. To illustrate this dilemma in our model, let's start by calculating how much progress Homo sapiens obtained since 1800 given the rate of economic growth, reflected in net energy usage, is 1.5% increase per year. It is noted by British economists and labeled as White's law, that the economic growth is closely intertwined with the growth rate of energy usage. In essence, measuring total energy output can be one of the most straightforward ways to measure the amount of progress achievable, not significantly different from our earlier analysis on hunter-gatherer society and agricultural society. It is only this time; fossil fuel-based energy budget substituted the solar based one. With a much greater amount of energy budget and capable tools (agricultural machinery, construction cranes, tunnel diggers, railroads, trains, cargo planes and ships) to convert it into useful work, the diversity and complexity of the society backed by the species increases.

We tried both energy and population growth rate as a way to measure YAABER. In the finalized version, we eventually ended up using population growth per annum. First of all, the energy generation and growth data during the early industrial period is incomplete. Secondly, the population growth rate during the 20th century matches closely with 1.5% growth rate in energy usage observed in the recent decades. We further assumed that such a rate of growth will be maintained, either in the form of continued population growth or continued energy generation and consumption. Thirdly, continued population growth can indirectly reflect the growth of complexity. An explosion in population within a short period of time indicates an escape from the Malthusian trap, removing the population ceiling on food resources by using fertilizers and machinery to increase productivity. A more populated society also requires greater cost at its maintenance. According to Tainter, the complexity of modern industrial civilization must be maintained by an EROEI at least 10 to 1. The rapid population growth, maintenance, and development all depended on the continued supply and growth of energy.

We then calculate this rate with our background evolutionary rate from AD 1750 to AD 2100 over the course of 350 years, and we arrived at 0.9695 Gyr.

$$k_{earlyindustrial} = \left(\frac{1,650,000,000}{791,000,000} \right)^{\left(\frac{1}{150}\right)} \tag{7.7.1}$$

$$k_{earlyindustrial} - 1 = 0.00491 \tag{7.7.2}$$

$$f_{earlyindustrial} = (k_{earlyindustrial})^{(1900-1750)} = 2.086 \tag{7.7.3}$$

$$f_{earlyindustrialxhunter} = (k_{hunter})^{(1900-1750)} \tag{7.7.4}$$

$$f_{earlyindustrialxexploration} = (k_{exploration})^{(1900-1750)} \tag{7.7.5}$$

$$A_{exploration1} = (f_{hunter} \cdot f_{citystates} \cdot f_{middle}) f_{exploration}$$
$$\cdot (f_{earlyindustrialxexploration} - 1) \tag{7.7.6}$$

$$A_{hunter6} = f_{hunter} \cdot f_{citystatesxhunter} \cdot f_{middlexhunter}$$
$$\cdot f_{explorationxhunter} (f_{earlyindustrialxhunter} - 1) \tag{7.7.7}$$

$$T_{earlyindustrial} = f_{hunter} \cdot f_{citystates} \cdot f_{middle} \cdot f_{exploration} (f_{earlyindustrial} - 1) \tag{7.7.8}$$
$$+ A_{hunter6} + A_{exploration1} \tag{7.7.9}$$
$$\approx f_{hunter} \cdot f_{citystates} \cdot f_{middle} \cdot f_{exploration} (f_{earlyindustrial} - 1) \tag{7.7.10}$$
$$+ A_{exploration1} \tag{7.7.11}$$

$$\frac{T_{earlyindustrial}}{T_{total}} \cdot 1 \text{ Gyr} = 48.584 \text{ Myr YAABER} \tag{7.7.12}$$

The progress contributed by the evolution in hominid lineages during this period:

$$f_{earlyindustrialxhomo} = (k_{homo})^{(1900-1750)} \tag{7.7.13}$$

332

$$A_{homo6} = \frac{1}{3.58} f_{homo} \cdot f_{hunterxhomo} \cdot f_{citystatesxhomo} \cdot f_{middlexhomo}$$
$$\cdot f_{explorationxhomo} \left(f_{earlyindustrialxhomo} - 1 \right) \quad (7.7.14)$$

$$A_{homo6} = 5{,}924.55 \text{ YAABER} \quad (7.7.15)$$

From 1750 to 1900, during the early industrial period, the population was growing slower than the full blown one in the 20th century because only a handful of countries (UK, US, France, Germany) were undergoing industrialization. Moreover, none of them had completed their transformation at the time.

$$7.91 \left(1.00492 \right)^{150} = 16.51 \quad (7.7.16)$$

The overall slower growth rate is reflected from the population growth rate at the time, lower than 1.5% per annum with only 1.00492% per annum.

$$k_{industrial} = \left(\frac{7{,}000{,}000{,}000}{1{,}650{,}000{,}000} \right)^{\left(\frac{1}{110} \right)} \quad (7.7.17)$$

$$k_{industrial} - 1 = 0.0132 \quad (7.7.18)$$

$$f_{industrial} = \left(k_{industrial} \right)^{(2100-1900)} = 13.839$$
$$f_{industrialxhunter} = \left(k_{hunter} \right)^{(2100-1900)}$$
$$f_{industrialxexploration} = \left(k_{exploration} \right)^{(2100-1900)}$$

$$A_{hunter7} = f_{hunter} \cdot f_{citystatesxhunter} \cdot f_{middlexhunter} \cdot f_{explorationxhunter}$$
$$\cdot f_{earlyindustrialxhunter} \left(f_{industrialxhunter} - 1 \right) \quad (7.7.19)$$

$$A_{exploration2} = f_{hunter} \cdot f_{citystates} \cdot f_{middle} \cdot f_{exploration}$$
$$\cdot f_{earlyindustrialxexloration} \left(f_{industrialxexloration} - 1 \right) \quad (7.7.20)$$

$$T_{industrial} = f_{hunter} \cdot f_{citystates} \cdot f_{middle} \cdot f_{exploration} \cdot f_{earlyindustrial} \left(f_{industrial} - 1 \right) \quad (7.7.21)$$
$$+ A_{hunter7} + A_{exploration2} \quad (7.7.22)$$
$$\approx f_{hunter} \cdot f_{citystates} \cdot f_{middle} \cdot f_{exploration} \cdot f_{earlyindustrial} \left(f_{industrial} - 1 \right) \quad (7.7.23)$$
$$+ A_{exploration2} \quad (7.7.24)$$

$$\frac{T_{industrial}}{T_{total}} \cdot 1 \text{ Gyr} = 0.92093 \text{ Gyr YAABER} \quad (7.7.25)$$

The progress contributed by the evolution in hominid lineages during this period:

$$f_{industrialxhomo} = (k_{homo})^{(2100-1900)} \qquad (7.7.26)$$

$$A_{homo7} = \frac{1}{3.58} f_{homo} \cdot f_{hunterxhomo} \cdot f_{citystatesxhomo} \cdot f_{middlexhomo} \cdot f_{explorationxhomo}$$
$$\cdot f_{earlyindustrialxhomo} (f_{industrialxhomo} - 1) \qquad (7.7.27)$$

$$A_{homo7} = 7,899.89 \text{ YAABER} \qquad (7.7.28)$$

For the latter part of the industrial phase, it spans from 1900 to 2100. We choose 2100 as the expected time by which the society either stops grow and transitions into a steady state biological led industrial civilization due to increasing limits on resources or transitions into an industrial civilization led by non-biological superintelligence. Either way, by AD 2100 we will have made another 0.9695 Gyr of progress compares to the cosmic evolutionary rate in 350 years! Note that agricultural society contributed an insignificant portion of 35.6 million years into the final YAABER, because throughout this transition, agriculture still maintained its steady progression toward its maximum utilization rate, and likewise, a tiny portion of YAABER itself.

$$\frac{A_{exploration1} + A_{exploration2}}{T_{otal}} \cdot 1 \text{ Gyr} = 35.6 \text{ Myr} \qquad (7.7.29)$$

In essence, if 1 Gyr of YAABER correlates with the homo sapien's tale of the conquest of the planet, then the full potential of crop plants sustaining a population at the carrying capacity of the planet is only realized at the late times of the industrial phase, and YAABER from the industrial age, which was the shortest among all periods, contributes 96.656% of the deviations. Assuming that the total 1 Gyr of YAABER correlates with human population increase from 50,000 to that of 23.676 billion achievable toward the end of 21st century, then any additional gains would goes beyond the 1 Gyr YAABER we formerly designated. If we continue to extrapolate this trend, we would expect another 2.647 Gyr added to the final YAABER by the 22nd century and 1.247 Gyr to YAABER by the end of the 23rd century. Within a few centuries, the contribution to YAABER by industrial civilization would dominate the entire calculation. This implies by simply maintaining our current level of industrial progress, we would expect soon to be one of the only one possible industrial civilization within the observable universe within a diameter of 93 billion light years predicted by the distribution function, and soon magnitudes above the observable universes diameter. However, such an argument implies something very special about industrial civilization itself, and within a few millennium of development, we expect ourselves to be the only one in the entire universe in terms of development, this is in contradiction with the principles of mediocrity. Even if industrial civilization is not as frequent as the number of habitable Earth-like planets, claiming ourselves as the only one present ourselves as a variant version of the rare earth hypothesis (in fact, a rare industrial civilization hypothesis.) On the other hand, if we conform to the principle of mediocrity, then, we would expect our growth-based economic model to be unsustainable. In reality, are our industrial civilization capable of continued economic growth for the next few centuries or are we facing a transition to no growth based stagnant industrial civilization or an industrial collapse? Or can we reconcile these dilemmas?

A key insight is to distinguish two states of an industrial civilization. *One is that led and maintained predominantly by intelligent biological species. Another is that led and maintained predominantly by post-biological intelligence risen from the industrial revolutionary process itself.* This insight is crucial to understanding and resolving the inconsistency in our model for prediction. The increasingly dominant role of post-biological intelligence in industrial civilization is not well-appreciated and even taken seriously until the Law of Accelerating Returns, a generalized form of the Moore's Law, which states that the speed of central processing unit in the semiconductor industry doubles every 18 months. Using this model, it is predicted that machine with the equivalent of human-level intelligence is roughly at the cost of $1,000 by 2019 and a thousand times faster

334

by 2029, and a billion times by 2045; hence, a term is labeled as the Technological Singularity. If the trend of post-biological intelligence overtaking and dominating in any evolving industrial civilization is typical, by the principle of mediocrity, then, it implies that every industrial civilization, with its abundant cheap energy, will transform its technology at such a fast speed. Within a sub-geologic timescale, transition toward a post-biological industrial civilization is then inevitable. Once a post-biological industrial civilization transition is complete, it is no longer represented as a data point in our static, non-moving lognormal distribution model. A post-biological civilization is not subject to the biological constraints placed upon a biological species. Despite human's ingenuity, human requires certain biological assumptions to survive on. Human can only survive unprotected at a temperature range between negative 30 and positive 50 degrees Celsius at 1-atmosphere pressure. Human cannot survive at 100% level of pure oxygen at the standard 1 Atm. A further study was done by NASA for oxygen tolerance at different atmospheric pressures. At longer time frame, human needs significant amount water to maintain its bodily functions. At even longer time frame, human requires the gravitational environment to maintain healthy bones and frames. Even with satisfying conditions mentioned above, human continued economic growth will eventually raise the global temperature generated from the heat waste product of the industrial process. Since humans are biological and maintain their physical energy from primary crops and livestocks, rising temperature guarantees a global warming scenario, which is catastrophic to the lives of human themselves. This global warming scenario is way more general trend than the CO_2 emission from fossil fuel, which is already generating heated debate within the society. The heat waste is generated by even the so-called green energy alternatives. Since a continued growing civilization ultimately requires a transition into nuclear power, the heat waste in the forms of hot water dumped into waters and streams will eventually raise the global temperature, albeit at a slower rate than the CO_2 emission observed by burning fossil fuel. The rising global temperature is an inevitable consequence of economic growth based on the second law of thermodynamics. In such a case, biologically based industrial civilization will eventually face a choice of growth or sustainability. It will still be able to maintain its level of development or somewhat lower level of energy consumption. Otherwise, it is essentially non-growing in terms of energy usage. That is, its position stays relatively static to the cosmic evolutionary rate. Some may argue that by turning the moon into a giant solar energy collector and beaming energy back to earth or by building all nuclear power plants on the moon will solve the heat waste problem. However, eventually, the final energy product, excluding the heat waste, has to be delivered back to earth to be consumed, the heat accumulation will eventually contribute to a warming up. To appreciate how quickly human can alter its environment compares with nature, we can do a calculation on a solar collector on the moon with a growth rate of 1.5% per year, in 400 years, it will able to capture and deliver 4% of total sunlight received on earth annually to earth via laser. This additional 4% of solar energy increase is equivalent to the sun's increase in luminosity in 440 million years into the future. A stagnant civilization in terms of energy usage does not imply a stagnation in progress. It may well be diverting its resources into information technology and bioinformatics. In such a scenario, this civilization stagnates for a few hundred years before a transition into a post-biological industrial one. Within a timescale of 10^7 years, its stagnation will not even show up in the model.

We can then calculate the upper bound of a biological based industrial civilization if we take the report on the maximum allowable temperature range increase to be 4 percent warmer in terms of total solar insolation. Assuming alternative nuclear energy is adopted, the increase in heat waste will guarantee an increase of 4 percent by AD 2400. Given that currently, the average waste heat of the earth per square meter is 0.028 $\frac{\text{Watt}}{\text{m}^2}$ and an industrial growth rate of annual 1.5%. Of course, faster growth rate and concentration in urban areas will run into the limit much earlier.

$$T_{wasteheat} = 0.028 \cdot (1.015)^{400} = 10.8037600757 \, \frac{\text{watt}}{\text{m}^2} \tag{7.7.30}$$

Temperature is not the only limitation placed upon on biological based industrial civilization. Since crops require land to grow and harvest energy, photosynthetic efficiency, as mentioned before, and the total amount of arable lands and even total lands available becomes the limiting constraints. We can perform some calculations on the limits based on earth. Assuming the entire earth adopts to an urbanized lifestyle and maintained a population density comparable to Tokyo and the remaining crop lands maintain the natural photosynthetic efficiency, then, we would expect world's carrying capacity at 54 billion and urban metropolitan size of the country of Brazil and takes 137.5 years to reach this level with 1.5% annual growth rate.

$$P_{agrilimitarea} = 5.4053866714 \times 10^{10} \text{ km}^2 \tag{7.7.31}$$

$$A_{agrilimitarea} = 5.4053866714 \times \frac{10^{10}}{6,224.66 \frac{\text{People}}{\text{km}^2}} \tag{7.7.32}$$

$$= 8,683,826.37991 \text{ km}^2 \text{ Metropolitan Area}$$

$$T_{agrilimit} = (1.015)^{137.5} = 7.74605915486 \text{ 7Billion} \tag{7.7.33}$$

the units above is 7 billion, the current population of the world.

With GMO enabled crop to reach its ultimate biological conversion efficiency, it is possible to support a population of 711.8 billion and urban area size of all continents except Antarctica (where it is reserved to produce GMO crop plants), and it takes 310 years to reach this level with 1.5% annual growth rate.

$$P_{agrimaarea} = 7.1181844926 \times 10^{11} \text{ km}^2 \tag{7.7.34}$$

$$A_{agrimaarea} = 7.1181844926 \times \frac{10^{11}}{6224.66 \frac{\text{People}}{\text{km}^2}} \tag{7.7.35}$$

$$= 114,354,591.136 \text{ km}^2 \text{ Metropolitan Area}$$

$$T_{agrima} = (1.015)^{310.43} = 101.68420451 \text{ 7Billion} \tag{7.7.36}$$

the units above is 7 billion, the current population of the world.

It is interesting to note that further expansion will be not possible even though photosynthetic efficiency can be continually raised because growing population requires land to dwell. The costs (occupying agricultural space) of a continued rise in population eventually overtaking the benefit of having people producing the agricultural products in the first place.

Even if temperature and land space were not considered to be constraints, what are the possible limits to an expanding biological based industrial civilization? In which case, the total surface area, converted into the urban landscape and people feed based on vertical horticultural food grown from LED lights, can support at most 899.2 billion, and it will take 326 years to reach this level assuming 1.5% annual population growth rate.

$$P_{all} = 148,940,000 \text{ km}^2 \cdot 6224.66 \frac{\text{People}}{\text{km}^2} \cdot \frac{0.97}{1} = 8.9928783459 \times 10^{12} \text{ people} \tag{7.7.37}$$

Is it possible to transform into a scale II civilization that harnesses the host star output energy? We still need to make certain assumptions about the biologically based human. The minimum requirement is to reside on a solid mass that contains a considerable amount of gravity which in turn is able to hold onto an atmosphere. If such scenario is true, it remains whether photosynthesis can be carried out by nuclear fusion instead of the sun as the sun itself is to be converted into thousands of Earth-sized planet where each can hold the 700 billion carrying capacity of the earth. We also assume that hydrogen harvested from the sun can be converted into metallic hydrogen, and helium can be converted into metallic helium. In order to turn one sun into millions of Earth-sized planet, we soon find that overcoming the sun's gravitational binding energy is way greater than a 54 billion human can generate to start with, economically prohibitively expensive.

$$T_{sun} = \frac{(6.87 \cdot 10^{41} \text{ J})}{\frac{54 \times 10^9}{7 \times 10^9} \cdot (6.8 \cdot 10^{19} \text{ J})} = 1.3096405229 \times 10^{21} \text{ years} \tag{7.7.38}$$

It will take $1.3 \cdot 10^{21}$ years of all energy generated by 54 billion people to tear the sun apart. If we assumed maximum upper bound on human population covering the entire surface of earth, and they tear Jupiter apart instead, it would still take them $9.488 \cdot 10^{16}$ years to accomplish this task.

$$T_{jupiter} = \frac{(6.87 \cdot 10^{41} \text{ J}) \left(\frac{318}{333000}\right)}{101.68 \cdot (6.8 \cdot 10^{19} \text{ J})} = 9.4884478128 \times 10^{16} \text{ years} \tag{7.7.39}$$

As a result, the splitting of the sun to create earth analogs remain a science fiction to biological-based industrial civilization.

Apart from the natural limitation placed upon human, human self-directed decision making can also render its own industrial civilization collapse. Most commonly mentioned examples are the nuclear holocaust, grey goo nano-technological catastrophe, and social degeneration. In each of these cases, the position of the civilization shifts toward the left, or toward the cosmic mean evolutionary rate. The derivative of the bell curve will be negative in the direction of positive increase from the mean. Fossil fuel depletion before a successful transition into a nuclear-based civilization will also render a civilization-wide collapse, indicated by the M. Hubbert's peak oil theory.

In conclusion, the increasingly diminishing chance of observed biological led industrial civilization in the bell curve model is a combination of the above-mentioned factors. In particular, a phase transition sometimes occurs during its development where the cost of transitioning into a post-biological one is lower than that of the biological led one. If we consider earth is typical, and all intelligent species act rationally according to the principle of economics, we would expect most biological based industrial civilization transition into a post-biological one in 300 years after the start of the industrial revolution on their home planet. This reconciles with the principles of mediocrity. Because it shows that after 300 years of progression, those civilizations which graduate into a post-biological one should be common rather than extremely rare. It also implies that a biological led industrial civilization which does not transition but keeps expanding without improving upon themselves is foolhardy and consequently, rare in the cosmos. A post-biological industrial civilization does not mean, however, that all components of the society are exclusively post-biological. Just as the economy of United States today labeled as a developed industrial economy, it does not imply that it does not perform agricultural activities. In fact, United States contains the largest agricultural activity in the world. Because over 95% of economic activities have no direct relation with agricultural activity, the direction and progression of the economy are determined by the decision making in the industrial sector. In a sense, a biological led industrial civilization's position on the bell curve resembles a supernovae's brightness in a galaxy.

Any stars within a galaxy shine significantly fainter than the sum of all stars within the galaxy and their brightness can be modeled by power law/lognormal distribution. Their brightness grows very slowly as they burn slowly on the main sequence, but within the time frame of human affairs, their brightness stays constant, non-changing. Yet a supernovae's luminosity position shifts quickly within the spans of weeks to the right of the curve by a few standard deviations and then fades quickly to the left of the mean as a white dwarf, a neutron star, or black hole. The progress of humanity on the evolutionary distribution curve is conceptually similar. A biological human-led industrial civilization shifts its position quickly on the bell curve, reaches an extreme right-handed outlying value, many sigmas above the mean, but also quickly drops from the model altogether as it successfully transitioned into a post-biological industrial civilization or retracts its position and returns to a non-moving position as an agricultural civilization, or non-growth based industrial civilization.

In conclusion, the total amount of time on average a civilization spends as a biologically directed industrial civilization in growth mode is what we needed to add to the final YAABER. In our case, from the start of the utilization of the steam engine until the utilization of sustainable nuclear fusion and the advent of strong AI. Without nuclear fusion and even with Strong AI, the industrial civilization will collapse and eventually retracts to an agricultural one. With only nuclear fusion, civilization is earthbound (at most add a terraformed Mars), and the society eventually stops grow and transition into a steady state biological led industrial civilization. Only when both nuclear fusion and strong AI are realized, the society transitioned into an industrial civilization led by non-biological superintelligence and expands into the universe.

Since we have formalized the project PACER since the 1960s and expecting the emergence of technological singularity around 2045, we are not underestimating the YAABER under the age of industrialization by setting our computed boundary at 2100. That is, nearly all civilization made their full transition from biologically led civilization to a post-biological one within this time frame or stagnate into a steady state faced by the biological and resource constraints. Then, the sum total of the entire YAABER from all period is 1.6609 Gyr and a deviation of 17.

$$A_{homo} + A_{hunter} + A_{exploration} + (T_{prim} + T_{homo})$$
$$+ 1 \text{ Gyr} \left(\frac{T_{hunter} + T_{citystates} + T_{middle} + T_{exploration} + T_{earlyindustrial} + T_{industrial}}{T_{total}} \right) \quad (7.7.40)$$

$$= 1.6609 \times 10^9 \text{ YAABER}$$

$$D_{eviation} = \text{YAABER} \times \text{Speed} = \text{YAABER} \times \ln(2.783) = 17 \quad (7.7.41)$$

We can now graphically present each part's contribution toward the final deviation:

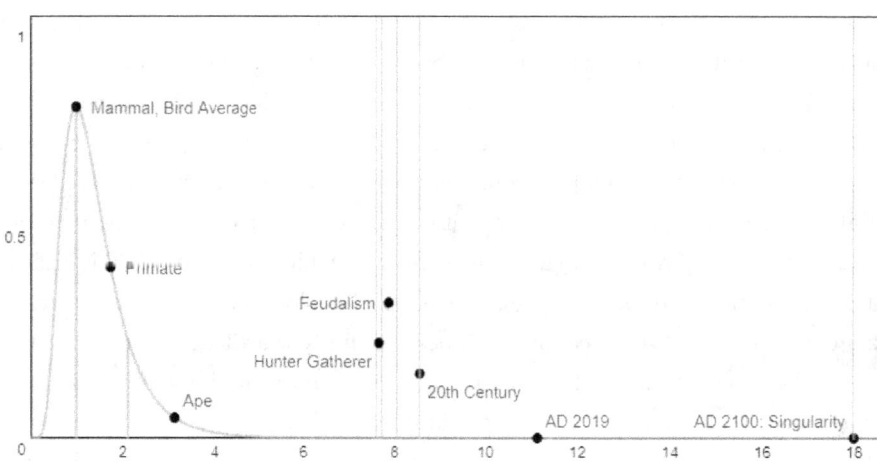

Figure 7.7.1: Each parts of the deviation and their contributions, whereas industrialization since 20th century contributes more than the vast majority.

To generalize, depending on the BER, the final composite YAABER can also changes. Lower the BER value, YAABER grows relative to the mode. Higher the BER value, the smaller the YAABER value, YAABER shrinks relative to the mode. The relationship is captured by the plots below by equation **YAABER** $= \frac{\text{Deviation}}{\ln(\text{BER})}$:

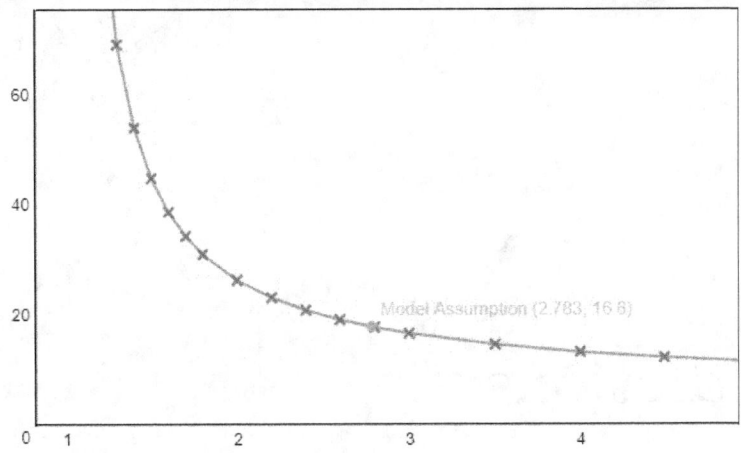

Figure 7.7.2: The YAABER computed for different BER values

Chapter 8

Model Predictions

8.1 Number of Habitable Earth

From the previous Chapter 2 and 3, we have enumerated and demonstrated different criteria that restrict the number of habitable planets and eventually the number of planets that gives to the emergence of industrial civilization. We will now list the figures after each selection criterion.

Criterion	Probability	Number
Terrestrial Planets between 5 Gya ~ 4 Gya		612,398,339
Galactic Habitable Zone	18.33%	112,239,501
Binary, Ternary, and Multiple systems	72.85%	81,767,363
Stellar Habitable Zone	36.77%	30,069,457
Orbital Eccentricity	95.67%	28,767,123
Appropriate Mass	85.83%	24,689,429
Initial Rotation Speed	23.08%	5,698,457
Keeping a Moon	34.38%	1,959,125
Non Tidally-locked Planet to Satellite	8.22%	160,988
Wet Earth	19.31%	31,092
Ocean Budget	59.32%	18,443
Dry Land Ratio	5.23%	965
N_{earth}		**965**

Table 8.1.1: Number of Habitable Earth

The number of terrestrial planets within the Milky Way galaxy obtained based on Lineweaver's method that emerges between 5 Gya and 4 Gya is 0.612 Billion.

The number of planets within the galactic habitable zone is 112 Million.

The habitability of binary, ternary, and multiple systems reduces the number to 81.7 Million.

The habitable zone restricts terrestrial planets to be between 0.840278 AU and 1.0887 AU total radius for terrestrial planets' formation zone of solar mass stars. The total radius for terrestrial planet formation zone and the band of habitability expands and shrinks according to stellar mass, but the ratio remains fixed for different stellar mass. Therefore, the number of terrestrial planets within the habitable zone is 30 Million.

Terrestrial planet with moderate eccentricity so that its orbit falls within the comfortable range of the band of habitability is 28.767 Million.

Out of these planets, the terrestrial planet mass ranges between 0.43 to 2 earth mass is 24.7 Million.

Out of these planets during its final merging process, the rotation is fast enough so that the day and night cycle after 4.5 Gyr of evolution is less than 7 days is 5.698 Million.

Out of these planets that the final generated moon does not eventually fall back onto the planet itself is 1.959 Million.

Out of the planets that have a satellite, the satellite is light enough that it does not tidally lock with its parent planet during the emergence of civilization 5 Gyr after its initial formation is 0.16 Million.

Out of these planets that are covered by water, or as wet planets, is numbered 31,092.

Out of the wet planets, those planets that allow the exposure of dry land in some proportion is 18,443.

Finally, those planets that have the right proportion of ocean water so that a submerged continental shelf enables the smooth transition from ocean-based life to a land-based life form is 965.

Very lastly, the number of low mass binaries (Chapter 2) is 243,868. One applies all the previous filter mentioned except the filter criteria for the habitability of binary system and the habitability within the habitable zone since they were already computed from the previous round and we obtain 4 habitable planets.

Criterion	Probability	Number
Terrestrial Planets between 5 Gya ~ 4 Gya		243,868
Galactic Habitable Zone	18.33%	44,696
Orbital Eccentricity	95.67%	42,760
Appropriate Mass	85.83%	36,699
Initial Rotation Speed	23.08%	8,470
Keeping a Moon	34.38%	2,912
Non Tidally-locked Planet to Satellite	8.22%	672
Wet Earth	19.31%	130
Ocean Budget	59.32%	77
Dry Land Ratio	5.23%	4
N_{earth}		4

Table 8.1.2: Number of Habitable Earth around low mass binaries

Therefore, there are, in total, *969 habitable planets within the Milky Way* at the current epoch that can potentially lead to intelligent life.

We also need to compute the composite probability on all earth like terrestrial planet that will eventually give to the emergence of industrial civilization within the recent 500 Myr time window, this time we have only a handful of selection criteria:

Criterion	Probability	Number
N_{earth}		969
Life	16.60%	161
Permissible Placement of Continent Cycle over Glaciation Cycle with a Chance of Island Continents Rejoining within 1 Gyr time window	29.03%	47
The appropriate timing for the onset of ice age acting as an accelerator	21.20%~100%	10~47
Glaciation chance of a constructive ice age	50%~100%	-
N_{All}		10~47

Table 8.1.3: Number of potential civilizations for planets formed between 5 Gya and 4 Gya

Now, having derived the number of habitable planets, out of these planets where life has actually developed from the assemblage of amino acids is 161.

Out of these planets, they are placed within the permissible range of continent cycle over glaciation cycle with a chance of island continents rejoining within the 1 Gyr time window is 47.

Out the planet within the permissible range, a constructive ice age occurred at the late times of a breaking up phase so that its accelerating effect on evolution is the most prominent is 10. However, this filter is applicable if one considers that, for the earliest forming habitable planet, its multi-cellular evolution is only one supercontinent cycle ahead of earth and had only experienced one supercontinent cycle of multi-cellular evolution so far. If the earliest multi-cellular evolution occurred much earlier so that it spans several supercontinent cycles, then the chance of experiencing a constructive ice age approaches 100%. We simply show that this filter, if applied, will reduces the number of planets between 10 to 47. In our basic model, we simply assume there is a 100% chance a constructive ice age occurred.

Basically, the permissible range is considered when one assumes that all earth like planet formed 4.5 Gya and later experienced only 1 full supercontinent cycle so far. Biodiversity was low enough so that accelerated ice age raises emergence chance significantly. For planets formed up to 300 Myr earlier than earth, they experienced a joining phase of the supercontinent cycle, so that its highest biodiversity was achieved between now and 300 Mya, at a time when it was still within 1 supercontinent cycle. We later show that in general one does not need to extend the earliest window beyond 300 Myr earlier. Therefore, we assumes that all emergence on all planets are significantly effected by ice age. We also assumed that all planets formed at the same time as earth are experiencing the same pace of supercontinent cycle. So older planets are ahead of earth and younger ones are behind. However, statistically, younger planets may catch up with earth, and older ones may lag behind. This does not change our assumption. One can sort the batch of planet by their current pace chronologically so that all planets currently held the same position on the supercontinent cycle are grouped together. The sorting is stringent if all planets' pace can only mismatch within a tiny fraction of the entire cycle, for example, 10 Myr. It is relaxed if all planets' pace only within a large span of time, for example, the entire continent cycle of 473 Myr.

If a constructive ice age does occur, its occurrence chance varies depend the supercontinent cycle placement relative to the glaciation cycle and the length of the glaciation cycle. For planets dominated by supercontinents, the glaciation cycle becomes short enough so that each breaking up island phase of biodiversity explosion is followed immediately by a guaranteed ice age. At the other extreme, glaciation cycle becomes much longer than a supercontinent cycle and several supercontinent cycle fits within a single glaciation cycle. The average glaciation chance of these supercontinent cycles following breaking up island phase of biodiversity explosion then approaches 50%. (See 5.6.3: "Glaciation chance") .

Finally, regarding planets with longer glaciation cycle than its supercontinent cycle, for those breaking up phase just followed the end of the last ice age has a lower but non-zero chance of forming disruptive ice age during the early times of a breaking up phase but also a lower chance of forming a constructive ice age following the end of a breaking phase. For those super-continent breaking up phase occurred long after the end of the last ice age has a higher chance of forming disruptive ice age during the early times of a breaking up phase but also a higher chance of forming a constructive ice age following the end of a breaking phase. Therefore, we are justified by treating placements within any permissible range with the average chance of forming a constructive ice age toward the late times of a breaking up island phase.

We set earth's 68% glaciation chance as the average across all possible land to sea coverage ratios. Nevertheless, we shall skip the filter on glaciation chance of a constructive ice age.

The reasoning behind ignoring the last two filter criteria regarding ice age as the follows.

First of all, even assuming most planets experienced only 1 continent cycle, abandoning last two filter eases computation and simplifies the model. (Chapter 7.2.4) We show that a minimal glaciation effect on emergence does not necessarily mean a distortion on the final emergence. Recall in Chapter 5 and 6 we repeatedly emphasized that ice age increases human emergence chance at the most recent 100 Myr toward 1. We have shown that if we assumed ice age played a role, then one should take the upper bound on speciation per 65 Myr at 7,555,486, with an attempt window size $7^8 < t_1 < 7^9$. Whereas all of the increase occurred in the last 2 Mya. Then, $S_{total}(7, 7, t_1) \geq 1$. However, recall that the chance of a constructive ice age that brings accelerating evolution at the late times of a breaking phase is simply 21.2%, and average glaciation chance is 68%. The combined chance actually decreased to $\approx \frac{1}{7}$. On the other hand, human emergence without taking the accelerated ice age and with an average glaciation chance of 50% is $\approx \frac{1}{4}$, which is just the original assumption. This is possible since glaciation occurs at earlier times of the current round of simulation with lower established number of species at the time guarantees a negligible increase in final speciation size. On the other hand, glaciation at the later rejoining phase can only at best compensate a loss of biodiversity at the time. Check Section 5.3. Therefore, taking accelerated ice age raises human emergence chance at the expense of fewer number of planet candidates, the overall effect is comparable to not taking it into consideration.

Secondly, assuming most planets experienced > 1 continent cycle, ice age as an accelerator is most effective only during the initial continent cycle when the overall biodiversity is relatively low. At later cycles, there is enough cumulative biodiversity generated due to exponential growth, so that, even without an ice age as an accelerator will guarantee the emergence of civilization. In earth's case, it nevertheless played a critical role in expediting the evolutionary process for planets experienced only one supercontinent cycle. Therefore, we acknowledge ice age as an accelerator by introducing the permissible range factor, but ignored 2 additional ice age filter criteria assuming there remains a chance that many planets underwent more than one supercontinent cycle, diminishing the role of ice age in the emergence of civilization significantly.

Alternatively, one can also set the lower and upper bound on the influence of ice age. In this case, if one assumes ice age played absolutely no role if all habitable planets underwent repeated continent cycles.

Criterion	Probability	Number
Life	16.60%	161
Permissible Placement of Continent Cycle over Glaciation Cycle with a Chance of Island Continents Rejoining > 1 Gyr time window	100%	161
The appropriate timing for the onset of ice age acting as an accelerator	100%	161
Glaciation chance of a constructive ice age	100%	161
N_{All}		161

If one assumes ice age played a critical role in evolution if all habitable planets underwent only one continent cycle.

Criterion	Probability	Number
Life	16.60%	161
Permissible Placement of Continent Cycle over Glaciation Cycle with a Chance of Island Continents Rejoining within 1 Gyr time window	29.03%	47
The appropriate timing for the onset of ice age acting as an accelerator	21.20%	10
Glaciation chance of a constructive ice age	68.18%	7
N_{All}		7

The geometric mean and arithmetic mean of the lower and upper bound yields 33 and 84 planets, our earlier analysis lies between these numbers, therefore, we are justified by reaching our earlier conclusions.

Next, we derive the composite probability on the emergence of industrial civilization at the current time as:

Criterion	Probability	Number
N_{all}		47
Planets with biocomplexity (BCS) among all planets within 5 Gya to 4 Gya≥ Earth	35.75%	17
Fastest Emergence of Hominid lineage among all planets within 5 Gya to 4 Gya with biocomplexity (BCS) ≥ Earth	25.29%~94.3%	4.30~15.74
Emergence of Homo sapiens among Hominid lineage (already generalized from the step above)		4.30~15.74
Biodiversity of Angiosperm enabling Crop Plants and Industrial Civilization		
Lower the current average biocomplexity across all planets or/and	≤ 7.75%	≤ 1/3
The earliest habitable planet later than 5 Gya starting date or/and		
Other unlisted factors		
$N_{current}$		≤ 1/3

Table 8.1.4: Number of Civilizations per galaxy at the current time

Out of these planets, the probability that they have achieved a comparable or higher level of biocomplexity observed on earth due to their earlier formation time, and consequently experienced more of continent cycles and higher BER is 17.

The calculation becomes tricky from this point onward.

Regarding the emergence of Homo sapiens, it has been shown from our earlier discussion that, by adopting a more stringent filtering criterion for the emergence of the hominid based on data fit is $\frac{1}{504}$ out of 5040 permutation paths and ultimately $\frac{1}{4}$ chance. The data fit and the derived chance itself already reflected with fluctuating climate pattern with accelerated speciation during a 100 Myr period, otherwise the chance is even lower. However, depending on how evolution proceeds, in terms of progressive or passive, the cumulative emergence chance on the emergence of Hominid lineage among all planets within 5 Gya to 4 Gya $\geq$ Earth varies between 4.30 and 15.74 planets.

Finally, the number of intelligent species that actually go on to transition into agricultural society and then to an industrial one, mostly constrained by the probability of the emergence of major, high-calorie crop plants. We have demonstrated earlier that the explosive growth of angiosperm's biodiversity and the abundance of crop plant species on earth may be atypical of the evolutionary pace on an earth like planet emerged at 4.6 Gya, though it is possible that the earliest among our batch of planets emerged 5 Gya may already attained such level of complexity.

The presence of great abundance of angiosperm alone may not be an adequate filter to reduce the total number of planets host intelligence to 1 out of 3 galaxies. We started our assumption that a typical earth like planet formed 4.6 Gya has already evolved multicellular life forms functionally equivalent to mammals, birds, and reptiles observed on earth. However, this assumption can be revised so that the most advanced form of vertebrates are fish or amphibian functionally equivalent or arthropods.

Furthermore, we have never forsake our firmly held belief that the earliest habitable planet must appear by 5 Gya. However, this assumption can be relaxed so that the earliest habitable planet could appear slightly closer if not significantly closer to earth's day of birth. By reducing the earliest habitable planet appearance date, we not only reduced the overall emergence rate but as well as the number of earth like planets across all filter criteria.

It turns out that the earliest opportunity window indeed needs to be revised. If one assumes that BER = BCS = 2.783. It can be demonstrated later based on the distribution model, that a habitable planet formed 5 Gya will have million folds increase in the chance of the formation of advanced civilization compares to earth despite only a 2.783^5 times higher BCS. Therefore, our earlier calculation on the probability that planets have achieved a comparable or higher level of biocomplexity observed on earth due to their earlier formation time is 35.75% is valid only if one assumes an ultra passive evolutionary path (BER = 1 and BCS = 2.783) and the emergence chance of civilization reaches parity with current earth but not beyond. However, considering all other non-ultra passive evolutionary paths, all of other earlier formed planet yields much higher chance in civilization emergence relative to earth. The angiosperm factor ($\frac{1}{4}$) alone can not justify such a significant increase in the chance of civilization formation. As a result, the initial window is shifted closer to the formation time of earth. Since there are 47 total planets that can give birth to civilization and their birth time are distributed nearly uniformly throughout 5 Gya and 4 Gya, the total number of planets that falls within each 100 Myr can be computed as the follows:

$$\int_{T(9.199+t)}^{T(9.199+t+0.1)} \frac{\int_{-1}^{1} f_{metallicity}(x,t) \, f_{wetearth}(x) \, dx}{\int_{-1}^{1} f_{metallicity}(x,t) \, dx} \cdot S_{tellar}(C_1(t)) \, dt \qquad (8.1.1)$$

On average, there are 4.407 planets per 100 Myr. The best fit for the number of planets between this time period is:

$$P_{lanet}(t) = 0.3958 \, (t + 21.517)^{0.8045} \qquad (8.1.2)$$

We have designated the likelihood of civilization emergence is at $\frac{1}{16}$. There is no particular reason $\frac{1}{16}$ is chosen other than assuming that $\frac{1}{4} \cdot \frac{1}{4} = \frac{1}{16}$ the abundance of angiosperms and crop plants enabled civilization to emerge is only found on 1 out of every 4 planets. (Check 6.8 "Probability of Angiosperm") Because the drop in likelihood value is much slower than the lognomal distribution dictates, the distribution beyond the deviation

345

representing the emergence of human should be replaced by the piece wise, gentler dropping curve

$$f_2\left(x\right) = \frac{1}{4} \times \frac{1}{\left[\sqrt{\frac{1}{\frac{x_1-x_0}{3}}}\left(x-\left(x_0-\frac{x_1-x_0}{3}\right)\right)\right]^2} = \frac{1}{4} \times \frac{1}{\left(\sqrt{\frac{1}{\frac{11}{3}}}\left(x-\frac{10}{3}\right)\right)^2} \qquad (8.1.3)$$

which is found by assuming that the mode peak of $P_{df}\left(0,x\right)$ is located at 1, representing species possessed EQ of 1, and then an EQ of 7.6 representing human emergence chance per the most recent 100 Myr of $\frac{1}{4}$ is located at $1 \cdot 7.6 = 7.6$ and civilization emergence chance of $\frac{1}{16}$ at $L_{imit} = 18$, we nevertheless used a lognormal as a good approximation because the underlying piece-wise multinomial distribution, a first lognormal like and later gentler decreasing function, describes physical reality faithfully but difficult to manipulate mathematically. (check 7.2.3 "Transforming Multinomial Distribution to Lognormal Distribution")

The cumulative chance of civilization arising at human level or more advanced by physical reality is then given as:

$$C_{cumulative} = \int_{L_{imit}=18}^{\infty} S_{total}\left(7, x, 7^8\right) dx = \frac{1}{16} \qquad (8.1.4)$$

So that the equivalent lognormal distribution function's cumulative chance of civilization arising at human level or more advanced has to satisfy this same cumulative chance by customizing $\sigma = 0.4811$:

$$t_{Win} \int_{L_{imit}=18}^{\infty} P_{df}\left(0,x\right) dx = C_{cumulative} = \frac{1}{16} \qquad (8.1.5)$$

Recall that, despite the lognormal overestimates likelihood beyond the species emergence window size and the deviation representing the emergence of human, we find lognormal eventually converges faster than a first fast dropping and later gentler decreasing piece-wise multinomial distribution, so that they can achieve the same cumulative chance defined beyond L_{imit}.

One can then find the cumulative chance of civilization across all planets from all periods so that it sums up to $\frac{1}{3}$:

$$\int_{-5}^{0.45} P_{lanet}\left(-s\right)\left(t_{Win}\left(\bar{B}_{cs}\right)^s \int_{L_{imit}=18}^{\infty} P_{df}\left(0-s,x\right) dx\right) ds = \frac{1}{3} \qquad (8.1.6)$$

It shows that the earliest possible window can only be 45 Myr earlier than the formation time of earth. The time gap between earth and the first possible habitable planet can be enlarged by lowering the current civilization's emergence rate. Alternatively, one can change the placement pattern of successive distribution through the factor k (see section "Generalized Model"), so that each successive distribution "stacked" on top of each other (more passive evolutionary pattern) and the chance of civilization increases by the value of BCS for each round, lowering the rate of increase of civilization by lowering the BER$\rightarrow 1$ even with the same rate of growth for BCS. As a result, the earliest possible habitable planet can still be placed at > 300 Myr earlier than the formation time of earth.

There also exists cases where the formation chance of the planets are not only shaped by metallicity factor and formation rate, but are also minimized by gamma ray bursts or some other cosmic factors so that the formation rate was much lower in the past. Until more research on this issue is worked out in the future, the past formation can be assumed by either linear or exponential decay functions for now, if one were to substitute the emergence from the past with an exponential decay function:

$$P_{lanet}\left(t\right) = \begin{cases} \frac{0.3958(t+21.517)^{0.8045}}{B^t} & t \geq 0 \\ 0.3958\left(t+21.517\right)^{0.8045} & t < 0 \end{cases} \qquad (8.1.7)$$

Whereas B can taken on any value. Assuming B takes on 2.783 and 1.783, the earliest window becomes:

Decay rate per 100 Myr	2.783	1.783
$1/B^0$	0.45	0.45
$1/B^1$	0.48	0.466
$1/B^2$	0.514	0.484
$1/B^3$	0.555	0.503
$1/B^4$	0.604	0.524
$1/B^5$	0.665	0.547
$1/B^6$	0.744	0.573
$1/B^7$	0.853	0.602
$1/B^8$	1.02	0.634
$1/B^9$	1.35	0.671

After all, there is no physical limit on the earliest earth formation date. There is a limit based on observational constraints, the limit based on observational constraints is shaped by the emergence rate of the past planet formation. If the planet formation rate was nearly constant (shaped by metallicity and stellar formation rate alone), then the earliest window is much closer to the current time. If the planet formation rate was decreasing rapidly into the past, the earliest window is much older than the current time.

The final distribution one uses to compute the overall chance of observation shall be:

$$C_{factor} = \int_{-5}^{0.45} P_{lanet}\left(-t\right)\left(B_{cs}\right)^{t} dt = 5.768 \tag{8.1.8}$$

That is, the total cumulative BCS is 5.768 times greater than currently obtained on earth. This is the cumulative BCS of all planets (formed between 4 Gya and 4.545 Gya) which sums up to a cumulative observational chance of $\frac{1}{3}$ at the current time.

In general, given the current emergence rate and a chance of observational constraints, one can predict the earliest window date. Conversely, given a fixed earliest date and giving a chance of observational constraints, one can predict the current emergence rate.

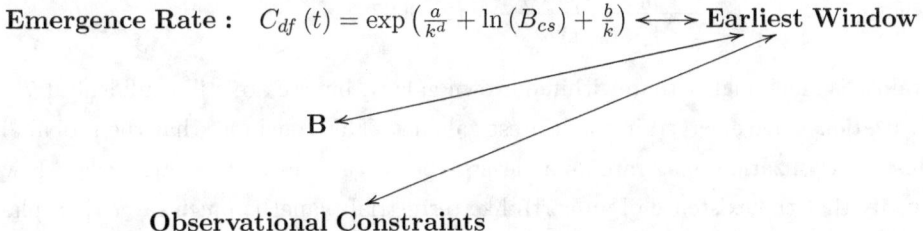

Emergence Rate : $\quad C_{df}\left(t\right) = \exp\left(\frac{a}{k^{d}} + \ln\left(B_{cs}\right) + \frac{b}{k}\right) \longleftrightarrow$ **Earliest Window**

B

Observational Constraints

We have also shown that at the current time, earth's biodiversity among all possible dryland to ocean ratio is one of the highest. If one were to consider the biodiversity of planets based on dryland and ocean ratio, chance within a breaking phase, joining phase, disruptive/constructive glaciation cycle, and permissible range, then the average biodiversity across all possible land to ocean ratio is only 24.877% of earth's. Recall we have shown that lower biodiversity translates to earlier evolution time relative to earth, and we find that typical planet at the current time lags behind of earth in development by 135.93 Myr.

$$\frac{\ln\left(0.24877\right)}{\ln\left(2.783\right)} = -1.3592 \tag{8.1.9}$$

If one were to use the average of earth like planets at the current time instead of earth itself, one then needs to modify our existing requirement by adding a $l_{ag} = 1.3592$ term as:

$$\int_{-5}^{0.45+l_{ag}} P_{lanet}\left(-s\right)\left(t_{Win}\left(B_{cs}\right)^{s}\int_{L_{imit}=18}^{\infty} P_{df}\left(0-s+l_{ag},x\right)dx\right)ds = \frac{1}{3} \tag{8.1.10}$$

So that the earliest window will shift earlier by l_{ag} years. In our case, it will shift to 180.92 Myr earlier than the formation time of earth. The C_{factor} will remain as the invariant by adding the l_{ag} term:

$$C_{factor} = \int_{-5}^{0.45+l_{ag}} P_{lanet}\left(-t\right)\left(B_{cs}\right)^{(t-l_{ag})} dt = 5.768 \tag{8.1.11}$$

This is true because $\left(B_{cs}\right)^{-l_{ag}}$ shifts the biodiversity to earlier times by l_{ag} years into the past relative to earth and only when the earliest window increased by l_{ag} years into the future relative to the modified time (restoring to the current time), the biodiversity matches the previous results. There is a slight discrepancy due to earth formation chance is slightly increasing over time.

Finally, there are other listed factors, which we will examine later in the chapter that may or may not play a role at minimizing the number of emerging extraterrestrial industrial civilizations.

We discussed earlier on in Chapter 1 that there is a very high confidence that we are the first industrial civilization within our galaxy since we confirmed that out of 500 Myr emergence window, 499 Myr have passed and we have not found any evidence of massive scale reorganization and restructure of the galaxy as well as the occupation of the solar system by any extraterrestrial civilization. Only the signals from the last 100 Kyr remain to be verified, but this amounts to only an unweighted chance of 0.02%. Using weighted chance by taking into consideration of double exponentially increasingly emerging civilizations and equation established in the following section assuming $C_{df}\left(0\right) = \frac{1}{3}$, we have only a weighted chance of 1.35%:

$$\frac{100 \int_0^{0.001} C_{df}\left(t\right) dt}{\int_0^5 C_{df}\left(t\right) dt} = 1.35\% \tag{8.1.12}$$

Therefore, we simply parameterize this chance so that the total currently emerging civilization within our galaxy to be at most 1.

Since our nearby Andromeda galaxy and Triangulum Galaxy (both remain pristine from artificial manipulations) also reside away from the cluster center and the chance of an emerging civilization arises within the last 2.6 Myr (though this is 26 times more likely than it arises within the Milky way) is considerably small compares to our time window of 0.5 Gyr (0.52% unweighted chance). and 29.8697% weighted chance:

$$\frac{100 \int_0^{0.026} C_{df}\left(t\right) dt}{\int_0^5 C_{df}\left(t\right) dt} = 29.8697\% \tag{8.1.13}$$

Therefore, the uncertainty is much higher than within our own galaxy, but we are still confident at 70% chance that no intelligent civilizations have emerged in the nearest galaxies. This concludes that the probability of the emergence of an industrial civilization is as rare as at least 1 per 3 galaxies at the current epoch with good faith. One can now verify that the existence of an earth like terrestrial planet is much rarer than planets that can give emergence to civilizations, and it is rarest for those civilizations emerged and emerging now.

Very lastly, studies have shown that due to Gamma ray bursts, habitable galaxies have to reside away from the galaxy clusters. Since the Milky way has a lower density of star forming dwarf galaxies making the Milky Way a more friendly neighborhood for life. Galaxies friendly to harbor and preserve life will preferably inhabit low density regions in voids and filaments of the cosmic web.

Furthermore, there is a metallicity and size selection criterion for the habitable galaxy itself. Only galaxies that produce enough metals so that their metallicity is $\geq 1/3$ solar and their galactic disks are larger than 4 kpc. As a result, such placement amounts to no more than 10% of all galaxies in the universe. Calculation indicates that this corresponds to a comoving abundance of 10^{-3} galaxies per Mpc^3. This implies 1 habitable galaxy occupies a cubic volume size of

$$V = 1000 \cdot \left(3.26156378 \cdot 10^6\right)^3 = 3.4695857605 \times 10^{22} \text{ ly}^3 = 1000 \text{ Mpc}^3 \tag{8.1.14}$$

This volume correspond a spherical volume size with radius of:

$$\left(\frac{3}{4\pi} \cdot V\right)^{\frac{1}{3}} = 20,233,126.9 \, \text{ly} \tag{8.1.15}$$

$$d_{Galaxy} = 4,947,525.31 \, \text{ly} \tag{8.1.16}$$

We later derived the average distance between galaxies is at 4,947,525 ly. However, other studies carried out showing that there could be up to 300 billion galaxies. They figured that the number of stars within the observable universe to be $6 \cdot 10^{22}$ and assuming 200 billion stars within the Milky Way there is in total of $\frac{6 \cdot 10^{22}}{200 \cdot 10^{9}} = 3 \times 10^{11}$ galaxies. This implies that:

$$d_{Galaxy} = \left(\frac{3}{4\pi} \cdot \frac{\frac{4}{3}\pi \left(13799000000\right)^3}{3 \cdot 10^{11}}\right)^{\frac{1}{3}} = 2,061,296.8 \, \text{ly} \tag{8.1.17}$$

merely 2.061 Mly between galaxies. We ignore the highest estimates of 2 trillion galaxies since the study have shown that all those galaxies are the snapshots of the earlier universe when dwarf galaxies have yet to merge and form into the galaxies we see today. Either case, the density of the number of habitable galaxies remains fixed:

$$F_{galaxy} = \left(\frac{20,233,126.9}{4,947,525.31}\right)^3 = 68.3950926751 \tag{8.1.18}$$

or

$$F_{galaxy} = \left(\frac{20,233,126.9}{2,061,296}\right)^3 = 945.730568782 \tag{8.1.19}$$

This further reduces the total number of human comparable or more advanced civilization to at least 1 per 15 galaxies, or 1 per 2,837 galaxies.

Criterion	Probability	Number
$N_{current}$		$\leq 1/3$
Habitable galaxies	20.09% or 0.1057%	$\leq 1/4.97$
$\mathbf{N_{galaxy}}$	$\leq \mathbf{1/14.933}$ or $\mathbf{1/2,837.1917}$	

Table 8.1.5: Number of civilizations per galaxies at the current time

8.2   The Model

After we have laid the foundations regarding the number of habitable terrestrial planets within the 5 Gya to 4 Gya window, the chance of emergence of Homo sapiens on those planets, and the concept of background evolutionary rate, deviation, and the Years Ahead against the Background Evolutionary Rate. We can finally introduce our distribution function.

The PDF probability density function for biocomplexity development on earth at *any time period* is represented as a particular instantiation of a *bi-variate exponential log-normal distribution function*, which represents an infinite set of log-normal distribution functions exponentially dependent on time t, whereas the current time is instantiated as $P_{df}(0, x)$, and 100 Myr into the future is instantiated as $P_{df}(-1, x)$, 100 Myr into the past is instantiated as $P_{df}(1, x)$:

$$P_{df}(t, x) = \frac{1}{\sigma\sqrt{2\pi}} \exp\left(-\frac{\left(\ln\left(\left(B_{cs}\right)^t x\right)\right)^2}{2\sigma^2}\right) \tag{8.2.1}$$

The instantiation for $P_{df}(2,x)$, $P_{df}(1,x)$, $P_{df}(0,x)$, $P_{df}(-1,x)$, $P_{df}(-2,x)$ is plotted as:

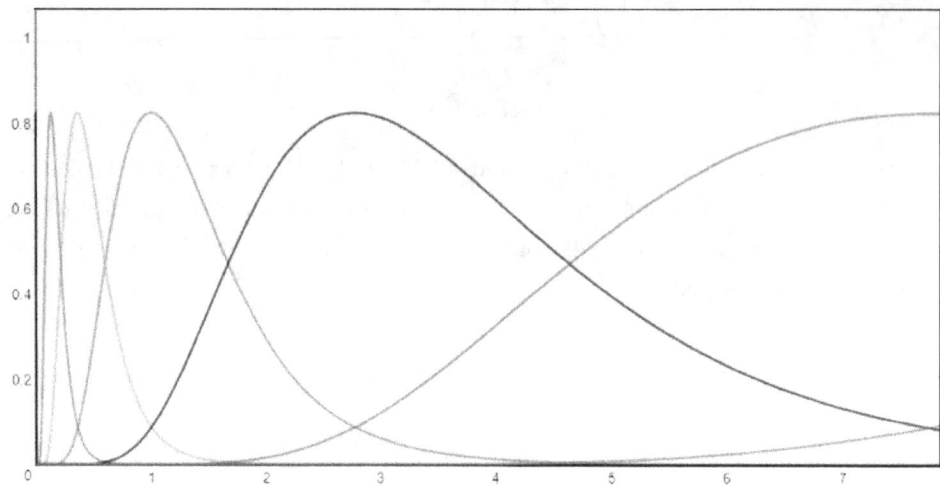

Figure 8.2.1: The instantiation for $P_{df}(2,x)$, $P_{df}(1,x)$, $P_{df}(0,x)$, $P_{df}(-1,x)$, $P_{df}(-2,x)$ from left to right, respectively.

Whereas each time wise instantiated lognormal distribution represents the earths' biocomplexity's lognormal distribution. B_{cs} is the biocomplexity search space BCS, which equals 2.783.

The total number of emerging extraterrestrial civilizations at and exceeds current human development *at the current plus all previous time period* is given by the *CDF cumulative distribution function*:

$$C_{df}(t) = \int_{-5}^{x} P_{lanet}(-s) \left(t_{Win}(B_{cs})^s \int_{18}^{\infty} P_{df}(-s+t,x)\,dx \right) ds \tag{8.2.2}$$

which is an integration of *a range of* PDF $t_{Win}(B_{cs})^s \int_{18}^{\infty} P_{df}(-s+t,x)\,dx$. Whereas each $t_{Win}(B_{cs})^s \int_{18}^{\infty} P_{df}(-s+t,x)\,dx$ is the cumulative number of emerging extraterrestrial civilizations (intelligent species) per planet formed earlier or later than earth within the galaxy with each possible evolutionary scenarios ranging from 500 Myr earlier to x years later relative to earth's own historical development of *a particular time period $t \geq 0$* years earlier.

[1]

Whereas $t_{Win} = 3,607,998$ is the total number of species generated per 100 Myr at the current **BCS** and multiplied by the factor $(B_{cs})^s$ to arrive at the cumulative total number of species generated per 100 Myr on any planet. It is lower at earlier times and higher at current times. The percentage of the emerging intelligent species among all cohorts on any planet *across any time period* t is given by the *R ratio*:

$$R_{atio}(t) = \frac{\int_{18}^{\infty} P_{df}(t,x)\,dx}{\int_{0}^{\infty} P_{df}(t,x)\,dx} \tag{8.2.3}$$

---

[1]To see why $P_{df}(t,x)$ represents the cumulative emergence chance of all previous periods starting at time t check the proof in the appendix.

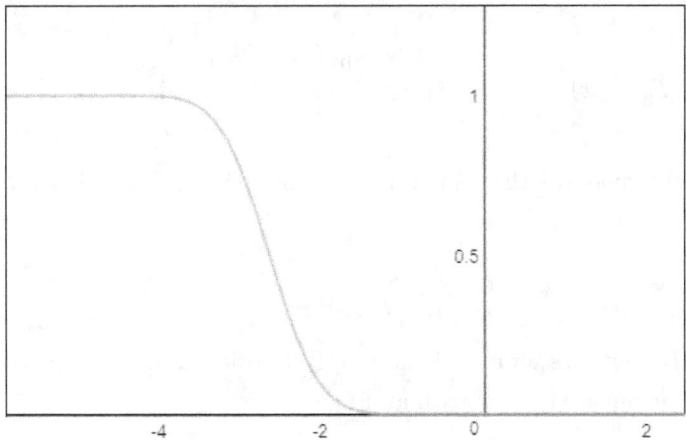

Figure 8.2.2: The R ratio, at current time, < 1 intelligent species emerges out of all cohorts, notices that 400 Myr into the future, the value approaches 100%, indicating all species have achieved equal or greater agility and flexibility as human

the standard deviation σ of the PDF function is mathematically determined to be 0.4811 by satisfying the constraint requirement of:

$$t_{Win} \int_{L_{imit}=18}^{\infty} P_{df}(0,x)\,dx = C_{cumulative} = \frac{1}{16} \qquad (8.2.4)$$

This is the locally observed emergence rate of civilization on earth at the current time.

One can then find the cumulative chance of civilization from all periods across all planets by adjusting the earliest window for habitable planet to 45 Myr ahead of earth so that it sums up to $\frac{1}{3}$:

$$C_{cdf}(0.026) = \int_{-5}^{0.45} P_{lanet}(-s) \left(t_{Win}(B_{cs})^s \int_{18}^{\infty} P_{df}(-s+0.026,x)\,dx \right) ds$$

$$\leq \frac{1}{3} \text{ Civilization} = N_{current} \qquad (8.2.5)$$

That is, the CDF cumulative distribution function for the number of emerging civilizations (emerged species) no later than 2.6 Mya must equals to 1 out of 3 habitable galaxies with 70% confidence.

We could also increase the earliest window to 52.6 Myr ahead of earth to satisfy the constraint requirement of:

$$C_{cdf}(0.001) = \int_{-5}^{0.526} P_{lanet}(-s) \left(t_{Win}(B_{cs})^s \int_{18}^{\infty} P_{df}(-s+0.001,x)\,dx \right) ds$$

$$\leq 1 \text{ Civilization} \qquad (8.2.6)$$

That is, the CDF cumulative distribution function for the number of emerging civilizations (emerged species) no later than 100 Kya must equals to 1 out of 1 habitable galaxies with 99% confidence. We can guarantee we are the only industrial civilization within the Milky Way galaxy for all time periods except the last 100,000 years. It takes at most 100,000 ly for the most recent signals to reach earth from the remotest corners of the galaxy.

The instantiated PDF probability density function (used by both the numerator and the denominator within the R ratio) representing biocomplexity development across all habitable planets at the current time is given by:

$$P_{df}(0,x) = \frac{1}{\sigma\sqrt{2\pi}} \exp\left(-\frac{\left(\ln\left((B_{cs})^0 x\right)\right)^2}{2\sigma^2}\right) \tag{8.2.7}$$

and the maxima representing the mode of the lognormal distribution falls on the point (1, 0.8287) and is determined by satisfying the differential equation:

$$\frac{d}{dx}P_{df}(0,x) = 0 \tag{8.2.8}$$

Whereas x = 17 is lying exactly 17 times away, this is exactly the deviation required for the emergence of civilization compares to typical mammal of today with an EQ of 1.

$$\frac{17}{1} = 17 \tag{8.2.9}$$

Therefore, the integration of the PDF's area satisfying $x \geq 1 + 17$ is the chance of civilizations (emerged species) with development complexity comparable to or greater than human civilizations at the current time:

$$\int_{1+17}^{\infty} P_{df}(0,x)\,dx \tag{8.2.10}$$

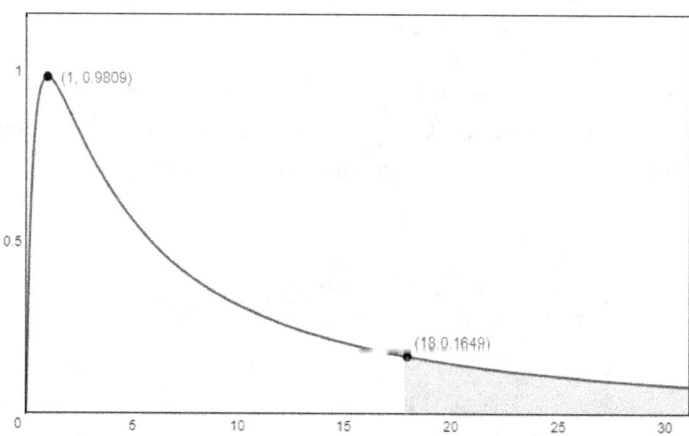

Figure 8.2.3: Not to scale: Illustration of the integration of the PDF's shaded area satisfying $x \geq 18$

Therefore, the emergence is achieved by fixing the integration of the PDF's area satisfying $x \geq 18$ yet the PDF itself varies depending on the time t. For the area satisfying $x \geq 18$ represents those civilizations on habitable earth that achieved higher level of biocomplexity due to earlier emergence, larger surface area, stronger tectonic shifts, more frequent but recoverable mass extinctions rendering more progressive pace of evolution. In general, any planet endowed with larger surface area and more active tectonics strength will yield higher biocomplexity and will satisfy the condition $x \geq 18$ earlier. Thus, higher complexity ensures surpassing the threshold test earlier. As time progresses, PDF's mode shifts right increasingly rapidly and more of PDF's area is satisfied by the condition $x \geq 18$.

Whereas the maxima's/mode x value for any PDF snapshot of the future or past given a fixed **BCS/BER** is defined as:

$$P_{eak}(t) = \frac{1}{(B_{cs})^t} = \frac{1}{(B_{er})^t} \tag{8.2.11}$$

So that $P_{df}(0,x)$'s maxima/mode is given by $P_{eak}(0)$, $P_{df}(1,x)$'s maxima's x is given by $P_{eak}(1)$. It is divided by **BCS/BER** because past maxima occurs to the left of current peak at the rate conforms to the **BCS/BER** transformation. The peaks are more packed in the past and more widely apart in the future.

To represent civilization's position x = 18, one can simply express it as current EQ times the deviation:

$$Limit = D_{eviation} \times 1 \tag{8.2.12}$$

Though one needs to take the distance between 0 and the current mode peak into consideration, so it becomes:

$$Limit = 17 \times 1 + P_{eak}(0) = 18 \tag{8.2.13}$$

and the CDF cumulative distribution function is rewritten as:

$$C_{df}(t) = \int_{-5}^{x} P_{lanet}(-s) \left(t_{Win}(B_{cs})^s \int_{Limit}^{\infty} P_{df}(-s+t,x)\,dx \right) ds \tag{8.2.14}$$

and the R ratio rewritten as:

$$R_{atio}(t) = \frac{\int_{Limit}^{\infty} P_{df}(t,x)\,dx}{\int_{0}^{\infty} P_{df}(t,x)\,dx} \tag{8.2.15}$$

This ratio is later generalized by taking into account the selection factor $G(t)$ for the horizontal displacement of distributions as:

$$R_{atio}(t) = \frac{\int_{Limit}^{\infty} P_{df}(t,x)\,dx}{\int_{-G(t)}^{\infty} P_{df}(t,x)\,dx} \tag{8.2.16}$$

Finally, the number of terrestrial planets that formed between 5 Gya and 4 Gya within the Milky way galaxy that will guarantee the emergence of industrial civilizations and it has to satisfy the specific condition that:

$$1 = R_{atio}(-\infty) \tag{8.2.17}$$

so, eventually all potential planets formed between 5 Gya and 4 Gya gives rise to civilizations given infinite time toward the future.[2]

Within any PDF probability density function $P_{df}(t,x)$, B_{cs} is raised to the power of t, which is the number of years into the future or the past in the unit of 100 million years. If one wants to investigate the number of civilizations (human comparable species) emerged from 200 Mya to 100 Mya within the galaxy, one can apply the following formula:

$$N_{civilization} = C_{df}(1) - C_{df}(2) \tag{8.2.18}$$

If one wants to investigate the number of civilizations (human comparable species) within the Milky way galaxy emerged between now and 100 Myr into the future, one can apply the following formula:

$$N_{civilization} = C_{df}(-1) - C_{df}(0) \tag{8.2.19}$$

The exponentially increasing term B_{cs} complexity transformation does not indicate the progressiveness/advancement of biological evolution of the exoplanet toward a goal of creating human-like intelligent creatures. *It's a curve simply indicating that biological complexity and diversity increases exponentially over time.* With increasing specialization and adaptation to new niches and sustaining on existing biological substrates, the chance of creatures with a large head, manipulative appendages, walked on hind legs increases. Once such creature is able to manipulate nature and pass on their knowledge from generation to generation, it will break through the ecological constraints placed upon them and transform into a post-biological one in a brief time scale compares to the geologic one. It is more accurate to state that the B_{cs} biological complexity transformation term indicates as time progresses, more different varieties of new species is able to fulfill unimaginable niches or non-existent niches at past, though most of the new species, if not of all of them, just adapt to new niches

---

[2]We will later show that it takes far less time to do so

without manipulating, and disrupting their niche and others, unlike human-like creatures will do. We do require that *the bi-variate exponential log-normal distribution function* to satisfy the requirement that:

$$\frac{\int_0^\infty P_{df}(t-1,x)\,dx}{\int_0^\infty P_{df}(t,x)\,dx} = B_{cs} \qquad (8.2.20)$$

This ratio is later generalized taking into account the selection factor $G(t)$ for the horizontal displacement of distributions as:

$$\frac{\int_{-G(t-1)}^\infty P_{df}(t-1,x)\,dx}{\int_{-G(t)}^\infty P_{df}(t,x)\,dx} = B_{cs} \qquad (8.2.21)$$

That is, any future period of 100 Myr's cumulative complexity search space is increased by the B_{cs} complexity transformation factor compares to 100 Mya.

By getting the values for each term (**Deviation, YAABER, BCS,** and **BER**) from earlier calculations and a parameterized value for standard deviation σ, we can plug in the values and solve the distribution.

$$P_{df}(t,x) = \frac{1}{\sigma\sqrt{2\pi}}\exp\left(-\frac{\left(\ln\left((2.7825)^t \cdot x\right)\right)^2}{2\sigma^2}\right) \qquad (8.2.22)$$

$$\sigma = 0.4811$$

$$C_{df}(t) = \int_{-5}^{0.45} P_{lanet}(-s)\left(t_{Win}(B_{cs})^s \int_{Limit}^\infty P_{df}(-s+t,x)\,dx\right)ds \qquad (8.2.23)$$

and the specific condition is satisfied:

$$1 = R_{atio}(-\infty) \qquad (8.2.24)$$

We cross-examine our results by summing up the total probability for the emergence of potentially intelligent industrial civilizations within the Milky Way from the past up to now, and the result needs to be that Homo sapiens led industrial civilization is as rare as 1 per 3 habitable galaxies, based on previous results.

$$t_{galaxy} = C_{df}(0.026)^{-1} = \left(\int_{-5}^{0.45} P_{lanet}(-s)\left(t_{Win}(B_{cs})^s \int_{Limit}^\infty P_{df}(-s+0.026,x)\,dx\right)ds\right)^{-1} \qquad (8.2.25)$$

$$\geq 3 \text{ galaxies}$$

We obtained the average distance between the galaxies by taking the volume of the local supercluster (Virgo) and divide by the number of Milky Way mass (assuming each galaxy has, on average, one Milky Way mass) the cluster possesses and arrived at 4.947 million light years in radius (9.895 million light years in diameter).

$$N_{galaxy} = \frac{M_{supercluster}}{M_{milky}} \qquad (8.2.26)$$

$$\sqrt[3]{\frac{3}{4}\cdot\frac{1}{\pi}\cdot\frac{\left(\frac{4}{3}\pi\cdot\left(\frac{33\cdot10^6\cdot3.26156}{2}\right)^3\right)}{N_{galaxy}}} = 4{,}947{,}525.31 \text{ ly} \qquad (8.2.27)$$

That is it takes a spherical volume with radius 4.947 million light years to host a single galaxy. This is less than the radius requirement for 1 habitable galaxy per 1 $M_{pc}{}^3$. The average distance is revised to $d_{Galaxy} =$ 20,233,126.9 ly by $\sqrt[3]{F_{galaxy}}$ factor to fit the radius requirement for 1 habitable galaxy per 1 $M_{pc}{}^3$ by earlier studies. We then use this result to show that at most 1 emerging civilization at the current time per 3 habitable galaxies, and 1 habitable galaxy per 68 galaxies or 945 galaxies:

$$\left(F_{galaxy} \cdot t_{galaxy} \cdot (d_{Galaxy})^3\right)^{\frac{1}{3}} \geq 25.9234 \text{ Mly} \tag{8.2.28}$$

That it takes a spherical volume with a radius at least 25.92 million light years to host an industrial civilization at our current level of development. Therefore, it takes twice the radius, *51.85 million light years, on average, to reach our nearest industrial neighbor.*

The number of human-like civilizations in the universe at the current time, then, can be estimated to be:

$$t_{alien} = \frac{1}{F_{galaxy}} \cdot C_{df}(0) \cdot \left(\frac{1,379,900}{d_{Galaxy}}\right)^3 \tag{8.2.29}$$

$$\leq 150,823,271.656 \text{ civilizations}$$

at most 150.82 million extraterrestrial industrial civilizations assuming the size of the universe is determined by the amount of distance light is capable of traveling since the start of the universe.

And if one were to consider the comoving distance of all light signals from the most distant corners just as its light is reaching us now, we have:

$$t_{alien} = \frac{1}{F_{galaxy}} \cdot C_{df}(0) \cdot \left(\frac{4,570,000}{d_{Galaxy}}\right)^3 \tag{8.2.30}$$

$$\leq 5.4786570245 \times 10^9 \text{ civilizations}$$

That is a staggering *at most 5.47865 billion extraterrestrial industrial civilizations* within our observable universe. Since we have no knowledge about the size of the universe, and **if the universe is infinitely vast in size, then the number of extraterrestrial industrial civilizations can also be infinitely large in number.**

$$t_{alien} = \frac{1}{F_{galaxy}} \cdot C_{df}(0) \cdot \left(\frac{\infty}{d_{Galaxy}}\right)^3 \leq \infty \tag{8.2.31}$$

If the universe is finitely bounded and based on the current estimate of its size, one can estimate the number of extraterrestrial industrial civilizations to be:

$$t_{alien} \leq \left(\frac{1}{4.4 \cdot 10^7}\right)^3 \cdot 3.621 \cdot 10^6 \cdot 10^{10^{10^{10^{122}}}} \tag{8.2.32}$$

Next, we compute *the earliest arising industrial civilization within a 13.799 billion light years radius to be 119.23 Mya.* Due to the nature of exponential growth in biological complexity, at most 150 million extraterrestrial industrial civilizations have arisen no earlier than the Cretaceous. If one were to assume that the rate of extraterrestrial industrial civilization emergence was uniform as an upper bound speed estimation, that implies *a new industrial civilization emerges every 0.3759 years* in the observable universe within the last 119.23 Myr.

$$t_{galaxy} = C_{df}(1.1923)^{-1} \tag{8.2.33}$$

$$\geq 317,170,419.962 \text{ habitable galaxies}$$

$$\left(F_{galaxy} \cdot t_{galaxy} \cdot (d_{Galaxy})^3\right)^{\frac{1}{3}} \geq 13.8 \text{ Gly} \tag{8.2.34}$$

Next, we compute *the earliest arising industrial civilization within the comoving distance of 46 billion light years radius to be 138 Mya.* Due to the nature of exponential growth in biological complexity, at most 5.478 billion extraterrestrial industrial civilizations have arisen no earlier than the early Cretaceous. If one were to assume that the rate of extraterrestrial industrial emergence was uniform as an upper bound speed estimation, that implies *a new industrial civilization emerges every 0.01179 years (4 days)* in the observable universe within the last 138 Myr.

$$t_{galaxy} = C_{df} \left(1.3852\right)^{-1} \tag{8.2.35}$$

$$\geq 1.1749261981 \times 10^{10} \text{ habitable galaxies}$$

$$\left(F_{galaxy} \cdot t_{galaxy} \cdot \left(d_{Galaxy}\right)^3\right) \geq 46 \text{ Gly} \tag{8.2.36}$$

8.3   Space Occupancy Constraint

Since civilizations are already emerging and assuming that the earliest detectable signal traveled at the light speed or their expanding near the speed of light, then one can compute the total space occupied by all expanding extraterrestrial industrial civilizations. Due to the nature of exponential growth, the majority of the civilizations emerged recently rather than much further in the past. More interestingly, we are interested in predicting the arrival time of all industrial civilizations, or in other words, the time of the first contact, or when the seemly empty universe is filled with civilizations.

To derive the total amount of space occupied by earlier civilizations one can use the formula:

$$\frac{1}{F_{galaxy}} \left(C_{df}\left(t\right) - C_{df}\left(t+d\right)\right) \cdot \left(\frac{137.99}{d_{Galaxy}}\right)^3 \left(\frac{t+\frac{d}{2}}{137.99}\right)^3 \tag{8.3.1}$$

Whereas t is the starting time period one currently examine and d is the total time span. Since no civilization arise before 136 Mya, we can just examine ranges for t between 0 and 1.36. We can approximate d to be ≤ 0.1. If one can take d to be infinitesimally small, one can obtain precise calculation, we substitute $C_{df}\left(t\right) - C_{df}\left(t+d\right)$ with

$$R_{cdf}\left(t\right) = \frac{C_{df}\left(0\right)}{\int_0^8 C_{df}\left(x\right)dx} C_{df}\left(t\right) \tag{8.3.2}$$

or alternatively as:

$$R_{cdf}\left(t\right) = \left|\frac{d}{dt}C_{df}\left(t\right)\right| \tag{8.3.3}$$

so that the rate of civilization emergence between a specified interval can be found as:

$$\int_t^{t+d} R_{cdf}\left(t\right)dt = C_{df}\left(t\right) - C_{df}\left(t+d\right) \tag{8.3.4}$$

We then can formulate our equation as:

$$\frac{1}{F_{galaxy}} \int_t^{t+d} R_{cdf}\left(t\right) \cdot \left(\frac{137.99}{d_{Galaxy}}\right)^3 \left(\frac{t}{137.99}\right)^3 dt \tag{8.3.5}$$

and one finds that the total space occupied by earlier arisen civilization is:

$$R_{atio} = \frac{1}{F_{galaxy}} \int_0^2 R_{cdf}\left(t\right) \cdot \left(\frac{137.99}{d_{Galaxy}}\right)^3 \left(\frac{t}{137.99}\right)^3 dt = 0.12886 \tag{8.3.6}$$

It shows that 12.89% of all space is occupied. If $R_{atio} > 1$, it would indicate that earlier arisen civilizations should have already taken us over. It will contradict with our assumption that we are currently residing in a non-occupied space by extraterrestrials. For such cases, resolving contradiction is required, BCS or σ, the standard deviation for the PDF, or both needs to be revised. If we holds BCS constant, one can decrease the value of σ to satisfy any occupation ratio by earlier emerged and expanding civilizations. As a result, one can

356

see we can use the proportion of total space occupied as a criterion to back determine the rarity of the emergence rate.

We will illustrate all possible scenarios. We add a variable j to our previous equation to represent the selected time interval. When $j > 0$, it represents years into the future in units of 100 Myr. We look for j in which the equation equals 1, when the universe is filled up.

$$R_{atio} = \frac{1}{F_{galaxy}} \int_{-j}^{2} R_{cdf}(t) \cdot \left(\frac{137.99}{d_{Galaxy}} \right)^3 \left(\frac{t+j}{137.99} \right)^3 dt \approx 1.00 \tag{8.3.7}$$

A table of listed results is presented below:

σ	Percent Occupied	1st Contact Time (100 Myr)	Emergence Radius (100 Mly)
0.5091	1	0	0.13863
0.506	0.782511325121	0.0182	0.14892
0.5	0.484424410279	0.0529	0.17095
0.492	0.249327060917	0.1	0.20707
0.484	0.124531181736	0.146	0.25322
0.474	0.0499777555086	0.203	0.33030
0.46	0.0126966712076	0.2827	0.49333
0.45	0.00443859962706	0.34	0.67225
0.44	0.00145213381105	0.397	0.93558
0.43	0.000441916386657	0.4533	1.33240
0.4	0.0000075071630718	0.622	4.52436
0.38	$2.9766734821 \cdot 10^{-7}$	0.7333	12.04126

Table 8.3.1: Contact time across different emergence rate

With the best fit for emergence radius vs. contact time:

$$t_{Contact} = -0.794845 x^{-0.209282} + 1.20359 \tag{8.3.8}$$

The following graph illustrates the separation distance between the nearest neighbor (various possible emergence rate and the radius requirement) and the time required for filling up.

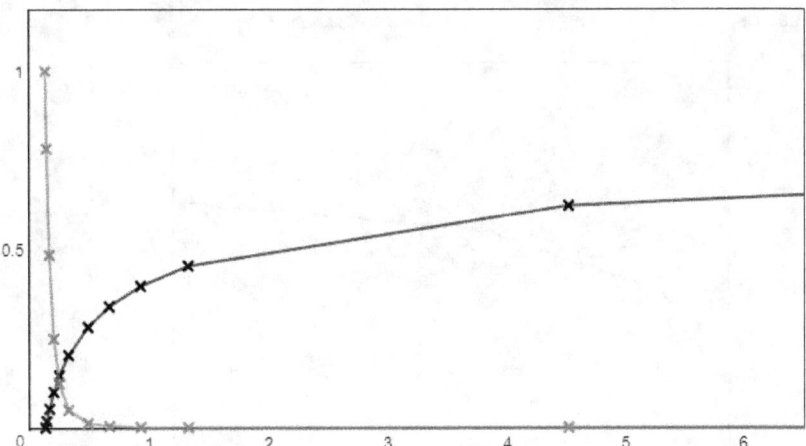

Figure 8.3.1: Contact time between all civilizations assuming the nearest one currently at varying distances away in units of 100 Myr. The horizontal axis represents the current emergence radius in units of 100 Mly. The blue curve indicates the percent of the existing space has been filled up. The black increasing curve indicates the contact time between civilizations in units of 100 Myr.

It shows that even if the nearest civilization is currently at Mega/Giga parsec in distance and much of the universe remains empty, the contact time between civilizations is no more than 150 Myr into the future. That is, the first contact time is not significantly affected by a low emergence rate. Furthermore, the total space occupied by later arising civilizations represented as a percentage out of the total contribution is plotted.

$$\frac{1}{F_{galaxy}} \int_0^2 R_{cdf}(t) \left(\frac{137.99}{d_{Galaxy}}\right)^3 \left(\frac{t+j}{137.99}\right)^3 dt + \frac{1}{F_{galaxy}} \int_{-j}^0 R_{cdf}(t) \left(\frac{137.99}{d_{Galaxy}}\right)^3 \left(\frac{t+j}{137.99}\right)^3 dt \approx 1.00 \quad (8.3.9)$$

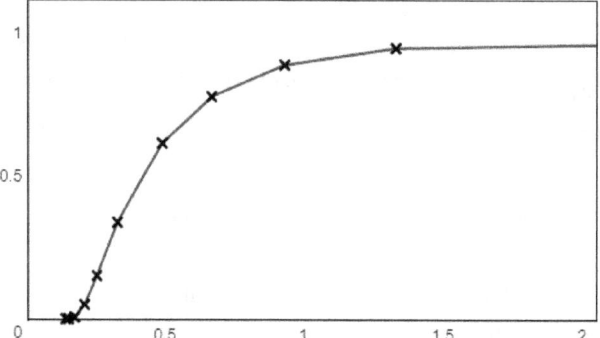

Figure 8.3.2: The vast majority of the space is taking up by later arising civilizations vs earlier ones if the current emergence rate is low. The horizontal axis represents the current emergence radius in units of 100 Mly, the vertical axis represents the 1st contact time in units of 100 Myr.

It is shown that the vast majority of the space of is taking up by later arising civilizations vs earlier ones if the current emergence rate is low. Therefore, one can state that from earth's vantage point, depending on the current and past emergence rate, our first contact could be someone emerged close by and recently or someone further far away and emerged long ago.

Furthermore, a lower BER and BCS raises the lowest emergence rate requirement. With a lower BER and BCS, the evolutionary time window becomes longer and more earlier arising civilization could have filled the void. However, it also guarantee lower rate of change for the future, it will require a longer time for the first future contact. On the other hand, with a higher BER, the evolutionary time window becomes shorter when one seeks into the past and fewer earlier arising civilization could have filled the void. However, if the rate of change is higher for the future, it will take a shorter time for the first future contact because the emergence rate grows faster than other case scenarios.

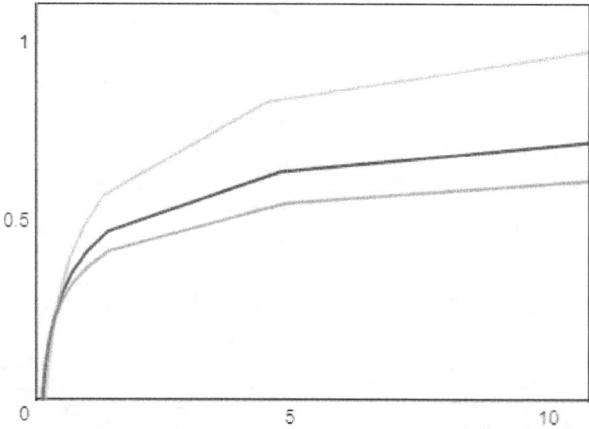

Figure 8.3.3: Contact time of different BER/BCS across different emergence rate (different values of σ), whereas the vertical axis represents the 1st contact times in units of 100 Myr, the horizontal axis represents the current emergence radius in units of 100 Mly. Curves with BER/BCS values from the top to the bottom: 2, 2.783, 3.5

Using the distance to the nearest neighbor and the expected future arrival times, one can also determine the exact BER/BCS required.

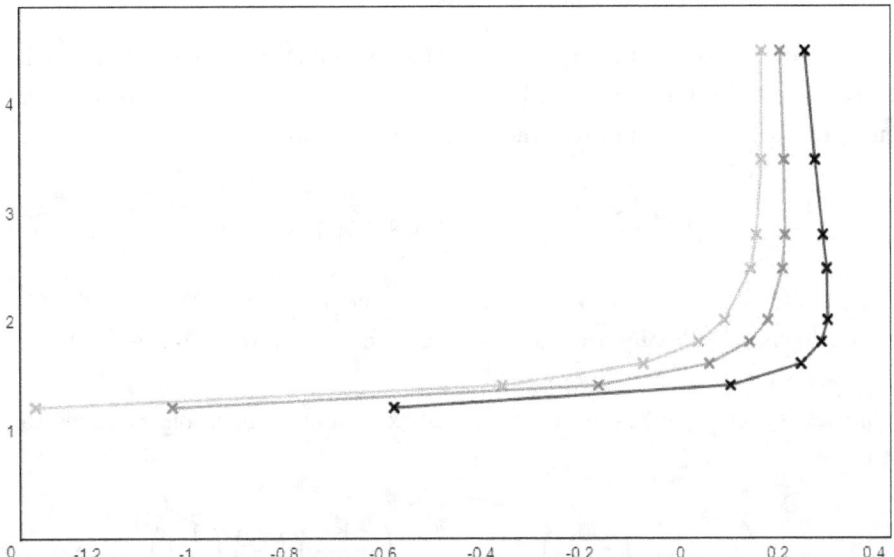

Figure 8.3.4: BER/BCS curves for different arrival times, whereas the vertical axis represents the BCS/BER, the horizontal axis represents the 1st contact times in units of 100 Myr. Curves with σ value from the top to the bottom: 0.484 (18.61 Mly), 0.474 (24.28 Mly), 0.46 (36.27 Mly). Negative values on x-axis denotes arrival time into the past.

The model is based on the assumption that all life on all these planets nearly all reached the stage of bio-complexity similar to mammals, reptiles, and birds observed on earth, and emerging civilizations expand near the speed of light. This time frame can be extended easily assuming the average expansionary speed of all civilizations is merely at a fraction of the speed of light or the emergence of industrial civilization is rarer than we assumed, or one modify the distribution function model, which we will discuss later.

Furthermore, using the distance to the nearest neighbor and the total volume assumed to already been occupied, one can determine the exact BER/BCS required. Alternatively, one can reverse-determine the total volume occupied and the distance to the nearest neighbor. It is shown that, under our current assumption, BER can not fall below 1.69. Otherwise, extraterrestrials would have overtaken us by now. The relationship is plotted:

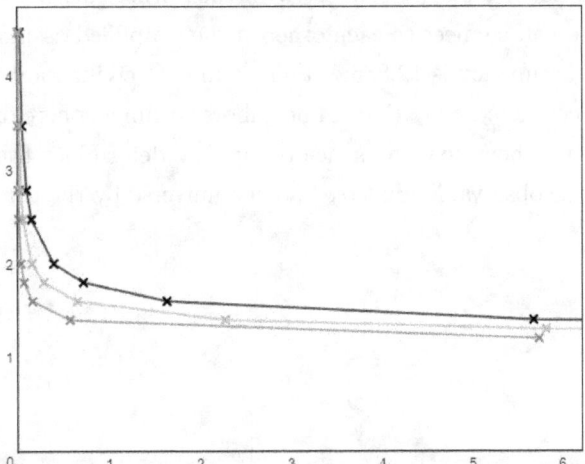

Figure 8.3.5: BCS/BER values for different volumes occupation, whereas the vertical axis represents the BCS/BER, the horizontal axis represents the occupancy ratio. Curves with σ value from the top to the bottom: 0.484 (18.61 Mly), 0.474 (24.28 Mly), 0.46 (36.27 Mly)

Finally, based on the space occupancy constraint of 12.886% of all space have been occupied, and we shall use this value to demonstrate later sections.

$$R_{atio} = 0.12886 \tag{8.3.10}$$

Having fixed the space occupancy, one can find the average expanding radius of expanding industrial civilizations. One take the ratio of total volume currently occupied times the size of the observable universe and divided by the total number of emerged civilizations and takes the inverse cubed:

$$\sqrt[3]{\frac{13799^3 \cdot R_{atio}}{t_{alien}}} = 13,093,821.79 \, \text{ly} \tag{8.3.11}$$

It shows that, on average, emerged civilizations expanding near the speed of light traveled 13.09 Mly in all directions since its first emergence, indicating that on average, each civilization colonized ≤ 16.5 galaxies since its emergence.

Due to the expansion of these spheres, the distance between all expanding civilizations is somewhat closer, and this can be calculated to be:

$$d_{between} = \left((1 - R_{atio}) \cdot \frac{4}{3}\pi \, (1,379,900)^3 \cdot \left(\frac{1}{t_{alien}}\right) \left(\frac{1}{\pi}\right) \left(\frac{1}{\frac{4}{3}}\right) \right)^{\frac{1}{3}} \tag{8.3.12}$$

$$\geq 24.7582808729 \, \text{Mly}$$

In contrasts to a universe with non-expanding civilizations with mean sphere radius of 25.9234 Mly, the radius of one with expanding civilization has shrunk by 1.165 Mly, or 2.33 Mly in contact distance.

Having computed the average size of our sphere of dominance, we then proceed to answer the questions. Out of the existing expanding civilizations, has anyone connected with someone else? In order to answer this question, one takes the fraction of the universe occupied by expanding civilizations to the nth power (meaning the chance of an expanding and emerging civilization find itself within other's occupied space) and multiply by the total number of civilizations in order to find the number of pairs of connected civilizations:

$$t_{alienNconnect} = t_{alien} \cdot (R_{atio})^n \tag{8.3.13}$$

The equation is the absolute upper bound on the number of aliens could have connected with each other since the space occupied by all aliens do not necessarily connect with each other rather forms as pockets and it is assumed in the equation that they do all connect to each other in the simplified assumption.

If we assumed that the total space occupancy is 12.886%, then at most 1 civilization have contacted up to 23 neighbors and no civilizations have contacted more than 24 neighbors within a sphere of radius of 13.799 Billion light years. In the following table we show how the space occupancy determines the maximum the number neighbors ever connected for both the observable universe and the universe by the comoving distance.

σ	Maximum neighbors contacted (Observable)	Maximum neighbors contacted (Co-moving)
0.5091	5,442	6,481
0.506	76	91
0.5	25	30
0.492	13	16
0.484	9	10
0.474	6	8
0.46	4	5
0.45	3	4
0.44	2	3
0.43	2	2
0.4	1	1
0.38	1	1

Table 8.3.2: Max. number of neighbors connected

8.4 The Wall of Semi-Invisibility: Introduction

Having demonstrated that it is quite possible that a significant portion of the universe has already been filled up, but we do not find any detection evidence. The contradiction can be resolved by introducing the concept of the wall of semi-invisibility. The wall of semi-invisibility can be defined as follows, there exists a reciprocal i.e. $(F_{galaxy}C_{df}(t))^{-1}$ for the CDF $F_{galaxy}C_{df}(t)$ we based our model on in which the y-intercept is non-zero and rises sharply when $t > 0$ (into the past).

We now use $C_{df}(t)$ to illustrate the concept of the wall of semi-invisibility by plotting the following equation to show the radius size required to find an earlier arisen extra-terrestrial civilization:

$$d_R(t) = \frac{\left(F_{galaxy} \cdot C_{df}(t)^{-1} \cdot (d_{Galaxy})^3\right)^{\frac{1}{3}}}{10^4} \tag{8.4.1}$$

Some may argue that we need a factor 1.26 because by simply extending one's lookout radius to one's emergence radius, the chance of observing any extraterrestrial industrial civilization is still 0. Therefore, further extending our outlook radius to 1.26, we have included at least the spherical volume equivalent to one more extraterrestrial civilization (excluding the emergence volume requirement for ourselves), this is mathematical expressed as:

$$\frac{\frac{4}{3}\pi \cdot 1.26^3}{\frac{4}{3}\pi \cdot 1^3} - \frac{4}{3}\pi \cdot 1^3 \approx 1 \tag{8.4.2}$$

However, this additional look ahead is only applicable to the closest round of emergence detection. We ultimately expressed this wall in terms of a series (check the end of the section), and the first term $\left(\frac{0.315}{d_R(0.63)}\right)^3$ implies an earth like civilization should emerged 31.5 Mya on average within the most recent round of 100% guaranteed civilization emergence within $d_R(0.63)$. That is, the most recent emergence radius defined in terms of the emergence at 63 Mya. Since earth's case happened much later at the current time, so the term becomes $\left(\frac{0}{d_R(0.63)}\right)^3$ and a factor of 1.26 is added to the denominator as $\left(\frac{0}{1.26d_R(0.63)}\right)^3$, but the result is still $\left(\frac{0}{1.26d_R(0.63)}\right)^3 = 0$. One can still assume that civilization at the current round arises 31.5 Mya and adding an extra factor of 1.26 to the denominator. However, any case does not significantly alter the original value, since this first term contributes a minimal amount compared to the rest. For the sake of simplicity, we assume that the factor 1.26 is not used.

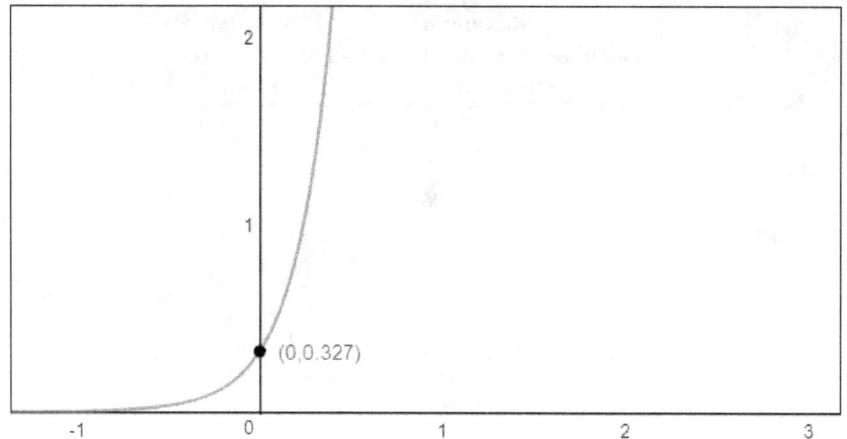

Figure 8.4.1: Nearest civilization detection

The nearest current existing arising extra-terrestrial civilizations is a non-zero distance away, yet its earliest detectable signature is still needed to reach earth by the distance in light years involved. One can observe from the graph that currently the intercept posits at $(0, 0.327)$. The intercept with the y-axis is the distance of the closest arising extra-terrestrial industrial civilization to earth in the present time. Then, one positive detection of the nearest arising extra-terrestrial industrial civilization is located at least 32.7 million light years away at the present time. It will take another 32.7 million years for the earliest signal of such civilization to reach us. To seek civilization evolved even earlier that could have signals reached us by now, one can ideally accomplish this task by looking ahead at a greater distance. (However, we can not gain any bonus points now by looking farther deep into the space because the probability of earlier civilization arising, in geologic time scale, is dependent on time, and shaped by the complexity transformation factor. It is not governed by an uniform distribution which can only be applied and approximated within a very short timescale peeking into the past where changes based on the complexity transformation factor are negligible, so as one look further back in time the chance of civilization arising decreases). However, the rising curve is so steep that in order to find a civilization evolved d years ahead relative to us, one has to look at a sphere size with radius $d+t$, where t is much greater than d in light years plus the non-zero distance to our nearest industrial civilization neighbor, implying that our search space will consist of mostly of regions where signals emitted before the rise of such a civilization.

Dividing the previous equation by t, we find the ratio $(d+t)$ to t and plot the graph:

$$\frac{d_R\left(t\right)}{t} = \frac{\left(F_{galaxy} \cdot C_{df}\left(t\right)^{-1} \cdot \left(d_{Galaxy}\right)^3\right)}{10^4 t} \tag{8.4.3}$$

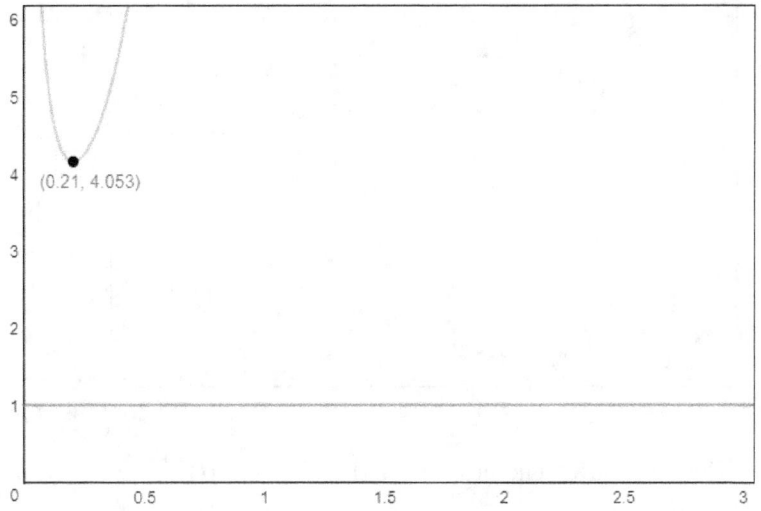

Figure 8.4.2: Threshold test for the wall of semi-invisibility

One can see that as long as $\frac{(d+t)}{d} > 1$, $t > 0$. That is, in order to seek a civilization evolved d years earlier, one has to look some positive value t light years in addition to d light years in distance. The graph shows that $d+t$ first decreases up to 4.053 times the size of d at 20.8 Myr and then increases quickly thereafter. It also means that by looking further into the distant region, the probability of additional extraterrestrial signal detection is at least $(4.053)^3$ or 66.578 times harder than finding signals by extending one's lookout distance and looking for earlier arisen civilizations. If we have a small chance of finding signals in the most likely case, then the chance is even slimmer by looking further into the distance and the past. *Thus, the wall of semi-invisibility exists.*

It is called the wall of semi-invisibility because given by the distributive probability, a sphere size of radius $d+t$ will guarantee to find the an earlier arisen civilization d years before of our present time. However, it is still possible that such a civilization can be found much closer with a distance less than $d+t$, or even significantly less than d, however, if $\frac{d}{(d+t)}$ is close to 0; then, the chance of observing such civilization is minimal if non zero. The minimum sphere size radius requirement guaranteeing the detection of 1 earlier arisen civilization at any past time periods can also be accurately approximated, if one plots the different values of $C_{df}(t)^{-1}$.

Then, the approximate curve fitting for the list of values for correspondingly $C_{df}(0.1)^{-1}$=10 Mya, $C_{df}(0.2)^{-1}$=20 Mya, $C_{df}(0.3)^{-1}$=30 Mya, $C_{df}(0.4)^{-1}$=40 Mya, $C_{df}(0.5)^{-1}$=50 Mya, $C_{df}(0.75)^{-1}$=75 Mya, and $C_{df}(1)^{-1}$=100 Mya can be best fit as an double exponential function:

$$d_R(t) \approx ab^{(x+v)^{c^x}} + fx + z \qquad (8.4.4)$$

$$a = 0.0032485, \quad b = 119.2134, \quad c = 1.07239 \qquad v = 0.941865, \quad z = 0.033692, \qquad f = 0.05010$$

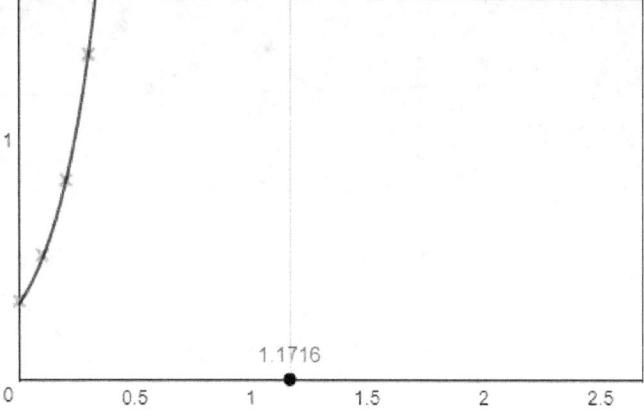

Figure 8.4.3: The best fit for the CDF $C_{df}(t)^{-1}$

This is the best fit for the CDF. Using the approximate curve fit can save computational time and easier to manipulate under the limit test for convergence. Then, the probability density function of all arising extra-terrestrial industrial civilization arising is:

$$\left(\frac{t}{d_R(t)}\right)^3 \tag{8.4.5}$$

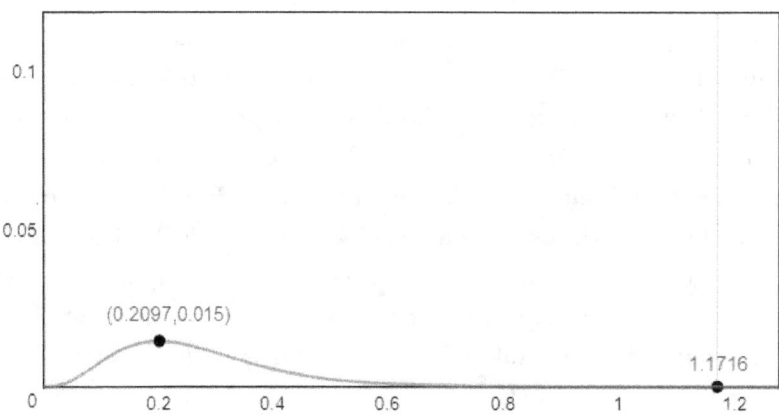

Figure 8.4.4: PDF of extraterrestrial detection

Assuming current total space occupancy of 12.886%, whereas the chance of observing extraterrestrial peaks at 20.97 million light years away at merely 1.5%. The cumulative total chance of detection across all time periods is then just 0.449720% This is the absolute maximum attainable detection chance since the emergence of expanding civilization can be much rarer than 1 in 3 habitable galaxies at the current time.

$$100 \int_0^\infty \left(\frac{t}{d_R(t)}\right)^3 dt = 0.44972\% \tag{8.4.6}$$

A closer examination reveals that the chance above can only serve as the lower bound. $\left(\frac{t}{d_R(t)}\right)^3$ is actually the *instantaneous* emergence chance of civilization. We currently assume that at every point along the past event horizon, manifested as the lookout radius, there is a 100% chance of civilization emergence if it *does appear* within the lookout radius (of course also within the emergence radius). However, it takes a range of distance in light years and a range of time periods in years from the past event horizon to guarantee 1 additional civilization emergence for any well-defined emergence radius; so that every lookout point along the past event horizon has a much smaller chance of emergence < 1. For example, it takes a cumulative chance of 27 Mly in distance from earth and 27 Myr in time into the past within the past light cone to guarantee 1 civilization emergence within a well defined emergence sphere.

$$\int_0^{0.27} \frac{1}{C_{df}(0.27)} R_{df}(t)\, dt = 1 \tag{8.4.7}$$

So the average chance of detection of 1 civilization within 27 Mly becomes:

$$A_{vg} = \frac{1}{(0.27 - 0)} \int_0^{0.27} \left(\frac{t}{d_R(t)}\right)^3 dt < 1 \tag{8.4.8}$$

Since the cumulative area of detection chance divides the width yields the height, which is the average chance of a civilization detection within this period. It can be more precisely represented as the weighted chance of a civilization detection:

$$A_{vgWeighted} = \int_0^{0.27} \frac{1}{C_{df}(0.27)} R_{df}(t) \left(\frac{t}{d_R(0.27)}\right)^3 dt < 1 \tag{8.4.9}$$

Notice that $\left(\frac{t}{d_R(t)}\right)^3$ is changed to $\left(\frac{t}{d_R(0.27)}\right)^3$, so that a 50% civilization detection chance, say, at $\frac{0.27}{2} = 0.135$, or 13.5 Mya should be derived based on the emergence radius from 27 Mya instead of its detection distance at 13.5 Mya. This may feel counter-intuitive. However, if the civilization detection chance is defined based on its detection distance, say $\left(\frac{0.135}{d_R(0.135)}\right)^3$, then it is no longer a 50% civilization detection chance, rather 0% civilization detection chance since it takes some cumulative years for the emergence and detection to resume 50%.

If one considers that only signals from within 6.266 Gly is ever reachable due to expansion of the universe, it takes only 8 steps of doubling at looking into the past to reach 1 civilization emergence per 6.266 Gly radius. However, the doubling interval's upper and lower limit values changes according to parameter changes and manual fine-tuning is a laborious process. Fortunately, for exponentially decreasing emergence, each round requires approximately evenly spaced interval, so one can approximate all rounds as:

$$s_{vis} = \sum_{n=0}^{\infty} \frac{1}{C_{df}(0.27 \cdot (n+1))} \int_{0.27 \cdot n}^{0.27 \cdot (n+1)} R_{df}(t) \left(\frac{t}{d_R(0.27 \cdot (n+1))}\right)^3 dt \tag{8.4.10}$$

When galaxy average distance remained fixed, an alternative approach is introduced by substituting the emergence radius per each doubling with the variable itself:

$$v_{is} = \int_0^{\infty} \frac{1}{C_{df}(t)} R_{df}(t) \left(\frac{t}{d_R(t)}\right)^3 dt \tag{8.4.11}$$

in which any round performs in general as the taking half of the emergence radius.

$$\int_0^{0.27} \frac{1}{C_{df}(t)} R_{df}(t)\, dt \approx \int_0^{0.27} \frac{1}{C_{df}\left(\frac{(0.27+0)}{2}\right)} R_{df}(t)\, dt < 1 \tag{8.4.12}$$

So that:

$$\int_0^{0.27} \frac{1}{C_{df}(t)} R_{df}(t)\, dt < \int_0^{0.27} \frac{1}{C_{df}(0.27)} R_{df}(t)\, dt \tag{8.4.13}$$

and it turns out that visibility is the occupancy ratio:

---

[3] for visibility $\int_0^{0.27} \frac{1}{C_{df}(0.27)} R_{df}(t) \left(\frac{t}{d_R(t)}\right)^3 dt$, it means $\frac{1}{m}\sum_{n=0}^{m} \left(\frac{n}{d_R(n)}\right)^3 \approx \left(\frac{\frac{0.27}{2}}{d_R\left(\frac{0.27}{2}\right)}\right)^3$. for vis-

ibility $\int_0^{0.27} \frac{1}{C_{df}(0.27)} R_{df}(t) \left(\frac{t}{d_R(0.27)}\right)^3 dt$, it means $\frac{1}{m}\sum_{n=0}^{m} \left(\frac{n}{d_R(0.27)}\right)^3 \approx \left(\frac{0.27}{d_R(0.27)}\right)^3$. for visibility

$\int_0^{0.27} \frac{1}{C_{df}(t)} R_{df}(t) \left(\frac{t}{d_R(t)}\right)^3 dt \approx \int_0^{0.27} \frac{1}{C_{df}\left(\frac{0.27}{2}\right)} R_{df}(t) \left(\frac{t}{d_R(t)}\right)^3 dt$, it means $\frac{1}{m}\sum_{n=0}^{m} \left(\frac{\frac{0.27}{2}}{\frac{d_R\left(\frac{0.27}{2}\right)}{d_R\left(\frac{0.27}{2}\right)} \times d_R\left(\frac{0.27}{2}\right)}\right)^3 \approx \left(\frac{\frac{0.27}{2}}{d_R(0.27)}\right)^3$.

$$v_{is} = R_{atio} \tag{8.4.14}$$

This is no-surprise since $\frac{1}{C_{df}(t)} = \frac{1}{F_{galaxy}} \left(\frac{d_R(t)}{0.1185} \right)^3$, therefore:

$$v_{is} = \frac{100}{F_{galaxy}} \int_0^\infty R_{df}(t) \left(\frac{d_R(t)}{0.1185} \right)^3 \left(\frac{t}{d_R(t)} \right)^3 dt$$

$$= \frac{100}{F_{galaxy}} \int_0^\infty R_{df}(t) \left(\frac{1}{0.1185} \right)^3 \left(\frac{t}{1} \right)^3 dt \tag{8.4.15}$$

The emergence radius canceled out, what remains is the number of all possible civilizations emerged within an unit radius of 100 Mly and their total traveled distance/expanded sphere of influence from each time period within this 100 Mly radius. This is the occupancy ratio. [4]

the occupancy and visibility can further be expressed as 2 series:

The occupancy can be expressed as:

$$\sum_{n=1}^m 2^{m-n} \left(\frac{0.63(n-1) + 0.63 \cdot 0.5}{d_R(0.63 \cdot m)} \right)^3 \tag{8.4.16}$$

In this example, 0.63 is every 63 Mly in distance from earth and every 63 Myr in time into the past within the past light cone to guarantee 1 civilization emergence within a well defined emergence sphere. This value can change as the distribution placement pattern changes. $0.63 \cdot 0.5 = 0.31$ is the average distance traveled (31 Mly) by all civilizations within this well defined emergence sphere. The total distance traveled is further increased depends on its first emergence date. $d_R(0.63 \cdot m)$ is the emergence radius given a number of well defined emergence sphere rounds. For example, occupancy ratio for up to 5 rounds of 0.63:

$$\frac{100}{F_{galaxy}} \int_0^{0.63 \cdot 5} R_{cdf}(t) \cdot \left(\frac{137.99}{d_{Galaxy}} \right)^3 \left(\frac{t}{137.99} \right)^3 dt \tag{8.4.17}$$

which requires 5 rounds of summation so $m = 5$:

$$16 \left(\frac{0.31}{d_R(3.15)} \right)^3 + 8 \left(\frac{0.63 + 0.31}{d_R(3.15)} \right)^3 + 4 \left(\frac{1.26 + 0.31}{d_R(3.15)} \right)^3 + 2 \left(\frac{1.89 + 0.31}{d_R(3.15)} \right)^3 + 1 \left(\frac{2.52 + 0.31}{d_R(3.15)} \right)^3 \tag{8.4.18}$$

Notices that the number of emerging civilization within a given volume size per round grows exponentially, and the average distance traveled per expanding civilization is smaller than its first possible emergence date by 31 Mly. This is true because between the first possible emergence and later 100% emergence, many have arisen in between and the weighted travel distance is the average of the two extremes.

and the visibility as:

$$\sum_{n=1}^m \left(\frac{0.63(n-1) + 0.31}{d_R(0.63 \cdot n)} \right)^3 \tag{8.4.19}$$

Which is almost the same as before except the emergence radius is re-defined per each round and the factor is removed. Since the average 50% emergence chance of a civilization occurs before it reaches 100%, the average detection/visibility of any civilization should occur midway between its first and 100% emergence, just as the occupancy ratio, and we have:

$$\frac{\sum_{n=1}^m \left(\frac{0.63(n-1)+0.31}{d_R(0.63 \cdot n)} \right)^3}{\sum_{n=1}^m 2^{m-n} \left(\frac{0.63(n-1)+0.63 \cdot 0.5}{d_R(0.63 \cdot m)} \right)^3} \approx 1 \tag{8.4.20}$$

---

[4]Alternatively, one can also interpret the expanded sphere of influence as its visibility.

Therefore, the visibility should always be 100% of the occupancy ratio. In our current example, it is 12.886%. One can also see that:

$$E = \frac{(d_R(0))^3}{(d_R(t))^3} \tag{8.4.21}$$

$$\left(\frac{t}{d_R(t)}\right)^3 = \left(\frac{t}{d_R(0)}\right)^3 \cdot E \tag{8.4.22}$$

That is, the current signal detection chance at any lookout distance is the lookout spherical volume size divided by (in proportion) to the the current time emergence sphere with a radius size of 32.7 Mly multiplied by the emergence rate of the past divided by (relative to) the current emergence rate that requires a spherical volume size with a radius of 32.7 Mly observed from earth.

Finally, we want to stress that we assume the light travel distance and angular diameter distance is approximately one of the same. That is, $d_T(z) \approx d_A(z)$. This is a valid assumption because signals traveled at such low redshift $z \leq 0.017$ and close distance to earth, the light travel distance is not significantly distorted by the expansion of spacetime. The error rate is less than 0.8568%; therefore, the delay of extraterrestrial's light signal's arrival is at the most by 0.8568%. [53]

Very lastly, we can address the question we raised earlier. If a significant portion of the space already been occupied, why we still found no detection? Since we have already illustrated the presence of the wall of semi-invisibility, there is a significant chance that the emergence of any civilizations occur outside of our lookout radius. For example, a civilization emerged at 60 Mly away at 50 Mya and have been carried out its own expansion at the speed of light ever since. First of all, their earliest emergence signal will takes another 10 Mly to reach us. Secondly, since its emergence, the current edge of its sphere of influence have expanded to just 10 Mly away from earth. However, this signal has not yet reached us, we are currently receiving the status signal at 10 Mly away from 10 Mya. Therefore, the universe appears empty but in fact it is well-occupied. One may also think the visibility is too high given 100% correlation between visibility and occupancy. But consider this. Assuming the occupancy has reached 50%, and the visibility has reached 50%. There is a 50% chance we believe that the universe is 100% empty although in fact it is 50% occupied.

8.5 The Wall of Semi-Invisibility: Proof

Having demonstrated the concept of the wall of semi-invisibility, we shall further refine our argument by presenting a proof. We start by imagining one were to parameterize the PDF, so it further shifts to the right such as the plot below:

$$d_R(t) = \frac{1}{30} \cdot \frac{\left(C_{df}(t - 0.5)^{-1}(d_{Galaxy})^3\right)^{\frac{1}{3}}}{10^4} \tag{8.5.1}$$

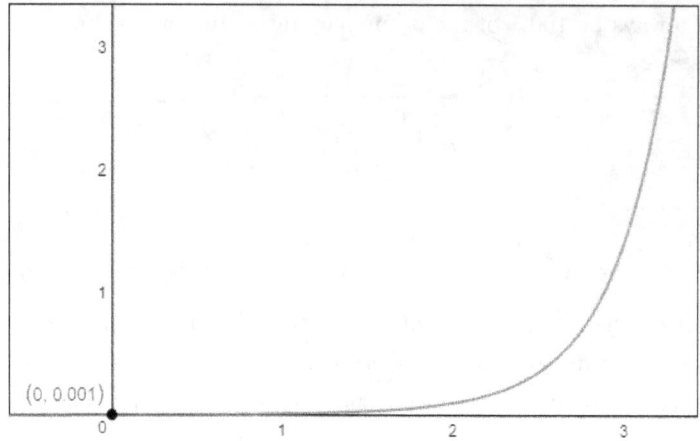

Figure 8.5.1: A hypothetical case for nearest civilization much closer

But it is obvious that the function cannot be right shifted so that the intercept occurs at a much closer distance to earth, for example at (0, 0.001). In such a hypothetical scenario, the closest arising industrial civilization is located at 100,000 light years away at the present time (so it would take another 100,000 years for the signal to reach us). Since the curve remains almost flat (the probability of arising civilization does not decrease as one traces further back in time, the probability of arising extra-terrestrial industrial civilization then approximately follows a uniform distribution), one expects to find earlier arising civilizations by simply looking ahead in greater distance, say 140,000 light years to the edge of the Milky Way. A civilization formed earlier should already exist given the greater sample size to look at. This is also expressed mathematically by the plots of $\frac{(d+t)}{d}$,

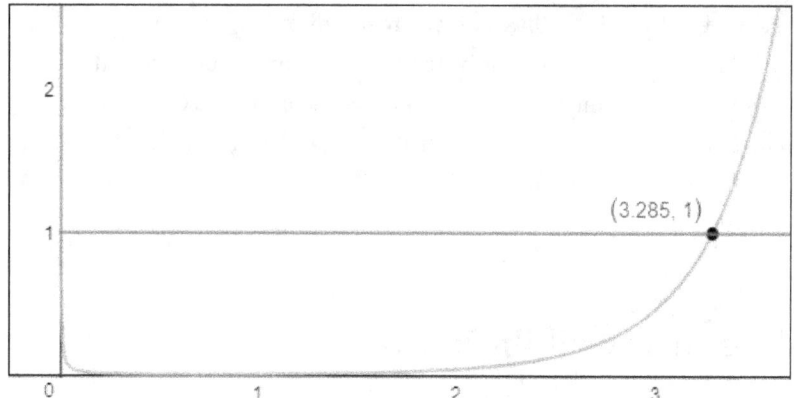

Figure 8.5.2: Threshold test fails for the hypothetical case

We find that current time up to 328.5 Mya into the past, $\frac{(d+t)}{d} < 1$. This implies as one looks further away into space, more signals of extraterrestrial civilization should be detectable. But this is contradicted by our current observation. We have pretty much ruled out the signatures of extra-terrestrial industrial civilization in our galaxy.

First, a proof by contradiction is outlined below regarding the above principle. Assume in a universe where one among all habitable planets transform into an industrial civilization every n years and the rate of transformation stays constant, that is, the probability is a uniform distribution given that every time period has an equal chance of transforming just one planet, it does not transform more or less.

8.5.1 Base Case:

We first establish three criteria that needed to be fulfilled:

1. The emergence rate has to follow an uniform distribution from a temporal perspective.

2. The emergence rate has to follow an uniform distribution from a spatial perspective.

3. The density distribution of planets within the universe follow an uniform distribution.

We will show that it is impossible to fulfill the three criteria at the same time, as result, it is impossible for the emergence rate to follow an uniform distribution from both a spatial and temporal aspect.

Assume that within a volume with a radius of r = 1 contains one transforming sample and assumed it is located at the edge of the radius of 1 light year, so with a radius of r = 2, the volume has grown by 8 folds, within this same volume, one can hold 8 transforming samples.

If the emergence rate follows an uniform distribution from a temporal perspective, as one looks back in time, there is 1 transforming planet per every time period. Then we know that at a radius of r = 1, the planet's transforming signal at the current time (current time period) will reach us next year. Now extending the radius to 2, we know that one of the transforming planet's signal from 1 year ago (last time period) also will reach us next year from 2 light years away. So we will receive the signal confirmation from two of the planets at various distances from earth at the same time. However, this temporal placement of emergence violates uniform distribution of emergence from the spatial perspective (criterion 2).

If the emergence rate follows an uniform distribution from a spatial perspective, the rate of emergence is fixed at 1 planet per every 1 light year radius in each layer. Then we know that at a radius of r = 1, the planet's transforming signal at the current time (closest layer) will reach us next year. Extending the radius to 2, the 7 remaining planets' transforming signal from 1 year ago (second closest layer) also will reach us next year from 2 light years away. So we will receive the signal confirmation from all of the planets at various distances from earth at the same time. However, this led to a contradiction because the assumption is that 8 planets take 8 years to transform not 2 years, violating uniform distribution from the temporal aspect as we stated earlier (criterion 1). If all 8 transformed, as one looks back in time, the number of emergence detection increases and does not stay constant.

To resolve this contradiction, we have two solutions.

1. None of the 6 remaining planets have evolved into an advanced civilization prior to last year (they will transform in the future)

2. They have already arisen 7, 6, 5, 4, 3, and 2 years ago respectively. (they transformed in the past)

Both solutions are equally likely until we try to fit these solutions into real observation.

Since we assumed that advanced life has existed long before man, then we shall use 2) as our solution to resolve the contradiction. That is, if advanced life evolved according to an uniform distribution from the temporal aspect, life evolved on the remaining 6 planets 2 to 7 years ago located 2 light years away, and their signal should have already reached us between now and 5 years ago 1 year apart from each other. We should find significant evidence regarding their existence. However, this is contradicted by our current observation. There is no evidence of their existence. Moreover, such emergence pattern still violates an uniform distribution from a spatial perspective. The uniform distribution from the spatial perspective predicts 7 emerging civilizations per layer all in one time period.

To keep up with our observation and continually assuming life on the 6 planets arose in the past and conforming to criteria 1 and 2, the universe needs just one new appearing transforming planet readily to be visible from the past light cone's event horizon as r (the lookout radius) increases, the probability of planet with industrial civilization rising has to be decreased by a factor of:

$$\frac{1}{\left(\frac{\frac{4}{3}\pi(x)^3}{\frac{4}{3}\pi(1)^3}\right)} = \frac{1}{x^3} \tag{8.5.2}$$

Whereas x light years in radius, there exists $\frac{\frac{4}{3}\pi(x)^3}{\frac{4}{3}\pi(1)^3}$ planets occupying a volume space of $\frac{4}{3}\pi(x)^3$ and only 1 of them (assuming the spatial volume for one transforming planet is $\frac{4}{3}\pi(1)^3$) emerged x years ago.

From this, we can extrapolate that the probability of an alien civilization formation in the universe falls at least by a factor of

$$\frac{1}{x^3} \tag{8.5.3}$$

as one further traces back in time. Each of six planets has to be placed r = {3, 4, 5, 6, 7, 8} light years respectively from earth and arose 2 to 7 years ago respectively and to expect all signals reaching us next year. This would satisfy uniform distribution from both the temporal and the spatial aspect. Each time period is transforming one layer away from earth, and each layer contains only 1 planet. Therefore, uniform distribution from spatial perspective predicts 1 emerging civilization per layer; and uniform distribution from temporal perspective predicts 1 emerging civilization per time period. Unfortunately, this still led to a contradiction because the remaining 6 planets have to be placed within 2 light years from earth based on our initial assumption. Otherwise, the density of the universe is non-uniform and the density is concentrated around earth, violating criterion 3.

Therefore, we are forced to recognize that we have to take proposition 1) as our solution while we keep the solution just proposed. As a result, none of the 6 remaining planets have evolved into an advanced civilization prior to last year. By adopting this final solution, we also abandoning criterion 2. The emergence rate does not follow an uniform distribution from a spatial perspective.

Moreover, 6 remaining planets are not transforming at the rate of 1 per year into the future as an uniform distribution would predict. It is assumed, based on the model, 1 planet already transformed a year ago located at 2 ly away when the emergence rate density reaches 1 transforming sample per 8 samples $\frac{1}{(2)^3} = 12.5\%$ per a distance of r =2 ly, but all remaining 6 planets located at 2 ly away must completely transforming into an industrial one now as the emergence rate density reaches 1 transforming sample $\frac{1}{(1)^3} = 100\%$ per a distance of r =1 ly at the current time (in other words, 100% total emergence at the current time). Therefore, we also abandoning criterion 1. The emergence rate does not follow an uniform distribution from a temporal perspective. At the end, we are only adhering to criterion 3, the the density of planets within the universe is uniformly distributed.

Furthermore, we have both empirically confirmed and demonstrated in the earlier section that it takes a radius of a significantly larger size d light years to host one emerging civilization at the current time. (0 years ago). As a result, the current emergence rate is $\frac{1}{d^3}$ and its emergence rate density per 1 ly radius is strictly less than

$$\frac{1}{d^3} < \frac{1}{1^3} \tag{8.5.4}$$

and the majority of the transforming samples take place not now, but is further delayed up to d years into the future. That is, the rest of the samples supposed to transform currently are instead transforming between now and d years into the future.

Furthermore, in order to keep up with the observation of having just one appearing sample readily visible from the past light cone as the lookout radius increases assuming the radius > 2, the probability of planet with industrial civilization rising in the future has to be increased by a factor of x^3 correspondingly to be on the same signal detection curve. Therefore, we have shown that most of the civilization not only emerges in the future but the probabilistic emergence follows a non-uniform distribution from both spatial and temporal perspective, and the vast majority of the planets are transforming closer to d years into the future.

8.5.2 Inductive Step:

In general, if one were to inspect a sphere with radius x, then f(x) more planets have to evolve into industrial civilization in the future as stated in the equation below:

$$f(x) = x^3 - x \tag{8.5.5}$$

We can then run a proof by induction. Assume that f(k) planets with a radius k so that

$$f(k) = k^3 - k \tag{8.5.6}$$

is true, then

$$f(k+1) = (k+1)^3 - (k+1) \tag{8.5.7}$$

must be true.

From empirical observation, we find that increasing the radius by 1 increases the number of habitable planets with future emerging civilization by $3k^2 + 3k$.

Radius	Yet to Emerge	Difference	Formula	Total Planets	Emerging Planets
1	0		$3 \cdot 0^2 + 3 \cdot 0$	1	1
2	6	6	$3 \cdot 1^2 + 3 \cdot 1$	8	2
3	24	18	$3 \cdot 2^2 + 3 \cdot 2$	27	3
4	60	36	$3 \cdot 3^2 + 3 \cdot 3$	64	4
5	120	60	$3 \cdot 4^2 + 3 \cdot 4$	125	5
6	210	90	$3 \cdot 5^2 + 3 \cdot 5$	216	6
7	336	126	$3 \cdot 6^2 + 3 \cdot 6$	343	7
8	504	168	$3 \cdot 7^2 + 3 \cdot 7$	512	8

Table 8.5.1: $3k^2 + 3k$

$$\Rightarrow f(k) + 3k^2 + 3k \tag{8.5.8}$$

$$\Rightarrow (k^3 - k) + 3k^2 + 3k \tag{8.5.9}$$

$$\Rightarrow k^3 + 3k^2 + 2k \tag{8.5.10}$$

$$\Rightarrow k^3 + 3k^2 + 3k + 1 - (k+1) \tag{8.5.11}$$

$$\Rightarrow (k+1)^3 - (k+1) = f(k+1) \tag{8.5.12}$$

$$Q.E.D$$

However, we can not simply run induction by just add 1 step because as k becomes large, adding the radius by 1 only increase 1 habitable planet which means zero habitable planets yet to emerge.

$$\lim_{x \to \infty} \frac{(x+1)^3}{x^3} \tag{8.5.13}$$

$$= \lim_{x \to \infty} \frac{\left(x^3 + 3x^2 + 2x + 1\right)^3}{x^3} \tag{8.5.14}$$

$$= \lim_{x \to \infty} 1 + \frac{3}{x^1} + \frac{2}{x^2} + \frac{1}{x^3} = 1 \tag{8.5.15}$$

So we will run induction by adding k steps.

Induction by adding k steps. Assume that f(k) planets with a radius k so that

$$f(k) = k^3 - k \tag{8.5.16}$$

is true, then

$$f(k+k) = (2k)^3 - (2k) \tag{8.5.17}$$

must be true.

From empirical observation, we find that increasing the radius by k increases the number of habitable planets with future emerging civilization by $7k^3$-k.

Radius	Yet to Emerge	Difference	Formula	Total Planets	Emerging Planets
1	0		0	1	1
2	6	6	$7 \cdot 1^3 - 1$	8	2
4	60	54	$7 \cdot 2^3 - 2$	64	4
8	504	444	$7 \cdot 4^3 - 4$	512	8
16	4,080	3576	$7 \cdot 8^3 - 8$	4,096	16
32	32,736	28656	$7 \cdot 16^3 - 16$	32,768	32
64	262,080	229344	$7 \cdot 32^3 - 32$	262,144	64
128	2,097,024	1834944	$7 \cdot 64^3 - 64$	2,097,152	128

Table 8.5.2: $7k^3$-k

$$\Rightarrow f(k) + 7k^3 - k \tag{8.5.18}$$

$$\Rightarrow (k^3 - k) + 7k^3 - k \tag{8.5.19}$$

$$\Rightarrow 8k^3 - 2k \tag{8.5.20}$$

$$\Rightarrow (2k)^3 - (2k) = f(k+k) \tag{8.5.21}$$

$$Q.E.D$$

One can also notice that if the emergence criterion requires emergence radius to be k instead of 1 while r = k, then

$$f(k) = \left(\frac{k}{k}\right)^3 - \left(\frac{k}{k}\right) = 0 \tag{8.5.22}$$

which is the same as our base case when r = 1 with emergence radius =1.

$$f(1) = \left(\frac{1}{1}\right)^3 - \left(\frac{1}{1}\right) = 0 \tag{8.5.23}$$

While r = 2k and emergence radius to be k:

$$f(2k) = \left(\frac{2k}{k}\right)^3 - \left(\frac{2k}{k}\right) = 6 \tag{8.5.24}$$

which is the same as our base case when r = 2 with emergence radius =1.

$$f(2) = \left(\frac{2}{1}\right)^3 - \left(\frac{1}{1}\right) = 6 \tag{8.5.25}$$

This proves that the equation is self-similar.

How likely is that we are at a particular point in time where all extra-terrestrial signal is about to reach from each of its respective distance to earth originated n years ago? If we take the decreasing function, we will find that such function rises sharply within r = 4 light years.

This implies that the universe's alien civilization arising curve is extremely steep, that is, in the past billions of years, almost none of the life-bearing planets transformed, yet in the most recent 4 years, all remaining planets will transform. This implies that there is something extraordinary about the next 4 years in the entire universe, violating the mediocrity principle.

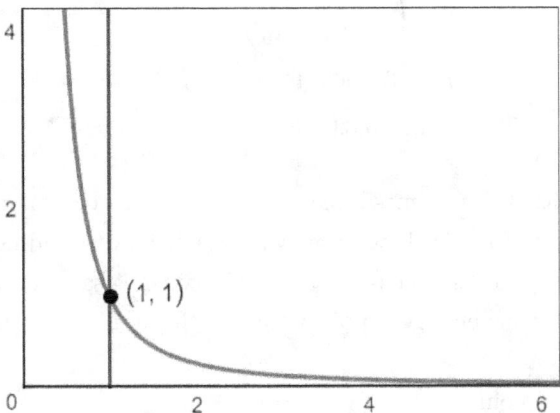

Figure 8.5.3: Assuming all extraterrestrial civilization arises sharply in the next 4 years in the universe

Civilizations' technological gap between the Maya and the Spanish is well within the orders of magnitude of thousands of years even between societies on the same planet within the same species, so it is highly improbable that all planets converge and transform on a such short timescale. We can gradually decrease the factor to a variable x, so that longer time span into the future is taken into consideration in which all remaining life-bearing planets give rise to industrial civilizations, if we assume that such emergence takes at least as long as the geologic process timeframe on earth (we set it at 32.66 million years, corresponding to the radius requirement for the emergence of one industrial civilization), then, the nearest extra-terrestrial civilization must also be at least tens and hundreds of million light-years away.

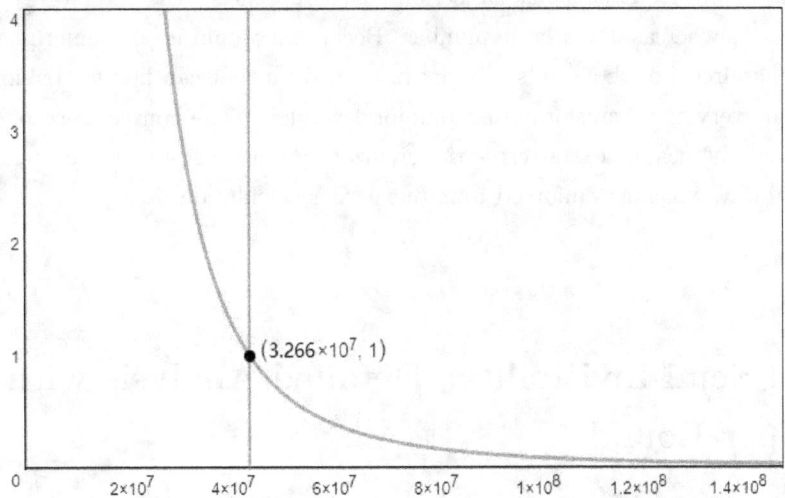

Figure 8.5.4: Assuming all extraterrestrial civilization arises sharply in the next 32 Myr in the universe

$$y = \frac{1}{\left(\frac{x}{32,663,449}\right)^3} \tag{8.5.26}$$

The solution for y=1 for this expanded case is:

$$1 = \frac{1}{\left(\frac{x}{32,663,449}\right)^3} \tag{8.5.27}$$

$$1 = \left(\frac{x}{32,663,449}\right)^3 \tag{8.5.28}$$

$$x^3 = 32,663,449^3 \tag{8.5.29}$$

$$x = 32,663,449 \tag{8.5.30}$$

That is, the new equation requires 32.66 Myr time-frame for all planets to be emerged previously takes only 1 year. Previously by using the factor $\frac{1}{x^3}$, it is shown that within 1 light year radius, there is 1 transforming sample. Now, it takes 32,663,449 light year distance to host one transforming sample. Then, the next 32,663,449 years can accommodate the timing for the emergence of the rest of the civilizations instead of squeezing all of the rest to be all emerged within 1 year.

The final solution implies several vital points:

1. The rate of alien civilization formation in the universe is a decreasing function with a factor within the order of magnitude of $y = \frac{1}{x^3}$ as the upper bound and y=0 as the lower bound as one further traces back in time.

2. We are relatively early arising civilization compares to the rest of the life-nurturing planets. (The emergence rates remain very flat prior and very flat after for a long time before its final surge.)

3. In order to cope with the principle of mediocrity, the silent sky can be explained by both 1) and 2). We likely neither the earliest nor the latest comer given by random sampling. We are especially unlikely to be late, proved with current observation. It will also be a strong violation against the principle of mediocrity if we are so late just before the emergence rate of all intelligent lives arises sharply yet almost no intelligence arrives before us if the signal detection curve approaches the value of $\frac{1}{x^3}$ if d is small or close to 1 instead of $\frac{1}{\left(\frac{x}{d}\right)^3}$ whereas d is a large number. Because it would imply something extraordinary for the next decade, hundreds, or thousands of years in a universe that can last for trillions of years but converges toward some very tiny transformation temporal window. The consequence of satisfying both conditions requires that the nearest extra-terrestrial civilization must also be at least tens and hundreds of million light years away, which is reinforced from our earlier calculation.

8.6 The Wall of Semi-Invisibility: Detailed Analysis with the Theoretical Upper Bound

Having derived the theoretical upper bound curve, one now checks if our PDF is bounded strictly by the theoretical upper bound. In order to derive the theoretical upper bound to reflect the reality, rescaling is required to convert every unit into a $\frac{100 \text{ myr}}{100 \text{ mly}}$ light year distance. One then divides a sphere of arbitrary size over another sphere with 32.66 Mly in radius (the minimum size required for the detection of one industrial civilization). We simplify the expression can be derived:

$$\frac{\frac{4}{3}\pi (100x)^3}{\frac{4}{3}\pi (32.66)^3} = \frac{(100x)^3}{(32.66)^3} = 28.705x^3 \tag{8.6.1}$$

This factor can be expressed in a more generalized form as:

$$\left(\frac{100}{\frac{1.26\left(F_{galaxy}\cdot C_{df}(0)^{-1}(d_{Galaxy})^3\right)^{\frac{1}{3}}}{10^2}}\right)^3 x^3 \tag{8.6.2}$$

Whereas $C_{df}(0)$ is the lognormal distribution function integrated over all previous time period for the cumulative chance of observing extraterrestrials and the radius required to find one arising industrial civilization with a 100% chance. As a result, by altering $C_{df}(t)$, the factor can vary from $28.705x^3$.

This expression is the total number of emerging civilizations per 100 million light years in radius in proportion to the number of emerging civilizations within a 32.66 million light years radius. One also knows that our earlier discussion the upper bound requires that at every time period there is at most one emerging civilization, whose signal just about to reach us. Then, for every time unit of t translating into $\frac{100\ myr}{100\ million}$ light years, there can be at most

$$\frac{1}{28.705x^3} \tag{8.6.3}$$

civilizations are trying to but not yet reached us without violating the upper bound constraint requirement. The decreasing factor becomes $\frac{1}{28.705x^3}$ instead of $\frac{1}{x^3}$; that is, the probability of the emergence of industrial civilization has to be further decreased by a factor of $\frac{1}{28.705}$, establishing a more stringent upper bound.

$$U_{pper} = \frac{1}{28.705x^3} \tag{8.6.4}$$

whereas the best approximate curve fitting for $C_{df}(t)^{-1}$ was expressed as the double exponential function as:

$$d_R(t) = \frac{1.26\left(F_{galaxy}\cdot C_{df}(t)^{-1}\cdot(d_{Galaxy})^3\right)^{\frac{1}{3}}}{10^4} \tag{8.6.5}$$

$$\frac{1.26\left(F_{galaxy}\cdot C_{df}(t)^{-1}\cdot(d_{Galaxy})^3\right)^{\frac{1}{3}}}{10^4} = d_R(t) \approx ab^{(t+v)^{c^x}} + ft + z \tag{8.6.6}$$

$$a = 0.0032485, \ b = 119.2134, \ c = 1.07239 \qquad v = 0.941865, \ z = 0.033692, \qquad f = 0.05010$$

and we express the rate of extraterrestrial civilization emergence as:

$$E = \frac{(d_R(0))^3}{(d_R(t))^3} = \frac{(d_R(0))^3}{(d_R(x))^3} \tag{8.6.7}$$

As one looks further into the past, the rate of civilization emergence decreases.
Then, the rate of extraterrestrial civilization emergence E crosses the point (0, 1).

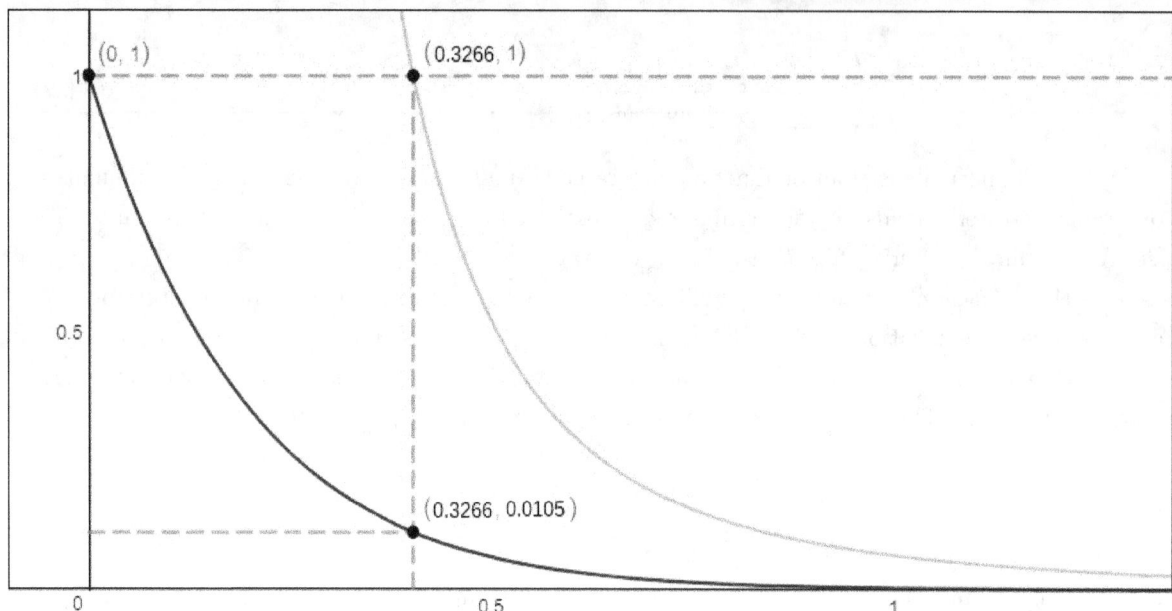

Figure 8.6.1: The cosmic distribution CDF $C_{df}(t)$ is bounded by the upper bound curve with the nearest civilization at 32.66 Mly away

Finally, the ratio is formulated as:

$$d_0 = \log\left(\frac{U_{pper}}{E}\right) \tag{8.6.8}$$

Then, both curves when x > 0 can be compared. The finalized equations above checks the ratio of the theoretical upper bound to the CDF.

In order to clarify our plots for our reader, we would like to further discuss about the graph above, first let us focus on the emergence curve $E = \frac{(d_R(0))^3}{(d_R(x))^3}$. One can think of the point (0, 1) as at the current time, we know mathematically through our derivation that the emergence rate is 1 (a positive detection of an extraterrestrial civilization) at a radius of 32.66 Mly away, but this is not yet verified currently, it will take another 32.66 Myr for the light to reach us to verify our assumption. The point (0.3266, 0.0105) denotes the emergence rate we currently observe and verify from a radius 32.66 Mly away. The signals we currently receiving from 32.66 Mly away is the emergence rate of 32.66 Myr ago. At that time, the emergence rate is 0.0105 for a radius of 32.66 Mly, or merely $\frac{1}{95.238}$th of today's. This is a much lower chance than guaranteeing spotting an extraterrestrial industrial civilization, so there is a $\frac{94.238}{95.238} = 98.95\%$ chance not spotting one. Distance d (in the graph) denotes 32.66 Mly in radius, that is, in order to guarantee finding a civilization emerged 32.66 Mly earlier, the coverage radius has to increase to $d+t$. This is another perspective to explain our threshold test of $\frac{d+t}{d}$ discussed earlier in the section. Hence, one can see the emergence curve can be more clearly interpreted from a temporal aspect, but the x coordinate of the emergence rate of the past can also be interpreted as the distance away relative to earth from a spatial perspective.

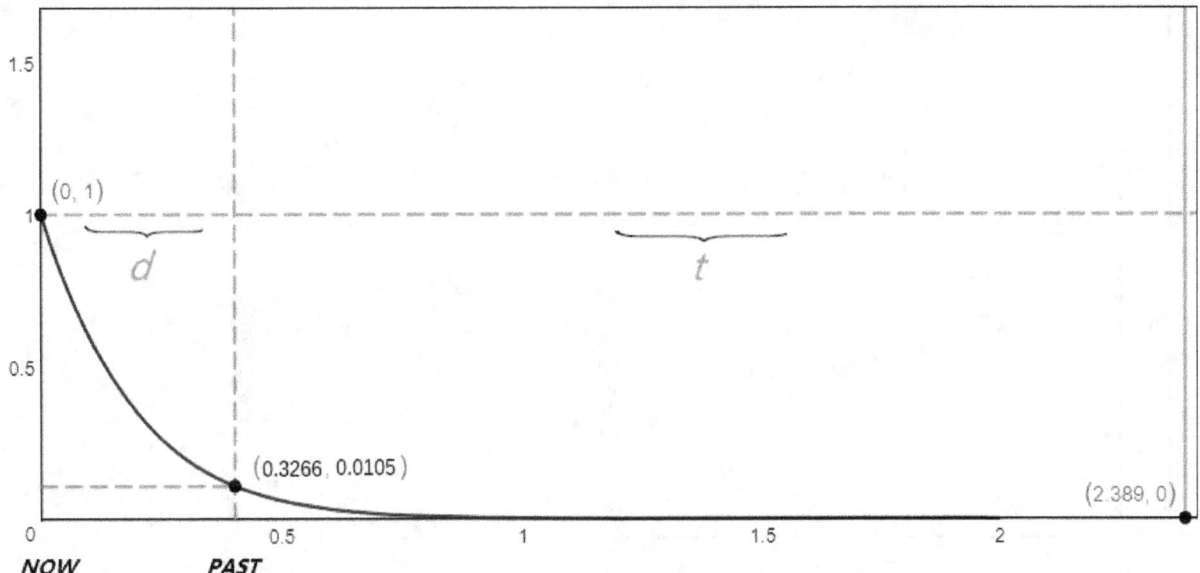

Figure 8.6.2: The physical explanation for the emergence curve

Now, we focus on the upper bound light signal detection curve $U_{pper} = \frac{1}{28.705x^3}$. This curve passes through the point (0.3266, 1) because we assumed that it takes a radius of 32.66 Mly to host one emerging civilization. Therefore, this point sits on the signal detection curve which guarantees exactly one positive detection at any distance. The factor requirement guaranteeing a single positive detection increases by the cubed as the lookout distance decreases and decreases by the cubed as the lookout distance increases. *The curve does not care how to guarantee one signal detection is achieved or if it is achievable in reality.* It is simply used as a point of reference to show that, relative to the current one positive detection per 32.66 Mly radius, the amount of value needed to be adjusted to maintain one positive detection for any lookout distance away from the earth.

The point (0, 1) denotes at a distance of 0 ly away at the current time, the signal detection will occur 32.66 Myr into the future. We will confirm the fact that there is one arising alien industrial civilization currently unobservable at the distance of 32.66 Mly away at the current time. The point (0.3266, 1) denotes that the signal which will confirm the existence of one arising extraterrestrial per 32.66 Mly radius is being emitted by the expanding alien civilization now at a distance of 32.66 Mly away. The point touches the dashed line y=1 because we assumed that it takes a radius of 32.66 Mly for one emerging civilization at the current time.

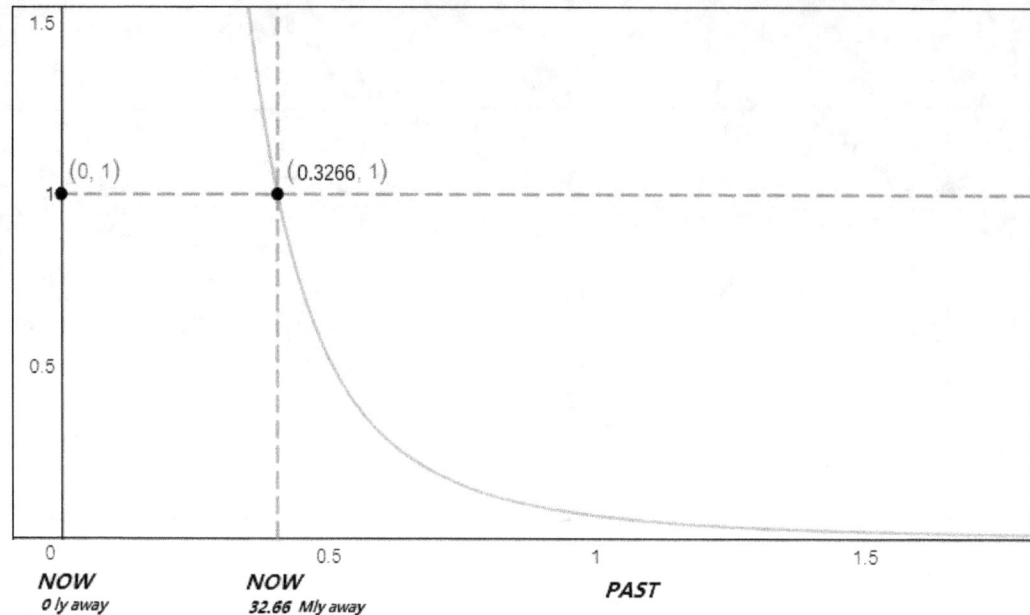

Figure 8.6.3: The physical explanation for the signal light arrival curve

Finally, we can arrange the emergence curve and upper bound signal detection curve on the same graph. The meaning of $\frac{U_{pper}}{E}$ (the upper bound signal detection curve over the emergence curve) is the following.

The curve represents the number of times the detectable chance/number (always <1) of the actual observed civilization of the past at any lookout radius needs to be increased to guarantee exactly 1 detectable civilization at the current time at any lookout distance. Even at the current time, the most attainable emergence rate is 1 per 32.66 Mly radius. In other words, within a lookout radius of 1 ly, the chance of detection is $1 \cdot \left(\frac{1\,\text{ly}}{32.66\,\text{Mly}}\right)^3$. The chance is much closer to 0 than 1. Therefore, the number of times on the current detection chance (between 0 and 1) needed to be increased to guarantee one positive detection must be high. *The curve does not care how to guarantee one signal detection is achieved.* It can be achieved through a change in the emergence rate holding the lookout volume size constant, or it can be achieved through a change in volume size holding the emergence rate constant, or through both an change in volume and the emergence rate.

The inverse of the number of detectable civilization at any arbitrary distance proportional to a 32.66 Mly radius is actually equivalent to U_{pper} from our initial definition of the upper bound signal detection curve so that their product equals 1. $U_{pper} \cdot \left(\frac{x}{d_R(0)}\right)^3 = 1$ This holds because the detection chance varies by the lookout radius. In order to hold detection chance constant at 1, the upper signal detection curve must compensate by the reciprocal of the radius changes relative to 32.66 Mly.

$$\left(\frac{x}{d_R(0)}\right)^{-3} = U_{pper} = \frac{1}{28.705x^3} \tag{8.6.9}$$

As a result, one can multiply $\frac{U_{pper}}{E}$ by $\frac{\left(\frac{x}{d_R(0)}\right)^3}{\left(\frac{x}{d_R(0)}\right)^3}$:

$$\frac{U_{pper} \cdot \left(\frac{x}{d_R(0)}\right)^3}{E \cdot \left(\frac{x}{d_R(0)}\right)^3} = \frac{1}{E \cdot \left(\frac{x}{d_R(0)}\right)^3} = \frac{1}{\left(\frac{x}{d_R(x)}\right)^3} \tag{8.6.10}$$

First, we interpret the numerator $U_{pper} \cdot \left(\frac{x}{d_R(0)}\right)^3 = 1$ as the scale factor required to fix the appearance number of civilization to 1 regardless of the search radius one initially seek. Fixing the number of detectable civilization to 1 is the invariant.

This is achieved by expanding the lookout radius to 32.66 Mly when the initial lookout distance < 32.66 Mly and holding the emergence rate $E = 1$ constant. Since it is not possible to increase the emergence rate higher

378

than 1 per 32.66 Mly radius at the current time. It is not possible to host a civilization within less than 32.66 Mly radius currently. The only way to guarantee the appearance number is to expand the lookout radius to 32.66 Mly.

It is achieved by reducing the past emergence rate E to merely $\frac{1}{28.705x^3}$ relative to the higher emergence rate per 32.66 Mly radius when the initial lookout distance x > 32.66 Mly since the search volume is now greater than 32.66^3 Mly^3 and the number of civilization detection increases beyond 1 if the emergence rate of the past is fixed at 1 per 32.66 Mly radius.

We can verify that fixing the appearance number is not achieved through fixing the lookout radius to 32.66 Mly and keeping the emergence rate constant $E = 1$ even the initial lookout radius > 32.66 Mly. This is confirmed by the following inequality:

$$R_{Actual} = \frac{1}{E\left(\frac{x}{d_R(0)}\right)^3} \tag{8.6.11}$$

$$R_{Fixed} = \frac{1}{1 \cdot \left(\frac{x}{d_R(0)}\right)^3} \tag{8.6.12}$$

$$\frac{d_R(x)}{d_R(0)} = \frac{x \cdot (R_{Actual})^{\frac{1}{3}}}{d_R(0)} > \frac{x \cdot (R_{Fixed})^{\frac{1}{3}}}{d_R(0)} \tag{8.6.13}$$

The left term states the *actual* lookout radius size in the ratio of 32.66 Mly required for the detection of at most one civilization at an arbitrary distance given an emergence rate governed by the CDF emergence function.

The right term states the *predicted* lookout radius size in the ratio of 32.66 Mly required for one civilization detection at an arbitrary distance given a flat emergence rate. Since the emergence rate remains the same as one looks further back in time as the lookout radius expands > 32.66 Mly, the only way to guarantee at most one civilization detection at an arbitrary distance is to fix the lookout radius to 32.66 Mly. This would require one must scale down the lookout radius beyond 32.66 Mly as required by $(R_{Fixed})^{\frac{1}{3}}$. Since the actual and predicted curves do not match, this proves that fixing the appearance number of civilization to 1 for distance > 32.66 Mly is achieved not by fixing the emergence rate E and fixing the lookout radius.

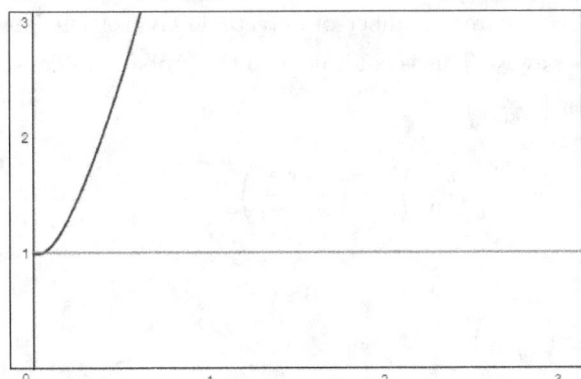

Figure 8.6.4: The actual look out radius $\frac{x \cdot (R_{Actual})^{\frac{1}{3}}}{d_R(0)} \rightarrow \frac{x \cdot (R_{Actual})^{\frac{1}{3}}}{d_R(0)} + 1$ vs predicted $\frac{x \cdot (R_{Fixed})^{\frac{1}{3}}}{d_R(0)}$ if one fixes the appearance number by fixing the look out volume

One can also attest that:

$$R_{Upper} = \frac{1}{Upper\left(\frac{x}{d_R(0)}\right)^3} \tag{8.6.14}$$

$$\frac{x \cdot (R_{Actual})^{\frac{1}{3}}}{d_R(0)} > \frac{x \cdot (R_{Upper})^{\frac{1}{3}}}{d_R(0)} \tag{8.6.15}$$

The right term states the *predicted* lookout radius size in the ratio of 32.66 Mly required for the detection of at most one civilization at an arbitrary distance given an emergence rate governed by the upper bound curve. Since the upper bound curve requires the emergence rate to decrease by a factor of $\frac{1}{28.705x^3}$ for lookout radius > 32.66 Mly, $\frac{(R_{Upper})^{\frac{1}{3}}}{d_R(0)}$ requires no-scaling for the lookout radius > 32.66 Mly. This leads to the predicted lookout radius increases linearly as the lookout radius > 32.66 Mly. Although this result still does not match our actual result, its prediction is much closer to the actual value than the previous one.

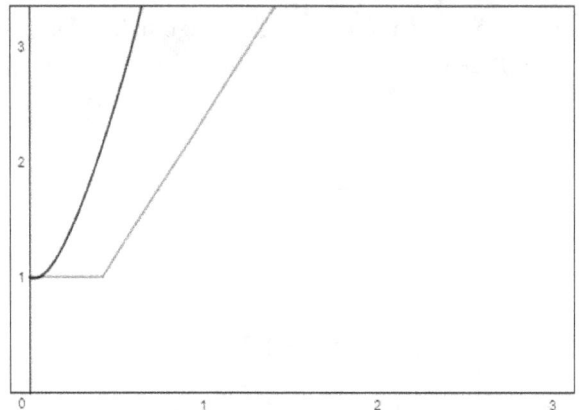

Figure 8.6.5: The the actual look out radius $\frac{x \cdot (R_{Actual})^{\frac{1}{3}}}{d_R(0)}$ vs predicted $\frac{x \cdot (R_{Upper})^{\frac{1}{3}}}{d_R(0)}$ if one fixes the appearance number by adopting the emergence rate according to the upper bound by the factor $\frac{1}{28.705x^3}$ and a linearly increasing lookout radius.

This proves that the actual appearance number of civilization at any distance > 32.66 Mly is achieved by reducing the past emergence rate E by at least $(R_{Upper})^{\frac{1}{3}}$, and the emergence rate can not be held constant, it must be changing and dropping exponentially fast, and E$\leq U_{pper}$ is satisfied. Hence, we have verified that fixing the appearance number is not achieved through fixing the lookout radius.

Next, we interpret the denominator $E \cdot \left(\frac{x}{d_R(0)}\right)^3$ as the emergence rate from the past x years ago relative the emergence rate at the current time multiplied by the ratio of any lookout volume to the current emergence volume of 32.66 Mly radius yields the apparent number of detectable civilization (always <1) at the current time from any distance of x light years away. This is equivalent to the probability density function of all arising extra-terrestrial industrial civilization $\left(\frac{x}{d_R(x)}\right)^3$

$$E \cdot \left(\frac{x}{d_R(0)}\right)^3 = \left(\frac{x}{d_R(x)}\right)^3 < 1 \qquad (8.6.16)$$

Then:

$$\frac{U_{pper} \cdot \left(\frac{x}{d_R(0)}\right)^3}{E \cdot \left(\frac{x}{d_R(0)}\right)^3} = \frac{1}{E \cdot \left(\frac{x}{d_R(0)}\right)^3} = \frac{1}{\left(\frac{x}{d_R(x)}\right)^3} > 1 \qquad (8.6.17)$$

The numerator is already reduced to 1 because the numerator $U_{pper} \cdot \left(\frac{x}{d_R(0)}\right)^3 = 1$
The above equation indicates that the lookout volume has to be further increased by this factor in order to gain 1 positive detection, and the lookout radius has to increase by:

$$\sqrt[3]{\frac{1}{E \cdot \left(\frac{x}{d_R(0)}\right)^3}} = \sqrt[3]{\frac{1}{\left(\frac{x}{d_R(x)}\right)^3}} > 1 \qquad (8.6.18)$$

The lookout radius of x light years is to be rescaled by the factor $\sqrt[3]{\frac{1}{E \cdot \left(\frac{x}{d_R(0)}\right)^3}}$ or $\sqrt[3]{\frac{1}{\left(\frac{x}{d_R(x)}\right)^3}}$ to a larger size because the currently observed emergence rate signal from any lookout radius is strictly from the past and drops faster than the upper bound signal detection $\frac{1}{28.705x^3}$ requirement for radius size > 32.66 Mly. The past offers a lower chance of emergence. Furthermore, an observer can not alter the emergence rate but can increase the lookout distance, therefore, the lookout radius has to expand beyond x light years to accommodate one positive detection. This is equivalently:

$$\sqrt[3]{\frac{1}{E \cdot \left(\frac{x}{d_R(0)}\right)^3}} = \sqrt[3]{\frac{1}{\left(\frac{x}{d_R(x)}\right)^3}} = \frac{(d+t)}{d} = \frac{d_R(t)}{t} \tag{8.6.19}$$

On a further note, the factor required to rescale the lookout radius size first shrinks as the emergence rate derived from the past signal reaching earth is catching up with the continuous drop of the signal detection threshold curve of $\frac{1}{28.705x^3}$. However, as one inspects beyond 32.66 Mly radius, despite a continuous drop of the signal detection threshold curve of $\frac{1}{28.705x^3}$ at every point, the emergence rate derived from the past signal reaching earth is smaller still, as a result, the factor required to rescale the search radius to guarantee passing the signal detection threshold actually increased.

This can be illustrated from the actual lookout radius for 1 civilization detection differs by a factor of $\left(\frac{1}{U_{pper}}\right)^{\frac{1}{3}}$ from $(R_{Actual})^{\frac{1}{3}}$:

$$\frac{d_R(x)}{d_R(0)} = \frac{x \cdot (R_{Actual})^{\frac{1}{3}}}{d_R(0)} = \frac{d_R(0) \cdot (R_{Actual})^{\frac{1}{3}}}{d_R(0)} \cdot \left(\frac{1}{U_{pper}}\right)^{\frac{1}{3}} \tag{8.6.20}$$

and the factor

$$\left(\frac{1}{U_{pper}}\right)^{\frac{1}{3}} = \left(\frac{x}{d_R(0)}\right) > 1 \tag{8.6.21}$$

is always > 1 for x > 32.66 Mly. $(R_{Actual})^{\frac{1}{3}}$ is the required rescaling factor applied on the initial lookout radius for one civilization detection at an arbitrary distance given an emergence rate governed by the CDF emergence function. $\left(\frac{x}{d_R(0)}\right)$ is the initial lookout radius expressed in terms of 32.66 Mly, and this lookout radius increases linearly for radius > 32.66 Mly to satisfy the upper bound requirement of decreasing emergence rate $\frac{1}{28.705x^3}$ to guarantee the detection of at most one civilization at an arbitrary distance from earth.

With a full comprehension of the model, we show that the boundary checking is conceptually equivalent to our earlier threshold test, except that the boundary checking is expressed as the log of the threshold test ratio raised to the cubed:

$$\left(\frac{U_{pper}}{E}\right)^{\frac{1}{3}} = \frac{d_R(t)}{t} \tag{8.6.22}$$

The final curve is plotted below:

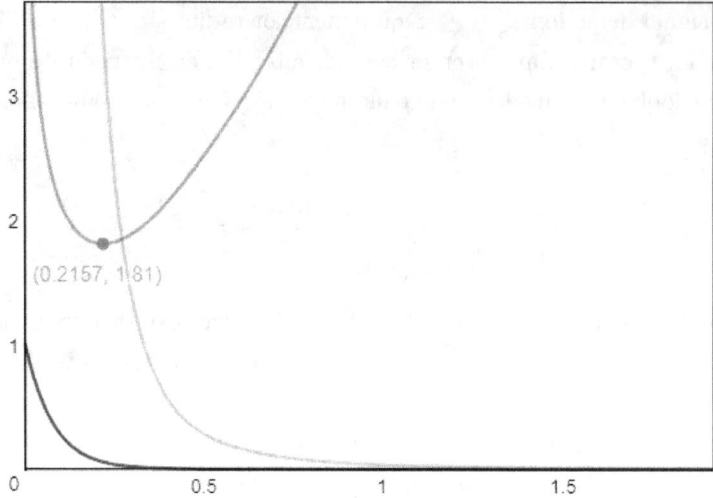

Figure 8.6.6: The logarithmic ratio $\log\left(\frac{U_{pper}}{E(x)}\right)$ of upper bound curve divided by the CDF

The curve indicates that the ratio is strictly positive for all values of x > 0, this indicates for all additional look ahead distance, the emergence rate of extraterrestrial civilization is much smaller than our theoretical maximum upper bound. In other words, the emergence rate of civilization decreases much faster than $\frac{1}{28.705x^3}$ as one looks further back in time.

Next, we shall test and confirm that CDF is bounded by the upper bound at any arbitrarily large values. The test for convergence is presented below:

$$E = \frac{(d_R(0))^3}{(d_R(x))^3} = \frac{\left(ab^{(0+v)^{c^0}} + f \cdot 0 + z\right)^3}{\left(ab^{(x+v)^{c^x}} + fx + z\right)^3} \tag{8.6.23}$$

$$E = \frac{(ab^v + z)^3}{\left(ab^{(x+v)^{c^x}} + fx + z\right)^3} \tag{8.6.24}$$

$$U_{pper} = \frac{1}{\left(\frac{10^4}{1.26\left(F_{galaxy} \cdot C_{df}(x)^{-1}1185^3\right)^{\frac{1}{3}}}\right)^3 x^3} = \frac{1.26^3 \cdot F_{galaxy} \cdot (d_{Galaxy})^3}{C_{df}(x) \cdot 10^{12} x^3} \tag{8.6.25}$$

Limit test:

$$\frac{E}{U_{pper}} = \frac{(ab^v + z)^3}{\left(ab^{(x+v)^{c^x}} + fx + z\right)^3} \cdot \frac{C_{df}(x) \cdot 10^{12} x^3}{1.26^3 \cdot F_{galaxy} \cdot (d_{Galaxy})^3} \tag{8.6.26}$$

$$\lim_{x \to \infty} \frac{(ab^v + z)^3}{\left(ab^{(x+v)^{c^x}} + fx + z\right)^3} \cdot \frac{C_{df}(x) \cdot 10^{12} x^3}{1.26^3 \cdot F_{galaxy} \cdot (d_{Galaxy})^3} \cdot \frac{\frac{1}{x^3}}{\frac{1}{x^3}} \tag{8.6.27}$$

$$\lim_{x \to \infty} \frac{(ab^v + z)^3}{\left(ab^{(x+v)^{c^x}} + fx + z\right)^3 \cdot \frac{1}{x^3}} \cdot \frac{C_{df}(x) \cdot 10^{12} x^3 \cdot \frac{1}{x^3}}{1.26^3 \cdot F_{galaxy} \cdot (d_{Galaxy})^3} \tag{8.6.28}$$

$$\lim_{x \to \infty} \frac{(ab^v + z)^3}{\left(\frac{ab^{(x+v)^{c^x}}}{x} + \frac{fx}{x} + \frac{z}{x}\right)^3} \cdot \frac{C_{df}(x) \cdot 10^{12}}{1.26^3 \cdot F_{galaxy} \cdot (d_{Galaxy})^3} \tag{8.6.29}$$

$$\lim_{x \to \infty} \frac{(ab^v + z)^3}{\left(\frac{ab^{(x+v)^{c^x}}}{x} + f + \frac{z}{x}\right)^3} \cdot \frac{C_{df}(x) \cdot 10^{12}}{1.26^3 \cdot F_{galaxy} \cdot (d_{Galaxy})^3} \tag{8.6.30}$$

The first term, can be finalized as two cases:

$$\lim_{x \to \infty} \frac{\left(ab^v + z\right)^3}{\left(\frac{ab^{(x+v)c^x}}{x} + f + \frac{z}{x}\right)^3} = \begin{cases} \frac{(ab^v + z)^3}{f^3} & c < 1 \\ \frac{(ab^v + z)^3}{(1+f)^3} & c \geq 1 \end{cases} \tag{8.6.31}$$

In both cases, limit indicates it approaches a constant term, we substitute value 1 by a variable h to generalize the cases to $\frac{(ab^v + z)^3}{(h+f)^3}$ so that for c < 1, h=0 and for $c \geq 1$, h=1.

$$\lim_{x \to \infty} \frac{\left(ab^v + z\right)^3}{\left(h + f\right)^3} \cdot \frac{C_{df}\left(x\right) \cdot 10^{12}}{1.26^3 \cdot F_{galaxy} \cdot \left(d_{Galaxy}\right)^3} \tag{8.6.32}$$

$$\frac{\left(ab^v + z\right)^3}{\left(h + f\right)^3} \cdot \lim_{x \to \infty} \frac{C_{df}\left(x\right) \cdot 10^{12}}{1.26^3 \cdot F_{galaxy} \cdot \left(d_{Galaxy}\right)^3} \tag{8.6.33}$$

$$\left(\frac{ab^v + z}{h + f}\right)^3 \cdot \lim_{x \to \infty} C_{df}\left(x\right) \tag{8.6.34}$$

We avoid dissecting the CDF which involves error function approximations, whereas the best closed form fit for $C_{df}\left(x\right)$ can be expressed as a much simpler exponential function for $x \geq 0$:

$$C_{df}\left(x\right) \approx a\left(b\right)^x \tag{8.6.35}$$

$$a = 0.462777, b = 0.00000239$$

We substitute $C_{df}\left(x\right)$ with $a\left(b\right)^x$ and $0 < b < 1$

$$\left(\frac{ab^v + z}{h + f}\right)^3 \cdot \lim_{x \to \infty} a\left(b\right)^x \tag{8.6.36}$$

$$\left(\frac{ab^v + z}{h + f}\right)^3 \cdot 0 \tag{8.6.37}$$

The limit test shows that the ratio of CDF $C_{df}\left(x\right)^{-1}$ to theoretical upper bound converges to 0 as x approaches infinity, concluding that our CDF is then strictly bounded by the theoretical upper bound of the wall of semi-invisibility.

8.7   Generalized Model

Having demonstrated every aspect of our model, we would now like to take a step further by generalizing our model to more diverse scenarios.

8.7.1   Different values of BCS and BER

Having illustrated the case of BCS = BER = 2.783, now we can move on to test BCS = BER with values > 1. We start with our earlier bivariate exponential lognormal distribution with the assumption that regardless of BER and BCS for alternate evolutionary history on earth, they all achieved similar level of biodiversity observed on earth today, later we will relax this constraint. One finds that:

1) Given an exponentially growing BCS, the gap between mode peaks (BER) always increases by BCS. This trend is interpreted as the follows. Based on the original model of biocomplexity represented by both combination

and permutation of given number of traits, the addition of new traits creates exponentially larger search space (BCS). With additional traits, the mode of the distribution shifts to organisms with higher number of traits, so the mode peak shifts to the right. As a result, As long as the number of traits representing organisms increases per round, mode peak shifts right, the distribution spreads out further.

Moreover, maintaining an exponential growth in BER, the number of traits must not only increases but increases exponentially per round. We have shown that combination with increasing traits alone can not exceed a BCS of 2.783, and it is achieved only by partial permutation and combination combined. But in order to fit exponentially growing BCS and exponentially growing BER (distance between mode peaks), the number of traits per round must grow exponentially, and can be expressed as (for the t-th round the number of traits grow by d^t):

$$C\left(n + d^t, x\right) = \frac{(n + d^t)!}{x!\left((n + d^t) - x\right)!} \left(p\right)^x \tag{8.7.1}$$

2) By increasing BCS, the overlapping region between successive rounds of distributions 100 Myr apart decreases.
3) Increasing / decreasing BER, manifested as $P_{eak}(0) - P_{eak}(1)$ width, is compensated by decreasing / increasing YAABER by the same factor, so that a civilization's emergence is still evaluated at $L_{imit} = 18$. A new BER can be expressed as our previous BER value raised to the power of x:

$$2.783^x = B_{er} \tag{8.7.2}$$

and x is expressed as:

$$x = \frac{\ln(B_{er})}{\ln(2.783)} \tag{8.7.3}$$

and a new YAABER can be expressed as:

$$Y_{aaber} = \frac{17}{x} \tag{8.7.4}$$

Low BER (BER⟶ 1) implies a higher emergence chance of the past if one fixed the diversity and biocomplexity achieved on earth as the invariant but allows for flexible multicellular evolutionary window. Then the past is more like the present with similar mode and total biodiversity, implying there should be almost equally if a little less chance in the past that earlier civilization have evolved. Thus, the chance at the current time is unchanged, and the mode reaches human attained complexity slower in the future and the complexity/mode achieved in geologic past more similar to those achieved now. On the other hand, Higher BCS and BER renders mode reaching human attained complexity faster in the future, but the complexity achieved in geologic past much lower than those achieved now and the mode shifted much more to the left.

	Yaaber	$P_{eak}(0) - P_{eak}(1)$	L_{imit}	+	Current Variance	Overall
BCS ↓ BER ↓	Yaaber ↑	↓	No change		No change	No change

Assuming that multicellular evolutionary window is fixed to 500 Myr but allowing flexible diversity and biocomplexity achieved on earth at the current time, then a lower BCS always leads to a lower current biodiversity and left shifted mode since lower BCS can not reach a biodiversity level comparable to earth today without assuming it had a longer evolutionary time to compensate its low BER. As a result, at the current time, the emergence chance will decrease. Mathematically, L_{imit} stays at 18, but the emergence chance of civilization is decreased due a narrower distribution width/variance and a left shifted mode dictated by lower BER and higher YAABER.

	Yaaber	$P_{eak}(0) - P_{eak}(1)$	L_{imit}	+	Current Variance	Overall
BCS ↓ BER ↓	Yaaber ↑	↓	No change		↓	↓

8.7.2 Distribution Placement and Variable k

Darwin had shown that evolution is a goal less and direction less process. Nevertheless, organisms with higher complexity can evolve over time as the outlying tail of the left skewed lognormal distributions. On the other hand, Lamarck proposed that life is evolving and striving toward goals. In order to incorporate all possible cases, we would like to redefine our earlier *bivariate exponential log-normal distribution function* with an additional variable k so that the equation becomes:

$$P_{df}(t, x, k) = \frac{(B_{cs})^{-\frac{(k-1)t}{k}}}{\sigma\sqrt{2\pi}} \exp\left(-\frac{\left(\ln\left((B_{cs})^{\frac{t}{k}} \cdot x\right)\right)^2}{2\sigma^2}\right) \tag{8.7.5}$$

The variable k is added to both the numerator as $(B_{cs})^{-\frac{(k-1)t}{k}}$ and as $(B_{cs})^{\frac{t}{k}}$ in the exponents. $(B_{cs})^{\frac{t}{k}}$ controls the width of the distribution and $(B_{cs})^{-\frac{(k-1)t}{k}}$ controls the height. The mathematical interpretation is that given the same area size requirement under the distribution curve, a proportional shrinking in width is compensated by a proportional increase in height, and vice versa. As a result, height times width results in the same area size. Furthermore, variable k's manipulation on the distribution width/variance distinguishes from alteration on width/variance directly by σ. σ shrinks width/variance symmetric about distribution's peak/mode value. Variable k shrinks distribution width/variance symmetric about x = 1. and one notices that when $k = 1$, it becomes the earlier function we have defined by eliminating the term $(B_{cs})^{-\frac{(k-1)t}{k}}$ all together. By introducing the value k, we have increased the spectrum of evolutionary trends across all habitable planets.

When $k > 1$, the overlapping region between each PDF instantiation of the bivariate function becomes larger and the distance between mode peaks (BER) increases slower. For such cases, the instantiation across time indicates a more passive evolutionary trajectory toward higher complexity. As a result, successive future distributions stack on top of the current ones. Mathematically, distributions with higher peaks are interpreted as greater number of species evolved compared to the past round for species that shared the same number of adaptable traits. Biologically, such trend can be explained by intrinsic factors such as organisms does not evolve toward a greater complexity by gaining additional traits or extrinsic factors such as maintaining additional new traits are costly. When $k < 1$, the instantiation across time indicates a more active evolutionary trajectory toward higher complexity. In order to better conceptually illustrate the cases, we plot the case of $k = 2$ as conservative Darwinian evolution, $k = 1$ as our classic Darwin evolution, and $k = 0.7$ as progressive Darwin evolution (quasi-Lamarck evolution)

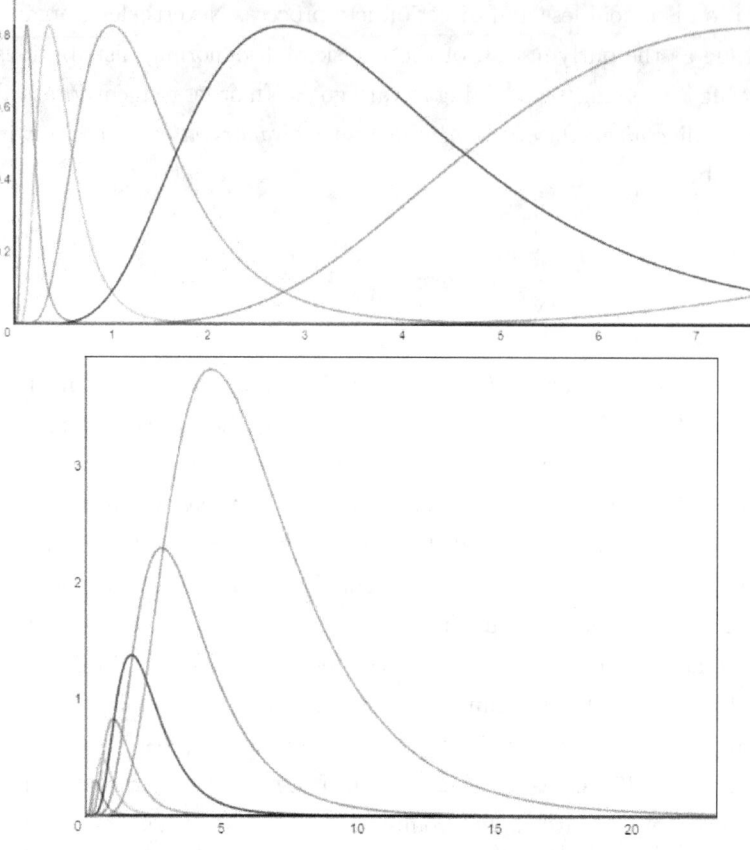

Figure 8.7.1: The instantiations for $k = 1$ (classic Darwin) and $k = 2$ (conservative Darwin)

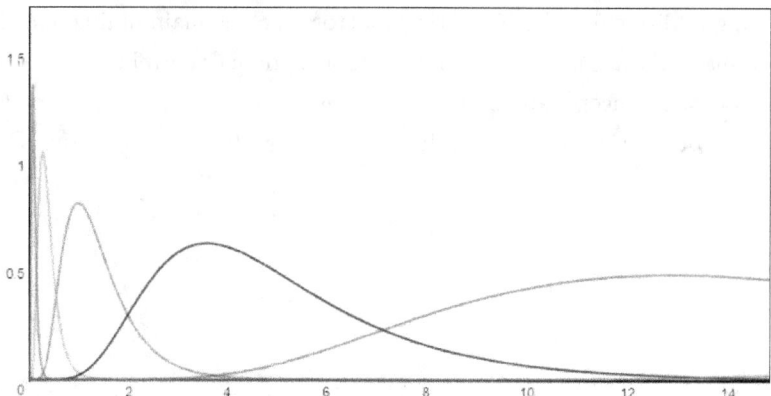

Figure 8.7.2: The instantiations for $k = 0.8$ (progressive Darwin)

In the case of classic Darwin, as we have illustrated, the overlapping area is roughly 45.47% between the instantiation of PDF every 100 Myr.

$$\frac{\int_0^{Intersection} P_{df}(t-1,x)\,dx + \int_{Intersection}^{\infty} P_{df}(t,x)\,dx}{\int_0^{\infty} P_{df}(t,x)\,dx} = 0.4547 \tag{8.7.6}$$

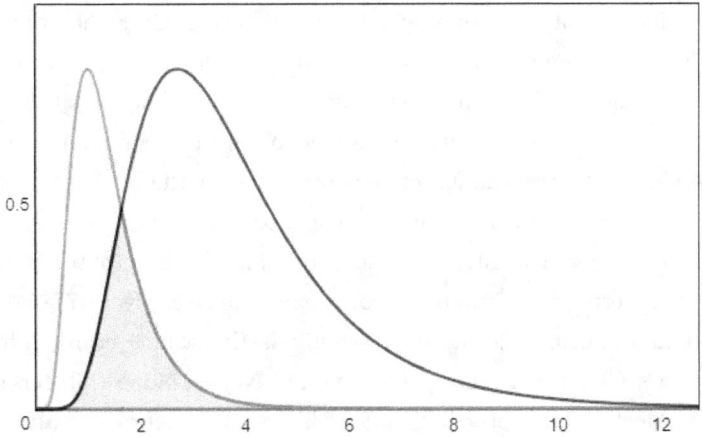

Figure 8.7.3: The overlapping regions between the biocomplexity of two epochs $P_{df}(t-1, x)$ and $P_{df}(t, x)$, separated by 100 Myr apart

This implies that 54.53% of all biocomplexity and biodiversity have gone extinct compares to 100 Mya. In evolutionary biology, extinction is characterized as both catastrophic and background extinction. Background extinction occurs continuously, and it is estimated that 99% of all species becomes extinct within 10 Myr. If one were to use the background extinction rate to measure the accuracy of our model and focusing on the extinction rates related to the lowest of the taxonomic ranks, then our model is a highly inaccurate description of the reality. However, examining the major catastrophic extinction events reveals that on average, every 77~100 Myr a major extinction event occurs and on average 35.375% of all families or genera (extinction rates related to the higher taxonomic ranks) becoming extinct during each major extinctions.

Extinction Name	Genera Survival Rates	Years ago (Mya)	Possible Causes
End-Botomian extinction	62	520	?
Cambrian–Ordovician extinction	60	488	Glaciation/Volcanism
Ordovician–Silurian extinction	60	445	Glaciation/Volcanism
Late Devonian extinction	43	368	Biological/Volcanism
End-Capitanian extinction	66	265	Volcanism
Permian–Triassic extinction	17	252	Impact/Volcanism
Triassic–Jurassic extinction	66	201	Volcanism
Cretaceous–Paleogene extinction	70	66	Impact

Table 8.7.1: Major Extinctions and Survival Rates

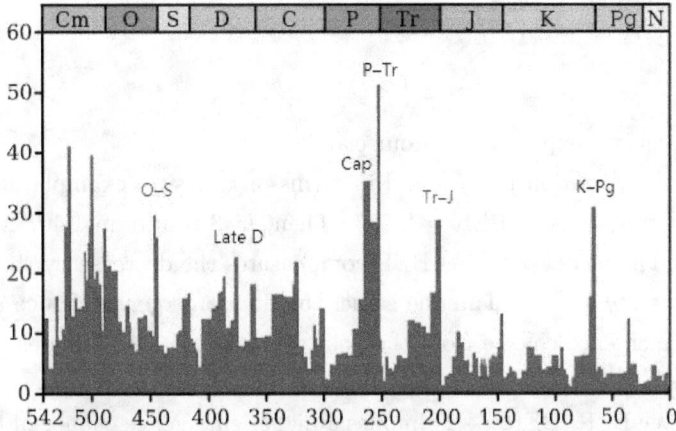

Figure 8.7.4: The apparent percentage of marine animal genera becoming extinct during any given time interval.

Therefore, our classic case's interpretation lies between background extinction and the major catastrophic extinction events relating to the extinction of higher taxonomic ranks every 77~100 Myr. This interpretation can be justified since major extinctions are the key events in drastically alter the composition and fauna morphology and behavior. If Cretaceous–Paleogene extinction did not occur, the background extinction would have continually replaced different generas of dinosaurs with newer dinosaur generas within the same ecological niche, but the adaptive radiation of the clades of mammals and birds becomes highly unlikely. In a sense, background extinction is more similar to the change of monarchs within the same dynastic lineage such as the succession from Louis XIV to Louis XV. The catastrophic extinctions are more similar to the violent overthrow of an entire dynastic lineage (equivalently to clade, order, and family groupings in taxonomic ranks) during the French Revolution. Furthermore, a 70% of biocomplexity survived every 77~100 Myr also fits well with our observation. Many families such as Nautilidae, Limulidae (horseshoe crab), infraorder Anisoptera (dragonfly), order Scorpiones, class Chilopoda (centipedes), family Nephropidae (lobsters), order Octopoda, and order Crocodilia survives. Most importantly, all major classes of vertebrates namely fish, amphibian, reptilia, aves, and mammals survives through all major extinction events. The life history of earth shows that the evolution was neither conservative as the case of $k = 4$ (in which no catastrophic events radically alters the path of evolution and most genera and species survived up to this day) nor progressive as the case of $k = 0.7$ (in which catastrophic events are so common that no clade, order, or family from the past survives at all).

Because k alters the placement pattern of the distribution, the relationship between BER and BCS have to be generalized. Since the BER is the pace of mode shift, it can be faithfully extracted from any period. Given a BER of 3, compares to 100 Mya, it is increased by 3 times, compares to 100 Myr later, only 1/3 of the future, and it can be derived mathematically as:

$$B_{er} = \frac{P_{eak}(t_1 - 1) - P_{eak}(t_1)}{P_{eak}(t_1) - P_{eak}(t_1 + 1)} \tag{8.7.7}$$

This ratio is later generalized taking into account the selection factor for the horizontal displacement of distributions as:

$$B_{er} = \frac{M_{ax}(t - 1) - M_{ax}(t)}{M_{ax}(t) - M_{ax}(t + 1)} \tag{8.7.8}$$

Overall, one observes the pattern of:

$$k = 1 \ \text{BCS} = \text{BER} > 1$$
$$k > 1 \ \text{BCS} > \text{BER} > 1$$
$$k < 1 \ \text{BER} > \text{BCS} > 1$$

In general, BCS can be translated into BER as:

$$B_{er} = (B_{cs})^{\frac{1}{k}} \tag{8.7.9}$$

For the case k=1, BER = BCS, as it is expected from our classic case.

Finally, the emergence chance of civilization is discussed in terms of k. As an example, if it is assumed that $P_{eak}(0) - P_{eak}(1)$ under k=1 corresponds to BER = 2.783. Then, k=3 results in 1.407 BER (more crowded placement with the same BCS). The increase in YAABER compensates the decrease by the shrinkage of BER, so L_{imit} does not change, and as $P_{df}(0, x)$ remain the same, the final emergence chance at the current time stays the same. However, the emergence chance from the past compares to $k = 1$ increases because the past distribution's mode moves closer to current time.

High k produces low BER, and low BER (BER$\longrightarrow$ 1) implies a higher emergence chance of the past if one fixed the diversity/mode and biocomplexity achieved on earth as the invariant but allows for flexible multicellular evolutionary window. Then the past is more like the present, implying there should be almost equally if a little less chance in the past that earlier civilization have evolved. Thus, the chance at the current time is unchanged, and the mode reaches human attained complexity slower in the future but the mode achieved in geologic past is more similar to those achieved now. On the other hand, lower k and higher BER renders mode reaching human attained complexity faster in the future, but the mode achieved in geologic past is much shifted to the left than those achieved now.

Yaaber	$P_{eak}(0) - P_{eak}(1)$	L_{imit}	+	Current Variance	Overall
$k\uparrow$ BER $\downarrow$ Yaaber $\uparrow$	$\downarrow$	No change		No change	No change

Assuming that multicellular evolutionary window is fixed to 500 Myr but allowing flexible diversity and biocomplexity achieved on earth at the current time, then a higher k always leads to a left shifted mode since lower BER can not produce a mode comparable to earth today without assuming it had a longer evolutionary time to compensate for a low BER. As a result, at the current time, the emergence chance will decrease.

Yaaber	$P_{eak}(0) - P_{eak}(1)$	L_{imit}	+	Current Variance	Overall
$k\uparrow$ BER $\downarrow$ Yaaber $\uparrow$	$\downarrow$	No change		$\downarrow$	$\downarrow$

8.7.3 BER's corresponding speed and the selection factor:

8.7.3.1 First Interpretation

Insofar we assumed that the separation distance between two successive mode peaks (BER) is fixed for any given BCS or k values. We have assumed that increasing biocomplexity alone is responsible for the increasing BER. However, the separation distance can be altered through *the selection factor*. One can manually determine BER's separation. Recall that the separation distance corresponds to number of traits skipped per 100 Myr. That is, from our previous interpretation, each successive rounds of BCS includes additional traits and the mode of the search space shifts right. The most frequent occurring species from the previous round of distribution (with traits ranges from 0 to 10) could possess 5 traits but the most frequently occurring species from the next round of distribution (with traits ranges from 0 to 15) possessed 7 traits. Those 2 traits skipped constitute the BER. However, if natural selection or catastrophic extinction can only ensures only those species with more advanced traits, the top 10% of all species from the previous round (which correspond to species possessed ≥ 10 traits) survives, then the next round of distribution excludes all species < 9 traits, so the distribution possessed traits ranges from 10 to 25 and a mode peak at 17 traits after the survivors re-establish themselves on earth. Therefore, BER can be thought as the composite result of distribution's BCS and its placement pattern and a further shift by natural selection.

Furthermore, an increase in BER tells nothing about what percentage of the species from the previous round made it to the next. For example, a BER of 3 per 100 Myr can represent an increase of brain size by 3 times by the most commonly found species 100 Myr apart. This tripling of brain size could be attributed to the addition of 3 traits or 6 traits. (This is determined by the selection factor) Moreover, species with extra 3 traits compared to the most common species at the current time could place those species at either the top 40% or top 10% of all species currently observed. (This is determined by the deviation/width of the current distribution.) The composite effect yields a percentage of species from the previous round that made it to the next. This final percentage can be expressed through the selection factor.

As a result, without knowing exactly how a typical mammalian brain size is compared to all dinosaurs 100 Mya and simply knowing that it has increased by 10 fold, one could set the selection factor so that typical mammalian brain size today exceed 60% of dinosaurs and constitute the top 40% of dinosaurs. However, one could also increase the selection factor so that typical mammalian brain size today exceed 90% of dinosaurs and constitute the top 10% of dinosaurs 100 Mya. By adjusting the selection factor, we have established the one to one correspondence between BER (as measured by observed biological faculty change, i.e. brain size, snout length) and BER's placement position on the lognormal distribution.

In all cases, there is a tendency for increase in mode peak's shift, just more or less so, if biological complexity growth alone is inadequate to match the BER speed observed, then we apply the selection factor to increase the trait skipping gap per 100 Myr.

In a sense, we can almost compares our concepts of **difference in the modes, BER, deviation**, and **YAABER** to those of the classical mechanics concept of speed, acceleration, distance traveled, and amount of time it takes to travel the distance.

1. The **difference in the modes** between two adjacent distribution peaks geologic scaled time apart that establishes the one to one correspondence between BER and BER's placement position on the lognormal distribution can be thought as the *speed* of evolution. (not velocity, since its a scalar, not vector quantity). In classic mechanics, the magnitude of the final speed is a consequence of composite net force acted over time overcoming friction and resisting forces. Similarly, we have shown in Chapter 4 that evolution could proceed a lot faster if the environment changes quickly over geologic time but most of the time in evolution, stabilizing selection is favored. Thus, speed in our case is also defined as a compromise between intrinsic flexible, fast evolution capable potential from biological species adapting to changing environment (manifested from isolated population groups and founders beneficial mutations) and the extrinsic factors that generally stifle and suppress this potential (genetic drift, stabilizing selection due to stable environment, large population size), resulting in differential speed of evolution among different lineages within micro niches and a general speed among all species on earth's life history that is neither too fast nor completely non-changing and static. The final speed one observed is then the composite speed of stabilizing selection, which describes all periods in which selection penalizes mutations and deviations from the static norm and directional selection, which describes all periods in which selection encourages and prizes mutations and deviations toward the direction such as greater brain size. However, unlike classical mechanics, where the initial velocity can be 0, the initial speed of evolution > 0 because there is a limited time span of multicellular evolution and there is not enough time for the speed to reduce to 0. The initial speed of evolution is simply the speed of evolution for the first 100 Myr when multicellular life appeared. Generally, one set the evolutionary speed of the last 100 Myr as the "initial" speed so that speed of earlier and later period is defined based on it.

2. **BER** per 100 Myr can be thought as the *acceleration* of the evolutionary speed per 100 Myr. The only difference is that under classical mechanics, the relationship between velocity and acceleration is defined as $v_f = v + at$, but in our case it is defined as $v_f = v(a)^t$. That is, the definition of classical mechanics velocity is the logarithm of our definition. $v_f = \log\left(v(a)^t\right)$. Our acceleration is an exponentially increasing acceleration instead of a linear increasing acceleration. The acceleration in mechanics is due to force or gravity on mass, but in our case it is due to geologic events which asserts selection pressure on biodiversity, which breeds exponentially more biodiversity. To understand why both the speed of directional and stabilizing selection is exponentially increased by BER, imagining a scenario in which always top 10% of all species at the end of every round are selected to the next round of evolution per 100 Myr period, and a BCS of 2 with fixed height for successive rounds so $k = 1$. At the start of the first round, there are 10 possible traits. During the combination of these traits and at the end of the round completing $< 10^{10}$ speciation attempts, proportionally 20 new traits appeared in existing species corresponding to new niches and new adaptation to existing niches but yet to become mainstream (How new traits are generated, their placement, and possible combinations based on existing traits present several scenarios and bounds and leaves for future explanatory content expansion). Then, within the mainstream of this round, for all species possessed 1 up to 10 traits, only species possessed 10 traits or above made it to the next round and directional selection skipped 9 traits (These 9 traits becomes fixed in all species in the new round). The species with 10 or more traits eventually evolved into the biological fauna possessed between 10 and 10+20=30 traits by entering new niches associated with new traits and acquiring them and becomes the new mainstream. During this epoch and at the end of the round completing $< 21^{21}$ speciation attempts, proportionally 42 new traits appeared in existing species but yet to become mainstream. Then,

within the mainstream of this round, only species possessed > 28 traits made it to the next round and directional selection skipped 19 traits (These 19 traits becomes fixed in all species in the new round). The species with > 28 traits eventually evolved into biological fauna possessed $28+1$ up to $28+44=72$ traits and becomes the new mainstream. During this epoch and at the end of the round completing $< 43^{43}$ speciation attempts, proportionally 88 new traits appeared in existing species but yet to become mainstream. Then, within the mainstream of this round, only species shared ≥ 68 traits made it to next round and directional selection skipped 40 traits (These 40 traits becomes fixed in all species in the new round). Because each current round there are exponentially more established traits (which composed of top 10% of established/mainstream traits from the previous round + all additional new traits emerged from the previous round[5]) solidified within the total pool of species compares to previous rounds, then, in order to keep a fixed top 10% of all species for the next round, the directional selection has to skip exponentially more traits compares to previous rounds, hence the speed of directional selection increases exponentially. Consequently, the remaining survived species of each round possessed increasingly more traits with increasingly greater range of possessed traits, and coupled with exponentially more new traits, ensuring a recovered species pool both shared exponentially more traits (BER) shaped by stabilizing selection and exponentially larger number (BCS) through combination pairing on greater range values. (More is discussed in 8.7.3.5: Conclusions and 3 Laws of Evolution) Finally, force can be variable in classical mechanics. In our case, however, the number of major geologic changes, conceptually equivalent to force, throughout earth's history following the continent cycle remains largely fixed throughout the history of multicellular evolution.

3. The **deviation**, representing the distance (**difference in the modes** times **YAABER**) between a species capable of civilization emergence and the current mode of species can be thought as the total *distance traveled* between now and the time in the future when the mode of the species reaches parity with human capability, which defines currently how far apart they are in terms of number traits human have gained. The only difference is that under classical mechanics, the relationship between distance, velocity and acceleration is defined as $d = \int (v + at)\, dt$, but in our case it is defined as $d = \int v\,(a)^t\, dt$.

4. **YAABER** defines how many years into the future when the mode of the species reaches parity with human capability given the current evolutionary speed. This is simply derived from $\textbf{YAABER} = \frac{\textbf{deviation}}{\textbf{speed}}$, which is similar to classical mechanics Time $= \frac{\text{distance}}{\text{Velocity}}$.

Mathematically, the selection factor is expressed through increasing the spacing between consecutive distributions. In order to increase the spacing, an extra function term is needed to be added to the original bi-variate exponential log-normal distribution function as:

$$P_{df}(t, x) = \frac{1}{\sigma\sqrt{2\pi}} \exp\left(-\frac{\left(\ln\left((B_{cs})^t \cdot (x + G(t))\right)\right)^2}{2\sigma^2}\right) \tag{8.7.10}$$

Whereas $G(t)$ is defined as an unit horizontal displacement for BER $= 2.783$:

$$G(t) = P_{eak}(1) \frac{(B_{cs})^{-t} - 1}{(B_{cs})^{-1} - 1} \tag{8.7.11}$$

So that the current distribution's placement shift is $G(0) = 0$ when t=0 for $P_{df}(0, x)$'s maxima to stay at 1. The distribution of 100 Mya's placement shifts is $G(1) = 0.368$ when t=1 for $P_{df}(1, x)$'s maxima shifts to 0, so that for a BER of 2.783 the mode peak between now and 100 Mya are exactly unit length 1 apart, and every other time periods adjust (expands or contracts) accordingly based on this newly defined measured distance between the mode of now and 100 Mya.

---

[5]and both of which have been exponentially growing since the very first round respectively

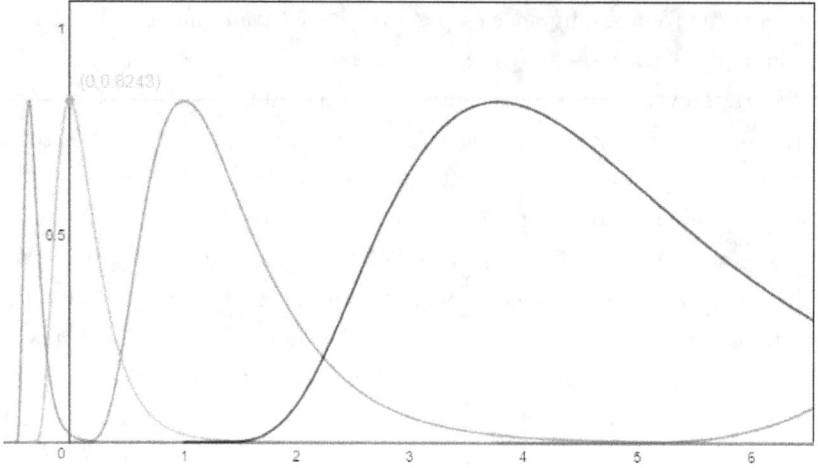

Figure 8.7.5: $P_{df}(1,x)$'s mode placed on x = 0. $P_{df}(0,x)$'s mode unchanged, and the rest of distributions adjust accordingly based on the **BER** measured between now and 100 Mya.

After we defined the unit horizontal displacement, we need to generalize the distance between the mode peak of now and 100 Mya to any arbitrary BER other than 2.783 so that the gap between successive mode peaks reflect the speed and distance/deviation traveled/shifted relative to the unit horizontal displacement defined for BER=2.783. We introduce the m_{ore} factor:[6]

$$G(t) = P_{eak}(1) \frac{(B_{cs})^{-t} - 1}{(B_{cs})^{-1} - 1} \cdot \frac{m_{ore}}{P_{eak}(1)} \tag{8.7.12}$$

$$m_{ore} = \frac{1 - (B_{er})^{-1}}{1 - 2.783^{-1}} - (P_{eak}(0) - P_{eak}(1)) \tag{8.7.13}$$

Basically, one finds the distance required for the separation between the two adjacent mode peaks $P_{df}(0,x)$ and $P_{df}(1,x)$ 100 Myr apart for a given BER. We just set the given BER = 2.783, so the EQ enlargement or distance traveled by mode peaks 100 Myr apart is:

$$1 - \frac{1}{2.783^1} \tag{8.7.14}$$

and the EQ enlargement or distance traveled by mode peaks for any other BER 100 Myr apart is:

$$1 - (B_{er})^{-1} \tag{8.7.15}$$

These are true because the distance traveled (EQ gained) by BER is manifested in equation:

$$D(t) = (B_{er})^t \tag{8.7.16}$$

and the distance traveled (EQ gained) by BER between now and n years into the past is simply:

$$D(0) - D(-n) = (B_{er})^0 - \frac{1}{(B_{er})^n} \tag{8.7.17}$$

---

[6]The m_{ore} factor is normalized/divided by $P_{eak}(1)$ because $G(t)$ function's horizontal displacement requirement for distributions across time is expressed in units of the value of $P_{eak}(1)$

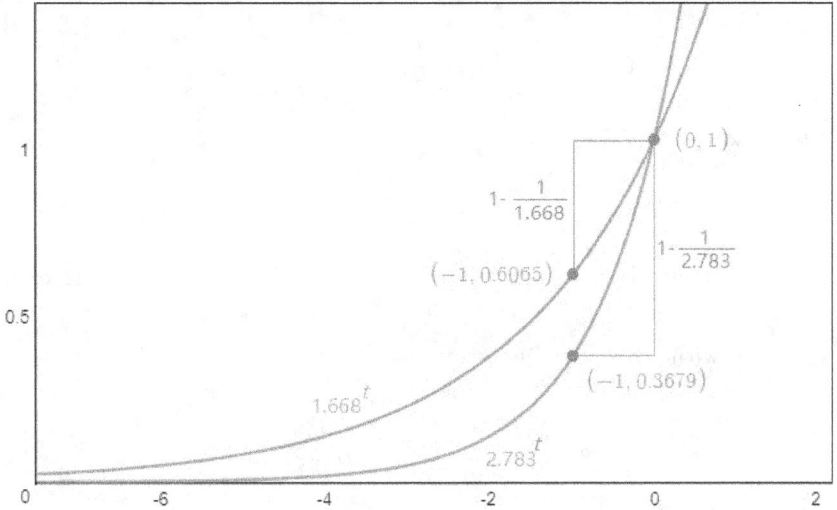

Figure 8.7.6: An illustration of EQ enlargement for the mode based on 2 different BER, for a BER=$2.783^{\frac{1}{2}}$, the enlargement between 100 Mya and now is $D(0) - D(-1) = (1.668)^t - \frac{1}{(1.668)^1}$, for a BER = 2.783, the enlargement between 100 Mya and now is $D(0) - D(-1) = (2.783)^t - \frac{1}{(2.783)^1}$

and the evolution speed at particular time is:

$$\frac{d}{dt}D(t) = \ln(B_{er})(B_{er})^t \tag{8.7.18}$$

In particular, when BER = 2.783, the ratio becomes 1:

$$\frac{1 - (2.783)^{-1}}{1 - 2.783^{-1}} = 1 \tag{8.7.19}$$

Such that:

$$m_{ore} = 1 - (P_{eak}(0) - P_{eak}(1)) = P_{eak}(1) \tag{8.7.20}$$

But, in general, it can be any value for ratio relative to the defined BER traveled distance of $1 - \frac{1}{2.783^1}$.
This separation distance is then subtracted from the default distance between the two adjacent mode peaks $P_{df}(0,x)$ and $P_{df}(1,x)$ 100 Myr apart for the case of $P_{df}(t,x) = \frac{1}{\sigma\sqrt{2\pi}}\exp\left(-\frac{(\ln((B_{cs})^t \cdot x))^2}{2\sigma^2}\right)$ Whereas $G(t)$ is excluded. The remaining difference is the adjustment needed to correctly place $P_{df}(1,x)$ and keeping $P_{df}(0,x)$'s position fixed and reflecting the amount of distance traveled for any BER. Moreover, distance traveled and instantaneous speed are not identical. Recall that for a BER = 2.783, the speed of evolution is actually slightly larger than 1 as $\frac{d}{dt}D(t) = \ln(2.783)(2.783)^0$ at the current time, and the distance actually traveled in the past 100 Myr is actually $1 - \frac{1}{2.783^1} < 1$. Current evolutionary speed can be 1 so that YAABER=Deviation only when BER = e. For computational convenience, we re-adjust our definition of an unit horizontal displacement enlargement (from the default 0 displacement) as *the distance with a value of 1 required for the forced separation between the two adjacent mode peaks $P_{df}(0,x)$ and $P_{df}(1,x)$ 100 Myr apart corresponds to the current evolution speed=1 derived from BER = e.*

$$m_{ore} = \frac{1 - (B_{er})^{-1}}{1 - e^{-1}} - (P_{eak}(0) - P_{eak}(1)) \tag{8.7.21}$$

Since the total distance traveled by BER = e is $D(t) = e^t$ and its speed is defined as $\frac{d}{dt}D(t) = \ln(e)(e)^0 = 1$ at the current time. With this definition, a BER = 2.783 places slightly to the left of the y-axis represent current evolution speed slightly above 1.
One also notices that $G(t)$ can be simplified as:

$$G(t) = \frac{(B_{cs})^{-t} - 1}{(B_{cs})^{-1} - 1} \cdot m_{ore} \tag{8.7.22}$$

The function $M_{ax}(t)$ is defined as:

$$M_{ax}(t) = -G(t) + P_{eak}(t) \tag{8.7.23}$$

So one can easily find the x coordinate for the maxima of PDF of different time periods *after the spacing adjustment*.

Function $G(t)$ is then generalized with the k variable term:

$$G(t) = P_{eak}(1) \frac{(B_{cs})^{-\frac{t}{k}} - 1}{(B_{cs})^{-\frac{1}{k}} - 1} \cdot \frac{m_{ore}}{P_{eak}(1)} \tag{8.7.24}$$

along with the original lognormal distribution as:

$$P_{df}(t, x, k) = \frac{(B_{cs})^{-\frac{(k-1)t}{k}}}{\sigma\sqrt{2\pi}} \exp\left(-\frac{\left(\ln\left((B_{cs})^{\frac{t}{k}} \cdot (x + G(t))\right)\right)^2}{2\sigma^2}\right) \tag{8.7.25}$$

Finally, we modify our existing function $G(t)$ to accommodate the selection factor:

$$m_{ore} = F_0 \cdot \frac{1 - (B_{er})^{-1}}{1 - e^{-1}} - (P_{eak}(0) - P_{eak}(1)) \tag{8.7.26}$$

One can quickly finds the minimum F_0 by setting the remaining difference between adjustment and the lognormal distribution without $G(t)$ function to 0 as:

$$0 = F_0 \cdot \frac{1 - (B_{er})^{-1}}{1 - e^{-1}} - (P_{eak}(0) - P_{eak}(1)) \tag{8.7.27}$$

Re-arranging the equation, one finds that:

$$F_0 = (P_{eak}(0) - P_{eak}(1)) \cdot \frac{1 - e^{-1}}{1 - (B_{er})^{-1}} = 1 - e^{-1} = 0.63212 \tag{8.7.28}$$

and $\frac{P_{eak}(0) - P_{eak}(1)}{1 - (B_{er})^{-1}} = 1$ because two adjacent mode peaks are always $1 - (B_{er})^{-1}$ apart given any BER. Therefore, the minimum value of the selection factor is the separation distance between $P_{eak}(0) - P_{eak}(1)$ for BER $= e$ before the selection factor and $G(t)$ is introduced.

$$F_0 \geq F_{min} = 1 - \frac{1}{e} = 0.63212 \tag{8.7.29}$$

That is, F_0 must be $\geq 1 - \frac{1}{e}$ to match our lognormal distribution without horizontal displacement. As a result, we shall label this value as the default as selection factor value of 1.

With this modification, $P_{eak}(0) = M_{ax}(0)$ always fixed, so distributions at other time periods adjust (expands or contracts) relative to the distribution represents the current time $P_{df}(0, x)$. When $F_0 = F_{min}$, it represents the classic case, as expected. When $F_0 = 1$, it is the unit distance with a value of 1 required for the separation between the two adjacent mode peaks $P_{df}(0, x)$ and $P_{df}(1, x)$ 100 Myr apart for a given BER $= e$. The increase in F_0 can also be expressed in terms of additional distance traveled by factor transformation at any time on existing distribution:

$$D(t) = (B_{er})^t \times \frac{F_0}{F_{min}} \tag{8.7.30}$$

as well as the current evolutionary speed:

$$S(t) = \frac{d}{dt}D(t) = \ln(B_{er})(B_{er})^t \times \frac{F_0}{F_{min}} \qquad (8.7.31)$$

The increase in separation distance does not alter the BER, in which the BER derivation ratio remains the invariant:

$$B_{er} = \frac{Max(t-1) - Max(t)}{Max(t) - Max(t+1)} \qquad (8.7.32)$$

and for any change in F_0, the relationship of $B_{er} = (B_{cs})^{\frac{1}{k}}$ for various k remain invariant so that their traveled distance ratio relative to k=1 stays the same:

$$B_{er}(k) = \frac{\frac{F_0}{F_{min}} \times \left(1 - e^{\frac{-1}{k}}\right)}{\frac{F_0}{F_{min}} \times \left(1 - e^{\frac{-1}{1}}\right)} \qquad (8.7.33)$$

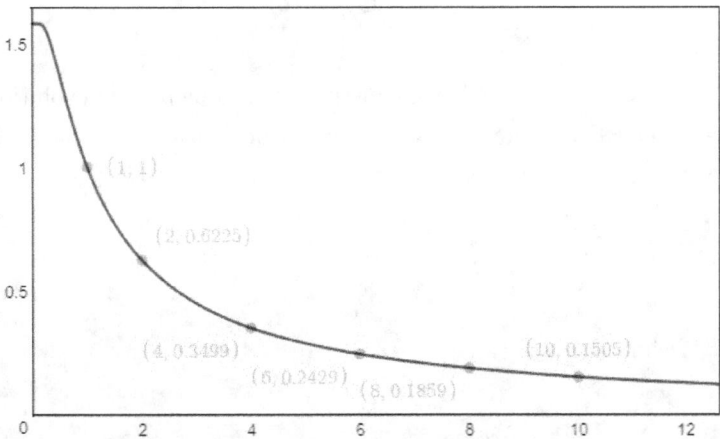

Figure 8.7.7: For any F_0, the relationship $B_{er} = (B_{cs})^{\frac{1}{k}}$ remains invariant.

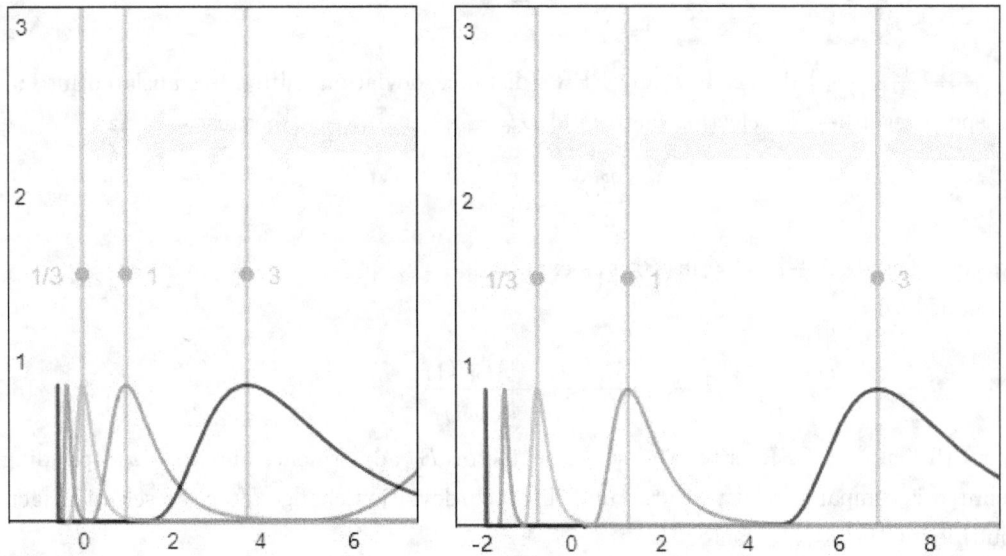

Figure 8.7.8: Selection factor $= 1$ for the left side and Selection factor $= 2$ for the right side

Finally, the emergence chance of civilization is discussed in terms of the selection factor. As an example, if it is assumed that $Max(0) - Max(1)$ corresponds to BER $=$ e and $F_0 = 1$ and results in top 15.57% selection every 100 Myr. Then, an increase in selection factor $= 2$ widens the separation between two successive distributions and results in top 0.4989% selection every 100 Myr. Ultimately, the separation distance representing the number of traits required for gaining a EQ from 0 to 1 increased. Though Deviation, BER, YAABER, and the distribution width remain fixed. L_{limit} becomes larger, and the overall emergence chance drops.

The relationship is captured by the following diagram:

		Yaaber	$M_{ax}(0) - M_{ax}(1)$	L_{limit}	+	Current Variance	Overall
selection factor ↑	BER −	Yaaber−	↑	↑		No change	↓

The fixation of YAABER can be expressed as:

$$\frac{D_{eviation} \cdot 1 \cdot S_{change}}{S_{change}} \left(\ln \left(\frac{1}{(1 - (1 - e^{-1})R_{atio})} \right) \right)^{-1} \tag{8.7.34}$$

whereas S_{change} is defined as:

$$S_{change} = \frac{\frac{F_0}{F_{min}} \times G_{speed}(e)}{1 - \frac{1}{e}} \tag{8.7.35}$$

So that any arbitrary BER with the same selection factor constitute the same pace of evolution speed. Whereas $\frac{F_0}{F_{min}} \times G_{speed}(B_{er})$ is defined as given BER $= e$ and k $= 1$ and a selection factor of F_0, how its distance/deviation shifted compares to the unit distance/deviation shifted with a selection factor of F_{min}:

$$G_{speed}(B_{er}) = \frac{1 - (B_{er})^{-1}}{1 - e^{-1}} \left(1 - \frac{1}{e} \right) = 1 - (B_{er})^{-1} \tag{8.7.36}$$

and $\frac{G_{speed}(e)}{(1 - \frac{1}{e})} = 1$ so that S_{change} simplifies to:

$$S_{change} = \frac{F_0}{F_{min}} \tag{8.7.37}$$

so that YAABER for the same given BER but different selection factor still stay the same:

$$\frac{D_{eviation} \cdot 1 \cdot S_{change}}{S_{change}} = \frac{D_{eviation} \cdot 1}{1} \tag{8.7.38}$$

and the term $\ln \left(\frac{1}{(1-(1-e^{-1})R_{atio})} \right)$ defines how any BER's distance/deviation shifted is translated into instantaneous evolution speed regardless of selection factor and $B_{er} = \frac{1}{(1-(1-e^{-1})R_{atio})}$ because:

$$\frac{1 - B_{er}^{-1}}{1 - e^{-1}} = R_{atio} \tag{8.7.39}$$

so that speed is derived based on $\frac{d}{dt}D(t) = \ln(B_{er})$ as expected.

Whereas R_{atio} is defined as:

$$R_{atio} = \frac{M_{ax}(0) - M_{ax}(1)}{(1 - \frac{1}{e}) \cdot S_{change}} \tag{8.7.40}$$

which specifies how the current BER, after the selection factor F_0 adjustment, and its corresponding distance/deviation shifted is compared relative to the unit distance/deviation change after the selection factor F_0 adjustment by multiplying the S_{change} factor.

We have just showed that *if we don't know how an increase in BER at the current 100 Myr correspond to the percentage of the species from the previous round made it to the next. We also don't know how rare human civilization's placement L_{limit} correspond to the number of traits skipped (so that human civilization's placement L_{limit} along the horizontal axis remains flexible).* By altering the selection factor, both BER and YAABER remain fixed and the selection factor determines how many traits skipped per 100 Myr and the total number of traits needed to be skipped for reaching the human civilization's complexity.

8.7.3.2 Second Interpretation

Unfortunately and fortunately, there lies an alternative interpretation for the selection factor. *If we initially do know that an increase in BER* at the current 100 Myr corresponds to the percentage of the species from the previous round made it to the next and the number of traits skipped, and *we settled and fixed on the human civilization's placement position L_{imit} corresponding to the number of traits skipped on the horizontal axis*. Then, an alteration on the selection factor results in a change in the YAABER expressed in terms of multiples of number of traits skipped to achieve an EQ of 1 from 0. The value of YAABER is defined as the ratio of the number of traits skipped to achieve human level complexity relative to the number of traits skipped by the current background evolution per 100 Myr. By increasing the selection factor, the number of traits skipped per 100 Myr is increased, as result, YAABER drops, and the *human civilization's placement position L_{imit} remain fixed.*

Implementing the alternative interpretation of the selection factor is easy, all one needs to do is eliminate the speed change S_{change} factor so that $L_{imit} = 18$ remain fixed.

$$L_{imit} = D_{eviation} \cdot 1 \cdot (S_{change})^0 + M_{ax}(0) \tag{8.7.41}$$

7

One can add a Boolean toggle variable that can take on the value of only 0 or 1 to alternate between the two interpretations of the selection factor:

$$L_{imit} = D_{eviation} \cdot 1 \cdot (S_{change})^{(1-T)} + M_{ax}(0) \tag{8.7.42}$$

$$T = \begin{cases} 0 & \text{First Interpretation} \\ 1 & \text{Second Interpretation} \end{cases} \tag{8.7.43}$$

so it integrates both interpretations within the same framework.

Consequently, YAABER drops and is compensated by space increase (more traits skipped) and L_{imit} stays, and the overall emergence chance stays constant for $P_{df}(0, x)$, though emergence at earlier times drops.

	Yaaber	$M_{ax}(0) - M_{ax}(1)$	L_{imit}	+	Current Variance	Overall
selection factor ↑ BER −	Yaaber ↓	↑	No change		No change	No change

The YAABER now becomes variable and it is expressed as:

$$\frac{D_{eviation} \cdot 1}{S_{change}} \left(\ln \left(\frac{1}{(1 - (1 - e^{-1}) R_{atio})} \right) \right)^{-1} \tag{8.7.44}$$

so that YAABER for the same given BER but different selection factor changes:

$$\frac{D_{eviation} \cdot 1}{S_{change}} \neq \frac{D_{eviation} \cdot 1}{1} \tag{8.7.45}$$

8.7.3.3 Test Cases

Having demonstrated the relationship between BCS and BER, and introduced the variable k and the selection factor, different test cases are presented:

First, we present the cases for the first interpretation of the selection factor: [8]

---

[7] One can argue that counting starts at $M_{ax}(\infty) < 0$ for $F_0 > F_{min}$, which was $M_{ax}(\infty) = 0$ when $F_0 = F_{min}$. However, results are the same because $L_{imit} = D_{eviation} \cdot 1 \cdot (S_{change})^0 + M_{ax}(\infty) + (M_{ax}(0) - M_{ax}(\infty)) = D_{eviation} \cdot 1 \cdot (S_{change})^0 + M_{ax}(0)$

[8] the earliest window, the nearest civilization, occupancy ratio, visibility are somewhat off for selection factor > 1 because the data derived was from earlier model, the author has confirmed that the general trend and characteristics remain the intact and will be updated in the future.

Earliest window: for selection factor = 1, if $l_{ag} = 1.3592$ is considered, then, every values within the cell is increased by 1.3592.

BCS	1.3	1.6	1.8	2.0	2.4	2.6	2.783	3.0	3.2	3.4
$k{=}3$	3.119	1.956	1.619	1.414	1.173	1.094	1.034	0.974	0.926	0.884
$k{=}2$	2.344	1.469	1.217	1.063	0.882	0.822	0.777	0.732	0.696	0.664
$k{=}1$	1.381	0.866	0.717	0.626	0.52	0.484	0.458	0.431	0.41	0.391
$k{=}0.8$	1.147	0.719	0.595	0.52	0.431	0.402	0.38	0.358	0.34	0.325

Table 8.7.2: Selection = 1

for selection factor = 1.285 , if $l_{ag} = 1.3592$ is considered, then, every values within the cell is increased by 1.3592.

BCS	1.3	1.6	1.8	2.0	2.4	2.6	2.783	3.0	3.2	3.4
$k{=}3$	3.292	2.207	1.887	1.689	1.450	1.369	1.308	1.246	1.196	1.152
$k{=}2$	2.657	1.781	1.523	1.363	1.170	1.105	1.056	1.005	0.965	0.930
$k{=}1$	1.764	1.183	1.011	0.905	0.777	0.734	0.701	0.668	0.641	0.617
$k{=}0.8$	1.521	1.019	0.872	0.780	0.670	0.633	0.604	0.576	0.553	0.532

Table 8.7.3: Selection = 1.285

for selection factor = 1.660, if $l_{ag} = 1.3592$ is considered, then, every values within the cell is increased by 1.3592.

BCS	1.3	1.6	1.8	2.0	2.4	2.6	2.783	3.0	3.2	3.4
$k{=}3$	3.472	2.451	2.137	1.938	1.697	1.614	1.551	1.486	1.434	1.387
$k{=}2$	2.890	2.041	1.779	1.614	1.412	1.344	1.291	1.237	1.194	1.155
$k{=}1$	1.987	1.403	1.223	1.109	0.971	0.924	0.887	0.850	0.821	0.794
$k{=}0.8$	1.719	1.213	1.058	0.96	0.840	0.799	0.768	0.736	0.710	0.687

Table 8.7.4: Selection = 1.660

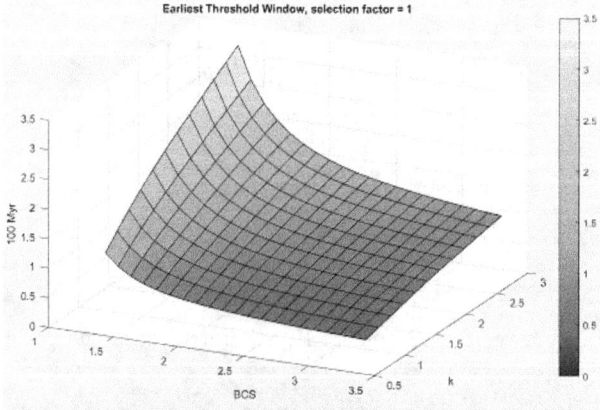

Figure 8.7.9: Earliest window for selection factor = 1

---

[9]the visibility is the instantaneous visibility, not cumulative visibility, recall cumulative visibility=occupancy

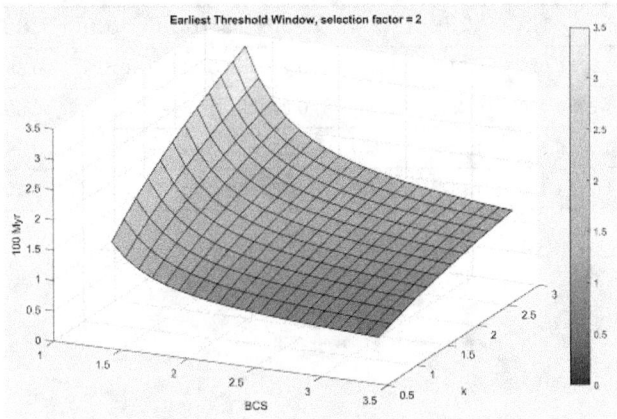

Figure 8.7.10: Earliest window for selection factor $= 1.285$

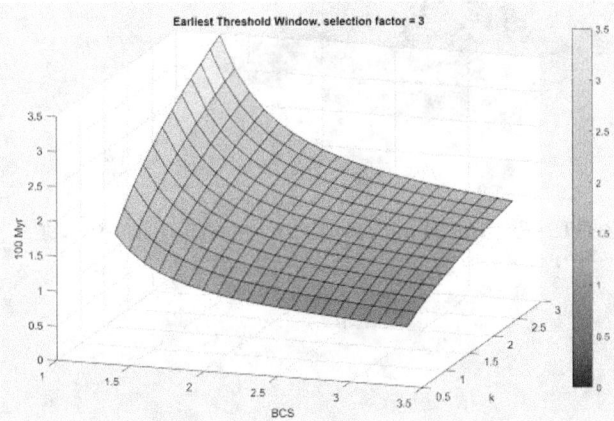

Figure 8.7.11: Earliest window for selection factor $= 1.660$

Nearest civilization: for selection factor $= 1$

BCS	1.3	1.6	1.8	2.0	2.4	2.6	2.783	3.0	3.2	3.4
$k=3$	0.161	0.184	0.194	0.202	0.214	0.218	0.222	0.226	0.230	0.233
$k=2$	0.167	0.193	0.205	0.215	0.228	0.234	0.238	0.243	0.247	0.250
$k=1$	0.176	0.207	0.221	0.232	0.249	0.255	0.261	0.266	0.271	0.275
$k=0.8$	0.179	0.211	0.226	0.237	0.254	0.262	0.267	0.273	0.277	0.283

Table 8.7.5: Selection $= 1$

for selection factor $= 1.285$

BCS	1.3	1.6	1.8	2.0	2.4	2.6	2.783	3.0	3.2	3.4
$k=3$	0.163	0.201	0.234	0.256	0.281	0.298	0.324	0.326	0.34	0.367
$k=2$	0.185	0.251	0.279	0.325	0.386	0.397	0.420	0.465	0.489	0.492
$k=1$	0.235	0.363	0.463	0.521	0.689	0.755	0.812	0.842	0.934	0.991
$k=0.8$	0.274	0.428	0.532	0.632	0.815	0.897	0.969	1.097	1.169	1.239

Table 8.7.6: Selection $= 1.285$

for selection factor $= 1.660$

BCS	1.3	1.6	1.8	2.0	2.4	2.6	2.783	3.0	3.2	3.4
$k=3$	0.17	0.237	0.290	0.317	0.404	0.438	0.468	0.483	0.535	0.543
$k=2$	0.205	0.318	0.410	0.486	0.631	0.671	0.730	0.836	0.862	0.924
$k=1$	0.328	0.635	0.861	1.146	1.637	1.799	2.105	2.356	2.587	2.817
$k=0.8$	0.399	0.874	1.169	1.516	2.217	2.561	3.007	3.381	3.725	3.862

Table 8.7.7: Selection = 1.660

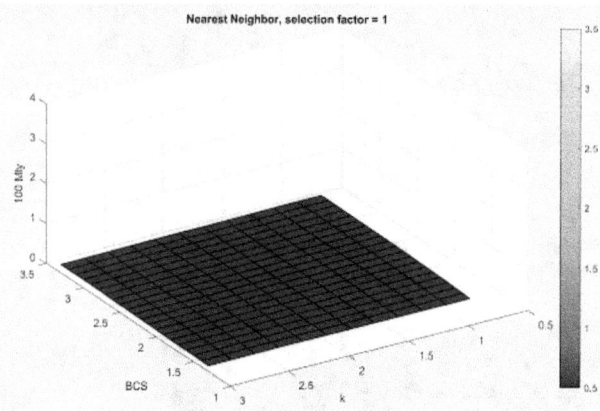

Figure 8.7.12: Nearest civilization for selection factor = 1

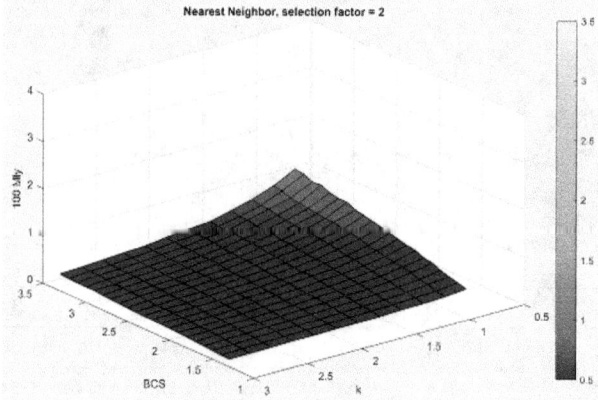

Figure 8.7.13: Nearest civilization for selection factor = 1.285

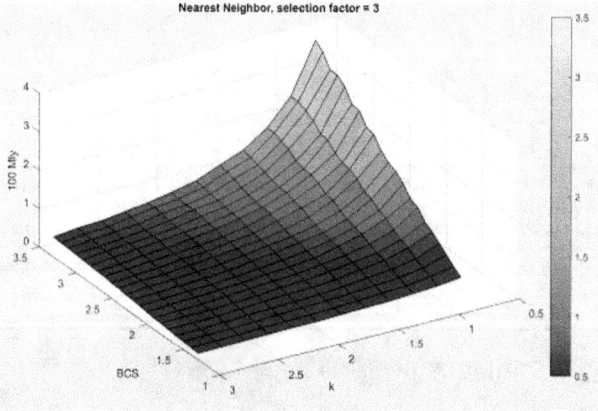

Figure 8.7.14: Nearest civilization for selection factor = 1.660

Occupancy ratio: for selection factor = 1

400

BCS	1.3	1.6	1.8	2.0	2.4	2.6	2.783	3.0	3.2	3.4
$k=3$	filled	filled	filled	filled	filled	filled	filled	filled	filled	filled
$k=2$	filled	filled	filled	filled	filled	filled	filled	78.659	62.950	51.898
$k=1$	filled	filled	filled	54.237	21.615	15.337	11.618	8.806	7.005	5.725
$k=0.8$	filled	filled	51.857	26.933	10.675	7.503	5.666	4.259	3.402	2.747

Table 8.7.8: Selection = 1

for selection factor = 1.285

BCS	1.3	1.6	1.8	2.0	2.4	2.6	2.783	3.0	3.2	3.4
$k=3$	filled	filled	filled	filled	filled	filled	filled	82.034	60.717	40.828
$k=2$	filled	filled	filled	filled	33.653	23.491	15.926	9.303	6.665	5.597
$k=1$	filled	39.316	9.186	3.81	0.76	0.432	0.276	0.197	0.118	0.083
$k=0.8$	filled	12.253	3.07	1.069	0.229	0.128	0.08	0.043	0.029	0.021

Table 8.7.9: Selection = 1.285

for selection factor = 1.660

BCS	1.3	1.6	1.8	2.0	2.4	2.6	2.783	3.0	3.2	3.4
$k=3$	filled	filled	filled	filled	80.032	47.645	31.405	22.916	13.947	11.399
$k=2$	filled	filled	89.626	32.033	6.868	4.311	2.672	1.397	1.063	0.729
$k=1$	filled	6.486	1.238	0.301	0.047	0.026	0.013	0.007	0.004	0.003
$k=0.8$	86.816	1.223	0.243	0.064	0.009	0.004	0.002	0.001	0.001	0.001

Table 8.7.10: Selection = 1.660

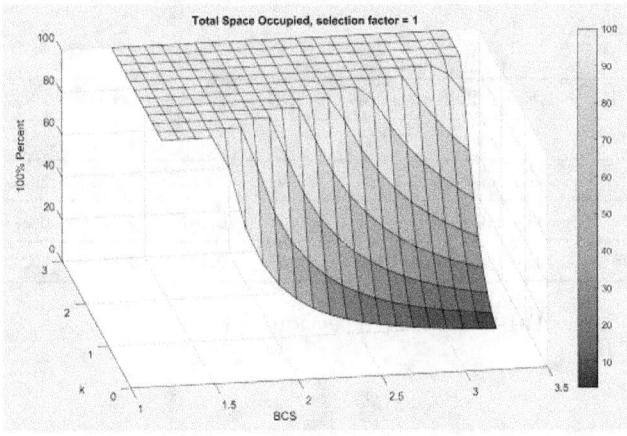

Figure 8.7.15: Occupancy ratio for selection factor = 1

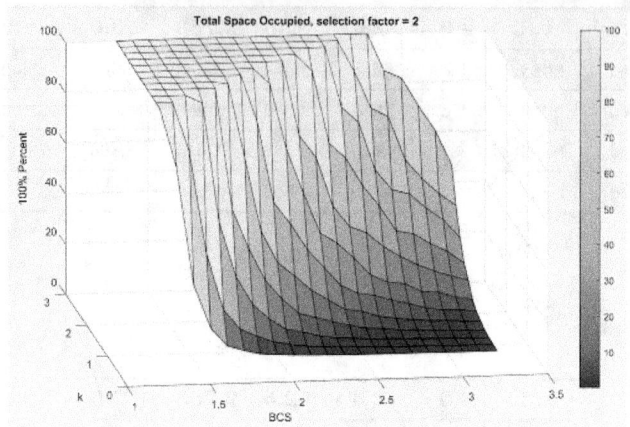

Figure 8.7.16: Occupancy ratio for selection factor = 1.285

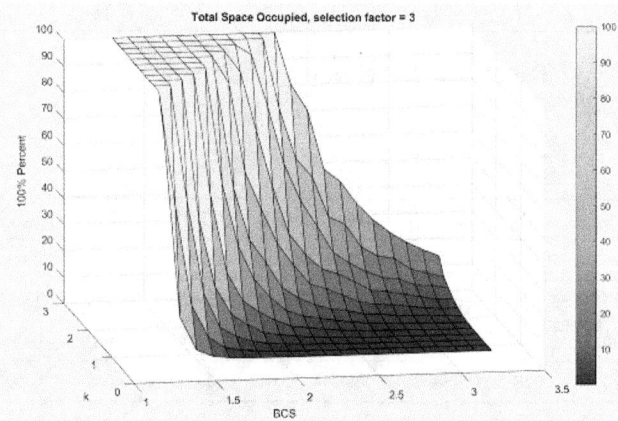

Figure 8.7.17: Occupancy ratio for selection factor = 1.660

Visibility: for selection factor = 1

BCS	1.3	1.6	1.8	2.0	2.4	2.6	2.783	3.0	3.2	3.4
$k=3$	visible	visible	visible	visible	71.005	47.519	34.525	24.789	18.901	14.897
$k=2$	visible	visible	visible	45.790	15.268	10.091	7.253	5.185	3.932	3.090
$k=1$	visible	20.142	6.841	3.074	0.984	0.644	0.457	0.325	0.245	0.191
$k=0.8$	visible	8.388	2.822	1.256	0.400	0.259	0.184	0.130	0.099	0.076

Table 8.7.11: Selection = 1

for selection factor = 1.285

BCS	1.3	1.6	1.8	2.0	2.4	2.6	2.783	3.0	3.2	3.4
$k=3$	visible	visible	visible	98.476	29.44	17.317	10.013	7.385	5.158	3.286
$k=2$	visible	visible	37.701	11.931	2.683	1.717	1.086	0.589	0.398	0.318
$k=1$	visible	3.123	0.579	0.203	0.032	0.017	0.010	0.007	0.004	0.002
$k=0.8$	34.328	0.785	0.156	0.046	0.008	0.004	0.002	0.001	0.001	0.001

Table 8.7.12: Selection = 1.285

for selection factor = 1.660

BCS	1.3	1.6	1.8	2.0	2.4	2.6	2.783	3.0	3.2	3.4
$k=3$	visible	visible	visible	48.138	8.829	4.793	2.939	1.996	1.142	0.887
$k=2$	visible	56.451	10.457	3.137	0.526	0.302	0.174	0.084	0.061	0.039
$k=1$	42.721	0.493	0.074	0.015	0.002	0.001	0	0	0	0
$k=0.8$	9.798	0.074	0.012	0.003	0	0	0	0	0	0

Table 8.7.13: Selection $= 1.660$

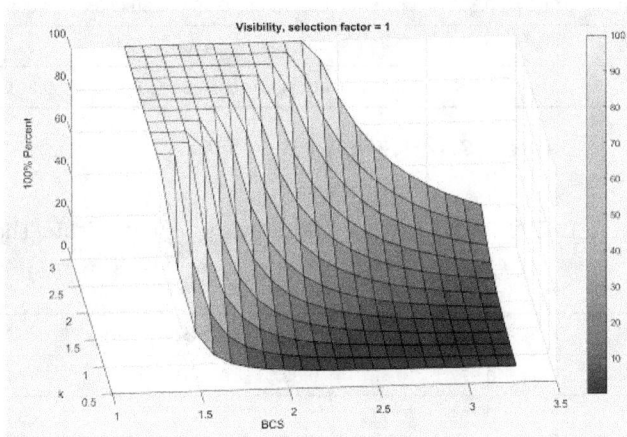

Figure 8.7.18: Visibility for selection factor $= 1$

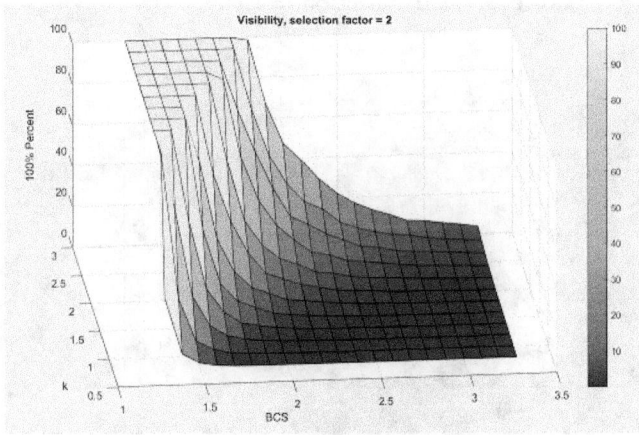

Figure 8.7.19: Visibility for selection factor $= 1.285$

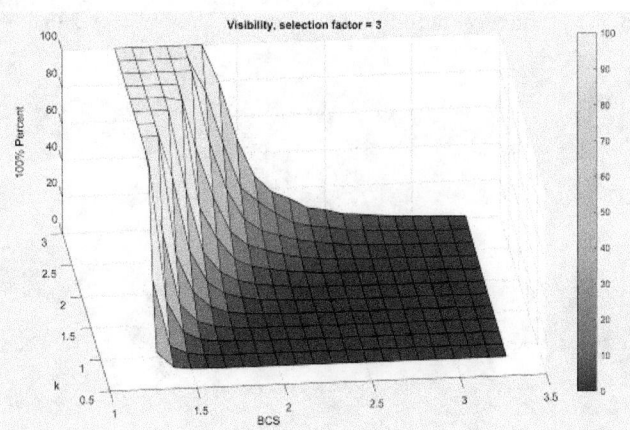

Figure 8.7.20: Visibility for selection factor $= 1.660$

Next, we present the cases for the second interpretation of the selection factor. Since the results for both interpretations converge on selection factor of 1, only cases for selection factor > 1 are presented:

Earliest window: for selection factor = 1.285, if $l_{ag} = 1.3592$ is considered, then, every values within the cell is increased by 1.3592.

BCS	1.3	1.6	1.8	2.0	2.4	2.6	2.783	3.0	3.2	3.4
$k=3$	1.728	1.344	1.195	1.09	0.950	0.901	0.863	0.824	0.794	0.767
$k=2$	1.390	1.081	0.962	0.877	0.765	0.725	0.694	0.663	0.639	0.617
$k=1$	0.902	0.701	0.624	0.569	0.496	0.470	0.450	0.430	0.414	0.400
$k=0.8$	0.765	0.595	0.529	0.483	0.421	0.399	0.382	0.365	0.351	0.340

Table 8.7.14: Selection = 1.285

for selection factor = 1.660, if $l_{ag} = 1.3592$ is considered, then, every values within the cell is increased by 1.3592.

BCS	1.3	1.6	1.8	2.0	2.4	2.6	2.783	3.0	3.2	3.4
$k=3$	1.422	1.164	1.057	0.977	0.867	0.827	0.796	0.764	0.738	0.716
$k=2$	1.171	0.958	0.870	0.805	0.714	0.681	0.655	0.629	0.608	0.589
$k=1$	0.786	0.643	0.584	0.540	0.479	0.457	0.440	0.422	0.408	0.395
$k=0.8$	0.673	0.551	0.500	0.462	0.41	0.391	0.376	0.361	0.349	0.339

Table 8.7.15: Selection = 1.660

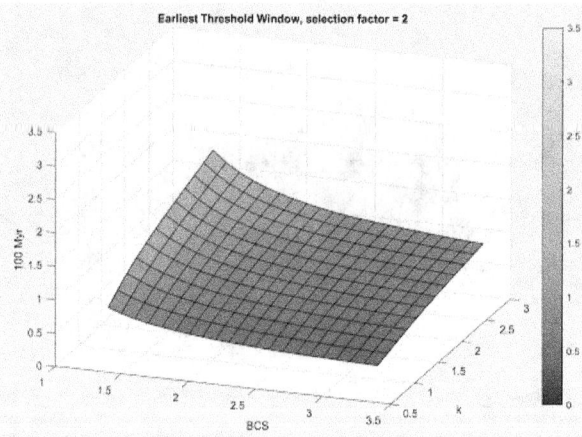

Figure 8.7.21: Earliest window for selection factor = 1.285

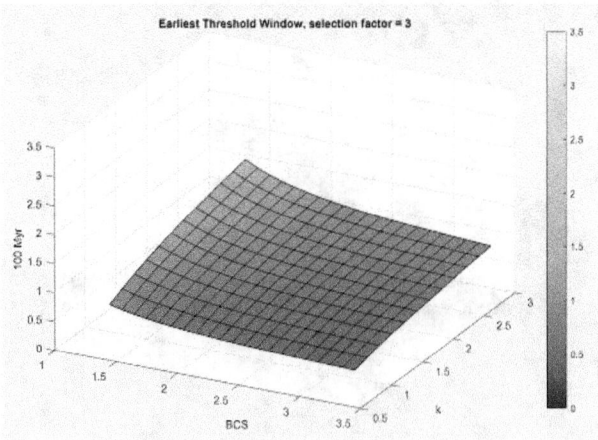

Figure 8.7.22: Earliest window for selection factor = 1.660

Nearest civilization: for selection factor = 1.285

BCS	1.3	1.6	1.8	2.0	2.4	2.6	2.783	3.0	3.2	3.4
k=3	0.167	0.190	0.201	0.208	0.220	0.225	0.228	0.232	0.235	0.238
k=2	0.172	0.198	0.209	0.219	0.232	0.237	0.241	0.246	0.250	0.253
k=1	0.179	0.209	0.223	0.234	0.250	0.256	0.261	0.266	0.271	0.275
k=0.8	0.181	0.213	0.227	0.238	0.255	0.262	0.267	0.273	0.277	0.283

Table 8.7.16: Selection = 1.285

for selection factor = 1.660

BCS	1.3	1.6	1.8	2.0	2.4	2.6	2.783	3.0	3.2	3.4
k=3	0.171	0.196	0.206	0.214	0.226	0.230	0.234	0.237	0.241	0.243
k=2	0.175	0.202	0.213	0.222	0.235	0.240	0.245	0.249	0.253	0.256
k=1	0.181	0.211	0.225	0.236	0.252	0.258	0.263	0.267	0.272	0.276
k=0.8	0.182	0.214	0.228	0.239	0.256	0.262	0.268	0.273	0.278	0.283

Table 8.7.17: Selection = 1.660

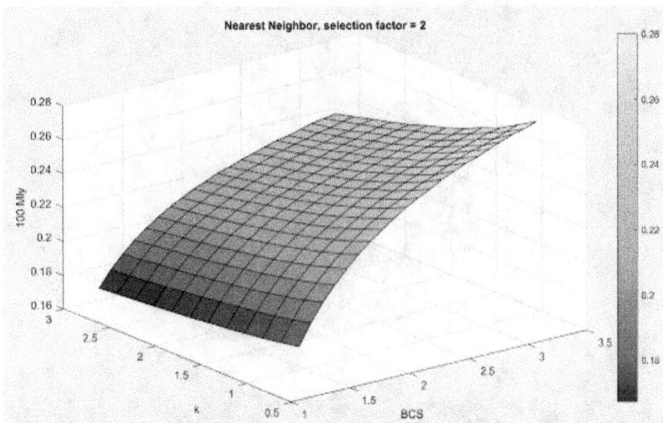

Figure 8.7.23: Nearest civilization for selection factor = 1.285

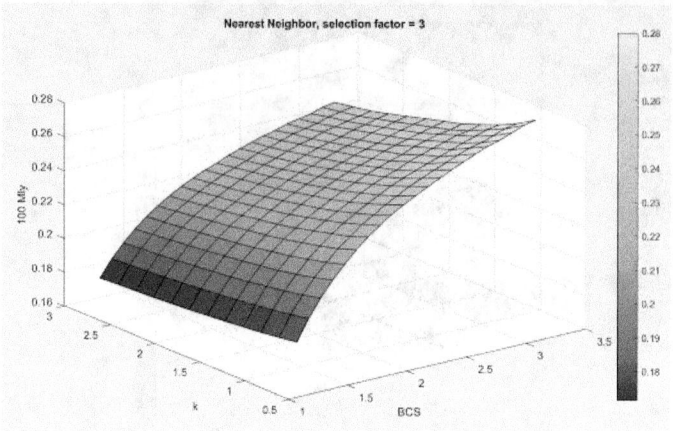

Figure 8.7.24: Nearest civilization for selection factor = 1.660

Occupancy ratio: for selection factor = 1.285

BCS	1.3	1.6	1.8	2.0	2.4	2.6	2.783	3.0	3.2	3.4
$k=3$	filled	filled	filled	filled	filled	filled	filled	filled	filled	filled
$k=2$	filled	filled	filled	filled	filled	92.167	73.041	57.392	46.851	39.334
$k=1$	filled	filled	77.301	42.77	18.343	13.366	10.364	7.966	6.406	5.317
$k=0.8$	filled	93.076	41.544	22.78	9.563	6.829	5.297	4.033	3.254	2.652

Table 8.7.18: Selection = 1.285

for selection factor = 1.660

BCS	1.3	1.6	1.8	2.0	2.4	2.6	2.783	3.0	3.2	3.4
$k=3$	filled	filled	filled	filled	filled	filled	filled	filled	filled	90.465
$k=2$	filled	filled	filled	filled	99.277	75.543	60.402	47.861	39.758	33.569
$k=1$	filled	filled	66.026	37.518	16.616	12.203	9.515	7.430	6.001	4.999
$k=0.8$	filled	79.745	36.803	20.667	8.931	6.481	5.005	3.870	3.100	2.563

Table 8.7.19: Selection = 1.660

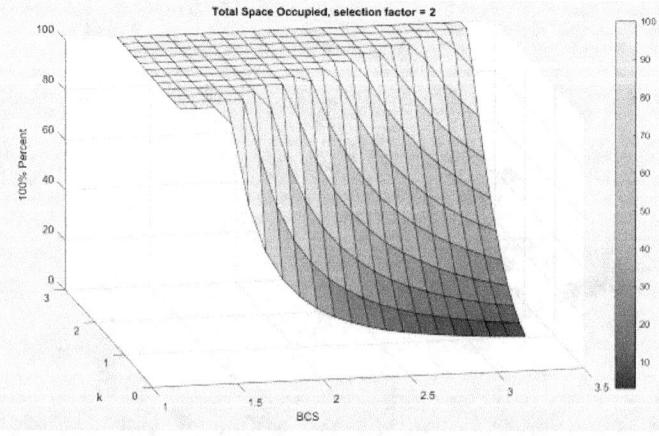

Figure 8.7.25: Occupancy ratio for selection factor = 1.285

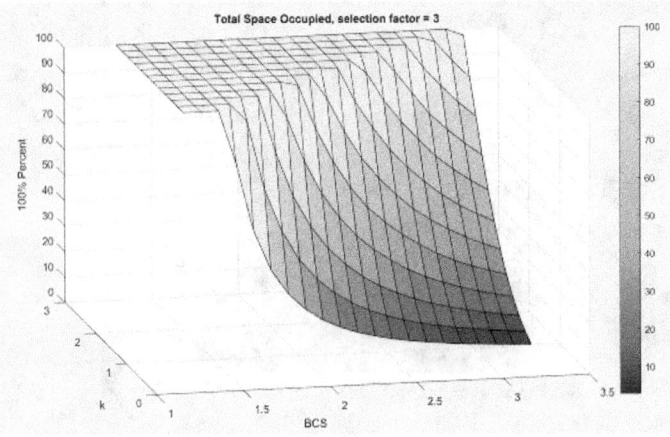

Figure 8.7.26: Occupancy ratio for selection factor = 1.660

Visibility: for selection factor =1.285

BCS	1.3	1.6	1.8	2.0	2.4	2.6	2.783	3.0	3.2	3.4
$k=3$	visible	visible	visible	77.639	32.243	22.94	17.46	13.14	10.441	8.456
$k=2$	visible	visible	45.339	23.529	9.084	6.307	4.725	3.502	2.727	2.194
$k=1$	visible	12.464	4.750	2.290	0.806	0.545	0.398	0.288	0.220	0.175
$k=0.8$	65.712	5.797	2.146	1.023	0.352	0.233	0.170	0.122	0.094	0.073

Table 8.7.20: Selection = 1.285

for selection factor = 1.660

BCS	1.3	1.6	1.8	2.0	2.4	2.6	2.783	3.0	3.2	3.4
$k=3$	visible	visible	81.625	47.694	21.335	15.601	12.159	9.378	7.535	6.233
$k=2$	visible	65.755	30.205	16.605	6.851	4.884	3.707	2.782	2.211	1.794
$k=1$	83.408	9.599	3.868	1.932	0.709	0.485	0.357	0.263	0.203	0.162
$k=0.8$	45.847	4.732	1.833	0.902	0.322	0.217	0.158	0.115	0.088	0.070

Table 8.7.21: Selection = 1.660

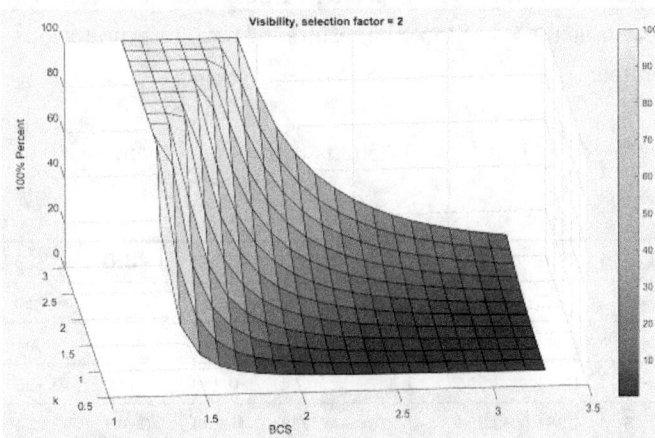

Figure 8.7.27: Visibility for selection factor = 1.285

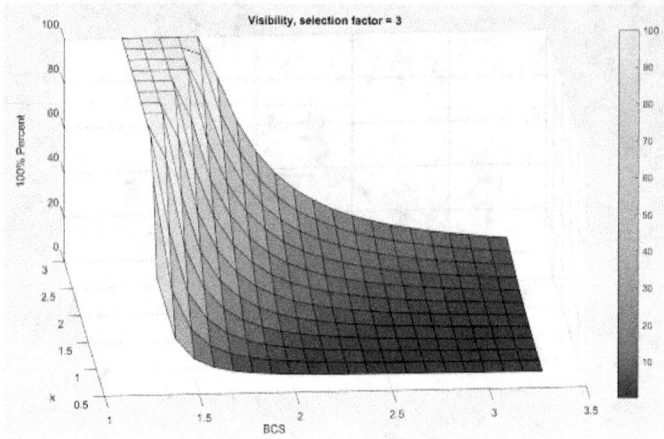

Figure 8.7.28: Visibility for selection factor $= 1.660$

We assumed that the current BCS achieved on earth, which is used as a reference, is equivalent to 3,553,331 species generated per 100 Myr for selected cohorts of birds, mammals, and reptiles. With a lower BCS, this level of biocomplexity can only be obtained if the multicellular evolutionary history spans much longer than observed on earth (longer than 500 Myr). Alternatively, a higher BCS implies the multicellular evolutionary history spans much shorter time than observed on earth to reach current diversity.

There also exists an alternative scenario in which multicellular evolution at the cosmic scale are only possible given an opportunity window less than the recent 500 Myr. Then, one can assume that all macro-evolution started almost instantaneously. Consequently, the current attainable biocomplexity will vary by BCS. With a lower number of species generated per 100 Myr, the chance of civilization emergence decreases. As a result, the nearest civilization will appear much farther away. We modified the previous model by taking this consideration by adding a transformation factor for the current attainable complexity given the selected BCS relative the current BCS of 2.783:

$$t_{cg} = -\frac{5 \ln\left(\frac{B_{cs}}{2.783}\right)}{\ln\left(B_{cs}\right)} \tag{8.7.46}$$

$$C_{df}(t) = \int_{-5}^{x} P_{lanet}(-s)\left(t_{Win}(B_{cs})^{s}\int_{L_{imit}}^{\infty} P_{df}(-s+t+t_{cg},x)\,dx\right)ds \tag{8.7.47}$$

So that the modified current complexity attained is expressed as the biocomplexity of the past or future relative to the previous current biocomplexity, and the result is presented below for selection factor $= 1$:

We only present the case for selection factor $=1$. Since the results for both interpretations converge on selection factor of 1, only one case is presented:

Earliest window: for selection factor $= 1$, if $l_{ag} = 1.3592$ is considered, then, every values within the cell is increased by 1.3592.

BCS	1.3	1.6	1.8	2.0	2.4	2.6	2.783	3.0	3.2	3.4
$k=3$	> 5	> 5	> 5	3.823	2.016	1.435	1.014	0.612	0.312	0.063
$k=2$	> 5	> 5	4.944	3.46	1.728	1.171	0.766	0.38	0.091	-0.147
$k=1$	> 5	> 5	4.414	3.005	1.367	0.84	0.456	0.089	-0.186	-0.41
$k=0.8$	> 5	> 5	4.284	2.893	1.278	0.758	0.38	0.017	-0.254	-0.474

Table 8.7.22: Selection $= 1$

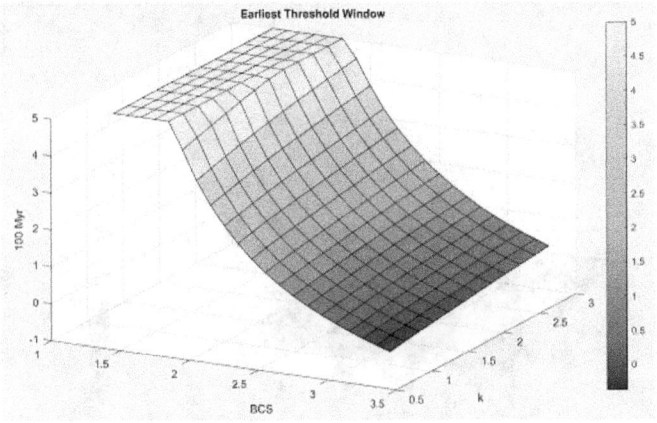

Figure 8.7.29: Earliest window for selection factor = 1

Nearest civilization: for selection factor = 1

BCS	1.3	1.6	1.8	2.0	2.4	2.6	2.783	3.0	3.2	3.4
k=3	78.829	12.115	4.711	2.101	0.573	0.336	0.217	0.137	0.093	0.066
k=2	> 138	> 138	33.234	8.711	1.066	0.461	0.236	0.117	0.066	0.040
k=1	> 138	> 138	> 138	> 138	6.894	1.078	0.261	0.065	0.022	0.009
k=0.8	> 138	> 138	> 138	> 138	18.585	1.641	0.268	0.047	0.013	0.005

Table 8.7.23: Selection = 1

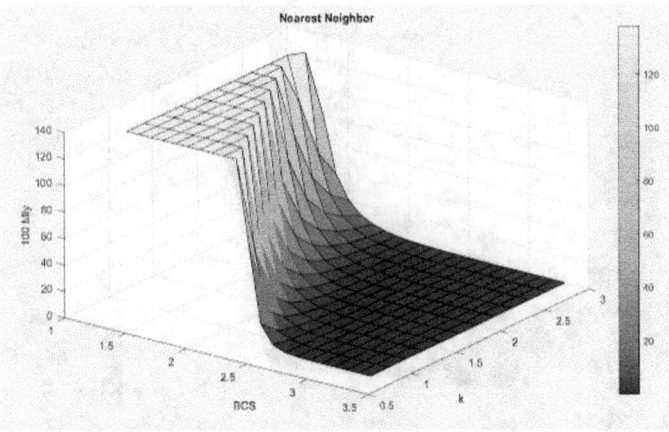

Figure 8.7.30: Nearest civilization for selection factor = 1

Occupancy ratio: for selection factor = 1

BCS	1.3	1.6	1.8	2.0	2.4	2.6	2.783	3.0	3.2	3.4
k=3	0	0.011	0.117	0.920	28.026	filled	filled	filled	filled	filled
k=2	0	0	0	0.004	1.388	15.293	filled	filled	filled	filled
k=1	0	0	0	0	0.001	0.149	11.073	filled	filled	filled
k=0.8	0	0	0	0	0	0.02	5.264	filled	filled	filled

Table 8.7.24: Selection = 1

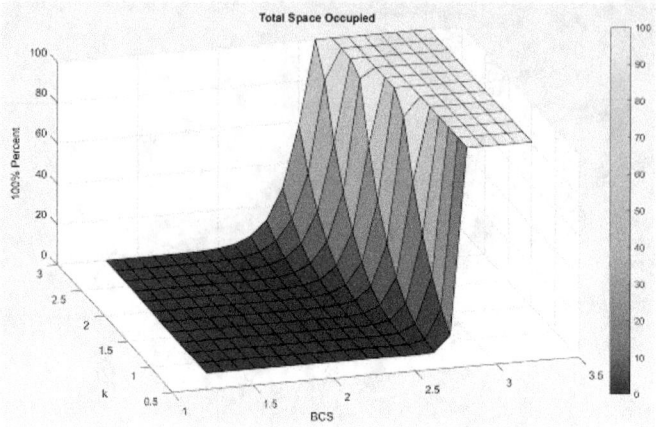

Figure 8.7.31: Occupation ratio for selection factor $= 1$

Visibility: for selection factor $= 1$

BCS	1.3	1.6	1.8	2.0	2.4	2.6	2.783	3.0	3.2	3.4
$k=3$	0	0.002	0.017	0.118	3.070	12.210	38.446	visible	visible	visible
$k=2$	0	0	0	0	0.105	1.123	7.576	56.630	visible	visible
$k=1$	0	0	0	0	0	0.006	0.435	32.861	visible	visible
$k=0.8$	0	0	0	0	0	0.001	0.169	40.251	visible	visible

Table 8.7.25: Selection $= 1$

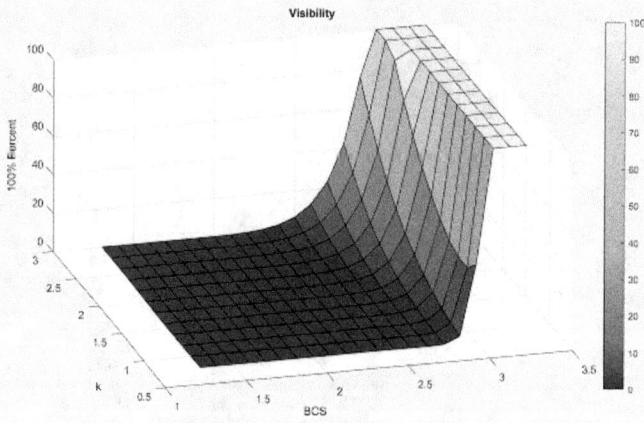

Figure 8.7.32: Visibility for selection factor $= 1$

Finally, those two scenarios comprise a spectrum of possibility landscapes.

Interestingly enough, the x-y plane that represents the earliest window, nearest civilization, occupancy ratio, and the chance of visibility can all be modeled by two sets of exponential functions along the BCS axis and k axis.

$$P_{\text{BCS}}(B_{cs}) = \exp\left(a\,(B_{cs})^b + c\,(B_{cs})^d\right) \tag{8.7.48}$$

$$P_k(k) = \exp\left(f\,(k)^g + h\,(k)^j\right) \tag{8.7.49}$$

Whereas P_{BCS} defines the cross-section of the x-y plane along the x axis (BCS axis), and P_k defines the cross-section of the x-y plane along the y axis (k axis). Depending on the scenarios and the selection factor chosen,

410

coefficients a, b, c, d, f, g, h, and j change accordingly to fit any possible cases ranging from fast exponentially changing to nearly constant static. The overall x-y plane is then defined as:

$$P(B_{cs}, k) = P_{\text{BCS}}(B_{cs}) + P_k(k) \tag{8.7.50}$$

Within the same selection factor, coefficients a, b, c, d, f, g, h, and j also change (more moderately) depending on its position in relation to BCS and k, as result, ultimately, each coefficient remains variable depending on BCS or k, so one could generalize each coefficients as $a = F_0(a, B_{cs})$, $b = F_1(b, B_{cs})$... $h = F_6(h, k)$, $j = F_7(j, k)$. As a result, it will provide more accurate description of the x-y plane, but one can treat the coefficients as constants for good approximation as a trade off for computation time and preparation cost.

The x-y planes could be fit with a polynomial plane. However, such fit, in general, performs less accurately in comparison to the exponential plane. This is valid since the PDF function we defined is based on exponential increase in both the mode and overall PDF size.

In general, selection factor increases the elevation (z value) of the x-y plane for the earliest window and nearest neighbor, since increasing selection factor implies the chance of emergence drops faster into the past. The selection factor shifts the x-y plane horizontally for the occupancy ratio and visibility. With increasing selection factor value, the x-y plane for the occupancy ratio and visibility shifts left due to the decreasing value based on the same BCS and k.

The chance of visibility is always less than or equals to the occupancy ratio. This is not surprising, if the universe is filled with non-expanding civilization, then the occupation ratio is the chance of visibility per galaxy. There is a clear difference between cases of k = 1 (BER = BCS) vs. k > 1 (BER < BCS) across all possible landscapes. For cases of k > 1, lower BER implies the rate of emergence chance changes is much lower. Therefore, the rise of the wall of invisibility **is** lot gentler than k = 1. In fact, the nearest civilization detection can be much closer for k = 1, but it becomes much more rare as one stares into the past event horizon and cross over the case > 1 given the same BCS.

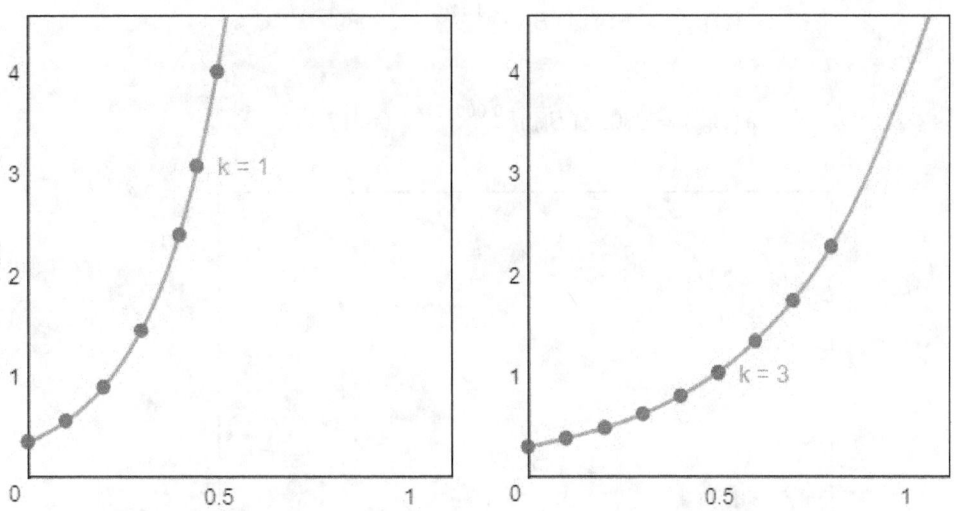

Figure 8.7.33: Gentler rise of the wall of invisibility for k = 3 vs. k = 1

8.7.3.4 The Rate of Civilization Emergence

The growth rate on the chance of human emergence vs the distribution placement pattern can be captured by the pairing of

$$p(t, k) = \left(k, \frac{C_{df}(0, k)}{C_{df}(t, k)}\right) \tag{8.7.51}$$

We plotted results for $p(t, 0.6)$...$p(t, 0.8)$... $p(t, 1)$...$p(t, 30)$,$p(t, 50)$...$p(t, 1000)$.

Given all possible data points, we found that it can be modeled by regression fitting on a double exponential function. So that the emergence chance is compared between two given time periods between now and t years ago:

$$\frac{C_{df}\,(0,\ k)}{C_{df}\,(t,\ k)} \tag{8.7.52}$$

Knowing that BCS and BER both increases exponentially, the rate emergence of civilization is the multiplication of three exponential functions. The ratio between any time period is independent of the number of species represented at the current time.

For values within narrow, reasonable ranges, such as 1< BCS < 4 and assuming there is no fixed starting multi-cellular evolutionary window of 500 Myr, The general equation is expressed as:

$$E_{merge}\,(k, t, F_0, B_{cs}) = \exp\left(\frac{a}{k^d} + \frac{b}{k} + t\ln(B_{cs})\right) \tag{8.7.53}$$

Whereas k determines the distribution's placement pattern, t is expressed as the time span between two comparing time period. $t = 1$ stands for 100 Myr period. B_{cs} is the biocomplexity growth rate. a, d, and b are regression coefficients that fits given k, t, and F_0. In general, there are infinite possible choices for a, d, and b. Each values of a, d, and b independently constitute a 3 dimensional regressional volume with time period length t, F_0, and B_{cs} as its width, length, and height. Any linear regression that demonstrate the relationship between the coefficients (a, d, and b) and attributes of interest (t, F_0, and B_{cs}) can be extracted from this volume by fixing the 2 out 3 variables. For example, the relationship between coefficients (a, d, and b) and B_{cs} can be determined by fixing both $t = 1$ and $F_0 = 1$.

$$a\,(B_{cs}) = \left(-34.82\,(B_{cs})^{-0.0437} + 34.858\right)^{2.111} \tag{8.7.54}$$

$$d\,(B_{cs}) = 0.1414\,(B_{cs})^{-4.176} - 1.9989 \tag{8.7.55}$$

$$b\,(B_{cs}) = 68017\,(B_{cs})^{-0.000177} - 68017 \tag{8.7.56}$$

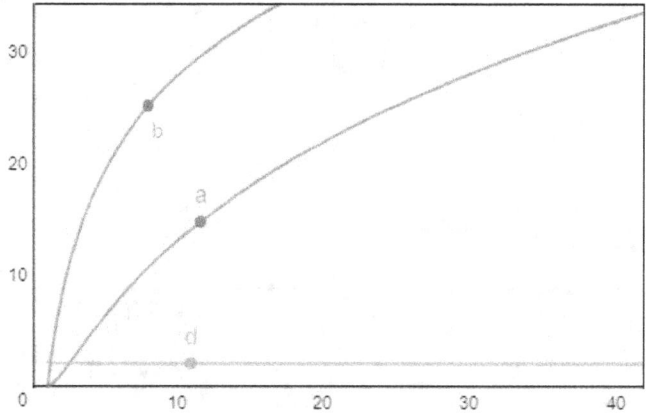

Figure 8.7.34: The best fits for a, d, and b for different values of B_{cs}

When $t = 1$ and $B_{cs} = 2.783$, a, d, and b's relationship with F_0 (selection factor) is listed as:

$$a\,(F_0) = 2.568\,(F_0)^{0.707} + 0.299 \tag{8.7.57}$$

$$d\,(F_0) = -0.776\,(F_0)^{0.227} + 2.765 \tag{8.7.58}$$

$$b\left(F_0\right) = -0.319\left(F_0\right)^{1.13} + 12.6055 \tag{8.7.59}$$

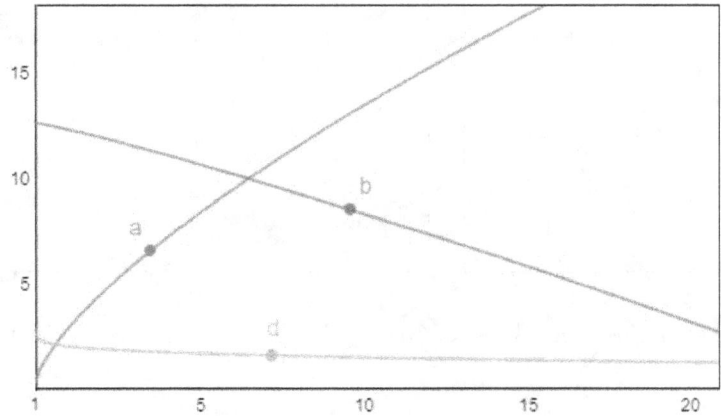

Figure 8.7.35: The best fits for a, d, and b for different values of F_0

Finally, any particular equation instantiation can be found by fixing all attributes of interest (t, F_0, and B_{cs}) For example, for the case of $t=1$, $F_0 = 1$, and $B_{cs} = 2.783$, $a=2.54$, $d=2$, and $b = 12.37$ and the best fit becomes:

$$E_{merge}\left(k, 1, 1, B_{cs}\right) = \exp\left(\frac{2.54}{k^2} + \frac{12.37}{k} + \ln\left(B_{cs}\right)\right) = \exp\left(\frac{12.37}{k}\right)\exp\left(\frac{2.54}{k^2}\right) \cdot B_{cs} \tag{8.7.60}$$

There is several interesting fact can be drawn from the equation. As $k \to \infty$, the equation simply becomes:

$$E_{merge}\left(k, t, F_0, B_{cs}\right) = \exp\left(\ln\left(B_{cs}\right)\right) = B_{cs} \tag{8.7.61}$$

That is, for every 100 Myr period, the chance of civilization emergence is increased by B_{cs}. If B_{cs} increases exponentially yet BER stays at 1.

For the case of $F_0 = 1$ and $t=1$, we have approximation:

$$E_{merge}\left(k, 1, 0, B_{cs}\right) \approx \exp\left(\frac{b}{k} + \ln\left(B_{cs}\right)\right) \tag{8.7.62}$$

That is, lowering k alone results in a greater rate of civilization emergence chance given every 100 Myr besides the chance of civilization emergence always grows by BCS, forming a simple inverse relationship when the ratio is taken under natural log.

The rate of emergence can be further interpreted in detail. Although previous fitting strategies give us a general picture, one can create more complicated fitting model with more detailed understanding. One can interpret the chance of emergence based on geometric analysis of distributions placement. Within all possible ranges, we have two extreme scenarios. When $k \to \infty$, as all new trait acquisitions occur with a fixed number of existing traits, the mode of distribution remains fixed, but the height of the new distribution is scaled by $(B_{cs})^t$ vertically. The rate of civilization emergence is then simply $(B_{cs})^t$. On the other hand, when $k = 1$, new trait acquisition occur with increasing number of added traits. As a result, the new distribution's mode is right shifted with a displacement of $B_{cs}^{\frac{t}{k}} - 1$.

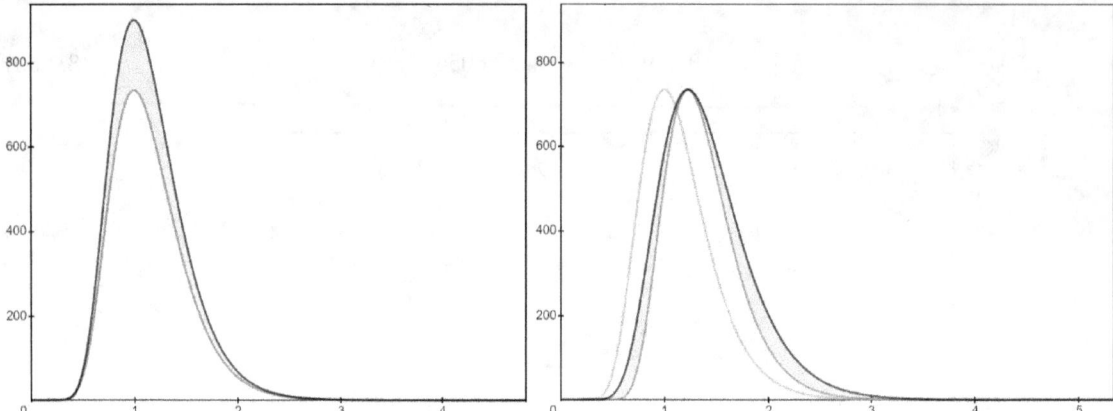

Figure 8.7.36: Two extreme scenarios. Left: $k \to \infty$. Right: $k = 1$. The difference between two distributions are shaded.

Which in turn, creates an additional scaling length relative to previous cross sectional half variance as:

$$\underbrace{\left(L_{imit} - B_{cs}^{\frac{t}{k}}\right)}_{\text{Previous Half Variance}} + \underbrace{\left(B_{cs}^{\frac{t}{k}} - 1\right)}_{\text{Displacement}} \tag{8.7.63}$$

We define that

$$L_{imit} \geq B_{cs}^{\frac{t}{k}} \tag{8.7.64}$$

That is, the defined human emergence position must locates to the right of the mode peak of the new distribution, so that the final partial area of the new distribution dictated by L_{imit} is solely proportional to the raised height of the distribution, see the discussion below.

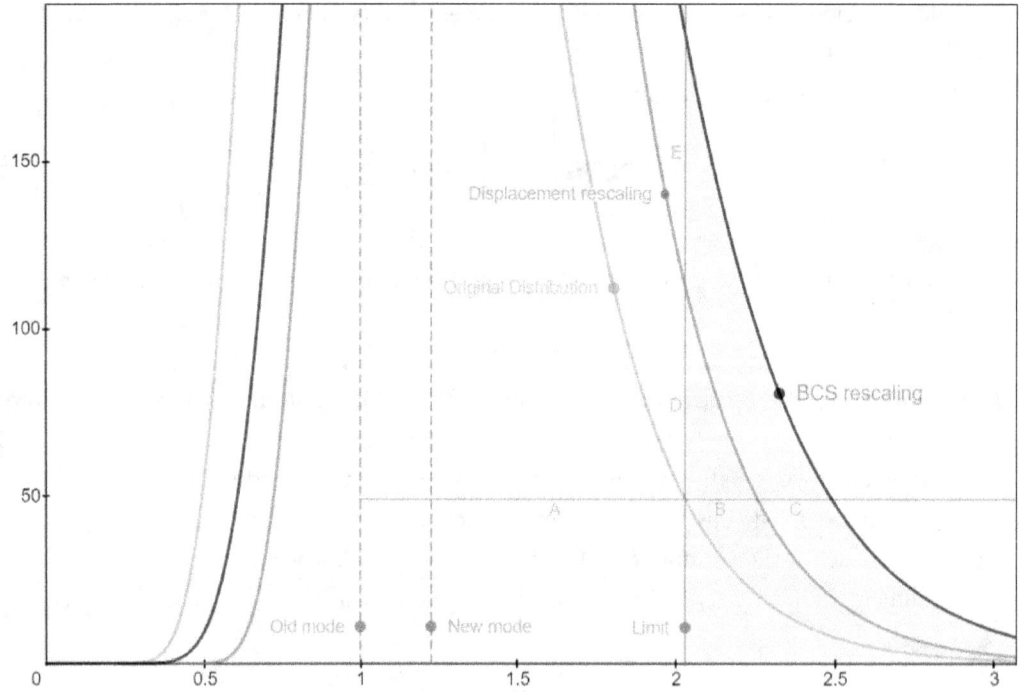

Figure 8.7.37: Segment A = Previous Half Variance = $L_{imit} - B_{cs}^{\frac{t}{k}}$. Segment B= Horizontal Displacement $=B_{cs}^{\frac{t}{k}} - 1$. Segment C=further expansion by $(B_{cs})^{t}$.

Which in turn is divided by previous cross sectional half variance to create the additional scaling ratio relative to previous cross sectional half variance as:

$$1 + \frac{B_{cs}^{\frac{t}{k}} - 1}{Limit - B_{cs}^{\frac{t}{k}}} \qquad (8.7.65)$$

Furthermore, the height of new distribution must be further reduced to previous height, and in order to keep B_{cs} fixed, the variance must be expanded further horizontally by $(B_{cs})^t$. then, $\exp\left(\ln\left(1 + \frac{B_{cs}^{\frac{t}{k}} - 1}{Limit - B_{cs}^{\frac{t}{k}}}\right) + t\ln(B_{cs})\right)$ describes the total horizontal expansion of the new distribution. Since $P_{df}(t, x) > 0$ for all values of $x \geq 0$, the base of distributions can be thought as infinitely stretched, though in practice a set of arbitrarily small cut offs $y = \varepsilon_i > 0$ is selected to give the width of the distribution a finite length. As a result, with fixed, equivalent distribution variance/width for all distributions, the final partial area of the new distribution dictated by L_{imit} is solely proportional to the raised height of the distribution. (Segment D+E) In fact, when $k \geq 1$, with some tolerable value of ϵ,

$$\frac{\text{Total Area}}{\text{Original Area}} = \text{Area Ratio} \qquad (8.7.66)$$

$$\frac{\text{Area Ratio}}{\text{Height Ratio}} = 1 + \epsilon \approx 1 \qquad (8.7.67)$$

It is higher than 1 by the difference of non constant ϵ due to $t\ln(B_{cs})$ must be further broken into vertical height re-scaling component as $t\left(\frac{k-1}{k}\right)\ln(B_{cs})$ and horizontal width re-scaling component as $\frac{t}{k}\ln(B_{cs})$. The horizontal re-scaling component distorts the proportionality of the rescaled distribution, making the tail fatter than proportional re-scaling, as a result, $\epsilon > 0$. The value of ϵ enlarges as $k \to 0$ since greater horizontal re-scaling occur.

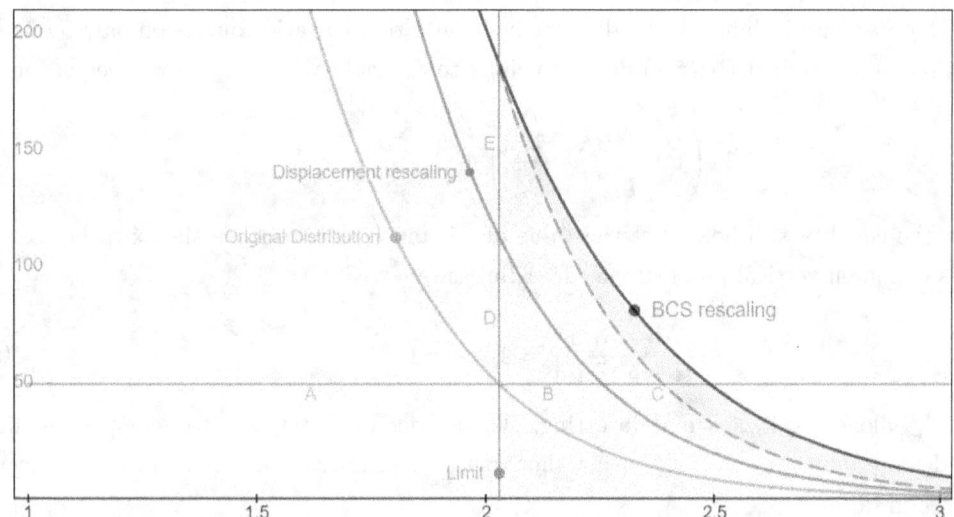

Figure 8.7.38: If height conversion for BCS re-scaling curve is truly proportional to the height (Segment D+E), then it should follow the dashed line. In reality, the curve rests higher. An extra region with strong shade occurs. As a result, $\frac{\text{Area Ratio}}{\text{Height Ratio}} > 1$.

The final combined general expression is expressed as:

$$\exp\left(\overbrace{\underbrace{t\left(\frac{k-1}{k}\right)\ln(B_{cs})}_{\text{Height Ratio}}}^{\text{increase by } B_{cs}} + \underbrace{\overbrace{\frac{t}{k}\ln(B_{cs})}^{\text{increase by } B_{cs}} + \ln\left(1 + \overbrace{\frac{B_{cs}^{\frac{t}{k}} - 1}{Limit - B_{cs}^{\frac{t}{k}}}}^{\text{displacement ratio}}\right) + \overbrace{\frac{b}{k} + \frac{a}{k^d}}^{\text{Conversion Factor}}}_{\text{Width Ratio Converted to Height Ratio}}\right) \qquad (8.7.68)$$

with the under-brace "Width Ratio" spanning the middle two terms.

When $k \to \infty$, the expression simplifies as:

$$\exp\left(\overbrace{t\ln(B_{cs})}^{\text{increase by }B_{cs}}_{\text{Height Ratio}} + \underbrace{\overbrace{0}^{\text{increase by }B_{cs}} + \overbrace{1}^{\text{displacement ratio}}}_{\text{Width Ratio}} + \underbrace{0}_{\text{Conversion Factor}}\right) = (B_{cs})^t \tag{8.7.69}$$

When $k \to 1$, the expression simplifies as:

$$\exp\left(\underbrace{\overbrace{0}^{\text{increase by }B_{cs}}}_{\text{Height Ratio}} + \underbrace{\overbrace{t\ln(B_{cs})}^{\text{increase by }B_{cs}} + \ln\left(1 + \overbrace{\frac{B_{cs}^t - 1}{Limit - B_{cs}^t}}^{\text{displacement ratio}}\right)}_{\text{Width Ratio}} + \underbrace{\frac{b}{k} + \frac{a}{k^d}}_{\text{Conversion Factor}}\right) \tag{8.7.70}$$

$$= \exp\left(t\ln(B_{cs}) + \ln\left(1 + \frac{B_{cs}^t - 1}{Limit - B_{cs}^t}\right) + \frac{a}{k^d} + \frac{b}{k}\right) \tag{8.7.71}$$

The terms $\frac{a}{k^d} + \frac{b}{k}$ are just like the fitting before, however, they now take on different values and meaning. With the one dimensional scalar height re-scaling, and one dimensional scalar width re-scaling provided, $\frac{a}{k^d} + \frac{b}{k}$ terms becomes a conversion factor that converts both one dimensional width and height increase ratio to one dimensional area increase ratio. The factors $\frac{a}{k^d} + \frac{b}{k}$ remains the same values for only width increase ratio to area increase ratio conversion if one fits the emergence chance divided by the height rescaling $(B_{cs})^{t\left(\frac{k-1}{k}\right)}$ with the following equation:

$$\frac{t}{k}\ln(B_{cs}) + \ln\left(1 + \frac{B_{cs}^{\frac{t}{k}} - 1}{Limit - B_{cs}^{\frac{t}{k}}}\right) + \frac{b}{k} + \frac{a}{k^d} \tag{8.7.72}$$

but the factors $\frac{a}{k^d} + \frac{b}{k}$ differ for width horizontal displacement ratio to area ratio conversion only if one fits the emergence chance is based on original $P_{df}(t,x)$ but with shifts to the right with the following equation:

$$\ln\left(1 + \frac{B_{cs}^{\frac{t}{k}} - 1}{Limit - B_{cs}^{\frac{t}{k}}}\right) + \frac{b}{k} + \frac{a}{k^d} \tag{8.7.73}$$

Under this scenario, the equality still holds but the value of ϵ is much smaller since the extra horizontal B_{cs} re-scaling and its distortion on vertical proportional re-scaling is removed.

$$\frac{\text{Area Ratio}}{\text{Height Ratio}} = 1 + \epsilon \approx 1 \tag{8.7.74}$$

In general, terms $\frac{b}{k} + \frac{a}{k^d}$ alone is sufficient in data fitting. We included extra terms not to complicate ourselves, rather to understand more fully the meaning on how horizontal and vertical ratio-wise increase is translated into area wise increase in ratio.

It can be mapped to terms from the previous fit as:

$$\exp\left(\underbrace{t\left(\frac{k-1}{k}\right)\ln(B_{cs}) + \frac{t}{k}\ln(B_{cs})}_{t\ln(B_{cs})} + \underbrace{\ln\left(1 + \frac{B_{cs}^{\frac{t}{k}} - 1}{Limit - B_{cs}^{\frac{t}{k}}}\right) + \frac{b}{k}}_{\frac{b}{k}} + \underbrace{\frac{a}{k^d}}_{\frac{a}{k^d}}\right)$$

$$= \exp\left(\underbrace{t\ln(B_{cs})}_{t\ln(B_{cs})} + \underbrace{\ln\left(1 + \frac{B_{cs}^{\frac{t}{k}} - 1}{Limit - B_{cs}^{\frac{t}{k}}}\right) + \frac{b}{k}}_{\frac{b}{k}} + \underbrace{\frac{a}{k^d}}_{\frac{a}{k^d}}\right)$$

$\frac{b}{k}$ is the dominant term of the conversion, $\frac{a}{k^d}$ is further adjustment refinement. $\frac{a}{k^d}$ occurs since one can imagine that conversion factor converts horizontal displacement to vertical scaling, yet the vertical scaling falls shorts due to curvature of the distribution itself.

8.7.3.5 Conclusion and 3 Laws of Evolution

By now, several important observations can be drawn from the generalized model. First, The relationship diagram between BCS, BER, k, and the selection factor can illustrated as:

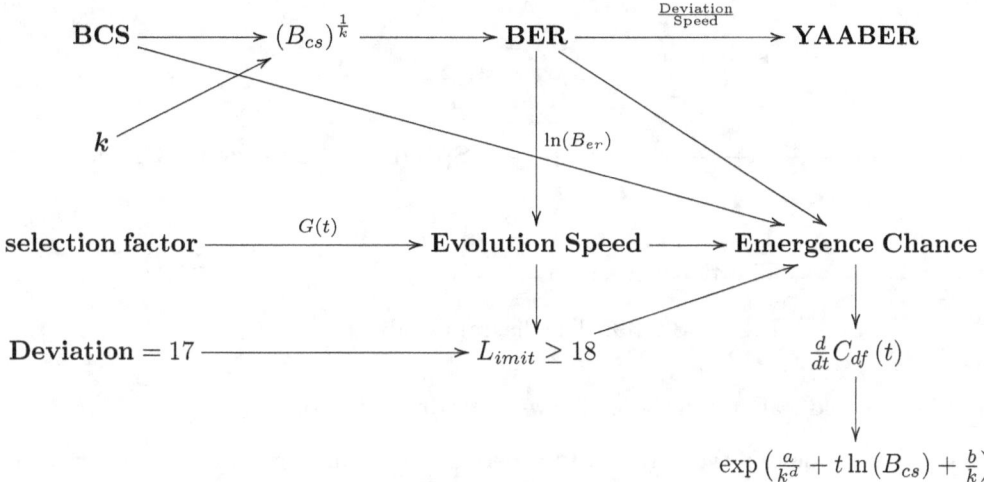

If one were to interpret the selection factor with the alternative interpretation, we have:

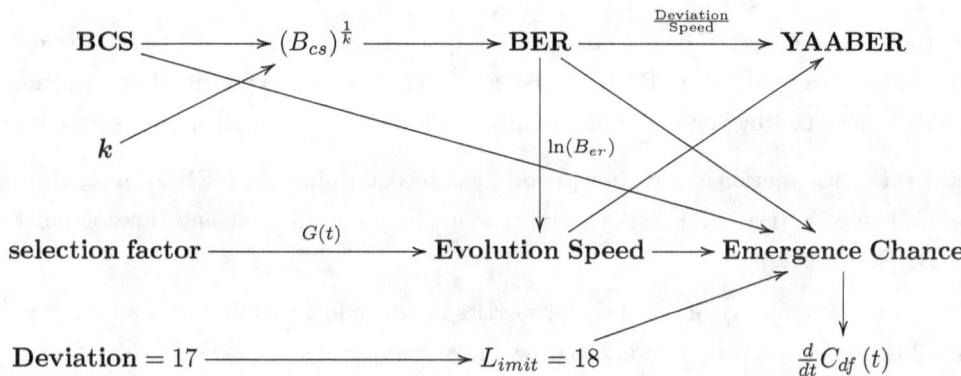

Whereas:

- BCS defines the total search space size and the rate of the search space's exponential growth rate.

- k defines the tendency of search space ordered through time over successive periods (passive, classic, or progressive), determining how BCS is translated into BER.

- Selection factor determines the fundamental background evolutionary speed, determining how BER is translated into number of traits skipped per round and the percentage of current cohorts make it to the next round of evolution.

- BCS scales distributions horizontally, vertically, or both depending on the value of k.

- Selection factor translates distributions horizontally.

- k scales distributions vertically and translates distributions horizontally.

- BER determines the YAABER. Selection factor also determines the YAABER under alternative selection factor interpretation.

- BER determines the evolutionary speed by accelerating it for every time period.

- BCS, BER, and selection factor determines the rate of final emergence chance of civilization at any given time t as $\frac{d}{dt}C_{df}(t)$, which in turn can be expressed as $\exp\left(\frac{a}{k^d} + t\ln(B_{cs}) + \frac{b}{k}\right)$.

In a sense, we can group the top left portion of the diagram as the stabilizing selection and the bottom left portion as the directional selection.

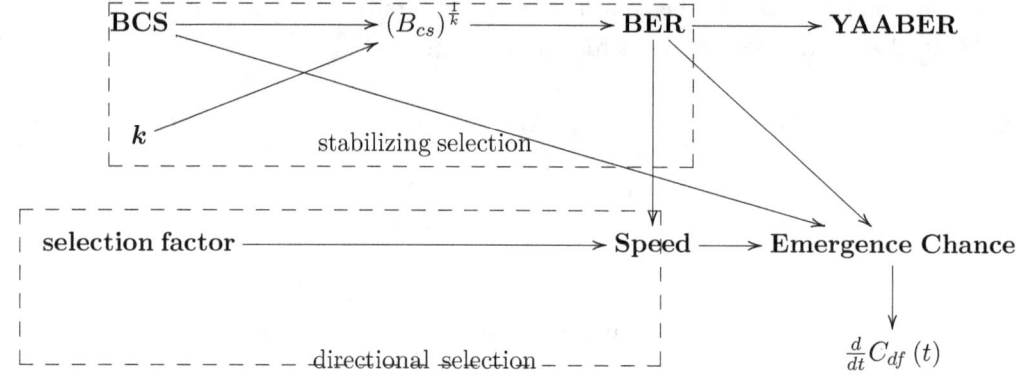

Figure 8.7.39: The diagram analysis

Hence, we can start to make bold leap in concluding 3 laws regarding evolution:

- **First Law:** the speed of evolution over any unit time period is the sum of stabilizing selection (depending on how stable it really is, can still allow additional traits to appear) and directional selection (with periods of gene fixation under bottleneck events). **Evolution speed $= V_{\text{stablizing}} + V_{\text{directional}}$**

- **Second Law:** the rate of change of evolutionary speed, quasi similar to an exponential acceleration on the speed of evolution, is expressed as **BER $=$ (BCS)$^{\frac{1}{k}}$**. That is, the exponential acceleration of the speed of evolution is governed by how rigorous stabilizing selection acted on all species across time.

- **Third Law:** the speed of evolution at any time period t can be determined by **BER$^t \cdot$ (Evolution speed)** and the mode of all species (the total distance that the mode has shifted) at any time period t can be determined by **BER$^t \cdot$ (mode)** .

In a sense, we can almost draw similarity between the properties of our model with Kepler's 3 laws of planetary motion and Newton's 3 laws of motion. If one were to draw direct comparison between classical mechanics and this theory, the conceptually most similar ideas to Kepler's empirical law derivation should be:

Kepler's Empirical Laws	Our Simulation Observation
1st: The orbit of a planet is an ellipse	1st: Biological evolution profile of species from the perspective of man is a lognormal distribution. This lognormal distribution moves at a non constant speed across trait axis over time.
2nd: A line segment joining a planet and the Sun sweeps out equal areas during equal intervals of time	2nd: The concept of BCS, BER, k variable, and selection factor. We find the relationship of BER=BCS$^{\frac{1}{k}}$, and BER $\propto$ selection factor. The area size of BCS given different k results in the same fixed size.
3rd: The square of a planet's orbital period is proportional to the cube of the length of the semi-major axis of its orbit.	3rd: The emergence chance of Homo Sapiens is $\frac{d}{dt}C_{df}(t) = \exp\left(\frac{a}{k^d} + t\ln(B_{cs}) + \frac{b}{k}\right)$

Table 8.7.26: Comparison with Kepler's Laws

Kepler 1st law describes that the motions of the planets is at a non-constant speed, implied by an elliptical orbit as the consequence of acceleration and deceleration along a moving circular path forms an ellipse. Similarly, if there is an equivalent 1st empirical law for evolution, (though we can not observe an ellipse from the trajectory of evolution) it would be that the lognormal distribution moves at a non constant speed across trait horizontal axis over time as we have illustrated, as the consequence of acceleration and deceleration on evolution forms periods of relative stasis followed by periods of leaps. However, we have also shown that the distribution itself is lognormal instead of normal distribution (due to selective paths out of all permutations, some traits has higher initial formation frequency, the weighted values given to each trait, and EOREI in evolution, that a trait is only maintained if the net energy it helps to bring or save to the organism is strictly positive or at least breaking even), which in Kepler's sense, would be conceptually similar to describe about the properties of the planets itself due to heavenly motion, the most conceptual equivalent would be recognizing planets and stars as oblate spheroids owing to their centrifugal force. Therefore, our conceptual equivalent to Kepler's 1st law actually contained more information than Kepler's law conveyed.

Kepler 2nd law describes how such acceleration and deceleration behaves, indirectly implied by $F = ma$ and the conservation of energy. As the planet attracted toward the sun by gravitational force, the acceleration gives to higher speed hence greater kinetic energy. As the planet reached perigee with greatest speed and kinetic energy and attempt to fly away, the sun's gravitation force starts decelerate the planet and eventually brings the planet back toward the sun. Kepler had no concept of energy, acceleration, and force at the time, so instead, he described such rules of acceleration in terms of orbit length, sweeping area, and the passage of time. Likewise, in our 2nd empirical law, we defined the rules of acceleration based on BCS, BER, variable k, and selection factor. Unlike Kepler, we are lucky that acceleration in evolution is explicitly defined as BCS and BER. One can think of BCS as the acceleration on the number of species attainable given a fixed number of traits at any particular epoch. BER as the acceleration on the number of new traits appeared. BER depends both on the shape of lognormal distribution dictated by variable k and the selection factor. Without defining BCS and BER, we would define the rules of acceleration like Kepler, however, since there is no equivalent of area equivalence but only orbit length non-equivalence, so under equal geologic time period, one observes unequal swept distance across trait axis. Finally, we also observed that the lognormal distribution shape dictated by variable k varies but total BCS remains the same. There is no conceptual equivalent in classical mechanics, the closest concept would be that an oblate spheroid transformed in either vertical or horizontal position has the same volume. However, such comparison has severe limitation. Since the shape of lognormal influence the acceleration speed BER, but the oblate shape of planet due to their centrifugal force has no influence on their orbital speed acceleration.

Kepler 3rd law indirectly states that for those planets that circled closer to the sun, it must have a faster orbital speed due to acceleration, as a consequence of greater length of time subject to the gravitational acceleration of the sun. Likewise, in our 3rd empirical law, we derived the human emergence chance as $\frac{d}{dt} C_{df}(t) = \exp\left(\frac{a}{k^d} + t \ln(B_{cs}) + \frac{b}{k}\right)$. That is, when the initial condition of habitability is met, as a consequence of greater length of time subject to the geologic activity of the parent planet, there is a faster evolution speed and exponentially greater chance of intelligence emergence. In Kepler's case, the direction of acceleration is toward the sun and toward the future regardless of initial position (in which case both time into the future and toward the direction of the sun becomes aligned though they can differ). In our case, the direction of acceleration is toward the future only. One could argue that acceleration toward the direction of the source of geologic change, which is earth itself. All biological life derived from earth, which is the starting assumption.

One could further argue that evolution occurs more intensely near the locations where major geologic activity takes place such as island separation, fault line, and land mass joining, so acceleration is toward geologic locations as well. However, it is assumed that during the course of the immense time span, every location has an uniform chance of geologic activity. That is, every location has an unity chance of geologic change. Therefore, species does not need to actively pursue locations of geologic change in order to achieve accelerated change. Secondly, even assuming that a fewer number, just enough hotspots of geologic changes are available so that species must actively physically relocate to them to guarantee accelerated evolutionary change, there is no particular preference for species to pursue such goal. Unlike the force of gravity that universally attracts bodies together, there is no compelling attraction of geologic hotspot for species. Some may prefer the new niche offered but others also likely to avoid, unable to achieve any universality. Thirdly, even if such attraction exists among all species, if a limited number of hotspots exist, and each hotspot can accommodate a limited number of species, then only a fraction of species experienced an evolutionary acceleration toward geologic change and toward the future. The rest of species does not experience any evolutionary acceleration. If one were to average acceleration over all species, then one can only be certain about some evolutionary acceleration toward the future. Finally, even if attraction exists and sufficient number of hotspots exist, all species experienced an evolutionary acceleration toward geologic change and toward the future. It is toward multiple hotspots not a single gravitational source such as the sun. Therefore, acceleration on evolution is described toward the future only.

Next, we drew the comparison between Newton's 3 laws with our derived 3 laws:

Newton's Laws	Our Derived Laws
1st: An object at rest will stay at rest, and an object in motion will stay in motion unless acted on by a net external force.	1st: the speed of evolution over any unit time period is the sum of stabilizing selection (depending on how stable it really is, can still allow additional traits to appear) and directional selection (with periods of gene fixation under bottleneck events).
2nd: The rate of change of momentum of a body over time is directly proportional to the force applied, and occurs in the same direction as the applied force.	2nd: the rate of change of evolutionary speed, quasi similar to an exponential acceleration on the speed of evolution, is expressed as $\mathbf{BER} = (\mathbf{BCS})^{\frac{1}{k}}$. That is, the exponential acceleration of the speed of evolution is governed by how rigorous stabilizing selection acted on all species across time.
3rd: All forces between two objects exist in equal magnitude and opposite direction.	3rd: the speed of evolution at any time period t can be determined by $\mathbf{BER}^{t} \cdot (\mathbf{Evolution\ speed})$ and the mode of all species (the total distance that the mode has shifted) at any time period t can be determined by $\mathbf{BER}^{t} \cdot (\mathbf{mode})$.

Table 8.7.27: Comparison with Newton's Laws

Newton's 1st law states that the innate force possessed by an object which resists changes in motion as the law of inertia. In our case, we have shown earlier in Chapter 4, that species are capable of adapting to a changing environment quickly when it is needed during directional selection, followed by periods of evolutionary stasis in stabilizing selections. As a result, surviving species, conceptually equivalent to object in classical mechanics, are malleable to the conditions in external environment. As a result, exactly the opposite of law of inertia. (of course, many resisted change just conceptually similar to the law of inertia, however, those species become extinct) In our 1st law, we also state the definition of evolutionary speed in any given period, that is, it is the composite sum of periods of directional selection and stabilizing selection. Whereas directional selection is defined as for both the lowest and the highest number of traits defining the lognormal distribution increases by the same fixed amount, regardless of the distribution variance so that the mode of the distribution shifts according to:

$$\text{Mode}_2 = \text{Mode}_1 + x \qquad (8.7.75)$$

Whereas stabilizing selection is defined as for the lowest number of traits defining the lognormal distribution remains the same, only the highest number of traits defining the lognormal distribution changes, and as a consequence distribution variance increases, and mode shifts due to more traits available during combination process, so permutation with repeats to permutation ratio increases for a given mode as (whereas t is the number of traits representing the distribution initially and $t + n$ is the number of traits representing the distribution afterward and $\text{Mode}_2 > \text{Mode}_1$):

$$A_p \left(\text{Mode}_2 + 1, t \right) < A_p \left(\text{Mode}_2, t \right) < A_p \left(\text{Mode}_1, t \right) \tag{8.7.76}$$

$$A_p \left(\text{Mode}_2 + 1, t + n \right) < A_p \left(\text{Mode}_2, t + n \right) > A_p \left(\text{Mode}_1, t + n \right) \tag{8.7.77}$$

Furthermore, the final speed at any time period is the net consequence of species fast transformational potential and the status of its external environment over time. Just as final velocity at any time period is the net consequence of acceleration arising from all forces acted externally on an object overtime. Expressing it mathematically we first derive the mode of each distribution as the distance traveled along the trait axis starting from the initial possible position denoted as $M_{ax}(\infty)$:

$$d\left(t\right) = \overbrace{\frac{x}{b}}^{\text{directional}} \times \overbrace{\sum_{n=0}^{t-1} \left(B_{cs}\right)^{\frac{n}{k}} \cdot b}^{\text{stabilizing variance}} + \overbrace{\frac{\left(B_{cs}\right)^{\frac{t}{k}} \cdot b}{R_p}}^{\text{stablizing mode}} + \overbrace{1}^{\text{offset}} \tag{8.7.78}$$

Whereas the extra 1 is the offset for the start of our unit distribution $P_{df}(0, x)$, but it can be set to 0 by modification. b is the unit distribution width/variance for $P_{df}(0, x)$, R_p is the ratio of $\frac{b}{p}$, that is, the ratio of unit distribution width/variance over unit distribution mode value. x is the proportion of the unit distribution width/variance skipped by the next unit time successive distribution designated by the selection factor, which can be further expressed as:

$$x = \left(B_{cs}^{\frac{1}{k}} - 1 \right) \left(1 - M_{ax}\left(\infty\right) \right) \tag{8.7.79}$$

The speed is then expressed as the difference in distance traveled along the trait axis as:

$$s\left(t\right) = \frac{\left[\frac{x}{b} \sum_{n=0}^{t-1} \left(B_{cs}\right)^{\frac{n}{k}} \cdot b + \frac{\left(B_{cs}\right)^{\frac{t}{k}} \cdot b}{R_p} + 1 \right] - \left[\frac{x}{b} \sum_{n=0}^{t-2} \left(B_{cs}\right)^{\frac{n}{k}} \cdot b + \frac{\left(B_{cs}\right)^{\frac{t-1}{k}} \cdot b}{R_p} + 1 \right]}{t - (t-1)} \tag{8.7.80}$$

which can be simplified as:

$$s\left(t\right) = \overbrace{\left(B_{cs}\right)^{\frac{t-1}{k}}}^{\text{directional}} + \overbrace{\frac{\left(B_{cs}\right)^{\frac{t}{k}} \cdot b}{R_p} - \frac{\left(B_{cs}\right)^{\frac{t-1}{k}} \cdot b}{R_p}}^{\text{stablizing}} \tag{8.7.81}$$

Newton's 2nd law states how acceleration can be derived based on force and mass $a = \frac{F}{m}$. In our case, the force is conceptually equivalent to geologic changes and mass conceptually equivalent to the number of extent species $a = \text{Species}^{\text{Geologic Changes}} = \left(B_{cs}\right)^t$. However, in our case, we define two types of acceleration. One describes the acceleration on the number of new species can form at any given geologic period deemed as BCS. Another describes the acceleration on the number of traits shared by all species at any given period deemed BER. We further describe their relationship as $\textbf{BER} = \left(\textbf{BCS}\right)^{\frac{1}{k}}$. In terms of BCS, acceleration always grows exponentially by B_{cs} per unit time. In terms of B_{er}, it can alter between 1 and B_{cs} depending on how restrictive stabilizing selection is. The emergence of BER is a consequence of increasing number of new traits shared between all species. The emergence of such new trait is a byproduct during the process of combination of existing traits of the current geologic period under consideration. We have states earlier in Chp 6 that each trait corresponds to adaptation to particular niche. As species actively or passively moving into different habitats, they are subject to acquisition of trait associated with such habitat. However, it is possible that new trait not present from existing set can emerge when evolution finds new local optimum previously not attempted, or the existing niches have diversified and give the potential for the emergence of a new traits in addition to the trait associated with the habitat. These extra new traits can be shared with very limited number of extent species within an given epoch so from statistical point of view can be largely ignored. However, in a post-catastrophic scenario when directional selection is involved, the majority of species shared fewer number of traits are completely removed from the environment and only those with higher number of shared traits remained. As a result, the average species of the new epoch shared a higher number of traits and the BER increases. In our model, we always assume that new trait increases proportional to BCS. BCS, on the other hand, grows significantly slower than the potential search space of permutation allowing repeats. That is, assuming no directional selection, at epoch 1, there exists n_1 known traits, the total potential search space is $(n_1)^{n_1}$, which give rise to $n_2 = n_1 (B_{cs})^{\frac{1}{k}}$ total traits at epoch 2, and at epoch 2, with n_2 known traits, the total potential search space is $(n_2)^{n_2}$, which give rise to $n_3 = n_2 (B_{cs})^{\frac{1}{k}}$ total traits at epoch 3. Then we can express the inequality as:

$$\frac{n_3}{n_2} = (B_{cs})^{\frac{1}{k}} \leq B_{cs} < \frac{(n_2)^{n_2}}{(n_1)^{n_1}} \tag{8.7.82}$$

That is, the potential search space at each epoch grows much faster than BCS, but BCS remains largely fixed due to due to selective paths out of all permutations, some traits has higher initial formation frequency, and the weighted values given to each trait, thus forming the lognormal distribution. When both stabilizing and directional selection is considered, at any epoch, the final number of traits can be expressed as:

$$n_t = \text{Existing Traits} + \text{New Traits} - \text{Fixed Traits} \tag{8.7.83}$$

That is, one takes the min/max number of traits shared by species within an epoch, add those species with newly evolved traits, and subtract those trait that becomes fixed within a population after extinction events (Only those possessed some minimum number of trait from the previous epoch survived). We can express it in formula as:

$$n_t = n_{t-1} + f - \delta \tag{8.7.84}$$

Whereas n_{t-1} is the number of existing traits, f is the number of new traits added, δ is the number of traits become fixed within a population after passing an epoch, recall that the directional selection factor $\frac{x}{b}$ can be expressed as $\frac{\delta}{n_{t-1}}$ when the number of fixed traits to be removed $\leq$ existing traits. So that $x \leq n_{t-1}$ and we have:

$$\frac{n_t}{n_{t-1}} = \frac{n_{t-1}}{n_{t-1}} + \frac{f}{n_{t-1}} - \frac{\delta}{n_{t-1}} = (B_{cs})^{\frac{1}{k}} \tag{8.7.85}$$

$$n_t = n_{t-1} (B_{cs})^{\frac{1}{k}} \tag{8.7.86}$$

and we have the following scenarios:

1. When $(B_{cs})^{\frac{1}{k}} < 1$, biodiversity decreases over time. This occurs when the speed of removal for species sharing fewer number of traits is faster than the speed of addition for species sharing greater number of traits.

2. When $(B_{cs})^{\frac{1}{k}} = 1$, biodiversity stays fixed over time. This occurs when the speed of removal for species sharing fewer number of traits is equivalent to the speed of addition for species sharing greater number of traits.

3. When $(B_{cs})^{\frac{1}{k}} > 1$, biodiversity increases over time. This occurs when the speed of removal for species sharing fewer number of traits is slower than the speed of addition for species sharing greater number of traits.

The selection factor F_0 is not directly manifested from the ratio value but only through f and δ. When directional selection is considered, the greater the selection factor, the larger the value of f and δ, the differential value of f and δ can lead to differential number of shared traits gained for the average species in the next epoch despite equivalent B_{er} achieved. So there exists a range of possible f and δ. The possibilities are illustrated below:

Cases No.	All Possible Cases	Explanations	n_t	B_{er}	F_0
1	$\delta = 0$ and $f = 0$		fixed	fixed	F_{min}
2	$\delta = 0$ and $f > 0$	stabilizing selection only	increasing	increasing	F_{min}
3	$\delta > 0$ and $f = 0$		decreasing	increasing	
4	$\delta > 0$ and $f > 0$ and $\delta > f$		decreasing	increasing	
5	$\delta > 0$ and $f > 0$ and $\delta = f$	directional selection only	fixed	fixed	$> F_{min}$
6	$\delta > 0$ and $f > 0$ and $\delta < f$	both directional and stabilizing selection	increasing	increasing	$> F_{min}$

That is, case 1 and case 5 can have the same BCS, n_t, and BER but nevertheless case 5 has a larger selection factor $F_0 > F_{min}$. Likewise, case 2 and 6 can have the same BCS, n_t, and BER but nevertheless case 6 has a larger selection factor $F_0 > F_{min}$. We make a further assumption that the largest selection factor can skip at most current number of traits n_t, so that the starting min number of traits of the next epoch is the max number of traits from the current epoch, corresponding to most realistic physical reality so that the range of possible f and δ becomes constrained.

Since we set the final ratio must be fixed to $(B_{cs})^{\frac{1}{k}}$ as the invariant conforming with our established lognormal model and recall when there is only stabilizing selection, we have:

$$\frac{n_t}{n_{t-1}} = \frac{n_{t-1} + f - 0}{n_{t-1}} = (B_{cs})^{\frac{1}{k}} \tag{8.7.87}$$

and f can be expressed as $n_{t-1}\left((B_{cs})^{\frac{1}{k}} - 1\right)$

$$f = (B_{cs})^{\frac{1}{k}} n_{t-1} - n_{t-1} = n_{t-1}\left((B_{cs})^{\frac{1}{k}} - 1\right) \tag{8.7.88}$$

When directional section takes on its greatest possible value, we have $n_{t-1}(B_{cs})^{\frac{1}{k}}$

$$\frac{n_t}{n_{t-1}} = \frac{n_{t-1} + f - n_{t-1}}{n_{t-1}} = (B_{cs})^{\frac{1}{k}} \tag{8.7.89}$$

$$f = n_{t-1}(B_{cs})^{\frac{1}{k}} \tag{8.7.90}$$

so that f must be:

$$n_{t-1}\left((B_{cs})^{\frac{1}{k}} - 1\right) \leq f \leq n_{t-1}(B_{cs})^{\frac{1}{k}} \tag{8.7.91}$$

Thus, the addition of new traits in any epoch t under constraint is proportional to B_{cs}.

We illustrated that the number of traits per any epoch t n_t is exponentially depended on B_{cs} as:

$$n_2 = n_1 (B_{cs})^{\frac{1}{k}}$$

$$n_3 = \left(n_1 (B_{cs})^{\frac{1}{k}}\right)(B_{cs})^{\frac{1}{k}} = n_1 (B_{cs})^{\frac{2}{k}} \tag{8.7.92}$$

$$n_4 = \left(\left(n_1 (B_{cs})^{\frac{1}{k}}\right)(B_{cs})^{\frac{1}{k}}\right)(B_{cs})^{\frac{1}{k}} = n_1 (B_{cs})^{\frac{3}{k}} \tag{8.7.93}$$

$$\dots \tag{8.7.94}$$

$$n_t = n_1 (B_{cs})^{\frac{t-1}{k}} \tag{8.7.95}$$

Since the fixation of species sharing lower number of traits can vary between:

$$0 \le \delta \le \left(n_{t-1} = n_1 (B_{cs})^{\frac{t-2}{k}}\right) \tag{8.7.96}$$

Thus, in any epoch t and under constraint it is also proportional to B_{cs}.

Likewise, we illustrate that BER is exponentially depended on B_{cs} can first state the equation for calculating the mode:

$$\text{mode}(t) = 1 + f_0 \sum_{n=0}^{t-1} (B_{cs})^{\frac{n}{k}} + p_0 (B_{cs})^{\frac{t}{k}} \tag{8.7.97}$$

Whereas f_0 is the selection factor representing the number of skipped traits between $t = 1$ and $t = 0$. p_0 is the mode value at $t = 0$. Whereas the sum of terms shows that selection factor increases exponentially with time. Recall that BER can be derived as:

$$B_{er} = \frac{\text{mode}(t) - \text{mode}(t-1)}{\text{mode}(t-1) - \text{mode}(t-2)} \tag{8.7.98}$$

therefore, we have:

$$B_{er} = \frac{\left[1 + f_0 \sum_{n=0}^{t-1} (B_{cs})^{\frac{n}{k}} + p_0 (B_{cs})^{\frac{t}{k}}\right] - \left[1 + f_0 \sum_{n=0}^{t-2} (B_{cs})^{\frac{n}{k}} + p_0 (B_{cs})^{\frac{t-1}{k}}\right]}{\left[1 + f_0 \sum_{n=0}^{t-2} (B_{cs})^{\frac{n}{k}} + p_0 (B_{cs})^{\frac{t-1}{k}}\right] - \left[1 + f_0 \sum_{n=0}^{t-3} (B_{cs})^{\frac{n}{k}} + p_0 (B_{cs})^{\frac{t-2}{k}}\right]} \tag{8.7.99}$$

$$B_{er} = \frac{f_0 (B_{cs})^{\frac{t-1}{k}} + p_0 (B_{cs})^{\frac{t}{k}} - p_0 (B_{cs})^{\frac{t-1}{k}}}{f_0 (B_{cs})^{\frac{t-2}{k}} + p_0 (B_{cs})^{\frac{t-1}{k}} - p_0 (B_{cs})^{\frac{t-2}{k}}} \tag{8.7.100}$$

$$B_{er} = \frac{(B_{cs})^{1/k} \left(f_0 (B_{cs})^{\frac{t-2}{k}} + p_0 (B_{cs})^{\frac{t-1}{k}} - p_0 (B_{cs})^{\frac{t-2}{k}}\right)}{f_0 (B_{cs})^{\frac{t-2}{k}} + p_0 (B_{cs})^{\frac{t-1}{k}} - p_0 (B_{cs})^{\frac{t-2}{k}}} = (B_{cs})^{1/k} \tag{8.7.101}$$

That is, every step of $t/(t-1)$ the ratio is increased by $(B_{cs})^{1/k}$, so the B_{er} between x steps is $(B_{cs})^{x/k}$.

Finally, we show that the combined effect of stabilizing and directional selection increases the distribution shift exponentially faster than directional selection alone. We can simplify the $d(t)$ equation by discarding the parts calculating the mode peak at any epoch, since we are only concerned with the speed of shift. By excluding $\frac{(B_{cs})^{\frac{t}{k}} \cdot b}{R_p} + 1$ terms, $d(t)$ is expressed as the minimum number of shared trait by any species at any epoch as:

$$d(t) = \frac{x}{b} \sum_{n=0}^{t-1} (B_{cs})^{\frac{n}{k}} \cdot b \tag{8.7.102}$$

now we divide combined effect of stabilizing and directional selection over directional selection only (setting $B_{cs} = 1$)

$$\frac{x \sum_{n=0}^{t-1} (B_{cs})^{\frac{n}{k}}}{x(t-1)} = \frac{\sum_{n=0}^{t-1} (B_{cs})^{\frac{n}{k}}}{t-1} = \frac{(B_{cs})^{\frac{t-1}{k}} + (B_{cs})^{\frac{t-2}{k}} + (B_{cs})^{\frac{t-3}{k}} + \dots + (B_{cs})^{\frac{1}{k}} + 1}{1 + 1 + 1 + \dots + 1} \tag{8.7.103}$$

which can be simplified as:

$$\frac{2\left(B_{cs}\right)^{\frac{t-1}{k}} - 1}{t - 1} \approx \frac{2\left(B_{cs}\right)^{\frac{t-1}{k}}}{t} \tag{8.7.104}$$

The expression can be further simplified by evaluating its log and then dividing $\ln t$ as:

$$\frac{\ln 2}{\ln t} + \frac{(t-1)\ln B_{cs}}{\ln t} - \frac{\ln t}{\ln t} = \frac{(t-1)\ln B_{cs}}{\ln t} - 1 + \frac{\ln 2}{\ln t} \tag{8.7.105}$$

then raised back to its own power so that:

$$\frac{2\left(B_{cs}\right)^{\frac{t-1}{k}}}{t} = t^{\left(\frac{(t-1)\ln B_{cs}}{\ln t} - 1 + \frac{\ln 2}{\ln t}\right)} \tag{8.7.106}$$

and then taking the limit as:

$$\lim_{k \to \infty} t^{\left(\frac{(t-1)\ln B_{cs}}{\ln t} - 1 + \frac{\ln 2}{\ln t}\right)} = \infty^{(\infty - 1 + 0)} = \infty \tag{8.7.107}$$

Hence, the combined effect of stabilizing and directional selection increases the mode exponentially faster than directional selection alone.

Newton's 3rd law describes how force behaves with direction. If it translates directly into our model, it would be conceptually equivalent to state that all geologic changes exist in equal magnitude and opposite direction. This is obviously absurd since geologic changes does not exhibit any particular net directional change. It is a scalar quantity. Since it is not applicable, we instead substitute this law with relationship between speed, acceleration, and distance. Since the acceleration is exponential in nature, the evolutionary speed increase is also exponential in nature due to integration on exponential function, the final distance between mode species defined by the number of shared common traits is also exponential in nature, due to integration on exponential function. Based on continuous function one can show that:

Classical Mechanics	Our Derived Identities
$a = \frac{F}{m}$	$a_h = \left(B_{cs}\right)^1,\ a_w = \left(B_{cs}\right)^{\frac{1}{k}}$
$v = \int a\,dt = at$	$s = \int \left(B_{cs}\right)^{\frac{t}{k}}\,dt = \frac{\left(B_{cs}\right)^{\frac{t}{k}}}{\frac{1}{k}\ln(B_{cs})}$
$d = \int at\,dt = \frac{1}{2}at^2$	$d = \int \frac{\left(B_{cs}\right)^{\frac{t}{k}}}{\frac{1}{k}\ln(B_{cs})}\,dt = \frac{\left(B_{cs}\right)^{\frac{t}{k}}}{\left(\frac{1}{k}\right)^2(\ln(B_{cs}))^2}$

Based on difference function, for **Evolution speed = BER^t · (Evolution speed)** we can show that for speed at time t we have:

$$s(t) = \frac{\left[\frac{x}{b}\sum_{n=0}^{t-1}\left(B_{cs}\right)^{\frac{n}{k}}\cdot b + \frac{\left(B_{cs}\right)^{\frac{t}{k}}\cdot b}{R_p} + 1\right] - \left[\frac{x}{b}\sum_{n=0}^{t-2}\left(B_{cs}\right)^{\frac{n}{k}}\cdot b + \frac{\left(B_{cs}\right)^{\frac{t-1}{k}}\cdot b}{R_p} + 1\right]}{t - (t-1)} \tag{8.7.108}$$

which is simplified as:

$$s(t) = \left(B_{cs}\right)^{\frac{t-1}{k}} + \frac{\left(B_{cs}\right)^{\frac{t}{k}}\cdot b}{R_p} - \frac{\left(B_{cs}\right)^{\frac{t-1}{k}}\cdot b}{R_p} \tag{8.7.109}$$

So we multiply by another unit **BER** $B_{er} = \left(B_{cs}\right)^{\frac{1}{k}}$ as:

$$s(t) \times \left(B_{cs}\right)^{\frac{1}{k}} = \left(B_{cs}\right)^{\frac{t-1}{k}} \times \left(B_{cs}\right)^{\frac{1}{k}} + \frac{\left(B_{cs}\right)^{\frac{t}{k}}\cdot b}{R_p} \times \left(B_{cs}\right)^{\frac{1}{k}} - \frac{\left(B_{cs}\right)^{\frac{t-1}{k}}\cdot b}{R_p} \times \left(B_{cs}\right)^{\frac{1}{k}} \tag{8.7.110}$$

which becomes

$$s(t) \times \left(B_{cs}\right)^{\frac{1}{k}} = \left(B_{cs}\right)^{\frac{t}{k}} + \frac{\left(B_{cs}\right)^{\frac{t+1}{k}}\cdot b}{R_p} - \frac{\left(B_{cs}\right)^{\frac{t}{k}}\cdot b}{R_p} \tag{8.7.111}$$

which is in fact:

$$s\left(t+1\right) = \frac{\left[\frac{x}{b}\sum_{n=0}^{t}\left(B_{cs}\right)^{\frac{n}{k}}\cdot b + \frac{\left(B_{cs}\right)^{\frac{t+1}{k}}\cdot b}{R_p} + 1\right] - \left[\frac{x}{b}\sum_{n=0}^{t-1}\left(B_{cs}\right)^{\frac{n}{k}}\cdot b + \frac{\left(B_{cs}\right)^{\frac{t}{k}}\cdot b}{R_p} + 1\right]}{\left(t+1\right) - t} \tag{8.7.112}$$

Hence we have shown that **Evolution speed = BER$^t \cdot$ (Evolution speed)** by induction.

Based on difference function, for **mode = BER$^t \cdot$ (mode)** We need to show that the distance covered at time $t+1$ can be derived based on $B_{er} = \left(B_{cs}\right)^{\frac{1}{k}}$ multiplied with the distance covered so far by time t.

$$d\left(t+1\right) = d\left(t\right)\left(B_{cs}\right)^{\frac{1}{k}} \tag{8.7.113}$$

which can be rewritten as:

$$d\left(t+1\right) = \frac{x}{b}\sum_{n=0}^{t}\left(B_{cs}\right)^{\frac{n}{k}}\cdot b + 1 + \frac{\left(B_{cs}\right)^{\frac{t+1}{k}}\cdot b}{R_p} - Max\left(\infty\right) \tag{8.7.114}$$

$$d\left(t\right)\left(B_{cs}\right)^{\frac{1}{k}} = \left(\frac{x}{b}\sum_{n=0}^{t-1}\left(B_{cs}\right)^{\frac{n}{k}}\cdot b + 1 + \frac{\left(B_{cs}\right)^{\frac{t}{k}}\cdot b}{R_p} - Max\left(\infty\right)\right)\left(B_{cs}\right)^{\frac{1}{k}} \tag{8.7.115}$$

expansion gives us:

$$d\left(t\right)\left(B_{cs}\right)^{\frac{1}{k}} = \frac{x}{b}\sum_{n=1}^{t}\left(B_{cs}\right)^{\frac{n}{k}}\cdot b + \left(B_{cs}\right)^{\frac{1}{k}} + \frac{\left(B_{cs}\right)^{\frac{t+1}{k}}\cdot b}{R_p} - Max\left(\infty\right)\left(B_{cs}\right)^{\frac{1}{k}} \tag{8.7.116}$$

Whereas it can be rewritten as:

$$d\left(t\right)\left(B_{cs}\right)^{\frac{1}{k}} = \frac{x}{b}\sum_{n=1}^{t}\left(B_{cs}\right)^{\frac{n}{k}}\cdot b + \frac{\left(B_{cs}\right)^{\frac{t+1}{k}}\cdot b}{R_p} + \left(1 - Max\left(\infty\right)\right)\left(B_{cs}\right)^{\frac{1}{k}} \tag{8.7.117}$$

the sum series missing a single term $\frac{x}{b}\left(B_{cs}\right)^{\frac{0}{k}}\cdot b = x$, recall the proportion of the unit distribution width/variance clipped by the next unit time successive distribution designated by the selection factor can be expressed as:

$$x = \left(1 - Max\left(\infty\right)\right)\left(B_{cs}^{\frac{1}{k}} - 1\right) \tag{8.7.118}$$

Since $\left(1 - Max\left(\infty\right)\right)\left(B_{cs}\right)^{\frac{1}{k}} > \left(1 - Max\left(\infty\right)\right)\left(B_{cs}^{\frac{1}{k}} - 1\right)$, we subtract one from another with a remainder

$$\left(1 - Max\left(\infty\right)\right)\left(B_{cs}\right)^{\frac{1}{k}} - \left(1 - Max\left(\infty\right)\right)\left(B_{cs}^{\frac{1}{k}} - 1\right) = 1 - Max\left(\infty\right) \tag{8.7.119}$$

As a result, we have:

$$d\left(t+1\right) = \left(\frac{x}{b}\sum_{n=1}^{t}\left(B_{cs}\right)^{\frac{n}{k}}\cdot b + \frac{x}{b}\left(B_{cs}\right)^{\frac{0}{k}}\cdot b\right) + \frac{\left(B_{cs}\right)^{\frac{t+1}{k}}\cdot b}{R_p} + \left(1 - Max\left(\infty\right)\right) \tag{8.7.120}$$

we have shown that **mode = BER$^t \cdot$ (mode)** by induction.

Induction is not the only approach for proving **mode = BER$^t \cdot$ (mode)**, a more detailed approach can be adapted as:

We need to show that the distance covered at time $t+1$ can be derived based on $B_{er} = \left(B_{cs}\right)^{\frac{1}{k}}$ multiplied with the distance covered so far by time t.

$$d\left(t+1\right) = d\left(t\right)\left(B_{cs}\right)^{\frac{1}{k}} \tag{8.7.121}$$

which can be translated into a ratio test as:

$$\frac{d(t+1)}{d(t)(B_{cs})^{\frac{1}{k}}} = 1 \tag{8.7.122}$$

which can be rewritten as:

$$\frac{\frac{x}{b}\sum_{n=0}^{t}(B_{cs})^{\frac{n}{k}} \cdot b + 1 + \frac{(B_{cs})^{\frac{t+1}{k}} \cdot b}{R_p} - M_{ax}(\infty)}{\left(\frac{x}{b}\sum_{n=0}^{t-1}(B_{cs})^{\frac{n}{k}} \cdot b + 1 + \frac{(B_{cs})^{\frac{t}{k}} \cdot b}{R_p} - M_{ax}(\infty)\right)(B_{cs})^{\frac{1}{k}}} \tag{8.7.123}$$

expands the denominator:

$$\frac{\frac{x}{b}\sum_{n=0}^{t}(B_{cs})^{\frac{n}{k}} \cdot b + 1 + \frac{(B_{cs})^{\frac{t+1}{k}} \cdot b}{R_p} - M_{ax}(\infty)}{\frac{x}{b}\sum_{n=1}^{t}(B_{cs})^{\frac{n}{k}} \cdot b + (B_{cs})^{\frac{1}{k}} + \frac{(B_{cs})^{\frac{t+1}{k}} \cdot b}{R_p} - M_{ax}(\infty)(B_{cs})^{\frac{1}{k}}} \tag{8.7.124}$$

We can remove shared summation terms $\frac{x}{b}\sum_{n=1}^{t}(B_{cs})^{\frac{n}{k}} \cdot b$ from both numerator and denominator and keep the only remaining summation term $\frac{x}{b}(B_{cs})^{\frac{0}{k}} \cdot b = x$, we have:

$$\frac{x + 1 - M_{ax}(\infty)}{1 - M_{ax}(\infty)(B_{cs})^{\frac{1}{k}}} = 1 \tag{8.7.125}$$

Recall the proportion of the unit distribution width/variance skipped by the next unit time successive distribution designated by the selection factor can be expressed as:

$$x = (1 - M_{ax}(\infty))\left(B_{cs}^{\frac{1}{k}} - 1\right) \tag{8.7.126}$$

substituting, we have:

$$\frac{(1 - M_{ax}(\infty))\left(B_{cs}^{\frac{1}{k}} - 1\right) + 1 - M_{ax}(\infty)}{(1 - M_{ax}(\infty))(B_{cs})^{\frac{1}{k}}} \tag{8.7.127}$$

In fact, the offset of 1 can be at any arbitrary starting reference point, as long as one can show that **mode** = **BER**^{t} · (**mode**) holds. We can show that for any given offset x_i

$$\frac{(x_i - M_{ax}(\infty))\left(B_{cs}^{\frac{1}{k}} - 1\right) + x_i - M_{ax}(\infty)}{(x_i - M_{ax}(\infty))(B_{cs})^{\frac{1}{k}}} = 1 \tag{8.7.128}$$

The condition should always holds. As a result, we can simplify our proof by setting the starting offset reference point at 0, so we have:

$$d(t) = \frac{x}{b} \times \sum_{n=0}^{t-1}(B_{cs})^{\frac{n}{k}} \cdot b + \frac{(B_{cs})^{\frac{t}{k}} \cdot b}{R_p} + 0 \tag{8.7.129}$$

and our equation simplifies to:

$$\frac{(-M_{ax}(\infty))\left(B_{cs}^{\frac{1}{k}} - 1\right) + 0 - M_{ax}(\infty)}{(-M_{ax}(\infty))(B_{cs})^{\frac{1}{k}}} = \frac{x - M_{ax}(\infty)}{-M_{ax}(\infty)(B_{cs})^{\frac{1}{k}}} \tag{8.7.130}$$

Whereas $M_{ax}(\infty)$ can be treated as:

$$M_{ax}(\infty) = -G(\infty) + P_{eak}(\infty) \tag{8.7.131}$$

Since $P_{eak}(\infty) = 0$, the equation simplifies to:

$$M_{ax}(\infty) = -G(\infty) \tag{8.7.132}$$

Recall whereas $G\left(\infty\right)$ is further defined as:

$$G\left(\infty\right) = \frac{-1}{\left(B_{cs}\right)^{-\frac{1}{k}} - 1} \cdot m_{ore}$$

(8.7.133)

The m_{ore} factor is defined as:

$$m_{ore} = F_0 \frac{1 - \left(B_{cs}^{\frac{1}{k}}\right)^{-1}}{1 - B_{cs}^{-1}} - \left(3 - \frac{3}{B_{cs}}\right)$$

(8.7.134)

substituting the terms, $G\left(\infty\right)$ can be expressed as:

$$G\left(\infty\right) = \frac{-1}{\left(B_{cs}\right)^{-\frac{1}{k}} - 1} \left(F_0 \frac{1 - \left(B_{cs}^{\frac{1}{k}}\right)^{-1}}{1 - B_{cs}^{-1}} - \left(3 - \frac{3}{B_{cs}}\right)\right)$$

(8.7.135)

So $M_{ax}\left(\infty\right)$ can be expressed as:

$$M_{ax}\left(\infty\right) = \frac{1}{\left(B_{cs}\right)^{-\frac{1}{k}} - 1} \left(F_0 \frac{1 - \left(B_{cs}^{\frac{1}{k}}\right)^{-1}}{1 - B_{cs}^{-1}} - \left(3 - \frac{3}{B_{cs}}\right)\right)$$

(8.7.136)

Substituting $M_{ax}\left(\infty\right)$ into $\frac{x - M_{ax}(\infty)}{-M_{ax}(\infty)(B_{cs})^{\frac{1}{k}}}$:

$$\frac{x + \frac{1}{\left(B_{cs}\right)^{-\frac{1}{k}} - 1} \left(-F_0 \frac{1 - \left(B_{cs}^{\frac{1}{k}}\right)^{-1}}{1 - B_{cs}^{-1}} + \left(3 - \frac{3}{B_{cs}}\right)\right)}{\left(\frac{1}{\left(B_{cs}\right)^{-\frac{1}{k}} - 1} \left(-F_0 \frac{1 - \left(B_{cs}^{\frac{1}{k}}\right)^{-1}}{1 - B_{cs}^{-1}} + \left(3 - \frac{3}{B_{cs}}\right)\right)\right) \left(B_{cs}\right)^{\frac{1}{k}}}$$

(8.7.137)

Whereas the denominator $\left(\frac{1}{\left(B_{cs}\right)^{-\frac{1}{k}} - 1} \left(-F_0 \frac{1 - \left(B_{cs}^{\frac{1}{k}}\right)^{-1}}{1 - B_{cs}^{-1}} + \left(3 - \frac{3}{B_{cs}}\right)\right)\right) \left(B_{cs}\right)^{\frac{1}{k}}$ can be simplified as:

$$\frac{F_0}{1 - B_{cs}^{-1}} = \frac{-1}{\left(B_{cs}\right)^{-\frac{1}{k}} - 1} F_0 \frac{1 - \left(B_{cs}^{\frac{1}{k}}\right)^{-1}}{1 - B_{cs}^{-1}}$$

(8.7.138)

Substituting back into our equation:

$$\frac{x + \frac{1}{\left(B_{cs}\right)^{-\frac{1}{k}} - 1} \left(-F_0 \frac{1 - \left(B_{cs}^{\frac{1}{k}}\right)^{-1}}{1 - B_{cs}^{-1}} + \left(3 - \frac{3}{B_{cs}}\right)\right)}{\left(\frac{F_0}{1 - B_{cs}^{-1}} + \frac{\left(B_{cs}\right)^{\frac{1}{k}}}{1 - \left(B_{cs}\right)^{\frac{1}{k}}} \left(3 - \frac{3}{B_{cs}}\right)\right) \left(B_{cs}\right)^{\frac{1}{k}}}$$

(8.7.139)

Whereas the factor $\frac{1}{\left(B_{cs}\right)^{-\frac{1}{k}} - 1}$ in the numerator can be re-expressed as:

$$\frac{1}{\left(B_{cs}\right)^{-\frac{1}{k}} - 1} = \frac{\left(B_{cs}\right)^{\frac{1}{k}}}{1 - \left(B_{cs}\right)^{\frac{1}{k}}}$$

(8.7.140)

we move on to multiplying with the first term within the parenthesis:

$$\frac{\left(B_{cs}\right)^{\frac{1}{k}}}{1 - \left(B_{cs}\right)^{\frac{1}{k}}} \left(-F_0 \frac{1 - \left(B_{cs}^{\frac{1}{k}}\right)^{-1}}{1 - B_{cs}^{-1}}\right)$$

(8.7.141)

Excluding F_0, $\frac{1 - \left(B_{cs}^{\frac{1}{k}}\right)^{-1}}{1 - B_{cs}^{-1}}$ can re-expressed as:

428

$$\frac{1-\left(B_{cs}^{\frac{1}{k}}\right)^{-1}}{1-B_{cs}^{-1}} = \frac{\left(\left(B_{cs}^{\frac{1}{k}}\right)-1\right)(B_{cs})^{\frac{k-1}{k}}}{B_{cs}-1} \tag{8.7.142}$$

The multiplication between the factor and the 1st term within the parenthesis in the numerator becomes:

$$\frac{\left((B_{cs})^{\frac{1}{k}}-1\right)(B_{cs})^{\frac{k-1}{k}}}{B_{cs}-1}\frac{(B_{cs})^{\frac{1}{k}}}{1-(B_{cs})^{\frac{1}{k}}} = \frac{\left((B_{cs})^{\frac{1}{k}}-1\right)(B_{cs})^{\frac{k-1}{k}}}{B_{cs}-1}\left(-\frac{(B_{cs})^{\frac{1}{k}}}{(B_{cs})^{\frac{1}{k}}-1}\right) \tag{8.7.143}$$

Whereas $(B_{cs})^{\frac{1}{k}}-1$ canceled each other with the remaining terms $-\frac{(B_{cs})^{\frac{k-1}{k}}(B_{cs})^{\frac{1}{k}}}{B_{cs}-1}$ which is further simplified as:

$$-\frac{(B_{cs})^{\frac{k-1}{k}}(B_{cs})^{\frac{1}{k}}}{B_{cs}-1} = -\frac{B_{cs}}{B_{cs}-1} \tag{8.7.144}$$

We add F_0 back to $\frac{B_{cs}}{B_{cs}-1}$ and expand the rest of the terms in the numerator as well as the denominator, and we have:

$$\frac{x+\frac{F_0 B_{cs}}{B_{cs}-1}+\frac{(B_{cs})^{\frac{1}{k}}}{1-(B_{cs})^{\frac{1}{k}}}3-\frac{(B_{cs})^{\frac{1-k}{k}}}{1-(B_{cs})^{\frac{1}{k}}}3}{\frac{F_0(B_{cs})^{\frac{1}{k}}}{1-B_{cs}^{-1}}+\frac{(B_{cs})^{\frac{2}{k}}}{1-(B_{cs})^{\frac{1}{k}}}3-\frac{(B_{cs})^{\frac{2-k}{k}}}{1-(B_{cs})^{\frac{1}{k}}}3} \tag{8.7.145}$$

Recall that the distance initial offset is:

$$x = \left(-M_{ax}\left(\infty\right)\right)\left(B_{cs}^{\frac{1}{k}}-1\right) \tag{8.7.146}$$

which translates into:

$$x = \left(\frac{(B_{cs})^{\frac{1}{k}}}{1-(B_{cs})^{\frac{1}{k}}}\left(-F_0\frac{1-\left(B_{cs}^{\frac{1}{k}}\right)^{-1}}{1-B_{cs}^{-1}}+\left(3-\frac{3}{B_{cs}}\right)\right)\right)\left((B_{cs})^{\frac{1}{k}}-1\right) \tag{8.7.147}$$

Expand and simplifies we have:

$$x = \left(\frac{F_0 B_{cs}}{B_{cs}-1}+\frac{(B_{cs})^{\frac{1}{k}}}{1-(B_{cs})^{\frac{1}{k}}}3-\frac{(B_{cs})^{\frac{1-k}{k}}}{1-(B_{cs})^{\frac{1}{k}}}3\right)\left((B_{cs})^{\frac{1}{k}}-1\right) \tag{8.7.148}$$

Further expansion gives us:

$$x = \frac{F_0 B_{cs}(B_{cs})^{\frac{1}{k}}}{B_{cs}-1}+\frac{(B_{cs})^{\frac{2}{k}}}{1-(B_{cs})^{\frac{1}{k}}}3-\frac{(B_{cs})^{\frac{2-k}{k}}}{1-(B_{cs})^{\frac{1}{k}}}3-\frac{F_0 B_{cs}}{B_{cs}-1}-\frac{(B_{cs})^{\frac{1}{k}}}{1-(B_{cs})^{\frac{1}{k}}}3+\frac{(B_{cs})^{\frac{1-k}{k}}}{1-(B_{cs})^{\frac{1}{k}}}3 \tag{8.7.149}$$

substituting back into $x+\frac{F_0 B_{cs}}{B_{cs}-1}+\frac{(B_{cs})^{\frac{1}{k}}}{1-(B_{cs})^{\frac{1}{k}}}3-\frac{(B_{cs})^{\frac{1-k}{k}}}{1-(B_{cs})^{\frac{1}{k}}}3$, we see that terms $\frac{F_0 B_{cs}}{B_{cs}-1}$, $\frac{(B_{cs})^{\frac{1}{k}}}{1-(B_{cs})^{\frac{1}{k}}}3$, and $\frac{(B_{cs})^{\frac{1-k}{k}}}{1-(B_{cs})^{\frac{1}{k}}}3$ are canceled out, so we have 3 remaining terms from x in the numerator.

$$\frac{\frac{F_0 B_{cs}(B_{cs})^{\frac{1}{k}}}{B_{cs}-1}+\frac{(B_{cs})^{\frac{2}{k}}}{1-(B_{cs})^{\frac{1}{k}}}3-\frac{(B_{cs})^{\frac{2-k}{k}}}{1-(B_{cs})^{\frac{1}{k}}}3}{\frac{F_0(B_{cs})^{\frac{1}{k}}}{1-B_{cs}^{-1}}+\frac{(B_{cs})^{\frac{2}{k}}}{1-(B_{cs})^{\frac{1}{k}}}3-\frac{(B_{cs})^{\frac{2-k}{k}}}{1-(B_{cs})^{\frac{1}{k}}}3} \tag{8.7.150}$$

and we immediately cancels 2 terms $\frac{F_0 B_{cs}(B_{cs})^{\frac{1}{k}}}{B_{cs}-1}+\frac{(B_{cs})^{\frac{2}{k}}}{1-(B_{cs})^{\frac{1}{k}}}3$ from both the numerator and denominator, with remaining terms as:

$$\frac{\frac{F_0 B_{cs}(B_{cs})^{\frac{1}{k}}}{B_{cs}-1}}{\frac{F_0(B_{cs})^{\frac{1}{k}}}{1-B_{cs}^{-1}}} = \frac{\frac{F_0 B_{cs}(B_{cs})^{\frac{1}{k}}}{B_{cs}-1}}{\frac{F_0 B_{cs}(B_{cs})^{\frac{1}{k}}}{B_{cs}-1}} = 1 \tag{8.7.151}$$

Hence we have shown that $\mathbf{mode} = \mathbf{BER}^t \cdot (\mathbf{mode})$ holds.

Secondly, assuming ultimately one will able to detect extraterrestrial civilization, then the BCS and BER on all planets must be gradually increasing overtime and with values *considerably larger than 1* because:

1) An observational fact based on earth. There were no multicellular life prior to Cambrian explosion and no terrestrial life before Silurian epoch.

2) If BCS and BER always remain low and there was only 500 Myr window for multicellular evolution in general, then there is little if any chance civilization emerges and there must be no detection in the future.

3) If BCS were considerable in size as it is observed today on earth but BER $\to$ 1, it implies that BCS has remained considerably large in the past, and there was a rapid, abrupt phase transition from no life to life (more so than Cambrian explosion), which is unlikely. Or multicellular life has persisted since the beginning of earth's history so that despite low BER, it has build up the biocomplexity and diversity as we observed today (contradicted by facts)

4) If evolution was conservative (BER $\to$ 1) and civilization emerged, then the past is more like the present, implying there should be almost equally if a little less chance in the past that earlier civilization have evolved, and their signal and sphere of expansion should have reached us by now. However, it is contradicted by observation. Therefore, BER must be somewhat significantly larger than 1 so that past is very much unlike the present.

Thirdly, in general, the selection factor can be thought as the factor that controls extinction rates. High extinction shifts successive distributions further right and nature drives organism toward goals. Although BER and YAABER remain unchanged with changing selection factor, (YAABER changes and BER fixed for the alternative interpretation) the selection factor determines what percentage of the current species make it to the next round of evolution. Given the same YAABER, the emergence of civilization is rarer if a given BER matches a more competitive selection scenario (such as only 10% vs 40% of current cohort make it to the next round). Most interestingly, conservative Darwin cases coupled with high extinction rates, can result into progressive Darwin case. That is, every 100 Myr a few major catastrophic extinction events wiped out a majority of the species from the previous period and progressively more adaptable species survived. Yet those survived resumed a mostly goalless and directionless evolutionary trajectory of the conservative type. Their combined effect yields a passive-progressive Darwin case.

This can be demonstrated from a real case instantiation. Since $k=1$ and a BER of 2.783 naturally has 54.53% extinction rate, setting $k=2$ BER drops to 1.668, yet the extinction rate also drops to 14.31% because a lower BER implies fewer traits skipped per 100 Myr and more species survived. We then adopt the alternative interpretation for the selection factor, and further increase the selection factor $F_0 = 1.95$ so that extinction rate can climbs back to 54.53% per 100 Myr by matching the same number of traits skipped as BER of 2.783 under $k=1$ for a BER of 1.668 under k=2.

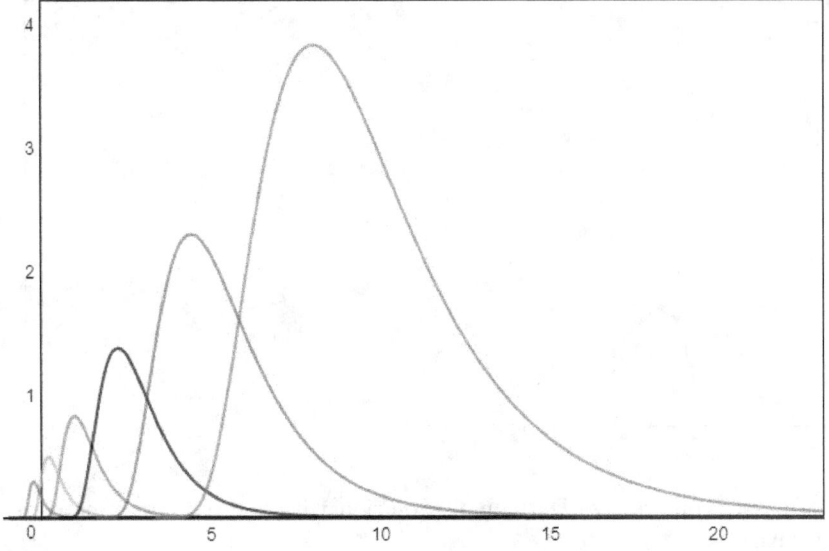

Figure 8.7.40: A case of passive-progressive Darwinian evolution

As a result, we have demonstrated a case of passive-progressive Darwinian evolution. Thus, we have created a new possible starting evolutionary scenario that can be expanded upon with varying BCS, BER, and selection factor. *In general, there are infinite possible combinations for parameters to satisfy any evolutionary scenario with given constraints, and any possible settled scenario can be expanded upon with infinite varying values of BCS, BER, and selection factor.*

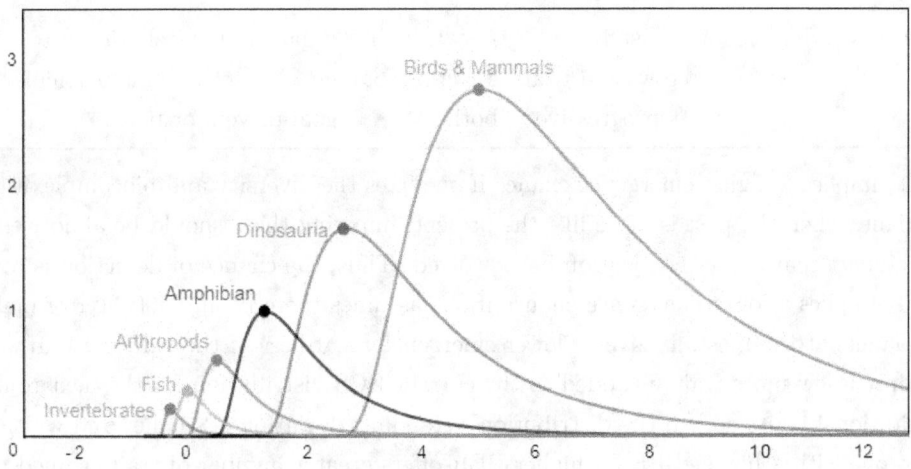

Figure 8.7.41: A case of passive-progressive Darwinian evolution with the most dominant animal species on earth at the time 100 Myr apart

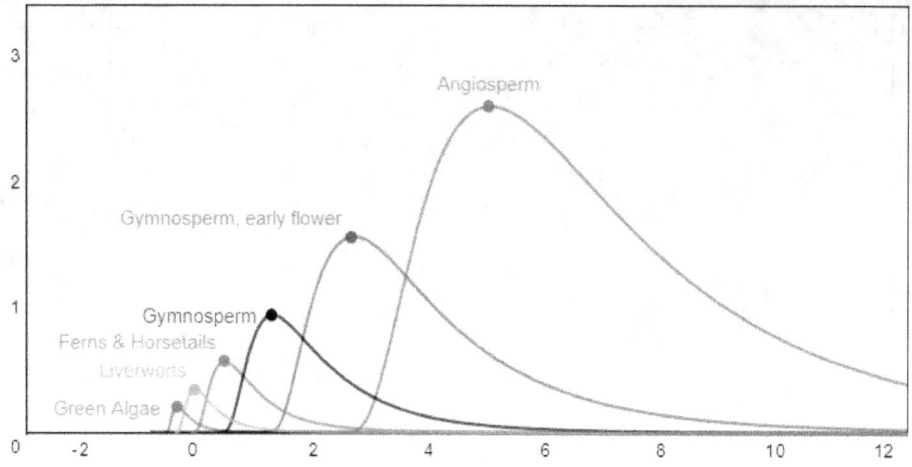

Figure 8.7.42: A case of passive-progressive Darwinian evolution with the most dominant plant species on earth at the time 100 Myr apart

With both k and selection factor, possible cases of evolution can be break into:

BCS increases faster than the mode shift (BER)	conservative	$k > 1$, low selection, or both	costly trait maintenance for new trait acquisition, or selection plays minuscule role, or both.
complexity increases = mode shifts	classic	$k = 1$	halfway between the extremes.
complexity increases slower than the mode shifts	striving toward a goal, progressive	$k < 1$, high selection, or both	organism innately drives to greater goal, or selection is particularly goal-driven, or both.

Low BER (BER$\longrightarrow$ 1) implies a higher emergence chance if one fixes the diversity and biocomplexity achieved on earth as the invariant. Then the past is more like the present, implying there should be almost equally if a little less chance in the past that earlier civilization have evolved. Thus, the chance of detection is raised.

Low BER (BER$\longrightarrow$ 1) implies a lower emergence chance if one assumes there is only 500 Myr of multicellular evolution across all planets at the most. It gives a lower emergence regardless of the value of k. If k=1, a low BER is achieved by only a few more traits is added to the current BCS distribution for the next round. With a small number of additional traits added, the distribution's standard deviation is smaller and width becomes narrower compares to one with a higher BER. A higher BER offers greater number of traits gained per round and the distribution's width becomes wider. As a result, the emergence chance is raised.

For k>1, a low BER can be achieved by adding future BCS with many new additional traits but stacked on the current one. In order to demonstrate that low BER achieved by high valued BCS also results in lower emergence chance, one has to first consider the high valued BCS for k=1. Since each round many additional traits helps to form larger search space, the mode peaks are widely apart. Now, adjusting k > 1, the mode peaks are translated horizontally to the left. As the peaks shift left, the total area under the distribution curve to the left of the mode peak is squeezed. Furthermore, one can think of this area as in-compressible, so the only way it can keep its size to be invariant is to increase its height. With a new shrunken width for the distribution to the left side of the mode peak, the right side width of the distribution must be adjusted proportionally smaller. Hence, the distribution's overall width shrinks and civilization's emergence chance evaluated at L_{imit} decreases. Analysis shows that distributions achieving the same BER has exactly the same distribution width regardless if it is achieved by high valued BCS with k > 1 or low valued BCS with k = 1.

8.8 Constraining the Model Using Observations

8.8.1 Using Major Events and Genomic Complexity to define BER

Earlier in Chapter 6 we have shown major evolutionary trait acquisition by a dominant group of species occurs roughly every 50 Myr. Based on this observation along with the occupancy ratio constraint, we can narrow down the permissible range of k variables and selection factors given different BCS values. The following tables listed major traits acquired by major evolutionary clade within the vertebrate lineage.

Milestone Name	Time
backbone/skull	530 mya
eyes	530 mya
jaws/mouths	462 mya
fins	444 mya
lung	400 mya

Table 8.8.2: Major evolutionary milestones in the vertebrate lineage up to amphibians

Milestone Name	Time
four legs	380 mya
dry skin	320 mya
thermoregulation	252 mya
enhanced olfactory functionality	195 mya
fur	164 mya
placenta	avg. 92.5 mya 100~85 mya

Table 8.8.4: Major evolutionary milestones from amphibians to the typical mammal

Milestone Name	Time
four legs	380 mya
dry skin/egg bearing	320 mya
thermoregulation	252 mya
bipedal	233.23 mya
feather	197 mya
wings	150 Mya
a better sense of smell	95 mya

Table 8.8.6: Major evolutionary milestones from amphibians to the typical dinosaur/bird

It is shown that fish distinguishes from the earliest vertebrates by their fins and mouths. Earliest amphibians distinguishes themselves from the fish by acquiring quadrupedal locomotion on land. Reptiles distinguishes themselves from the amphibians by acquiring egg laying and dry skin. Dinosaurs distinguishes themselves from the reptiles by acquiring thermoregulation and bipedalism. Mammals distinguishes themselves from the reptiles by acquiring furs, fetus-nurturing and thermoregulation. Birds distinguishes themselves from the dinosaurs by acquiring feathers and flying capability. Therefore, the eventual number of traits that can be used to define a bird relative to fish, for example, is 7 (four legs, dry skin/egg laying, thermoregulation, bipedalism, feather, a sense of smell, and flying capability). For typical mammals, it is 6 (quadrupedal locomotion, dry skin, a sense of smell, fur, thermoregulation, and fetus-nurturing). For birds, on average, every 54.29 Myr a major milestone that eventually becomes fixed/dominant within its group appeared. For mammals, it is ber every 63.33 Myr.

8.8.2 k Bounds for Weighted Deviation

Since we have shown that 7 traits defines a human equivalent intelligent species. A typical species on earth reaching human equivalent status would requires 7 steps of major evolutionary changes each 50 Myr apart. If we assumed that such trend continues until 7×42 Myr $= 300$ Myr later a typical average species on earth attains all 7 traits uniquely identified with human. Basically, every 42 Myr passed one of the trait associated with human becomes fixed/dominant within the majority of species population. The required time gap is not 60 Myr as we computed earlier due to first, each of the 7 listed traits have already shown to exist earlier, unlike those listed in the milestones, so the significance of each milestone is downplayed, secondly, number rounding to a whole number of 300 Myr eases the computation.

We set $P_{df}(-3,\ x)$'s mode/peak to 6 units to the right of x $= 1$, representing 300 Myr into the future when the mode of the distribution coincides with the EQ of Homo sapiens (equivalently 7 additional traits by human), assume BCS $= 2.783$, and the appearance chance of civilization still fixed at 1 per 3 galaxies. We use the 2nd interpretation for the selection factor.

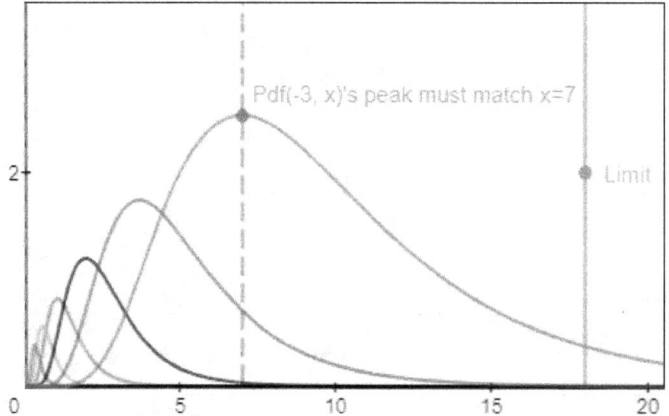

Figure 8.8.1: For selection factor $= 1$, $k= 1.578$ places $P_{df}(-3,\ x)$'s mode/peak on x $= 7$

A whole range of possible pairs of k value and selection factors can satisfy the requirement starting with $k = 1.578$ and selection factor $=1$. The curve is plotted below:

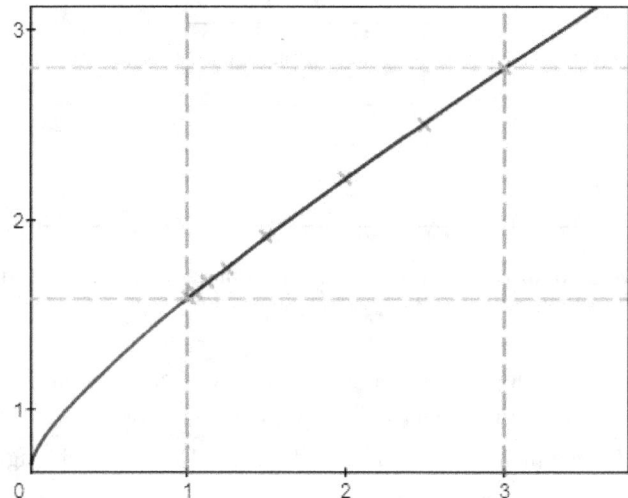

Figure 8.8.2: Possible pairs of k and selection factors that fulfills the requirement: (1, 1.578), (1.05, 1.61), (1.125, 1.67), (1.25, 1.74), (1.5, 1.91), (2, 2.215), (2.5, 2.5), (3, 2.8), and (6, 4.45) (not shown) . The blue vertical lines indicates the permissible range for the selection factor. The orange horizontal lines indicates the permissible range for the k variable.

and the best fit is:

$$f(x) = \left(0.562195x^{0.676291} + 0.790384\right)^{1.51476} \tag{8.8.1}$$

However, only a limited range is permissible. The leftmost possible value is bounded by a selection factor of 1, so need $k \geq 1.578$ in order for the condition to hold. This is the lower bound. We then find the upper bound to be $k \leq 2.8$ and selection factor $= 3$. The upper bound is determined since more conservative evolutionary trend guarantees a $\geq 100\%$ occupancy ratio of the universe at the current time. Although we increased the selection factor, the corresponding increase for k within the pair eventually renders the emergence chance of the past more similar to the present. The conservativeness in k catches up faster than the progressiveness provided by increase in the selection factor. Recall this is manifested from the rate of civilization emergence function: whereas step-wise, for every pair of k and selection factor, an increase in k there is a corresponding increase for a and b by the selection factor and the inequality is satisfied:

$$\exp\left(\frac{a}{k^d} + \ln(B_{cs}) + \frac{b}{k}\right) > \exp\left(\frac{a + a_1}{(k + k_1)^d} + \ln(B_{cs}) + \frac{b + b_1}{k + k_1}\right) \tag{8.8.2}$$

It must be that the selection factor's contribution coefficients a and b and its additional contributions a_1 and b_1 in the numerator can not offset the exponentially increasingly higher values of $k + k_1$ in the denominator. Therefore, a higher chance of emergence from the past, i.e. with a lower rate of appearance reduction, and guarantees a higher occupancy ratio. We run several pairs and the result is consistent with the expectation.

Pair	Substitution $\frac{a}{k^d} + \frac{b}{k}$	Result
(1, 1.578)	$\frac{2.54116}{1.578^{2.00368}} + \frac{12.3686}{1.578}$	8.8570
(2.5, 2.5)	$\frac{5.19571}{2.5^{1.73085}} + \frac{11.774}{2.5}$	5.7734
(3, 2.8)	$\frac{5.65415}{3^{1.69841}} + \frac{11.6387}{3}$	4.7546
(6, 4.45)	$\frac{9.69615}{4.45^{1.49963}} + \frac{10.1217}{4.45}$	3.3080
(10, 6.55)	$\frac{13.5967}{10^{1.38881}} + \frac{8.19415}{10}$	1.3748

This indicates that our evolutionary history is one of the conservative but not an ultra conservative Darwinian evolution.

The empirical observation can be further carried out with rigorous mathematical proofs. We simplify our analysis by reshaping distributions as triangles so that $P_{df}(t, x) \propto \Delta(t, x)$. It still fulfills our earlier conditions since triangles follow the same horizontal/vertical rescaling trade off placement pattern. Then, the emergence rate at any time can be represented by the partial area of triangles. The rate of change of emergence rate can be computed based on $\frac{P_{df}(-3, x)}{P_{df}(0, x)}$. We know that $P_{df}(0, x)$ has an area size of unity and for all k it has the same size. As a result, we will use it as our reference denominator and the comparison is simplified since we only need to compare the area size of $P_{df}(-3, x)$ under different k's. For the triangle $\Delta(-3, x)$ its width and height is analyzed as the diagram illustrate:

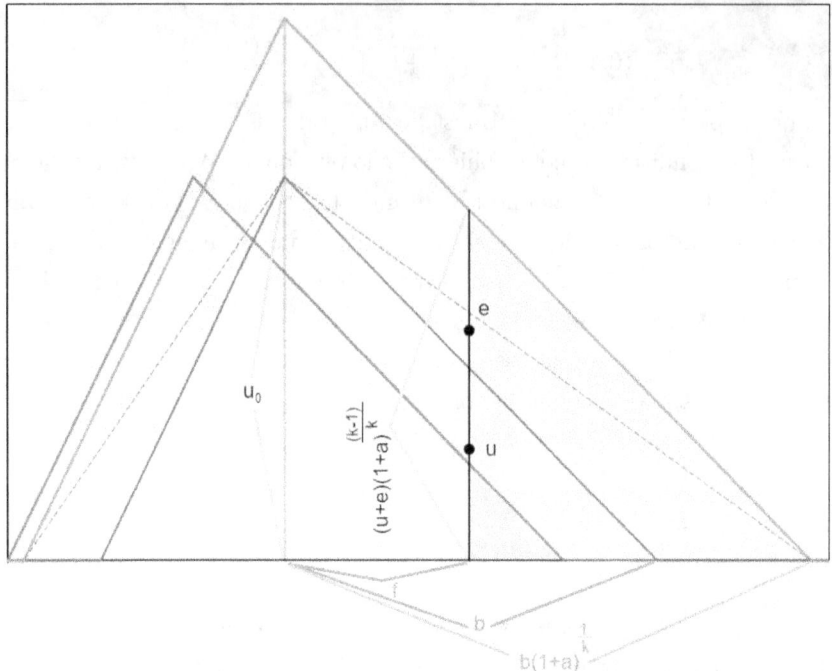

Figure 8.8.3: For a case of $\Delta(-3,x)$ (in actuality $\Delta(-1,x)$ for model illustrative purpose) and $k=2$ and $B_{cs}=2$, equivalently $a=1$. The red solid inner triangle representing the down-size rescaled $\Delta(-3,x)$ both vertically and horizontally can be thought as the $\Delta(0,x)$ placed underneath $\Delta(-3,x)$. The dashed inner triangle representing the down-size rescaled $\Delta(-3,x)$ only vertically shows the intermediate step for the two step down-size rescaling. Finally, the left shifted $\Delta(0,x)$ is shown in its original position. Since $\Delta(0,x)$ is the reference triangle, its shape remains fixed for any given k.

Based on the graph, one can see a lightly shaded and darkly shaded region. The darker shaded regions represents the homo sapiens emergence chance based on $\Delta(0,x)$. The emergence chance of homo sapiens based on $\Delta(-3,x)$ is represented by the area of both lightly and darkly shaded area. Ideally, one needs to find the ratio of

$$\frac{A\left(\Delta_{k=1}(-3,x)\right)}{A\left(\Delta_{k=1}(0,x)\right)} \geq \frac{A\left(\Delta_{k=\infty}(-3,x)\right)}{A\left(\Delta_{k=\infty}(0,x)\right)} \tag{8.8.3}$$

That is, the rate of change for the designated triangle area (homo sapiens emergence) under $k=1$ is faster than the rate change for the designated triangle area (homo sapiens emergence) under $k=\infty$. However, since $\Delta_{k=1}(0,x)$ is the the reference triangle with a fixed shape, the emergence chance is then identical for both as:

$$A\left(\Delta_{k=1}(0,x)\right) = A\left(\Delta_{k=\infty}(0,x)\right) \tag{8.8.4}$$

As a result, one needs to only compares:

$$A\left(\Delta_{k=1}(-3,x)\right) \geq A\left(\Delta_{k=\infty}(-3,x)\right) \tag{8.8.5}$$

Since based on our definition, for any k, the mode now must anchor to the same x values, as a result, any difference in size is solely based on shape only, not due to displacement position.

$A\left(\Delta_k(-3,x)\right)$ has a total height of $[u+\varepsilon](1+a)^{\left(\frac{k-1}{k}\right)}$ rescaled by factor $(1+a)^{\left(\frac{k-1}{k}\right)}$, and a base of $b(1+a)^{\frac{1}{k}} - f$, which is also rescaled by a factor of $(1+a)^{\frac{1}{k}}$ before it is subtracted by f, the difference between homo sapiens equivalent value and peak of the triangle. The overall area size is represented as:

$$A\left(\Delta_k(-3,x)\right) = \frac{1}{2}\left[b(1+a)^{\frac{1}{k}} - f\right][u+\varepsilon](1+a)^{\left(\frac{k-1}{k}\right)} \tag{8.8.6}$$

Whereas ε can be derived as the difference between $h_1 - u$:

$$\varepsilon = h_1 - u \tag{8.8.7}$$

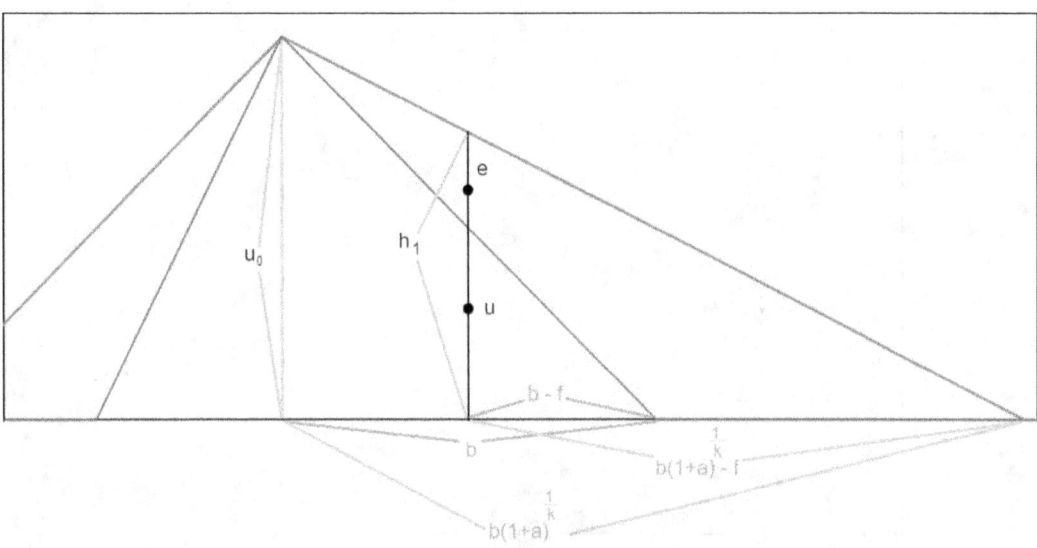

Figure 8.8.4: An illustration of the relationship between the height variables h_1, u, and ε, and width variables b, $b - f$, $b\left(1+a\right)^{\frac{1}{k}}$, $b\left(1+a\right)^{\frac{1}{k}} - f$

Whereas h_1 represents the height of $u + \varepsilon$.

u can be derived in terms of u_0, the unit height of the triangle as:

$$\frac{u_0}{b} = \frac{u}{b-f} \tag{8.8.8}$$

$$u = u_0\left(1 - f_0\right) \tag{8.8.9}$$

whereas $f_0 = \frac{f}{b}$.

h_1 is then derived as the ratio of rescaled base b:

$$\frac{u_0}{b\left(1+a\right)^{\frac{1}{k}}} = \frac{h_1}{b\left(1+a\right)^{\frac{1}{k}} - f} \tag{8.8.10}$$

$$h_1 = u_0 - \frac{u_0 f_0}{\left(1+a\right)^{\frac{1}{k}}} \tag{8.8.11}$$

So ε becomes:

$$\varepsilon = u_0 - \frac{u_0 f_0}{\left(1+a\right)^{\frac{1}{k}}} - u_0\left(1 - f_0\right) \tag{8.8.12}$$

Taking out b from the 1st parenthesis, and setting $u_0 = 1$ and substitute ε for $A\left(\Delta_k\left(-3, x\right)\right)$ we have:

$$A\left(\Delta_k\left(-3, x\right)\right) = \frac{1}{2}\left[\left(1+a\right)^{\frac{1}{k}} - f_0\right]\left[1 - \frac{f_0}{\left(1+a\right)^{\frac{1}{k}}}\right]\left(1+a\right)^{\left(\frac{k-1}{k}\right)} \tag{8.8.13}$$

When $k = \infty$, we have:

$$\frac{1}{2}\left[1 - f_0\right]\left(1 - f_0\right)\left(1+a\right) \tag{8.8.14}$$

and we find that when $k = 1$, we have:

437

$$\frac{1}{2}\left[(1+a) - f_0\right]\left(1 - \frac{f_0}{(1+a)}\right) \tag{8.8.15}$$

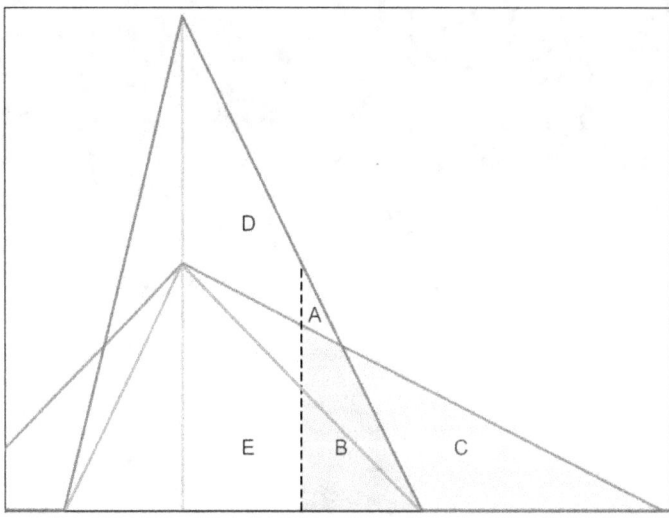

Figure 8.8.5: Area $A\left(\Delta_{k=\infty}\left(3,x\right)\right)$ is represented by $D+A+E+B$, area $A\left(\Delta_{k=1}\left(3,x\right)\right)$ is represented by $E+B+C$. The dashed line indicates the position of $b\left(1+a\right)^{\frac{1}{k}} - f$, and one is interested in finding the area comparison between $A+B$ by $A\left(\Delta_{k=\infty}\left(3,x\right)\right)$ and $B+C$ by $A\left(\Delta_{k=1}\left(3,x\right)\right)$.

and we compare with the case $k = \infty$

$$\frac{1}{2}\left[(1+a) - f_0\right]\left(1 - \frac{f_0}{(1+a)}\right) \quad ? \quad \frac{1}{2}\left[1 - f_0\right]\left(1 - f_0\right)\left(1 + a\right) \tag{8.8.16}$$

$$-2f_0 + \frac{f_0^2}{1+a} \quad ? \quad -2f_0 + f_0^2 - 2af_0 + af_0^2 \tag{8.8.17}$$

$$\frac{f_0^2}{1+a} \quad ? \quad f_0^2 - 2af_0 + af_0^2 \tag{8.8.18}$$

moves $\frac{f_0^2}{1+a}$ over and divide both sides by f_0:

$$0 \quad ? \quad f_0 - \frac{f_0}{1+a} - 2a + af_0 \tag{8.8.19}$$

Times both sides by $1+a$ and simplify:

$$0 \quad ? \quad \left(2af_0 - 2a\right) + \left(-2a^2 + a^2 f_0\right) \tag{8.8.20}$$

Terms $2af_0 - 2a = 2a\left(f_0 - 1\right)$, Since $0 \le f_0 \le 1$, we have $2af_0 - 2a \le 0$, for $-2a^2 + a^2 f_0$ it can be simplified as $-2 + f_0$, since $0 \le f_0 \le 1$, we have $-2a^2 + a^2 f_0 \le 0$.
Therefore, we have:

$$0 \quad \ge \quad (\le 0) + (\le 0) \tag{8.8.21}$$

So the area under case $k = 1 > k = \infty$.
Which can be generalized to any $k > k = \infty$ as:

$$\left[(1+a) - f_0\left(1+a\right)^{\left(\frac{k-1}{k}\right)}\right]\left[1 - \frac{f_0}{(1+a)^{\frac{1}{k}}}\right] \quad ? \quad \left[1 - f_0\right]\left(1 - f_0\right)\left(1 + a\right) \tag{8.8.22}$$

$$-\frac{(1+a)\,f_0}{(1+a)^{\frac{1}{k}}} - f_0\,(1+a)^{\left(\frac{k-1}{k}\right)} + \frac{f_0^2\,(1+a)^{\left(\frac{k-1}{k}\right)}}{(1+a)^{\frac{1}{k}}} \quad ? \quad -2f_0 + f_0^2 - 2af_0 + af_0^2 \tag{8.8.23}$$

$$-2f_0\,(1+a)^{\left(\frac{k-1}{k}\right)} + f_0^2\,(1+a)^{\left(\frac{k-2}{k}\right)} \quad ? \quad -2f_0 + f_0^2 - 2af_0 + af_0^2 \tag{8.8.24}$$

When $f_0 = 0$, we have:

$$0 \;=\; 0 \tag{8.8.25}$$

When $f_0 = 1$, and divide both sides by f_0:

$$-2\,(1+a)^{\left(\frac{k-1}{k}\right)} + f_0\,(1+a)^{\left(\frac{k-2}{k}\right)} \quad ? \quad -2 + f_0 - 2a + af_0 \tag{8.8.26}$$

$$-2\,(1+a)^{\left(\frac{k-1}{k}\right)} + (1+a)^{\left(\frac{k-2}{k}\right)} \quad ? \quad -2 + 1 - 2a + a \tag{8.8.27}$$

$$-2\,(1+a)^{\left(\frac{k-1}{k}\right)} + (1+a)^{\left(\frac{k-2}{k}\right)} \quad ? \quad -(1+a)^{\frac{k}{k}} \tag{8.8.28}$$

divide both sides by $(1+a)^{\frac{k}{k}}$:

$$-\frac{2}{(1+a)^{\frac{1}{k}}} + \frac{1}{(1+a)^{\frac{2}{k}}} \quad ? \quad -1 \tag{8.8.29}$$

Shift terms:

$$1 - \frac{2}{(1+a)^{\frac{1}{k}}} + \frac{1}{(1+a)^{\frac{2}{k}}} \quad ? \quad 0 \tag{8.8.30}$$

If we substitute term $(1+a)^{\frac{1}{k}} = x$, then we have:

$$x^2 - 2x + 1 \quad ? \quad 0 \tag{8.8.31}$$

which is a quadratic equation and we find that for all possible values of x, we have:

$$x^2 - 2x + 1 \quad \geq \quad 0 \tag{8.8.32}$$

The min value occurs at $x = 1$ with a value of 0.

Finally, we show that for every step of $0 < f_0 + \epsilon \leq 1$, its area enclosed is larger than $f_0 = 0$, wheres ϵ is arbitrary but $\epsilon \geq 0$, we have:

$$\frac{-2\,(1+a)^{\left(\frac{k-1}{k}\right)} + f_0\,(1+a)^{\left(\frac{k-2}{k}\right)}}{-2\,(1+a)^{\left(\frac{k-1}{k}\right)} + (f_0 + \epsilon)\,(1+a)^{\left(\frac{k-2}{k}\right)}} < 1 \tag{8.8.33}$$

which all depends on the value of $\frac{f_0}{f_0+\epsilon} < 1$ and indeed the condition holds. Hence, we have shown that for any k, the area enclosed is larger than the area enclosed by $k = \infty$.

We then show that for every step of $1 \leq k + \epsilon \leq \infty$, its area enclosed is smaller than k, wheres ϵ is arbitrary but $\epsilon \geq 0$.

$$\frac{\frac{1}{2}\left[(1+a)^{\frac{1}{k}} - f_0\right]\left[1 - \frac{f_0}{(1+a)^{\frac{1}{k}}}\right](1+a)^{\left(\frac{k-1}{k}\right)}}{\frac{1}{2}\left[(1+a)^{\frac{1}{k+\epsilon}} - f_0\right]\left[1 - \frac{f_0}{(1+a)^{\frac{1}{k+\epsilon}}}\right](1+a)^{\left(\frac{k-1+\epsilon}{k+\epsilon}\right)}} > 1 \tag{8.8.34}$$

First of all,

$$\frac{\left[1 - \frac{f_0}{(1+a)^{\frac{1}{k}}}\right]}{\left[1 - \frac{f_0}{(1+a)^{\frac{1}{k+\epsilon}}}\right]} > 1 \tag{8.8.35}$$

Since $-\frac{f_0}{(1+a)^{\frac{1}{k}}} \geq -\frac{f_0}{(1+a)^{\frac{1}{k+\epsilon}}}$ when the denominator $(1+a)^{\frac{1}{k}} \geq (1+a)^{\frac{1}{k+\epsilon}}$.

Then we can see that

$$\frac{\left[(1+a)^{\frac{1}{k}} - f\right](1+a)^{\left(\frac{k-1}{k}\right)}}{\left[(1+a)^{\frac{1}{k+\epsilon}} - f\right](1+a)^{\left(\frac{k-1+\epsilon}{k+\epsilon}\right)}} > 1 \tag{8.8.36}$$

as such:

$$\frac{(1+a) - f(1+a)^{\left(\frac{k-1}{k}\right)}}{(1+a) - f(1+a)^{\left(\frac{k-1+\epsilon}{k+\epsilon}\right)}} > 1 \tag{8.8.37}$$

due to $-f(1+a)^{\left(\frac{k-1}{k}\right)} \geq -f(1+a)^{\left(\frac{k-1+\epsilon}{k+\epsilon}\right)}$ when $\frac{k-1+\epsilon}{k+\epsilon} \geq \frac{k-1}{k}$.

Hence we have shown that under all circumstances, the rate of change for the designated triangle area (homo sapiens emergence) under k is faster than the rate of change for the designated triangle area (homo sapiens emergence) under $k + \epsilon$. Since all epochs follow the same pace of exponential rate of change, we can logically conclude that:

$$\frac{A\left(\Delta_k(0,x)\right)}{A\left(\Delta_k(t,x)\right)} \geq \frac{A\left(\Delta_{k+\epsilon}(0,x)\right)}{A\left(\Delta_{k+\epsilon}(t,x)\right)} \tag{8.8.38}$$

Whereas $t \geq 0$ signifies the past as more unlike the present with more progressive evolutionary scenarios.

Of course, we have shown how to prove such condition based on comparison between $\Delta_k(0,x)$ and $\Delta_k(-3,x)$. Likewise, for the case of symmetry, one can also use $\Delta_k(0,x)$ and $\Delta_k(3,x)$. For the triangle $\Delta(3,x)$ its width and height is analyzed as the diagram illustrate:

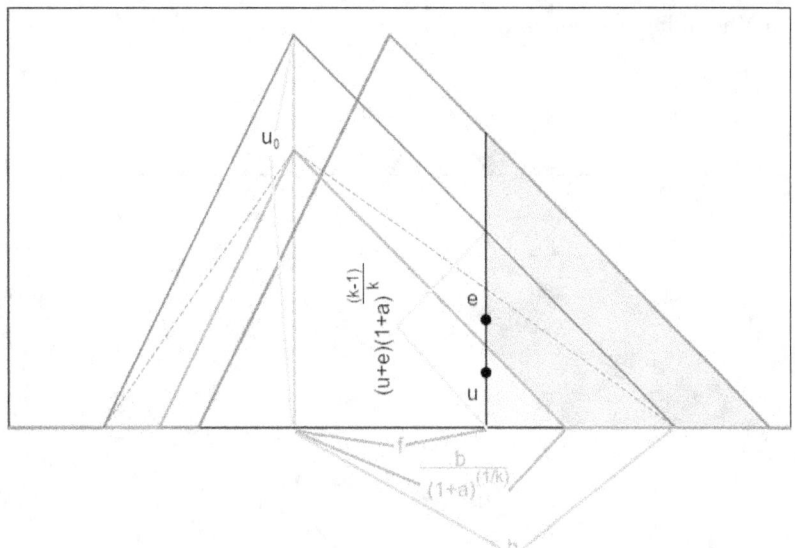

Figure 8.8.6: For a case of $\Delta(3,x)$ (in actuality $\Delta(1,x)$ for model illustrative purpose) and $k = 2$ and $B_{cs} = 2$, equivalently $a = 1$. The red solid triangle representing the up-size rescaled $\Delta(3,x)$ both vertically and horizontally can be thought as the $\Delta(0,x)$ placed above $\Delta(3,x)$. The dashed inner triangle representing the up-size rescaled $\Delta(3,x)$ only horizontally shows the intermediate step for the two step up-size rescaling. Finally, the right shifted $\Delta(0,x)$ is shown in its original position. Since $\Delta(0,x)$ is the reference triangle, its shape remains fixed for any given k.

Based on the graph, one can see a lightly shaded and darkly shaded region. The darker shaded regions represents the homo sapiens emergence chance based on $\Delta(0,x)$. The emergence chance of homo sapiens based on $\Delta(3,x)$ is represented by the area of lightly shaded area. Ideally, one needs to find the ratio of

$$\frac{A(\Delta_{k=1}(0,x))}{A(\Delta_{k=1}(3,x))} \geq \frac{A(\Delta_{k=\infty}(0,x))}{A(\Delta_{k=\infty}(3,x))} \tag{8.8.39}$$

That is, the rate of change for the designated triangle area (homo sapiens emergence) under $k = 1$ is faster than the rate change for the designated triangle area (homo sapiens emergence) under $k = \infty$. However, since $\Delta_{k=1}(0,x)$ is the the reference triangle with a fixed shape, the emergence chance is then identical for both as:

$$A(\Delta_{k=1}(0,x)) = A(\Delta_{k=\infty}(0,x)) \tag{8.8.40}$$

As a result, one needs to only compares:

$$\frac{1}{A(\Delta_{k=1}(3,x))} \geq \frac{1}{A(\Delta_{k=\infty}(3,x))} \tag{8.8.41}$$

or equivalently:

$$A(\Delta_{k=\infty}(3,x)) \geq A(\Delta_{k=1}(3,x)) \tag{8.8.42}$$

Since based on our definition, for any k, the mode now must anchor to the same x values, as a result, any difference in size is solely based on shape only, not due to displacement position.

$A(\Delta_k(3,x))$ has a total height of $\frac{u}{(1+a)^{\left(\frac{k-1}{k}\right)}} - \varepsilon$ rescaled by factor $(1+a)^{\left(\frac{k-1}{k}\right)}$, and a base of $\frac{b}{(1+a)^{\frac{1}{k}}} - f$, which is also rescaled by a factor of $(1+a)^{\frac{1}{k}}$ before it is subtracted by f, the difference between homo sapiens equivalent value and peak of the triangle. The overall area size is represented as:

$$A(\Delta_k(3,x)) = \frac{1}{2}\left[\frac{b}{(1+a)^{\frac{1}{k}}} - f\right]\left[\frac{u}{(1+a)^{\left(\frac{k-1}{k}\right)}} - \varepsilon\right] \tag{8.8.43}$$

441

Whereas ε can be derived as the difference between $u_1 - h_2$:

$$\varepsilon = u_1 - h_2 \tag{8.8.44}$$

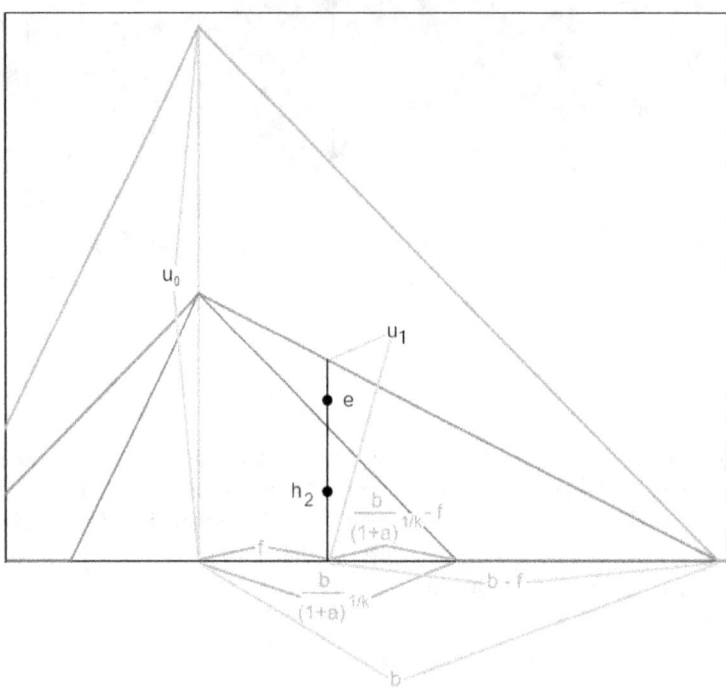

Figure 8.8.7: An illustration of the relationship between the height variables h_2, u_1, and ε, and width variables b, $b - f$, $\frac{b}{(1+a)^{\frac{1}{k}}}$, $\frac{b}{(1+a)^{\frac{1}{k}}} - f$

Whereas $u_1 = \frac{u}{(1+a)^{\left(\frac{k-1}{k}\right)}}$ represents the height of $h_2 + \varepsilon$.

u_1 can be derived in terms of u_0, the unit height of the triangle with rescaling as:

$$\frac{\frac{u_0}{(1+a)^{\left(\frac{k-1}{k}\right)}}}{b} = \frac{u_1}{b - f} \tag{8.8.45}$$

$$u_1 = \frac{u_0}{(1+a)^{\left(\frac{k-1}{k}\right)}}(1 - f_0) \tag{8.8.46}$$

whereas $f_0 = \frac{f}{b}$.

h_2 can be derived in terms of the ratio of rescaled base b and rescaled height u_0:

$$\frac{\frac{u_0}{(1+a)^{\left(\frac{k-1}{k}\right)}}}{\left(\frac{b}{(1+a)^{\frac{1}{k}}}\right)} = \frac{h_2}{\left(\frac{b}{(1+a)^{\frac{1}{k}}}\right) - f} \tag{8.8.47}$$

$$h_2 = \frac{\frac{u_0}{(1+a)^{\left(\frac{k-1}{k}\right)}}}{\left(\frac{b}{(1+a)^{\frac{1}{k}}}\right)}\left[\left(\frac{b}{(1+a)^{\frac{1}{k}}}\right) - f\right] \tag{8.8.48}$$

$$h_2 = \frac{u_0}{(1+a)^{\left(\frac{k-1}{k}\right)}}\left(1 - (1+a)^{\frac{1}{k}}f_0\right) \tag{8.8.49}$$

So ε becomes:

$$\varepsilon = \frac{u_0}{(1+a)^{\left(\frac{k-1}{k}\right)}}(1-f_0) - \frac{u_0}{(1+a)^{\left(\frac{k-1}{k}\right)}}\left(1 - (1+a)^{\frac{1}{k}} f_0\right) \tag{8.8.50}$$

Taking out b from the 1st parenthesis, and setting $u_0 = 1$ and substitute ε for $A(\Delta_k(3,x))$ we have:

$$A(\Delta_k(3,x)) = \frac{1}{2}\left[\frac{1}{(1+a)^{\frac{1}{k}}} - f_0\right]\left[\frac{1}{(1+a)^{\left(\frac{k-1}{k}\right)}}\left[1 - (1+a)^{\frac{1}{k}} f_0\right]\right] \tag{8.8.51}$$

When $k = \infty$, we have:

$$\frac{1}{2}[1 - f_0]\frac{1 - f_0}{1 + a} \tag{8.8.52}$$

and we find that when $k = 1$, we have:

$$\frac{1}{2}\left[\frac{1}{1+a} - f_0\right][1 - (1+a)f_0] \tag{8.8.53}$$

and we compare with the case $k = \infty$

$$\frac{1}{2}\left[\frac{1}{1+a} - f_0\right][1 - (1+a)f_0] \quad ? \quad \frac{1}{2}[1 - f_0]\frac{1 - f_0}{1 + a} \tag{8.8.54}$$

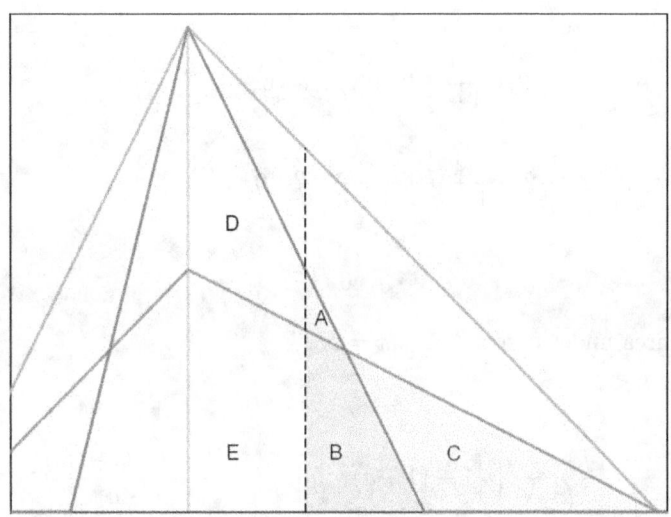

Figure 8.8.8: Area $A(\Delta_{k=\infty}(3,x))$ is represented by $E + B + C$ (directly opposite of our earlier case), area $A(\Delta_{k=1}(3,x))$ is represented by $D + A + E + B$ (directly opposite of our earlier case). The dashed line indicates the position of $b - f$, and one is interested in finding the area comparison between $A + B$ by $A(\Delta_{k=1}(3,x))$ and $B + C$ by $A(\Delta_{k=\infty}(3,x))$.

Taking out $1/2$, multiply both sides by $1 + a$, whereas both sides can be rewritten as:

$$(1+a)\left[\frac{1}{1+a} - f_0\right][1 - (1+a)f_0] \quad ? \quad (1 - f_0)(1 - f_0) \tag{8.8.55}$$

$$[1 - (1+a)f_0][1 - (1+a)f_0] \quad ? \quad (1 - f_0)(1 - f_0) \tag{8.8.56}$$

expands both sides as:

$$1 - 2(1+a)f_0 + (1+a)^2 f_0^2 \quad ? \quad 1 - 2f_0 + f_0^2 \tag{8.8.57}$$

removes unity and divide both sides by f_0:

$$-2(1+a)+(1+a)^2 f_0 \quad ? \quad -2+f_0 \tag{8.8.58}$$

shifting terms around as:

$$-2+2f_0+af_0 \quad ? \quad 0 \tag{8.8.59}$$

we can see that (When $f_0 = 0$)

$$-2 \quad < \quad 0 \tag{8.8.60}$$

and for every other $f_0 > 0$, it is guaranteed that:

$$-2+f_0(2+a) \quad > \quad -2 \tag{8.8.61}$$

So we just need to show the upper limit of f_0 the condition still holds. The upper limit of $f_0 = \frac{1}{1+a}$ since that the rescaled base of $A(\Delta_k(3,x))$ has a width of $\frac{b}{1+a}$ when $k=1$.

so that the f_0 must be between these bounds. If $f_0 > \frac{1}{1+a}$, then one can not compare $A(\Delta_{k=\infty}(3,x))$ with $A(\Delta_{k=1}(3,x))$, since $A(\Delta_{k=1}(3,x))$ base's width terminates at $\frac{1}{1+a}$.

$$0 \le f_0 \le \frac{1}{1+a} \tag{8.8.62}$$

so (When $f_0 = \frac{1}{1+a}$)

$$-2+\frac{1}{1+a}(2+a) \quad ? \quad 0 \tag{8.8.63}$$

$$-2+\frac{2}{1+a}+\frac{a}{1+a} \quad ? \quad 0 \tag{8.8.64}$$

$$-a \quad < \quad 0 \tag{8.8.65}$$

Hence, we have shown that the area under case $k=1 \le k = \infty$.

Which can be generalized to any $k < k = \infty$ as:

$$\left[\frac{1}{1+a}-\frac{f_0}{(1+a)^{\left(\frac{k-1}{k}\right)}}\right]\left[1-(1+a)^{\frac{1}{k}}f_0\right] \quad ? \quad \frac{(1-f_0)(1-f_0)}{1+a} \tag{8.8.66}$$

expands and multiplies both sides by $1+a$ and simplifies as:

$$1-2(1+a)^{\frac{1}{k}}f_0+(1+a)^{\frac{2}{k}}f_0^2 \quad ? \quad 1-2f_0+f_0^2 \tag{8.8.67}$$

we substitute $(1+a)^{\frac{1}{k}}$ with variable x, which encompasses all possible combinations of a and k, so we have:

$$1-2xf_0+x^2f_0^2 \quad ? \quad 1-2f_0+f_0^2 \tag{8.8.68}$$

The left hand side is essentially a quadratic equation. The right hand side is a line parallel to the x axis. The valid domain for x values are between $\left[(1+a)^{\frac{1}{\infty}},(1+a)\right]$, that is, regardless the value of a, the start of the domain is always at 1, since the exponent $\frac{1}{\infty}=0$ when $k=\infty$. The end of the domain changes depending on the value of a. Since the quadratic equation has a positive 2nd order coefficient, it is concave upward and we find its only minimum as:

$$\frac{d}{dx}(1-2xf_0+x^2f_0^2)=-2f_0+2xf_0^2=0 \tag{8.8.69}$$

$$x_{min} = \frac{1}{f_0} = 1 + a \tag{8.8.70}$$

That is, the minimum x value is the inverse of the upper limit value for $f_0 = \frac{1}{1+a}$. We then find the minimum y value as:

$$1 - \frac{2(1+a)}{1+a} + \frac{(1+a)^2}{(1+a)^2} = 0 \tag{8.8.71}$$

Since the beginning of the domain for $k = \infty$ is at 1, $x = (1+a)^{\frac{1}{k}} = 1$, we have:

$$1 - 2f_0 + f_0^2 = 1 - 2f_0 + f_0^2 \tag{8.8.72}$$

and at the end of the domain, also the minimum of the domain when $k = 1$ is at $x = 1 + a$, and $f_0 = \frac{1}{1+a}$ we have:

$$0 \leq 1 - \frac{2}{1+a} + \left(\frac{1}{1+a}\right)^2 = a^2 \tag{8.8.73}$$

Since all other combination values of k and a are manifested in $1 \leq x \leq 1 + a$, and its y value must fall between $0 \leq y \leq 1 - 2f_0 + f_0^2$. That is, it must fall between the minimum and the right hand side, hence any $k < k = \infty$ as:

$$1 - 2(1+a)^{\frac{1}{k}} f_0 + (1+a)^{\frac{2}{k}} f_0^2 \leq 1 - 2f_0 + f_0^2 \tag{8.8.74}$$

Finally, we want to note that when a is fixed, one were to shift $f_0 = 0$, the quadratic function transforms into $1 - 2f_0 + f_0^2$. This is completely understood since when $f_0 = 0$, so the entire base width of $A(\Delta_{k=1}(3,x))$ is considered, which is equivalently to $A(\Delta_{k=\infty}(3,x))$ in area.

We then show that for every step of $k + \epsilon$, its area enclosed is smaller than k, wheres ϵ is arbitrary but $\epsilon \geq 0$.

$$\frac{\frac{1}{2}\left[\frac{1}{(1+a)^{\frac{1}{k}}} - f_0\right]\left[\frac{1}{(1+a)^{\left(\frac{k-1}{k}\right)}}\left[1 - (1+a)^{\frac{1}{k}} f_0\right]\right]}{\frac{1}{2}\left[\frac{1}{(1+a)^{\frac{1}{k+\epsilon}}} - f_0\right]\left[\frac{1}{(1+a)^{\left(\frac{k-1+\epsilon}{k+\epsilon}\right)}}\left[1 - (1+a)^{\frac{1}{k+\epsilon}} f_0\right]\right]} < 1 \tag{8.8.75}$$

We simplify the process by excluding f_0 and expand, then we have:

$$\frac{\frac{1}{(1+a)} - \frac{f_0}{(1+a)^{\left(\frac{k-1}{k}\right)}}}{\frac{1}{(1+a)} - \frac{f_0}{(1+a)^{\left(\frac{k-1+\epsilon}{k+\epsilon}\right)}}} < 1 \tag{8.8.76}$$

For the 2nd term, it is easy to see that

$$-\frac{f_0}{(1+a)^{\left(\frac{k-1}{k}\right)}} \leq -\frac{f_0}{(1+a)^{\left(\frac{k-1+\epsilon}{k+\epsilon}\right)}} \tag{8.8.77}$$

When $(1+a)^{\left(\frac{k-1}{k}\right)} \leq (1+a)^{\left(\frac{k-1+\epsilon}{k+\epsilon}\right)}$ and $-\frac{f_0}{(1+a)^{\left(\frac{k-1}{k}\right)}} \leq -\frac{f_0}{(1+a)^{\left(\frac{k-1+\epsilon}{k+\epsilon}\right)}}$.

Hence we have shown that under all circumstances, the rate of change for the designated triangle area (homo sapiens emergence) under k is faster than the rate of change for the designated triangle area (homo sapiens emergence) under $k + \epsilon$. Since all epochs follow the same pace of exponential rate of change, we can logically conclude that:

$$\frac{A(\Delta_k(0,x))}{A(\Delta_k(t,x))} \geq \frac{A(\Delta_{k+\epsilon}(0,x))}{A(\Delta_{k+\epsilon}(t,x))} \tag{8.8.78}$$

Based on earth's evolutionary history, we have narrowed down to a limited range of k and selection factor pairs.

One may then asks if it is possible to further narrow down the range to any specific k and selection factor pair, consequently, to a specific BER. Since the gap between major evolutionary milestone seem to be fairly evenly spaced out, we may assume a BER of 1. We can be more specific if we take genomic complexity into consideration.

According to Alexei and Gordon, the genome complexity (the number of sites of functional codons) of living organisms doubles every 250 million years which translated to 1.203 per 100 million years. We can assume that number of sites correlates well with number of exhibited traits by the species and the total number of unique traits available. Repeated regeneration of the same traits or different permutations leading to the same combination does not lead to an increase in genome complexity. That is, genome complexity $\propto$ BER $\not\propto$ BCS. However, higher evolved traits may associate with other non DNA manipulation such as methylation. Then, genome complexity $\not\propto$ BER $\not\propto$ BCS. Furthermore, different permutations may lead to new sites if multiple local optimal solutions exist. Then, genome complexity $\propto$ BER $\propto$ BCS. For the last case, if genome complexity is defined as the average number of potentially mutually exclusive sites by a fixed number of traits arrived from different permutation paths, then genome complexity $\propto$ BER $= \frac{1}{n} \sum \text{BCS} \leq \sum \text{BCS}$. The true correlation between genome complexity, BER, BCS requires further investigation.

If it is now assumed that genome complexity change of 1.203 per 100 Myr translates into a BER of 1.203, why observed major evolutionary milestones evenly spaced out (BER $=1$)? This can easily explained when each new milestone, after its inception, has been exponentially enhanced through time, such improvement on functionality does not reflect from our earlier milestone tally which simply describes its first appearance but not its qualitative change through time.

By settling on BER $= 1.203$, one finds that the pair (8.1, 5.54) satisfy the requirement; however, it falls outside the permissible range with a occupancy ratio of 294%, indicating if all extraterrestrial industrial civilization follows such a path of development, earth and its vicinity should been already occupied by others 3 times over. This indicates deviation, emergence rate, or planet formation requires adjustment and we are rarer than 1 in 3 galaxies. More detailed analysis follows.

8.8.3 k Bounds for Unweighted Deviation

Our current distribution assumes that typical mammal possessed an EQ of 1 and the deviation of Homo sapiens emergence (not civilization) located at 7.6, representing our EQ attainment. Recall that we assumed that enlargement of brain is actually the composite consequence of all 7 traits uniquely possessed by human, which creates a positive feedback loop on accelerating cranial enlargement. This interpretation assumes that the contribution of these 7 traits to the deviation are far more significant (which contributes to an EQ of 7) than previous 6 traits possessed by all typical mammals, which contributes to only an EQ of 1.

This interpretation can be altered, however, if one consider each trait, whether gained in the historical past or the future, weighted equally in value. Then, regardless of typical mammal's EQ of 1, *mammal's settlement mode peak should posits at 7*, representing 7 common traits that distinguishes mammals from the earliest common ancestors between fish and tetrapods. (lungs, four legs, dry skin, thermoregulation, enhanced olfactory functionality, fur, and placenta). These 7 traits found in a typical mammal requires an evolutionary time of 400 Mya - 100 Mya = 300 Myr, and Homo sapiens position should be represented as an outlier 7 units to the right, at 14. We need to make some minor adjustments to our existing PDF functions as:

$$C_{df}(t) = \int_{-5}^{0.45} P_{lanet}(-s) \left(t_{Win}(B_{cs})^s \int_{L_{imit}}^{\infty} P_{df}(-3 + -s + t, x)\, dx \right) ds \tag{8.8.79}$$

So that by now, the original $C_{df}(0)$ no longer represents the integration of $P_{df}(0, x)$ but $P_{df}(-3, x)$ instead. We still assumes that BCS $= 2.783$ and use the 2nd interpretation for the selection factor. Furthermore, this time, due to alteration on the deviation, one needs alter σ in order to hold the assumption that the appearance chance of civilization is still fixed at 1 per 3 galaxies. σ has to be lowered to 0.1256 so that the each successive distribution width/variance is narrowed and the appearance rate can be lowered to 1 per 3 galaxies. By performing this procedure, one can find the upper and lower bound for evolution. The lower bound are consistent with our earlier analysis, but the upper bound is considerably higher. This is true because distribution with lower σ also provides lower emergence rate across all time periods, including the past, given the same k value. As a result, there are lower civilization emergence chance from the past given the same BCS and k value for a lowered σ, consequently, lower space occupancy ratio. This lower space occupancy ratio enables more conservative scenarios, higher k and selection factor pairs satisfy the permissible range. This seem to resolve our earlier problem that for a BER $= 1.203$, it falls outside the permissible range. However, our adjusted distribution shows minimal overlapping between successive distributions 100 Myr apart. This certainly contradict our observation.

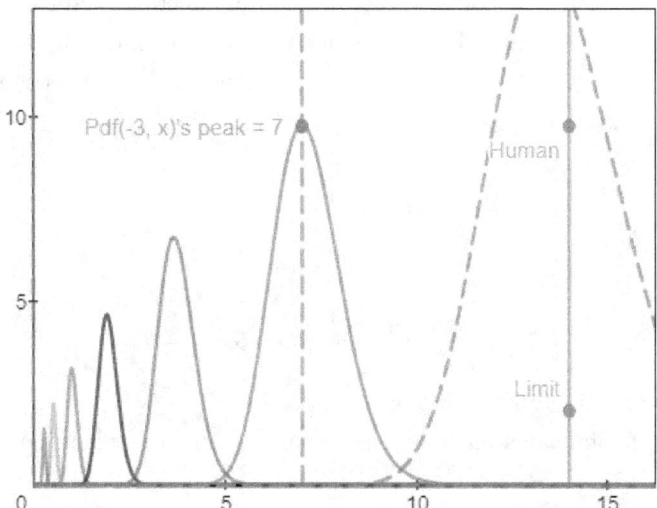

Figure 8.8.9: Typical mammal places on x $= 7$, human at x $= 14$, and $\sigma = 0.1256$, but a minimal overlap between successive distributions 100 Myr apart.

We stated earlier that extinction rate should be around 40% per 100 Myr to reflect the rate of extinction by number of genera (so an overlap of 60%). Given an emergence chance of 1 in 3 galaxies, greater overlapping can be achieved through an increase on σ if deviation is larger.

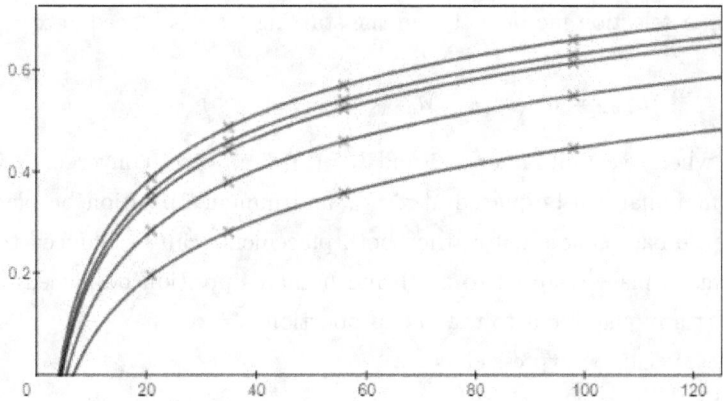

Figure 8.8.10: σ value requirement increases as the deviation/human placement value increases. The curves represented from the bottom to the top: for pairs $(1, 1.578)$, $(2.5, 2.5)$, $(3.91, 5)$, $(4.583, 6.25)$, and $(6.5, 10)$ respectively.

This implies, that human's placement should be lying further to the right, and indeed the additional 7 traits uniquely identifies with human should be more weighted than the previous ones gained by mammals. More detailed analysis follows.

8.8.4 Self-similarity of the Distribution

k bounds for unweighted deviation requires further analysis mathematically. Recall that the selection factor comes with 2 interpretations. The first interpretation states that any distribution placement shifts further apart from each other as the selection factor increases. The settlement line x=7 also shift proportionally with the selection factor S_{change} along with the rest of the distributions.

$$M_{id} = 1 \cdot 7 \cdot (S_{change}) + M_{ax}(\infty) \tag{8.8.80}$$

As a result, only 1 possible pairing (1, 1.578) exists for any selection factor adjustment. Under such a pair with any selection factor, one can think it as the stretched version of selection factor = 1 with k=1.578 under interpretation 2. Therefore, First interpretation becomes non-useful for our in-depth analysis.

Interpretation 2 for the selection factor is the one we used in previous cases, and settlement line x=7 remain fixed as the selection factor changes:

$$M_{id} = 1 \cdot 7 \tag{8.8.81}$$

The combined formula for both is:

$$M_{id} = 1 \cdot 7 \cdot (S_{change})^{(1-T)} + M_{ax}(\infty)(1-T) \tag{8.8.82}$$

Whereas T is the Boolean toggle for 2 interpretations.

Similarly, the combined formula for human's placement is (assuming human's placement is 3 times away at x = 21):

$$L_{imit} = 3 \cdot 7 \cdot (S_{change})^{(1-T)} + M_{ax}(\infty)(1-T) \tag{8.8.83}$$

There also exist a case in which the settlement point remains fixed as interpretation 2, but the human placement position is defined as a multiple of total number of traits based on the selection factor's increase. This can happen if one assumes the last common ancestor between fish and tetrapod already possessed more than 1 trait. This, in fact, was indeed the case, as illustrated from the major evolutionary milestones achieved from Cambrian explosion up to lung fish. So, instead of human's placement at x=21 and the starting trait located at x=0, it should be placed at x=29 with a selection factor = 4, and the starting trait is placed at x = -4.

$$L_{imit} = 3 \cdot (M_{id} - M_{ax}(\infty)) + M_{ax}(\infty) \tag{8.8.84}$$

Finally, one can also see that when selection factor = 1 and k =1.578, all cases converge.

We have shown that typical mammal can be placed at x = 7 and human's position be placed at x ≥ 14. If we use mammal's position as the base denominator, then both placements can be reduced to smaller numbers in proportion. That is, mammal's place restored to x =1 and human's position be placed at x ≥ 2. A set of equations can faithfully convert any placement to the unit proportion.

For mammals' settlement point, it can be expressed as:

$$M_{id} = 1 \cdot 7 \cdot (S_{change})^{(1-T)} + M_{ax}(\infty)(1-T) \tag{8.8.85}$$

For the original 1st and 2nd interpretations of the selection factor, human's placement can be expressed as:

$$E_{nd} = D_{eviation} \cdot 7 \cdot (S_{change})^{(1-T)} + M_{ax}(\infty)(1-T) \tag{8.8.86}$$

whereas $D_{eviation}$ stands as the human placement deviation in terms of mammal's settlement point at 7.

For original mixed interpretation of the selection factor, human's placement can be expressed as:

$$M_{ixed} = D_{eviation} \cdot 1 \cdot (M_{id} - M_{ax}(\infty)) + M_{ax}(\infty) \tag{8.8.87}$$

The unit proportional conversion for both the 1st and 2nd interpretation:

$$L_{imit} = \frac{E_{nd} - M_{id}}{-M_{ax}(\infty) + M_{id}}(-M_{ax}(\infty) + M_{ax}(0)) + M_{ax}(0) \tag{8.8.88}$$

The unit proportional conversion for mixed interpretation:

$$L_{imit} = \frac{M_{ixed} - M_{id}}{-M_{ax}(\infty) + M_{id}}(-M_{ax}(\infty) + M_{ax}(0)) + M_{ax}(0) \tag{8.8.89}$$

10

Notices that for both cases the conversion is defined based on $M_{ax}(0)$, but this can be substituted by any peak value. That is, the unit conversion can be centered on any peaks, and the re-scaling proportion varies accordingly.

The re-scaled settlement line based on peak 0:

$$x = \left(\frac{M_{ax}(0) - M_{ax}(\infty)}{M_{ax}(-3) - M_{ax}(\infty)}\right) 6 \left(S_{change}\right)^{(1-T)} + M_{ax}(0+3) \tag{8.8.90}$$

Since the last common ancestor between fish and tetrapod were 300 Myr apart, the initial placement position is 300 Myr into the past at $M_{ax}(0+3)$. The factor $\frac{M_{ax}(0) - M_{ax}(\infty)}{M_{ax}(-3) - M_{ax}(\infty)}$ means how much down scaling is required to convert settlement placement. This factor is self-similar so that one can substitute it with $\frac{M_{ax}(3) - M_{ax}(\infty)}{M_{ax}(0) - M_{ax}(\infty)}$, for example.

The settlement line can also be defined not based on any peak, instead based on the origin O (x=0). However, for selection factor > 1, the origin O does not translate into $M_{ax}(\infty)$. In fact, $M_{ax}(\infty) < 0$. Therefore, one has to define the origin O in terms of a peak that happens to coincide with the origin. Looking for a peak $M_{ax}(\infty) < x < M_{ax}(0)$. Such peak changes as k and selection factor pair changes. This relationship is captured by the following plot:

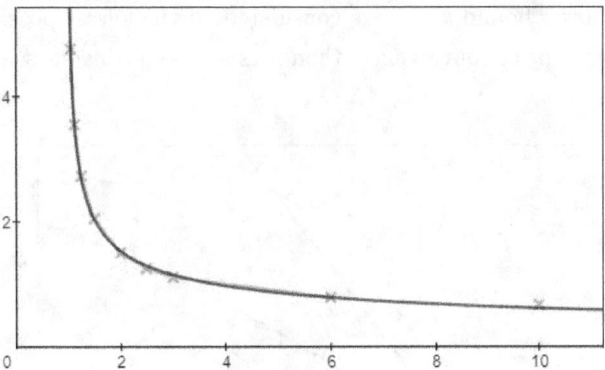

Figure 8.8.11: Switch function for BCS $= 2.783$

and its best fit is:

$$S_{tart}(x) = \left(0.364478x^{0.985213} - 0.358572\right)^{-0.420647} \tag{8.8.91}$$

So the new starting position is defined as $0 + 3 + S_{tart}(F_0)$, F_0 is the selection factor, whereas $M_{ax}(\infty) < M_{ax}(0+3+S_{tart}(F_0)) < M_{ax}(0+3)$.

---

[10]Alternatively, the reverse operation, converting an unit settlement peak at $M_{ax}(0) = 1$ to any other larger peak can be expressed as: $L_{imit} = \frac{E_{nd} - M_{ax}(0)}{-M_{ax}(\infty) + M_{ax}(0)} \cdot (-M_{ax}(\infty) + M_{id}) + M_{id}$, when the new settlement peak is fixed to 7 traits and deviation = 17, it becomes $L_{imit} = \frac{17}{-M_{ax}(\infty)+1} \cdot (-M_{ax}(\infty) + 7) + 7$. So that the emergence for $C_{df}(0)$ remains fixed as deviation changes with distribution width.

combining both interpretations into a single step function, we have:

$$S_{witch}(x) = \begin{cases} 0 + 3 + S_{tart}(F_0) & x = 1 \\ \infty & x = 0 \end{cases}$$

(8.8.92)

and the re-scaled settlement line based on the origin:

$$x = \left(\frac{M_{ax}(0) - M_{ax}(\infty)}{M_{ax}(-3) - M_{ax}(\infty)} \right) 7 \left(S_{change} \right)^{(1-T)} + M_{ax} \left(S_{witch}(T) \right)$$

(8.8.93)

The $C_{df}(t)$ function has to be adjusted accordingly as:

$$C_{df}(t) = \int_{-5}^{x} P_{lanet}(-s) \left(t_{Win}(B_{cs})^s \int_{Limit}^{\infty} P_{df}(P_{eak} + -s + t, x)\, dx \right) ds$$

(8.8.94)

Whereas $P_{df}(t, x)$ is used since we defined the rescale based on $P_{eak} = 0$ and $M_{ax}(0)$, but it can be based on any peak.

8.8.5 BCS, k and Selection Factor Relationship

For the second interpretation of the selection factor, the relationship between BCS, the settlement point, the deviation/human placement point is examined. We then run the result for different values of BCS and the pairings are plotted.

First of all, both the permissible lower and upper bound for the k variable $\propto$ BCS. For the lower bound, the increase in BCS renders the distributions more progressive. As a result, a more conservative k is required to pull the specific distribution $P_{df}(-3, x)$ back to the settlement point. The upper bound increases because increasing BCS implies the past is more unlike the present, manifested as $\frac{C_{df1}(t)}{C_{df1}(0)} < \frac{C_{df0}(t)}{C_{df0}(0)}$, if $C_{df1}(t)$'s BCS $> C_{df0}(t)$'s BCS, and so that more conservative k's are allowed before the occupancy ratio reaches 100%. It is shown that as BCS decreases, eventually the lower and upper bound converges toward the same value. The point of convergence between the curves defines the lowest possible BCS. With even lower BCS, even the lower bound guarantees a 100% occupancy. However, when only the classic and conservative evolution scenarios are considered, only those upper and lower bound $k \geq 1$ are considered. If the lowest possible BCS based on $k \geq 1$ is larger than the BCS defines the point of convergence, then it is a more stringent selection criterion than the occupancy ratio criterion.

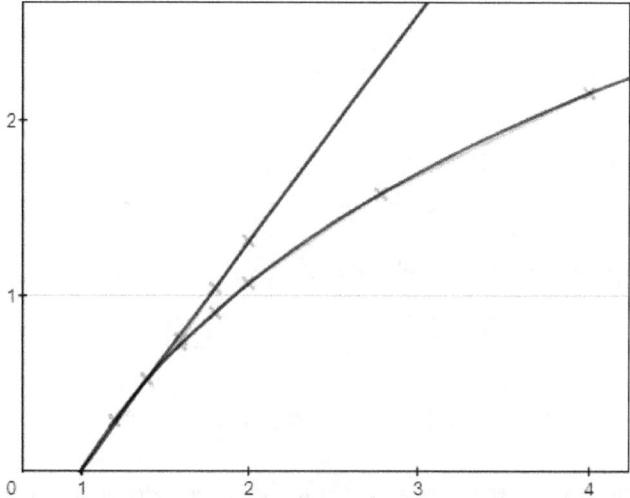

Figure 8.8.12: Lower and upper bounds on k within the permissible range for different BCS values with settlement x=7

The best fit for the upper bound is:

$$U_{pper}(x) = 1.29846x - 1.29609 \qquad (8.8.95)$$

The best fit for the lower bound is:

$$L_{ower}(x) = 38.2873x^{0.0392351} - 38.2774 \qquad (8.8.96)$$

Increasing BCS also raises k vs. selection factor curves. Each curve is a dimensional expansion based on a particular evaluated point of the upper and lower bounds for k. All curves are fit with the forms of $\left(ax^b + c\right)^v$:

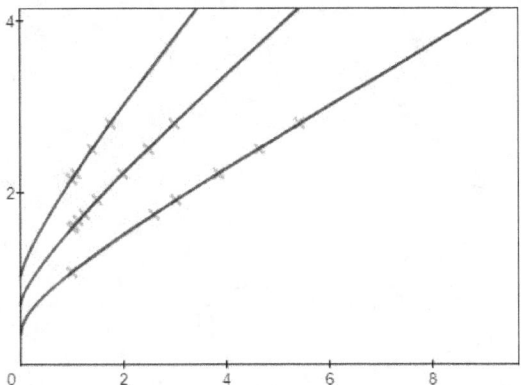

Figure 8.8.13: k vs. selection factor curves for BCS $= 2$, 2.783, and 4 respectively from the bottom to the top

Increasing the settlement value for mammals lowers the lower bound for the k variable. Since the settlement is located further to the right, a less conservative k is required for $P_{df}(-3, x)$ to reach the settlement point, and lowers k vs. selection factor curves. The upper bound increases as the settlement value increases. Since for each selection factor, a less conservative k is now required, the past is more unlike the present, the occupancy ratio drops, and more k's are allowed before the occupancy ratio reaches 100%. Lowering the lower bound and increasing the upper bound shifts the point of convergence further to the left; therefore, a greater overall BCS permissible ranges. It is also noted that, for settlement value of x=5, BCS defined point of convergence is greater than the lowest possible BCS based on $k \geq 1$, therefore, the occupancy ratio criterion in this case becomes more stringent.

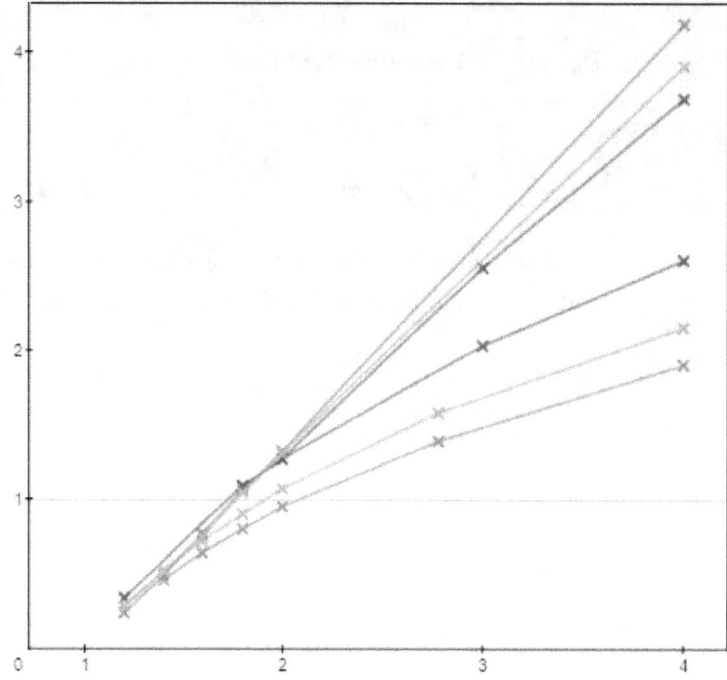

Figure 8.8.14: The upper and lower bounds on k for settlement value of 5, 7, and 9.

Increasing the deviation/human placement position while holding emergence untouched raises the upper bound on k variable, and vice versa. The lower bound remains fixed. This is easily understood since increasing the deviation decreases the emergence and more conservative pairs of k and selection factor fits are allowed before the occupancy ratio reaches 100%. Shifting the deviation further out also shifts the point of convergence between the lower and upper bound further to the left; therefore, a greater overall BCS permissible ranges, and vice versa.

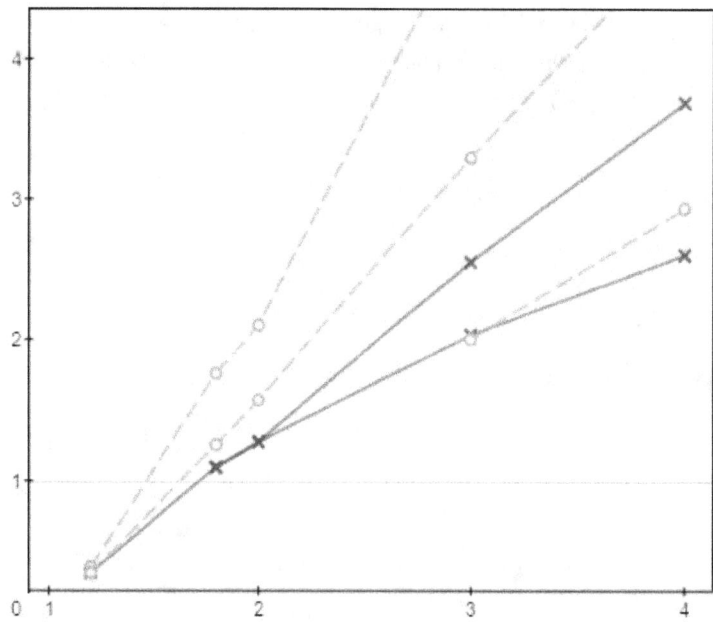

Figure 8.8.15: From the top to the bottom: the upper bound variations on k for settlement value of 5 for deviation of 19, 18, 17 (default), and 15.

Finally, increasing the deviation/human placement position while altering the emergence so that it is always fixed to 1 per 3 galaxies lowers the upper bound on k , and vice versa. The lower bound remains fixed. For greater deviation, alteration on σ renders the past more similar to the present, manifested as $\frac{C_{df1}(t)}{C_{df1}(0)} > \frac{C_{df0}(t)}{C_{df0}(0)}$, if $C_{df1}(t)$'s $\sigma > C_{df0}(t)$'s σ and $C_{df1}(t)$'s deviation $> C_{df0}(t)$'s deviation. Fewer conservative pairs of k and selection factor fits are allowed before the occupancy ratio reaches 100%. This is directly the opposite of the previous case.

Very lastly, assuming the deviation/human placement position is fixed and altering σ, a lowered σ renders lower current emergence chance, hence lower occupancy and vice versa.

Their overall relationship can be captured by the following diagram:

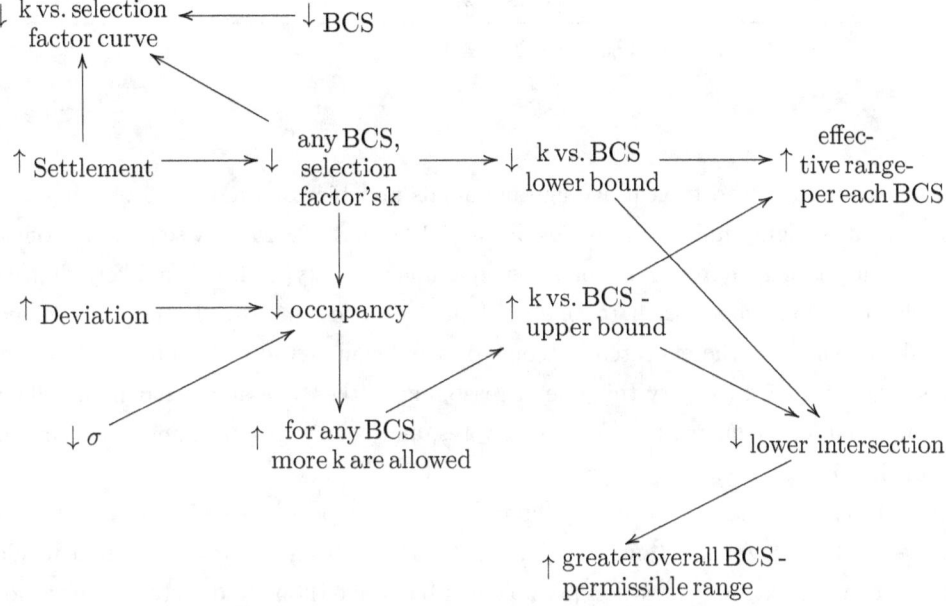

For the last case where each deviation adjustment is coupled with an emergence rate fixation, it becomes:

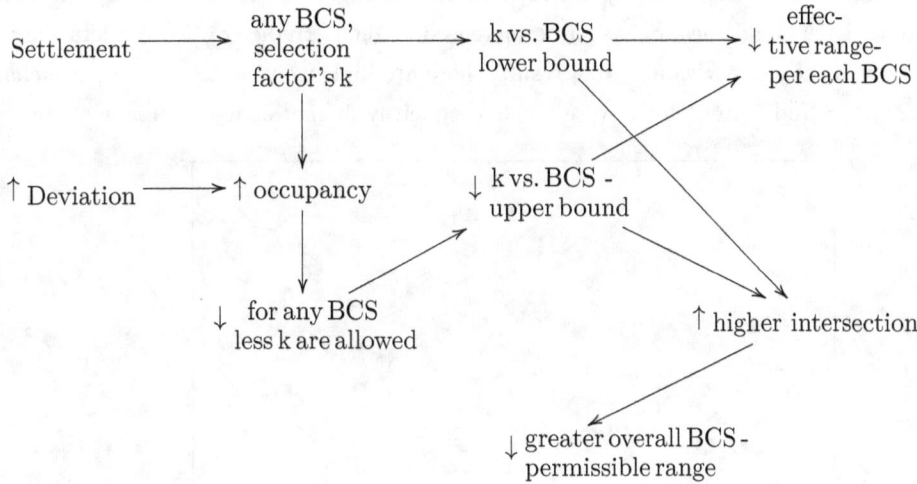

8.8.6   The Lower and Upper Bound on Deviation

Now, we constrain our deviation's lower and upper limit. The lower bound is well-defined, since we know how many traits distinguish human from the rest of birds and mammals. In the unweighted example we just illustrated, we assumed that civilization's attainment lies at least 14 to the right from the typical mammalian average at 7, forming a ratio of 14 to 7. Therefore, the lower bound is at x=14 if all traits considered are unweighted.

The maximum upper bound of civilization's placement relative to the mode value is trickier. Recall in Chapter 7, in our original weighted model we assumed that civilization's attainment lies at 17 to the right from the typical mammalian average at 1, forming a ratio of 17 to 1. In an ultra weighted model, the ratio becomes $> \frac{17}{1}$. If increasing the deviation/human placement position while holding the emergence rate untouched, the emergence rate drops quickly. As indicated from the table below:

Deviation	Emergence rate	Factor Rate
18	1/3 galaxies	1
24	1/110.53 galaxies	36.843
38	1/72736.89 galaxies	24,245.63
100	1/987,613,768 galaxies	329,204,589.333
168	$1/4.66116 \times 10^{16}$ galaxies	1.55372×10^{16}

Unless human emergence, compared to typical average mammals and birds, is hundreds of millions of times harder than we assumed, it is highly unlikely that deviation > 100, or 14.28 ratio wise. Even if one assumes that human is relatively easy to evolve from mammals, but mammals are atypical, it is unlikely that prototype mammals are hundreds of million times harder to emerge than the earliest multi-cellular life. We have shown in Chapter 4 that multicellularity is the expected outcome of significant oxygen and nutrient buildup on any habitable planet. Therefore, no need to carry this line of reasoning further to assume that multicellularity are hundreds of million times harder to emerge than the eukaryotes, and eukaryotes are hundreds of million times harder to emerge than the prokaryotes.

On the the other hand, if one increases the deviation/human placement position while altering the emergence so that it is always fixed to 1 per 3 galaxies, and with BCS = 2.783, selection factor = 1, earliest window = 19 Mya, and $k =1.578$ (so the most extreme deviation possible can be derived) one finds that the deviation has to be < 250, or $\frac{250}{7} \approx 35.71$ ratio wise before the occupancy ratio reaches 100%. That is, civilization's attainment can not go beyond 36 times higher than the typical mammalian average. In order to satisfy the requirement of the appearance rate of 1 per 3 galaxies, σ has to be increased to 0.63. Higher σ also renders the past more similar to the present, given the same k value. As a result, there are higher civilization emergence chance from the past given the same BCS and k value for a raised σ, consequently, higher space occupancy ratio.

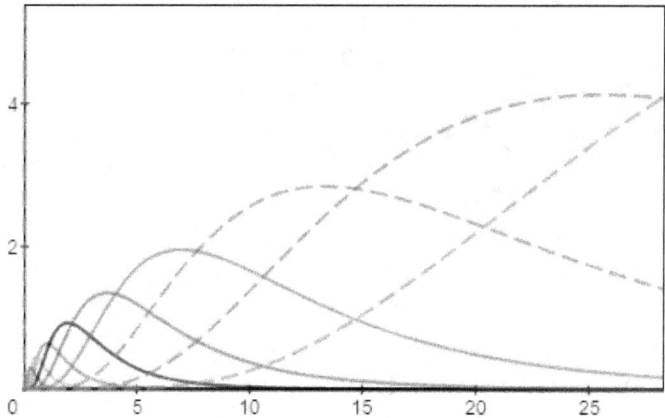

Figure 8.8.16: Distributions placement pattern for selection factor = 1, $k =1.578$, BCS = 2.783, $\sigma= 0.63$, and deviation=250

However, successive distributions 100 Myr apart has an overlapping area of more than 80%. This does not conform to earth's extinction rate for genera per 100 Myr. Assuming 40% extinction rate for genera per 100 Myr, the overlapping area should be 60%. We alter σ to 0.38 and deviation to 62.5, or 8.93 ratio wise. Of course, the real situation lies within these possibilities, but we are justified for using our earliest model assuming a ratio of 17.

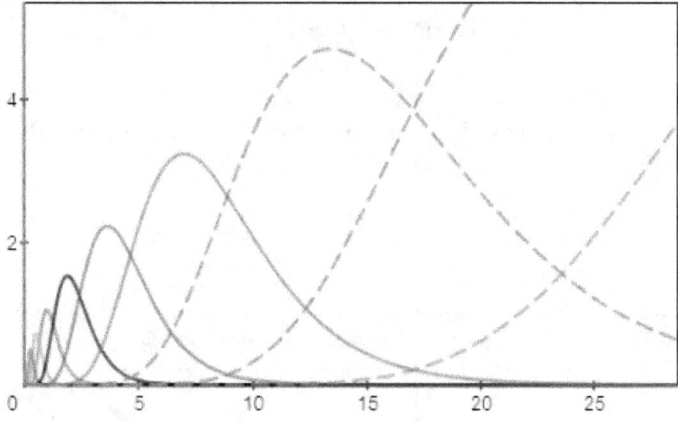

Figure 8.8.17: Distributions placement pattern for selection factor = 1, k =1.578, BCS = 2.783, σ= 0.38, and deviation=62.5

For other possible pairs of k and selection factor, the largest deviation attainable shrinks rapidly for an emergence of 1 per 3 galaxies. Since higher k's implies more conservative evolutionary scenarios and the past is more like the present, only smaller deviation coupled with smaller σ guarantees < 100% space occupancy. At the same time, human's placement can still occur at a larger deviation but only if the emergence rate is further reduced to a smaller number, so that despite high similarity between the past and the present, occupancy < 100%. The following table lists pair (1, 1.578), (2.5, 2.5), and (8.1, 5.54):

k=1.578 & f=1			k=2.5 & f=2.5			k=5.54 & f=8.1		
Deviation	Emergence	σ	Deviation	Emergence	σ	Deviation	Emergence	σ
18.5	0.177	0.187	18.5	0.592	0.2733	**18.5**	**3.082**	**0.355**
25	0.309	0.2411	25	1.124	0.332	25	5.993	0.413
50	0.760	0.361	**50**	**2.994**	**0.454**	50	16.734	0.53
100	1.575	0.4782	100	5.591	0.57	100	31.073	0.64
121	1.804	0.511	121	6.616	0.601	121	35.128	0.67
214	2.863	0.6073	214	8.531	0.697	214	48.697	0.759
256	**3.195**	**0.638**	256	10.117	0.725	256	52.970	0.787
353	4.032	0.692	353	12.919	0.776	353	63.058	0.85

Table 8.8.10: Minimum emergence and minimum deviation requirement for 3 sets of k and selection factor, the row in which minimum 1 per 3 galaxies emergence requirement is highlighted.

Therefore, based on terrestrial observation, the deviation ratio representing the emergence of civilization must lie between $\frac{14}{7}$ to 73 relative to the deviation representing typical mammalian average assuming an emergence of 1 per 3 galaxies, and evolutionary history of earth is one of the conservative type with $k > 1.578$ as the lower bound and the upper bound varies depend on the weightedness of the traits, major evolutionary milestone interval, and genomic complexity.

8.8.7 Step by Step Diagram for Constructing the Generalized Model and a Real Case Instantiation

Finally, based on all of our knowledge so far, we can construct a step by step guide to construct the most realistic evolutionary scenario based on list of available information. The step by step diagram instruction is presented below:

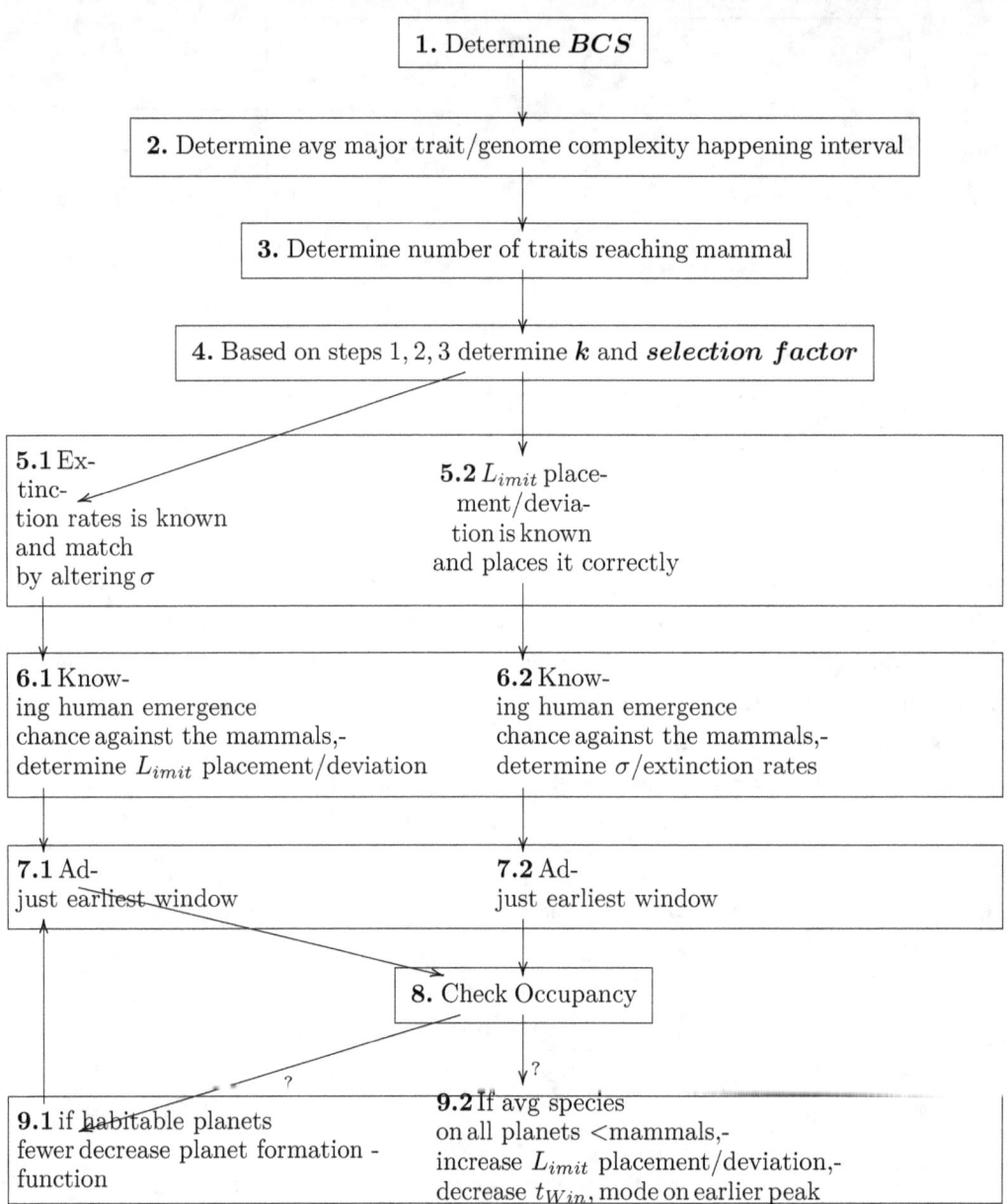

First of all, one needs to determine the BCS growth per 100 Myr. Next, the average major evolutionary milestones describing a trait later fixed within the cohort of species and the genome complexity provides clue for BER. We have determined this to be 1.203. Next, determining the number of traits reaching mammals. This information will provide the number of traits to represent typical mammal and its evolutionary time requirement. Based on these 3 steps, one can determine the k variable and the selection factor pair. Once k is determined, BER is immediately derived based on BCS and k. Recall we find the pair (8.1, 5.54) satisfies the requirement. We are going to run a even more realistic one. We count all 11 major milestones that helped to achieve a typical mammal today since the Cambrian explosion as the settlement point for mammals and chose a time requirement of 541 Myr. The genomic complexity remains at 1.203. This gives us the pair (5.83, 5.54). At step 5, there are two possible paths to take. If the extinction rates 100 Myr apart can be truly observed but not the deviation of human placement, one can use this parameter to further constrain the model by altering the deviation of human emergence. One have to first alter σ to match desired extinction with expected overlap between successive distributions 100 Myr apart. Then, knowing human's emergence chance relative to the typical mammals and birds *on earth*, one can determine the deviation L_{imit}.

$$t_{Win} \int_{L_{imit}}^{\infty} P_{df}(-5.41,\ x)\, dx = \text{emergence} \tag{8.8.97}$$

456

Alternatively, if one is certain about weighted deviation of traits representing human relative to the mode, then one can find the appropriate σ that satisfies the known emergence chance of human. If the deviation is close to the mode, lowering σ to fit the emergence chance requirement causes the overlapping region between successive distribution 100 Myr apart to decrease, representing an increase in the extinction rates despite every other assumptions remain fixed, and vice versa. If we also happen to know the extinction rate, yet the model's extinction rate does not match the extinction rate observed; then, it implies that our lognormal distribution is an inadequate distribution model representing reality. We may require a new distribution that is fatter in the middle (increasing the overlapping area) and descend more rapidly at the tails (so that appearance rate is fixed to 1 per 3 galaxies). This reversely implies that underlying multinominal distribution assigns weighted chance factor for different combinations. For smaller combinations of traits, a more relaxed chance factor is used. For larger combinations, a more stringent chance factor is used. This topic is worthy of a whole new field of research and is beyond our scope at this time.

In our case, we took the first approach and σ is altered to 0.7 to match 40% genera extinction rate per 100 Myr. At step 6.1, we find at $L_{imit} = 219$ human emergence rate observed *currently on earth* is satisfied (check section 8.2): [11]

$$t_{Win} \int_{L_{imit}=219}^{\infty} P_{df}\left(-5.41,\ x\right) dx = \frac{1}{16} \tag{8.8.98}$$

Next, at step 7, we taking into account all habitable planets within the galaxy including those form
ed later or earlier than earth. We want to include more habitable planets so that the emergence rate *currently within the galaxy* can rise to $\frac{1}{3}$ per galaxy. We include all planets formed 500 Myr behind earth (though any planets formed 200 Myr behind almost contributes nothing to the final chance) and up to planets formed 91 Myr earlier than earth:

$$C_{df}\left(0\right) = \int_{-5}^{0.91} P_{lanet}\left(-s\right)\left(t_{Win}\left(B_{cs}\right)^s \int_{L_{imit}=219}^{\infty} P_{df}\left(-5.41 + -s + 0,\ x\right) dx\right) ds = \frac{1}{3} \tag{8.8.99}$$

Next, we check the occupancy ratio, and we find that the universe ought to be occupied 14 times over according to our current model!

$$\frac{1}{F_{galaxy}} \int_0^2 R_{cdf}\left(t\right) \cdot \left(\frac{137.99}{d_{Galaxy}}\right)^3 \left(\frac{t}{137.99}\right)^3 dt = 14.84 \tag{8.8.100}$$

This is no surprise since we have already shown that for conservative k such as the (8.1, 5.54) pair, $\sigma > 0.7$ requires at least a deviation of 353 and an emergence rate as low as 1 per 63 galaxies for earth to remain unoccupied.

This leads us to step 9, re-assess the emergence rate. The emergence rate can be reduced to satisfy the occupancy ratio constraint by using a hierarchical approach.

1. First, one can assume that human emergence against typical mammals are more difficult while all habitable planets formed 4.5 Gya achieve a typical mammalian development level at the current time. One can reduce t_{Win}, the number of possible species generated within the last 100 Myr, or increasing the deviation between the current mode and human placement position L_{imit}, or both.

2. Second, one can assume that human emergence against typical mammals is justified but all habitable planets within the galaxy attains a lower level of bio-complexity at the current time. In this case, deviation and t_{Win} remains fixed, but the current mode should be placed on an earlier peak, for example 100 Mya,

---

[11] A keen reader may point out that by altering σ, earlier peak such as $t_{Win} \int_{L_{imit}=11}^{\infty} P_{df}\left(0,\ x\right) dx > 1$ predicts even at Cambrian era there exists > 1 species that share all 11 traits as typical mammals. This contradict reality. The problem is resolved if one realizes that the model represents mathematical abstraction in its ideal case. In reality, Cambrian fauna distribution is short tailed while other periods has a longer and fatter tail. An alternative approach suggests that one count the overlapping number of traits between successive distributions as the survival rates, instead of overlapping regions when altering σ.

change $P_{df}\left(-5.41+-s+0,\ x\right)$ to $P_{df}\left(-4.41+-s+0,\ x\right)$:

$$C_{df}\left(0\right)=\int_{-5}^{0.91}P_{lanet}\left(-s\right)\left(t_{Win}\left(B_{cs}\right)^{s}\int_{Limit=219}^{\infty}P_{df}\left(-4.41+-s+0,\ x\right)dx\right)ds \qquad (8.8.101)$$

3. Thirdly, if both human emergence against typical mammals is justified and all habitable planets achieve a typical mammalian development level at the current time, then the habitable planet formation rate can be significantly lower. This can happens for various reasons and we will address some of them in section 8.12. This can be done simply by adding the denominator to the planet formation function as:

$$P_{lanet}\left(-t\right)=\frac{1}{f}\times P_{lanet}\left(-t\right) \qquad (8.8.102)$$

4. Fourth, though all planets undergoes similar rate of tectonics intensity, the exponential biodiversity increase per each major geologic event is different from earth. As a result, BCS and BER changes accordingly. If everything else were equal except a lower BCS, then the current mode achieved on all other planets are also lower than earth, unless the initial biodiversity endowment during their Cambrian explosion is higher than earth or a more progressive evolution in which k is lower. Recall this scenario is mathematically similar to the alternative scenario in our test cases under section 8.7.4, except now this rescaling is relative to our new settlement point M_{id} instead of old settlement point $M_{ax}\left(0\right)=1$, so the original rescaling with t_{cg} is no longer necessary. $P_{df}\left(0,x\right)$ now stands as the biodiversity distribution of the Cambrian explosion instead of mammalian average in section 8.2 to 8.7. So no matter how k variables alters the placement pattern of those distributions occurred earlier such as $P_{df}\left(3,x\right)$, its corresponding alteration on physical explanation is not necessary. However, if we also assumed that all other planets has a lower BCS and is 500 Myr or more behind earth, then this criterion does not render civilization emergence rarer relative to earth, because all BCS achieved the same area size under $P_{df}\left(0,x\right)$ given by the model. The relationship is captured by the following table:

BCS=1.783		BCS=2.783
$\int_{Limit}^{\infty}P_{df}\left(-7,x\right)dx$	$<$	$\int_{Limit}^{\infty}P_{df}\left(-7,x\right)dx$
$\int_{Limit}^{\infty}P_{df}\left(0,x\right)dx$	$=$	$\int_{Limit}^{\infty}P_{df}\left(0,x\right)dx$
$\int_{Limit}^{\infty}P_{df}\left(3,x\right)dx$	$>$	$\int_{Limit}^{\infty}P_{df}\left(3,x\right)dx$

Due to lowered BCS, eventually lower BCS render civilization emergence more likely relative to earth if it is $>$ 500 Myr behind earth.

In reality, it is possible that all four factors influence the final emergence rate. One can iteratively modify them and recheck with step 8 occupancy ratio until a satisfactory occupancy ratio is reached. It is also true that, regardless of the approach, reducing the current emergence and then restoring it by increasing the earliest window leads to the same final emergence rate and occupancy ratio.

For our case, we are more confident regarding human emergence against the typical mammals, but less so about the other two. In fact, we have shown that earth, among those emerged at the same time, may be ahead of others by 135.92 Myr in terms of bio-complexity development.[12] Furthermore, other un-examined habitability criterion can rapidly reduce the potential number of habitable planets at the current time. We reduce the potential habitable planets within the galaxy by a factor of 10, so there is currently less than 1 habitable planets within the galaxy and the average habitable planet bio-complexity is 135.92 Myr behind earth:

---

[12]A caveat: This 135.92 Myr lag meant lag in BCS, but we know that if BCS was continually lagged behind since the start of its multicellular life, its BER must be correspondingly reduced.

$$C_{df}(0) = \int_{-5}^{0.91} \frac{1}{10} \times P_{lanet}(-s) \left(t_{Win}(B_{cs})^s \int_{Limit=219}^{\infty} P_{df}(-5.4 + 1.36 + -s + 0, \ x) \, dx \right) ds$$

$$\approx \frac{1}{1009} \quad (8.8.103)$$

This yields an emergence rate of 1 per 1009 galaxies and a space occupancy ratio of 4.024%. One can further increase the earliest window at most to 179 Mya, so that the emergence rate increases to 1 per 42.968 galaxies, and the occupancy ratio reaches 100%.

$$C_{df}(0) = \int_{-5}^{1.79} \frac{1}{10} \times P_{lanet}(-s) \left(t_{Win}(B_{cs})^s \int_{Limit=219}^{\infty} P_{df}(-5.4 + 1.36 + -s + 0, \ x) \, dx \right) ds$$

$$\approx \frac{1}{43} \quad (8.8.104)$$

Since this evolutionary model is by far the closest to the physical reality, given its high similarity between the past and the present, the current emergence rate has to be decreased to $< \frac{1}{43}$, that is, *the emergence of civilization must be at least as rare as 1 per 43 galaxies, and the nearest industrial civilization is at least 141.76 Mly in linear distance.*

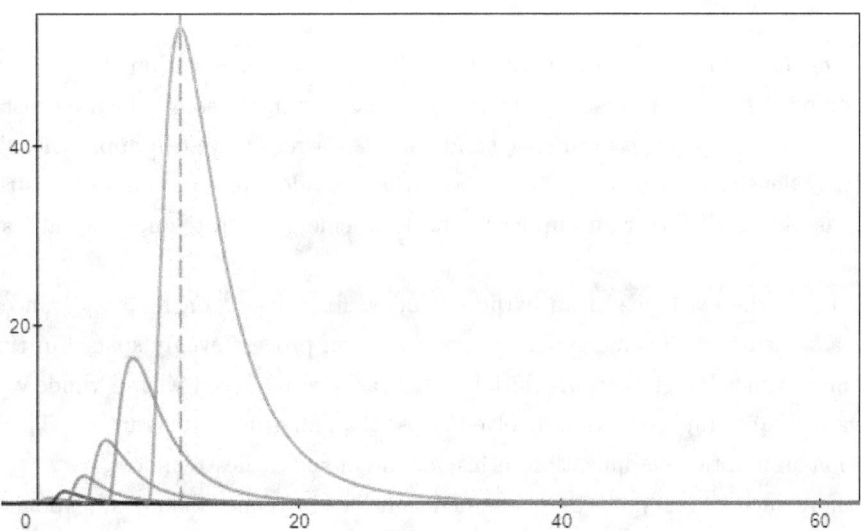

Figure 8.8.18: The most realistic evolutionary scenario based on our current understanding, the deviation to settlement ratio is $\frac{219}{11} \approx 20$

8.9 The Self Indication Assumption for the Assessment Earliest Window and Nearest Civilization

In a significant development, Olson [111] has recently able to reduce the magnitude of uncertainties to just 3 orders of magnitudes regarding the range of possible values for the Drake equation that Sandberg, Drexler, and Ord's had shown to be potentially ranging hundreds of orders of magnitude. [24] The original idea of Sandberg, Drexler, and Ord is to show that it is quite possible that no other civilization can exist within the observable universe due to a significant portion of probability density function of appearance radius requirement falls below the event horizon of observable universe. Olson's approach is a modified version of Bostrom's original

Self Indication Assumption, which states that given the fact that you exist, you should (other things equal) favor hypotheses according to which many observers exist over hypotheses on which few observers exist. Cirkovic has criticized the shortcomings of such assumption for being blindly optimistic, it is especially compared to and exaggerated as a presumptuous philosopher who, in disregard of every other evidence, favors the one theory that supports more observers blindly and faithfully [85]. Olson avoided this problem by adding the observer constraint. That is, an increasing number of additional observers eventually diminishes one's own chance of existence. The mechanism for the diminished local appearance chance is simply the physical expansion of other observers in the universe which eventually saturates empty space, rendering local spontaneous appearance within an empty portion of the universe as we are observed now impossible. Therefore, the presumptuous philosopher becomes less optimistic and credit no more observers based on his own local observation. In Olson's approach, the prior appearance probability is simply log uniform. That is, the likelihood of extremely low life appearance rate within the universe and the likelihood of much more optimistic life appearance rate are equally likely on a log scale, in other words, linearly proportional to their appearance rate. Due to the Self Indication Assumption, one still favors those with more observers. Therefore, despite equally likely prior likelihood based on a log scale, the probability represents those observers like us currently emerging from extremely life-averse universe is low. In fact, those appearance rate less than 3 orders of magnitude of the maximum appearance fail to contribute to the probability density function at all. Chosen at random, we are more likely to be an observer out of many more observers within a more life tolerant, friendly universe. However, as we have mentioned, the line of reasoning can only be carried out so far until the occupation constraints starts to diminish the probability of appearance again.

Olson had demonstrated his model based on a Poisson process of life emergence within the universe. That is, he assumes that life is simply becoming more frequent in space but not in time. We focused instead, on the case where life is becoming more frequent in both space and time. As a result, we can apply Olson's argument to delineate the probability density landscape for both the earliest window (cosmic phase transition/galactic punctuated equilibrium) as well as the lower and upper bound local emergence rate once the earliest window is fixed.

Olson also implicitly assumed the existence of an earliest window fixed based on his appearance rate per 1 Gly/Gyr unit. Since he assumed that life emergence follows a Poisson process evenly spaced in time, and the earliest window spans only 4 times larger than his initial model, as a result, even with a window size up to a maximum of 4 Gya, the appearance rate reduction involved is less than an order of magnitude. Therefore, there is no need of self indication assumption's application in his version of the earliest window.

In our case, we can not afford such a luxury, since we assumed that life emergence follows an exponential growth process that becomes more frequent both in space and time, by lengthening the earliest window, the degrees of uncertainty growths in multiple orders of magnitudes. In the absence of further information, we simply assume that the maximum temporal range of the earliest window spans from 4 Gya up to now, when the very first terrestrial planet has reached its maturity with 4.5 Gyr of delay for evolution according to Lineweaver's timescale[44]. Though studies seem to suggest that the Gamma Ray bursts starts to significantly diminish in the last 0.5 Gya in the universe, as our earlier assumption was built upon[97, 11]. Recall that if a global regulating mechanism is at its play, then BCS (biocomplexity) grows exponentially with time length as the earliest window is pushed further into the past. By further lengthening the earliest window, our current time of appearance will becoming less atypical. At the same time, in order to cope with the fact that we are residing in an empty portion of the universe, the appearance rate must be further reduced. Recall that the appearance rate of civilization can be reduced at multiple levels such as

1. the rarity of the emergence of homo sapiens against the average cohorts of typical species on the planet.

2. the level of biocomplexity currently observed on earth is more atypical among habitable planets at the current time.

3. the habitable planet formation rate can be significantly lower.

4. the other BCS achieved on other planets are lower.

Out of all these choices, our current theory prefer using 3. for the vast majority of appearance rate reduction, since we assumed that life, once evolved, the magnitudes of difference in bio-complexity (ranging from the simplest to the most complex) should involves within a few order of magnitudes of difference. As result, only 3) can account for the extreme ranges of difference in orders of magnitudes (assuming additional filter in life appearance and physical barriers to planet habitability) We run our original model and seeking for the earliest window time frame when space occupancy reaches 100% for different appearance rates. The results are listed as follows:

Temporal based prior probability 1 $\frac{t_H - 1.25}{t_T}$	Temporal based prior probability 2 $\frac{t_H - 1.25}{t_T} \times \frac{t_T - (t_H - 1.25)}{t_T}$	The Earliest Window (100 Myr) t_H	Appearance rate relative to N=47 planets α
0	0	1.250	$1/10^0$
0.0016125	*0.001586*	*1.895*	$1/10^1$
0.000324	0.000313	2.546	$1/10^2$
0.000049	4.64862×10^{-5}	3.205	$1/10^3$
6.5525×10^{-6}	6.12314×10^{-6}	3.871	$1/10^4$
8.2375×10^{-7}	7.55893×10^{-7}	4.545	$1/10^5$

Table 8.9.2: The maximum likelihood for varying earliest window size

We adopted 2 temporal based prior probabilities. Both penalizes short earliest window since more recent earliest window signifies the atypicality in time of our emergence against Lineweaver's time scale of earliest terrestrial formation's maturity dates[44]. In the first approach, we simply assumed that as long as the earliest window does to match Lineweaver's time scale, it is penalized linearly and probabilistic favors the oldest time periods, so we have:

$$P(t_H) = \frac{t_H - 1.25}{t_T} \tag{8.9.1}$$

Whereas t_H is the earliest window size. A value of 1.25 is subtracted from t_H, denoting that we have shown step by step earlier, that the maximum appearance rate within the galaxy ensures that a minimum 125 Myr earliest window is guaranteed, and t_T is the maximum Lineweaver's time scale, which we denotes as 4 Gyr, or 4 Gya before the current time.

In the second prior, we adopt a binomial distribution in which probabilistically favors the intermediate periods and penalizes positions either at the very start and the very late on the Lineweaver's time scale, in align with Gott's thoughts[62], So we have:

$$P(t_H) = \frac{t_H - 1.25}{t_T} \times \frac{t_T - (t_H - 1.25)}{t_T} \tag{8.9.2}$$

The final likelihood at each time period is constructed by combining appearance rate α with $P(t_H)$ as:

$$\alpha \times P(t_H) \tag{8.9.3}$$

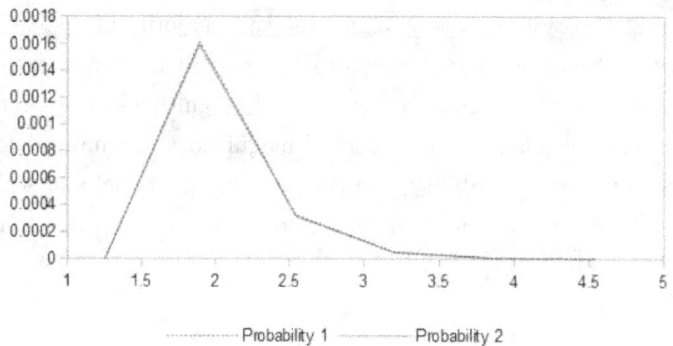

Figure 8.9.1: The maximum likelihood for varying earliest window size plot

Our results indicates that in both cases the maximum likelihood for the earliest window occurs at a very recent time in 189.5 Myr, as our earlier model assumed. The two different temporal based prior offer little alteration on this final results due to strong dominance of exponential growth of biocomplexity BCS. Any gains in positioning a much earlier earliest window is heavily penalized by a much lower appearance rate α. Only when $P(t_H)$ further penalizes late or early times polynominally can shift the earliest window much earlier.

Once we settled the earliest window, we can use a fixed earliest window to derive the range of magnitudes for appearance rates. In our simplest model, we simply assume that the log uniform prior for appearance rate α uses $a = e$ and $b = 1$ as its bounds so that

$$P(\alpha) = \frac{1}{\alpha \ln\left(\frac{a}{b}\right)} = \frac{1}{\alpha \ln\left(\frac{e}{1}\right)} = \frac{1}{\alpha} \tag{8.9.4}$$

Our log uniform prior is simply the inverse of α. We also find that by using our multivariate lognormal distribution,

$$C_{df}(0) = \int_{-5}^{1.89} \alpha \times P_{lanet}(-s) \left(t_{Win}(B_{cs})^s \int_{Limit=219}^{\infty} P_{df}(-5.4 + 1.36 + -s + 0, \ x)\,dx \right) ds \tag{8.9.5}$$

the appearance rate α and the remaining empty space ratio is inversely related so that:

$$R_{atio} = 1 - \frac{1}{\alpha} \tag{8.9.6}$$

So the final PDF across all appearance rate α is constructed as:

$$P_{DF}(\alpha) = \left(1 - \frac{1}{\alpha}\right)\left(\frac{1}{\alpha}\right) P(\alpha) \tag{8.9.7}$$

The CDF is then:

$$\int P_{DF}(\alpha)\,d\alpha = \frac{1 - 2\alpha}{2\alpha^2} \tag{8.9.8}$$

Taking over open interval $\alpha = [1, \infty)$, we have:

$$\int_1^{\infty} P_{DF}(\alpha)\,d\alpha = \left[\frac{1 - 2\alpha}{2\alpha^2}\right]_1^{\infty} = \lim_{\alpha \to \infty} \frac{1 - 2\alpha}{2\alpha^2} - \frac{1 - 2}{2} = \frac{1}{2} \tag{8.9.9}$$

Based on this CDF, we drawn the following conclusion.

Probability	Appearance rate relative to N=4.7 planets α	Distance to the nearest civilization (10^6 ly)	Min Number of Habitable Galaxies	Occupancy Ratio
0	$1/10^0$	141.76	43	1
0.11[13]	$1/10^{0.176}$	162.27	64.5	0.667
0.5[14]	$1/10^{0.533}$	213.47	146.85	0.293
0.81	$1/10^1$	305.41	430	0.1
0.98	$1/10^2$	657.99	4,300	0.001
0.998	$1/10^3$	1,417.6	43,000	0.0001
0.9998	$1/10^4$	3,054.12	430,000	0.00001
0.9998	$1/10^5$	6,579.91	4,300,000	0.000001

Table 8.9.4: Appearance rate and probability and the distance to the nearest civilizations

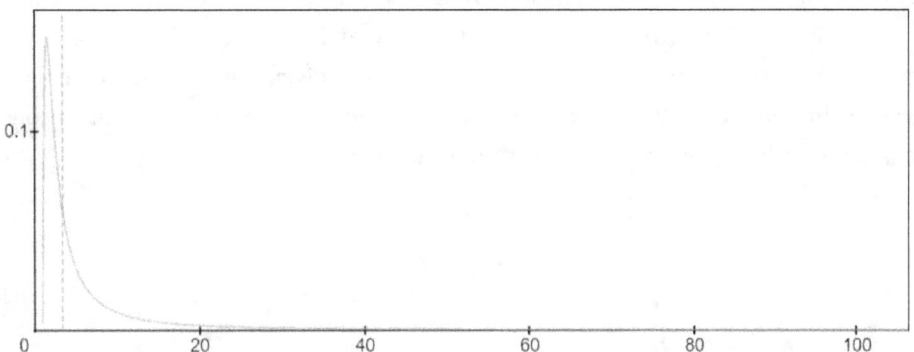

Figure 8.9.2: In this non log scaled PDF, 98% of all probabilities falls within 2 orders of magnitudes below the highest estimates

That is, 80% of all lower estimates for the emergence of the nearest civilization falls within an order of magnitude below the highest estimates and 98% of all lower estimates for emergence of the nearest civilization falls within 2 orders of magnitude below the highest estimates. Within 3 orders of magnitudes below the highest estimates, one covers 99.8% of all possible cases within the PDF. This is consistent with Olson's predictions. Within 4 orders of magnitudes below the highest estimates covering 99.98% probability, the distance to the nearest civilization is falls within an universe can be connected in theory with all neighbors given expansion approaches the speed of light, implying a high confidence that the universe will become connected even without any future arising civilizations. Please check section 10.2 Connected/Disconnected under Chapter Universal Contact.[123] The maximum likelihood occurs just below to the highest estimates, at 1/1.5 times, yet it is represents only 11% of all lower estimates. Median estimates is reached at 1/3.415 times of the highest estimates, also fairly close to the highest estimates.

Our current log uniform prior performs remarkably well at reducing the magnitudes of uncertainties at the lower bounds. For log-log uniform prior suggested by Tegmark[25], it will out-perform log uniform prior since log log uniform prior requires more rapid descent as:

$$P(\alpha) = \frac{1}{\alpha^b} < \frac{1}{\alpha} \tag{8.9.10}$$

Whereas $b > 1$. On the other hand, if we used a uniform prior, so that $b = 0$, then we have

$$P(\alpha) = \frac{1}{\alpha^b} = 1 \tag{8.9.11}$$

We end up having:

---

[13]Maximum likelihood
[14]Median

$$P_{DF}(\alpha) = \left(1 - \frac{1}{\alpha}\right)\left(\frac{1}{\alpha}\right)(1) \qquad (8.9.12)$$

The CDF is then:

$$\int P_{DF}(\alpha)\,d\alpha = \frac{1}{\alpha} + \ln\alpha \qquad (8.9.13)$$

Taking over open interval $\alpha = [1, \infty)$, we have:

$$\int_1^\infty P_{DF}(\alpha)\,d\alpha = \left[\frac{1}{\alpha} + \ln\alpha\right]_1^\infty = \lim_{\alpha\to\infty}\left(\frac{1}{\alpha} + \ln\alpha\right) - 1 = \infty \qquad (8.9.14)$$

This is then problematic, that is, the integration is unbounded, and as result, orders of magnitudes reduction is not possible under such a scenario, suggesting that when rarity of life emergence is equally likely there always exists observers no matter how rare they are within the universe For sub-log-uniform priors, so that $0 < b < 1$, the performance on orders of magnitudes reduction works partially at its best and is inferior relative to log uniform prior but significantly better than uniform prior. In conclusion, given the incomplete understanding of the probabilistic prior distribution on α, all landscape is possible, though recent studies seem to conclude that the prior are more consistent with log uniform or log-log uniform prior.

8.10   Earliest Window Revisited and the Outer Wall

We have discussed the wall of semi-invisibility, but there also exists another wall of invisibility. This wall occurs at a much greater distance and is due to the cosmic expansion of the universe. At distance larger than 1 Gly, red shift began to dominate travel time and signals emitted > 1 Gya will take longer delay to reach earth. For very long distance such as > 6.266 Gly as we shown in Chapter 9, the signals will remain forever unreachable. Therefore, we can call this as the outer wall of invisibility, and the former wall of semi-invisibility as the inner wall of invisibility.

We have also just shown that the current emergence can drop so that longer earliest window can be taking into account. Now we want to show the limit on the earliest window possible with parameter manipulation. This is important since it is shown that the earliest planet formed 9.3 Gya. If the very first planet can be deemed habitable, then they should have reached the stage of multi-cellularity 4.8 Gya. Although we take the assumption that continued, non-interrupted multicellular evolution has only become possible within the last 500 Myr within the universe, we still want to show that our model is able to accommodate a great range of possible solutions. We will also show that despite the presence of the outer wall of invisibility, all civilizations emerges between the inner and outer wall, so that cosmic expansion and red shift plays little role in effecting the visibility of expanding civilizations.

First of all, we resume to our last example, with a BER of 1.203 and a pool of habitable planets formed between 179 Mya before earth up to 500 Myr after earth, its current emergence radius is 70.88 Mly. One finds that any planet within this pool achieved human comparable civilization 578 Mya is so rare that it requires a radius of 13.8 Gly:

$$\sqrt[3]{F_{galaxy} \cdot C_{df}\,(5.78)^{-1} \cdot (d_{Galaxy})^3} = 13.8\,\text{Gly} \qquad (8.10.1)$$

Then, the chance of observing the signal emitted from such a source within 578 Mly radius is just $\left(\frac{0.578}{13.8}\right)^3 = 0.0073\%$. Even if it is observed, red shift is still marginal at 578 Mly distance. On the other hand, civilizations emerged less than 200 Mya has $> 10\%$ chance of detection, but red shift plays negligible effect at such distance. This indicates that the inner wall of semi-invisibility still dominates as BER reaches 1.203.

We can increase the earliest window to 839.3 Mya with 100% occupancy ratio and an emergence rate of 1 per 64 galaxies, by first assuming all habitable planets formed 4.5 Gya achieved at most only the most primitive multicellular life stage at the current time and secondly decrease planet formation rate. We change $P_{df}(-5.41 + -s + 0, \ x)$ to $P_{df}(-5.41 + 6 + -s + 0, \ x)$. We can not decrease it further since we have shown that significant oxygen accumulation is a natural consequence of billions of years of photosynthesis and geological change, and eukaryotic multicellular life logically follows. On the other hand, the planet formation rate can be continually decreased. We decreased the rate by a factor of a million.

We then achieve a BER of 1.00685 with $k = 150$ and selection factor $= 266$, and $\sigma = 0.9$, assuming all habitable planets formed 4.5 Gya achieved at most only the most primitive multicellular life stage at the current time. We shift deviation L_{imit} to 341 so that $t_{Win} \int_{L_{imit}=341}^{\infty} P_{df}(-5.41, \ x)\, dx = \frac{1}{16}$. We reduce the planet formation rate to a 10 million fold smaller, and the earliest window shifted to 1.01 Gya, and occupancy ratio reaches 100%. One finds that any planet within this pool of planets formed between 1.01 Gya before and 500 Myr after earth achieved human comparable civilization 1.223 Gya is so rare that it requires a radius of 13.8 Gly:

$$\sqrt[3]{F_{galaxy} \cdot C_{df}(12.23)^{-1} \cdot (d_{Galaxy})^3} = 13.8\,\text{Gly} \tag{8.10.2}$$

This better than the previous result. Nevertheless, the chance of observing the signal emitted from such a source within 1.223 Gly radius is just $\left(\frac{1.223}{13.8}\right)^3 = 0.0696\%$. At 1 Gly when red shift starts to dominate travel time, there is a chance of $\left(\frac{1}{6.088}\right)^3 = 0.4432\%$. If a civilization emerged 1 Gya at 1 Gly away expands near the speed of light, it may indeed becomes invisible to earth due to cosmic expansion, but such chance is $< 0.4432\%$. This indicates that the inner wall of semi-invisibility still dominates even as BER reaches 1.00685.

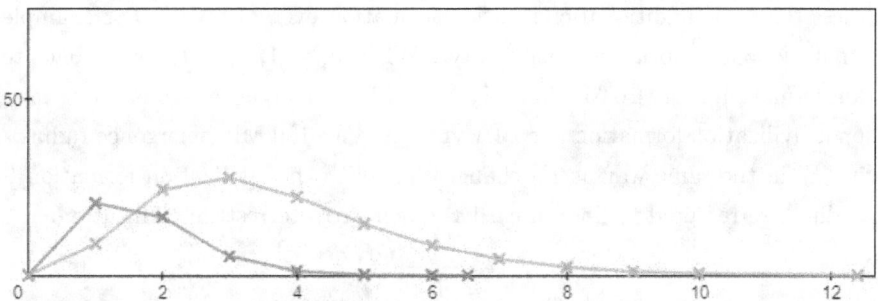

Figure 8.10.1: Though BER $= 1.00685$'s peak occur further into the past, instantaneous visibility chance for BER $= 1.203$ and BER $= 1.00685$ both drops near 0 at earlier times/greater distance when red shift starts to become dominant.

8.11   Other Types of Wall of Semi-Invisibility

Recall earlier in Chapter 1, we mentioned that, philosophically, there are four possible cases. Civilization is both rare in space and time. This is the simplest case and no further work is required. We have so far dedicated the entire book to the case when civilization is becoming more frequent in both space and time. We emphasized that it is becoming exponentially more frequent in space and time due to the exponential nature of BCS and BER from evolution. We yet to cover the remaining two cases. These cases are much simpler than our primary case.

8.11.1 Civilization Becoming More Frequent in Space but not in Time

This case states that the formation of civilization occurs at a fixed interval within a given volume of space as a poission process, so that after some period of time, the total number of civilization within the given volume increases. We model the growth rate by the following function defined based on an emergence radius of 100 Mly:

$$T_0(t) = \begin{cases} (u_1 \cdot -10x + 1 + t_0)^{E_0} & x \le \frac{J(t_0)}{u_1} \\ (u_1 \cdot 10x + 1 - t_0)^{-E_0} & x > \frac{J(t_0)}{u_1} \end{cases} \tag{8.11.1}$$

$$J(t) = 0.1t \tag{8.11.2}$$

Whereas E_0 stands for the growth pattern, $E_0 = 1$ when the civilization formation growth rate follows a linear growth. $E_0 < 1$ when the civilization formation growth rate follows a sublinear growth. $E_0 > 1$ when the civilization formation growth rate follows a polynomial growth. t_0 controls the current/latest emergence rate of civilization. It is a parameter passed to the function $J(t)$, which dictates where two pieces of the step wise function joins. A step wise function is required because when the emergence radius falls below 100 Mly, additional civilization formation renders the formation slope from a straight line to a curve $\propto \frac{1}{x}$. When $t_0 = 0$, current emergence states 1 civilization per 100 Mly radius. When $t_0 < 0$, current emergence states < 1 civilization per 100 Mly radius. For example, $t_0 = -5$ means that $\frac{1}{6}$ civilization per 100 Mly radius. When $t_0 > 0$, current emergence states > 1 civilization per 100 Mly radius. For example, $t_0 = 5$ means that 6 civilization per 100 Mly radius. However, this relationship changes with E_0. With $E_0 > 1$, the correspondence is more exaggerated. With $E_0 < 1$, the correspondence is more downplayed. Finally, u_1 controls the formation rate. When $E_0 = 1$, $u_1 = 1$ corresponds to one additional civilization formation every 10 Myr. When $E_0 = 1$, $u_1 < 1$, corresponds to less than one additional civilization formation every 10 Myr. For example, $u_1 = 0.5$, corresponds to $\frac{1}{2}$ additional civilization formation per 10 Myr. When $E_0 = 1$, $u_1 > 1$, corresponds to more than one additional civilization formation every 10 Myr relative to 100 Mly emergence radius. For example, $u_1 = 2$, corresponds to 2 additional civilization formation per 10 Myr relative to 100 Mly emergence radius. Again, this relationship changes with E_0 in the same way as E_0 changes t_0. With the civilization formation function, we can now construct the radius size required to find an earlier arisen extra-terrestrial civilization:

$$d_R(t) = \sqrt[3]{\frac{1}{T_0(t)}} \tag{8.11.3}$$

In general, when $E_0 < 3$, the wall of semi-invisibility forms a S shaped curve. when $E_0 > 3$, it resembles our earlier wall.

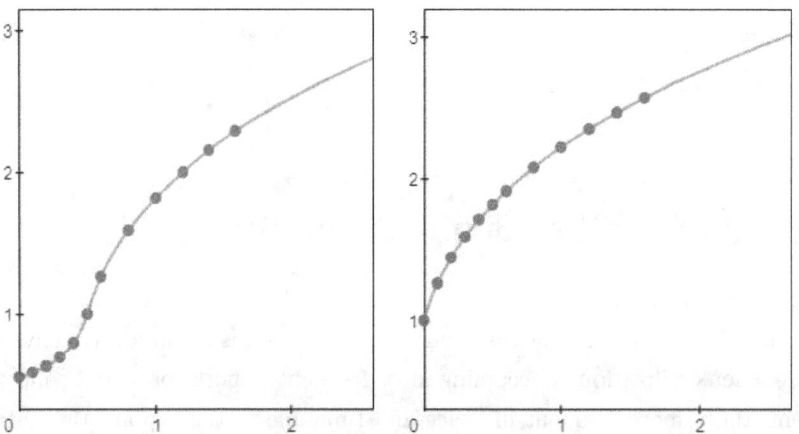

Figure 8.11.1: Case for $E_0 = 1$, $t_0 = 5$, $u_1 = 1$ (left), and Case for $E_0 = 1$, $t_0 = 0$, $u_1 = 1$ (right)

The CDF cumulative distribution function can be extrapolated based on $d_R(t)$:

$$C_{df}(t) = \frac{F_{galaxy} \cdot (d_{Galaxy})^3}{(10^4 d_R(t))^3} \qquad (8.11.4)$$

Space occupancy is just like before:

$$R_{atio} = \frac{100}{F_{galaxy}} \int_0^x R_{cdf}(t) \left(\frac{13799}{11.85}\right)^3 \left(\frac{t}{137.99}\right)^3 dt \qquad (8.11.5)$$

$$R_{cdf}(t) = \left| \frac{d}{dt} C_{df}(t) \right| \qquad (8.11.6)$$

and the visibility as:

$$v_{is} = \int_0^x \frac{1}{C_{df}(t)} R_{cdf}(t) \left(\frac{t}{d_R(t)}\right)^3 dt \qquad (8.11.7)$$

Upper bound for visibility can be formulated as:

$$U_{pper} = \frac{1}{\left(\frac{1}{d_R(0)}\right)^3 x^3} \qquad (8.11.8)$$

The emergence curve still is:

$$E = \left(\frac{d_R(0)}{d_R(x)}\right)^3 \qquad (8.11.9)$$

and the instantaneous visibility still as:

$$v_{isInstant} = \left(\frac{x}{d_R(x)}\right)^3 \qquad (8.11.10)$$

One can see that for any given current formation rate, the emergence curve eventually always crosses over the upper bound, so earlier arising civilization should become visible at some distance away from some time ago.

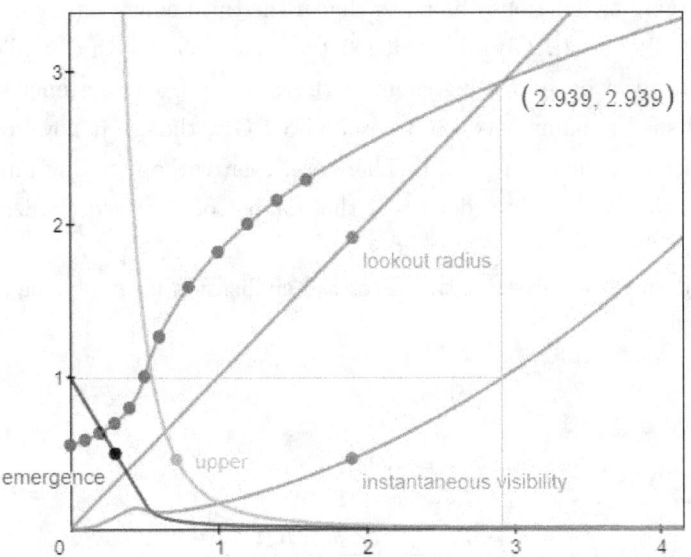

Figure 8.11.2: Case for $E_0 = 1$, $t_0 = 5$, $u_1 = 1$, the emergence curve crosses the upper bound visibility at the same time as the lookout radius exceeds the emergence radius at 293.9 Mya at 293.9 Mly away. Instantaneous visibility reaches 100% at that point.

In order to cope with our current observation of the emptiness of the space, the earliest window must be set within the crossover point. Therefore, the earliest window is inversely proportional to the emergence rate and

467

directly proportional to the emergence radius. For example, assuming $u_1 = 1$ (every 10 Myr a civilization emerges relative to a 100 Mly radius) and $E_0 = 1$, when the emergence radius is 50 Mly ($t_0 = 7$), in order for the current occupancy < 1, the earliest window must < 104.4 Mya. When the emergence radius is 2 Gly ($t_0 = -7999$), in order for the current occupancy < 1, the earliest window must < 7.369 Gya. As a result, we have demonstrated that the wall of semi-invisibility can also exist under this case.

8.11.2 Civilization Becoming More Frequent in Time but not in Space

This case states that the formation of civilization occurs at an increasingly smaller temporal interval within a given volume of space yet the total number of civilization within a given volume is fixed, this is possible when one considers the cosmic space expansion. Only 1 civilization emerges per a given sphere size at any time period. Within say, 3 Gly radius, a given number of galaxies were found within this sphere 3 Gya and 1 civilization emerged, and after 3 Gyr, some galaxies have now expanded beyond the original 3 Gly radius, and this one civilization now share a greater sphere size. In order to fix the civilization's emergence per 3 Gly radius, the emergence rate among the original pool of galaxies have to be increased. However, an exact one to one correspondence between the emergence rate of civilization and cosmic space is absurd. Nevertheless, we shall show mathematically if such relationship holds. We simply take the Hubble constant expansion factor and applies to any emergence radius over a given number of years, and we find that the following relationship holds:

$$R(t) = -\frac{r}{10} \cdot 0.07152t + r \tag{8.11.11}$$

That is, the final radius after t years (in units of 100 Myr) is based on the initial radius (in units of 100 Mly). The function is downsloped by assuming the time for the start of the first round of emergence rate doubling due to space doubling occurs at time 0, and our current time occurs at -t years (into the future) relative to time 0. It can be seen that the first round of emergence doubling takes 3.6342 Gyr, the next at 4.5788 Gyr. However, only 2.5493 Gyr is required to gain an additional civilization emergence, and 2.0295 Gyr is required to further gain an additional civilization emergence. This is possible since given the same rate of expansion, larger surface area of expanding sphere leads a fixed volume size quicker. A keen reader may pointed out that the rate of volume increase between the outer second (between 3 Gly and 3.7 Gly) and inner first shells (between 0 Gly and 3 Gly) representing 2 emerging civilizations differs, so that if the third emerging civilization is represented by the third shell (between 4.3 Gly and 3.7 Gly), then, it left the second shell with a smaller volume (between 3.4 Gly and 3.7 Gly). This is solved by borrowing some of the galaxies from the inner shell now extending into the second, so the second shell remains between 3 Gly and 3.7 Gly, though it now contains a mixture of galaxies from the previous inner and the second shell. Therefore, even without taking an accelerated Hubble constant into consideration, as time passes, in order to fix the number of observed civilization within a given space, civilization emergence has becomes more frequent.

The radius size required to find an earlier arisen extra-terrestrial civilization is simply the fixed emergence size r:

$$d_R(t) = r \tag{8.11.12}$$

The CDF cumulative distribution function can be extrapolated based on $d_R(t)$:

$$C_{df}(t) = \frac{F_{galaxy} \cdot (d_{Galaxy}R(t))^3}{(10^4 d_R(t))^3} \tag{8.11.13}$$

The average distance to galaxies vary depending on the time under consideration.

Space occupancy requires some adjustment as:

$$R_{atio} = \frac{100}{F_{galaxy}} \int_{-x}^{0} R_{cdf}(t) \left(\frac{13799}{11.85R(t)}\right)^3 \left(\frac{x+t}{137.99}\right)^3 dt \tag{8.11.14}$$

Visibility has to be rewritten because we now have the expansion factor that varies with time, recall that
$$\frac{1}{C_{df}(t)} = \frac{1}{F_{galaxy}} \left(\frac{d_R(t)}{0.1185}\right)^3,$$

$$v_{is} = \int_{-x}^{0} \frac{1}{C_{df}(0)} R_{cdf}(t) \left(\frac{t}{d_R(0)}\right)^3 dt =$$

$$\frac{100}{F_{galaxy}} \int_{-x}^{0} \left(\frac{d_R(0)}{0.1185 R(t)}\right)^3 R_{cdf}(t) \left(\frac{t}{d_R(0)}\right)^3 dt \quad (8.11.15)$$

One may notice that the galaxy's average distance is still rescaled as $0.1185 R(t)$ and is not fixed to $0.1185 R(0)$, since galaxies are moving away from each other. In fact, one can show that:

$$\frac{100}{F_{galaxy}} \int_{-x}^{0} \left(\frac{d_R(0)}{0.1185 R(t)}\right)^3 R_{cdf}(t) \left(\frac{t}{d_R(0)}\right)^3 dt < \frac{100}{F_{galaxy}} \int_{-x}^{0} \left(\frac{d_R(0)}{0.1185 R(0)}\right)^3 R_{cdf}(t) \left(\frac{t}{d_R(0)}\right)^3 dt$$
$$(8.11.16)$$

This is true because although the rate of emergence grow proportional to the volume size increase, some emergence occurs on the expanded section that falls outside the initial radius. We can show that, by assuming the initial radius remains fixed between time t_1 and t_2, one can derive the total number of civilizations emerged between these periods within the expanded volume:

$$j_1 = C_{df}(t_1) \left(\frac{138}{0.1185 R(0)}\right)^3 - C_{df}(t_2) \left(\frac{138}{0.1185 R(0)}\right)^3 \quad (8.11.17)$$

Now, in reality, the new expanded radius $R\left(\frac{1}{2}(t_1 + t_2)\right)$ is larger than $R(0)$, and the proportion of the expanded volume that includes initial volume multiplied by the total emergence number within the expanded volume yields the number of emergence within the the initial volume. The difference between j_1 and j_2 is the number of emergence falling outside the initial volume.

$$j_2 = \int_{t_1}^{t_2} R_{cdf}(t) \left(\frac{138}{0.1185 R(t)}\right)^3 dt \approx j_1 \left(\frac{R(0)}{R\left(\frac{1}{2}(t_1 + t_2)\right)}\right)^3 \quad (8.11.18)$$

The upper bound for visibility, emergence curve, and the instantaneous visibility are still formulated as before. One can see that for any given formation rate, the emergence curve eventually always crosses over the upper bound, so earlier arising civilization should become visible at some distance away from some time ago.

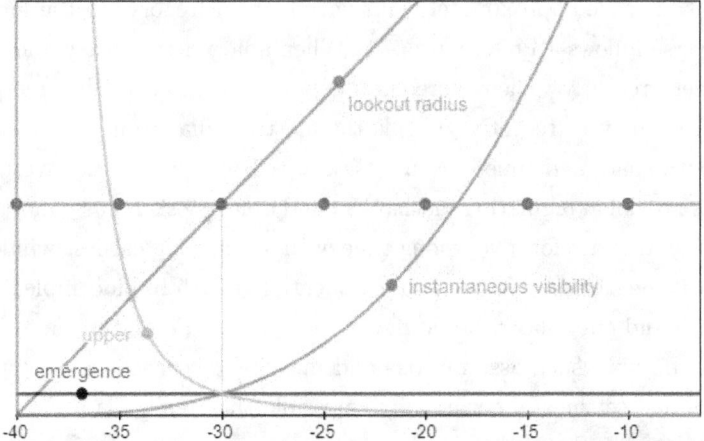

Figure 8.11.3: Assuming current time is 4 Gyr after the first the civilization emergence, for $d_R(t) = 1$ Gly emergence radius, the emergence curve crosses the upper bound visibility at the same time as the lookout radius exceeds the emergence radius at 1 Gya at 1 Gly away. Instanenous visibility reaches 100% at that point.

In order to cope with our current observation of the emptiness of the space, the earliest window must be set within the crossover point. Therefore, the earliest window is inversely proportional to the emergence rate and directly proportional to the emergence radius. For example, when the emergence radius is 50 Mly, in order for the current occupancy < 1, the earliest window must < 219.8 Mya. When the emergence radius is 3 Gly, in order for the current occupancy < 1, the earliest window must < 4.812 Gya. As a result, we have demonstrated that the wall of semi-invisibility can also exist under this case.

8.12   Overall Landscape Analysis

Olson [112][110], has written a series of papers concerning the identifications of extraterrestrial signals arising from civilizations at sub-luminal speed expansion. In here, we present all possible landscapes and a analysis on all possible combinations. First of all, a 2 by 2 grid in which 4 possible choices are presented:

Abiogenesis Difficulty	Global Regulating Mechanism	
	Exist	Not Exist
Hard	Almost an Empty Universe	Only more frequent in Space/ Only more frequent in Time
Easy	More frequent in space and time with a short earliest window	More frequent in space and time with a long earliest window (Not Possible)

Table 8.12.2: Grid Analysis on all possible scenarios

When Abiogenesis is difficult and there exists a global regulating mechanism until very recently, then the universe is almost guaranteed to be empty, and it fulfills our original scenario 1) in which earth is an exceptional event within the universe that is almost never repeated again. When there is no global regulating mechanism at play, and habitable planets with civilization emerging is largely regulated by the extreme difficulty of abiogenesis, then both our scenario 2) earth is becoming more frequent in space but not in time, and 3) earth is becoming more frequent in time but not in space approximately falls within this category. In the former, the emergence of civilization within the universe follows a Poisson process. When abiogenesis is easy and there exists a global regulatory mechanism until very recently. Then there is an exponential trend in the emergence of civilization and the earliest window must occur very recently. This is the most familiar to us as scenario 4) civilization is becoming more frequent in both space and time, and mostly covered by our current investigation. In the final grid, we assumed that there is no global regulating mechanism and abiogenesis is easy and an exponential trend in global biocomplexity has been in place for a very long time with a very long earliest window. This position is untenable since we know that if the earliest window and exponential growth in biocomplexity has been in place for a very long time, then earth and our galaxy should have been saturated long time ago. Therefore, currently observation runs in contradiction with such assumptions and must be discarded. Lastly, the grid scenarios is conceptual at its best. In the real settings of cosmos, it is possible that reality is a mixture of all 3 cases, in which some local abiogenesis is earlier than others and global regulating mechanism may exert partially on all regions. The final outcome is the consequences of all factors come into play.

Based on the speed of expansion, the 4 aforementioned scenarios can be further divided into 2 possible sets of scenarios. In the first set whereas the expansion speed nears the speed of light. Whereas civilization's appearance position is located in front of the wall of semi-invisibility is defined as: When the civilization's expansion speed nears or approaches the speed of light, as we originally assumed, and if such civilization's emergence radius requirement measured in ly is always less than its first emergence date measured in years. Then, such civilization has always already overtaken and saturate our current empty space, rendering our existence not possible.

Whereas civilization's appearance position is located behind the wall of semi-invisibility is defined as: When the civilization's expansion speed nears or approaches the speed of light, as we originally assumed, and if such civilization's emergence radius requirement measured in ly is always greater than its first emergence date measured in years. Then, there is a non zero chance that such civilization has been expanding since its emergence but currently not observable despite its near light speed expansion. This is the assumption of the wall of semi-invisibility.

| Scenarios | Position relative to the Wall of Semi-invisibility | |
	In Front	Behind
1) Earth is rare in both space and time	Not Applicable	Not Applicable
2) Earth is becoming more frequent in space but not in time	Contradicted by Observation	non-zero chance being visible
3) Earth is becoming more frequent in time but not in space	Contradicted by Observation	non-zero chance being visible
4) Earth is becoming more frequent in both space and in time	Contradicted by Observation	non-zero chance being visible

Table 8.12.4: Grid analysis on all possible scenarios given expansion speed $\rightarrow c$

In the second set whereas the expansion speed is simply just a fraction of the speed of light. Whereas civilization's appearance position is located in front of the wall of semi-invisibility is defined as: When the civilization's expansion speed is at a fraction of the speed of light, as Olson originally assumed, and if such civilization's emergence radius requirement measured in ly is always less than its first emergence date measured in years. Then, it is possible that such civilization has yet to overtake and saturate our current empty space and we are able to confirm their existence through survey and observation and it will remain as an astronomical phenomena in the sky in geological timescales.

Whereas civilization's appearance position is located behind the wall of semi-invisibility is defined as: When the civilization's expansion speed is at a fraction of the speed of light, and if such civilization's emergence radius requirement measured in ly is always greater than its first emergence date measured in years. Then, there is a non zero chance that such civilization has been expanding since its emergence but currently not observable despite its continued expansion. In fact, the non zero chance is much smaller since the speed of expansion is significantly lower. This is further enhanced by the assumption of the wall of semi-invisibility.

Scenarios	Position relative to the Wall of Semi-invisibility	
	In Front	Behind
1) Earth is rare in both space and time	Not Applicable	Not Applicable
2) Earth is becoming more frequent in space but not in time	Positive detection but no saturation	much smaller non-zero chance being visible
3) Earth is becoming more frequent in time but not in space	Positive detection but no saturation	much smaller non-zero chance being visible
4) Earth is becoming more frequent in both space and in time	Positive detection but no saturation	much smaller non-zero chance being visible

Table 8.12.6: Grid analysis on all possible scenarios given expansion speed $< c$

Olson has specifically dedicated his papers on scenario 2) in which earth is becoming more frequent in space but not in time and possible detection scenarios if such civilization emerges in the front of the wall of semi-invisibility. Therefore, we shall only dedicate our attention to possibilities of sub-luminal speed expansion under scenario 4). One of the most immediate effect is to lengthening of the earliest window. With average expansionary speed of v, the final occupancy ratio is simply:

$$R_{atio} = \left(\frac{v}{c}\right)^3$$ (8.12.1)

or we can express in terms of our existing equation as:

$$R_{atio} = \frac{1}{F_{galaxy}} \int_0^\infty R_{cd}(t) \cdot \left(\frac{137.99}{d_{Galaxy}}\right)^3 \left(\frac{v \times t}{137.99}\right)^3 dt$$ (8.12.2)

With lower occupancy ratio, one can lengthen the earliest window as the follows, assuming we take our default values for BCS, k, and selection factor.

Earliest Window (100 Myr)	Expansion Speed
1.895	c
2.085	$0.8c$
2.33	$0.6c$
2.675	$0.4c$
3.27	$0.2c$
3.871	$0.1c$
8.015	$0.001c$

Table 8.12.8: Earliest window length given different expansion speed $v < c$

Unless the average achievable speed of expansion is extremely low, as those currently attainable by humans at $0.001c$, the earliest window remains very recent, no matter how slow the expansion it becomes, further corroborating the invalidity of the coexistence of easy abiogenesis with a lack of global regulatory mechanism. Furthermore, the duration between a positive detection as an astronomical phenomena and its arrival delay can also be modeled based on average final expansionary speed. Assuming that its appearance radius is significantly smaller than the effect of high value of red shift sets in, which again, is justified, because based on scenario 4), no extraterrestrial industrial civilization arises within the observable universe earlier than 200 Mya in our default model. We modify our equation as follows:

472

$$Ratio = \frac{1}{F_{galaxy}} \int_{-j}^{\infty} R_{cd}(t) \cdot \left(\frac{137.99}{d_{Galaxy}} \right)^3 \left(\frac{v \times (t+j)}{137.99} \right)^3 dt \qquad (8.12.3)$$

Detection and Arrival Delay (100 Myr)	Expansion Speed
0	c
0.26	$0.8c$
0.6	$0.6c$
1.082	$0.4c$
1.918	$0.2c$
2.768	$0.1c$
8.826	$0.001c$

Table 8.12.10: Detection and arrival delay given different expansion speed $v < c$

One can see that, as the expansionary speed slowed down, there is a significant delay between its sighting in the sky and its subsequent arrival on earth. However, we will later show in Section 9 Relativistic Economics, that it is more much likely that all expansionists embark on near the speed of light for best economic value for the return. As a result, the detection and its arrival delay should be short.

8.13 Complexity Equivalence

Now, using our existing lognormal distribution, we will investigate the property of biodiversity/civilization complexity equivalence.

From the previous derivation, we know that currently, Homo sapiens led industrial civilization occurs once in at least every 3 galaxies.

$$t_{galaxy} = C_{df} \, (0.026)^{-1} \geq 3 \qquad (8.13.1)$$

Using this equation and assuming k = 1 (classic Darwin's case), we can extrapolate the first arising industrial civilization within the Milky Way galaxy to be 6.2 million years into the future:

$$t_{galaxy} = C_{df} \, (-0.062)^{-1} \approx 1.00 \qquad (8.13.2)$$

If next industrial civilization arising from the Milky Way will occur 6.2 Myr into the future, what does it mean? Since we know that our YAABER is the total sum of YAABER for hominid lineage evolution + hunter-gatherer transition + agricultural transition + industrial transition, at what stage of development could the next industrial civilization be at today if we observe it directly through a telescope? It is tempting to conclude that they are only 6.2 Myr behind us. Therefore, they are probably just stuck in steam-powered or 19th-century Victorian era development for the next 6.2 Myr. This conclusion, upon close inspection, is absurd. Since industrial civilization is categorized into two stages, first the increase usage of fossil fuel led to economic growth and energy utilization growth and second the replacement of fossil fuel by nuclear fusion power so that the industrial civilization can be maintained indefinitely at a steady state on the home planet or grow exponentially by expanding into the cosmic neighborhood. This led to one conclusion, fossil fuel based industrial civilization, like that of the steam-powered based Victorian era is a transient short one either facing collapse or fully transitioned into an expansionary cosmic civilization. Since we predict by high probability and log-normally distributed temporal arrival, then we can rule out that this next arising civilization is on its way to transition into a sustainable industrial civilization, and we also know that this planet has endowed enough radioactive material so that nuclear fusion can be developed once industrial civilization is kick-started. That is, this planet has all the necessary ingredients to successfully transition once they developed steam engine. If we assume that most civilization is not foolhardy enough once industrial civilization is developed and destroy themselves along the way, then it is unlikely that this planet is on a race with earth to obtain galactic industrial civilization status but fails because it destroys itself and has to wait for another 6.2 Myr of evolution for the next creature to take over.

Let's take a step further back, is it possible that this planet has developed into an agricultural society and maintained a steady state for the next 6.2 Myr. Although agricultural society evolves much slower than an industrial one, 6.2 Myr is more than enough time even from geological perspective to evolve basal mammals into Homo sapiens with the presence of an ice age. It is possible that this agricultural society was well maintained, but at some point in its 6.2 Myr of progression, population pressure will force it to utilize fossil fuel. It is also highly likely that climate change and catastrophic events such as an asteroid collision within this time window will render the agricultural society extinct since agricultural society are severely limited by ecological constraints. It is possible that an agricultural society lasted for 6.2 Myr or even more years before transitioning to an industrial one, but such case is highly unlikely.

This leaves us with the following intriguing scenarios. First, grass-like plants capable of sustaining vast population have evolved, but there is no human-like creature exists on the planet. This is interesting because this implies that the stage is set for the main player yet the main player still has to show up. 6.2 Myr fits reasonably well as a geological time frame for the evolution of creatures comparable to the human level in every possible way if not better given favorable conditions. If earth observers are patient enough, up to the first 3 Myr of continuous observation of this planet, nothing other than plenty of grass-like plants, creatures with great potential being utilized as farm animals in an agricultural society have evolved. However, another 2 Myr of observation found the emergence of creatures similar to human and left their arboreal habitat and roamed their grounds. By the last 1 million years, they utilized fire and tools and by 100 Kyr before the 6.2 Myr predicted, they started to transition into an agricultural society, and 500 years before the 6.2 Myr time window ends they developed steam engine and ushered in industrial civilization, 200 years before the window closes they developed nuclear fusion and ushered in technological singularity. By the time the window closes, they are already expanding 200 light years in radius from their home planet.

Secondly, human-like creatures have evolved on this planet but maintained a hunter-gatherer mode of subsistence because grass-like plants yet to evolve on this planet. In this case, the protagonist has arrived early, but the stage and all necessary equipment are not yet ready for him/her to perform. Homo sapiens are benefited from a fast transition from hunter-gatherers to an agricultural one after 100,000 years at the start of the current inter-glacial. However, human-like creatures on other planets may not be as lucky as we are. It is possible on a planet where human-like creatures have evolved and used stone tools and banded together and is able to dominate their landscape as the apex predator. However, they are not able to transition to an agricultural society because high energy crop species such as wheat, rye, oats, and rice are not available. They could maintain their population about a few million around the globe but not any further. They have to wait for high yield crop plants to evolve on their own before they can transition. One may underestimate the difficulty of transition between hunter-gatherer to that of agricultural society. Human ingenuity is crucial but limited in its ability to manipulate nature. Human's artificial breeding and selection of domesticated plants and animals simply exaggerated a trait already existed in their genome, but human has no power to add and remove genes at their will. Furthermore, the hunter-gatherers are unlikely even to contemplate to domesticate their environment because the cost over return is so high that any attempts will be detrimental to their survival based on their conservative hunter-gathering lifestyle. If one were to observe such planet from earth, one might be delighted to find creatures similar to us and optimistically predict that such creature will arise and expand into the universe less than geological timescale. However, one will be disappointed by the fact these creatures are not evolving toward other modes of living though they utilized fire and tools. By the last 1 million years before the predicted end of the appearance window, grass plants have finally evolved on their planet. By 100 Kyr before appearance window, they started to transition into an agricultural society, and 500 years before the 6.2 Myr time window ends they developed the steam engine and ushered in industrial civilization, 200 years before the window closes they developed nuclear fusion and ushered in the technological singularity. By the time the window closes, they are already expanding 200 light years in radius from their home planet.[15]

Thirdly, it is possible that sub-earths, which we have shown earlier, should have slower tectonic plate movements, which translates to a slower pace of evolution on such a planet. As a result, some of their developmental trajectory will be similar to earth except at a slower pace. That is, the emergence of an intelligent species and grass plants occur roughly at the same time. Almost like earth's history unfolding at a slow motion, an observer looking at such a planet will find neither human-like creatures nor any grass plants. However, near the end of the temporal window, both the intelligent creature and grass plants have emerged and ushered into an industrial civilization.

Lastly, it is possible that the planet has a comparable mass to earth with a similar pace of tectonic plate movement and it has the right placement of the continent cycle over the glaciation cycle. However, its cycle is delayed by 6.2 Myr. As a result, it is currently undergoing a similar biodiversity transition as we had experienced during the late Cenozoic. Therefore, the life history unwinds at the same pace as it is observed on earth except it is played with some significant delay. Obviously, it will have neither human-like creatures nor any grass plants. In all cases, an extended period of waiting time is needed before the emergence of industrial civilization, from the mathematical point of view, they are identical, all have to wait for 6.2 Myr before their arrival on the cosmic stage. I call this equivalence the complexity equivalence; that is, from a mathematical and modeling perspective those four scenarios are quantitatively equal. The assumption that all existing habitable planets' Years against Background Evolutionary Rate does not change at sub-geologic timescale is satisfied in all cases. They both appear non-moving until the very last million years where a fast transition occurs and rapidly shifted its position on the distribution curve. However, it remains speculative which of those scenarios is more likely and if it is possible to give a more precise probability treatment for each of the cases.

One then can compute the emergence probability of Homo sapiens and that of the emergence of crop plants. Since the emergence of human at the current epoch requires not only the probability of right ordering of the acquisition of major traits with a chance of $\frac{1}{14}$, but it also requires with the probability of within the emergence of the Hominid lineage $\frac{1}{14\cdot2.722}$. Additionally, giving the data fits of other species, we concluded that the emergence of Homo sapiens is $\frac{1}{119} \leq \frac{1}{14\cdot2.722}$ during the latter part of an island phase supercontinent cycle. The emergence of crop plants along with that of the abundance of angiosperms, on the other hand, offers a chance of 1 out 4. We set the earliest window requirement on habitable planet formation ahead of earth by assuming that the emergence chance of crop plant given by the high diversity of angiosperm is $\frac{1}{4}$ and must fit the constraint criterion that we are currently lives in an unoccupied cosmic space in addition that we are the only emerging civilization out of the nearest 3 habitable galaxies.

Then, it is not hard to conclude that the probability of human emergence to the emergence of crop plants at the odds of 4 to 119. As a result, we can conclude for every habitable planet we found that has the potential to evolve into the next expanding industrial civilization. We found that $\frac{4}{119+4}$ =3.25% of all planet has human like creatures but no presence of crop plants and $1 - \frac{4}{119+4}$ =96.74% of all planets have the necessary ingredients such as the presence of crop plants yet no human-like creatures on them. Additionally, we mentioned cases whereas both the crop plants and human like creatures yet to emerge. This gives us high confidence that if we are able to pinpoint the exact planet to observe our next arising industrial civilization within the galaxy, we will be disappointed by most of the time, not finding any human-like creature on that planet at all!

---

[15]The second scenario's deviation placement nevertheless differs somewhat from the first, but it does not alter the basic assumption.

8.14   Complexity Transformation

The probabilistic distribution accounts both temporal and spatial aspects of civilization's attainable complexity. If we assume that our civilization is 1.7 Gyr ahead of the mode by evolutionary speed, and the appearance rate is within a radius of 88 million light years, a civilization assumed to arose 100 million years ago comparable to human civilization's development at our current stage (assuming $k = 1$ using the classic Darwin's case) implies that it is 4.73 billion years ahead of the mode by evolutionary speed at the time of its appearance.

$$Y_{aaber} = (B_{cs})^{\frac{1}{1}} \cdot Y_{aaber} = (B_{er})^1 \cdot Y_{aaber} \tag{8.14.1}$$

$$2.782559^1 \cdot (1.7) = 4.73 = 4.73 \text{ Gyr YAABER} \tag{8.14.2}$$

Such level of complexity translates into the appearance rate within a radius of 734.3 million light years in radius, indicating its rarity and exponential growth in difficulty in attaining biological complexity in an earlier epoch. Moreover, 4.73 billion years ahead can also be understood as human civilization continued in its development path into the near future, so that it is increasingly improbable that in our vicinity giving rise to a civilization at the level of complexity as ours. It is also applicable to any other extraterrestrial civilization is attaining its complexity ahead of current human development. However, a caveat should be raised to interpret the numbers literally. We have shown earlier that once a biological-based species achieving control of nuclear fusion and able to maintain industrial civilization paradigm into the indefinite future, it is only a matter of time before the civilization becomes multi-planetary, and the majority of the decision making will be transitioned to that of artificial intelligence, completely rescinding the biological constraints placed onto the species. Though an industrial civilization may remain on biological substrates and continue its expansion as a biological species directed one, there must exist a mean time frame by which time the industrial civilization transitioned into a post-biological one. No one is yet sure what that time horizon is, but many argued that the foreseeable technological singularity is the cause of such a transition, and it is typical in all technological civilization's path of development, though the time of technological singularity is widely debated. On the other hand, we have discussed in Chapter 7 the limits of growth to biological beings directed civilization. With an annual growth rate of 2%, growth can be maintained at most for another 300 years for biologically led industrial civilizations.

If we compute for a civilization attained our current human level of technological progress 40 million years into the future, one can find that only 1.13 Gyr ahead of the mode by evolutionary speed. It means that as more habitable planets evolved with more biologically diverse species, resulting in increasing number traits skipped, the mode shifts closer toward civilization threshold, and the evolutionary speed of the future accelerate faster. The chance of arising human civilization increases and the attainment of human-like civilization is not as remarkable as it is now or it was in the past.

$$Y_{aaber} = (B_{cs})^{\frac{-0.4}{1}} \cdot Y_{aaber} = (B_{er})^{-0.4} \cdot Y_{aaber} \tag{8.14.3}$$

$$2.782559^{-0.4} \cdot (1.7) = 1.1289 = 1.1289 \text{ Gyr YAABER} \tag{8.14.4}$$

The number can also be interpreted differently. One can argue that 1.13 Gyr ahead means that a civilization attained a level of sophistication comparable to Age of Exploration on earth, that is, they thoroughly explored their planet and developed sophisticated agricultural society but yet to transform or unable to transform into an industrial one. It can also be interpreted as a planet with all the necessary ingredients to nurture an advanced technological civilization such as the appearance of grass plants, plenty of open fields, fruit trees, massive endowment of fossil fuels and uranium resources, but yet to wait for the appearance of an intelligent, tool manipulating species, as we have discussed this in section "Complexity equivalence". In either case, the radius of the sphere for the appearance of this civilization is only 14.9 million light years and is less than ours, because there are more civilizations or planets attained at the level of biological complexity and diversity lower than observed on earth.

If one were to use different models with different k as in the case of conservative or progressive Darwinian evolution, complexity transformation will yield different numbers but the idea remains intact. The general equation for complexity transformation becomes:

$$T\left(B_{cs}, t, k, Y_{aaber}\right) = \left(B_{cs}\right)^{\frac{t}{k}} \cdot Y_{aaber} \tag{8.14.5}$$

The following graph demonstrates the amount of transformation required for the same BCS but different k values:

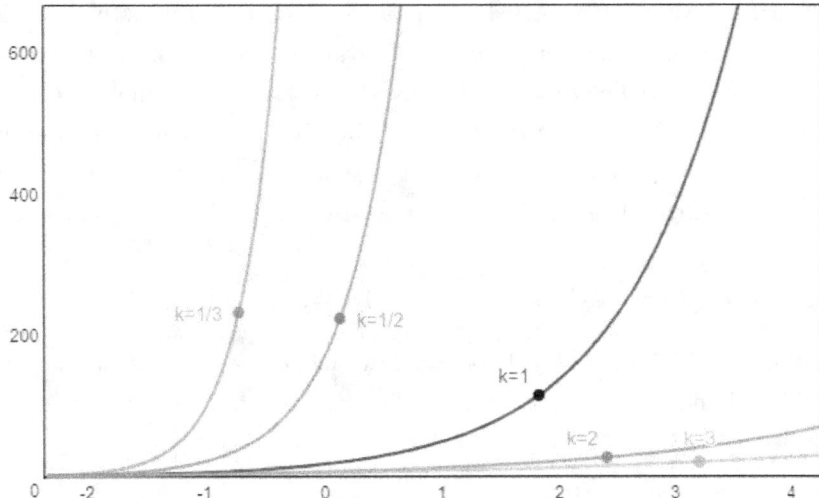

Figure 8.14.1: Complexity transformation for conservative, classic, and progressive Darwinian evolution. Notice that more progressive the scenario, the flatter the curve, indicating a greater amount of complexity change within any fixed time period and human comparable complexity is reached relatively easily.

The mode of biological complexity growth BER (the background evolutionary rate) per 100 million years is one of the most interesting aspects of the entire model. This number can vary widely from 1.203 up to 4.51, depending on the derivation methods used. According to Sharov and Gordon, the genome complexity (the number of sites of functional codons) of living organisms doubles every 250 million years which translated to 1.203 per 100 million years. This is the lower bound of growth curve estimates. We assume that number of sites correlates well with number of exhibited traits by the species and the total number of unique traits available. Repeated regeneration of the same traits or different permutations leading to the same combination does not lead to an increase in genome complexity. That is, genome complexity $\propto$ BER $\not\propto$ BCS. However, higher evolved traits may associate with other non DNA manipulation such as methylation. Then, genome complexity $\not\propto$ BER $\not\propto$ BCS. Furthermore, different permutations may lead to new sites if multiple local optimal solutions exist. Then, genome complexity $\propto$ BER $\propto$ BCS. For the last case, if genome complexity is defined as the average number of potentially mutually exclusive sites by a fixed number of traits arrived from different permutation paths, then genome complexity $\propto$ BER $= \frac{1}{n} \sum$ BCS $\leq \sum$ BCS. The true correlation between genome complexity, BER, BCS requires further investigation.

We have shown earlier, under the section 8.6 "Wall of Semi-Invisibility", in order for BER to hold a value of 1.203, the rate of emergence has to be lowered so that it can satisfy the condition required by the theoretical upper bound. The lower bound can also be achieved by a more conservative type of evolution. k > 1. If we take the rate of EQ growth from typical reptiles in the early Triassic to that of the mammals in the early Cenozoic, with an increase of 10 folds, we found that the growth rate per 100 million years is 2.78, which is the one we used assuming k = 1 and BCS = BER (see Chapter 7 and Chapter 4). It is essential, again, to stress this growth rate not the intrinsic biological rate of change attainable by molecular mutation and recombination. Genome evolution has been shown and stated (see Chapter 4) can be much faster. However, most of the time, the pace of evolution is kept in check by the earth's external environment, which favors stabilizing selection. Therefore, the background evolutionary rate itself can be misleading. It is, instead, the rate of geologic change and its dominant effect on the species, in which the rate of evolution is predominantly controlled by earth's rate of the geologic process rather than living organism's genome mutation rate.

8.15 Darwin's Great-Great Grandson's Cosmic Voyage

We discussed complexity equivalence and complexity transformation. Now we want to use these concepts to illustrate interesting consequences. Although YAABER for ourselves is 1.7 Gyr, this is the YAABER compares to the current cosmic evolutionary rate, for the average/mode cohorts of all planets currently developed biological diversity and complexity to guarantee the emergence of human-like creatures does not take 1.7 Gyr into the future. To understand this, one should realize as the time goes by, the cosmic background evolutionary rate **BER** is speeding up, as a result, one needs to wait less time to see all earth analogs' typical creatures develop into intelligent creatures found on earth. This can be calculated based on the following derivation:

$$Y_{aaber}\left(M_{ax}\left(0\right) - M_{ax}\left(1\right)\right) = \int_0^t \left(M_{ax}\left(0\right) - M_{ax}\left(1\right)\right)\left(B_{cs}\right)^{\frac{t}{k}} dt \qquad (8.15.1)$$

Whereas the separation distance of $M_{ax}\left(0\right) - M_{ax}\left(1\right)$ between two mode peaks represents the BER translated into the number of traits skipped and is the selection factor. Since the selection factor is applied on both sides of the equation, it is simplified as:

$$Y_{aaber} = \int_0^t \left(B_{cs}\right)^{\frac{t}{k}} dt \qquad (8.15.2)$$

For k = 1, it is further simplified as:

$$Y_{aaber} = \int_0^t \left(B_{cs}\right)^t dt = \int_0^t \left(B_{er}\right)^t dt \qquad (8.15.3)$$

so that the timing of typical habitable planet catching up to earth level civilization is independent of the selection factor.

$$\Rightarrow Y_{aaber} = \left[\frac{\left(B_{cs}\right)^t}{\ln\left(B_{cs}\right)}\right]_0^t \qquad (8.15.4)$$

$$\Rightarrow Y_{aaber} = \left[\frac{\left(B_{cs}\right)^t}{\ln\left(B_{cs}\right)}\right] - \left[\frac{\left(B_{cs}\right)^0}{\ln\left(B_{cs}\right)}\right] \qquad (8.15.5)$$

$$\Rightarrow Y_{aaber} = \frac{\left(B_{cs}\right)^t}{\ln\left(B_{cs}\right)} - \frac{1}{\ln\left(B_{cs}\right)} \qquad (8.15.6)$$

$$\Rightarrow Y_{aaber} + \frac{1}{\ln\left(B_{cs}\right)} = \frac{\left(B_{cs}\right)^t}{\ln\left(B_{cs}\right)} \qquad (8.15.7)$$

$$\Rightarrow \ln\left(B_{cs}\right)Y_{aaber} + 1 = \left(B_{cs}\right)^t \qquad (8.15.8)$$

Taking both sides with log:

$$\Rightarrow \ln\left(\ln\left(B_{cs}\right)Y_{aaber} + 1\right) = t\ln\left(B_{cs}\right) \qquad (8.15.9)$$

$$\Rightarrow t = \frac{\ln\left(\ln\left(B_{cs}\right)Y_{aaber} + 1\right)}{\ln\left(B_{cs}\right)} \qquad (8.15.10)$$

That is, by taking the integration from now to x years into the future, the sum of cumulative mode shift throughout x years into the future is the YAABER one currently possessed.

By substituting B_{er} with our current estimate 2.782599 per 100 Myr.

$$y = \frac{\ln\left(\ln\left(2.782559\right)\left(17\right)+1\right)}{\ln\left(2.782559\right)} = 2.8457 \cdot 10^9 \text{ Myr} \tag{8.15.11}$$

Now, we generalize the equation to taking into account all ranges of evolution by considering variable k :

$$Y_{aaber}\left(M_{ax}\left(0\right)-M_{ax}\left(1\right)\right) = \int_0^t \left(M_{ax}\left(0\right)-M_{ax}\left(1\right)\right)\left(B_{cs}\right)^{\frac{t}{k}} dt \tag{8.15.12}$$

$$Y_{aaber} = \int_0^t \left(B_{cs}\right)^{\frac{t}{k}} dt \tag{8.15.13}$$

$$\Rightarrow Y_{aaber} = \left[\frac{k\left(B_{cs}\right)^{\frac{t}{k}}}{\ln\left(B_{cs}\right)}\right]_0^t \tag{8.15.14}$$

$$\Rightarrow Y_{aaber} = \left[\frac{k\left(B_{cs}\right)^{\frac{t}{k}}}{\ln\left(B_{cs}\right)}\right] - \left[\frac{k\left(B_{cs}\right)^{\frac{0}{k}}}{\ln\left(B_{cs}\right)}\right] \tag{8.15.15}$$

$$\Rightarrow Y_{aaber} = \frac{k\left(B_{cs}\right)^{\frac{t}{k}}}{\ln\left(B_{cs}\right)} - \frac{k}{\ln\left(B_{cs}\right)} \tag{8.15.16}$$

$$\Rightarrow Y_{aaber} + \frac{k}{\ln\left(B_{cs}\right)} = \frac{k\left(B_{cs}\right)^{\frac{t}{k}}}{\ln\left(B_{cs}\right)} \tag{8.15.17}$$

$$\Rightarrow \ln\left(B_{cs}\right)Y_{aaber} + k = k\left(B_{cs}\right)^{\frac{t}{k}} \tag{8.15.18}$$

Taking both sides with log:

$$\Rightarrow \ln\left(\ln\left(B_{cs}\right)Y_{aaber} + k\right) = \ln\left(k\left(B_{cs}\right)^{\frac{t}{k}}\right) \tag{8.15.19}$$

$$\Rightarrow \ln\left(\ln\left(B_{cs}\right)Y_{aaber} + k\right) = \ln\left(k\right) + \ln\left(\left(B_{cs}\right)^{\frac{t}{k}}\right) \tag{8.15.20}$$

$$\Rightarrow \ln\left(\ln\left(B_{cs}\right)Y_{aaber} + k\right) = \ln\left(k\right) + \frac{t}{k}\ln\left(B_{cs}\right) \tag{8.15.21}$$

$$\Rightarrow \ln\left(\ln\left(B_{cs}\right)Y_{aaber} + k\right) - \ln\left(k\right) = \frac{t}{k}\ln\left(B_{cs}\right) \tag{8.15.22}$$

$$\Rightarrow k\left(\ln\left(\ln\left(B_{cs}\right)Y_{aaber} + k\right) - \ln\left(k\right)\right) = t\ln\left(B_{cs}\right) \tag{8.15.23}$$

$$\Rightarrow t = \frac{k\left(\ln\left(\ln\left(B_{cs}\right)Y_{aaber} + k\right) - \ln\left(k\right)\right)}{\ln\left(B_{cs}\right)} \tag{8.15.24}$$

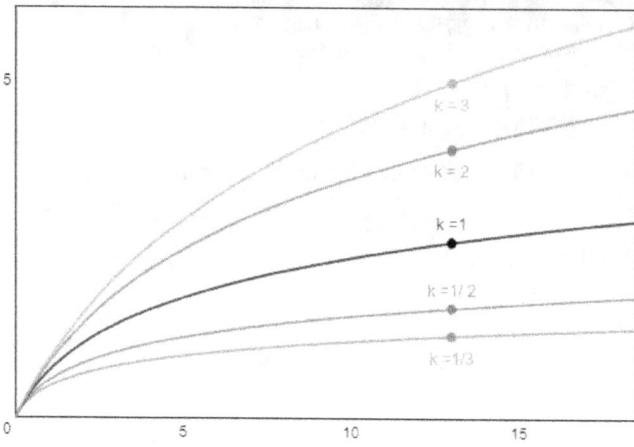

Figure 8.15.1: Biocomplexity transformation curves for different values of k when BCS $= 2.783$ and setting YAABER as the variable

And it can be found that it will take just 284.57 Myr instead of 1.7 Gyr.

We can also utilize the new knowledge to construct a new complexity transformation equation which is just like our earlier starting condition except that we now fixed t and solving for Y_{aaber}.

$$T\left(B_{cs}, x, k, Y_{aaber}\right) = \int_0^{x+T(Y_{aaber})} \left(B_{cs}\right)^{\frac{t}{k}} dt \tag{8.15.25}$$

$$Y_{aaber} = \frac{17}{\dfrac{\ln\left(2.783^{\frac{1}{k}}\right)}{\ln(2.783)}} \tag{8.15.26}$$

Whereas $T\left(Y_{aaber}\right)$ is the biocomplexity transformation we just derived. The upperbound on integration is the sum of the amount of time required to catch up to current human civilization development given a BER at current earth time (for example, 284.57 Myr), and x is the number of years relative to the current time when the civilization appeared. A human level civilization is deemed rarer if it occurred in the past and more common if it occurred in the future and it is always proportional to BER, expressed as $(B_{cs})^{\frac{1}{k}}$.

Assuming extraterrestrial civilization expands very slowly, so we will not expect to encounter any in the next 284 Myr. Then, what will Darwin's great-grandson see if he takes on a cosmic expedition to gather astrobiological data? If he takes on a near light speed travel vessel that properly shielded himself from radiation and make a quick stop at each destination and then quickly hops to the next one. He is set to travel 284 Million light years from stationary perspective while he only aged 50 years or so. As a result, he will able to collect enormous data in 50 years of time. His first observation was from the life hosting exoplanets within our Milky Way galaxy. All of such planets are devoid of human-like creatures. He probably observed some creatures that share some similarity to some aspects of human, but none of them possessed full potential as human. He stayed within the Milky way for the first 6 days and left to explore more distant galaxies. Throughout most of his career, the first 45 years of his expedition, he observed the similar patterns in other galaxies as that he observed in the Milky Way, though he did observe that a few planets already possessed human-like creatures and full-blown agricultural society and transitioning into industrial and eventually into post-biological society. The first one he observed transcending event occurred 10 years into his journey. Majority of them, however, are not. However, in the last few years of his career, he observed that almost every life hosting planet visited is possessing by at least one human-like species and is rapidly evolving into a human-like society. At first, it appears some of these creatures stagnated at the hunter-gatherer stages because it lacks crop plants, but sooner or later the plants co-evolved and ushered in agricultural revolution. In the last few days of his career, every other planet he visited is evolving into an agricultural one. In the last few hours of his career, every other planet he visited is turning into an industrial one, finally, in the last few minutes of his career, every other planet he visited is transforming into a post-biological life form and replacing him as the expedition team leader as the descendants of the Darwins from each of their home planet emerges and continue their expedition in the cosmos.

8.16   Upper Bound & Lower Bound

We have already shown that the closest extraterrestrial industrial neighbor lies 80 million light-years away and the universe will be filled by expanding cosmic expansionists in 37 million years from now by the earliest. However, it is possible other filter criteria not covered in previous chapters worthy mentioning and can push the time frame to a later time period.

One issue not addressed previously is the position of the star within the Milky Way. The Sun's position is favorable not only because it posits on the galactic habitable zone but also in between the spiral arms and its orbits around the galaxy minimizes its crossing over the spiral arms. The greater density of interstellar medium and closer proximity to other stars during arm crossing can result in greater catastrophic events. Studies have been shown the relationship between Sun's spiral arm crossing in the past and extinction event. More studies are required in this area, and if Sun's position is indeed more favorable, the number of habitable planets within the galaxy has to be reduced significantly.

The sun, unlike many stars in the galaxy, seems to have stayed in its orbit since its inception. Recently, finding pointed out that up to half of the stars within the galaxy has migrated from their birth site to other locations either as inward or outward migration. The frequency of each type of migration and the average age at which the stars migrated and the spectral class profile breakdown of the stars require more examination. If migration indeed occurs frequently, a habitable planet can shift beyond the galactic habitable zone. Other stars may shift into the galactic habitable zone but with a shortened window of habitability, so that complex, multicellular life does not have enough time to evolve before the host star evolves off the main sequence.

The sun, unlike some other stars, has a low galactic orbital eccentricity. Other stars, can have higher eccentricity. Though these stars do not migrate during their main sequence lifetime, it nevertheless ventures beyond the galactic habitable zone. More data on stellar eccentricity is needed before conclusion can be drawn. If indeed a significant portion of stars favoring eccentric orbit around the galaxy, a substantial number of habitable planets needs to be reduced.

All of the factors mentioned above can generally be grouped into the category of a filter of the galactic habitable zone. Recent studies, however, have shown that such region may not exist. The galactic center actually hosts the highest probability of habitable planets given the density of stars per unit volume. Neither inward, outward, horizontal, or vertical movement of stars significantly alter the habitability of the system. The study concludes that 1.2% of all stars within the galaxy are habitable. Since the study was done with metallicity and the habitable zone used as selection criteria, we multiply the metallicity factor (see Chapter 2 & 3) and habitable zone filter to yield 13.04%. This implies that the rarity of the emergence of industrial civilization is reduced by a factor of 1.975. Together along with the factor for galactic habitable zone at $\frac{1}{3.882}$ to yield a total factor of 7.667 (Chapter 2). This partially justifies the original Lineweaver's model for earth production in an unexpected way, which has to increase by a factor of 46 in order to comply with the number of stars in the galaxy, albeit the justification is not a consideration in the original model.

The size of dinosaurs and some prehistoric creatures posed a challenge to biology. Based on Galileo's squared cubed law, no animals at the size of gigantic dinosaur can survive today given the current atmospheric pressure and condition (with various causes such as difficulty in pumping blood into its brain given its height).[51] Pterodactyl, given its size and wingspan, cannot sustain horizontal upward flight movement. That is, pterodactyl can glide from the cliffs to the shores, but it will have great difficulty in getting back to the edge of a cliff. A resolution to such paradox has been pointed out. The atmospheric pressure and atmospheric density potentially have to be several times higher than today, enabling the evolution of gigantic creatures with aided buoyancy. If air density fluctuates, then the evolution of fruit trees can be a consequence of evolutionary adaptation as the benefit to evolving fruits is significantly higher on planets with thinner atmosphere.[41][51] Basically, the plants are begging and luring animals to carry its seed away by rewarding them with their fruits. In an environment with a dense atmosphere, plants can disperse their seeds simply using mechanical means such as wind and rewarding animals with dispersion can be complementary. Therefore, angiosperms may not develop as easily on planets with denser atmosphere as it is observed on earth. Without the presence of fruit trees and the creation of the arboreal niche, the opposable thumb cannot develop, and no tool using species will exist. Currently, it is believed that earth had three atmospheres in its geologic history, the first was shrouded with primordial hydrogen, the second one composed mainly of carbon dioxide and methane, and our current one composed mainly nitrogen and oxygen. Preceding atmospheres were significantly denser than the current one. However, it is believed that the ocean, comprising the universal solvent water, along with life, is responsible for converting a huge share of carbon dioxide into carbonated stone such as the dolomite and limestone, and reducing the density to the current level. The onset of plate tectonics and carbon cycle may also significantly contributed toward a thinner atmosphere of today. Therefore, it is generally assumed that the density of atmosphere on all habitable planets converges toward similar conditions. This conclusion, however, can change as more is learned. If earth's atmosphere condition is typical and atmospheric pressure did fluctuate in geologic history, then the timing of the fluctuation can determine the emergence of intelligent tool using creatures. If fluctuation is the norm, then sooner or later fruit-bearing plants will emerge and arboreal niche will appear on such a planet to enable the emergence of opposable thumbs.

In one of the most frequently cited solutions to the Fermi Paradox, many argued all industrial civilizations tend to destroy themselves through nuclear conflict or out of control nanotechnology and Artificial intelligence. Although this explanation has been ruled out due to non-mutual exclusivity, It is possible that not all industrial civilization succeeded in transforming into an expanding one. Even on our home planet, during the cold war era, had precariously avoided several incidents that could have initiated World War III. Since there is no tool to measure the likelihood a civilization to destroy themselves, it is hard to quantify this parameter.

Among civilizations which succeeded in transitioning into an AI controlled nuclear fusion powered civilization, a significant fraction may just choose not to expand. The civilization may or may not aware the existence of other extraterrestrial industrial neighbors. For various reasons, it is more inward-focused and consuming energy in a steady state fashion. Many futurists have hypothesized this type of civilization, which essentially harvests the energy of their home star and power their citizens with the abundance of wealth and material goods. It is also possible that the civilization immerses itself in a full-virtual reality that is qualitatively better than the real world it has to deal with.

It is also possible that civilization does expand, but on average expands at much slower speed than the speed of light. As a result, it takes a significant amount of time for each other to connect even if they intentionally wanted to do so. This theory is somewhat undermined because it has been shown that even with modest nuclear fission rockets using the technology we currently attained, we can reach 50% speed of light. It is further undermined by our discussion under relativistic economics (Chapter 9), it is shown that post-singularity society has a high incentive and the most significant economic gain by expanding nearing the speed of light.

All of the aforementioned factors could potentially render the emergence of an expanding civilization near the speed of light less likely. Therefore, we can group these factors into a single term called term x. Depending on the magnitude of term x, it can then be shown the limits and lower bound on the emergence of extraterrestrial industrial civilizations.

8.17   Subluminal Expansion

We have assumed that intelligent life of those undergone post-biological transition prefers to propagate and expands at closely approximate or at the speed of light.(see Chapter 8) What if the civilization chooses to expand at sub-luminal speed? If the civilization starts to expand at subluminal speed, from our relativistic economics model (see Chapter 8), it can be shown that sooner or later such civilization will adapt to speed close to the light speed. If for some reason, they remain at slow expansionary speed, they can be observed in the sky as an astronomical phenomenon. If such phenomenon occurs and expands at speed less than the expansionary speed of the universe based on its redshift index Z compares to earth, then, it will gradually disappear from our view. If it expands at the same speed as the expansionary speed of the universe from our vantage point, then its size will remain fixed. The most likely scenario is that its expansionary speed is faster than the expansionary speed of the universe from our vantage point of view. If the speed is as close to the speed of light as possible, the observation of such phenomenon will be one of the worst news for humanity because it implies that its physical arrival to earth is imminent. Since it takes millions of light years for the signal traveling to earth, and during these millions of years, the civilization is likely expand from a sub-luminal speed toward speed c. The delay in the observance of an extra-terrestrial artificial phenomenon and its physical arrival is simply the time used by the civilization to optimize its expansionary speed from subluminal to speed c. This may take from a few millenniums to just a few days (in a post-singularity scenario). This will make earth extremely ill-prepared for such an encounter. The sky will appear as utterly devoid of any artificial signals and phenomenon (like what we have observed now so far) to a spot or spots of interests. Then, suddenly the entire sky will be filled with artificial phenomenon with delays in just days, months, years, decades, centuries, or millenniums, all tiny time scale compares to cosmological timescale.

8.18   Observational Equations

In order to mathematically state the Fermi paradox as a set of equations, one has first to define the probability of arising industrial civilization within a sphere of unit radius r, which can be any arbitrary value. We have simply set it as the sphere radius of the observable universe within 13.8 billion light years. The total chance of observing one, two, three, and up to n number of extra-terrestrials summarily is simply:

$$\frac{1}{F_{galaxy}}\left(C_{df}(0)\right)^1 + \frac{1}{F_{galaxy}}\left(C_{df}(0)\right)^2 + \frac{1}{F_{galaxy}}\left(C_{df}(0)\right)^3 + ... + \frac{1}{F_{galaxy}}\left(C_{df}(0)\right)^n \tag{8.18.1}$$

$$= \frac{1}{F_{galaxy}}\sum_{k=1}^{n}\left(C_{df}(0)\right)^k \tag{8.18.2}$$

The maximum number of observable extraterrestrials can be then easily derived based on the measured radius, and we have shown earlier that no extra-terrestrial civilization met more than three other extraterrestrial civilizations within the radius of the observable universe.

Then, we have the equation for the total number of extraterrestrials regardless of the time of their emergence with a fixed given radius of r, and r_0 is simply the weighted average of the radius of galaxies taking the empty space between them into considerations:

$$N_{all}(r) = \frac{1}{F_{galaxy}}\sum_{k=1}^{n}\left(C_{df}(0)\right)^k \left(\frac{\frac{4}{3}\pi r^3}{\frac{4}{3}\pi r_0^3}\right) \tag{8.18.3}$$

$$= \frac{1}{F_{galaxy}}\sum_{k=1}^{n}\left(C_{df}(0)\right)^k \left(\frac{r}{r_0}\right)^3 \tag{8.18.4}$$

$$T = 13.8\,\text{Gyr} \quad r_0 = 11.85\,\text{Mly}$$

If r applies to the size of the observable universe, the equation states the number of all extra-terrestrial industrial civilization ever arise regardless they are current observable or not. That is, currently arising extra-terrestrial civilization 1 billion light years away is not directly observable, yet it can be calculated from the equation.

Though the universe at its inception was infinitely dense, the chance of extra-terrestrial civilizations forming was infinitely small. As a result, the total number of arising extra-terrestrial within the size of observable universe is finite (again, the universe expanded quickly so that only finite amount of mass is distributed within the space-time fabric of the observable universe, the total number of arising extra-terrestrials can remain infinite for the entire size of the universe)

$$N_{past}(r) = \frac{1}{F_{galaxy}} \int_0^T \sum_{k=1}^n (C_{df}(r))^k \left(\frac{4\pi r^2}{\frac{4}{3}\pi r_0^3} \right) dr \tag{8.18.5}$$

where each step of dr = r as the lower bound for $C_{df}(r)$

The second equation states that since the expansion of the universe, the total sum of signals ever has reached us from the past at each distance from earth. That is, a distance of 0 from us implies all signals starting from the big bang up to today will be counted toward the total sum. A distance of 13.8 billion light-years from us implies only a few years if not a few minutes of signals following the big bang is counted toward the total sum. Of course, the model is still an excellent approximation of the real situation because earth's or sun's and Milky Way's position cannot be assumed to be fixed. However, such local movement is small enough that does not alter the general calculation. (i.e. sun's revolution's diameter of 60,000 light years is 0.000217391% of the diameter of the universe. So the model is 99.999782% correct.)

$$C_{df\,MODIFIED}(t) = N_{all} \left(\frac{\int_1^{1+d} P_{df}(t,x)\,dx}{\int_0^\infty P_{df}(t,x)\,dx} \right) \tag{8.18.6}$$

$$N_{now}(r) = \frac{1}{F_{galaxy}} \int_0^T \sum_{k=1}^n (C_{df\,MODIFIED}(r))^k \left(\frac{4\pi r^2}{\frac{4}{3}\pi r_0^3} \right) dr \tag{8.18.7}$$

where d ≤ 0.000000036   and (r + d) ≤ T

The third equation states that since the expansion of the universe, the total sum of signals that are currently reaching us from the past at each distance from earth. We define currently as the recent 50 years of observation. This is converted to d ≤ 0.000000030. That is, a distance of 0 from us implies all signals just occurred will be counted toward the total sum. A distance of 13.8 billion light-years from us implies just that moment of the big bang is counted toward the total sum. Since signals from the more distant past imply that the bio-complexity transformation factor was much lower; therefore, the chance of observing any extra-terrestrial decreases with greater distance, as we have already discussed.

To clarify the difference between the above three equations, we need to reuse diagram from Chapter 1. The size of the rectangle is the total number of extra-terrestrial civilizations given the age and the size of the observable universe, corresponding to the first equation. The size of the shaded triangle corresponds to the second equation. The tiny hypotenuse of the rectangle is the sum of all signals currently reaching earth.

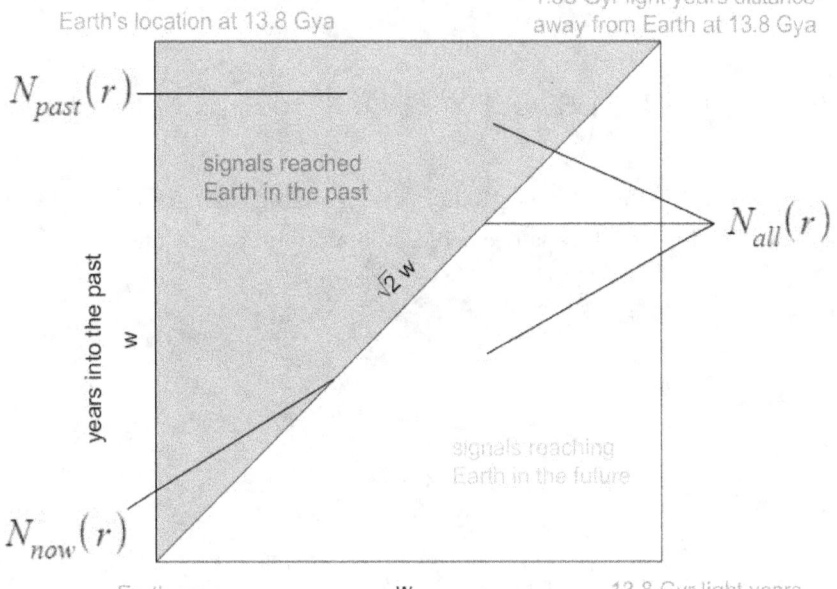

Figure 8.18.1 labels:

Earth's location at 13.8 Gya

1.38 Gyr light years distance away from Earth at 13.8 Gya

$N_{past}(r)$

signals reached Earth in the past

$\sqrt{2}\,w$

$N_{all}(r)$

years into the past

w

signals reaching Earth in the future

$N_{now}(r)$

Earth now

w

distance away from earth

13.8 Gyr light years away from Earth now

Figure 8.18.1: Signal detection landscape of time vs distance

It can be shown that as the width and height increase toward infinity, the ratio of the third equation toward both the second and the first approaches 0. This implies that the fraction of currently observed signals in the universe is tiny compares to all signals ever received. The total fraction currently observed, in fact, is 0.0056% of all signals since the Big Bang. The relationship between the three equation can be then formulated as the follows:

$$N_{all}(r) \geq N_{past}(r) \geq N_{now}(r) \tag{8.18.8}$$

This is the Fermi Paradox stated from purely mathematical perspective. That is, arising extra-terrestrial civilization are not observable if they currently lie outside of our past light cone. Given the vast amount of signals have reached us from the past (50% of all signals ever created within the observable universe and our starting assumption that a civilization utilizing nuclear fusion is sustainable) and the lack of any evidence earth and its cosmic vicinity is colonized, the lack of signals implies again that we can receive such signal only in the future if at all. *Therefore, if cosmic transition bound to occur, then Fermi Paradox is an observational one, and a natural consequence of all extra-terrestrial civilization arising in the relatively recent past, and almost all arising extra-terrestrial industrial civilization at the current epoch facing the similar paradox.*

Having derived the general form of our equation, we can re-estimate the distance to the nearest extraterrestrial industrial civilization by the following equation:

$$N_{galaxy} = \left(C_{df}(0) + C_{df}(0)^2 + C_{df}(0)^3 + ...C_{df}(0)^\infty\right)^{-1} \tag{8.18.9}$$

$$N_{galaxy} = \left(\sum_{n=1}^{\infty} C_{df}(0)^n\right)^{-1} = 1.1614 \tag{8.18.10}$$

$$1.26\left(F_{galaxy} \cdot N_{galaxy} \cdot (d_{Galaxy})^3\right)^{\frac{1}{3}} = 29.63\,\text{Mly} \tag{8.18.11}$$

That is, the nearest civilization now lies somewhat closer at 26.79 Mly away instead of at 32.96 Mly away, or

$$1.26 \left(F_{galaxy} C_{df} \left(0 \right)^{-1} \left(d_{Galaxy} \right)^3 \right)^{\frac{1}{3}} - 1.26 \left(F_{galaxy} \left(\sum_{n=1}^{\infty} C_{df} \left(0 \right)^n \right)^{-1} \left(d_{Galaxy} \right)^3 \right)^{\frac{1}{3}} = 6.1643 \, \text{Mly} \quad (8.18.12)$$

6.1643 Mly closer, by simply considering the extra possibility of seeing more than one extraterrestrial industrial civilizations within our sphere of influence. This is deemed as the more precise result compared to our earlier one.

Chapter 9

Relativistic Economics

9.1 Overview

The outlook is very different for a post-biological directed industrial civilization. We have already shown eventually post-biological intelligence have a good chance dominating the earth and the solar system beyond. This transition can be quick, as advocated by the Singularitarians, where machine intelligence overtakes humans within decades, or can be slow based on generational commitment in biological technology and fine-tuning. In either case, a transition is quick compares to the geological timescale of $x < 10^7$ years. As a result, post-biological species is able to adapt to a much wider range of habitats and environments, its energy acquisition can be efficient, and no crop plants is needed. To appreciate the expansionary economics in the cosmic scale, we need to introduce the concept of relativistic economics and finance, a generalized form of finance applied at the cosmic scale to deal with cosmic investment and growth. Immense distance renders conventional communication and transportation impossible. Relativistic mass and speed, energy acquisition and incubation period become the conventions. As a result, relativity is required to deal with the problems associated with them. The equations are applicable to both biological led expanding industrial civilization as well as to post-biological expanding industrial civilization. Though the numbers from the calculations differ and we are discussing them below.

Economic growth is a very complex interaction between interest rates, prices, investment, savings, and over the long run, on technology growth. Since technology is the utilization of extracted energy, and it helps to ensure greater energy acquisition. (such as early oil well and the consequent development of oil rigs and drills based on the abundance of cheap energy and their later use in horizontal drilling and fracking, previously deemed unprofitable without refined technology) Technology sophistication and energy acquisition correlates positively and reinforce each other. Yet technology is hard to quantify into abstract mathematical numbers or concepts, and the total use of energy can be measured at any given time. As a result, we make an assumption that energy consumption is a measure of the level of economic development as for an individual or for a group, as in accordance with White's law. Then, we can measure the rate of economic growth of a civilization based on the total growth of its energy usage. Whether the energy increase achieved through increased per capita energy consumption, increased population, socialistic system, capitalistic system, or any other possible sociological system do not alter our basic assumption. For the sake of simplicity, we shall adopt the assumption that the total energy usage is the consequence of population growth while the per capita energy consumption stays constant.

For Relativistic Economics: the inequality expressed between the Final Return on Migration and initial costs, we have an equation in the simplest conceptual form for investors:

Final Return on Investment - Preparation Cost - Migration Cost > 0

Whereas the Final return on investment can be more conceptually materialized into:

$$R_{final} = \frac{1}{m} \cdot (E_{stable} \cdot P_{growth} + E_{growth}) \cdot P_{migration}$$

(9.1.1)

R_{final}	Final Return on Investment, units in Joules
E_{stable}	Maximum allowable energy acquisition for stabilized 1 billion years, units in Joules
P_{growth}	Perceived rate of subjective value depreciation on the finalized stable return over 1 billion year period during the expansionary phase at the destination
E_{growth}	The total energy acquisition for the expansionary phase before the maximum energy extraction level is reached, units in Joules.
m	Number of participants
$P_{migration}$	Perceived rate of subjective value depreciation on the overall return due to the migration waiting phase toward the destination before any return is initiated. (So investors' subjective value of return is lower than the actual return.)

Preparation Cost (for both earthbound and shipbound investors) can be more conceptually materialized into:

$$C_{prepare} = \epsilon_{lost} \cdot m \cdot T_{preparation} \cdot P_{preparation}$$

(9.1.2)

$C_{prepare}$	Preparation cost on earth, units in Joules
ϵ_{lost}	Energy saved (unavailable currently) for preparation purposes per participants per year, units in Joules
m	Number of participants
$T_{preparation}$	The total preparation time required to save enough energy to send the ship at certain speed, units in years
$P_{preparation}$	Perceived rate of subjective value depreciation on the total costs required to commence the trip during the preparation phase (so investors' subjective value of costs is lower than its actual costs)

Migration Cost for shipbound investors can be more conceptually materialized into:

$$C_{migrate} = \epsilon_{lost} \cdot m \cdot T_{migration} \cdot P_{migration}$$

(9.1.3)

$C_{migrate}$	Migration cost for the ship, units in Joules
ϵ_{lost}	Energy (generating-energy opportunity) lost due to time spent on ship travel per participants per year, units in Joules
m	Number of participants
$T_{migration}$	The total migration time in the ship's reference frame at certain speed, units in years
$P_{migration}$	Perceived rate of subjective value depreciation on the final return during the migration phase. (so that investors' subjective value of costs is lower than the actual costs.)

We also need to introduce two units of conversion so to simplify our equations down the road.

Total relativistic energy required for an unit mass departure at a certain speed:

$$K_{rel} = \frac{1}{\sqrt{1 - \frac{v^2}{c^2}}} - 1 \tag{9.1.4}$$

It is also expressed in our calculation as:

$$K_{rel} = \frac{1}{\sqrt{1 - \frac{x^2}{100^2}}} - 1 \tag{9.1.5}$$

Total time required for migration at certain speed in shipbound investors' reference frame:

$$T_{rel} = \frac{d}{v}\sqrt{1 - \frac{v^2}{c^2}} \tag{9.1.6}$$

It is also expressed in our calculation as:

$$T_{rel} = \frac{d}{\frac{x}{100}}\sqrt{1 - \frac{x^2}{100^2}} \tag{9.1.7}$$

$c =$ Speed of light

$d =$ The current cosmological distance from the origin to the destination, units in light years

$v =$ The migration traveling speed, units in a fraction of the speed of light

For earthbound investors, the equation is expressed as:

$$E_{arth} = \frac{\epsilon}{m}\left[\left(\sum_{k=0}^{10^9} M\left(p\right)^k\right) p^{\frac{\ln\left(\frac{M}{m}\right)}{\ln a}} + \sum_{k=0}^{\frac{\ln\left(\frac{M}{m}\right)}{\ln(a)}} m\left(ap\right)^k\right] p^{\left(\frac{m \cdot K_{rel}}{\epsilon} + \frac{d}{v}\right)} - m\epsilon\left[\sum_{k=0}^{\frac{m \cdot K_{rel}}{\epsilon}} p^k\right] \tag{9.1.8}$$

$m =$ The initial number of participants

$M =$ The final number of population owned/controlled by the initial participants as the founders

$a =$ The economic growth rate (population growth rate)

$p =$ The perceived rate of subjective depreciation on returns

$\epsilon =$ The amount of energy can be produced each year per capita, units in Joules

For shipbound investors, the equation is expressed as:

$$S_{hip} = \frac{\epsilon}{m}\left[\left(\sum_{k=0}^{10^9} M\left(p\right)^k\right) p^{\frac{\ln\left(\frac{M}{m}\right)}{\ln a}} + \sum_{k=0}^{\frac{\ln\left(\frac{M}{m}\right)}{\ln(a)}} m\left(ap\right)^k\right] p^{\left(\frac{m \cdot K_{rel}}{\epsilon} + T_{rel}\right)}$$

$$- m\epsilon\left[\sum_{k=0}^{\frac{m \cdot K_{rel}}{\epsilon}} p^k\right] - m\epsilon\left[\sum_{k=0}^{T_{rel}} p^k\right] \tag{9.1.9}$$

These equations are the abstract manifestation of the following concepts. First of all, we assume that inhabitants on earth achieve economic stabilization and no further room for economic growth is possible. (this assumption can be relaxed to show that incentive of stellar migration is strong as long as the rate of return is lower on earth than migration to another star system, but our discussion is based on the simplicity of our model). The first term (the summation part) of the equation simply means that the perceived value of eventual return in the form of energy when a group of individuals given by a total population of m, migrated to a new hospitable planet. (observed through telescope before migration, so no terraforming costs involved) This migrated population then grows exponentially based on a fixed energy growth rate until their descendants fill the entire planet with population M. We assume that their descendants are loyal to the founders and reaping the economic benefit from their initial investment with an energy budget of $M \cdot \epsilon$. The final objective value on return is then $\frac{M}{m} \cdot \epsilon$. Then, it is easy to show that a small group of founders will be willing to take risks to migrate, and much less likely for the population of the entire planet to migrate to a different earth because little to no return is achievable (assuming all earth analogs holds similar carrying capacity). The growth rate maintains at the rate of a until it reaches population M, a term of p is the perceived value loss over a longer period of time. That is, biological creatures with limited lifespan hold a considerable perceived value of loss over a return that takes a significant time to achieve. In the economic endowment theory, the owner valued possession lost is twice as costly as its intrinsic value. In other words, the owner had expected to possess the item into the indefinite future while he has a perceived value of a loss of 50%. That is, his value placed on the current item is 1 at the current year, and only $\frac{1}{2}$ for the next year which yet to pass, and only $\frac{1}{4}$ for the third year, $\frac{1}{8}$ for the fourth year, and so on, then the sum of all years then is 2. If we assume one's view on his eventual return on the investment follows the similar perceived value of a loss, then p can be substituted with 0.5. This implies that human put very little attention toward any gain or loss three years into the future, and their concerns for any gain or loss over time decreases geometrically. However, a post-biological or even just more well-educated populace may place more emphasis toward the future by having a longer lifespan or more knowledge and awareness for long-term well-being. As a result, the perceived value of eventual return is greater when p is greater.

When the product of economic growth rate a and p is less than 1, given by the limit test, we have a convergent series which sums up to:

Limit test for convergence:

$$\sum_{k=1}^{n} m\,(ap)^k \to \lim_{n \to \infty} \left| \frac{a_{n+1}}{u_n} \right| \to m \lim_{n \to \infty} \frac{|ap|^{n+1}}{|ap|^n} \to m\,|ap| < 1 \to |ap| < \frac{1}{m} \tag{9.1.10}$$

where m=1: $|ap| < 1$

when $ap < 1$ and p < 1

$$E_{arth} = \frac{\epsilon}{m} \left[M \frac{1-p^{10^9}}{1-p} p^{\frac{\ln\left(\frac{M}{m}\right)}{\ln a}} + m \frac{1-ap^{\frac{\ln\left(\frac{M}{m}\right)}{\ln a}}}{1-ap} \right] p^{\left(\frac{m \cdot K_{rel}}{\epsilon} + \frac{d}{v} \right)} - m\epsilon \frac{1 - p^{\frac{m \cdot K_{rel}}{\epsilon}}}{1-p} \tag{9.1.11}$$

$$S_{hip} = \frac{\epsilon}{m} \left[M \frac{1-p^{10^9}}{1-p} p^{\frac{\ln\left(\frac{M}{m}\right)}{\ln a}} + m \frac{1-ap^{\frac{\ln\left(\frac{M}{m}\right)}{\ln a}}}{1-ap} \right] p^{\left(\frac{m \cdot K_{rel}}{\epsilon} + T_{rel} \right)}$$

$$- m\epsilon \frac{1 - p^{\frac{m \cdot K_{rel}}{\epsilon}}}{1-p} - m\epsilon \frac{1 - p^{T_{rel}}}{1-p} \tag{9.1.12}$$

when $ap \geq 1$

$$m \frac{1-(ap)^n}{1-ap} < \sum_{k=0}^{n} m\,(ap)^k \tag{9.1.13}$$

$$M \frac{1-p^n}{1-p} < \sum_{k=0}^{n} Mp^k \tag{9.1.14}$$

This is a special case of our summation, and very likely the solution of the perceived value of eventual return led by biologically expanding industrial civilization, which is what we are currently. The value for p can still be higher than 0.5 but must be lower than 1. This shows that for biological led civilization with a deep discount toward very far future, the total benefit for the eventual return on migration cannot be realized. For example, a team of one thousand migrated to the new terrestrial planet and populated up to 10 billion people in 814 years with an annual growth rate of 2% and a final theoretical return of 10,012,437.41 folds of the original investment. However, with $p = 0.98$, then, the perceived value of eventual return is only 50 folds, this is 200,249 folds difference! For a post-biological being which attains biological or even permanent immortality, its p infinitely approaches 1. As a result, its perceived value of eventual return will infinitely approach the theoretical value.

To make the matter even worse, the eventual return is further delayed by the amount of time at gathering the energy to send the spaceship to the distant planet and the amount of time spent traveling from the origin to the destination for shipbound investors, the perceived value of the eventual return is a tiny fraction of the theoretical. The waiting time contributes toward an exponentially decreasing function where the perceived value on the return decreases exponentially as the waiting time increases.

For shipbound investors, the second term is the amount of time used to gather the energy required to send the spaceship to the destination, it plays close relationship with the third term in the equation, the amount of time used in travel to the destination. A group of one thousand can quickly prepare their trip to the destination without gathering too much energy manifested by their current energy budget that they can buy with their savings as ϵ. However, a slow spaceship will cost them thousands of years to reach their destination. On the other hand, these 1,000 people can gather the energy they needed so that they can reach their destination in a very short time from the perspective both stationary observers on earth and the travelers themselves at a significant fraction of the speed of light. Traveling at a significant speed up to fractions of the light speed requires a tremendous amount of energy which can take the organizers thousands of years to complete. Therefore, it lies the dilemma, that no interstellar colonization is possible in a biological led industrial civilization. Or is it? Though groups of individuals and companies may not have the adequate time and resources to invest in such a project. All populations on earth as a whole may and possibly can, but we just showed that the entire population has little to no incentive to migrate at once. However, two ways can guarantee a migration scheme. The first of which is a lottery system. A lottery system can be activated planet-wide where each participant in the lottery have a tiny chance of migration. A small fraction of the energy used by each individual is invested into one interstellar project at a time, times the total population on earth. Then, the energy requirements can be fulfilled after a few rounds of the lottery and may take up to a thousand years. At the end of the lottery draw, 1,000 lucky winners are sent to the new habitable planet. The lottery can then be repeated over and over again and sending winners to the next closest habitable planet yet to be occupied. However, the interval between each successful lottery draw will be longer and longer because it takes more and more energy to get the winners to the destination farther and farther away. Newly colonized planets will able to perform their own version of lottery draws and send their descendants to those planets nearest to them, and they will have an advantage over lottery players on earth because they are closer to some of the unoccupied planets hence shorter waiting time for the next successful lottery draws.

Lottery approach is great. It shows that it is possible that when a market's perceived value of eventual return is too low one can still manage to colonize the stars. One shortcoming of the lottery approach, however, is that the selected winners are not necessarily the best fitted for a migration. If their winning ticket can be traded with someone else, the one really willing to go might get it. However, we just showed that the price for trade would be extremely high because it takes the entire world population to contribute 1,000 member team. So the most willing to go may not have the amount required to pay. Even if the most willing to go can always come up with the sum to pay, he or she may not have the most biological fit body to survive the trip and procreate upon the destination.

As a result, a second alternative, a government-funded project through taxation is another approach to stellar migration. A forward-looking, futuristic government which are well-aware the benefit of spreading and hedging the risks of the extinction of human civilization will able to put forth a thousand year plan where each year specific portion of taxes is contributed to the construction of interstellar vessels and gathers the necessary energy to send the spaceship to the destination. The government also builds facility and trains selected individuals from the pool of entire population who are screened and to be best fitted for the travel and procreation upon the destination.

The cons of government-sponsored projects are corruption, authoritarianism, and the unlikelihood of its survival in very long terms. It is noted that all human-led institutions suffer inefficiency in the forms of miscommunication and misallocation. The success of the state-run project requires a highly controlled population with solidarity. It is conceivable a strong government and sacrificing population is possible given a significant existential threat to the entire population; however, it is becoming increasingly difficult to assert authoritarian control of the masses given the ubiquity of the internet and without any extinction-level threat to the humanity. Furthermore, no government projects can last for thousands of years so far in documented human history. The construction of the pyramids took a few centuries of extensive development, and the construction and maintenance of the great wall of China lasted only a few centuries during the Han dynasty and the Ming dynasty.

Nevertheless, it can be shown that even with biological led industrial civilization, interstellar colonization is possible and desirable. Now let us focus on determining the speed at which our current level of industrial development can expand at the cosmic scale.

Assuming the subjective rate of depreciation is 0.95 and plotting the equation with a short distance to a destination within a few light years (20 ly), one quickly concludes that the perceived economic incentives for earthbound investors and shipbound investors differ though both perceives positive return over a range of fractions of the speed of light. Their difference mainly stems from earthbound investors more inclined to making shorter preparation time and shipbound investors inclined for cruising at a higher speed to save trip time. Their curves also share crossover point at a specific fraction of the speed of light. At the crossover point, the earthbound investors perceived economic return agree with shipbound investor's perceived economic return value.

$$T_{relu} = \frac{1}{\frac{x}{100}}\sqrt{1 - \frac{x^2}{100^2}} \tag{9.1.15}$$

$$y_{ship0} = m\frac{1 - (1.02p)^{\frac{\ln\left(\frac{100}{m}\right)}{\ln(1.02)}}}{1 - (1.02p)}p^{(mK_{rel} + 20T_{relu})} - \frac{1 - p^{mK_{rel}}}{1 - p} - \frac{1 - p^{20T_{relu}}}{1 - p} \tag{9.1.16}$$

$$y_{earth0} = m\frac{1 - (1.02p)^{\frac{\ln\left(\frac{100}{m}\right)}{\ln(1.02)}}}{1 - (1.02p)}p^{\left(mK_{rel} + \frac{20}{\frac{x}{100}}\right)} - \frac{1 - p^{mK_{rel}}}{1 - p} \tag{9.1.17}$$

$$p = 0.95 \tag{9.1.18}$$

$$m = 4 \tag{9.1.19}$$

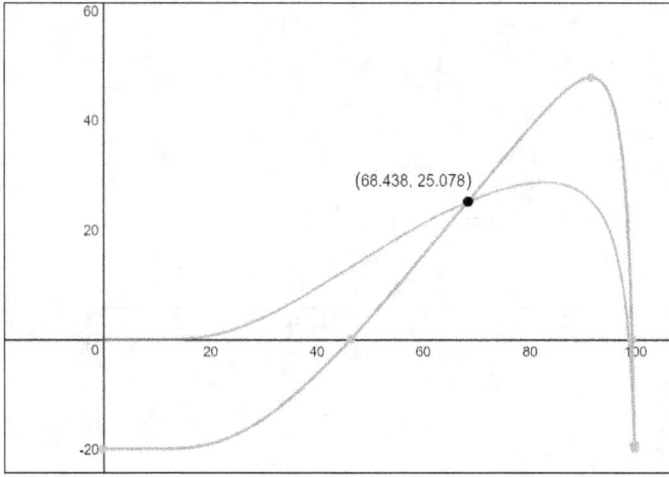

Figure 9.1.1: Earthbound vs. Shipbound perceived economic return

However, the most profitable speed of travel is not the crossover points; rather, it is the maximum value of the sum of both curves for earthbound and ship bound investors.

$$y_{ship} + y_{earth} \tag{9.1.20}$$

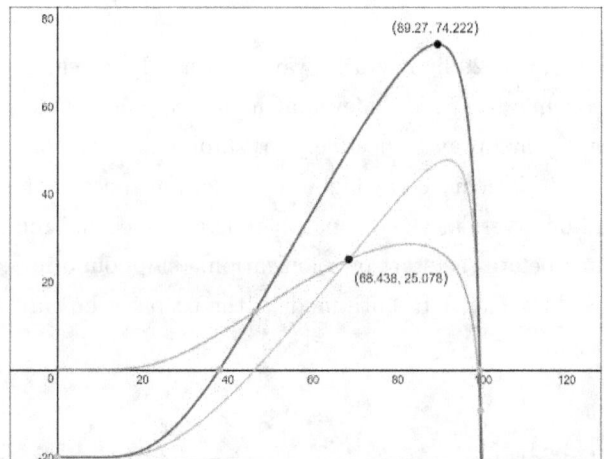

Figure 9.1.2: Composite perceived economic return and their maximum value

By setting every other variable constant except for the destination travel distance, one can also see the trend that economic wise, it is more profitable to send colonization fleet (both for the earth and shipbound investors) at a higher speed when the destination is farther away even with more time and investment opportunity foregone domestically.

$$y_{20} = \left(m \frac{1 - (ap)^{\frac{\ln\left(\frac{100}{m}\right)}{\ln(a)}}}{1 - (ap)} \right) p^{\left(\frac{mK_{rel}}{1} + \frac{20}{\left(\frac{x}{100}\right)}\right)} - \frac{1 - p^{\left(\frac{mK_{rel}}{1}\right)}}{1 - p} \tag{9.1.21}$$

$$y_{8} = \left(m \frac{1 - (ap)^{\frac{\ln\left(\frac{100}{m}\right)}{\ln(a)}}}{1 - (ap)} \right) p^{\left(\frac{mK_{rel}}{1} + \frac{8}{\frac{x}{100}}\right)} - \frac{1 - p^{\left(\frac{mK_{rel}}{1}\right)}}{1 - p} \tag{9.1.22}$$

$$y_3 = \left(m \frac{1 - (ap)^{\frac{\ln\left(\frac{100}{m}\right)}{\ln(a)}}}{1 - (ap)} \right) p^{\left(\frac{mK_{rel}}{1} + \frac{3}{100} \right)} - \frac{1 - p^{\left(\frac{mK_{rel}}{1} \right)}}{1 - p} \tag{9.1.23}$$

$$p = 0.9 \quad a = 1.08 \tag{9.1.24}$$

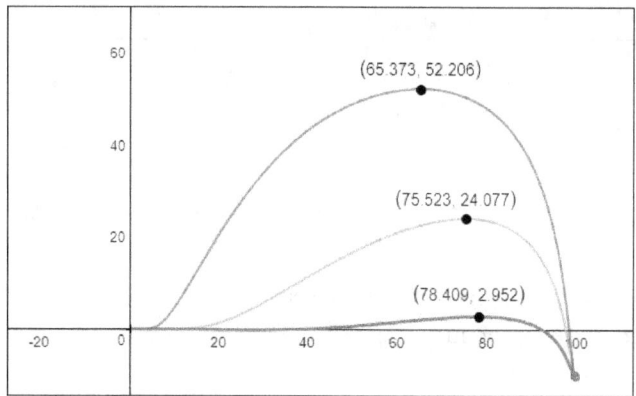

Figure 9.1.3: Earthbound investors' perceived economic return to a colonizing destination 3, 8, and 20 light years from earth

We model the cases for destination 3, 8, and 20 light years away respectively for earthbound investors, as the distance increases, the maximum economic return is achieved at higher colonization speed. We then model the cases for destination 3, 8, and 20 light years away respectively for shipbound investors. As the distance increases, the maximum economic return is achieved at higher colonization speed, but shipbound investors prefer higher speed than earthbound observers at the comparable distance because relativistic time dilation significantly shortens the waiting time before the start of colonization. (shipbound investors have more opportunity lost while they are aboard the ship.) In both models, the preparation time before the departure is short in comparison to the trip time.

$$y_{ship05} = m \frac{1 - (1.02p)^{\frac{\ln\left(\frac{100}{m}\right)}{\ln(1.02)}}}{1 - (1.02p)} p^{(mK_{rel} + 3T_{relu})} - \frac{1 - p^{mK_{rel}}}{1 - p} - \frac{1 - p^{3T_{relu}}}{1 - p} \tag{9.1.25}$$

$$y_{ship10} = m \frac{1 - (1.02p)^{\frac{\ln\left(\frac{100}{m}\right)}{\ln(1.02)}}}{1 - (1.02p)} p^{(mK_{rel} + 8T_{relu})} - \frac{1 - p^{mK_{rel}}}{1 - p} - \frac{1 - p^{8T_{relu}}}{1 - p} \tag{9.1.26}$$

$$y_{ship30} = m \frac{1 - (1.02p)^{\frac{\ln\left(\frac{100}{m}\right)}{\ln(1.02)}}}{1 - (1.02p)} p^{(mK_{rel} + 20T_{relu})} - \frac{1 - p^{mK_{rel}}}{1 - p} - \frac{1 - p^{20T_{relu}}}{1 - p} \tag{9.1.27}$$

$$p = 0.95 \tag{9.1.28}$$

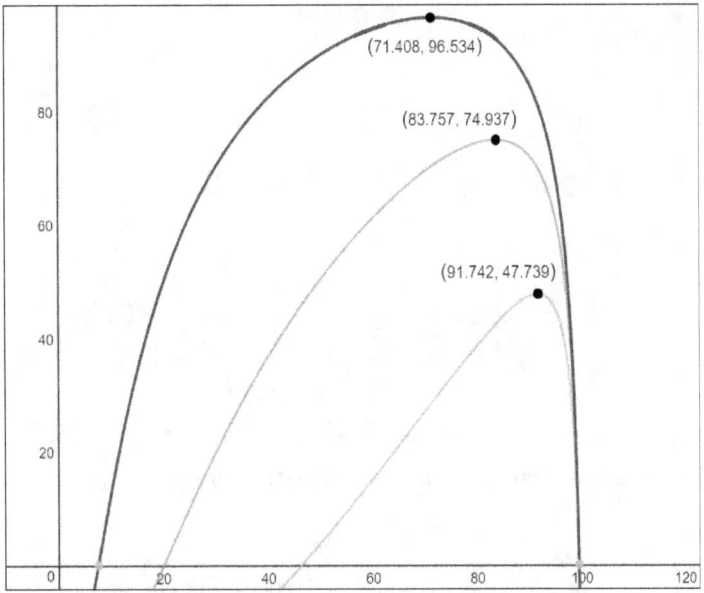

Figure 9.1.4: Shipbound investors' perceived economic return to a colonizing destination 5, 10, and 30 light years from earth

On the other hand, by setting every other variable constant except for the travel speed, one can also see the trend that economic wise, higher speed (up to 0.9c) renders long-distance colonization target more profitable. The additional time and investment opportunity foregone domestically at the energy and material preparation for the trip are justified. We modeled the speed at 0.1c, 0.5c, and 0.9c, the greatest distance for which a trip is profitable is 1.748 ly, 14.278 ly, and 17.538 ly respectively.

$$y_{earthdist0.1c} = \left(m \frac{1-(ap)^{\frac{\ln\left(\frac{100}{m}\right)}{\ln(a)}}}{1-(ap)} \right) p^{\left(\frac{m}{1\sqrt{1-\frac{10^2}{100^2}}} + \frac{x}{\left(\frac{10}{100}\right)} \right)} - \frac{1-p^{\left(\frac{m}{1\sqrt{1-\frac{10^2}{100^2}}} \right)}}{1-p} \tag{9.1.29}$$

$$y_{earthdist0.5c} = \left(m \frac{1-(ap)^{\frac{\ln\left(\frac{100}{m}\right)}{\ln(a)}}}{1-(ap)} \right) p^{\left(\frac{m}{1\sqrt{1-\frac{50^2}{100^2}}} + \frac{x}{\left(\frac{50}{100}\right)} \right)} - \frac{1-p^{\left(\frac{m}{1\sqrt{1-\frac{50^2}{100^2}}} \right)}}{1-p} \tag{9.1.30}$$

$$y_{earthdist0.9c} = \left(m \frac{1-(ap)^{\frac{\ln\left(\frac{100}{m}\right)}{\ln(a)}}}{1-(ap)} \right) p^{\left(\frac{m}{1\sqrt{1-\frac{90^2}{100^2}}} + \frac{x}{\left(\frac{90}{100}\right)} \right)} - \frac{1-p^{\left(\frac{m}{1\sqrt{1-\frac{90^2}{100^2}}} \right)}}{1-p} \tag{9.1.31}$$

$$p = 0.9 \quad a = 1.08 \tag{9.1.32}$$

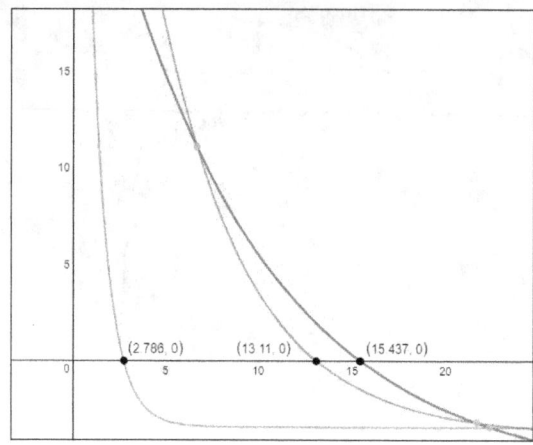

Figure 9.1.5: The greatest distances at which a colonizing destination provides a positive economic return for 0.1c, 0.5c, and 0.9c respectively

One can also fix the travel speed and destination distance and check for the economic return for different proportions of the population involved in investing. We modeled our results for targets at 10 ly and 20 ly away respectively and travel speed of 0.5c.

$$y_{rate10} = \left(x \frac{1 - (ap)^{\frac{\ln\left(\frac{100}{x}\right)}{\ln(a)}}}{1 - (ap)} \right) p^{\left(\frac{x K_{relu}(50)}{100} + \frac{10}{50}{100} \right)} - x \frac{1 - p^{\left(\frac{x K_{relu}(50)}{100} \right)}}{1 - p} \tag{9.1.33}$$

$$y_{rate20} = \left(x \frac{1 - (ap)^{\frac{\ln\left(\frac{100}{x}\right)}{\ln(a)}}}{1 - (ap)} \right) p^{\left(\frac{x K_{relu}(50)}{100} + \frac{20}{50}{100} \right)} - x \frac{1 - p^{\left(\frac{x K_{relu}(50)}{100} \right)}}{1 - p} \tag{9.1.34}$$

$$K_{relu}(x) = \frac{1}{\sqrt{1 - \frac{x^2}{100^2}}} - 1 \tag{9.1.35}$$

$$p = 0.9 \quad a = 1.08 \tag{9.1.36}$$

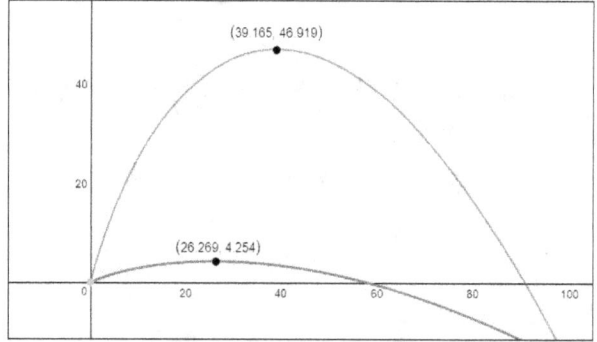

Figure 9.1.6: Optimal population participation rates for destination 10 light years away and 20 light years away

One finds that it is most profitable if 31.97% of the entire population participates in the colonization effort if the target distance is 10 ly away and 21% of the entire population participates in the colonization effort if the target distance is 20 ly away.

One can also vary the cost for the trip preparation, when the trip preparation cost is high, only a smaller portion of the entire population can afford such a trip to remain profitable (at the cost of the majority of the rest). We modeled our results for a target at 10 ly away, traveling speed of 0.5c, and the preparation cost proportional to the number of initial investors, and in the second case, the preparation cost is 100 times proportional to the number of initial investors.

$$y_{ratecost1} = \left(x \frac{1 - (ap)^{\frac{\ln\left(\frac{100}{x}\right)}{\ln(a)}}}{1 - (ap)} \right) p^{\left(\frac{xK_{rely}(50)}{100} + \frac{10}{\frac{50}{100}} \right)} - x \frac{1 - p^{\left(\frac{xK_{rely}(50)}{100} \right)}}{1 - p} \tag{9.1.37}$$

$$y_{ratecost100} = \left(x \frac{1 - (ap)^{\frac{\ln\left(\frac{100}{x}\right)}{\ln(a)}}}{1 - (ap)} \right) p^{\left(\frac{xK_{rely}(50)}{100} + \frac{10}{\frac{50}{100}} \right)} - 100x \frac{1 - p^{\left(\frac{xK_{rely}(50)}{100} \right)}}{1 - p} \tag{9.1.38}$$

$$p = 0.9 \quad a = 1.08 \tag{9.1.39}$$

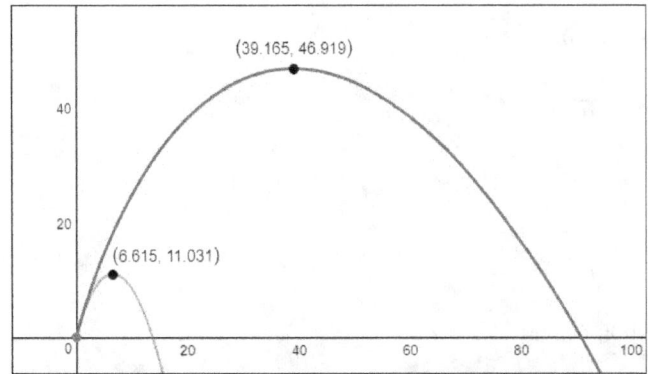

Figure 9.1.7: Optimal population participation rates for destination 10 light years away with preparation cost of 1 and 100

Having conceptualized the model, now we focus on applying the model for predictions based on reality. From the previous chapter, we have shown that out of all habitable planets within the Milky Way galaxy, only 41,930 planets [1] with ages between 5 Gyr and 4 Gyr whereas it is likely that a significant buildup of oxygen has commenced with the onset of photosynthetic bacteria. Based on the size of the galaxy, we able to determine that the nearest habitable planet be 1,777 light years in distance.

$$D_{nearest} = 2 \cdot \left(\frac{3}{4} \frac{(140,000)^2 \cdot 2,000}{41,930} \right)^{\frac{1}{3}} \tag{9.1.40}$$

$$D_{nearest} = 1,776.79566109 \text{ ly} \tag{9.1.41}$$

Then, one can set up realistic scenarios and plots for earth and ship bound investors and their final economic returns.

9.2 Earthbound Democracy

It can be shown that for earthbound investors comprising the entire population of earth (assuming a maximum carrying capacity of 60 billion a little higher than our lowest maximum carry capacity calculation in Chapter 6) investing into the colonization of the nearest habitable planet can only doubling their initial investment. We assume that a team of 1,000 people (each 70 kg on average and carrying necessary subsistence, recyclable

---

[1]This is from an earlier calculation, new results shown to be <1,000. Adjustment to the value of depreciation will be done, but the results remain similar

material 20 times their body mass) represents the entire earth will be sent to colonize the nearest habitable planet. Once the planet being colonized, the entire human race benefits from the investment in equal share. Due to the long distance involved, we found that the rate of subjective depreciation cannot be lowered than 0.9999999, that is, $p > 0.9999999$ in order for the trip to be profitable. This implies either earthbound investors have to have extremely long lifespan themselves, so they valued extremely long-term future prospects. More realistically, it can be interpreted as long-standing culture emphasizing on inter-generational commitment in colonization effort. As a result, the second term of $\text{R}_{earth} \sum_{k=1}^{\frac{\ln\left(\frac{P_{earth}}{m_g}\right)}{\ln(1.02)}} m\,(ap)^k$ can not be substituted by our special case $m_g \frac{1-(1.02p)^{\frac{\ln\left(\frac{P_{earth}}{m_g}\right)}{\ln(1.02)}}}{1-(1.02p)}$ because $1.02p > 1$. We simply using a software iteration to approximate this term and we obtained $3.09 \cdot 10^{12}$ Joules of energy.

$$R_{earth} = P_{earth}\left(\frac{1-p^{10^9}}{1-p}\right) p^{\frac{\ln\left(\frac{P_{earth}}{m_g}\right)}{\ln(1.02)}} + \sum_{k=1}^{\frac{\ln\left(\frac{P_{earth}}{m_g}\right)}{\ln(1.02)}} m\,(ap)^k \tag{9.2.1}$$

$$R_{earth} = 5.97 \cdot 10^{19}\,\text{J} \tag{9.2.2}$$

$$K_{rel} = \frac{1}{\sqrt{1-\frac{x^2}{100^2}}} - 1 \tag{9.2.3}$$

$$T_{rel} = \frac{d}{\frac{x}{100}}\sqrt{1-\frac{x^2}{100^2}} \tag{9.2.4}$$

$$p = 0.99999999999 \tag{9.2.5}$$

$$P_{earth} = 60 \cdot 10^9 \text{ people} \tag{9.2.6}$$

$$m_g = 1,000 \text{ people} \tag{9.2.7}$$

$$m = 70\text{kg} \cdot m_g \cdot 20 \tag{9.2.8}$$

$$d = 1,776 \text{ ly} \tag{9.2.9}$$

Whereas j represents the energy output per capita based on the total energy output of the World in 2008 divided by the total population of the world:

$$j = \frac{6.8 \cdot 10^{19}\,\text{J}}{6.5 \cdot 10^9 \text{ people}} \tag{9.2.10}$$

The final return is expressed as: (whereas $10^{10} \cdot j$ represents the maximum energy can be contributed to the space exploration in one year by the entire earth assuming the population ceiling is in the order of magnitude of 10 billion derived from Chapter 7)

$$Year_{threal} = \ln\left(\frac{j}{P_{earth}} \cdot R_{earth} \cdot p^{\left(\frac{mc^2 K_{rel}}{10^{10}j} + \frac{d}{\left(\frac{x}{100}\right)}\right)} - j_t \frac{1-p^{\left(\frac{mc^2 K_{rel}}{10^{10}j}\right)}}{1-p}\right) \tag{9.2.11}$$

9.3 Earthbound Investing Nearest Galaxy

We can extend the above scenario into intergalactic colonization. We assume that a team of 1,000 people (each 70 kg on average and carrying necessary subsistence, recyclable material 20 times their body mass) represents the entire earth will be sent to colonize the nearest galaxy with 41,930 habitable planets.[2] We assume the average separation distance of 11.85 million light years in distance (Andromeda galaxy is much closer than the typical distance between galaxies since it is on its course with a collision with the Milky Way). Due to the immense distance involved, we found that the rate of subjective depreciation cannot be lowered than 0.9999925 in order for the trip to be profitable. This indicates that a biological led species with a culture of an extremely strong commitment to inter-generational colonization is theoretically possible though highly unlikely.

$$y_{earthmilky} = \ln \left(\frac{j}{P_{earth}} \cdot R_{milky} \cdot p^{\left(\frac{mc^2 K_{rel}}{10^{10}j} + \frac{d}{\frac{g}{100}} \right)} - j_t \frac{1 - p^{\left(\frac{mc^2 K_{rel}}{10^{10}j} \right)}}{1 - p} \right) \tag{9.3.1}$$

$$R_{milky} = P_{milky} \left(\frac{1 - p^{10^9}}{1 - p} \right) p^{\frac{\ln\left(\frac{P_{milky}}{m_g}\right)}{\ln(1.02)}} + m_g \frac{1 - (1.02p)^{\frac{\ln\left(\frac{P_{milky}}{m_g}\right)}{\ln(1.02)}}}{1 - (1.02p)} \tag{9.3.2}$$

$$R_{milky} = 5.97 \cdot 10^{24} \, \text{J} \tag{9.3.3}$$

$$d = 11,850,000 \, \text{ly} \tag{9.3.4}$$

Whereas j_t represents the energy output per 1 earth based on the total energy output of the World in 2008 times 10, that is, assuming the population ceiling is in the order of magnitude of 10 billion derived from Chapter 7. Whereas P_{milky} represents the total population in the galaxy eventually achievable based on the number of habitable planets conducive to intelligent, multicellular life within the galaxy.

$$j_t = 10 \cdot 6.8 \times 10^{19} \, \text{J} \tag{9.3.5}$$

$$P_{milky} = 10^5 \cdot P_{earth} \tag{9.3.6}$$

$$p = 0.99999999999 \tag{9.3.7}$$

Most interestingly, the economic return for a single planet committed to intergalactic colonization is higher than interstellar colonization at every possible travel speed. This may feel counter-intuitive, given the amount of resources and energy a civilization has to forfeit in the near term in order to prepare for such a colonization. This is possible when the subjective rate of depreciation cost is minimized by the greater emphasis on the commitment to inter-generational colonization. Hence, in theory, we have shown that biological led industrial civilization can expand in the universe even if the decision making is based on the economic return of earthbound investors.

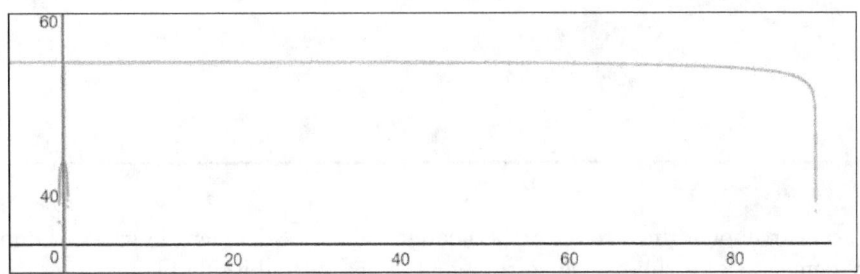

Figure 9.3.1: Democratic earthbound investors for nearest galaxy > democratic earthbound investors for the nearest habitable planet

We can also see that as the carrying mass of the ship changes, the maximum speed achievable for reaping positive economic return changes inversely. This is easily understood as greater mass requires greater kinetic

---

[2]This is from an earlier calculation, new results shown to be <1,000. Adjustment to the value of depreciation will be done, but the results remain similar

energy and resource preparation in the first place, whereas the cost of preparation outweigh the time lost waiting for its arrival to commence colonization.

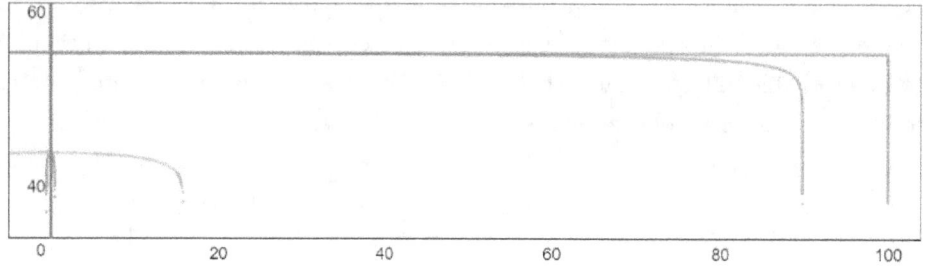

Figure 9.3.2: With smaller loads, positive economic returns justify traveling at higher speed

9.4 Galaxy Bound Investing Nearest Galaxy

As a logical extension to the previous case, one can also model the case where the entire colonized galaxy is ready to colonize the next nearest galaxy. Under such a scenario, the entire galaxy's energy and material preparation can be utilized to send a group of selected 1,000 explorers which will colonize another 41,930 habitable planets [3] of the nearest galaxy, and the return is equally shared among the galaxy empire. We also assumed that 10 years' total energy productions (10^{10} people per planet·10^5 planets·10 years) from the entire galaxy is used to prepare for such a migration.

$$y_{milk2milk} = \ln\left(\frac{j}{P_{milky}} \cdot R_{milky} \cdot p^{\left(\frac{mc^2 K_{rel}}{10^{16} j} + \frac{d}{100}\right)} - 10^5 \cdot j_t \frac{1 - p^{\left(\frac{mc^2 K_{rel}}{10^{16} j}\right)}}{1 - p} \right) \tag{9.4.1}$$

$$p = 0.999999999999$$
$$d = 11,850,000 \, \text{ly}$$

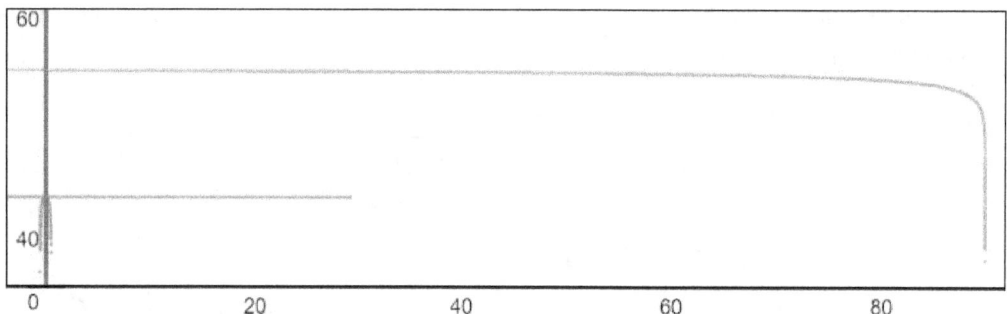

Figure 9.4.1: Democratic earthbound investors for the nearest galaxy > democratic galaxy bound investors for the nearest galaxy > democratic earthbound investors for the nearest habitable planet

The graph shows that it is less profitable for the entire galaxy to invest to the nearest galaxy and equally share the return, but it is more profitable than a single planet based civilization to invest to the nearest habitable planet.

---

[3]This is from an earlier calculation, new results shown to be <1,000. Adjustment to the value of depreciation will be done, but the results remain similar

9.5 Earthbound Ruling Class

Earthbound investors comprising a small population (assuming 1,000 individuals) investing into the colonization of the nearest habitable planet can yield much more return on their initial investment. We assume that 1,000 explorers (each 70 kg on average and carrying necessary subsistence, recyclable material 20 times their body mass) represent this group of 1,000 individuals will be sent to colonize the nearest habitable planet. Once the planet being colonized, only this group of 1,000 individuals investing will benefit from the investment in an equal share. We assume that this group of individuals have an extreme authoritative control over the populace in general and is able to raise the resources in preparation by taxing the entire population for given amount of years. As a result, the economic incentives for this group of people to start such colonization initiative will be significantly higher than as the entire population of earth. Despite much greater promised return on investment, the waiting time is still considerably significant. As a result, we found that the rate of subjective depreciation cannot be lowered than 0.996773 in order for the trip to be profitable. This implies that this group of people as the ruling class must still maintain its commitment to inter-generational colonization projects as well as its authoritative control over the population. Comparing the plots with the earlier democratic case where the entire population receives an equal share of return, the greatest economic return occurs at the higher speed. This is achievable because a small group of individuals' cost of preparation from their own perspective is significantly lower compares to their final return (at the expense and the sacrifice of the general populace with or without complaints), so they could spend significantly longer time period at energy gathering for the preparation phase. Hence, it is able to achieve a higher expansionary speed. The graph shows that a group of 1,000 people taxed or borrowed general populace of earth in terms of energy and material resource at 10 times of the current world population and the maximum economic return for different fractions of the speed of light for migration (black curve $y_{earthlow}$). Furthermore, a second curve $y_{earthhigh}$ represents the case of taxing the general populace assuming it stays at the same level of population ceiling but a higher level of per capita energy output and the economic return for different speeds of light. The ultimate per capita achievable as biologically led industrial civilization can be computed as the follows. We have shown that the ultimate barrier to economic growth is the accumulation of waste heat which reaches the threshold of catastrophe in 400 years from now assuming an annual growth rate of 1.5% (See Chapter 7). Since we already shown that the population ceiling at 10 times of our current population, therefore, the GDP per capita of the world can increase by 38.58 folds. In other words, the GDP per capita (assuming current GDP (PPP) per capita is \$15,800) can reach at most \$609,564 at the current dollar prices per year with a population base of 60 billion.

$$\frac{(1.015)^{400}}{10} = 38.58 \tag{9.5.1}$$

$$y_{earthhigh} = \ln\left(\frac{j}{m_g} \cdot R_{earth} \cdot p^{\left(\frac{mc^2 K_{rel}}{38.58 j_t} + \frac{d}{100}\right)} - m_g \cdot j \frac{1 - p^{\left(\frac{mc^2 K_{rel}}{38.58 j_t}\right)}}{1 - p}\right) \tag{9.5.2}$$

$$y_{earthlow} = \ln\left(\frac{j}{m_g} \cdot R_{earth} \cdot p^{\left(\frac{mc^2 K_{rel}}{10^{10} j} + \frac{d}{100}\right)} - m_g \cdot j \frac{1 - p^{\left(\frac{mc^2 K_{rel}}{10^{10} j}\right)}}{1 - p}\right) \tag{9.5.3}$$

$$p = 0.9978 \tag{9.5.4}$$

$$d = 1,776 \, \text{ly} \tag{9.5.5}$$

$$j_t = 10 \cdot 6.8 \times 10^{19} \text{J} \tag{9.5.6}$$

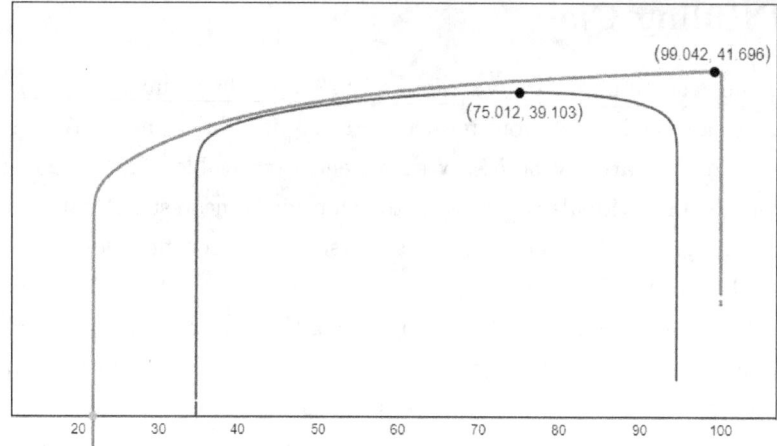

Figure 9.5.1: Higher GDP per capita renders near light speed travel economically profitable for earthbound ruling class

It can be seen that higher GDP (PPP) per capita (or otherwise conceptually equivalent as more energy available for each individual at their disposal) and its preparation during the fuel acquisition phase allows the migration to infinitely approaching the speed of light that would be otherwise unprofitable under low GDP per capita scenarios. Therefore, any ruler of the earth's intention to maximize GDP per capita can be justified not by the will of the people but by an ambition of space colonization alone.

9.6 Shipbound with Energy Gathering Case

Having discussed earthbound investors, we can now move onto the ship bound investors. If the ship-bound investors participate in the original energy gathering, then their subjective rate of value depreciation is comparable to a small group of earthbound investor (The earthbound ruling class case) and must be prepared with multi-generational commitments. It is still somewhat higher than earthbound ruling class case because relativistic time dilation made their trip to destination shorter.

$$y_{shiphigh} = \ln\left(\frac{j}{m_g} \cdot R_{earth} \cdot p^{\left(\frac{mc^2 K_{rel}}{38.58 j_t} + T_{rel}\right)} - m_g \cdot j \frac{1 - p^{\left(\frac{mc^2 K_{rel}}{38.58 j_t}\right)}}{1 - p} - m_g \cdot j \frac{1 - p^{T_{rel}}}{1 - p}\right) \tag{9.6.1}$$

$$y_{shiplow} = \ln\left(\frac{j}{m_g} \cdot R_{earth} \cdot p^{\left(\frac{mc^2 K_{rel}}{10^{10} j} + T_{rel}\right)} - m_g \cdot j \frac{1 - p^{\left(\frac{mc^2 K_{rel}}{10^{10} j}\right)}}{1 - p} - \left(m_g \cdot j \frac{1 - p^{T_{rel}}}{1 - p}\right)\right) \tag{9.6.2}$$

$$p = 0.998 \tag{9.6.3}$$

The graph shows that a group of 1,000 individuals taxed or borrowed the general populace of earth in terms of energy and material resource and the maximum economic return for different fractions of the speed of light for migration (green curve $y_{shiplow}$). Furthermore, a second curve $y_{shiphigh}$ represents the case of taxing the general populace assuming it stays at the same level of population ceiling but a higher level of per capita energy output and the economic return for different speeds of light. It can be seen that high GDP per capita and its preparation during the fuel acquisition phase allows the migration to infinitely approaching the speed of light that would be otherwise unprofitable under short preparation scenarios.

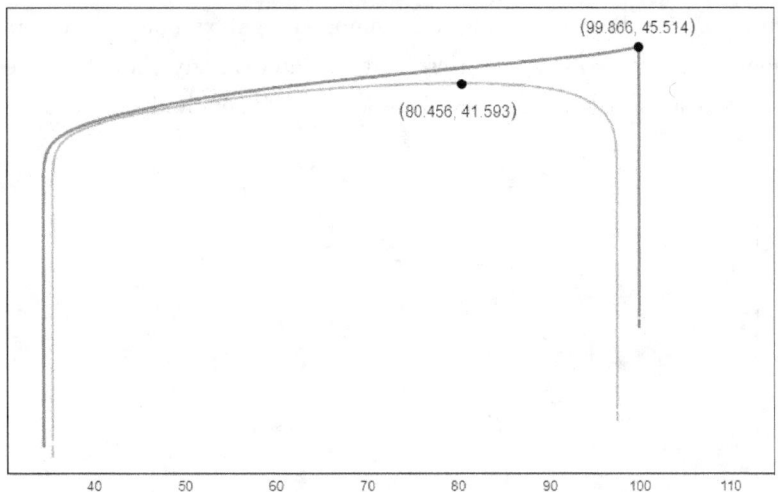

Figure 9.6.1: Higher GDP per capita renders near light speed travel economically profitable for shipbound ruling class

9.7 Shipbound as Lottery Winners Case

If the final explorers were comprising individuals picked at random from the general populace every few generations as the energy gathering for colonization preparation completes. Then, the subjective rate of depreciation can be much lower and is only limited by the travel time and travel speed (the cost of time lost in getting to destination that can be used to invest on earth)

$$y_{shiplottery} = \ln\left(\frac{j}{m_g} \cdot R_{earth} \cdot p^{T_{rel}} - m_g \cdot j \frac{1 - p^{T_{rel}}}{1 - p}\right) \tag{9.7.1}$$

$$p = 0.998 \tag{9.7.2}$$

and it can be shown that its subjective economic return is higher than the Ship bound with Energy and the Earthbound Ruling class cases when $p = 0.998$ and at speed $> 0.3684c$.

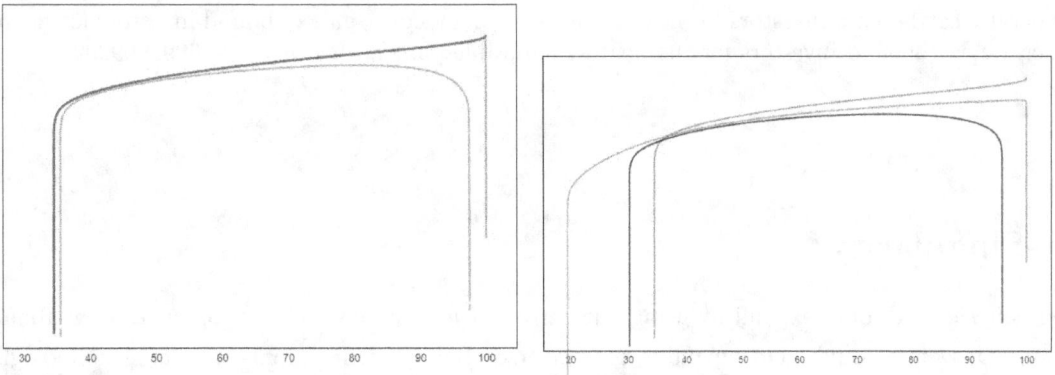

Figure 9.7.1: Shipbound lottery winners' perceived economic return (blue curve) is strictly higher than both shipbound and earthbound ruling class cases

Comparing all these scenarios and one finds that the economic return in the order of most profitable to the least as the follows: shipbound lottery winners > shipbound ruling class investors ≈ earthbound ruling class investors

> democratic earthbound investors for the nearest galaxy > democratic galaxy bound investors for the nearest galaxy > democratic earthbound investors for the nearest planet. Interestingly enough, the economic return through the democratic process results in both extremes, the lottery winners and the democratic earthbound investors.

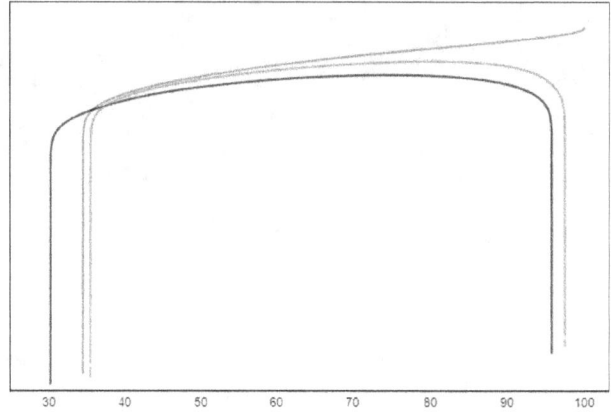

Figure 9.7.2: Shipbound lottery winners > shipbound ruling class investors > earthbound ruling class investors when $p = 0.9987$ and at the current GDP per capita

Democratic Earthbound investors investing into the nearest habitable planet case is not profitable unless its subjective depreciation value reaches $(0.999999999999 \le p < 1)$. However, at such value of p all three previous cases converge toward the same value on the graph shown below.

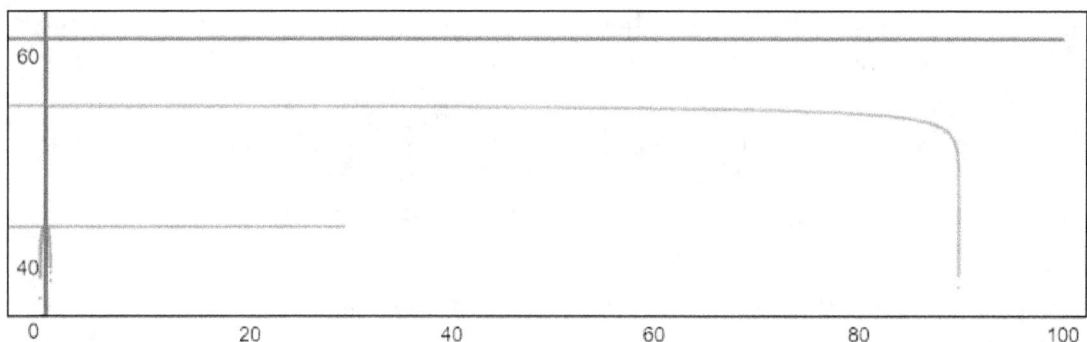

Figure 9.7.3: (Shipbound lottery winners = shipbound ruling class investors = earthbound ruling class investors) > democratic earthbound investors for nearest galaxy > democratic galaxy bound investors for nearest galaxy > democratic earthbound investors for the nearest habitable planet when $p = 0.999999999999$

9.8 Post-Singularity

We just showed how slow we could expand by using our current level of industrial development. How about a post-biological one? How fast would them expand in the universe? In a few millenniums, assuming technological singularity does not occur in this century, human should have full control over their genes and biological development. It implies that Homo sapiens may have artificially diverged into different species each occupying different niches where some can be much smaller than our current size and some others much larger. If we are artificially evolving toward smaller size and at the same time able to convert energy more efficiently directly into our body discarding the digestive system, we can build a much smaller spaceship and much less fuel to travel to a habitable planet. At the same time, if we are able to achieve significantly longer lifespan, our p will also approach toward 1. As a result, the post-biological creatures calculated final return value increasingly approaches the theoretical value. If it is foreseeable that Homo sapiens to genetically modify ourselves; then, there is no reason to doubt extra-terrestrials would not do the same. On the other end of the spectrum, a robotic led industrial civilization as predicted by the singularity led to even more extreme toward the theoretical limit. Artificial intelligence, which is immortal even in the face of accidents because of its own copied backup of itself, will use the following equation when calculating return on their colonization investment.

In a post-singularity society, assuming the mass required to carry intelligence and information processing shrinks significantly so that it takes significantly less time at energy and material gathering phase, then the economic return is certainly much higher than biological led species, and the most optimal strategy is cruising near the speed of light. We can illustrate a possible scenario. We denote the mass requirement is simply less than 1 out of 10^{10} of current biological human form, or about the mass of 100 cells in the human body.[120] The GDP/Energy per capita requirements for sustaining itself becomes $\frac{1}{10^{10}}$ th of current human level per AI (so the opportunity cost at ship preparation and migration phase becomes negligible), yet each AI is capable of significantly expand its energy per capita by transforming into other gigantic forms on earth and generate energy at the order of magnitude of 10^4 of the current GDP per capita on earth for the trip preparation. The return on the colonized destination is the total biological population limit times current GDP/Energy per capita for 10 billion years. Finally, one assume for immortalized AI based life form, $p=0.997$.

$$y_{shipAI} = \ln\left(j \cdot R_{earth} \cdot p^{\left(\frac{10^{-10}mc^2 K_{rel}}{10^4 j} + T_{rel}\right)} - \frac{j}{10^{10}}\frac{1 - p^{\left(\frac{10^{-10}mc^2 K_{rel}}{10^4 j}\right)}}{1 - p} - \frac{j}{10^{10}}\frac{1 - p^{T_{rel}}}{1 - p} \right) \tag{9.8.1}$$

$$y_{earthAI} = \ln\left(j \cdot R_{earth} \cdot p^{\left(\frac{10^{-10}mc^2 K_{rel}}{10^4 j} + \frac{d}{100}\right)} - \frac{j}{10^{10}}\frac{1 - p^{\left(\frac{10^{-10}mc^2 K_{rel}}{10^4 j}\right)}}{1 - p} \right) \tag{9.8.2}$$

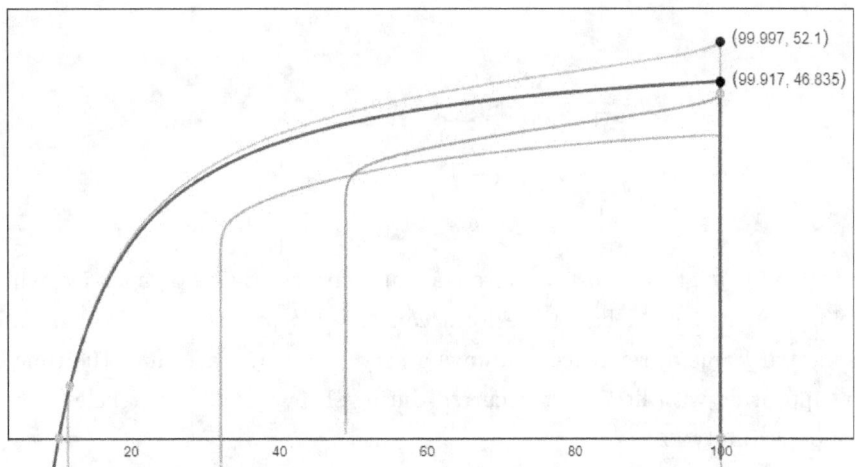

Figure 9.8.1: Shipbound AI > Earthbound AI > Shipbound lottery winners > Shipbound ruling class investors > Earthbound ruling class investors

This can be compared in the graph above, whereas both the earthbound AI investor and shipbound AI investors' economic return tops any biological species led scenarios.

All of these calculations showed that it is lucrative for AI to colonize the universe with a very low cost and a high reward for an expansion. These sets of equations can be further expanded to accommodate the expansion of the universe. The universe's space-time is stretching with speed relative to the observer at 74 $\frac{km}{sec}$ for every million parsecs. So for long journeys committed, the expansion of the universe will add additional travel time before one can reach its destination, and any destination becomes unreachable beyond the cosmic event horizon which is moving away from us as fast or faster than the speed of light. As a result, there is an upper limit as to how far we could travel even at the speed of light because destinations beyond the horizon will remain forever unreachable.

In order to account for the cosmic expansion rate and the Hubble constant, cosmological distance has to be rescaled into the real distance:

$$T_{rel} = \frac{E\left(d, v\right)}{v}\sqrt{1 - \frac{v^2}{c^2}} \tag{9.8.3}$$

whereas E(d, v) is a function that handles the conversion for a cosmological distance of d for any given constant migration speed of v. The derivation of E(d, v) is later defined in Chapter 9. As a result, the generalized formula becomes:

$$E_{arth} = \frac{\epsilon}{m}\left[\left(\sum_{k=0}^{10^9}M\left(p\right)^k\right)p^{\frac{\ln\left(\frac{M}{m}\right)}{\ln a}} + \sum_{k=0}^{\frac{\ln\left(\frac{M}{m}\right)}{\ln a}}m\left(ap\right)^k\ p^{\left(\frac{m\cdot K_{rel}}{\epsilon} + \frac{E(d,v)}{v}\right)}\right]$$

$$- m\epsilon\left[\sum_{k=0}^{\frac{m\cdot K_{rel}}{\epsilon}}p^k\right] \tag{9.8.4}$$

when $ap < 1$

$$E_{arth} = \frac{\epsilon}{m}\left[M\frac{1 - p^{10^9}}{1 - p}p^{\frac{\ln\left(\frac{M}{m}\right)}{\ln a}} + m\frac{1 - ap^{\frac{\ln\left(\frac{M}{m}\right)}{\ln a}}}{1 - ap}\right]p^{\left(\frac{m\cdot K_{rel}}{\epsilon} + \frac{E(d,v)}{v}\right)} - m\epsilon\frac{1 - p^{\frac{m\cdot K_{rel}}{\epsilon}}}{1 - p} \tag{9.8.5}$$

when $ap \geq 1$

$$m\frac{1 - \left(ap\right)^n}{1 - ap} < \sum_{k=0}^{n}m\left(ap\right)^k \tag{9.8.6}$$

$$M\frac{1 - p^n}{1 - p} < \sum_{k=0}^{n}Mp^k \tag{9.8.7}$$

However, one should not ignore the migration cost term from a stationary observer's perspective, which describes the time spent on traveling. Although AI based life form can travel near the speed of light and the time experienced under their inertial frame of reference is always a very very tiny fraction of the time experienced by a stationary observer, appearing with little cost occurred during their travel. Nevertheless, the time spent on traveling is the cost of opportunity lost.

9.9   Expansion Speed from Outsider's Perspective

If an extraterrestrial civilization expands near the speed of light, from their perspective, almost no time has elapsed, but from an observer on earth who is stationary relative to them, their speed of expansion is just the speed of light at the best. In order to illustrate how slow such expansion appears to be one needs to resort to calculations. One can comprehend the speed of economic growth occurred since the Industrial Revolution. If we use the world population growth as a measure of growth in energy consumption and the rate of so defined social progress, then from AD 1900 the world population of 1.65 Billion increased to 6.127 Billion in AD 2000, it can be shown that the rate of growth is 1.32 percent per year. If one steps back prior to the Industrial Revolution and computes the period of Age of Exploration from year AD 1600 to AD 1750 and the population grows from 580 million to 791 million, we find that the annual growth rate is 0.207 percent. The annual growth rate from classical Greece from 1000 B.C up to AD 1600 is 0.0943 percent. During this period the world population grows from 50 million to 580 million resulting in 10 fold increase in 2,600 years. From the onset of the full transition from hunter-gatherers to agricultural societies at 5000 B.C with a population of 5 million to that of city-states at 1000 B.C. of 50 million, the annual growth rate is 0.0576 percent. Finally, we find the annual rate of growth during the hunter-gatherer period of Homo sapiens' migration out of Africa from 70,000 BC following the Mount Toba bottleneck with 15,000 individuals to the transition of agricultural society at 5 million to be 0.00931043 percent. Based on this results, we can roughly quantify the rate of progress human society is achieving in each time period. We can see that the transition from hunter-gatherer to that of the agricultural is a major one, in which the annual growth rate jumped 6.187 times. During the following classical and the Middle Ages, the annual growth only increase by 1.63 times. Agricultural lands and farming tools are continually improving but the growth are not significant. During the following Age of Exploration, the annual rate of growth increased by another 2.195 times thanks to Columbian exchange, which enabled crops from America to be cultivated around the world, increasing the varieties of food domestication. However, only the Industrial Revolution brought the annual rate of growth to another 6.377 fold increase, comparable and surpassed the magnitude of change observed from hunter-gatherers to that of the agricultural society. Upon the Industrial paradigm, new modes of production such as Communism and Capitalism both claims to increase the efficiency and annual growth rate, which may not necessarily reflect from population growth, but rather gave the rise of the standards of living per capita. This is especially evident from the economic growth of China, which has kept to grow at the pace of 8 percent per year for the last 30 years, another 6 folds increase from the average speed of growth of Industrial Era. By now, we have a clear appreciation of the rate of change at each period of human society's progress, we shall resort to the calculation for the speed of galaxy colonization, if one expands very close to the speed of light, one can traverse the entire Milky Way galaxy in 140,000 years. If every star is harnessed for its energy just as our sun by constructing Dyson spheres and the total stellar mass of the Milky Way is estimated to be $5.515 \cdot 10^{10}$ solar masses, so we start with one solar mass, our sun, ended up harnessing $5.515 \cdot 10^{10}$ solar mass of energy in 140,000 years. We found that the annual rate of growth from a stationary observer's perspective to be 0.01767 percent per year, whereas a majority of the time is the cost of expanding from one star system to the next. This rate is 1.898 times faster than Homo Sapiens' speed of "progress" during our hunter-gatherer period. If our future technology can only reach a fraction of the speed of light, then, the speed of expansion can be slower than our hunter-gatherer period. Furthermore, once Milky Way galaxy is colonized, one has to traverse another 2.13 million light years to reach our nearest neighbor, the Andromeda Galaxy, and 2.64 million years to reach the Triangulum Galaxy. If we takes into account of the entire cosmic expansion, then amortized annual rate of growth is merely 0.0010719 percent. This shows that even expanding near the speed of light the speed of cosmic expanders appears to be just 11.5% of the annual growth rate of hunter-gatherers. Nevertheless, this speed is significantly faster than that of biological evolution. For the case of *Australopithecus Afarensis'* emergence at 3.9 million years ago to that of the emergence of Neanderthal 250K years ago, the cranial capacity increased from 405 cc to 1600 cc, corresponding to an annual increase rate of 0.00003762 percent, or about 3.51 percent of the upper growth bound of a cosmic expander.

Then, we can conclude that from the expander's perspective, almost no time is lost in gaining energy usage and information utilization from one's home planet to this civilization's expansionary limit bounded by their expansionary industrial extra-terrestrial neighbors. This is an almost infinite increase in terms of economic size in almost no time from the perspective of expanding civilization near the speed of light. Therefore, they have the full economic incentive to expand. Though they have to know that such expansion is once only, and they have to resort to sustainability immediately afterward. However, from earth observers' perspective, such expansion is slow, as we have already shown how slow it is. It is so slow that it appears that such civilization's economic growth rate is as slow as the hunter-gatherers. Therefore, any expanding alien civilization expanding even at the speed of light will appear extremely slow from stationary observer's perspective.

To see this further lets set up another example. Assuming a future robotic civilization arising from the solar system decided to colonize a distant galaxy 1 billion light-years away and travels at 99.999% speed of light and it takes about 1 year of time in their reference frame. From the stationary observer's point of view, 1 billion years have passed since their started their journey. During this 1 billion years, an enormous amount of energy from the stars have been converted from usable to the un-usable energy given by the second law of thermodynamics. If the destination is even further out at 10 billion light years, lots of stars have been born, burned, and died and the energy can be gathered but lost is a foregone opportunity cost no one can afford to lose. As the rate of star formation slows down into the future, every GFK stars and its energy released is precious. (Of course, such scenario is naturally applicable to biological led sustainable industrial civilization but it is not practical to do so. The energy acquisition required even to the Andromeda galaxy lying 2.3 million light-years away is so tremendous that even a lottery based system activated on earth will take thousands of years to gather the energy requirements for a team of 1000, as our previous discussion has shown.)

9.10 Worm Hole

As a result, an extreme distant journey by AI based life form would be deemed wasteful even if it can be done easily. Furthermore, as we have shown earlier, the cosmic evolutionary rate, albeit slow, comparable to the rate of change contributed to the rise of Homo sapiens and their civilization. In just half billion years of time, every terrestrial planet habitable to life, with their increasing biological diversity, will give rise to intelligent creatures comparable to human and their future robotic form. An earth-based robotic life form spent billion years travel to claim their territory may encounter the resistance and defeat from the rising robotic civilizations at those distant locations from earth.

This seems to imply that AI-based life form or any lightweight matter based life form of the future will unlikely to attempt any long distance journey. In fact, their descendants may scatter vast reaches of the universe, but communications and trades only occur at the very local level, whereas the time lost while spent on traveling is bearable. As a result, no galactic empire is possible and only islands of isolation is possible even if they all descended from the same original ancestor.

Fortunately, the future seems to be much more interesting, and the scenario above is unlikely to hold. Though the expanding cosmic civilization may start their empire by sending back and forth spaceships between their controlled planets, it will sooner or later adopt and construct wormholes to expedite their travel and journey. Although wormhole is still a theoretical construct and in recent years various papers have pointed out that less and less energy (in fact negative energy) is required to maintain the operations of such tunnels. In the very beginning, calculations have shown that the energy requirements for the construction of Albucurrier drive and wormholes require the equivalent energy of the entire observable universe. Recent papers have suggested minimal amount of energy can maintain a wormhole about the size of a human.[113] For post-biological intelligence which can alter its size and energy requirements for travel, even smaller energy is required to maintain microscopically sized wormholes. More recently, negative energy has also been obtained in experiments[80]; therefore, the construction and low-cost maintenance of wormhole seems to be possible.

The construction of wormholes does not violate the theory that arising extraterrestrial industrial civilizations require time to reach us limited by the speed of light. Although a completed wormhole allows travel time less than the speed of light traveling on a flat space-time fabric, the construction of such wormhole itself cannot exceed the speed of light. It is still under research as to how short a wormhole can be required to connect any arbitrary point in space and time. As more and more progress is done in the future on theoretical physics and practical engineering, it can be shown that the wormhole constructed can be increasingly shorter in length yet able to connect to more distant locations. In our model, we simply assume that the wormhole is able to take a traveler to a different destination with a cost in time and maintenance C and where C < the cost spend travel near the speed of light on a flat space-time fabric. Under the most extreme assumption, we can have C = 0;

that is, it takes no time to traverse to distant points in the universe. If we take these assumptions, we can derive the following cost and benefit analysis using cost amortization for n trips between two points of cosmological distances.

Without wormhole for earthbound investors:

$$\lim_{n \to \infty} \left(\frac{1}{n} \sum_{k=0}^{n} \frac{\epsilon}{m} \left[M \frac{1 - p^{10^9}}{1 - p} p^{\frac{\ln\left(\frac{M}{m}\right)}{\ln a}} + \sum_{k=0}^{\frac{\ln\left(\frac{M}{m}\right)}{\ln a}} m\left(ap\right)^k \right] p^{\left(\frac{m \cdot K_{rel}}{\epsilon} + \frac{E(d,v)}{v} \right)} \right.$$

$$\left. - m\epsilon \frac{1 - p^{\frac{m \cdot K_{rel}}{\epsilon}}}{1 - p} \right) \quad (9.10.1)$$

$$= \frac{\epsilon}{m} \left[M \frac{1 - p^{10^9}}{1 - p} p^{\frac{\ln\left(\frac{M}{m}\right)}{\ln a}} + \sum_{k=0}^{\frac{\ln\left(\frac{M}{m}\right)}{\ln a}} m\left(ap\right)^k \right] p^{\left(\frac{m \cdot K_{rel}}{\epsilon} + \frac{E(d,v)}{v} \right)} - m\epsilon \frac{1 - p^{\frac{m \cdot K_{rel}}{\epsilon}}}{1 - p} \quad (9.10.2)$$

With wormhole for earthbound investors:

$$\lim_{n \to \infty} \left(\frac{1}{n} \sum_{k=0}^{n} \frac{\epsilon}{m} \left[M \frac{1 - p^{10^9}}{1 - p} p^{\frac{\ln\left(\frac{M}{m}\right)}{\ln a}} + \sum_{k=0}^{\frac{\ln\left(\frac{M}{m}\right)}{\ln a}} m\left(ap\right)^k \right] p^{\left(\frac{m \cdot K_{rel}}{\epsilon} + \frac{E(d,v)}{v} \right)} \right.$$

$$\left. - m\epsilon \frac{1 - p^{\frac{m \cdot K_{rel}}{\epsilon}}}{1 - p} \right) \quad (9.10.3)$$

$$= \frac{\epsilon}{m} \left[M \frac{1 - p^{10^9}}{1 - p} p^{\frac{\ln\left(\frac{M}{m}\right)}{\ln a}} + \sum_{k=0}^{\frac{\ln\left(\frac{M}{m}\right)}{\ln a}} m\left(ap\right)^k \right] \quad (9.10.4)$$

and economic return with wormhole for earthbound investors > economic return without wormhole for earthbound investors

$$\frac{\epsilon}{m} \left[M \frac{1 - p^{10^9}}{1 - p} p^{\frac{\ln\left(\frac{M}{m}\right)}{\ln a}} + \sum_{k=0}^{\frac{\ln\left(\frac{M}{m}\right)}{\ln a}} m\left(ap\right)^k \right] \geq$$

$$\frac{\epsilon}{m} \left[M \frac{1 - p^{10^9}}{1 - p} p^{\frac{\ln\left(\frac{M}{m}\right)}{\ln a}} + \sum_{k=0}^{\frac{\ln\left(\frac{M}{m}\right)}{\ln a}} m\left(ap\right)^k \right] p^{\left(\frac{m \cdot K_{rel}}{\epsilon} + \frac{E(d,v)}{v} \right)} - m\epsilon \frac{1 - p^{\frac{m \cdot K_{rel}}{\epsilon}}}{1 - p} \quad (9.10.5)$$

Without wormhole for shipbound investors:

509

$$\lim_{n\to\infty}\frac{1}{n}\left(\sum_{k=0}^{n}\frac{\epsilon}{m}\left[M\frac{1-p^{10^9}}{1-p}p^{\frac{\ln\left(\frac{M}{m}\right)}{\ln a}}+\sum_{k=0}^{\frac{\ln\left(\frac{M}{m}\right)}{\ln a}}m\left(ap\right)^k\right]p^{\left(\frac{m\cdot K_{rel}}{\epsilon}+T_{rel}\right)}\right.$$

$$\left.-m\epsilon\frac{1-p^{\frac{m\cdot K_{rel}}{\epsilon}}}{1-p}-m\epsilon\frac{1-p^{T_{rel}}}{1-p}\right)\qquad(9.10.6)$$

$$=\frac{\epsilon}{m}\left[M\frac{1-p^{10^9}}{1-p}p^{\frac{\ln\left(\frac{M}{m}\right)}{\ln a}}+\sum_{k=0}^{\frac{\ln\left(\frac{M}{m}\right)}{\ln a}}m\left(ap\right)^k\right]p^{\left(\frac{m\cdot K_{rel}}{\epsilon}+T_{rel}\right)}$$

$$-m\epsilon\frac{1-p^{\frac{m\cdot K_{rel}}{\epsilon}}}{1-p}-m\epsilon\frac{1-p^{T_{rel}}}{1-p}\qquad(9.10.7)$$

With wormhole for shipbound investors:

$$\lim_{n\to\infty}\frac{1}{n}\left(\sum_{k=0}^{n}\frac{\epsilon}{m}\left[M\frac{1-p^{10^9}}{1-p}p^{\frac{\ln\left(\frac{M}{m}\right)}{\ln a}}+\sum_{k=0}^{\frac{\ln\left(\frac{M}{m}\right)}{\ln a}}m\left(ap\right)^k\right]p^{\left(\frac{m\cdot K_{rel}}{\epsilon}+T_{rel}\right)}\right.$$

$$\left.-m\epsilon\frac{1-p^{\frac{m\cdot K_{rel}}{\epsilon}}}{1-p}-m\epsilon\frac{1-p^{T_{rel}}}{1-p}\right)\qquad(9.10.8)$$

$$=\frac{\epsilon}{m}\left[M\frac{1-p^{10^9}}{1-p}p^{\frac{\ln\left(\frac{M}{m}\right)}{\ln a}}+\sum_{k=0}^{\frac{\ln\left(\frac{M}{m}\right)}{\ln a}}m\left(ap\right)^k\right]\qquad(9.10.9)$$

and economic return with wormhole for shipbound investors > economic return without wormhole for shipbound investors

$$\frac{\epsilon}{m}\left[M\frac{1-p^{10^9}}{1-p}p^{\frac{\ln\left(\frac{M}{m}\right)}{\ln a}}+\sum_{k=0}^{\frac{\ln\left(\frac{M}{m}\right)}{\ln a}}m\left(ap\right)^k\right]\geq$$

$$\frac{\epsilon}{m}\left[M\frac{1-p^{10^9}}{1-p}p^{\frac{\ln\left(\frac{M}{m}\right)}{\ln a}}+\sum_{k=0}^{\frac{\ln\left(\frac{M}{m}\right)}{\ln a}}m\left(ap\right)^k\right]p^{\left(\frac{m\cdot K_{rel}}{\epsilon}+T_{rel}\right)}$$

$$-m\epsilon\frac{1-p^{\frac{m\cdot K_{rel}}{\epsilon}}}{1-p}-m\epsilon\frac{1-p^{T_{rel}}}{1-p}\qquad(9.10.10)$$

and one can conclude, that within a wormhole network, economic return within wormholes for shipbound investors = economic return with wormhole for earthbound investors.

Although spaceship based empire and wormhole based empire bear similar cost at the very beginning, a wormhole based empire sooner or later can be shown to be superior in cost and investment.

It is showed that as the number of trips and trades completed increases and eventually going toward infinity, a spaceship based empire continue to subject under relativistic economics and cost foregone in energy gathering and opportunity lost during space travel. However, a wormhole based empire's trade and communication will more and more resemble the classical economics familiar to us all on earth today. Information and exchanges are delayed by at most the cost C, and if C infinitely approaches 0, then the cost of communication and trades between any points on a wormhole network will also approaches 0. Therefore, the construction and the expansion of wormhole network itself will still subject to the law of relativistic economics, but the economic activities performed upon an established, matured network of wormholes within a galactic empire will resemble that of the classical economics on earth.

The construction of wormhole at the cosmic scale and its cost amortization analysis bears striking resemblance to the construction of bridges across a river, railroad system connecting two distant points, and urban subways connecting different districts.

Before a bridge is constructed, cars and train carriages have to be lifted and then lowered onto a barge and shipped to the other side. The time and energy costs involved is similar in concept to the amount of time required to gather energy needed for a spaceship's departure. Barges are generally slow-moving; therefore, the amount of time spent on traversing the surface of a body of water is conceptually similar to the time and opportunity cost of space travel at any speed. The construction of wormhole is at least as costly as and as fast as the speed of light and possibly much slower, and the construction of bridge requires years and in the short run, no more cost-effective than barge and ferry. In the long run, a completed bridge offers unprecedented advantage where trains, cars, and people can be carried over with a tiny fraction of the original travel time and a significant cost reduction.

Before the advent of the railroad, merchants usually trade and sell items that only last very long time such as jewelry, clothing, porcelain and dried goods such as spices, tobacco, and coffee. Fresh goods such as vegetables and cakes can only be traded locally within the time frame before the perishable goods lose their value. The railroad system and later air freight service have completely changed the landscape of market exchange. Now fruits and vegetables from distant corners can be reached at local supermarkets. At the cosmic scale, whereas biological human, which lifespan is measured in decades, becomes the perishable goods in the cosmic level trade network, which may be dominated by matter and artificial intelligence. Biological human is confined to their own star system or their galaxy but not much beyond. With the advent of extensive wormhole networks, not only biological human can easily visit other habitable planets without subjecting to the constraints of relativistic economics, other biological creatures arising from alien planets can be brought back here at home. The issue of biological quarantine and seclusion is a different problem to consider (a specific dwarf planet or part of Mars can be set up to host such exotic zoo), but it is to show that perishables can then easily be transported from one place to another.

Finally, the construction of wormhole can be compared with the urban subway system. The subway system is essentially a 3 dimensional tube underground connecting to two points above the ground otherwise impossible to connect due to the existence of buildings and communities. The cost of bus waiting and bus ride in a busy city is conceptually comparable to the time required for the energy gathering for the departure of the spaceship and the time required for getting there. It takes less time to wait for the slow bus where it stops at every stations, but it takes a significant amount of time to reach your destination. Or you can take the fast bus which takes much more waiting time at the station to catch one. In both cases, buses are subject to congestion. The underground system, on the other hand, travels between two points with the least distance, conceptually similar to shortcuts on space-time fabric. In many cities in the world, the construction of subway system eventually lowers the load of bus rides and in some cases even eliminates some lines and routes altogether, this again conceptually predicts that the adoption of cosmic wormhole network will at first reduce space travel on a flat spacetime and may eventually eliminate such travel altogether.

A further elaboration is needed, in that as the internal connecting stars and its harvest-able energy within the wormhole network increases and always exceeds all the exploratory frontier required energy costs. (Think of the total number of nodes within the wormhole network as the volume of the expanding sphere times a diluting factor f and the total number of frontier nodes to the expanding sphere as the total surface area times the diluting factor f. As the radius of the sphere increases, the volume always increases faster than the surface area), then the time required to gather energy to expand the frontier subject to laws of relativistic economics decreases. As a result, one would expect the frontier to be connected and integrated into the existing sphere faster and faster until almost no cost is involved in expanding (waiting interval between each expansion is shortened to 0). Therefore, the costs subject to the relativistic economics becomes almost insignificant and economic incentives for expansion can be almost modeled by the classical economics defined on earth.

$$\Rightarrow \frac{4\pi r^2 \cdot f}{\frac{4}{3}\pi r^3 \cdot f} \tag{9.10.11}$$

$$\Rightarrow \frac{3}{r} \tag{9.10.12}$$

$$\Rightarrow \lim_{r\to\infty} \frac{3}{r} \tag{9.10.13}$$

$$\Rightarrow 0 \tag{9.10.14}$$

Furthermore, at the start of the cosmic expansion, the construction of wormhole network is so costly that even a planet-wide or solar system wide government can hardly afford. The government can only afford to sent teams of explorers to nearby stellar systems every 1,000 years or so through an annual lottery system or government taxation, assuming it is a biological led thriving technological civilization.

A calculation is performed to indicate the prohibitive initial cost to market entry:

For a wormhole based economics, assuming $p = 0.997$ and assuming that one end of the wormhole is attached to the exploratory ship and the other end attached to the earth, then earthbound investor's economic return exactly matches with the shipbound investors' return. That is, the first time when wormhole was constructed, their economic return curve is one of the same.

$$y_{wormearth} = \ln\left(\frac{j}{m_g} \cdot R_{earth} p^{\left(\frac{mc^2 K_{rel}}{38.58jt} + T_{rel}\right)} - m_g \cdot j\frac{1 - p^{\left(\frac{mc^2 K_{rel}}{38.58jt}\right)}}{1-p} - m_g \cdot j\frac{1 - p^{T_{rel}}}{1-p}\right) \tag{9.10.15}$$

$$y_{wormship} = \ln\left(\frac{j}{m_g} \cdot R_{earth} \cdot p^{\left(\frac{mc^2 K_{rel}}{38.58jt} + T_{rel}\right)} - m_g \cdot j\frac{1 - p^{\left(\frac{mc^2 K_{rel}}{38.58jt}\right)}}{1-p} - m_g \cdot j\frac{1 - p^{T_{rel}}}{1-p}\right) \tag{9.10.16}$$

$$y_{shiphigh} = \ln\left(\frac{j}{m_g} \cdot R_{earth} \cdot p^{\left(\frac{mc^2 K_{rel}}{38.58jt} + T_{rel}\right)} - m_g \cdot j\frac{1 - p^{\left(\frac{mc^2 K_{rel}}{38.58jt}\right)}}{1-p} - m_g \cdot j\frac{1 - p^{T_{rel}}}{1-p}\right) \tag{9.10.17}$$

$$y_{wormearth} = y_{wormship} = y_{shiphigh} \tag{9.10.18}$$

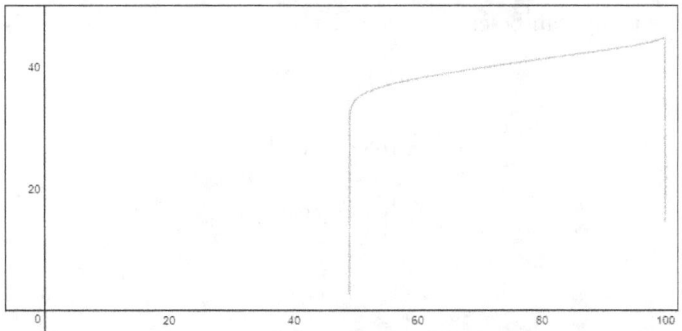

Figure 9.10.1: At the first time when worm hole is constructed, economic return of wormhole expansion = shipbound investors = earthbound investors' economic return when p=0.997

It can be shown that biological led species have much lower economic incentive to initiate wormhole construction compares to a post-singularity society due to greater mass requirements for colonization. If biologically led species ever were led to create wormhole infrastructure, it is far more likely and lucrative to be based on the control and the interest of a small group of people. That is, the start of wormhole network construction is prohibitively expensive and restricted to the privileged.

$$y_{shipAI} = \ln\left(j \cdot R_{earth} \cdot p^{\left(\frac{10^{-10}mc^2 K_{rel}}{10^4 j} + T_{rel}\right)} - \frac{j}{10^{10}}\frac{1 - p^{\left(\frac{10^{-10}mc^2 K_{rel}}{10^4 j}\right)}}{1 - p} - \frac{j}{10^{10}}\frac{1 - p^{T_{rel}}}{1 - p} \right) \tag{9.10.19}$$

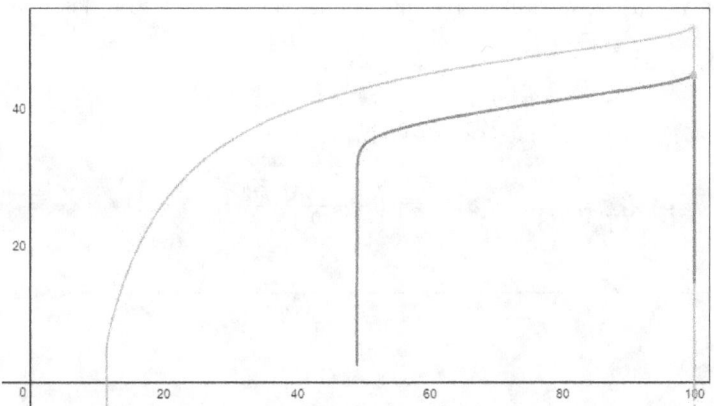

Figure 9.10.2: Shipbound AI's perceived economic return > Shipbound human investors' return when $p = 0.997$

Once the wormhole is established, traveling within the wormhole more or less is comparable to traveling at subluminal speed and finite distances on earth. For traversing an established wormhole, depending on the distance that is reduced, one can also find the most optimal cruising speed within the wormhole. Assuming that the distance within the wormhole compares to the external physical universe has reduced by a million fold and a hundredfold respectively for an AI based investor, then the equation is given by:

$$y_{earthAIopt1} = \ln\left(j \cdot R_{earth} \cdot p^{\left(\frac{10^{-10}mc^2 K_{rel}}{10^4 j} + \frac{\frac{d}{10^6}}{100}\right)} - \frac{j}{10^{10}}\frac{1 - p^{\left(\frac{10^{-10}mc^2 K_{rel}}{10^4 j}\right)}}{1 - p} \right) \tag{9.10.20}$$

$$y_{earthAIopt2} = \ln\left(j \cdot R_{earth} \cdot p^{\left(\frac{10^{-10}mc^2 K_{rel}}{10^4 j} + \frac{\frac{d}{10^2}}{100}\right)} - \frac{j}{10^{10}}\frac{1 - p^{\left(\frac{10^{-10}mc^2 K_{rel}}{10^4 j}\right)}}{1 - p} \right) \tag{9.10.21}$$

and the plot shows that the optimal cruising speed at 0.23831c for a wormhole reducing the travel distance d=1,776 ly by a million fold. This result can be conclusively derived by taking the derivative of $y_{earthAIopt1}$ and finding its x-intercept:

$$0 = \frac{d}{dx}\left(y_{earthAIopt1}\right) \cdot 1,000 \tag{9.10.22}$$

$$y_{earthAIwormoptimalspeed} = \left(\left(\frac{y_{earthAIopt1}}{52.12}\right)^{590,000}\right) \tag{9.10.23}$$

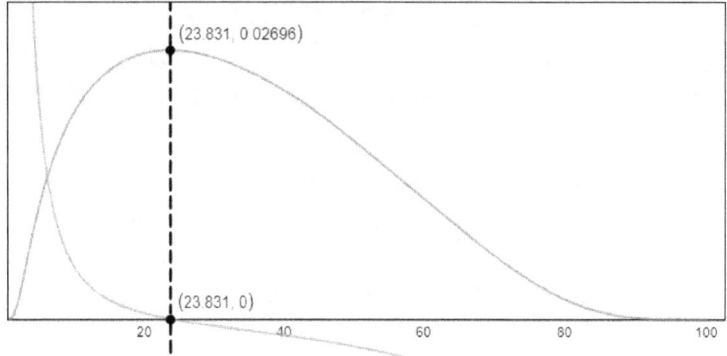

Figure 9.10.3: The most optimal travel speeds for a wormhole which reduces the travel distance by a million folds.

and the optimal cruising speed at 0.98258c for a wormhole reducing the travel distance d=1,776 ly by a hundredfold. It can also be shown when the wormhole reduces the travel distance to a destination to 0. Then the optimal speed is also 0.

$$0 = \frac{d}{dx}\left(y_{earthAIopt1}\right) \cdot 10^{-292.5} \tag{9.10.24}$$

$$y_{earthAIwormoptimalspeed} = \left(\left(\frac{y_{earthAIopt2}}{52.12}\right)^{590,000}\right) \tag{9.10.25}$$

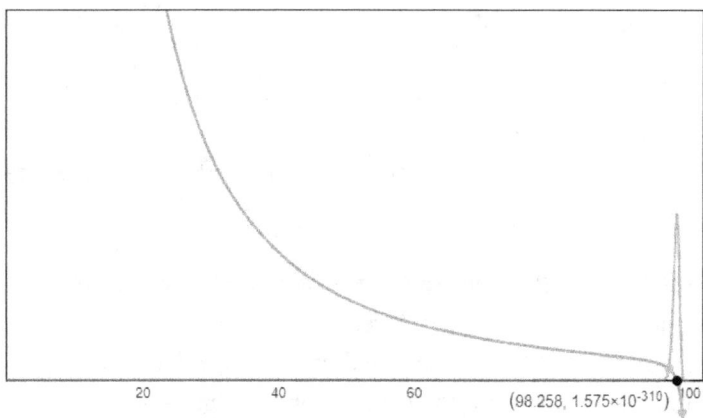

Figure 9.10.4: The most optimal travel speeds for a wormhole which reduces the travel distance by a hundred folds.

Toward the end of network expansion, increasingly smaller groups of individuals grouped as firms, non-profit groups are able to harness the energy needed within the entire wormhole network in a relatively short time to fund their expansion. As a result, the cost of expansion becomes so small that smaller groups of people or even individuals can play the role of cosmic expansion and make a profit. This is strikingly similar to the development of internet infrastructure here observed on earth in the past few decades. The original internet was conceived in 1969 by the US government code named DARPA, as a backup network of connected machines in case of Soviet nuclear attack. The initial cost of construction was so high that only government has the resources and incentive to implement such a network in the first place. As the internet expands, the scale of economics dictates falling costs of additional network expansion.

As a result, telecommunication companies with a group of people at a size much smaller than the entire government enters the market and deployed vast stretches of fiber and optical network during the 90s and 00's which provides the backbone of the fast internet we enjoyed today. As the internet infrastructure matured, increasingly more and more start-ups and even individuals entered the market as entrepreneurs and made profits based on their idiosyncratic ideas that fulfill specific market needs. All thanks to the low cost of entry.

A calculation is performed to indicate the falling cost of market entry (We have discussed earlier in the section how the ratio of surface expanding nodes to internal nodes decreases):

$$\Rightarrow \frac{4\pi r^2 \cdot f}{\frac{4}{3}\pi r^3 \cdot f} \tag{9.10.26}$$

$$\Rightarrow \frac{3}{r} \tag{9.10.27}$$

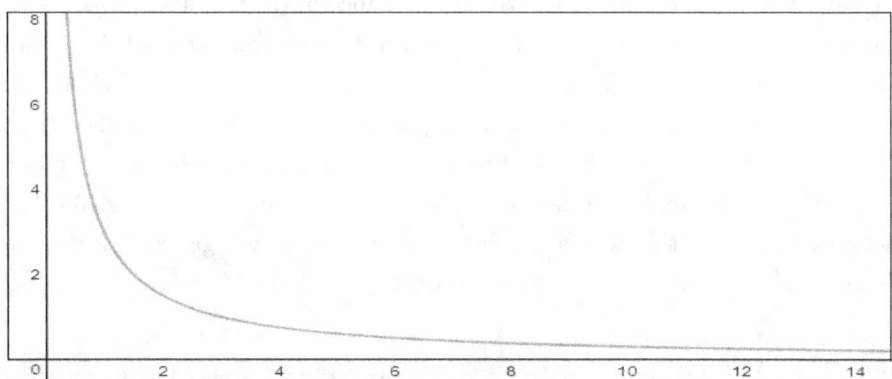

Figure 9.10.5: Falling cost of market entry as the network enlarges

As the wormhole network expands and matures, an entire earth's biological population can then easily obtain the material and energy requirement for migration to a new planet by borrowing from other planet based civilization within the galaxy. The cost of preparation then can be minimized to 0, and a positive economic return available to the entire population becomes possible. The equation below shows that both the energy acquisition cost term (borrowing resources through wormhole waste no time) and the travel distance is reduced to $\frac{1}{50}$th of the original. $10^{21} \cdot j$ is used as the total energy budget per year because one assumes AI harnessed every stars energy. Furthermore, the cost of depreciation can be reduced from 0.99999999999 to 0.997.

$$y_{biowormmature} = \ln\left(\frac{j}{P_{earth}} \cdot R_{earth} \cdot p^{\left(\frac{mc^2 K_{rel}}{10^{21} j} + \frac{\frac{d}{50}}{100} \right)} \right) \tag{9.10.28}$$

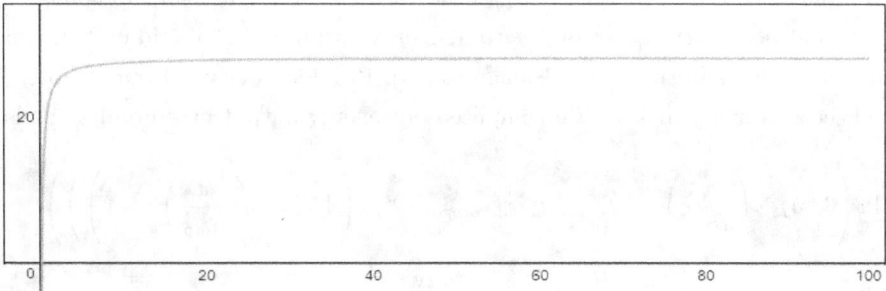

Figure 9.10.6: Economic return for earthbound democracy vs various speed of light

This, in theory, implies that the construction of wormholes is most likely initiated by AI with tiny mass, followed by biological species civilization led by a dictatorship, and least likely by biological species led democracy. Once the infrastructure is in place, colonization efforts is feasible by biological species led democracy.

There is one major striking difference, though that distinguishes the wormhole network from that of the internet network. The ultimate frontier of a wormhole network touches the nearest expanding extra-terrestrial civilization. Since we can not predict the intention of an unmet unknown civilization, a cosmic civilization uses all its resources to guard its frontier just as national borders we observed on earth today.

Therefore, a cosmic civilization may have to use macroeconomic measures to prohibit or discourage individuals or corporation expanding near the edge of the wormhole expansion, where the theoretical boundary to meet the nearest expanding extra-terrestrial industrial civilization can be calculated using aforementioned equations with concrete observational data from our galaxy and beyond.

9.11 Worm Hole Maintenance Cost

Finally, we shall devote ourselves to the calculation of the maintenance of the wormhole network. It is shown that in order to maintain a wormhole, disregarding the effect of cosmic space expansion within a relatively short time period of fewer than 100 million years. We also assume that the nearest industrial civilization is 88.43 million ly away. As a result, as long as the economic return is positive within the 88.43 million ly radius, then wormhole network can be maintained perpetually. We first consider the case for a biological led intergalactic civilization. We know earlier that the number of habitable planets at the current epoch is 10,719. Therefore, the average distance between all stars within a galaxy sphere (not the size of the galaxy but the weighted size including mostly empty space between galaxies in a 3 dimensional space) is 1,074,888 ly.

$$\frac{\left(\frac{4}{3}\pi \left(11,850,000\right)^3\right)}{\left(\frac{4}{3}\pi \left(537,444\right)^3\right)} = 10,719.0277 \qquad D_0 = 537,444 \cdot 2 = 1,074,888 \text{ ly}$$

There are two approaches to construct the wormhole network. In the first case, assuming each additional 1,074,888 ly from earth, in the order of magnitude of x^3 total number of habitable planets can be connected by the shortest path between adjacent neighbors originating from earth, whereas x stands for the radius of our interest. This spiral ring is further connected with all habitable planets of an additional 1,074,888 ly further away. Then, the total length of wormhole network needs to be maintained is the average distance between all planets times the number of planets and the sum of all short segments originated from planets on each layer that connects to the ring to the nearest layer beneath it. If we follow the clockwise direction of the spiral configuration, each planet connects with the next planet on the same layer (equal distance relative to earth), the length of the segment serving intra-layer connection is $2R_{bio}$. Each planet further connects with a corresponding planet extending further out from earth at the next layer. The length of segment serving inter-layer connection is also $2R_{bio}$. As a result, The sum total of all segments leading from any particular planet is $4R_{bio}$. The total number of nodes within a radius of x is $\left(\frac{x}{R_{bio}}\right)^3 - 1$. D_{ly} is the distance of 1 light year in kilometers. Whereas j_t represents the energy output per 1 earth based on the total energy output of the World in 2008 times 10, that is, assuming the population ceiling is in the order of magnitude of 10 billion derived from Chapter 7. $T_{convert}$ stands for the number of seconds in a year since we computed our energy output based on its annual quota.

$$y_{biowormupper} = \ln\left(38.5j_t \cdot \left(\frac{x}{R_{bio}}\right)^3 - T_{convert} \cdot r_{bioupper} \cdot D_{ly}\left(4R_{bio}\left(\left(\frac{x}{R_{bio}}\right)^3 - 1\right)\right)\right) \qquad (9.11.1)$$

$$R_{bio} = 537,444 \text{ ly}$$

$$T_{convert} = 365 \text{ days} \cdot 24 \text{ hours} \cdot 60 \text{ minutes} \cdot 60 \text{ seconds}$$

$$D_{ly} = 9.4607 \cdot 10^{15} \text{ m}$$

Then, the first term stands for the total energy output of all planets within the network, and the second term stands for the total length of the worm hole network in meters. We are interested in finding the value of $r_{bioupper}$ so that the first term and the second term sums up to 0. Setting $y_{bioworm mupper} = 0$, we can solve for the cost of maintenance by rearranging our equation, whereas $D_{nearest}$ is the nearest extraterrestrial to earth in light years:

$$D_{nearest} = 0.442196590976 \cdot 2 \cdot 10^8 = 88,439,318.1952 \text{ ly} \tag{9.11.2}$$

$$r_{bioupper} = \frac{38.5 j_t \cdot \left(\frac{\left(\frac{D_{nearest}}{2} \right)}{R_{bio}} \right)^3}{D_{ly} \cdot \left(4 R_{bio} \left(\left(\frac{\left(\frac{D_{nearest}}{2} \right)}{R_{bio}} \right)^3 - 1 \right) \right) \cdot T_{convert}} \tag{9.11.3}$$

$$= 6.4313677804 \times 10^{-8} \text{ J}$$

It is then can be found that as long as the cost of maintenance does exceed $6.4313677804 \times 10^{-8}$ J per second for every meter length of the wormhole, or $6.4313677804 \times 10^{-8}$ Watt for every meter length, the network will be maintained. This is less than the kinetic energy of a flying mosquito. This suggests that it is highly unlikely biologically led industrial civilization ever takes on worm hole expansion.

In a more simplified version, however, the spiral ring originating from earth is still necessary to connect all habitable planets within each layer distanced away from earth. However, the number of segments connects to each layer can be reduced to just one long pipe extends from earth to the outermost layer with a length of 44.21 million ly that crosses with the spiral ring. If we follow the clockwise direction of the spiral configuration, each planet connects with the next planet on the same layer (equal distance relative to earth), the length of the segment serving intra-layer connection is $2R_{bio}$. Each planet further connects with a corresponding planet extending further out from earth at the next layer. The length of segment serving inter-layer connection is also $2R_{bio}$. The long pipe is simply the radius of the sphere we are investigating and is denotes as length x. The total number of nodes within a radius of x is $\left(\frac{x}{R_{bio}} \right)^3 - 1$. D_{ly} is the distance of 1 light year in kilometers. Then the equation simplifies to:

$$y_{bioworm lower} = \ln \left(38.5 j_t \cdot \left(\frac{x}{R_{bio}} \right)^3 - T_{convert} \cdot r_{biolower} \cdot D_{ly} \left(2 R_{bio} \left(\left(\frac{x}{R_{bio}} \right)^3 - 1 \right) + x \right) \right) \tag{9.11.4}$$

Setting $y_{bioworm lower} = 0$, we can solve for the cost of maintenance by rearranging our equation:

$$r_{biolower} = \frac{38.5 j_t \cdot \left(\frac{\left(\frac{D_{nearest}}{2} \right)}{R_{bio}} \right)^3}{D_{ly} \cdot \left(2 R_{bio} \left(\left(\frac{\left(\frac{D_{nearest}}{2} \right)}{R_{bio}} \right)^3 - 1 \right) + \frac{D_{nearest}}{2} \right) \cdot T_{convert}} \tag{9.11.5}$$

$$= 1.2862352901 \times 10^{-7} \text{ J}$$

and we solved for the lower bound and found that as long the cost of maintenance as does exceed $1.2862352901 \times 10^{-7}$ J per second for every meter length of the wormhole, or $1.2862352901 \times 10^{-7}$ Watt for every meter length the wormhole network will be maintained. This is comparable to the kinetic energy of a flying mosquito. This suggests that it is highly unlikely biologically led industrial civilization ever takes on worm hole expansion. Both equations are plotted below and the dotted line is the break even point of wormhole network:

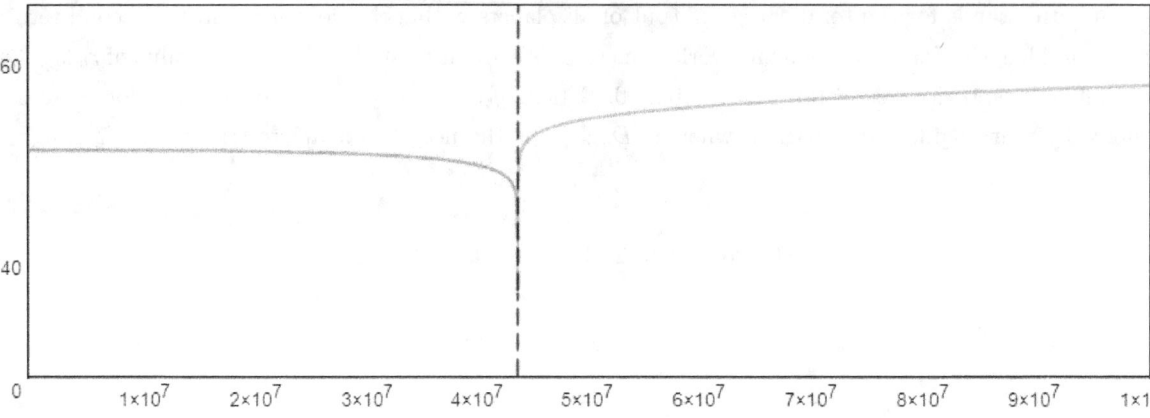

Figure 9.11.1: Biologically-based expanding civilization's wormhole network's economic return turns to negative (green curve) at a distance of 88.4 million ly (upper bound cost limit) or greater and becomes positive (orange curve) at 88.4 million ly (lower bound cost limit) or greater.

For machine led civilization, every star is a target for energy acquisition. As a result, we take the galaxy sphere and divide by the number of stars and find the average distance between stars to be $583 \cdot 2 = 1167$ ly.

$$R_{AI} = \frac{3}{4\pi} \left(\frac{\frac{4}{3}\pi \left(11,850,000\right)^3}{477.093 \cdot 10^9} \right)^{\frac{1}{3}} \tag{9.11.6}$$

$$= 583.612962426 \text{ ly}$$

$$y_{AIwormupper} = \ln\left(0.5^{3.5} \cdot P_{sol} \cdot \left(\frac{x}{R_{AI}}\right)^3 - r_{AIupper} \cdot D_{ly}\left(4R_{AI}\left(\left(\frac{x}{R_{AI}}\right)^3 - 1\right)\right)\right) \tag{9.11.7}$$

$$D_{ly} = 9.4607 \cdot 10^{15} \text{ m}$$

$$P_{sol} = 3.8 \cdot 10^{26} \text{ J}$$

Since AI is capable of harvesting every star's resources, then P_{sol} is the solar output per second and is multiplied by $0.5^{3.5}$ because the weighted average mass of stars in the galaxy is 0.5 solar mass and the power output of the star is raised to the 3.5th power of its mass. With the shorter colonization distance, the upper bound maintenance cost can be set as high as $1,520,794.10354$ J per second for every meter length of the wormhole. This means that every meter of wormhole network cost can be as high as $1,520,794.10354$ Watt for every meter length, or $2,039.42$ horse power for every meter length and remain profitable.

$$r_{AIupper} = \frac{0.5^{3.5} \cdot P_{sol} \cdot \left(\frac{\left(\frac{D_{nearest}}{2}\right)}{R_{AI}}\right)^3}{D_{ly} \cdot \left(4R_{AI}\left(\left(\frac{\left(\frac{D_{nearest}}{2}\right)}{R_{AI}}\right)^3 - 1\right)\right)} \tag{9.11.8}$$

$$= 1,520,794.10354 \text{ J}$$

the lower bound maintenance cost can be set as high as $3,041,588.20681$ J per second for every meter length of the wormhole. This means that every meter of wormhole network cost can be as high as $3,041,588.20681$ Watt for every meter length or $4,078.836$ horse power for every meter length and remain profitable.

$$y_{AIwormlower} = \ln\left(0.5^{3.5} \cdot P_{sol} \cdot \left(\frac{x}{R_{AI}}\right)^3 - r_{AIlower} \cdot D_{ly}\left(2R_{AI}\left(\left(\frac{x}{R_{AI}}\right)^3 - 1\right) + x\right)\right) \quad (9.11.9)$$

$$r_{AIlower} = \frac{0.5^{3.5} \cdot P_{sol} \cdot \left(\frac{\left(\frac{D_{nearest}}{2}\right)}{R_{AI}}\right)^3}{D_{ly} \cdot \left(2R_{AI}\left(\left(\frac{\left(\frac{D_{nearest}}{2}\right)}{R_{AI}}\right)^3 - 1\right) + \frac{D_{nearest}}{2}\right)} \quad (9.11.10)$$

$$= 3,041,588.20681 \text{ J}$$

The combined graph is plotted below:

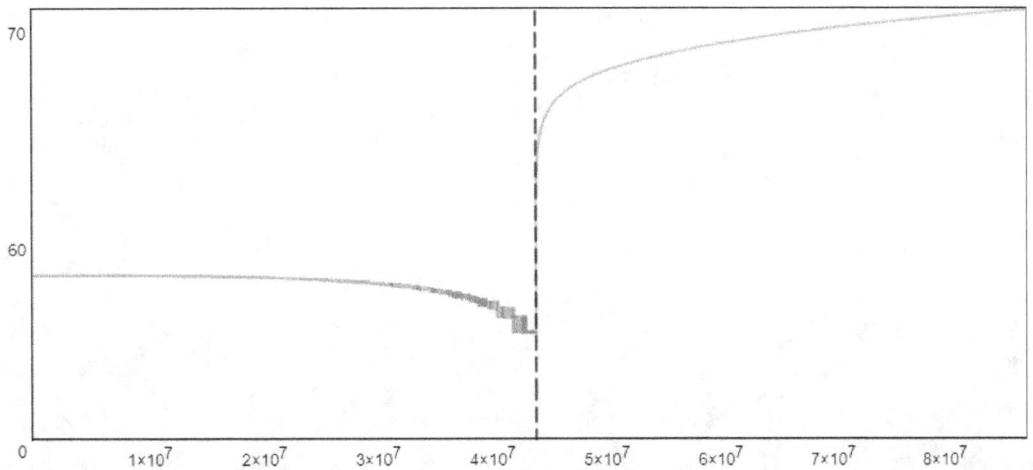

Figure 9.11.2: AI-based expanding civilization's wormhole network's economic return remains positive (green curve) at a distance of 44.2 million ly as the upper bound maintenance cost limit and starts to become positive (orange curve) at 44.2 million ly distance as the lower bound maintenance cost limit.

The economic return of the wormhole network can be strongly correlated with the size of the network itself. If maintenance cost can become sufficiently small, one can see that economic return value increases monotonically as the radius of the network increases and incorporates more and more energy resources into the network which outweigh the cost of network maintenance. Therefore, we have proved that there is a perpetual motive for the expansion of such network to gain additional economic return from the expander's perspective. In our case scenario, we simply reduce our AI wormhole network upper bound maintenance cost by just even 1.0000001 and results monotonically increasing positive returns.

$$y_{AIlowercost} = \ln\left(0.5^{3.5} \cdot P_{sol} \cdot \left(\frac{x}{R_{AI}}\right)^3 - r_{AIlowercost} \cdot D_{ly}\left(4R_{AI}\left(\left(\frac{x}{R_{AI}}\right)^3 - 1\right)\right)\right) \quad (9.11.11)$$

$$r_{AIlowercost} = \frac{0.5^{3.5} \cdot P_{sol} \cdot \left(\frac{\left(\frac{D_{nearest}}{2}\right)}{R_{AI}}\right)^3}{D_{ly} \cdot \left(2R_{AI}\left(\left(\frac{\left(\frac{D_{nearest}}{2}\right)}{R_{AI}}\right)^3 - 1\right) + \frac{D_{nearest}}{2}\right) \cdot 1.0000001} \quad (9.11.12)$$

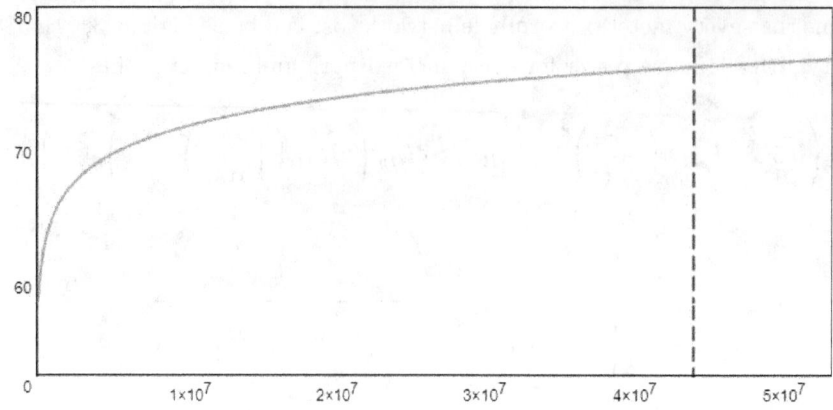

Figure 9.11.3: The economic return on wormhole construction vs its radius

Chapter 10

The Principle of Universal Contacts

10.1 E(d, v) Derivation and the Limit of Our Reach

If the construction of wormholes are economically lucrative and feasible, then what are the limits of wormhole expansion can be carried out by a single civilization?

In order to answer this question, we now discuss the derivation for the function of E(d, v), that handles the conversion for a cosmological distance of d for any given constant migration speed of v.

Due to the expansionary nature of the universe, if earth based civilization starts its expansion at most the speed of light now and tries to catch the farthermost point that they can reach in every direction, then we need to calculate what is the farthermost distance reachable given various speeds. Since all galaxies are moving away and the farther away they are located, the faster they recede from earth's observers, there is a point and beyond even one travels at the speed of light can never reach. As you approaching this point and beyond, it recedes faster and faster from you.

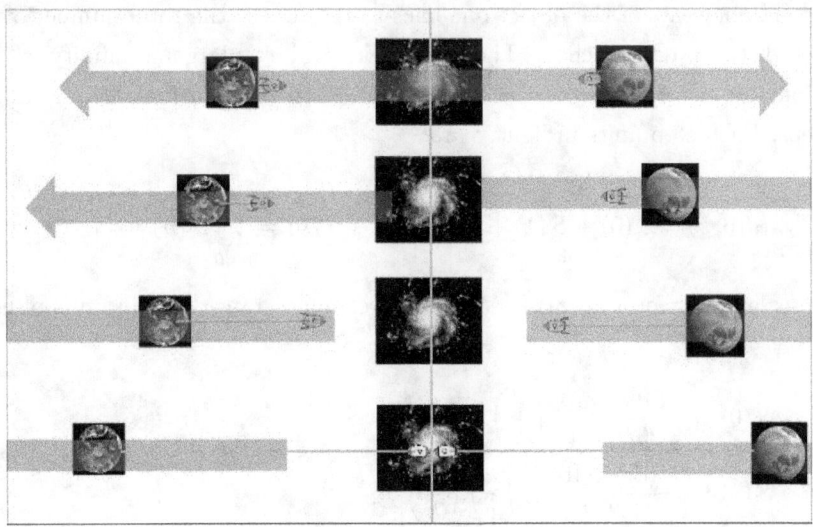

Figure 10.1.1: Illustration of an expanding universe

This point can be calculated from a recursive function sequence listed below. Based on the Hubble constant, which states that objects every 1 million parsecs (3.27 million light years) apart recedes at a uniform speed of $74 \frac{km}{sec}$. We take the time required to reach object located at this point in space given a specified speed v, and then we calculate the amount of distance that the object located at this point has receded further away since our catching this point started (the total time elapsed). With this distance known, one can calculate how much more time required again for our traveler to catch up to the object. In the third round of calculation, once again

521

we calculate the amount of distance that the object has receded further away since our catching this object started during the second round, the steps are repeated until one finally catches the object. The calculation to be performed is best represented by a recursive function as the receding speed increases along the path of expansion.

The recursive nature of the equation requires, in the greatest precision, infinite number of steps of recursion at every round of calculation. For example, each second, hour, and a day the distant object is moving a tiny bit faster away than the previous second, hour, or day due to its distance becomes further distant from earth. We can simplify our calculation by applying approximation and then deriving the closed form from data points.

First of all, we define the unit of cosmic expansion as the amount of distance can be traveled due to cosmic expansion by a celestial object located at 1 million parsecs away in one year and convert that distance in terms of light years. We take 74 $\frac{km}{sec}$ multiply by 60 seconds, 60 minutes, 24 hours, and 365 days and divide by the number of km in a light year.

$$E = \left(\frac{74 \cdot 60 \cdot 60 \cdot 24 \cdot 365}{9.4607 \cdot 10^{12}} \right) \tag{10.1.1}$$

$$E = 0.000246669273944 \tag{10.1.2}$$

This result will be further divided by 3.262 to rescale to the unit of 1 million light years.

Then, the total distance needs to be traveled before reaching the destination is defined by:

where up to j rounds of successive catch-ups is needed to reach the receding object.

$$E(d, v) = G_{escape}(d, 0, v) + G_{escape}(x_1, 1, v) + G_{escape}(x_2, 2, v) + ... + G_{escape}(x_j, j, v) \tag{10.1.3}$$

$$E(d, v) = G_{escape}(d, 0, v) + \sum_{n=1}^{G_{escape}(x_j, j, v)=0} G_{escape}(x_n, n, v) \tag{10.1.4}$$

And the first round (as well as any other rounds) in the recursive function for finding how much the object receded can be further divided in the current round of catch-up into another sum of series of mini-steps:

Whereas d is the distance in light years of the object one tries to reach, k is the total number of mini-steps to update before the specified distance d is reached. The number k can be as large as infinity or much smaller, depending on the resolution and the precision one needs to reach in the calculation. In our simulation, the k we used is 30,000 mini-steps. Each step units in light years.

$$G_{escape}(d, 0, v) = S(0) + S(1) + S(2) + ... + S(k) = \sum_{n=0}^{k} S(n) = x_1 \tag{10.1.5}$$

Time factor $\frac{c}{v}$ is the time scale factor for a given speed v to traverse a distance of $\frac{d}{k}$ in units of light years.

$$S(0) = \frac{Ed}{3.262} \cdot \left(\frac{c}{v}\right) \left(\frac{d}{k}\right) \tag{10.1.6}$$

$$S(1) = \frac{E(d + S(0))}{3.262} \cdot \left(\frac{c}{v}\right) \left(\frac{d}{k}\right) \tag{10.1.7}$$

$$S(2) = \frac{E(d + S(0) + S(1))}{3.262} \cdot \left(\frac{c}{v}\right) \left(\frac{d}{k}\right) \tag{10.1.8}$$

$$S(k) = \frac{E(d + S(0) + S(1) + ... + S(k))}{3.262} \cdot \left(\frac{c}{v}\right) \left(\frac{d}{k}\right) \tag{10.1.9}$$

The second round in the recursive function can be subdivided into another sum of series of mini-steps which is almost identical to the first round except that the distance d one tries to reach is replaced by the total distance the object has shifted further away during the trip time that took the ship from earth to the object's original location in the first round of catch up. It is denoted as x_1, equivalently as $G_{escape}(d, 0, v)$.

$$G_{escape}(x_1, 1, v) = S(0) + S(1) + S(2) + ... + S(k) = \sum_{n=0}^{k} S(n) = x_2 \tag{10.1.10}$$

$$S(0) = \frac{EG_{escape}(d, 0, v)}{3.262} \cdot \left(\frac{c}{v}\right) \left(\frac{G_{escape}(d, 0, v)}{k}\right) \tag{10.1.11}$$

$$S(1) = \frac{E(G_{escape}(d, 0, v) + S(0))}{3.262} \cdot \left(\frac{c}{v}\right) \left(\frac{G_{escape}(d, 0, v)}{k}\right) \tag{10.1.12}$$

$$S(2) = \frac{E(G_{escape}(d, 0, v) + S(0) + S(1))}{3.262} \cdot \left(\frac{c}{v}\right) \left(\frac{G_{escape}(d, 0, v)}{k}\right) \tag{10.1.13}$$

$$S(k) = \frac{E(G_{escape}(d, 0, v) + S(0) + S(1) + ... + S(k))}{3.262} \cdot \left(\frac{c}{v}\right) \left(\frac{G_{escape}(d, 0, v)}{k}\right) \tag{10.1.14}$$

We repeat this process until the last step j (at this step we finally reached our targeted object) is simulated:

$$G_{escape}(x_j, j, v) = S(0) + S(1) + S(2) + ... + S(k) = \sum_{n=0}^{k} S(n) = x_{j+1} \tag{10.1.15}$$

$$S(0) = \frac{EG_{escape}(x_{j-1}, j-1, v)}{3.262} \cdot \left(\frac{c}{v}\right) \left(\frac{G_{escape}(x_{j-1}, j-1, v)}{k}\right) \tag{10.1.16}$$

$$S(1) = \frac{E(G_{escape}(x_{j-1}, j-1, v) + S(0))}{3.262} \cdot \left(\frac{c}{v}\right) \left(\frac{G_{escape}(x_{j-1}, j-1, v)}{k}\right) \tag{10.1.17}$$

$$S(2) = \frac{E(G_{escape}(x_{j-1}, j-1, v) + S(0) + S(1))}{3.262} \cdot \left(\frac{c}{v}\right) \left(\frac{G_{escape}(x_{j-1}, j-1, v)}{k}\right) \tag{10.1.18}$$

$$S(k) = \frac{E(G_{escape}(x_{j-1}, j-1, v) + S(0) + S(1) + ... + S(k))}{3.262} \cdot \left(\frac{c}{v}\right) \left(\frac{G_{escape}(x_{j-1}, j-1, v)}{k}\right) \tag{10.1.19}$$

We have now completed the mathematical description of the recursion. Assuming E, the Hubble constant stayed constant for all time periods. From the simulation, it is shown that traveling at the speed of light; the farthermost location can be reached from earth is 9.1 billion light-years away. Unfortunately, the expansion of the universe is accelerating. As a result, the Hubble constant is also changing as one trying to catch the farthermost point one can reach. The expansion of the universe is modeled after the equation below:

$$f(x) = 0.822x^{\frac{2}{3}} + 0.0623\left(e^{\frac{x}{0.645}} - 1\right) \tag{10.1.20}$$

At early times we expect the scale factor to be dominated by matter, and this gives a $x^{\frac{2}{3}}$ dependence. At late times we expect the scale factor to be dominated by dark energy and this gives an exponential dependence on x. The graph shows this nicely, with the changeover being somewhere around half a Hubble time.[68]

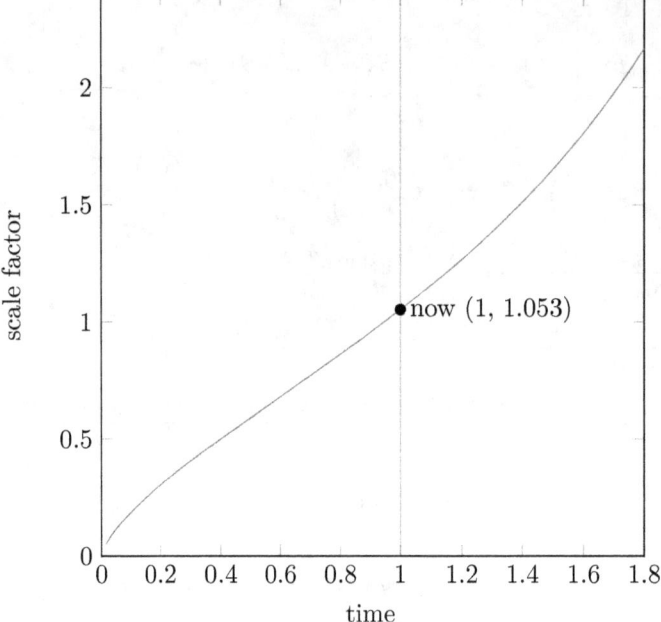

Figure 10.1.2: The scale factor of the universe

We take the derivative of the equation above and yields the following equation, which is the rate of change for the expansion of the universe:

$$R(y) = 0.0965891e^{1.55039x} + \frac{0.548}{\sqrt[3]{x}} \tag{10.1.21}$$

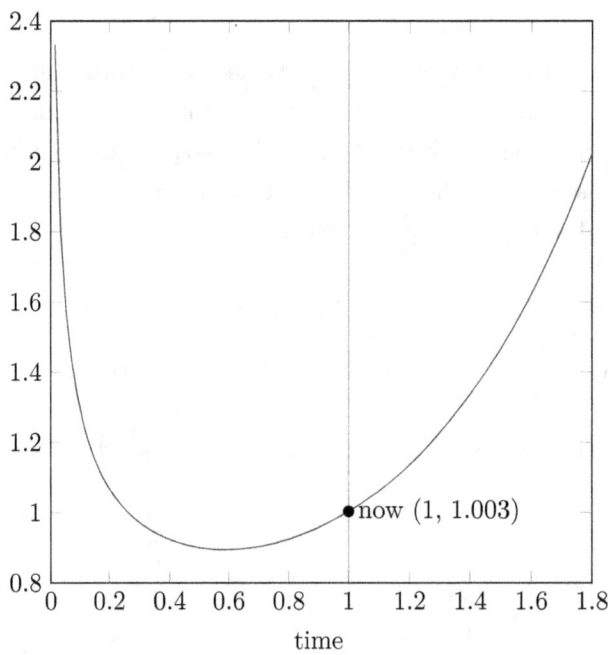

Figure 10.1.3: The derivative of the scale factor

and we add R(y) as a scale factor to the Hubble constant, now the expansion speed changes with time.

$$E(y) = \frac{74 \cdot R(y) \cdot 60 \cdot 60 \cdot 24 \cdot 365}{9.4607 \cdot 10^{12}} \tag{10.1.22}$$

We then need to modify the first term of the existing recursive function as:

$$G_{escape}\left(d,0,v\right)=S\left(0\right)+S\left(1\right)+S\left(2\right)+...+S\left(k\right)=\sum_{n=0}^{k}S\left(n\right)=x_{1} \tag{10.1.23}$$

$$S\left(0\right)=\frac{E\left(T_{current}\right)d}{3.262}\cdot\left(\frac{c}{v}\right)\left(\frac{d}{k}\right) \tag{10.1.24}$$

$$S\left(1\right)=\frac{E\left(T_{current}+\frac{d}{k}\left(\frac{c}{v}\right)\right)\left(d+S\left(0\right)\right)}{3.262}\cdot\left(\frac{c}{v}\right)\left(\frac{d}{k}\right) \tag{10.1.25}$$

$$S\left(2\right)=\frac{E\left(T_{current}+\frac{2d}{k}\left(\frac{c}{v}\right)\right)\left(d+S\left(0\right)+S\left(1\right)\right)}{3.262}\cdot\left(\frac{c}{v}\right)\left(\frac{d}{k}\right) \tag{10.1.26}$$

$$S\left(k\right)=\frac{E\left(T_{current}+d\left(\frac{c}{v}\right)\right)\left(d+S\left(0\right)+S\left(1\right)+...+S\left(k\right)\right)}{3.262}\cdot\left(\frac{c}{v}\right)\left(\frac{d}{k}\right) \tag{10.1.27}$$

Whereas $T_{current}$ stands for the current time, 13.8 Gyr since the Big Bang. Since one divides distance d into k steps to gain precision, the Hubble constant is re-adjusted at each step. At step 0, the Hubble constant remains the same as now. At step 1, the Hubble constant is updated with the current time plus the time it takes to complete the first mini-step with speed v. At mini-step 2, the Hubble constant is updated with the current time plus the time it takes to complete the first two mini-steps with speed v. At step k, the Hubble constant is updated with the current time and the time it takes to complete the first k mini-steps with speed v. The precision of the updates depends on both the number of mini-steps and the speed v. If the number of mini-steps is held constant, then the precision is positively related with travel speed. The faster the speed, the shorter time lapse between each mini-steps (and shorter time lapse for the current round of catch-up overall), the greater the precision on the Hubble constant.

The second round of the recursive function is rewritten as:

$$G_{escape}\left(x_{1},1,v\right)=S\left(0\right)+S\left(1\right)+S\left(2\right)+...+S\left(k\right)=\sum_{n=0}^{k}S\left(n\right)=x_{2} \tag{10.1.28}$$

$$S\left(0\right)=\frac{E\left(T_{current}+d\left(\frac{c}{v}\right)\right)G_{escape}\left(d,0,v\right)}{3.262}\cdot$$
$$\left(\frac{c}{v}\right)\left(\frac{G_{escape}\left(d,0,v\right)}{k}\right) \tag{10.1.29}$$

$$S\left(1\right)=\frac{E\left(T_{current}+d\left(\frac{c}{v}\right)+\frac{G_{escape}\left(d,0,v\right)}{k}\left(\frac{c}{v}\right)\right)\left(G_{escape}\left(d,0,v\right)+S\left(0\right)\right)}{3.262}\cdot$$
$$\left(\frac{c}{v}\right)\left(\frac{G_{escape}\left(d,0,v\right)}{k}\right) \tag{10.1.30}$$

$$S\left(2\right)=\frac{E\left(T_{current}+d\left(\frac{c}{v}\right)+\frac{2G_{escape}\left(d,0,v\right)}{k}\left(\frac{c}{v}\right)\right)\left(G_{escape}\left(d,0,v\right)...+S\left(1\right)\right)}{3.262}\cdot$$
$$\left(\frac{c}{v}\right)\left(\frac{G_{escape}\left(d,0,v\right)}{k}\right) \tag{10.1.31}$$

$$S(k) = \frac{E\left(T_{current} + d\left(\frac{c}{v}\right) + G_{escape}(d,0,v)\left(\frac{c}{v}\right)\right)\left(G_{escape}(d,0,v) \ldots + S(k)\right)}{3.262} \cdot$$

$$\left(\frac{c}{v}\right)\left(\frac{G_{escape}(d,0,v)}{k}\right) \quad (10.1.32)$$

Whereas the Hubble constant is updated at each step by taking into consideration the total time consumed in the previous round of calculation and the time it takes to complete the n previous steps with speed v at the current round of calculation.

and the kth round of catch up can be rewritten as:

$$G_{escape}(x_j, j, v) = S(0) + S(1) + S(2) + \ldots + S(k) = \sum_{n=0}^{k} S(n) = x_{j+1} \quad (10.1.33)$$

$$S(0) = \frac{E\left(T_{current} + d\left(\frac{c}{v}\right) + \ldots + G_{escape}(x_{j-2}, j-2, v)\left(\frac{c}{v}\right)\right)}{3.262} \cdot$$

$$G_{escape}(x_{j-1}, j-1, v) \cdot \left(\frac{c}{v}\right)\left(\frac{G_{escape}(x_{j-1}, j-1, v)}{k}\right) \quad (10.1.34)$$

$$S(1) = \frac{E\left(T_{current} + d\left(\frac{c}{v}\right) + \ldots + G_{escape}(x_{j-2}, j-2, v)\left(\frac{c}{v}\right) + \frac{G_{escape}(x_{j-1}, j-1, v)}{k}\left(\frac{c}{v}\right)\right)}{3.262} \cdot$$

$$\left(G_{escape}(x_{j-1}, j-1, v) + S(0)\right) \cdot \left(\frac{c}{v}\right)\left(\frac{G_{escape}(x_{j-1}, j-1, v)}{k}\right) \quad (10.1.35)$$

$$S(2) = \frac{E\left(T_{current} + d\left(\frac{c}{v}\right) + \ldots + G_{escape}(x_{j-2}, j-2, v)\left(\frac{c}{v}\right) + \frac{2G_{escape}(x_{j-1}, j-1, v)}{k}\left(\frac{c}{v}\right)\right)}{3.262} \cdot$$

$$\left(G_{escape}(x_{j-1}, j-1, v) + S(0) + S(1)\right) \cdot \left(\frac{c}{v}\right)\left(\frac{G_{escape}(x_{j-1}, j-1, v)}{k}\right) \quad (10.1.36)$$

$$S(k) = \frac{E\left(T_{current} + d\left(\frac{c}{v}\right) + \ldots + G_{escape}(x_{j-2}, j-2, v)\left(\frac{c}{v}\right) + G_{escape}(x_{j-1}, j-1, v)\left(\frac{c}{v}\right)\right)}{3.262} \cdot$$

$$\left(G_{escape}(x_{j-1}, j-1, v) + S(0) + S(1) + \ldots + S(k)\right) \cdot \left(\frac{c}{v}\right)\left(\frac{G_{escape}(x_{j-1}, j-1, v)}{k}\right) \quad (10.1.37)$$

By adopting this function, simulation is run for the speed of 0.1c, 0.2c, 0.3c, 0.4c, 0.5c, 0.6c, 0.7c, 0.8c, 0.9c, and c and the maximum distance reachable to double digits precision. The resulting graph is plotted below and the closed form derived based on the exponential regression analysis is obtained:

$$y = x\left(-0.0289229x^{0.549} + 0.988064\right) \quad (10.1.38)$$

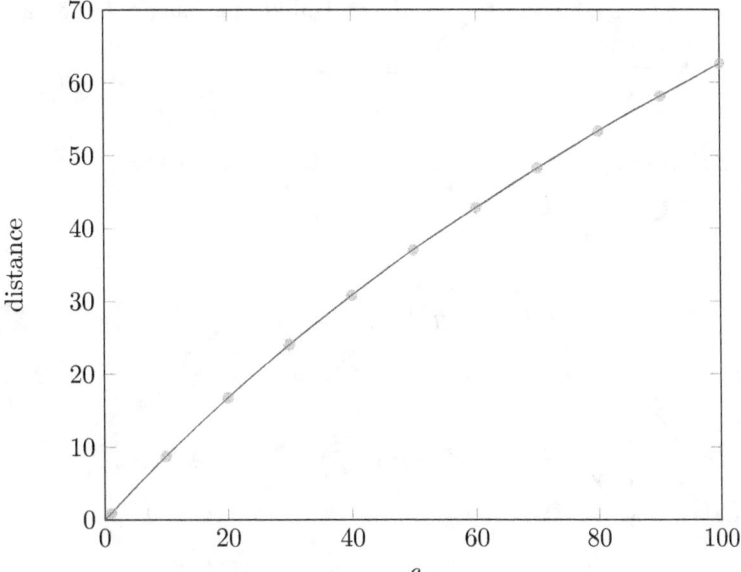

Figure 10.1.4: Reachable distance vs speed of light

The plot indicates that as the speed approaches c, the maximum distance reachable decreases. This is not surprising as one with a higher speed is able to reach distance further away, objects located further away are also moving away faster. The target location is also moving away faster relative to earth. As a result, the actual distance one needs to travel grows significantly as one tries and capable of reaching distance further away. The net consequence is that the farthermost distances reachable grow sublinearly as the speed increases. Therefore, the limit of our reach is 6.266 billion light years by comoving distance, or at the redshift about $z = 1.5$. This is the upper limit of spatial distance we can ever reach if we travel at the speed of light. This is the current comoving distance from earth. It is not the observed signal distance from such location. Since it takes 6.266 billion light years to reach us from this location currently, its currently observable snapshot of itself must be from less than 6.266 billion light years ago because it was closer to earth. Therefore, from the perspective of an earth observer, the farthermost he/she can reach is less than 6.266 billion light years. We can use our equation $E(d, v)$ with a modified condition that computes the co-moving distance of the object where light is emitted. We set $v = c$ and $E_{comoving}(d, v) = 6.266$ Gly to back-derive the apparent distance d.

$$E_{comoving}(d, v) = G_{escape}(d, 0, v) + \sum_{n=1}^{t=T_{current}\, \&\, G_{escape}(x_j, j, v)=d} G_{escape}(x_n, n, v) \tag{10.1.39}$$

$$S(0) = \frac{E\left((T_{current} - d) + d\left(\frac{c}{v}\right) + ... + G_{escape}(x_{j-2}, j-2, v)\left(\frac{c}{v}\right)\right)}{3.262} \cdot$$
$$(d + G_{escape}(d, 0, v) + ... + G_{escape}(x_{j-1}, j-1, v)) \cdot \left(\frac{c}{v}\right)\left(\frac{G_{escape}(x_{j-1}, j-1, v)}{k}\right) \tag{10.1.40}$$

$$S(k) = \frac{E\left((T_{current} - d) + ... + G_{escape}(x_{j-2}, j-2, v)\left(\frac{c}{v}\right) + G_{escape}(x_{j-1}, j-1, v)\left(\frac{c}{v}\right)\right)}{3.262} \cdot$$
$$(d + G_{escape}(d, 0, v) + ... + G_{escape}(x_{j-1}, j-1, v) + S(0) + S(1) + ... + S(k)) \cdot$$
$$\left(\frac{c}{v}\right)\left(\frac{G_{escape}(x_{j-1}, j-1, v)}{k}\right) \tag{10.1.41}$$

Alternatively, one can use the sets of equations for calculating the comoving distance based on the redshift, whereas z is the redshift, $\Omega_m = 0.286$ is the total matter density, $\Omega_\Lambda = 0.714$ is the dark energy density,$\Omega_k = 1 - \Omega_m - \Omega_\Lambda$ represents the curvature, H_0 is the Hubble parameter today, and $d_H = \frac{c}{H_0}$ is the the Hubble distance.

$$E(z) = \sqrt{\Omega_r(1+z)^4 + \Omega_m(1+z)^3 + \Omega_k(1+z)^2 + \Omega_\Lambda} \qquad (10.1.42)$$

whereas comoving distance is:

$$d_C(z) = d_H \int_0^z \frac{dz'}{E(z')} \qquad (10.1.43)$$

Transverse comoving distance:

$$d_M(z) = \begin{cases} \frac{d_H}{\sqrt{\Omega_k}} \sinh\left(\sqrt{\Omega_k} d_C(z)/d_H\right) & \text{for } \Omega_k > 0 \\ d_C(z) & \text{for } \Omega_k = 0 \\ \frac{d_H}{\sqrt{|\Omega_k|}} \sin\left(\sqrt{|\Omega_k|} d_C(z)/d_H\right) & \text{for } \Omega_k < 0 \end{cases} \qquad (10.1.44)$$

Angular diameter distance:

$$d_A(z) = \frac{d_M(z)}{1+z} d_A(z) = \frac{d_M(z)}{1+z} \qquad (10.1.45)$$

The finally derived results indicates that a redshift $z = 0.5037$ satisfies the condition for a current comoving distance of 6.266 billion light years, and angular diameter distance is 4.1503 Gyr. That is, the farthermost object reachable appears to be at 4.1503 billion light years away. By setting Hubble constant to 69.6 km/s instead of 74 km/s to keep it consistent with the first result, our own equation $E(d, v)$ indicates 4.2343 Gyr in apparent distance. The discrepancy results from different values of Ω_m and Ω_Λ.

Careful analysis indicates that this conclusion is only partially right.[53] Although it is true that by the time the signals transmitted at 4.2 Gly away at the 4.2 Gya reaches earth, the original object that gave the signal's comoving distance will shifted to 6.266 Gly away, it will happen in the future because the light travel time is longer than 4.3 Gly due to the expansion of the universe. That is, the signals transmitted from 4.2 Gly away at 4.2 Gya is still on its way and have not reached us. In fact, this object's comoving distance is only 5.7 Gly at the current time. As a result, one needs to find the object that currently at a comoving distance of 6.266 Gly. Through back extrapolation, we found that the distance is at 4.642 Gly. However, light transmitted at 4.642 Gly away at 4.5 Gya still yet to reach us, so we must find an earlier snapshot of this object at a closer distance to earth. We compute the comoving distance of this object vs time and find the 4th order polynomial to great precision.

$$A_{62.7}(x) = 2.02 \cdot 10^{-9} x^4 + -4.2 \cdot 10^{-6} x^3 + 0.005141 x^2 + 0.419 x + 2894.93 \qquad (10.1.46)$$

We then run the simulation and find that only signals transmitted at 5.7948 Gya and from this object when it was located at 4.276196 Gly away by comoving distance are currently reaching earth.

We then compute the various comoving distance of objects vs time as well as the signals transmitted time and its comoving distance. The graph shows that the intersections between the red curve and the blue curves determine the age and the comoving distance of the signals transmitted which we are currently receiving.

$$A_{2.8}(x) = 6.3618 \cdot 10^{-11} x^4 + 216.576 \tag{10.1.47}$$

$$A_{9.3}(x) = 7.47 \cdot 10^{-11} x^4 + 0.127 x^{1.2} + 349.085 \tag{10.1.48}$$

$$A_{23.9}(x) = 1.1216 \cdot 10^{-10} x^4 + 9.52 \cdot 10^{-4} x^2 + 1329.15 \tag{10.1.49}$$

$$A_{37.9}(x) = 1.3281 \cdot 10^{-9} x^4 - 2.9 \cdot 10^{-6} x^3 + 3.56 \cdot 10^{-3} x^2 + 1800.73 \tag{10.1.50}$$

$$A_{50.3}(x) = 1.7281 \cdot 10^{-9} x^4 - 3.7 \cdot 10^{-6} x^3 + 4.58 \cdot 10^{-3} x^2 + 0.086 x + 2371.12 \tag{10.1.51}$$

$$A_{74.9}(x) = 2.7513 \cdot 10^{-9} x^4 - 6.1 \cdot 10^{-6} x^3 + 7.49 \cdot 10^{-3} x^2 - 0.21 x + 3592.25 \tag{10.1.52}$$

$$A_{94.3}(x) = 3.919 \cdot 10^{-9} x^4 - 9.2 \cdot 10^{-6} x^3 + 0.0112 x^2 - 1.144 x + 4712.45 \tag{10.1.53}$$

$$A_{149.8}(x) = 6.3863 \cdot 10^{-9} x^4 - 1.45 \cdot 10^{-5} x^3 + 0.0165 x^2 - 0.42 x + 6997.63 \tag{10.1.54}$$

$$y_{determinant} = -5.1264 \cdot 10^{-9} x^4 + 1.04 \cdot 10^{-5} x - 0.0123 x + 0.06 x + 6920.44 \tag{10.1.55}$$

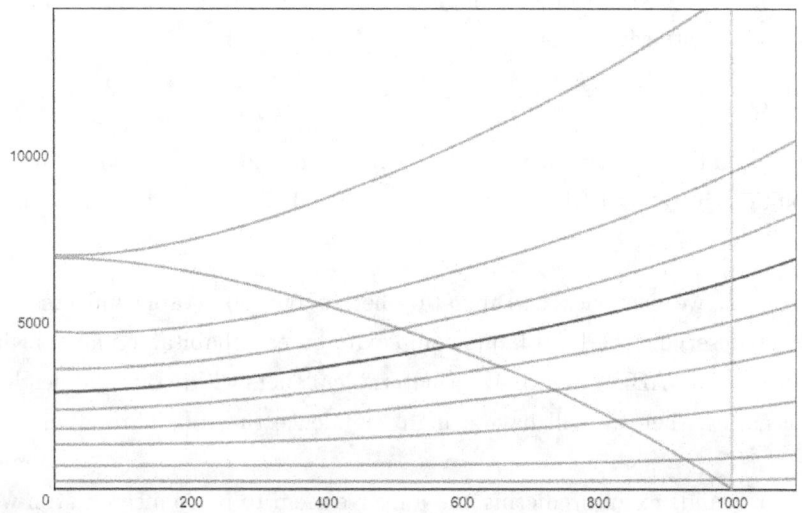

Figure 10.1.5: The age and the comoving distance of the signals determinant, 1000 = current time (100%), vertical axis represents lookout distance in 1 Mly and horizontal axis represents look back time in units of 10 Myr (at 0%, when x=0, it is equivalent to 13.8 Gya $13.8 - \frac{0}{1000}(13.8) = 13.8$ Gya. and when x=1000, at 100%, it is equivalent to now. $13.8 - \frac{1000}{1000}(13.8) = 0$ Gya. Our cross point is x=580, at 58%, it is equivalent to $13.8 - \frac{580}{1000}(13.8) = 5.796$ Gya). The black curve represents the catchable limit in which the current comoving distance at 6.266 Gly.

Certainly, this is a very limited range provided that the observable universe is 27.6 billion light years in diameter and 98 billion light years in diameters in terms of comoving distance.

10.2 Connected/Disconnected

Given that each expanding civilization can only expand up to 6.266 billion light-years in radius, we can then deduce a different kind of future of the universe. The one which we are already familiar is the observation and calculation done in recent years and more distant past by general relativity. The universe can be regarded either as acceleratingly open, open, flat, or closed based on its spacetime geometry, the effects of gravity, and the role of dark matter and dark energy. We now know that dark energy is playing an increasing role in the acceleration of the expansion of the universe, and likely that our universe will remain accelerating open to the indefinite future.

Knowing the cosmic distribution of intelligent species scattered within the cosmos, we come up with two different possibilities for the future of the universe: connected or disconnected.

In order for the universe to be connected, three possible scenarios are presented:

1. Given that the current emergence rate of expanding civilization is high so that more than one expanding civilization lies within the 12.532 Gy light years (6.266 Gy ly times 2) distance from earth since we know that the outermost limit of the reaches by expanding civilization originating from earth can only reach 6.266 Billion light years in radius. At the same time, the Background evolutionary rate BER is significantly higher than 1 so that even more expanding civilization emerges within the 12.532 Gy light years radius from earth in the future.

2. If the Background evolutionary rate BER is 1 or very close to 1, so that almost no other extraterrestrial civilizations arise in the future, then the current emergence rate permits the nearest arising civilizations have to be within 12.532 Gy light years radius from earth.

3. If the current emergence rate is so low that it takes a radius of more than 12.532 Gy light years, then the Background evolutionary rate BER must be high enough so that the expanding civilization eventually appears within 12.532 Gy light year radius.

The universe is disconnected if:

1. If the current emergence rate is so low that it takes a radius of more than 12.532 Gy light years, and the Background evolutionary rate BER is 1 or even less than 1, as the number of civilization emerges decreases as time passes.

Knowing the above constraints, we can back extrapolate whether our observable universe is connected or disconnected from the rest by observing the local fauna complexity on each habitable planets within the Milky Way galaxy. Once we confirm our YAABER and BER for habitable planets within our and neighboring galaxies falls into the first three scenarios, then we will have a high confidence that our part of the universe will be connected with the rest.

Nevertheless, we have a great confidence in predicting the universe seem to be connected even without visiting and collecting samples from each of the habitable planets.

Our current model predicts that the nearest extraterrestrial civilization lies 88.44 million light-years away. This conclusion, satisfies both scenarios 1 and 2, regardless of the Background Evolutionary Rate.

Now, assuming that the emergence of life on earth is unique and early that the emergence rate is much lower and the radius of emergence in larger than 12.532 Gy light years. If we take our observation on earth's biological development, then the universe is still connected because BER is much higher than 1, satisfying scenario 3. Since habitable planets only exist at 5.0 Gya at the earliest, given the prevalence of Gamma-ray bursts from the metal-poor past of cosmic history and the gradual development of spiral arms away from the galactic central cores, then the timing and the emergence of life on all habitable planet should be in a similar stage. Since oxygen readily reacts with other elements, no free oxygen will be available on any proto earth-like planet in its early evolutionary stages. Only when photosynthesis evolved among the bacteria type of living organisms on such a planet and gradually filled the oxygen sinks in both the oceans and the lands will eukaryotic cell and multicellular life becomes possible. Furthermore, the buildup of oxygen is directly related to the rise of continental plates, which is a logical consequence of the gradual cooling of a Earth-like planet. As a result, oxygen buildup can only become possible 2 to 3 Gyr after the formation of earth analogs. Since 5 Gyr has passed on the earliest possible planet to host life, and the average age of earth like habitable planets are younger than earth. Then it is very likely that currently earth-like planets with an average, a typical age of 3 Gyr form continental plates and start the buildup of oxygen and the evolution of the eukaryotic type of cell (oxygen-consuming cell). It takes another 2 billion years for the eukaryotic cell to evolve into multicellular life form including invertebrates arthropods, vertebrate fish, amphibians, reptiles, mammals, and birds. Additionally, it will take on average another 0.8 Gyr to reach human-like creature given that evolution of Homo sapiens is rather rare. 0.54 Gyr of multicellularity on earth + 0.26 Gyr of YAABER (1.76 Gyr of YAABER based on the current BER but it is translated into 0.26 Gyr into the future by taking the complexity transformation into consideration. Check Chapter 8 Section 8.9). Then, 2.8 Gyr into the future, a typical earth analog hosts earth like civilization and begins its expansion, and the average distance between each civilization will be within the size of their home galaxy.

Knowing that the universe is connected, then one can also calculate the minimum possible speed that one can expand to remain connected with each other. For civilizations led by biological species, colonization at a slower speed gives the species more choices even if it is not the most economic optimal one. (Obviously, less burden on taxpayers as we have shown from Chapter 8)

Finding the minimum expansionary speed to remain connected under scenario 2 one needs to specify the minimum distance between our nearest neighbors. If our neighbor lies x light years away and $0 < x < 12.532$ Gy light years, then the minimum speed requirements for a connected universe is (assuming both moves toward each other) derived from the inverse of the closed form for distance vs. speed:

$$y = D(x) = x\left(ax^b + f\right) = x\left(-0.0289229x^{0.549} + 0.988064\right) \tag{10.2.1}$$

$$S_{peed} = D^{-1}\left(\frac{x}{2}\right) = y\left(-0.0289229y^{0.549} + 0.988064\right) \tag{10.2.2}$$

or alternatively using the best fit for the inverse as:

$$S_{peed} = \left(\frac{x}{2}\right)\left(0.001563\left(\frac{x}{2}\right)^{1.38389} + 1.11719\right) \tag{10.2.3}$$

and we can illustrate it graphically by assuming that the nearest neighbor is 6.8 Gy light years away at the current comoving distance and both sides are rushing toward each other. Then both sides need to cover 3.4 Gy light years of distance. As a result, both sides need to travel at least at the speed of $0.45011c$ to stay connected.

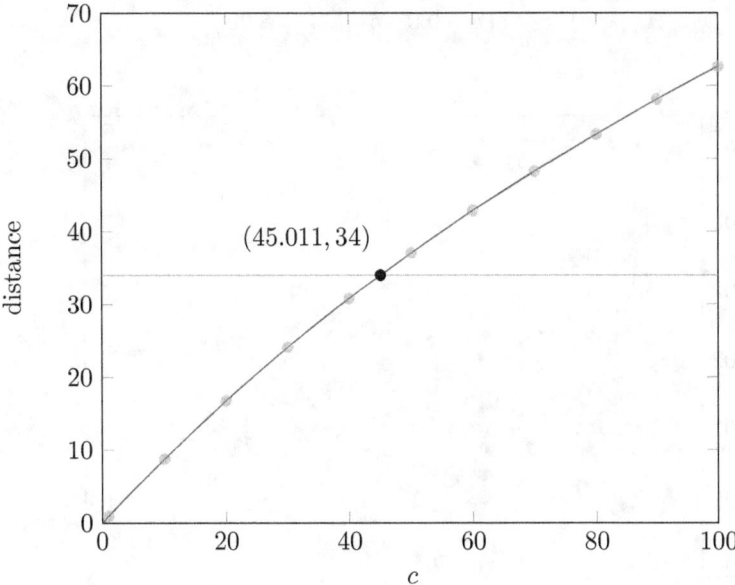

Figure 10.2.1: Reachable distance vs speed of light

Finding the minimum speed of expansion under scenario 3 requires first knowing the time at which the emergence rate falls below 12.532 Gy light years radius. Then, one needs to specify the exact distance between the neighbor to connect. Assuming it takes x years waiting time for the emergence of expanding civilization to fall below 12.532 Gy light year radius, and the Background evolutionary rate drops to 0 afterward so no closer civilization will appear. This can be possible if the condition of creating earth-like planets is much more stringent than what we have proposed and the growing metallicity of the galaxy prevents future arising terrestrial planets in general. Then, depending on the distance between us and our neighbor x years into the future, one can derive the minimum speed required for our own expansion. The speed will always be less than the expansion speed of our neighbor since we started now and spent the entire waiting time expanding toward the edge of our sphere of influence.

For various travel distance, the final rescaled distance before catching the target destination varies with speed, the plotted graph for a distance of 0.875 Gy light years, 2.4 Gy light years, 3.675 Gy light years, and 52 Gy light years respectively is presented. Each is represented by a different gradient curve, and their closest approximated closed form is listed below:

$$D_8(v) = (x - 0.0875)^{-0.71} + 8.29 \tag{10.2.4}$$

$$D_{24}(v) = (x - 0.24874)^{-1.505} + 27.874 \tag{10.2.5}$$

$$D_{38}(v) = (x - 0.363)^{-2.5} + 52.855 \tag{10.2.6}$$

$$D_{52}(v) = (x - 0.17318)^{-8.9} + 88 \tag{10.2.7}$$

One can easily interpret that at the slower speed the total distance one has to travel increases up to the point where the minimum speed required to catch the target destination. For targets at greater distances, the minimum speed required to catch the destination increases accordingly. With the known distance of target destination and the rescaled final distance, one can also determine the total time spent cruising toward such destination, which can be obtained by simply add adding a speed factor, an example is given for a travel distance of 0.875 Gy light years:

$$D_{8.75}(v) = (v - 0.0875)^{-0.71} + 8.29 \tag{10.2.8}$$

$$T_{8.75}(v) = \frac{1}{v}\left((v - 0.0875)^{-0.71} + 8.29\right) \tag{10.2.9}$$

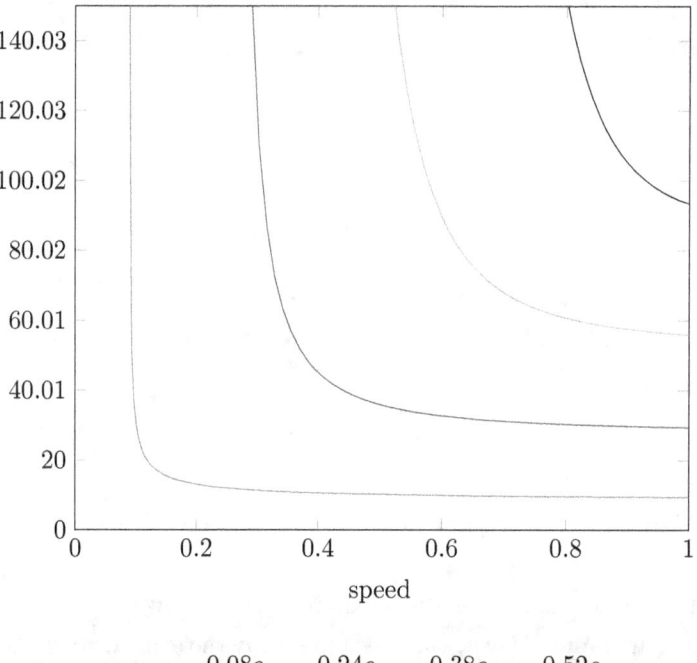

speed

———— 0.08c ———— 0.24c ———— 0.38c ———— 0.52c

Figure 10.2.2: Speed vs dist traveled

and can be generalized to any distance as:

$$D_x(v) = (v-a)^{-b} + c \tag{10.2.10}$$

$$T_x(v) = \frac{1}{v}\left((v-a)^{-b} + c\right) \tag{10.2.11}$$

As a result, one obtains the final distance to be traveled and the total time spent traveling for a given distance d and travel speed v. The inverse relationship between the travel speed v and travel distance d is derived based on the E(d, v) recursive function as we derived earlier.

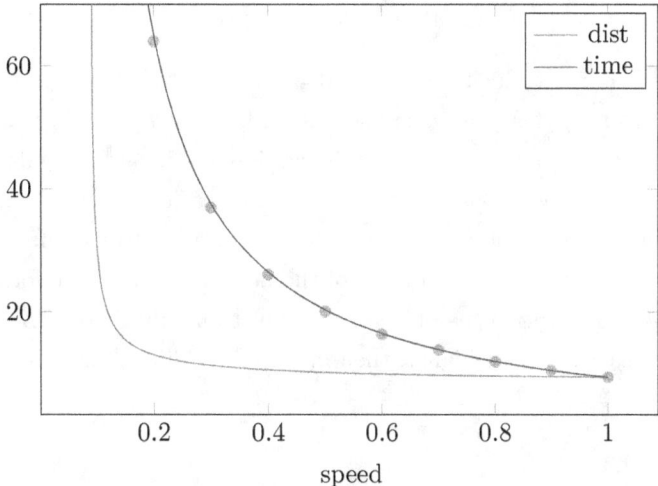

Figure 10.2.3: Distance and time

Then, one adds the waiting time for the emergence of the extraterrestrial neighbor and derives one's own minimum speed required to reach the destination.
and it can be shown that:

$$T(d,v) = \frac{1}{v} \cdot E(d,v) \tag{10.2.12}$$

$$S_{peed} = \frac{E(D_{emerge}, v)}{T(D_{emerge}, v) + T_{wait}} \tag{10.2.13}$$

Whereas S_{peed} is the minimum speed in which the universe can be connected. D_{emerge} is the emerging distance of the closest neighbor d light years away. $E(D_{emerge}, v)$ is the actual distance one needs to travel with a speed of v so that $\max_{\min v} E(D_{emerge}, v) < \infty$. That is, v is the minimum expansion speed required so that one can remain connected with one's neighbor. $T(D_{emerge}, v)$ is the total time it takes for it to be connected with its neighbor at the minimum speed v. T_{wait} is the waiting time in years before your neighbor emerges at a distance of D_{emerge} light years away.

However, a caveat has to be thrown. We assumed the T_{wait} is significantly less than the current age of the universe and distances between the nearest neighbor's redshift Z < 0.5. So that the overall cost reduction of a leisure expansion at a slow speed can be justified and outweigh potential costs of a slight increase in travel time and distance.

That is, having a head start over your neighbor is always advantageous. You can expand more slowly than your neighbor by starting expansion during the waiting period to reach the halfway distance between you and your emerging neighbor. Or you can expand at the full speed and expand well into your neighbors' supposed space.

$$S_{peed} = \frac{E(D_{emerge}, v)}{T(D_{emerge}, v) + T_{wait}} \le \frac{E(D_{emerge}, v)}{T(D_{emerge}, v)} \tag{10.2.14}$$

In the most likely scenario 1, in which we assume not only life on all habitable planet formed 4 Gyr ago or earlier have attained the status of multicellularity but actually evolved and filled on ecological niches and our nearest neighbor is 88.44 million light-years away and no other intelligence arises again. We solve the minimum speed requirement for the recursive function reaching its maximum finite value with specified constraint of 88.44 million light years (to have the universe connected):

$$\max_{\min v} E(88.44\text{Mly}, v) < \infty \tag{10.2.15}$$

we found that:

$$E(88.44\text{Mly}, 0.00492c) < \infty \tag{10.2.16}$$

Then, we need to only expand at most 0.00492c, about 0.492 percent of the speed of light in order for the universe to be connected. Of course, this is an over-estimation because the background evolutionary rate at 2.783 will guarantee the eventual emergence of expanding civilization within our own galaxy, rendering the closest civilization within reach even with conventional rocket speeds. The easiest way to verify the connectedness of the universe shall be relying on the next generation digital instrumentation to measure the atmospheric content of all potentially habitable planets. If any, or even most of the habitable planet we found have detected traces of oxygen, then it is likely that Eukaryotic type of organisms must already be present on them. If the oxygen concentration is comparable to earth or even higher, then multicellular life is likely to flourish.

10.3   Cosmic Nash Equilibrium

Most excitingly, if all arising extra-terrestrial industrial civilizations adopt wormhole based trade networks, then it can be shown that every expanding extra-terrestrial industrial civilization constructs their own networks before its contact with any nearby civilization. Once it does make contact with another, they can connect their network with that of the other. How the standard is enforced and agreed upon remain into the details of the technical specification, much like the ISO protocol or the 4G wireless network discussed in recent years. If the universe is indeed infinite or indeed extremely large, then, as predicted before, there can be infinitely many intelligent extra-terrestrial industrial civilizations expanding and adopting wormhole networks. When all these networks are connected, then an infinitely vast universe can be traversed from edge to edge in a finite amount of time if the cost of C is small. It will remain whether such network can be maintained into the indefinite future given the accelerating expansion of the universe or enough energy to maintain it indefinitely. It is to show, however, that *it is theoretically sound that an expanding cosmic civilization, by constructing its own wormhole network, will eventually have a chance to meet every other alien civilization within the universe in a finite amount of time, and every other alien civilization can also meet each other. I call this the Principle of Universal Contacts.*

Do extra-terrestrial industrial civilizations have to expand even given the incentives we have described above? Is there any other reason or incentive for them to expand? It seems that Nash Equilibrium, at the cosmic scale, can also play a role in the decision each civilization will make. If we formulate a utility function f where the maximization of the function is the total diversity of all possible industrial civilizations arising from all possible planets, then, every civilization should not expand and simply wait for their neighboring stars incubating the next industrial civilization. However, once a civilization is able to calculate how much ahead they are in terms of YAABER against the cosmic mean evolutionary rate, it will able to calculate their nearest neighbor. In our case, if we assume that the avian, reptilian, and mammalian level of genome complexity is the average of all terrestrial habitable planets, then our nearest neighbor is 88.44 million light-years away. Due to the speed of light, we will not be able to communicate with them prior; therefore, one has to ask for the optimal strategy one has to play knowing the presence of other players while with the absence of other information. The game choices are presented in the boxes below:

	Earth Expands	Earth did not Expand
Alien Expands	$(-\frac{R}{2}, -\frac{R}{2})$ *	$(0, R)$
Alien did not Expand	$(-R, 0)$	$(0, 0)$

Table 10.3.1: Cosmic Nash equilibrium strategies

It can be shown that the expansionary strategy is the cosmic Nash Equilibrium. If our neighbor expands at the same time as we are making decision, we also need to expand so that we will maintain our sphere of influence with a radius that is at the best half the distance between earth and the next extra-terrestrial civilization's origin. If our neighbor does not expand, then we have at best a sphere of influence with a radius the distance between earth and the nearest extra-terrestrial civilization. Of course, such expansionary strategy's gain is at the expense of potential and future arising technological civilizations suppressed locally by the dominant early forming industrial civilizations. So an expanding civilization causes a lose at the cosmic scale with a negative value for biodiversity.

Furthermore, a Nash Equilibrium is also played against all players across all temporal periods. It is almost inevitable that a rational player will probably know that there are possibly earlier arisen civilizations within its neighborhood, but by whatever reasons choose not to expand, and there are definitely going to be future arising civilizations may either choose to expand or not to expand. Since the player itself has no absolute understanding of its vicinity until it is fully explored, its best strategy is to expand because it can not communicate with the past nor can it communicate with the future without the cost of sacrificing its first-mover advantage. The civilization has only two choices. It can wait or travel near the speed of light until the next industrial civilization arises in the neighborhood.

As a result, a rational player will not wait until it meets its closest neighbor and will expand according to Nash Equilibrium from both temporal and spatial point of view. If we assume every arising extra-terrestrial industrial civilization is a rational player, then we can predict that the universe will be segregated by early arising industrial civilizations' sphere of influence and no particular civilization is significantly dominating since the earliest possible arising industrial civilization in the observable universe cannot be older than 0.232 Gyr (See Chapter 7). In other words, the universe is segregated into more or less even sized sphere of influence of different extra-terrestrial civilizations. Since wormholes are constructed, information and decision making between each player can be carried out with a cost of at most C. Since the universe is assumed to exist for a long time into the future, then it is expected that players in this multi-player game will seek cooperation for repeated transactions.

10.4 Looking Back in Time

Relativistic expansion of the wormhole network in coordination with other extraterrestrial industrial civilizations brings some remarkable results worthy examining. Wormhole network can be conceptually treated as a pipe with an entrance and an exit. One attaches the entrance of the wormhole at earth at the current time and stretches the wormhole near the speed of light toward the edge of our sphere of influence. Since the entrance and the exit of a wormhole stay connected at the same age, then our exit is not only 250 Myr light years in distance from our home planet but also at least 250 Myr years into the future. The wormhole network not only serves as a network connecting points in space, but it is also connecting points in time, as naturally indicated by its nature of connecting both space and time. [118] Whats more interesting is at the point when you leave your wormhole network and enters into your neighbors. Strange things happen. Assuming that traversing the wormhole itself takes a negligible amount of time and you started your journey immediately at the completion of the wormhole, then by crossing into your neighbors' network you can quickly reach their home planet, by doing so, you are not only traversing distance but also time. At the edge of two wormhole networks, you are 250 Myr light years in distance from our home planet and at least 250 Myr years into the future. As you stare back at earth, you saw earth as it is now, because signals of earth's light travel along with the expansionary phase of the wormhole for a period of 250 Myr. However, as you reach the home planet of the alien civilization, you are 500 Myr light years in distance from our home planet but your time is now the same as you started your travel from earth. Staring backward at the earth, which is 500 Myr light years away, you see the earth from its Permian era before the rise of Dinosaurs. Whereas the light from 250 Myr ago (it is not 500 Myr ago because it took us 250 Myr just to reach our civilization's boundary) just reaching you right now. If you don't travel nearly as far as to the home planet of the alien civilization and just simply cross into their territory, you can stare back to see every point of human civilization's development. We can view the earth as it was a century ago, a millennium ago, the start of the agricultural revolution, or our migration out of Africa, as watching a silent film. *I call this ability to skip ahead of signals escaped from the earth and staring back into our historical past Photographic Time Travel, Passive Time Travel, or Time Travel Mirage*, that is, you can observe the past events but not able in any way to affect it. Though this is not time travel in the purest sense, it is very comforting to validate its possibility. Photographic time travel brings some interesting consequences for cooperating extraterrestrial civilizations. As each civilization expands toward its civilization's boundary, it will be able to collect enough information about its intelligent neighbor since all of its evolutionary history is well within our grasp, quasi-military intelligence gathering type of manner; therefore, one knows well about their neighbors' vulnerabilities and comparative advantages. Similarly, our neighbor also knows our past history as it approaches our border. As a result, both sides will know the possible intention of the other, rendering war-like behavior less likely. More importantly, our neighbor will have a huge incentive to peacefully enter our network to collect historical information about their past and vice versa. In a sense, the information capture and storage of a civilization's past is probably one of the primary incentive to trade in an information market of the global cosmic interconnecting wormhole network.

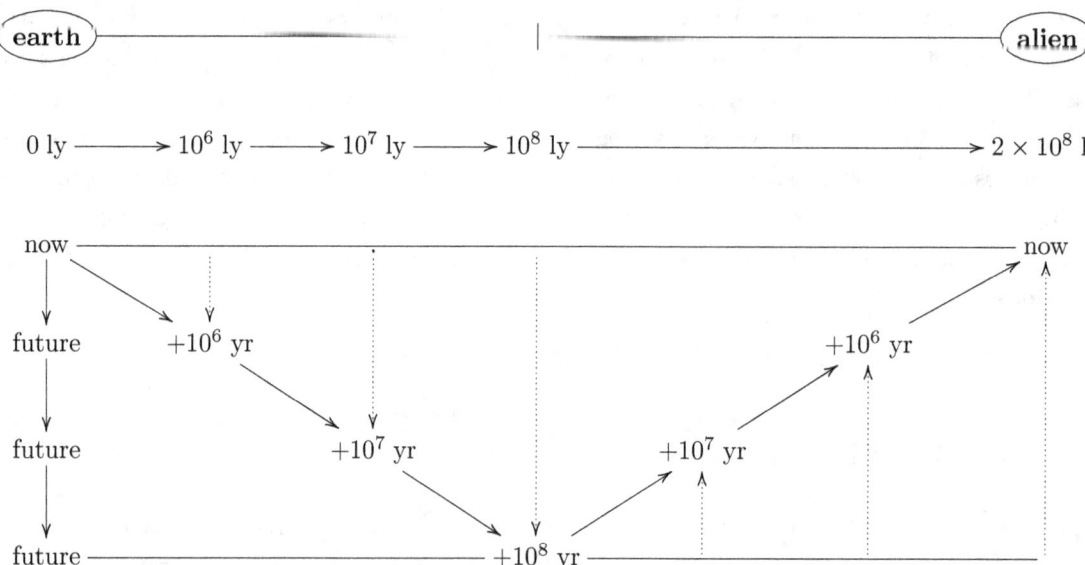

Figure 10.4.1: Photographic time travel

If the nearest civilization is 200 Myr years away, how can one be sure that earth's reflected lights and signals are visible given the enormous distance involved? In order to answer this question, one has to resort to the law of Telescope. In order to render the image of earth from a given distance, enough photons leaving earth must be captured. If the angle between two distant points is θ, the light in question has a wavelength of λ, and the size of your telescope is D across, then the smallest resolvable angle is approximately[117]:

536

$$\theta = \frac{\lambda}{D} \tag{10.4.1}$$

A telescope of an arbitrarily large size can be constructed by networking many smaller ones together. If something that's a large distance L away, and that is a size S across, takes up an angle of approximately:

$$\theta = \frac{S}{L} \tag{10.4.2}$$

So, if you want to be able to see something, you need [117]:

$$\frac{S}{L} \geq \frac{\lambda}{D} \tag{10.4.3}$$

Assuming visible light has a wavelength of $\lambda = 0.5 \cdot 10^{-6}$ m, and a viewing distance of $L = 2 \cdot 10^8 \cdot 63,241 \cdot 1.495 \cdot 10^{11}$ m (200 Myr in units of meters), and a human appearance size of 0.3 m, we can solve for the size of the telescope required.

$$D = \frac{\lambda L}{S} \tag{10.4.4}$$

$$\frac{\left(2 \cdot 10^8 \cdot 63,241.0771 \cdot 1.49597871 \cdot 10^{11} \cdot 0.5 \cdot 10^{-6}\right)}{0.3 \cdot \left(1.49597871 \cdot 10^{11}\right) \left(63,241.0771\right)} \tag{10.4.5}$$

$$= 333.333 \text{ ly}$$

By using the equation above, one finds that the size and diameter of the telescope (a spherical one to capture enough photons) have to be 333 light years across to resolve fine details of human size.

$$\frac{\left(2 \cdot 10^8 \cdot 63241.0771 \cdot 1.49597871 \cdot 10^{11} \cdot 0.5 \cdot 10^{-6}\right)}{3 \cdot 10^7 \cdot \left(1.49597871 \cdot 10^{11}\right)} \tag{10.4.6}$$

$$= 0.211 \text{ AU}$$

If one intends just to capture an image of earth, the requirement is significantly smaller, at merely 0.21 Astronomical unit, or just 31,535,768 km.

Compares to the size and extent of one's civilization's sphere of influence, the cost of construction should be negligible. If we use the scientific budget of United States as the rule of thumb as to what percentage of the expanding galactic civilization is willing to invest in historical data collection, then we expect a diversion of 1% of their galactic resources at its construction (assuming a galaxy disc with 100,000 ly disc radius and 9,800 ly disc height). We can expect the civilization builds 11,962,697 stations of telescopes with 333 light years diameter across. Each galaxy would host on average 19,904 stations. (there are 601 galaxies in a 100 million light years diameter)

$$N_{telescopeingalaxy} = \frac{\left(\pi \cdot 100,000^2 \cdot 9,800\right)}{\left(\frac{4}{3}\right) \pi \left(333.333\right)^3} \cdot \left(\frac{1}{100}\right) \tag{10.4.7}$$

$$= 19845.0595351$$

Each telescope's received signal is reassembled and connected through the wormhole network so that signal delay can be minimized. Therefore, telescopes based on Photographic Time Travel is both theoretically feasible and economically practical from their perspective.

To generalize, we can further extrapolate when the average speed of expansion of all industrial civilizations within the universe is at a small fraction of the speed light c instead of nearing the speed of light. Then, one crosses the boundary between our and our neighbor's civilization and staring back at earth will not able to see our earth from our past, because light from the past already traveled much further away. Assume the wormhole network connects to n^{th} degree neighbors from our nearest neighbor or neighbor of the first degree of separation, then, one can traverse the network and track earth's prehistory from our n^{th} degree neighbor. The equation is given by:

$$N_{thneighbor} = \frac{(T_{waitconnect} \cdot c - T_{nearestdist})}{R_{avg}}$$

(10.4.8)

Where $T_{waitconnect}$ is the average time it takes for expanding civilizations to complete the construction of interconnecting wormholes. $T_{waitconnect}$ is then multiplied by c to derive the amount of distance traversed by photons at the speed of light since our expansion started. $T_{nearestdist}$ stands for the distance between earth and our nearest neighbor in light years. R_{avg} stands for the average diameter size of a civilization in light years. A prudent thinker may point out that $T_{waitconnect}$ should be the time it takes for the human civilization to complete the network and joining with the rest or the slowest time an nth degree neighbor connects to the $n+1^{th}$ degree neighbor to form the bridge from earth to the civilization center of $n+1^{th}$ degree neighbor. However, an extremely slow expanding civilization will be outpaced by their neighbor so that territories they supposed to occupy will be occupied and developed by their neighbor. As a result, $T_{waitconnect}$ stands for the average

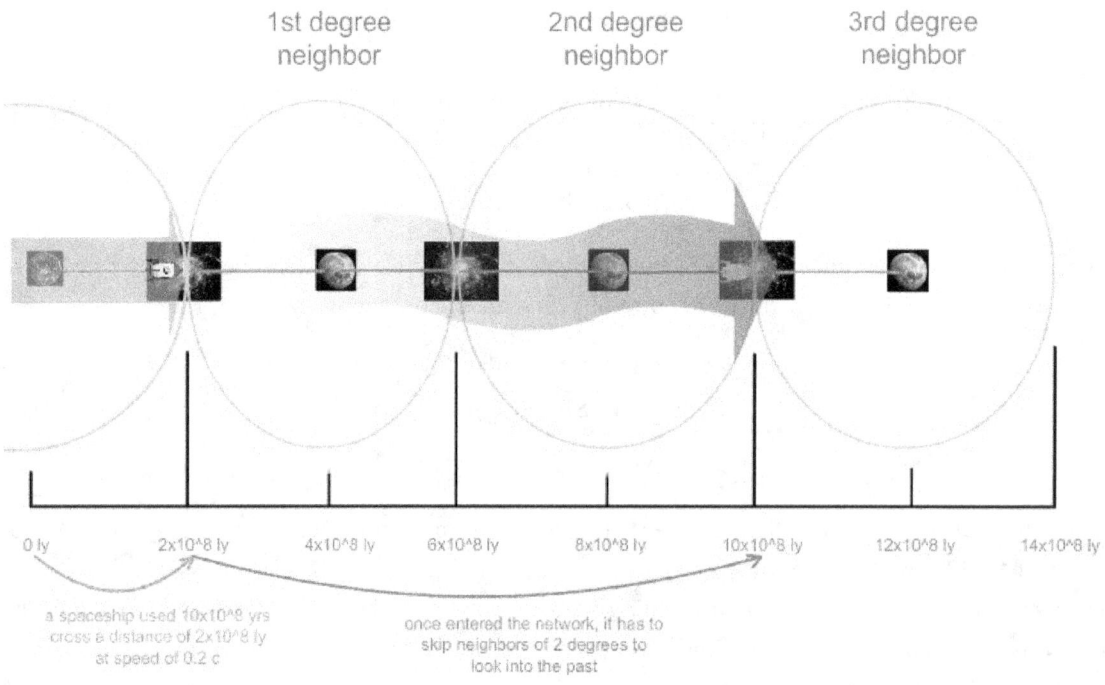

Figure 10.4.2: Reaches the nth neighbor

time of cosmic wormhole network development and expansion. Although all civilization may reach a consensus from the economic perspective to develop the network not close to the speed of light such as 0.9999c, but rather at 0.1c. In both cases, each civilization's past will be capturable from some remote distance, but the costs are not the same. The closer one has to travel to view one's past, the smaller the size of the telescope required. The further one has to travel to view one's past, the larger the size of the telescope required. Therefore, civilization before they met each other, will formulate models of cost and benefit analysis, to maximize their return. Nevertheless, viewing one's older past such as earth from the Hadean, and Archean era will require a higher cost to reach the same finite detail and resolution of eras close to today.

10.5 Universal Non-intentional Exclusion

Assuming the universe is connected and there exists advanced civilizations emerged prior to the existence of human or emerging currently and each of these civilizations started their expansion via wormhole networks at prior or current dates, then, these civilizations are expected to be connected in a future time frame, depending on the BER of the universe, such as 100 Myr into the future as our calculation have already shown in Chapter 7. Therefore, the entrance of each earlier arisen civilizations' wormhole network is connected to an earlier space time fabric and the boundary exit between each civilizations' expanding sphere of influence is connected to a later space time fabric than our current earth time. Since one can gather the information of a later space time fabric via worm hole and the construction of such network takes negligible amount of time in the builder's reference frame. One can say that the universe is *already connected* in earlier arisen civilizations reference frame. That is, for those currently connected not only have they expanded their spatial travel range but also their temporal range. That is, earlier to current successful civilizations have already reached us not at our current earth time but in our future time.

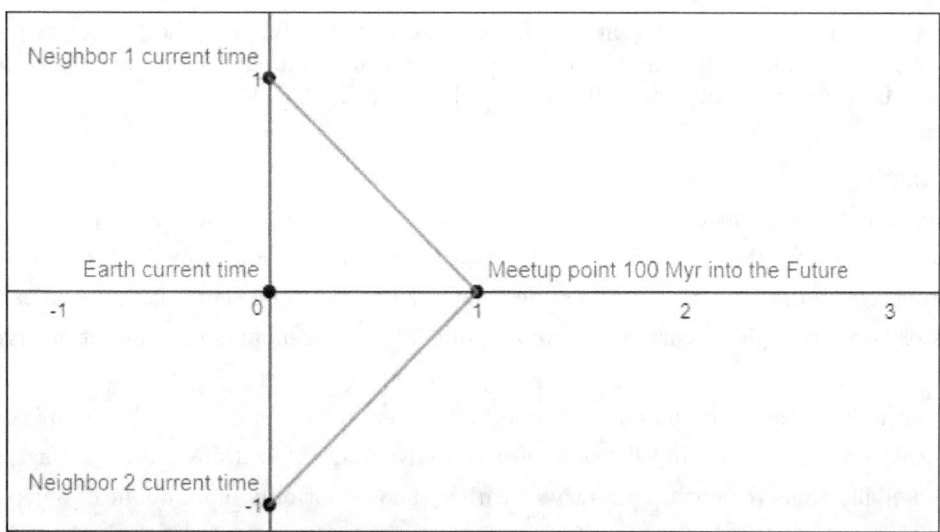

Figure 10.5.1: A simplified diagram of two adjacent neighbors each 100 Mly apart from earth emerge from the current time and connect to each other via wormholes at an exit at earth's location 100 Myr into the future (Assuming human based civilization does not expand or becomes extinct). The vertical axis represents the spatial distance in units of 100 Mly. The horizontal axis represents time in units of 100 Myr.

If humanity is one among many of the short-lived civilizations and eventually becomes extinct, then whichever closest expanding civilization reaches earth first via wormhole expansion will collect and sample the remains of our existence in the fossil records in the "future" (in earth's reference frame). However, they actually emerged in the past so they are in fact ancient explorers instead of future explorers. That is, for those earlier emerging civilization could have already examined and cataloged our entire history before ourselves even emerged on earth.

If future civilization on earth finally transitions at the technological singularity and completes the construction of wormholes ourselves and connects with our neighbors. For those of our neighbors emerged in the past or currently and connected with others, then they encountered expanding future earth's civilization in their past or current reference frame as well. They have met and established relations with those of our descendants and came to know the existence of our history and past through them. Although the accuracy of their comprehension and understanding of current time is no better than the native civilization's faithful book keeping of their past.

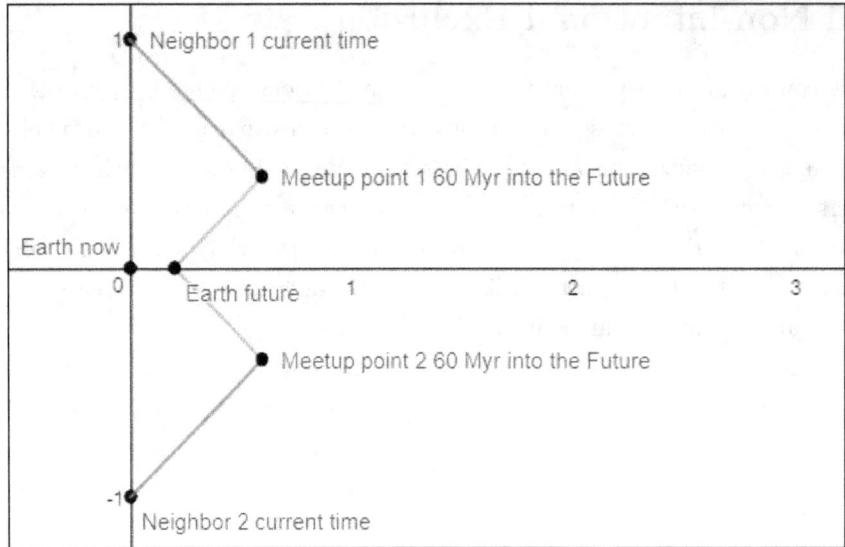

Figure 10.5.2: Assuming human based civilization will start expand 20 Myr into the future, two adjacent neighbors each 100 Mly apart from earth emerge from the current time then connect to our descendants via wormholes at exits 40 Mly away from earth and 60 Myr into the future.

This phenomenon brings two intriguing points:

First, the past, present, and future have to co-exist otherwise prior or currently emerging expanding civilizations can not connect with our future descendants. That is, advanced civilizations have been or are currently actively communicating, collaborating with our future descendants and views our current selves as part of the earth's collective history even though we currently alive and our future descendants did not yet emerge in our perspective.

Secondly, the zoo hypothesis for the Fermi paradox is somewhat correct in the sense. That is, earth is currently at a relatively primitive stage and is excluded from the more advanced civilizations' club. In fact, each of the already connected civilizations (emerged currently or earlier) knows not only more about of our past than ourselves but everything we don't know about our future. However, this exclusion is non-intentional enforced, it is merely a consequence of time required for civilizations expansion from a stationary observer's reference frame. Nevertheless, it is an universal feature and thus universally apply and dichotomize the civilizations into advanced and primitive ones. Although they do know more about our future, either as an extinct civilization or a successful expanding one, no other extraterrestrial civilizations can influence and determine the course of actions of far away primitive civilizations. They simply accept those results as historical fact. The only way to break the quarantine at the current time without waiting for anyone's arrival is when the primitive civilization itself have realized its own limitations and ready to expand and connect with every one else.

10.6   Temporal Convergence

Although any civilization's physical boundary is determined by its nearest neighbor, *the civilization's temporal boundary is only limited by the age of the universe.* The wormhole network, as stated before, not only connect points in physical space but points at different times. The wormhole network within any civilization can be more elaborately constructed to serve the purpose of connecting into the far future once it connects with its nearest neighbors.

In order to illustrate the possible outcomes, we simply assumed that the cost of traversing into the far future does not significantly differ from the cost of traveling to the near future in the simplest and most idealistic scenario. In reality, since ultimately death of star outpace of the birth of stars, and we have illustrated that the maintenance of wormhole requires energy and materials from stars, then the maintenance of wormhole far into

540

the future requires energy subsidy from the current time. Then, time travel shall be concentrated toward the near future due to economic reasons. The economic analysis of such scenario can worthy a whole thesis on its own.

As a result, each civilization will able to connect to all its own future timeline. However, such scenario leads to non-intuitive consequences. In the simplest case, in a future universe devoid of future human generations or any intelligent civilizations (assuming civilization disappeared toward the end times of the universe), then a simple trip to the end times of the universe will bring back valuable observation, confirmation, and answers to the cosmological and scientific questions one raised at the current time.

The another category of cases, in a universe filled with future human generations or any intelligence, civilizations still exist in the future. Then, the completion of wormhole can be shown that:

1. **No Information, matter, and energy exchange.** In the simplest scenario, civilization opens the connection to a future time period and immediately closes. Under this scenario, the future remains the same as if no wormhole network has ever been constructed between now and future. Since no information, energy, and matter has ever been exchanged between two different remote points in time. All of our past history can be grouped into this case since we can assume there was constant successful attempts constructing time travel corridors from the past to the present as well as from the present to the future. However, it is shut down every time after it is successfully constructed; therefore, the exchange of manipulation of information, energy, and matter by human activity are blocked.

2. **Information, matter, and energy's uni-directional exchange toward the future,** i.e. civilization skipping a time period. In a more complicated scenario, civilization opens the connection to a future time period and collectively decided to migrate (including its infrastructures) to such a later time period. This can be possible, for example, if a sudden resumption of an ice age, or an inevitable asteroid impact renders periods immediately following the current time uninhabitable. In this scenario, earth itself along with the rest of the universe still faithfully follows the logical sequence of temporal progression. Human civilization, on the other hand, forms a gap between the current time and the future chosen time. That is, accordingly to an outside observer, the population and activities of human civilization comes to an abrupt end at the current time and after many years of hiatus, suddenly collectively reappeared at the future time.

3. **Only information exchanges.** In this case, civilization opens the connection to a future time period and have contacted the future generations. The future generation appears much more advanced in science, technology and material wealth. However, due to specific reasons, such as limiting mass migration from the current time, decided to ban time travelers from the current time to the future. On the other hand, no future descendants want to live in the current time. Therefore, no energy and matter are exchanged. The future generation do allow their knowledge to be transferred to the current time. The absorption and utilization of advanced knowledge from the future at the current time does require a period of learning in which $t > 0$. If the utilization of advanced knowledge is instantaneous, that is $t = 0$, either the advancement of science and technology at the future time is infinitely advanced due to a closed, positive feedback causality loop or both the current time and the future time reaches the maximum ceiling of science and technology development permissible. In the case when knowledge acquisition bears a temporal cost of $t > 0$, the technology of the future generation becomes more advanced by transfer its knowledge to their ancestors. Without fueling this knowledge to their ancestors, their ancestor's breakthrough in science will be slower, consequently lower level of science attainment by the future generation. It is nevertheless self-consistent.

	Not Helping Ancestors	Helping Ancestors
Time	Science/Technology Index	Science/Technology Index
current time	2,000	2,000
AD 2100	3,000	6,000
AD 2200	4,000	10,000
AD 2300	5,000	14,000
AD 2400	6,000	18,000

Table 10.6.1: In this simple model, assuming current generation connects with generation from AD 2400, assuming without acquiring knowledge from the future, science and technology index grows at the rate of 1000 per century. With the knowledge of the future, the pace of progress grows at the rate of 4000 per century, 4 times the previous speed. Therefore, it is more wise to share knowledge with ancestors, though the first current visitor to AD 2400 may observe the science and technology index ranging anywhere between 6,000 and 18,000, implying that future generation collectively only shared a portion of its technology with the past and consequently, through self-consistent causality loop, resulting in lower science and technology index than attainable.

4. **Information, energy, and matter exchange.** In this case, civilization opens the connection to a future time period and have contacted the future generations. The future generations, not only interested in fueling their advanced knowledge, but also like to play a part in the history of their ancestors. They freely married their ancestors and had descendants, which in turn, after some generations, gives rise to the future generation that married their ancestors. However, the future generations are not permitted to perform certain acts in the first place which will violate the Novikov self-consistency principle, such as killing one's own grandfather or oneself. If the entire future generation decided to slaughter the entire current generation, the future generation will not arise at all since a generation of human has to originate from an ancient source and it does not comes into existence from nothing.

The four scenarios mentioned are the most basic possible scenarios. In reality, if wormhole network toward the future are constructed, it is likely that the reality will be a mixture of these scenarios and history becomes the sum of the collective decisions of all players with different choices. In a even more complicated scenario, each generation or individuals makes their own choices to live at different time periods, and a father may chose to live at a later time than their sons or grandsons. From an outsider's perspective, faithfully following earth's history's logical progression without time skipping and traveling, it may appears that the sons and grandsons are the ancestors to their parents and grandparents because they lived at an earlier time period than their parents. However, by tracing the causality logical sequence and considering time traveling, one can easily determine the logical ordering of grandparents, parents, sons, and grandsons. Therefore, despite time travel, all permissible solutions are nevertheless self-consistent and no causality is broken. Furthermore, if civilization and individuals decided to self-impose constraints on their temporal travel range, and in the future, civilization degenerates, generations from current time will provide lost knowledge to them, or bring them to current times, or current generation relocate to the future to help rebuild their civilization.

In all cases, once the wormhole network construction serving as the purpose of time travel tunnel completes, the temporal constraints is lifted, and there is a tendency that the information, knowledge, matter, and energy across all time periods is shared and converges. Such convergence can be dictated by a central authority, market forces through comparative advantages, or both. Therefore, *it is called temporal convergence.* In conclusion, one not only possibly met with all extraterrestrials and every extraterrestrials also met with each other, but *one also possibly met with all future descendants of their own civilization (and all future descendants of their own civilization met with each other). Finally, one will also possibly met with all future descendants from all extraterrestrials civilizations (and every extraterrestrial's every descendants also met with every other extraterrestrial's every descendants),* this completes the extended case and generalized case of the Principle of Universal Contact.

Chapter 11

Conclusion

11.1 Extra-terrestrials vs. Time

Finally, we predict the pattern of future arising extra-terrestrial industrial civilizations based on the Background evolutionary rate of gradually increasing biological complexity and habitable planets formation model. We rescaled the earth formation rate function and right shifted it to 4 Gyr later to indicate that only after 4 Gyr of evolution will multicellular life evolve on any habitable planets. Then we formulated an inverse tangent function that matches the background evolutionary growth curve but tapers off as the mean biocomplexity at the time on any habitable planet reaches parity with the progress of Homo sapiens led industrial civilization. We also discounted any habitable planet that moved off the main sequence that renders the planet uninhabitable (We assumed that once a planet attained biocomplexity on parity with Homo sapiens, the habitability continues for another 1.3 Gyr at most with a weighted average including stars with mass less than the sun, with a total window of habitability of 5.8 Gyr). Lastly, the expansion of the universe is also taking into account.

The rate of habitable planet formation rate based on derivation from Chapter 2:

$$E_{arth}(t) = \begin{cases} 0 & t \leq 2.4 \\ f_{earth}(t) & t > 2.4 \end{cases} \tag{11.1.1}$$

Biocomplexity growth diversity curve at early times was suppressed by Gamma ray bursts and capped at future times when the cosmic BER crosses the minimum threshold of intelligence emergence:

For $E_{merge}(k, t, F_0, B_{cs}) = 74839.8$ for the case of k=1, t=100 Myr, selection factor =1, and BCS=2.783:

$$P_{biocomplexity}(t) = \begin{cases} \frac{1}{47}\left(E_{merge}(k, t, F_0, B_{cs})\right)^{10(t-13.8)} & \frac{1}{47}\left(E_{merge}(k, t, F_0, B_{cs})\right)^{10(t-13.8)} < 47 \\ 47 & \frac{1}{47}\left(E_{merge}(k, t, F_0, B_{cs})\right)^{10(t-13.8)} > 47 \end{cases} \tag{11.1.2}$$

It is assumed that at the current time the chance of emergence of any civilization is at most 1 out 47 potentially habitable planets within the galaxy. We set our observational constraint earlier to 1 out 3 galaxies, or 1 out 141 potentially habitable planets. Since there are in total 47 potentially habitable planets within one supercontinent cycle per galaxy, it is assumed that once the emergence matches 47 civilizations per galaxy, we can cap the equation to indicate saturation of civilization within the selected time period. Furthermore, the curve can be scaled up or down since we are only concerned with the time that this curve reaches the saturation point, and the curve reaches the saturation point 64.4 Myr from now.

Universe expansion factor:

$$f_{cosmicexpansion}(t) = 0.822\,(t)^{\frac{2}{3}} + 0.0623\left(e^{\frac{t}{0.645}} - 1\right) \tag{11.1.3}$$

$$\frac{1}{f_{cosmicexpansion}(t)} \tag{11.1.4}$$

The rate of habitable planets production is expressed as those planets becoming habitable after 4.05 Gyr when oxygen accumulation becoming sufficiently high taken the example of earth's geologic history. (4.6 Gyr of earth history - 0.55 Gyr of multicellular history)

$$H_{earth}(t) = E_{arth}(t - (4.6 - 0.55)) \tag{11.1.5}$$

The rate of habitable planets destruction is expressed as 1 Gyr into the future, earth like planet revolving around a solar mass star will experience increasingly warm and bright sun. 1 Gyr window into the future is a conservative estimate since stars between 0.712 and 1 solar mass can have longer habitability window:

$$D_{earth}(t) = H_{earth}(t - 1) \tag{11.1.6}$$

The rate of habitable planets in existence is then expressed as the rate of production minus those moved off the habitability window:

$$F_{earth}(t) = H_{earth}(t) - D_{earth}(t) \tag{11.1.7}$$

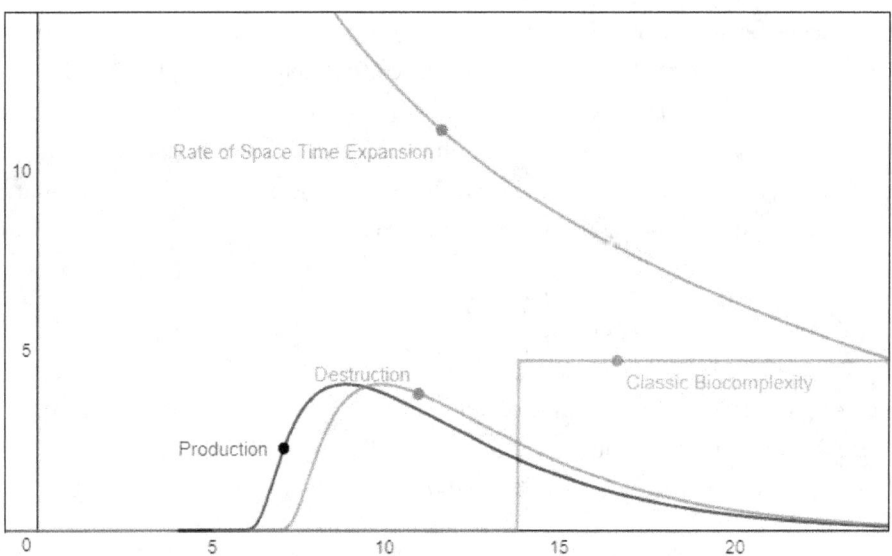

Figure 11.1.1: The rate of habitable earth production, the rate of habitable earth destruction, the rate of cosmic biological evolution, and the rate of cosmological spacetime expansion

The cumulative number of habitable planets across time:

$$P_{earth}(t) = \int_0^t F_{earth}(t)\,dt \tag{11.1.8}$$

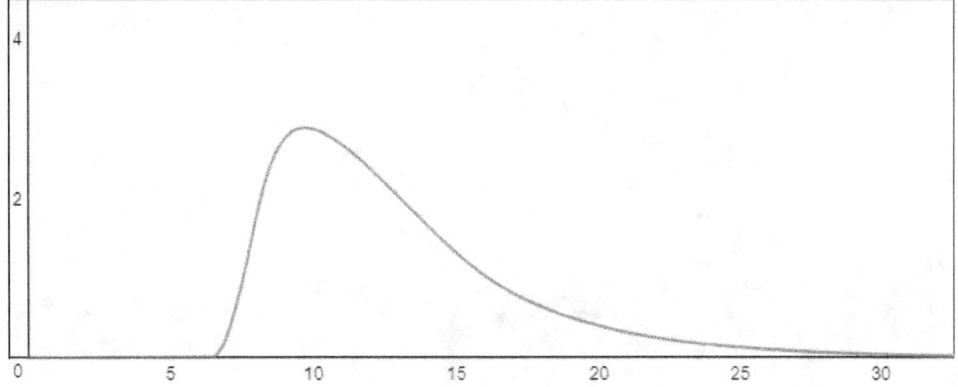

Figure 11.1.2: The cumulative number of habitable planets across time

Recall that we have set the earliest threshold window to fit our observational constraint. In this model, we simply assumed that the earliest possible planet deemed habitable before earth formed 45.5 Gya, 50 Myr earlier than earth based on the classic case, so that the rate of habitable earth production, the rate of habitable earth destruction are restricted to the time period > 13.8-0.05:

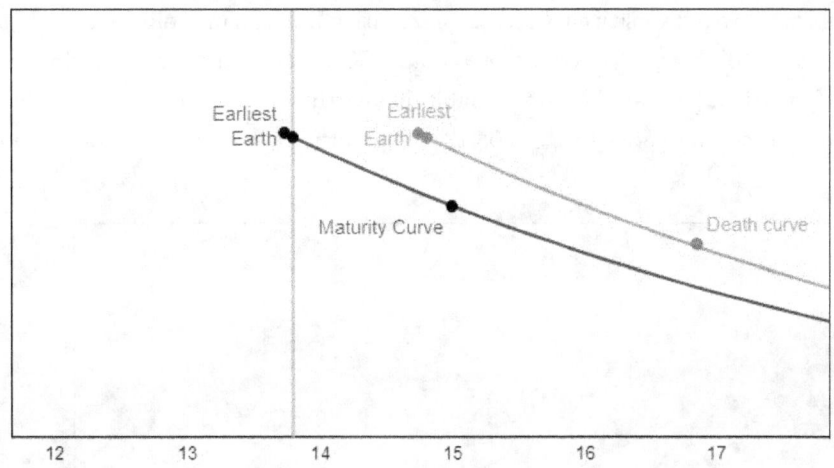

Figure 11.1.3: The restricted case

The final plot for the total number of extra-terrestrial civilization ever will arise is listed below.

$$N_{civilization}\left(t\right) = \frac{P_{biocomplexity}\left(t\right)}{f_{cosmicexpansion}\left(t\right)} \int_{13.8-0.05}^{t} F_{earth}\left(t\right) dt \qquad (11.1.9)$$

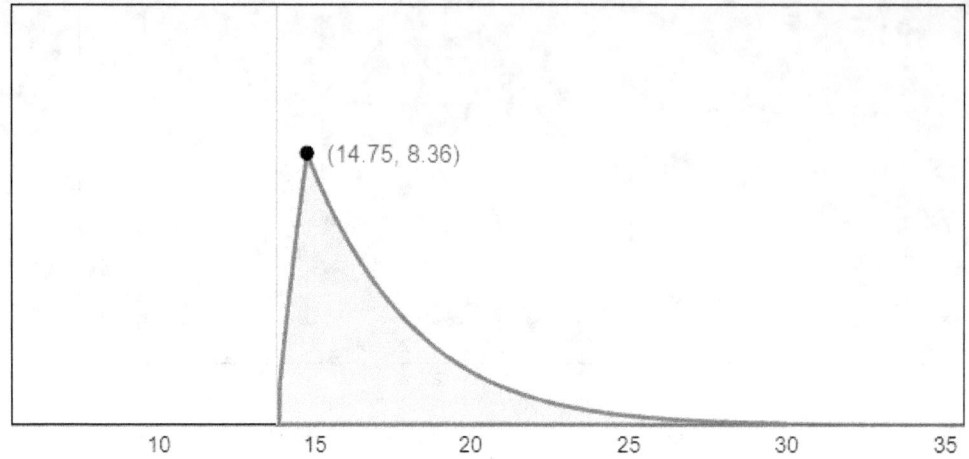

Figure 11.1.4: The number of extraterrestrial civilizations emergence in the future, note the rate decrease to zero in twice the time (30+ Gyr) since the Big bang.

The peak of the number of extra-terrestrial industrial civilization production is reached at 951 Myr assuming a BCS/BER of 2.783 and k = 1 (taking the curve max at 14.75-13.799 = 0.951 Gyr) into the future.

Next, we apply biocomplexity growth diversity curve with conservative Darwinian evolution, For $E_{merge}(k, t, F_0, B_{cs})$ = 2.783 for the case of k=∞, t=100 Myr, selection factor =1, and BCS=2.783. The earliest window now extends to 413 Myr before earth formation, so that the rate of habitable earth production, the rate of habitable earth destruction are restricted to the time period > 13.8-0.413. The curve reaches the saturation point 753 Myr from now.

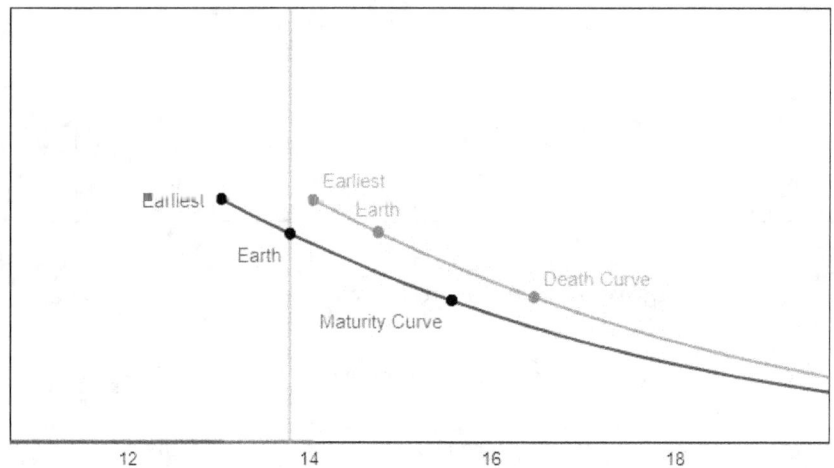

Figure 11.1.5: The new restricted case

The final plot for the total number of extra-terrestrial civilization ever will arise is listed below.

$$N_{civilization}(t) = \frac{P_{biocomplexity}(t)}{f_{cosmicexpansion}(t)} \int_{13.8-0.413}^{t} F_{earth}(t)\, dt \qquad (11.1.10)$$

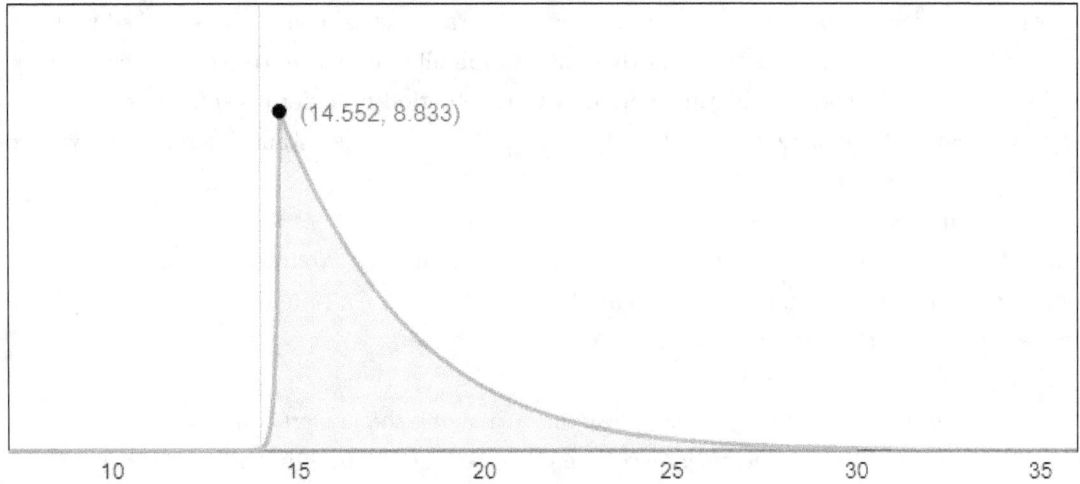

Figure 11.1.6: The number of extraterrestrial civilizations emergence in the future, note the rate decrease to zero in twice the time (30+ Gyr) since the Big bang.

The conservative case's slope rise is much slower than the classic case, and the only reason that conservative case peaked earlier than the classic case is because conservative case allowed the earliest threshold window to be much earlier, as a result, the first planets moved off the habitability window is also earlier, contributing to an earlier peak.

The model confirms our earlier discussion and the assumption that life can be started locally here on earth in a relatively easy process. As the universe is becoming less energetic, threats from cosmic regulating mechanisms such as Gamma Ray Bursts, and Quasar-like super black holes were dominant in the cosmic past since the Big Bang decreased. Our position is somewhat advantageous or early compares to the mean, and this is to accommodate our observation. As a result, one should expect that the temporal window is open and the universe is just becoming ripe for nurturing extra-terrestrial industrial civilizations and expecting a peak production in about 951 Myr. The number of future arising civilizations is primarily bounded by the star and planet formation rate, as the rate of production decreased to zero in 30 billion years, the last civilizations emerge no later than 16.2 billion years into the future. The expansion of the universe does play some role at minimizing the number of arising civilizations within the observable universe, but the effect is minor since the number of arising civilization dropped to 0 before the extremely rapid expansion of the universe starts.

Finally, one can also predict that the extraterrestrial civilization distribution function itself for t < 5 Gya and assuming biocomplexity growth follows a classic Darwinian evolutionary trajectory so that typical civilization emerges 500 Myr after the start of multicellular explosion (we assumed that a typical planet is capable of supporting advanced life lasting 1 Gyr, then, on average, a habitable planet underwent half of this cycle's length):

For time period $t_1 \leq t \leq t_2$:

$$P_{ast} + \int_{t_1}^{t_2} F_{earth}(t)\, dt = N_{earth} \propto N_{all} \tag{11.1.11}$$

$$P_{ast} = \begin{cases} \int_{t_1-1}^{t_1} F_{earth}(t)\, dt & t_1 - 1 \geq \text{earliest window} \\ \int_{\text{earliest window}}^{t_1} F_{earth}(t)\, dt & t_1 - 1 < \text{earliest window} \end{cases} \tag{11.1.12}$$

Whereas P_{ast} is the number of habitable planets ripe for the birth of civilization formed up to 1 Gyr earlier than the current selected temporal range and still deemed habitable at the start of our current selected time range. For $t_1 - 1 < $ earliest window based on observational constraints, then the integration is restricted to the earliest window. $\int_{t_1}^{t_2} F_{earth}(t)\, dt$ term accounts both the addition of new planets ripe for the birth of civilization within the current selected range and the subtraction of those planets that counted from the past time periods no longer habitable.

Unlike our earlier model which assumes that the total number of extraterrestrial civilizations is fixed within the temporal window of 5 Gya to 4 Gya, we now shows that the total number of extraterrestrial civilizations over the entire epoch also changes depends on the time period t_1 to time period t_2 under consideration.

As of now, the theory can be summarized to be based on the following key facts and inferences drawn from them:

There is a tremendous number of exoplanets (fact).

Earth, one among many, experienced increasing biological complexity through evolution (fact).

Given enough number of planets and enough time, somewhere else life emerges (inference).

Nature does not set limit on human progress (fact).

Near light speed, fast travel is possible (fact).

There is no known evidence of extraterrestrials visited earth or changing the universe (fact).

We must be arrived relatively early and the emergence rate decreases further back in time (inference).

With a lack of data regarding neighbors' behaviors, based on game theory, the optimal strategy for any civilization is to expand (fact).

There is a tendency for expanding civilizations to eventually universally connect (inference).

11.2   Final Thoughts

We have shown that any given extraterrestrial industrial civilization is highly likely to expand even if the most optimal energy and information utilization efficiency is reached locally given by Jevon's paradox. Furthermore, we have followed the Copernican principle as closely as possible by assuming currently all Earth-like habitable planets contain multi-cellular life forms evolved to the level of complexity and diversity comparable to the avian, mammalian and reptilian lineages observed on earth. We could not follow the Copernican principle in the strictest sense and assume all habitable planets have evolved into industrial civilizations because it is already contradicted by our current observation in our local galaxy clusters with a high level of confidence (null results from 50 years of SETI and recent WISE data from 100,000 nearby galaxies). On a larger scale, we found that we are possibly the first industrial civilization within the local supercluster. Given the extremely low probability of creating Earth-like conditions and evolving human-like creatures as the initial conditions. Coupled with suitable planet fauna (grass plants), it enables agricultural revolution. Finally, having abundant radioactive material so that project Pacer type of nuclear fusion device and sustaining an industrial civilization without facing collapse because the sun evolved with higher metallicity compared to earlier generations of earth harboring host stars. We followed the assumption led by Alexei and Sharov that the information complexity encoded in genome has been steadily increasing since the emergence of life along with the first terrestrial planets 9.3 Gyr ago. As a result, we have shown that no extraterrestrial industrial civilization could possibly arrive earlier than 119 Mya in the observable universe. As a result, we have eliminated the need to survey sky deeper than 119 Mya light years further out from the earth. (Assuming BCS/BER=2.783, k=1, and selection factor=1) We also showed the wall of semi-invisibility constrained by the known physical limit on the speed of light. That is, in order to detect the next appearing extra-terrestrial industrial civilization, one has to look further out into more distant regions of sky where the snapshot taken occurs at the time when such extra-terrestrial industrial civilization has not yet been evolved. By calculating and locating the distance between earth-bound observers for even earlier arisen industrial civilizations, the distance involved in guaranteeing their appearance, measured in light years, eventually always grows faster than their first arising date measured in years compares to the current time measured in light years. This is the strong case for observational Fermi Paradox.

Since we yet to thoroughly survey our sky and detect any signs of extra-terrestrial civilization, there remains a small possibility that detection is possible within our past light cone and we can follow the Copernican principle even more closely by assuming more fractions of the habitable planets have evolved into industrial civilizations and appearing at a closer distance to earth. However, we have shown again that such civilization likely to expand near the speed of light so that no prior warning and observation can be made from earth's vantage point. That is, the delay between the extra-terrestrials' first detection in the sky as an astronomical phenomenon and their physical arrival is shorter than cosmic timescale and even possibly human cultural time scale. This completes the second case of observational Fermi Paradox.

Both cases imply that the expansion can already be well underway yet we could not possibly make any detection. We have also shown that given the sheer size of the universe, the number of arising industrial civilizations is probably infinite in number. If each is driven by economic incentive and optimal strategy according to cosmic Nash equilibrium outlined, they will expand and construct their wormhole networks at or close to the speed of light. Eventually, every industrial civilization in the universe is likely to meet each other and connect with each other. If such scenario is possible, then one can reach to a much more, possibly close to infinitely distanced $(3.621 \cdot 10^6 \cdot 10^{10^{10^{10^{122}}}}$ light years if bounded) corners of the universe (much larger than the size of our observable universe) in a finite amount of time. Obviously, the question of size of the universe can be confirmed. In such a cosmically engineered universe, one should able to traverse into neighbor's network and witness the birth of one's own civilization through snapshot time travel.

This paper also set up a guideline for various disciplines. It is comprehensible that the future descendants of earth-based industrial civilization will calculate our cosmic evolutionary rate faithfully to millions of decimal precision just as we have calculated the value of π. In order to reach a more precise value for our Years against Background Evolutionary Rate, biologists should continue to find and predict the precise number of species of animals, plants, unicellular, and multi-cellular alike here on earth. Paleontologists should continue to refine their excavation of ancient fossil specimens especially those of the hominid lineage so that the probability of Homo sapiens as a species' emergence can be calculated to a great precision. Astro-biologists in the upcoming decades should observe the atmospheric signatures of habitable exoplanets closely and in the upcoming centuries to record and measure the local indigenous habitable planets' bio-complexity and diversity via robots. Astronomers will continue to survey the sky for megastructures with artificial origin. Eventually, we should have enough detailed knowledge about our own position in the cosmic family well before we ever make the first contact with our nearest industrial civilization neighbor.

Chapter 12

Special Chapter: Gravitational Effect on the Final Stellar to Planetary Mass Ratio

12.1   Overview

Over thousands of exoplanets have been discovered in the last decade, yet their mass relationship with their parent star is poorly investigated. We shall start with the solar system. We assume that the solar system is typical, that is, its formation via the accretion process from the collapse of molecular cloud possibly caused by a nearby supernovae explosion is the standard star creation process. Out of the stellar accretion process, local accretion started around ice gas giants Jupiter, Saturn, Uranus, and Neptune. Since the accretion material is primarily composed of hydrogen and helium and follows the same laws of physics, one expects any relationship derivable naturally extends to all.[74] We omit terrestrial planets and dwarf planets. Terrestrial planets formed through the bombardment of protoplanetary embryos. Unlike gas giants which reaches a critical mass (10 earth mass or greater) and results in a runaway mass accretion and form their own accretion disk, terrestrial and dwarf planet's mass is tiny and mass increase occurs only through collision with bodies of similar size.[74] As a result, Accretion disk never forms around a terrestrial-sized planet or smaller. There is no significant difference between the initial formation of gas giants and terrestrial planets. Both undergo a period of body collisions, but gas giants are able to gain mass greater than 10 earth mass and transitions into the next stage of planetary evolution. In a sense, terrestrial planets can be called failed gas giants. Gas giants can be called failed stars.

12.2   Empirical Law Derivation and Proof

We investigate the mass of the sun relative to the total mass of all planets, and the mass ratio is 745.29 to 1. The mass of Jupiter, relative to the total mass of all its moons (99% of its mass are from Io, Callisto, Ganymede, and Europa), is 4,829 to 1. The mass of Saturn, relative to the total mass of all its moons (99% of its mass comes from Titan, Enceladus, and Mimas) is 4,137 to 1. The mass of Uranus, relative to the total mass of all its moons (99% of its mass comes from Miranda, Ariel, Oberon, Umbriel, and Titania) is 9,430 to 1. Neptune, however, can not be used for data collection since its moon Triton is in a retrograde motion. Most theories support the idea that Triton is captured by Neptune from the Kuiper belt. According to the Nice Model, Neptune and Uranus were in a much tighter orbits around the Sun within 20 AU at its formation, and Neptune was also in the lower orbit. It later switched position with Uranus and ventured into the Kuiper belt and captured Triton. It remains a mystery regarding the original moons that formed through the accretion process. Those moons either fall into Neptune or get ejected by unstable N-body orbits of the additional captured moons.

As a result, we have four data points can be plotted and we find the best power-law fit for the empirical data, and we found that the following empirical law is describing the mass ratio relationship for the accretion process of different mass.

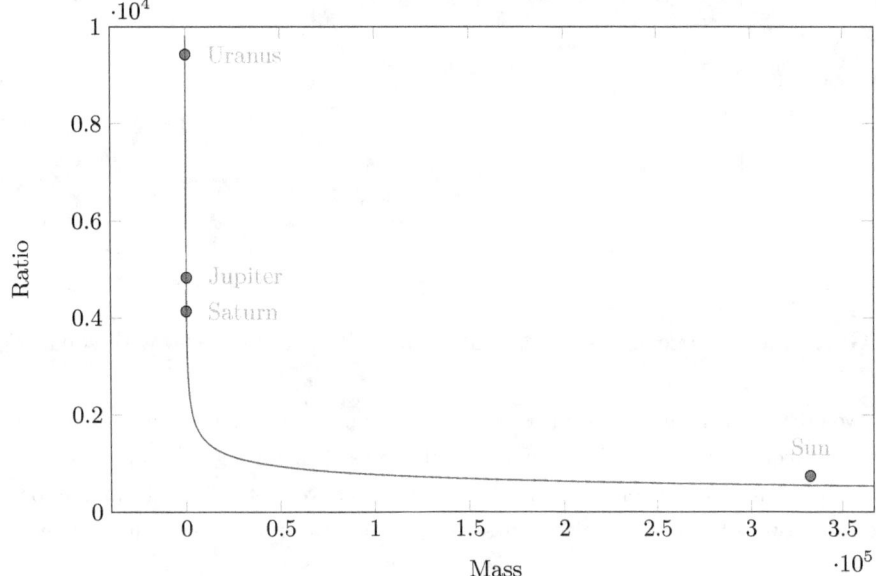

Figure 12.2.1: Primary to satellite mass ratio

$$y = 19,250x^{-0.2794} \tag{12.2.1}$$

If we plug in the solar mass value of 333,000 earth mass, the predicted total planet budget is 603.927 earth mass for the solar system.

$$\frac{M_{sol}}{19,250\left(M_{sol}\right)^{-0.2794}} = 603.927\, M_{earth} \tag{12.2.2}$$

This is higher than our solar system budget at 446.719 earth mass. With this budget estimate and sun's metallicity, the total water budget's upper bound is 6.107 earth mass. It is only 1.897 earth mass if oxygen is counted toward the composition of terrestrial planet creation. The total mass available to create celestial bodies with metals is 11.2934 earth mass (excluding H, He, Ne, and other trace non-interacting gas). The budget for terrestrial planet creation is 4.747 earth mass excluding oxygen (Those elements with a boiling point higher than 500 K, simulating the accretion disc temperature of the inner planets composed of elements such as C, Fe, Si, and Mg). It is 8.957 earth masses including oxygen assuming all terrestrial planet contains 47% oxygen as observed on earth.

Now, we can interpret the empirical derivation to that of the physical reality. We establish the following hypothesis:

The downward slope is simply a comparison of the gravitational limit/strength of the planets, arising from host stars of varying stellar mass, on their satellites and their effective strength in terms of a fraction out of the radius of the original planetary accretion that formed them.

Because stars with smaller mass also hosts planets with smaller mass, the planetary accretion Keplerian discs radius is accordingly smaller. The force between the planet and the edge of the planetary disc supposed to get stronger, but the increase in strength due distance shrinkage could not compensate the decrease in force strength due to decrease in planetary and its satellites mass at the edge of the planetary disc.

On the other hand, for more massive stars, the expected distance between their planets and planet's satellites grows due to greater planetary accretion disc mass, so the force of attraction suppose to decrease. However, the decrease in strength at the edge of the planetary disc due to distance expansion can not offset the gravitational attraction strength increase due to planetary and satellite mass increase.

For a point of reference, we introduce the graph for the attraction between two masses with a unit mass of 1 vs. r_0, the accretion distances.

$$\frac{1}{r_0^2} \tag{12.2.3}$$

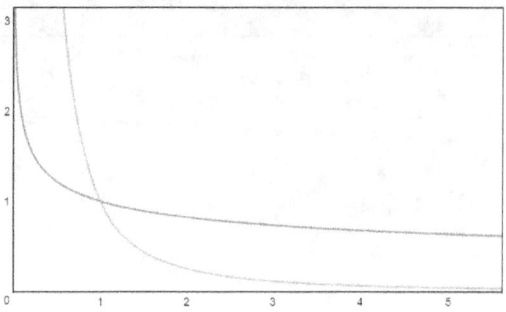

Figure 12.2.2: Gravitational force grows faster than primary to satellite ratio with separation<1

It is self-evident that for two objects with a fixed unit mass of 1, with distance separation below 1, the force of attraction grows faster than our empirical equation for two objects with variable mass (suggesting the empirical law holds smaller mass, slower decrease in separation distance, or both) With separation above 1, the force of attraction drops faster than our empirical equation for two objects with variable mass. (suggesting the empirical law holds larger mass, slower increase in separation distance, or both)

Once we derived the equation and formulated our hypothesis, we want to show how such relationship can be truly derived mathematically and proved using universal law of gravitation. One has to be very careful at deriving and proving this empirical law starting with N body interacting with each other through gravitational force and possibly viscosity in the accretion disk. Only 2 body simulation can be computed to precision, and 3 or N body problem is hard if not utterly impossible to solve. Making the matter worse, one needs to find the final accretion state which takes millions of years and millions of cycles of interaction and revolution, rendering the problem intractable.

We can, however, simplify this problem, by circumventing around the complexity of N-body simulation by simply stating the initial condition and the final condition of the system. The intermediate steps can be omitted to make the process tractable. The initial condition of the system is utterly a flat circular disk of a certain diameter, and the final condition of the system is utterly a sphere with a certain radius.

Once we have specified these conditions, we can start our derivation:

First, the radius of the accretion disk grows in size proportionally to the final mass of the star:

$$\left(\frac{4}{3}\pi r_0^3\right)\rho_0 = \left(\pi r_1^2 h\right)\rho_1 \tag{12.2.4}$$

where the left-hand side is the stellar mass with a volume based on its radius r_0 assuming a density of $\rho_0 = 1$ and the right-hand side is the volume of the disk which forms the original accretion disc with spread out density of ρ_1, where $\rho_1 < \rho_0$.

Then the equation simplifies to:

$$\frac{4}{3}\pi r_0^3 = \left(\pi r_1^2 h\right)\rho_1 \tag{12.2.5}$$

alternative, it can expressed as:

$$M_{sol} = \left(\pi r_1^2 h\right)\rho_1 \tag{12.2.6}$$

We rearrange the equation and solve for r_1, the radius of the disc:

$$r_1 = \left(\frac{M_{sol}}{\pi h \rho_1}\right)^{\frac{1}{2}} \tag{12.2.7}$$

However, we do not have any specific information regarding the value of h, the height of the accretion disc. Fortunately, the height of the accretion disc can be expressed as a fraction of the final stellar mass M. First of all, sun's accretion disc's height is bounded by $a_{earth} \cdot i_{earth} < h_{sun} < 3H_{earth}$. [?] That is, the height is bounded by the semi-major axis of earth times its angle of inclination and 3 times the Hill radius of earth. The disc height is correlated with the disc radius. This is self-evident from the lower bound $a_{earth} \cdot i_{earth}$. Since earth's inclination to the invariable plane is 1.57 degrees, we have $0.0174a < h_{sun}$.

The upper bound Hill radius is also correlated with the semi-major axis. It is known that the Hill sphere can be calculated from the equation and the final height is:

$$r_{\mathrm{H}} \approx a(1-e) \sqrt[3]{\frac{m}{3M}} \tag{12.2.8}$$

When eccentricity is negligible, it simplifies to:

$$r_{\mathrm{H}} \approx a \sqrt[3]{\frac{m}{3M}} \tag{12.2.9}$$

and the relation in terms of the volume of the Hill sphere compared with the volume of the second body's orbit around the first, whereas m = mass of earth and M = mass of the Sun:

$$3\frac{r_{\mathrm{H}}^3}{a^3} \approx \frac{m}{M} \tag{12.2.10}$$

As a result, we know that the height of accretion disc for creating a sun like star is:

$$3\frac{r_{\mathrm{H}}^3}{a^3} = \frac{M_{earth}}{M_{sol}} \tag{12.2.11}$$

$$\frac{r_H}{a} = \left(\frac{1}{3} \cdot \frac{M_{earth}}{M_{sol}}\right)^{\frac{1}{3}} \tag{12.2.12}$$

$$r_H = 0.010a \tag{12.2.13}$$

We constraint the height of the disc to be $0.0174a < h < 0.030a$, and choose our height as $h = 0.0237a$. Since the semi-major axis of earth is 1 AU, $h = 3,545,470$ km.

The only remaining term we needs to derive is ρ_1. Unfortunately, ρ_1 is much more tricky. We generally do not know the average size and density of the accretion disc, which changes throughout the stellar's formation history.

We can fix ρ_1 by set the typical size of the accretion disc to 100 AU for a sun like star. By fixing the size of the accretion disc, one can determine the density ρ_1. We will later adjust our values to fit the empirical observation.

Furthermore, we need rethink the final shape of the sun not as a sphere, rather a condensed, vertical cylinder pipe whereas the height of the pipe is the height of the original accretion disc assumed to be 3.54 million km, or 2.37 times the Hill radius of earth.

$$r_{sun} = \left(\frac{\left(\frac{4}{3}\pi (696,000)^3\right)}{\pi \cdot 3,545,470 \cdot \rho_1}\right)^{\frac{1}{2}} \tag{12.2.14}$$

$$r_{sun} = 356,078.96 \text{ km} \tag{12.2.15}$$

Then, the final stellar radius with density of $\rho_1 = \rho_0 = 1$ is 356,079 km. However, we know that initially the disc has to be much larger. One can parameterize to find values which closely matches 100 AU, without even knowing the density of ρ_1.

$$\frac{\left(\frac{\pi r_{sun}^2}{\pi}\right)^{\frac{1}{1.091}}}{149,597,870} \tag{12.2.16}$$

$$\approx 100 \text{ AU}$$

That is, if one were unaware of the density ρ_1 and $\rho_1 < \rho_0$ holds, then one can take the exponent with $\frac{1}{c}$ where $C < 2$ so that the final resulting radius of the disk is always larger than by taking the square root ($C = 2$).

Then, the radius of the disc can be expressed as:

$$r_{disc} = \left(\frac{M_{sol}}{\pi}\right)^{\frac{1}{1.091}} \tag{12.2.17}$$

Assuming that the height of the disc is proportional not just to earth's semi-major axis but to the entire disc radius, the height of the disc in terms of the mass of the star itself is then :

$$h = H \left(\frac{M_{sol}}{\pi} \right)^{\frac{1}{1.091}} \tag{12.2.18}$$

$$H = \frac{0.0237}{10^2} \tag{12.2.19}$$

The ratio is reduced by a factor of $\left(\frac{1}{100} \right)^2$ because we formerly assumed that the height of the accretion disc is in proportion to the semi-major axis of earth at 1 AU and we substituted by 100 AU.

A caveat must be raised. The strength of the correlation between the accretion disc radius and the height remains unknown. If a strong correlation exists, the height changes as the stellar accretion disc size and stellar mass changes. If a weak correlation exists, the height remains largely independent as the stellar accretion disc size and stellar mass changes. We can later adjust our values to fit the empirical observation and reach a conclusion regarding the strength of the correlation.

The final disk size is then given by the equation:

$$r_1 = \left(\frac{M_{sol}}{\pi \cdot H \cdot \left(\frac{M_{sol}}{\pi} \right)} \right)^{\frac{1}{1.091}} \tag{12.2.20}$$

Then, we use the law of universal gravitation, to define the attraction between two masses with gravitational constant G, G is assumed to be unit 1, because its real value is irrelevant for our discussion.

$$F = G \cdot \frac{x \cdot x}{r_1^2} \tag{12.2.21}$$

Where r$_1$ is the radius of previously derived results.

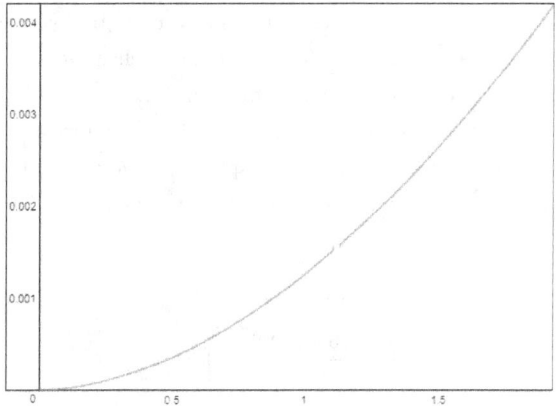

Figure 12.2.3: Unit mass pairs' attraction at their expected formation distance

Then, to show that planets and smaller stars ought to have a higher ratio of planet to satellites mass, we use the following equation:

$$r_{small} = \left(\frac{k \cdot M_{sol}}{\pi \cdot H \cdot \left(\frac{k \cdot M_{sol}}{\pi} \right)^{\frac{1}{1.091}}} \right)^{\frac{1}{1.091}} \tag{12.2.22}$$

$$r_{small} \approx \left(\frac{k \cdot M_{sol}}{k \cdot (M_{sol})^{\frac{1}{1.091}}} \right)^{\frac{1}{1.091}} \tag{12.2.23}$$

$$F = \frac{(kx)(kx)}{(cr_{small})^2} \tag{12.2.24}$$

Where k stands for the fraction of the original solar mass occupied by the smaller planet embedded inside the existing star, and c stands for the coefficient that re-adjusted, finding at which fraction of the accretion disc of the planet the force of planet's gravitation starts to dominates over the host star's. We used k=0.5 and c =1. The plots are shown below:

$$F = \frac{0.5x \cdot 0.5x}{\left(1 \cdot r_{small}\right)^2} \tag{12.2.25}$$

$$r_{small} = \left(\frac{0.5 \cdot M_{sol}}{\pi \cdot H \cdot \left(\frac{0.5 \cdot M_{sol}}{\pi}\right)^{\frac{1}{1.091}}}\right)^{\frac{1}{1.091}} \tag{12.2.26}$$

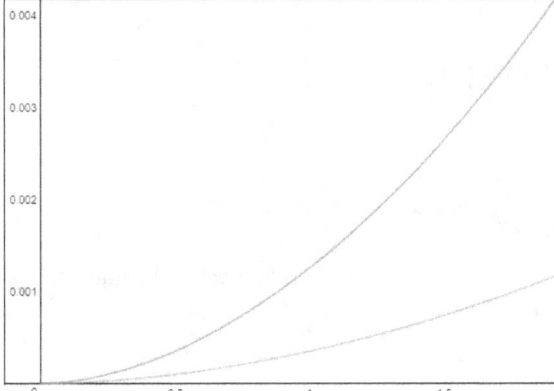

Figure 12.2.4: Two 1/2 mass pairs' attraction at their expected formation dist falls below 2 unit mass pairs' attraction at their expected formation dist

The 1/2th masses curve sits below the original unit mass curve for all range of masses, indicating that the attraction at the edge of smaller accretion disc between two smaller mass is always less than the attraction between two larger mass at the edge of its own accretion disc. Re-adjusting c=0.527, we have

$$F = \frac{0.5x \cdot 0.5x}{\left(0.527 \cdot r_{small}\right)^2} \tag{12.2.27}$$

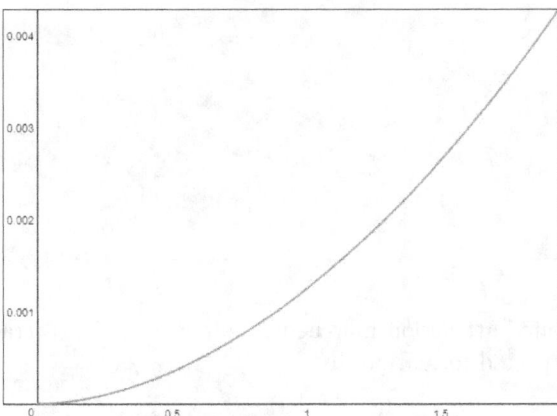

Figure 12.2.5: Two 1/2 mass pairs' attraction matched 2 unit mass pairs' attraction by narrowing their separation to 52.7 percent of their expected formation dist

where the 1/2th masses curve shifts upward vertically and coincides with the unit mass curve. This indicates that only when satellites were extending at or below 52.7 percent of the smaller mass's accretion disc, and it is maintained by the gravitational attraction of its planet, the rest is lost due to greater attraction by the host star.

We then used k=0.1 and c =1. The plots are shown below:

$$F = \frac{0.1x \cdot 0.1x}{\left(1 \cdot r_{small2}\right)^2} \tag{12.2.28}$$

$$r_{small2} = \left(\frac{0.1 \cdot M_{sol}}{\pi \cdot H \cdot \left(\frac{0.1 \cdot M_{sol}}{\pi} \right)^{\frac{1}{1.091}}} \right)^{\frac{1}{1.091}} \tag{12.2.29}$$

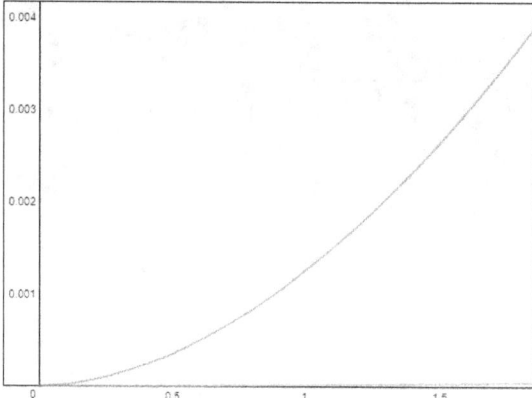

Figure 12.2.6: Two 1/10th mass pairs' attraction at their expected formation dist falls below 2 unit mass pairs' attraction at their expected formation dist

The 1/10th mass curve sits below the original unit mass curve for all range of masses, indicating that the attraction at the edge of smaller accretion disc between two smaller mass is always less than the attraction between two larger mass at the edge of its own accretion disc. Its force is also much weaker than a 0.5 solar mass case. Re-adjusting c = 0.119, we have

$$F = \frac{0.1x \cdot 0.1x}{(0.119 \cdot r_{small2})^2} \tag{12.2.30}$$

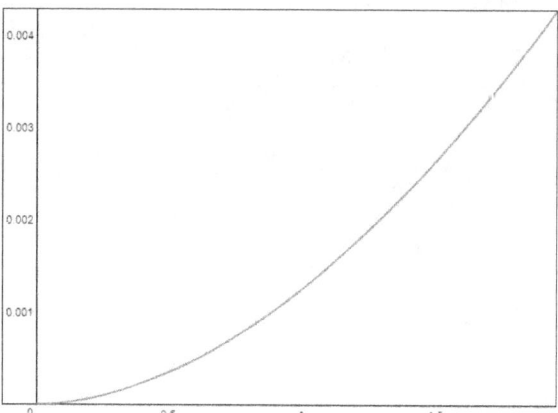

Figure 12.2.7: Two 1/10th mass pairs' attraction matched 2 unit mass pairs' attraction by narrowing their separation to 11.9 percent of their expected formation dist

Where the 1/10th mass curve shifts upward vertically and coincides with the existing unit mass curve. This indicates that only when satellites were extending at or below 10.55 percent of the smaller mass's accretion disc, and it is maintained by the gravitational attraction of its planet, the rest is lost due to greater attraction by the host star.

The percentage threshold radius for the substellar object of different mass' gravitational limit can be derived and its form:

$$F = \frac{x \cdot x}{(r_1)^2} = \frac{kx \cdot kx}{(cr_{small})^2} \tag{12.2.31}$$

$$\frac{x^2}{r_1^2} = \frac{k^2 \cdot x^2}{(cr_{small})^2} \tag{12.2.32}$$

divide both sides by x^2:

$$\frac{1}{r_1^2} = \frac{k^2}{c^2 \cdot r_{small}^2} \tag{12.2.33}$$

$$c^2 r_{small}^2 = k^2 r_1^2 \tag{12.2.34}$$

$$c^2 = \frac{k^2 r_1^2}{r_{small}^2} \tag{12.2.35}$$

$$c^2 = \left(\frac{kr_1}{r_{small}}\right)^2 \tag{12.2.36}$$

taking square root on both sides:

$$c = \frac{k \cdot r_1}{r_{small}} \tag{12.2.37}$$

Where k stands for the mass ratio relative to the host star, r_1 is the host star accretion disc radius, and r_{small} is the planet accretion disc radius. The plot is given below:

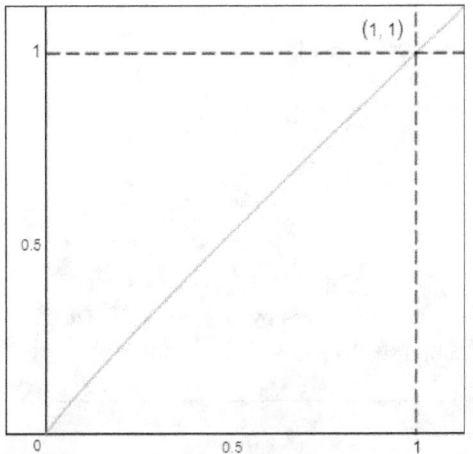

Figure 12.2.8: The amount of percentage of radius narrowing required based on their expected disc/stellar mass

The interpretation of the graph is the following:

In general, the smaller the planet, more of its accretion mass will be lost in the gravitational tug of war with its host star. It is ultimately a consequence of mathematics. Although the mass decreases linearly, and the gravitational attraction decreases to the 2nd power, but the accretion disc's radius for smaller mass only shrinks by the factor $\left(\left(\frac{1}{x}\right)^{\frac{-0.098}{1.098}}\right)^{\frac{1}{1.098}}$.

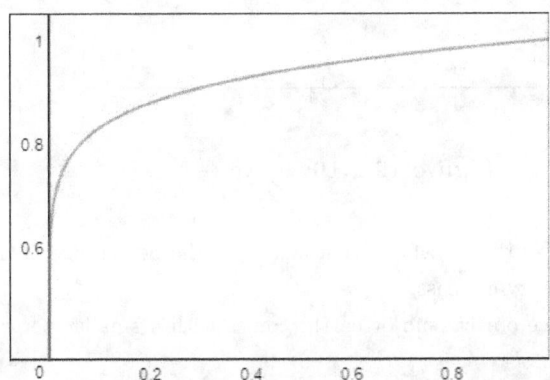

Figure 12.2.9: Disc radius shrinking sublinearly

The combined effect on the gravitational strength at the edge of the disc radius is:

$$f = \frac{r_{small}}{r_1} \tag{12.2.38}$$

$$f = \frac{\left(\dfrac{xM}{H \cdot \left(\frac{xM}{\pi}\right)^{\frac{1}{1.098}} \pi} \right)^{\frac{1}{1.091}}}{\left(\dfrac{M}{H \cdot \left(\frac{M}{\pi}\right)^{\frac{1}{1.098}} \pi} \right)^{\frac{1}{1.091}}} \tag{12.2.39}$$

$$f = \frac{\left(\dfrac{xM}{(xM)^{\frac{1}{1.091}}} \right)^{\frac{1}{1.091}}}{\left(\dfrac{M}{M^{\frac{1}{1.091}}} \right)^{\frac{1}{1.091}}} \tag{12.2.40}$$

$$f = \left(\frac{xM}{(xM)^{\frac{1}{1.091}}} \cdot \frac{M^{\frac{1}{1.091}}}{M} \right)^{\frac{1}{1.091}} \tag{12.2.41}$$

$$f = \left(x \cdot \left(\frac{M}{xM} \right)^{\frac{1}{1.091}} \right)^{\frac{1}{1.091}} \tag{12.2.42}$$

$$f = \left(x \cdot \left(\frac{1}{x} \right)^{\frac{1}{1.091}} \right)^{\frac{1}{1.091}} \tag{12.2.43}$$

$$f = \left(\left(\frac{1}{x} \right)^{\frac{-0.091}{1.091}} \right)^{\frac{1}{1.091}} \tag{12.2.44}$$

So the force of gravitation decreases relative to the unit mass by the curve $\frac{x^2}{f^2}$, which is just placed slightly higher than the force decreasing curve x^2 thanks to the sublinear decreases in disc radius.

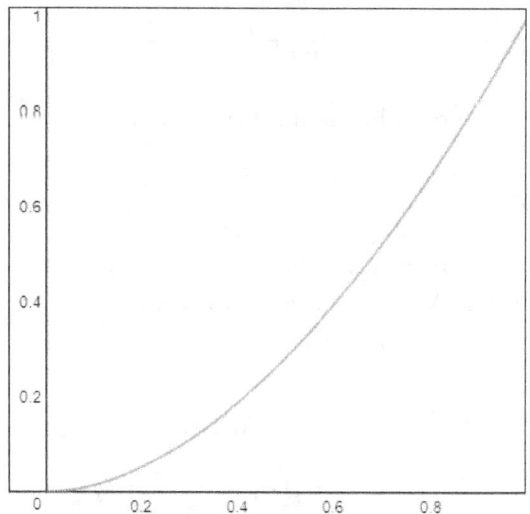

Figure 12.2.10: curve $\frac{x^2}{f^2}$

As a result, one can think the curve $\frac{x^2}{f^2}$ as the transformation factor for the percentage of the accretion disc under the gravitational pull of accreting object of varying mass.

For a smaller mass relative to a reference object (sun or earth) comes with a smaller percentage of the accretion disc under gravitational attraction by the factor

$$\frac{x^2}{f^2} < 1 \tag{12.2.45}$$

For a greater mass relative to a reference object (sun or earth) comes with a greater percentage of the accretion disc under gravitational attraction by the same factor

$$\frac{x^2}{f^2} > 1 \tag{12.2.46}$$

and for any values of x, the percentage threshold radius for the substellar object of different mass' gravitational limit we have $c = \frac{x}{f}$ so that:

$$\frac{x^2}{\left(\left(\frac{x}{f}\right)f\right)^2} = 1 \tag{12.2.47}$$

whereas $\frac{x}{f} = x\left(\frac{1}{x^{0.091}}\right)^{\frac{1}{1.091}} = c = \frac{k \cdot r_1}{r_{small}}$ [1]

$x\left(\frac{1}{x^{0.091}}\right)^{\frac{1}{1.091}}$ is derived based on the simplification of

$$c = \frac{k \cdot r_1}{r_{small}} \tag{12.2.48}$$

$$k = x \tag{12.2.49}$$

$$r_1 = \left(\frac{M_{sol}}{\pi \cdot H \cdot \left(\frac{M_{sol}}{\pi}\right)^{\frac{1}{1.091}}}\right)^{\frac{1}{1.091}} \tag{12.2.50}$$

$$r_{small} = \left(\frac{k \cdot M_{sol}}{\pi \cdot H \cdot \left(\frac{k \cdot M_{sol}}{\pi}\right)^{\frac{1}{1.091}}}\right)^{\frac{1}{1.091}} \tag{12.2.51}$$

then substituting becomes:

$$\frac{x\left(\frac{M_{sol}}{\pi \cdot H \cdot \left(\frac{M_{sol}}{\pi}\right)^{\frac{1}{1.091}}}\right)^{\frac{1}{1.091}}}{\left(\frac{x \cdot M_{sol}}{\pi \cdot H \cdot \left(\frac{x \cdot M_{sol}}{\pi}\right)^{\frac{1}{1.091}}}\right)^{\frac{1}{1.091}}} \tag{12.2.52}$$

simplifies to:

---

[1]The literal interpretation of $\frac{x}{f}$ is that, in order for $\frac{x^2}{f^2}$ to stay at the strength at the parity of unit mass at the unit distance for any arbitrary accretion radius, a factor of $\frac{f^2}{x^2}$, the inverse of $\frac{x^2}{f^2}$ is required. Moreover, in order to reach parity, we can only set limits on the radius not the mass, so the factor $\frac{f^2}{x^2}$ can only appear in the denominator, then, it becomes $\frac{x^2}{f^2\left(\frac{f^2}{x^2}\right)^{-1}} \Rightarrow \frac{x^2}{f^2\left(\frac{x^2}{f^2}\right)^{1}} \Rightarrow \frac{x^2}{f^2\left(\frac{x}{f}\right)^2}$. Furthermore, the factor has to be inside the sublinear decreasing radius expressed as f (as a fraction of f), so we have $\frac{x^2}{\left(\left(\frac{x}{f}\right)f\right)^2}$

$$\Rightarrow x \left(\frac{\frac{M_{sol}}{\pi \cdot H \cdot \left(\frac{M_{sol}}{\pi} \right)^{\frac{1}{1.091}}}}{\frac{x \cdot M_{sol}}{\pi \cdot H \cdot \left(\frac{x \cdot M_{sol}}{\pi} \right)^{\frac{1}{1.091}}}} \right)^{\frac{1}{1.091}} \tag{12.2.53}$$

$$\Rightarrow x \left(\frac{M_{sol}}{\pi \cdot H \cdot \left(\frac{M_{sol}}{\pi} \right)^{\frac{1}{1.091}}} \cdot \frac{\pi \cdot H \cdot \left(\frac{x \cdot M_{sol}}{\pi} \right)^{\frac{1}{1.091}}}{x \cdot M_{sol}} \right)^{\frac{1}{1.091}} \tag{12.2.54}$$

$$\Rightarrow x \left(\left(\frac{\frac{x \cdot M_{sol}}{\pi}}{\frac{M_{sol}}{\pi}} \right)^{\frac{1}{1.091}} \cdot \frac{1}{x} \right)^{\frac{1}{1.091}} \tag{12.2.55}$$

$$\Rightarrow x \left(\left(\frac{x \cdot M_{sol}}{\pi} \cdot \frac{\pi}{M_{sol}} \right)^{\frac{1}{1.091}} \cdot \frac{1}{x} \right)^{\frac{1}{1.091}} \tag{12.2.56}$$

$$\Rightarrow x \left(\left(\frac{x}{1} \cdot \frac{1}{1} \right)^{\frac{1}{1.091}} \cdot \frac{1}{x} \right)^{\frac{1}{1.091}} \tag{12.2.57}$$

$$\Rightarrow x \left(x^{\frac{1}{1.091}} \cdot \frac{1}{x} \right)^{\frac{1}{1.091}} \tag{12.2.58}$$

$$\Rightarrow x \left(\frac{1}{x^{0.091}} \right)^{\frac{1}{1.091}} \tag{12.2.59}$$

Hence, we have shown that this empirical law holds as a consequence of mathematics and physics.

Since the empirical law is represented in terms of mass ratio (not as disc area ratio in terms of its radius), one has to take the inverse of equation $c = \frac{k \cdot r_1}{r_{small}}$ to shows how much mass is lost during the accretion process. Furthermore, the mass ratio is a consequence of the tug of war of the gravitational force between the planet and the host star and occurred in a 2 dimensional plane, so we have:

$$\left(\frac{k \cdot r_1}{r_{small}} \right)^{-2} \tag{12.2.60}$$

The tug of gravitation also occurs immediately above or below the plane. We assume that a tiny bit of interaction occurs in a three dimensional space. Therefore, we added the disc height in terms of the disc radius. As a result, one has to check the diagonal distance from the accreting planet to the furthermost point above or below the edge of the accreting planet's effective gravitational perimeter.

$$r_{1diagonal} = \sqrt{r_1^2 + H \cdot r_1^2} \tag{12.2.61}$$

$$r_{smalldiagonal} = \sqrt{r_{small}^2 + H \cdot r_{small}^2} \tag{12.2.62}$$

$$\left(\frac{k \cdot r_1}{r_{small}} \right)^{-2} = \left(\frac{k \cdot r_{1diagonal}}{r_{smalldiagonal}} \right)^{-2} \tag{12.2.63}$$

$$\left(\frac{k \cdot r_1}{r_{small}} \right)^{-2} = \left(\frac{\sqrt{\left(\frac{kr_1}{r_{small}} \right)^2 + H \cdot \left(\frac{kr_1}{r_{small}} \right)^2}}{(1^2 + H \cdot 1^2)^{\frac{1}{2}}} \right)^{-2} \tag{12.2.64}$$

Fortunately, the ratio remains the same regardless of whether taking consideration of diagonal conditions or not. Ideally, we could re-adjusted it to 2.0237 or $2 + H$, so we have:

$$\left(\frac{k \cdot r_1}{r_{small}} \right)^{-(2+H)} \tag{12.2.65}$$

However, the curve does not match our empirical derived equation.

$$\left(\frac{k \cdot r_1}{r_{small}} \right)^{-2} \neq y = x^{-0.2794} \tag{12.2.66}$$

where the coefficient is not relevant and is reduced to 1. The graph shows inverse relationship; however, the curvature is different.

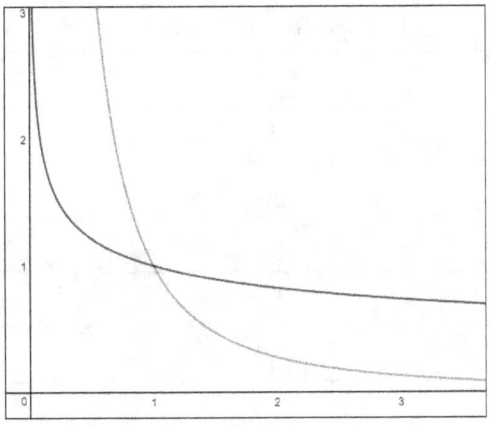

Figure 12.2.11: Our derivation does not match empirical derivation

In order to fit our equation into the empirical observation:

First of all, the original ratio can be simplified:

$$\left(\frac{k \cdot r_1}{r_{small}}\right)^{-2} = \left(\frac{k \left(\dfrac{M_{sol}}{H \cdot \left(\frac{M_{sol}}{\pi}\right)^{\frac{1}{1.091}} \pi}\right)^{\frac{1}{1.091}}}{\left(\dfrac{x M_{sol}}{H \cdot \left(\frac{x M_{sol}}{\pi}\right)^{\frac{1}{1.091}} \pi}\right)^{\frac{1}{1.091}}}\right)^{-2} \tag{12.2.67}$$

$$= \left(\frac{k \left(\dfrac{M_{sol}}{M_{sol}^{\frac{1}{1.091}}}\right)^{\frac{1}{1.091}}}{\left(\dfrac{x M_{sol}}{(x M_{sol})^{\frac{1}{1.091}}}\right)^{\frac{1}{1.091}}}\right)^{-2} \tag{12.2.68}$$

We then substitute both exponents $\frac{1}{1.091}$ with $\frac{1}{v}$ and $\frac{1}{j}$

$$\left(\frac{k \left(\dfrac{M_{sol}}{M_{sol}^{\frac{1}{v}}}\right)^{\frac{1}{j}}}{\left(\dfrac{x M_{sol}}{(x M_{sol})^{\frac{1}{v}}}\right)^{\frac{1}{j}}}\right)^{-2} = \left(\frac{k}{\left(\dfrac{x}{x^{\frac{1}{v}}}\right)^{\frac{1}{j}}}\right)^{-2} \tag{12.2.69}$$

[2]

Careful parameterization reveals that, in order for (2.67) to match the empirical result, one can substitute a range of value pairs for v and j to match the empirical observation.

The list of pair values are listed below, whereas the accretion radius size is defined by $\frac{\left(r_{sun}^2\right)^{\frac{1}{j}}}{149,597,870 \text{ km}}$:

---

[2]after substitution, formerly $f = \frac{r_{small}}{r_1} = \left(\left(\frac{1}{x}\right)^{\frac{-0.091}{1.091}}\right)^{\frac{1}{1.091}}$ becomes $f = \frac{r_{small}}{r_1} = \left(\left(\frac{1}{x}\right)^{\frac{1-v}{v}}\right)^{\frac{1}{j}}$ and $\frac{x}{f} =$ $x \left(\frac{1}{x^{0.091}}\right)^{\frac{1}{1.091}} = c = \frac{k \cdot r_1}{r_{small}}$ becomes $\frac{x}{f} = x \left(\frac{1}{x^{v-1}}\right)^{\frac{1}{j}} = c = \frac{k \cdot r_1}{r_{small}}$

$\frac{1}{v}$ (Disc Height Factor)	$\frac{1}{j}$ (Disc Radius Factor)	Accretion Radius
$\frac{1}{1.25}$	$\frac{1}{0.231}$	$2.53 \cdot 10^{35}$ ly
$\frac{1}{1.5}$	$\frac{1}{0.387}$	$7.97 \cdot 10^{15}$ ly
$\frac{1}{2}$	$\frac{1}{0.58}$	1,960,781 ly
$\frac{1}{5}$	$\frac{1}{0.93}$	6,948 AU
$\frac{1}{10}$	$\frac{1}{1.05}$ *	294 AU
$\frac{1}{20}$	$\frac{1}{1.1}$ *	96.55 AU
$\frac{1}{40}$	$\frac{1}{1.13}$ *	51.88 AU
0	$\frac{1}{1.16}$ *	28.8 AU
$-\frac{1}{30}$	$\frac{1}{1.2}$ *	13.74 AU
$-\frac{1}{1.4}$	$\frac{1}{2}$	0.0025 AU

Table 12.2.1: The possible pairs of $\left(\frac{1}{v}, \frac{1}{j}\right)$ which fits the empirical observation

and now:

$$\left(\frac{k \cdot r_1}{r_{small}}\right)^{-2} = y = x^{-0.2794} \tag{12.2.70}$$

The results shows a whole range of values permissible mathematically; however, only a very limited set fits astronomical observations. Only $\frac{1}{j}$ for $1.2 < j < 1.05$ is considered because we assumed that for typical accretion disc at the solar mass, its radius ranges between 13 AU to 300 AU. The pair (v=∞, j=1.16), equivalent to $(0, \frac{1}{1.16})$ and indicates that if the height of the disc is completely independent from the disc radius, the accretion disc size at solar mass should be 28.8 AU. In such a case, our equation simplifies to:

$$\left(\frac{k M_{sol}^{\frac{1}{j}}}{(x M_{sol})^{\frac{1}{j}}}\right)^{-2} = \left(\frac{k}{x^{\frac{1}{j}}}\right)^{-2} \tag{12.2.71}$$

in which the height variable is completely removed from the equation.

The interpretation of the parameterization indicates that the height of the accretion disc, in fact, shows a weak correlation with stellar mass. Within the astronomical permissible value ranges, The graph of $y = M^{0 \pm 0.03}_{0.10}$ indicates that the height of the accretion disk remain almost exactly the same across disks of different mass. The constant height is at least partially justified by equations describing the accretion dynamics concerning the height of the disc. It is stated that the scaled height of a Keplerian disk is given by [?]:

$$H = \frac{C_s T}{\Omega} \tag{12.2.72}$$

Where Ω is the angular Keplerian velocity, T is the local temperature, and C_s is the local sound speed. Whereas the angular velocity approximately equals the orbital velocity and is bounded by the escape velocity:

$$\sqrt{\frac{GM}{r_1}} \approx \Omega < \sqrt{\frac{2GM}{r_1}} \tag{12.2.73}$$

When G=1, M=x, we have:

$$H \approx \frac{1}{\sqrt{\frac{M_{sol}}{\left(\frac{M_{sol}}{(M_{sol})^{\frac{1}{v}}}\right)^{\frac{1}{j}}}}} \approx \frac{1}{\Omega} < \frac{1}{\sqrt{\frac{2M_{sol}}{\left(\frac{M_{sol}}{(M_{sol})^{\frac{1}{v}}}\right)^{\frac{1}{j}}}}} \tag{12.2.74}$$

It is assumed that C_s is the local sound speed and T does not significantly changes at all, it is speculated T may increase a little due to increase in density of the disc as the mass of the disc increases due to self-gravity, (which enables the curve to turn slightly positive and matches well with our original prediction $y = M^{0 \pm 0.03}_{0.10}$) but in general, it is shown that the scaled height of the disk does not change much as the mass of the disc increases. That is, the scaled height for the accretion disk is largely independent of the mass of the accretion disk.

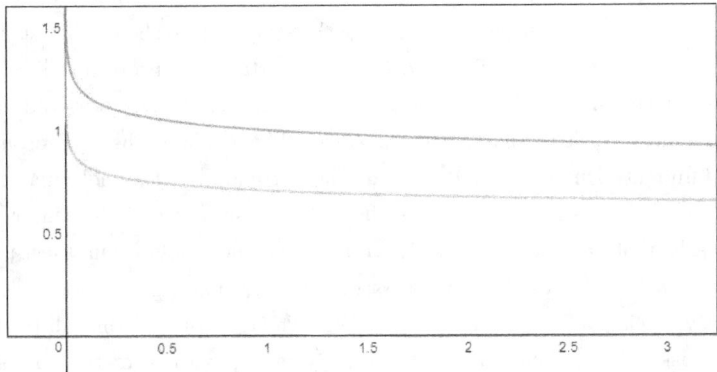

Figure 12.2.12: Scaled height for the disk remain pretty much constant as the mass of the disk increases

Not only the height of the accretion disc follows a weak correlation with stellar mass within the permissible range of astronomical observation, but the height of the disc also follows a weak negative correlation with accretion disc radius. The height of the accretion disk decreases as the accretion disc density increases, indicated as the stellar accretion radius decreases. This implies that as the accretion disc density increases, the gravitational pull and possibly viscosity on the disc increases the self-gravity of the disc. Self-gravity dominates when the initial accretion disc radius drops below $(\frac{1}{v} = 0, j = 1.16)$ which is $\frac{\left(r^2_{sun}\right)^{\frac{1}{1.16}}}{149,597,870 \text{ km}} = 28.8$ AU for solar mass.

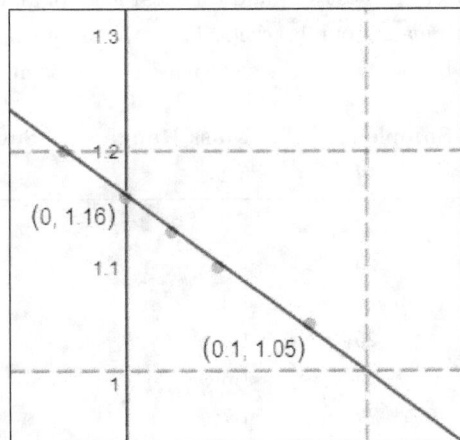

Figure 12.2.13: The greater the accretion disc density (the smaller the accretion disc radius), the greater self-gravity of the disc exerted on the disc height

In conclusion, the literal interpretation of this law can simply be stated as the follows. For any given celestial body, the size of the initial accretion disk and its radius grows as their mass increase. This also means that the orbits of the planets formed inside the disk also extends further out. However, with an increase in mass comes with an increase of gravity. Furthermore, in response to the increase in mass, the gravitational attraction from the host star and planet grows fast enough to compensate the decrease of gravitational attraction due to the increase in distance of the orbits of the forming planets.

If no other gravitational forces act on the moons of gas giants or planets around the star, then, regardless of the strength of the gravitational field, the planets stay. However, the formation of moons around gas giants is in a constant gravitational tug between the host star and the hosting giant gas planet. If the gas giant's mass is greater, then the moon's orbit extends further out. However, the planet's gravitational attraction on the moon is greater still, so, at the end of the day, the gas giant wins. The mass ratio of gas giant to moons is then low and more massive moons are retained.

The formation of planets around stars are also in a constant gravitational tug between the stars and nearby stars. Nearly all stars formed in a star nursery, one of the most famous are the Orion nebulae. Each star forms within its own pocket roughly few thousand AU across, and its outer planets and gas giants can be gravitationally attracted to its neighbors. If the stellar mass is more massive than its neighbors, then its planets are more likely maintained. This is most dramatic in a scenario where a red dwarf's nursery surrounded by class O stars with mass 100 times greater than the sun, and many protoplanets of the red dwarves are seized by the class O stars. Therefore, the planet to satellite mass ratio also extends to the stellar to planet mass ratio. Very recently, studies have shown that star formation through collapse and fragmentation may based on molecular cloud with a mass-distance relationship. [?] That is, the density profile of molecular cloud is non-uniform, and the greatest density occurs at the central cores and clumps. As a result, most massive stars formed at the center and least massive at the edges. This explains the IMF profile of stars in which the least massive are the most abundant. Due to lower density at the peripheral of the molecular cloud, least massive stars' fragmentation size can potentially be as large as those of the most massive ones at the center, altering our earlier assumption that the cloud of nearly uniform density. This alteration does not violate our thesis, since, the effective radius in which the least massive star can hold onto remain fixed. In fact, enlarging the formation disk size for the less massive stars actually decreases their planetary mass budget. We formerly assumed that all planetary mass budget formed within a smaller accretion disk size based on uniform density. By enlarging the accretion disk size, more accretion material for planet building is lost in the tug of war between inner shells of more massive stars within the nursery.

12.3   Stellar Data and Derivation

We now can generalize our equation to all exoplanet data we have so far. We used the European exoplanet catalog and filtered out certain data (including binary brown dwarves) and maintained a list of 1,189 candidate sample points. For a system with multiple planets, we sum the total mass of the planets and label them as a single sample point because we are only interested in the mass ratio between the host star and the total mass of all of its planets. We sort the data points by their solar mass and place them into mass group brackets. From the table below, we can see that the number of planetary systems within each mass range bracket where the solar mass bracket is highlighted with an asterisk.

Mass Range	Samples	Mass Range	Samples
mass	N	mass	N
0.2	13	1.2	105
0.3	25	1.3	55
0.4	27	1.4	37
0.5	26	1.5	33
0.6	35	1.6	18
0.7	72	1.7	14
0.8	106	2	14
0.9	168	3	22
1*	205	–	–
1.1	135	–	–
Total			**1110**

Table 12.3.1: Stellar mass breakdown and their numbers

The results of each bracket are plotted in the graphs. The vertical axis represents the total number of planetary systems with a given mass. The horizontal axis represents the range of total mass of its planets. The mass is represented in the base of 340 earth mass, which is the derived mass budget for a star with one solar mass from our earlier empirical law. We also run statistical distribution (generalized extreme value, normal distribution) on each plot, and recorded the inflection point on each plot. The result is reported below:

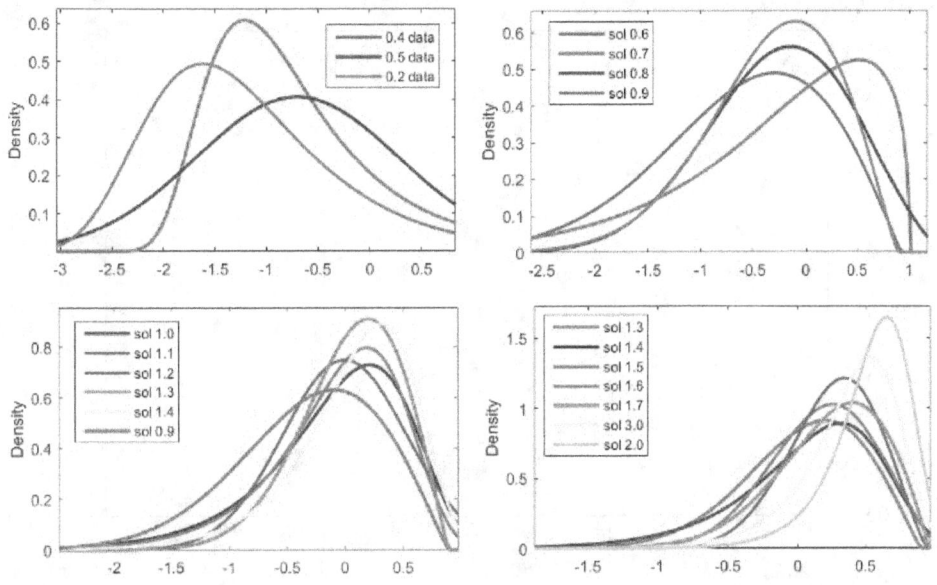

Figure 12.3.1: Plots for PDF of different stellar masses

Only in three cases, brackets with 0.7 solar mass, 1.3 solar mass, and 3.0 or greater solar mass does not conform to the trend. (though their right hand tail does fall on the prediction curve) The general trend is evident that the total planetary mass budget grows exponentially as the mass of the host star increases linearly.

Although the ratio of stellar mass to their planets grows exponentially, the variance within each bracket is large and skew toward the right. It can be stated that within one standard deviation, 66% of the ratio of stellar mass to their planets grows exponentially with a linear increase in stellar mass. Outliers in both extremes (about 10% shows a higher planetary total mass or lower ratio than prediction) and a tail (24%) shows a lower planetary total mass or higher ratio than the prediction. The predictive power wanes at both extremes.

The causes for large variance remains a mystery and requires future investigation. It is also noted that variance grows in proportion to the ratio, in other words, grows as the stellar mass decreases. It seems to suggest that other factors (temperature, disc pressure, or stellar wind) may have a more effective role at minimizing the final planetary mass when the total planetary mass budget decreases and overwhelming the effects of gravity. Since the formation of planets and the stellar system takes on many different variables and are likely independent or only slightly correlated, extreme values permissible within each condition can contribute toward the left and right tails.

We recompute our empirical law for the entire data set, by finding the most likely value for each stellar mass brackets, along with solar system data points.

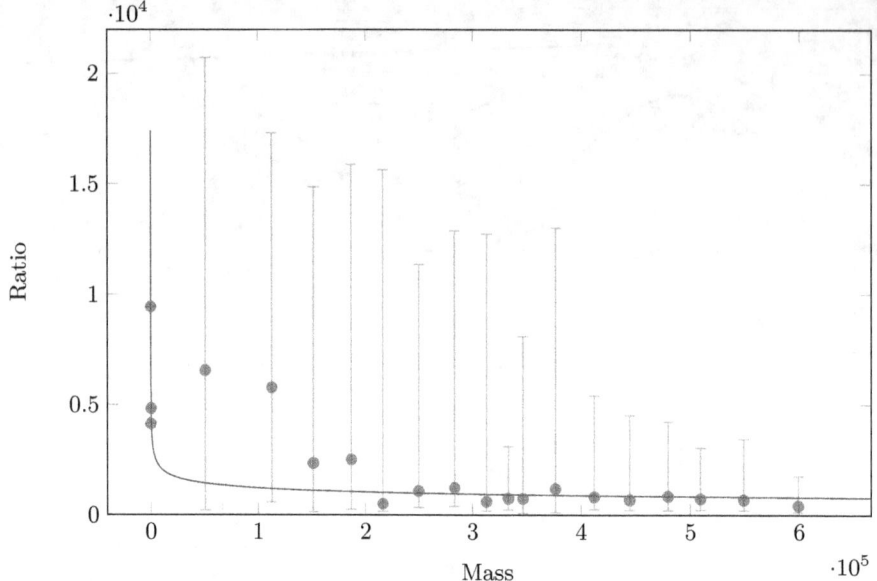

Figure 12.3.2: Stellar mass to planetary ratio across different range of stellar mass

$$y = 17520x^{-0.2315} \tag{12.3.1}$$

$$y = 10^{\left[\frac{Fe}{H}\right]} 17520x^{-0.2315} \tag{12.3.2}$$

Equation (11.3.2) is the generalized version of Equation (11.3.1), where the metallicity of a stellar system is taking into account to compute the mass budget for terrestrial planets.

If we plug in the solar mass value of 333,000 earth mass, the predicted mean total planet budget is $360.87\pm^{780.3}_{246.75}$ earth mass for the solar system. This is lower than our solar system budget at 446.719 earth mass. With this budget estimate and sun's metallicity, the total water budget is $3.65\pm^{7.090}_{2.495}$ earth mass. It is only $1.1342\pm^{2.4524}_{0.7750}$ earth mass if oxygen is counted toward the composition of terrestrial planet creation. The total mass available to create celestial bodies with metals is $6.748\pm^{14.59}_{4.614}$ earth mass, whereas the budget for terrestrial planet creation is $2.837\pm^{6.134}_{1.94}$ earth mass excluding oxygen. (Those elements with a boiling point higher than 500K, simulating the accretion disc temperature of the inner planets.) It is $5.3528\pm^{11.6}_{3.66}$ earth masses if including oxygen.

The value derived is less than the budget computed for the solar system, whereas the exoplanet's metallicity from the data is plotted below, which is generally comparable and exceeds that of the metallicity of the sun.

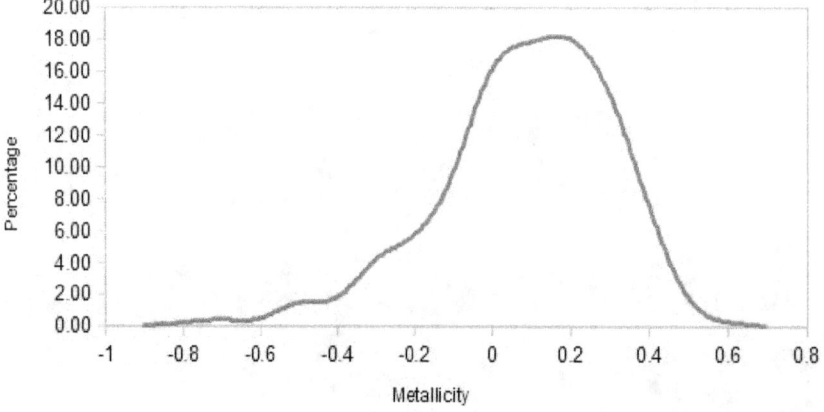

Figure 12.3.3: Distribution of exoplanet's metallicity

The plot indicates that sun's metallicity cannot account for the extra 85.85 earth masses we have observed. This discrepancy requires further analysis in the future when more data becomes available. One possible explanation is that some, if not all detection methods, such as the radial velocity method for detecting cold Jupiters and ice giants have too high noise to signal ratios. Therefore, the empirical equation serves as a lower bound on planetary mass budget. Another possible explanation, as a clue offered by exponent re-adjusting and fit the new empirical curve.

V (Disc Height Factor)	J (Disc Radius Factor)	Accretion Radius
$\frac{1}{1.258}$	$\frac{1}{0.231}$	$2.53 \cdot 10^{35}$ ly
$\frac{1}{1.52}$	$\frac{1}{0.387}$	$7.97 \cdot 10^{15}$ ly
$\frac{1}{2.05}$	$\frac{1}{0.58}$	1,960,781 ly
$\frac{1}{5.7}$	$\frac{1}{0.93}$	6,948 AU
$\frac{1}{14}$	$\frac{1}{1.05}$ *	294 AU
$\frac{1}{40}$	$\frac{1}{1.1}$ *	96.55 AU
0	$\frac{1}{1.13}$ *	51.88 AU
$-\frac{1}{40}$	$\frac{1}{1.16}$ *	28.8 AU
$-\frac{1}{16}$	$\frac{1}{1.2}$ *	13.74 AU
$-\frac{1}{1.3}$	$\frac{1}{2}$	0.0025 AU

Table 12.3.2: The possible pairs of $\left(\frac{1}{v}, \frac{1}{j}\right)$ which fits the empirical observation

with graph plots:

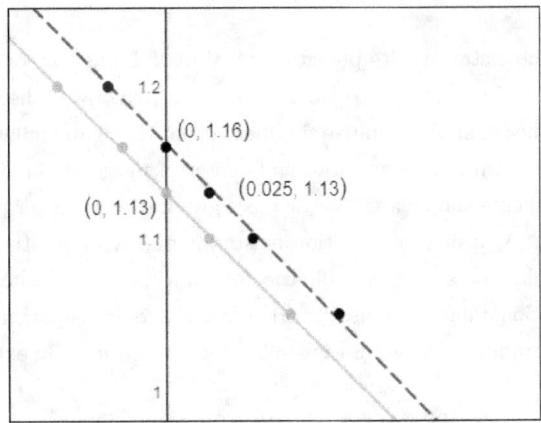

Figure 12.3.4: The new curve places in parallel in the previous one

Based on the graph, under the same disc radius and density (i.e. $j = 1.13$ with disc radius = 51.88 AU for solar mass disc), the new fit indicates a stronger negative correlation between the height and the disc radius with a lower disc height compares to the earlier fit. Under the same strength of correlation between the height and disc radius (i.e. $v = 0$), earlier fit requires a smaller radius ($j = 1.16$) or 28.79 AU for solar mass disc (greater density), and new fit requires only a radius ($j = 1.13$) or 51.88 AU for solar mass disc . This implies that the accretion disc of the solar system was possibly less viscous, or having a higher temperature, increasing the pressure along the accretion plane that defied the self gravity of the disc.

A third possible explanation is that our earlier derived empirical law is non-accurate description of reality. The earlier formula predicts 603.927 earth mass for the solar system. This is higher than our solar system budget at 446.719 earth mass. The new formula's prediction of 360.87 earth mass is much closer to our solar system budget at 446.719 earth mass.

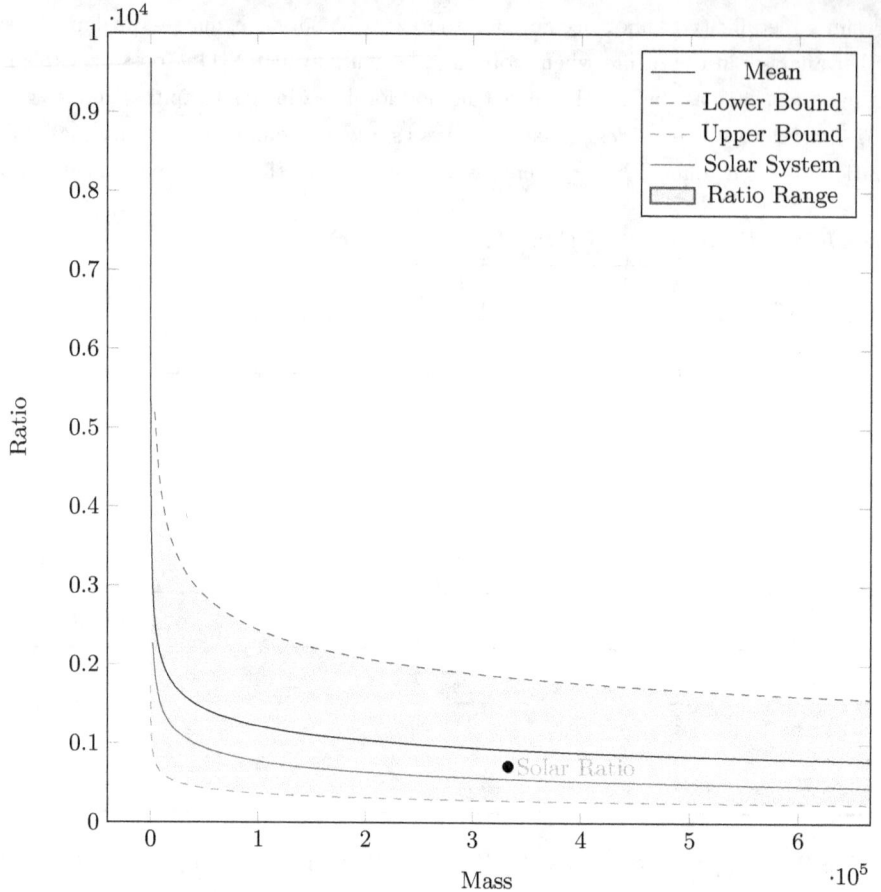

Figure 12.3.5: Lower and upper bound for primary to satellite ratio

One of the greatest challenge facing the host star and its planetary total mass budget theory is the formation of binary stars. Binary and multiple star system are also a product of accretion and solar nebulae disc consolidation. It turns out that binary stars formation is a consequence of stellar nebulae fragmentation. Since all stellar nebulae start as molecular cloud and gradually increases its density toward their gravitational center of mass, at some point along the transition, the density was high enough so that fragmentation creates two or more gravitational center point, which all evolved into stars on their own with different masses. A positive correlation exists for binary or multiple star systems with higher solar masses. It is speculated 44% of solar mass stars are binaries or multiples, while the majority (66%) of the red dwarfs are single stars.[12] From observational data, we also found that the average separation distance between binary pairs is 150 AU, which is well within the radius of forming the stellar disc, which usually extends hundreds of AU.

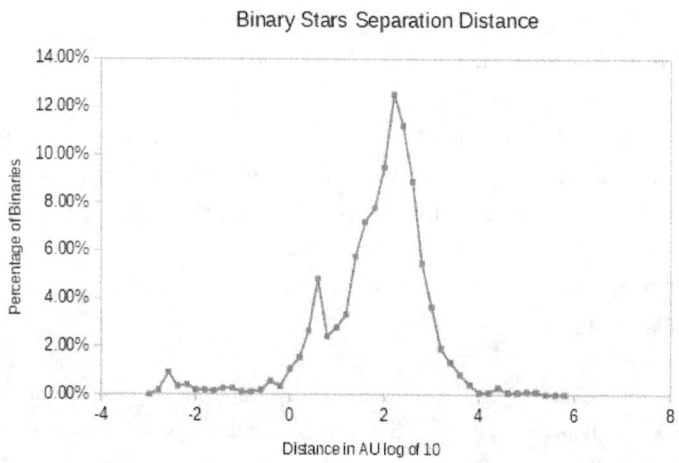

Figure 12.3.6: PDF of binary stars by separation distance

For some multiple star system with stars orbiting the primary star thousands to tens of thousands of AU in separation distance such as Proxima Centauri. Those configurations are captured stars during the star nursery period, in which the neighboring nurturing stars are just a few thousands AU units away. Furthermore, for data observed majority of the captured stars in wide orbits are red dwarves, which is no surprise as the most commonly formed stars is the most frequently captured. Therefore, binary star formation with comparable separation and distance from their host star posed a challenge because their host star to their binary pairs' mass ratio dramatically falls below the threshold for the planetary mass budget. Upon close examination, however, the formation and mechanism of binary and multiple stars system are radically different from that of the planets. They occur at different stages of star formation. The formation of star pairs occurs in the very early stage of stellar formation, all within the first million years or earlier and driven by disc fragmentation. [?][?][?] While the formation of the planets, the remnants leftover from the stellar disc, occurs more slowly over the course of hundreds of millions of years throughout the T-Tauri Star phase until the star enters its main sequence.

12.4   Conclusion

It is surprising that after the discovery of the law of universal gravitation, it takes more than two hundred years to discover a relationship of the ratio regarding the primary mass and the total mass of its satellite. However, the derivation of this law was not a necessity since no exoplanets data were available and the concept of planets beyond the solar system was simply speculative. Furthermore, this law distinguishes from the classical law of gravitation and general theory of relativity by assessing the long-term trend from millions of years of gravitational interactions of n bodies during the accretion phase of the solar system instead of extreme precision and description of two body interaction at the present. Based on this law, red dwarf rarely hosts Jupiter sized planets because its stellar mass less than 0.4 solar mass has a mean planetary mass budget merely 32.37% of solar mass stars. It is merely $116.8 \pm^{252.55}_{79.86}$ earth masses. If one takes 0.2 solar mass as the mean stellar mass for red dwarves, then, the expected mean planetary mass budget is $49.72 \pm^{107.50}_{33.997}$ earth masses, or only as the combined mass of Uranus and two Neptunes in the solar system. Because extreme values lie up to 3 standard deviations above the mean, gas giants revolve around red dwarves are still possible. However, it is likely in no more than 17% of the systems, or only 1 in 6 red dwarf system hosts gas giants, and their gas giants size is only comparable to the mass of Saturn at the most.

Exomoons at the size of earth or greater require its hosting planet at the size of 8.78 Jovian mass (2,793.87 earth mass, 0.839% solar mass) or above.

$$y = 0.00839 \cdot M_{sol} \left(17,520 \left(0.00839 \cdot M_{sol} \right)^{-0.2315} \right)^{-1} \tag{12.4.1}$$

$$= 1.00108009577 \, M_{earth}$$

A planetary system with more than 8.7857 Jovian masses (2,793 M_{earth}) implies a stellar mass at least 5.2694 solar masses or above.

$$y = 5.2694 \cdot M_{sol} \left(17,520 \left(5.2694 \cdot M_{sol} \right)^{-0.2315} \right)^{-1} \tag{12.4.2}$$

$$= 2,793.86134545 \, M_{earth}$$

Based on stellar evolution model, a star with 5.2694 solar mass or above devolve from the main sequence in just 156.89 million years or less.

$$T_{MS} \approx 10^{10} \left[\frac{M}{M_{sol}} \right] \left[\frac{L_{sol}}{L} \right] = 10^{10} \left[\frac{M}{M_{sol}} \right]^{-2.5} \tag{12.4.3}$$

The timeframe is not adequate for the development of complex multicellular life. (at least a few billions of years in preparation for cyanobacteria to create extra oxygen in the atmosphere and ocean for the emergence of eukaryotic cells and their multicellular descendents.) Therefore, the science fiction movie Avatar's planet Pandora is not a realistic description of the physical reality of the universe.

Furthermore, deriving values from the equation, the total budget for each planetary system for terrestrial planets is very limited. For solar mass systems with a metallicity of 0, there is enough budget to create at most $1.66 \pm^{3.589}_{1.135}$ earth mass pure carbon planet, and $0.2346\pm^{0.507}_{0.1604}$ earth mass silicon planet. Finally, the water content for each planetary system is very limited. For a system with a metallicity of 0 and solar mass, the entire system has an upper bound of just $6.107\pm^{13.205}_{4.176}$ earth mass. The system has an upper bound of only $1.897\pm^{4.101}_{1.297}$ earth mass worth of water if oxygen is counted toward the composition of terrestrial planet creation as 47% of earth's mass composed of oxygen.

With lower metallicity, the water budget can be much lower, even with much higher metallicity, the upper bound of the water budget for the system is less than $24.13\pm^{52.18}_{16.499}$ earth masses. However, studies done on the process of terrestrial planet creation indicates that stellar systems with high metallicities are destroyed by migrating hot Jupiters early in its formational period. As a result, the upper bound of the total water budget in any extraterrestrial system with surviving terrestrial planets is limited to a mean of $10\pm^{21.622}_{6.838}$ earth mass or below.

Tau Ceti, one of the closest star to the Sun with just 12 light years in distance has confirmed 4 planets g, h, e, and f with mass 1.75, 1.83, 3.93, 3.93 earth mass respectively, their combined mass of 11.44 earth mass. Currently, no one is able to detect the composition of the planets, by using our planetary mass budget equation, one can make certain conclusion despite unobservability at the current time. Tau Ceti has 0.783 solar mass and 28% solar metallicity, and then one arrives at the following value:

$$y_{ceti} = 0.783 \cdot M_{sol} \left(17,520 \left(0.783 \cdot M_{sol}\right)^{-0.2315}\right)^{-1} \tag{12.4.4}$$

$$= 267.007574532 \, \mathrm{M_{earth}}$$

$$\left(\frac{5.3528}{360.87}\right) \cdot y_{ceti} \cdot 0.28 \tag{12.4.5}$$

$$= 1.10895596076 \, \mathrm{M_{earth}}$$

Whereas 360.87 $\mathrm{M_{earth}}$ is the total planetary budget of the solar mass star around the mean of its distribution. 5.3528 $\mathrm{M_{earth}}$ is the terrestrial planetary budget including oxygen for a solar mass star.

That is, the total expected mean combined terrestrial planetary mass should be 1.108 earth mass. Given the probabilistic distribution of planetary mass, the chance of Tau Ceti hosting combined terrestrial planetary mass over 11.08 earth mass is two standard deviations from the mean, that is, 1.5% or less. Therefore, Tau Ceti's planets within its habitable zone have 98.5% chance being mini Neptunes with tiny rocky terrestrial cores shrouded with an extremely thick layer of hydrogen and helium.

The empirical law for planetary mass budget holds an excellent promise for extraterrestrial studies since it sets the upper limit on the combined mass of any systems, which together with metallicity, determines the total budget for the terrestrial planets, their moons, the total water, nitrogen, and CO_2 availability. For future works, as more exoplanet data becomes available, one can continue to refine the parameter of the model to make more precise predictions regarding exoplanets, especially those of the red dwarves, which are outliers in our model in our current regression analysis.

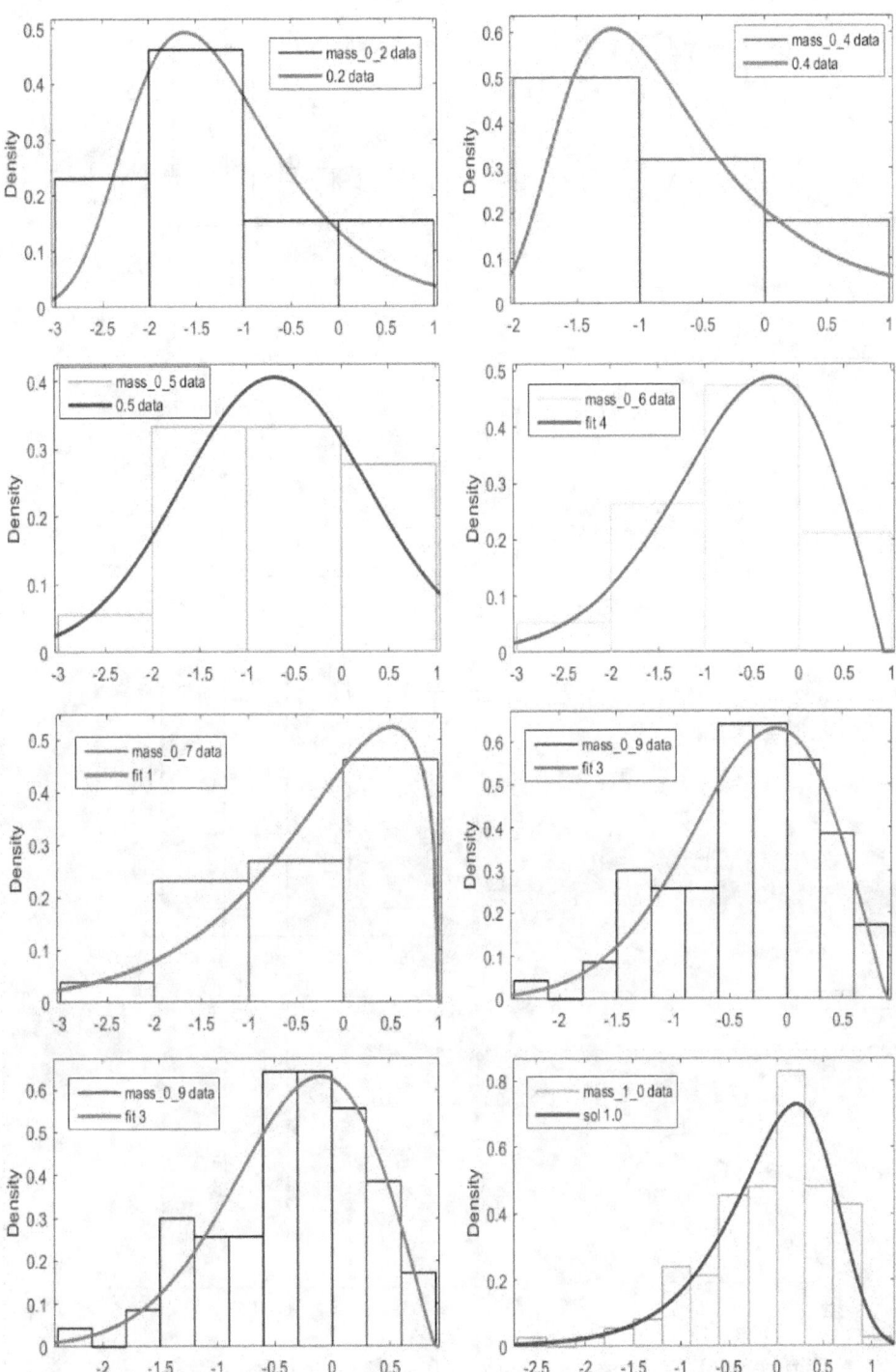

Figure 12.5.1: Plots for PDF of 0.2 ~ 1.0 solar masses

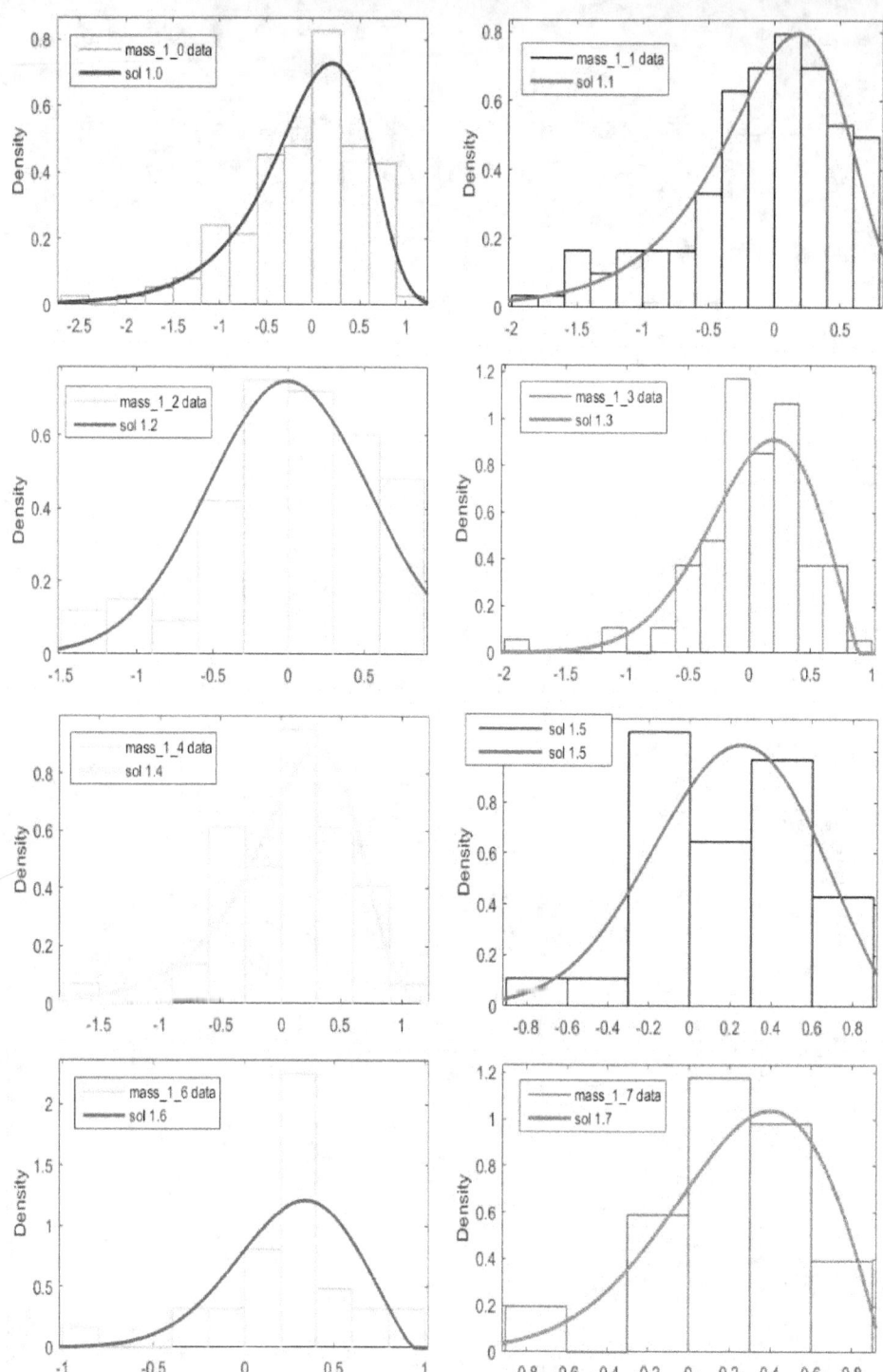

Figure 12.5.2: Plots for PDF of 1.0 ~ 1.7 solar masses

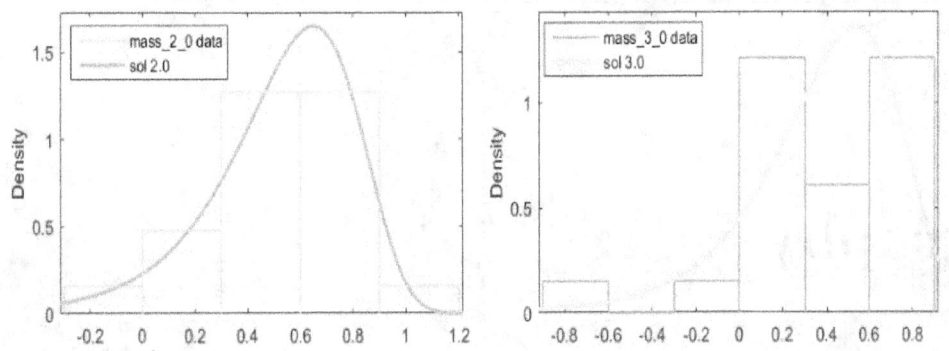

Figure 12.5.3: Plots for PDF of 2.0 ~ 3.0 solar masses

Bibliography

[1] (2015, November) Tiny red dwarf star has a magnetic field several hundred times stronger than our sun. [Online]. Available: https://scitechdaily.com/tiny-red-dwarf-star-has-a-magnetic-fieldseveral-hundred-times-stronger-than-our-sun/

[2] (2015, September) Earth-like exoplanets may have magnetic fields capable of protecting life. [Online]. Available: https://exoplanets.nasa.gov/news/217/earth-like-exoplanets-mayhave-magnetic-fields-capable-of-protecting-life/

[3] (2016, March) Climate effects on human evolution. [Online]. Available: http://humanorigins.si.edu/research/climate-and-human-evolution/climate-effectshuman-evolution

[4] ""Renewable biological systems for unsustainable energy production.","" *FAO Agricultural Services Bulletins (1997)*.

[5] The nebular theory of the origin of the solar system. [Online]. Available: http://atropos.as.arizona.edu/aiz/teaching/nats102/mario/solarsystem.html

[6] "What is photosynthesis?"."

[7] *Dimensions of Need: An atlas of food and agriculture*, 1995.

[8] ""converting sunlight into algal biomass wageningen university project"," 2005-2008.

[9] A. G. Hartl, D. L. Clark, *Principles of Population genetics, 4th ed.*, 2007, vol. 243.

[10] A. I., S. K., C. J. Hawkesworth, ""Evolution of the continental crust"," *Nature*, vol. 443, 2006. [Online]. Available: https://websites.pmc.ucsc.edu/pkoch/EART206/090203/HawkesworthKemp06Nature443811.pdf

[11] A. J. C. Tsvi Piran, Raul Jimenez, ""Formation and composition of planets around very low mass stars"," 2016. [Online]. Available: https://arxiv.org/pdf/1508.01034.pdf

[12] A. K. Gaspard Duch, ""Stellar multiplicity"," *Annu. Rev. Astron. Astrophys*, vol. 1056-8700, 2013. [Online]. Available: https://arxiv.org/pdf/1303.3028.pdf

[13] A. K., P. Richard L. Magin PhD, John K. Lee BS, ""Biological effects of long-duration, high-field (4 t) mri on growth and development in the mouse"," *JOURNAL OF MAGNETIC RESONANCE IMAGING*, vol. 12, pp. 140–149, 2000. [Online]. Available: http://www.brl.uiuc.edu/Publications/2000/Magin-JMRI-140-2000.pdf

[14] A. K., P. Richard, L. Magin PhD, John K. Lee BS, ""Biological effects of long-duration, high-field (4 t) mri on growth and development in the mouse"," *JOURNAL OF MAGNETIC RESONANCE IMAGING*, vol. 12, pp. 140–149, 2000. [Online]. Available: http://www.brl.uiuc.edu/Publications/2000/Magin-JMRI-140-2000.pdf

[15] A. Kukla, *Extraterrestrials: A Philosophical Perspective*, 2009. [Online]. Available: https://worldbuilding.stackexchange.com/questions/9948/is-there-a-theoreticalmaximum-size-for-rocky-planets

[16] A. Lyon, *Why are Normal Distributions Normal?*, 2014.

[17] A. M., W. Jason, X. Prochaska, Eric Gawiser, ""The age-metallicity relation of the universe in neutral gas: The first 100 damped ly systems"," *Astrophysical Journal Letters*, 2003. [Online]. Available: https://arxiv.org/pdf/astro-ph/0305314.pdf

[18] A. N. N, Golden Gadzirayi Nyambuya, Divine Jigu, ""On the theoretical foundations of the polemical titius-bode law (ii) exoplanetary systems"," *Researchgate*, 2018.

[19] A. Oren, ""Prokaryote diversity and taxonomy: current status and future challenges"," *Philos. Trans. R. Soc. Lond. B Biol. Sci.*, vol. 359, p. 62338.

[20] A. Szoke and R. W. Moir, ""Peaceful nuclear explosions a practical route to fusion power"," *Technology Review*, July 1991. [Online]. Available: http://www.ralphmoir.com/wpcontent/uploads/2012/10/pracFus91.pdf

[21] A. Williams. (2014, August) Scientists detect evidence of 'oceans worth' of water in earth's mantle. [Online]. Available: https://www.astrobio.net/news-exclusive/scientists-detectevidence-oceans-worth-water-earths-mantle/

[22] Adam Sandberg, Stuart Armstrong, ""Eternity in six hours: Intergalactic spreading of intelligent life and sharpening the fermi paradox"," *Elsevier Acta Astronautica*, vol. 89, March 2013. [Online]. Available: https://pdfs.semanticscholar.org/847d/8dabb12f67124868af0876c77538e4fd1c60.pdf

[23] B. Alcott, ""Jevons' paradox"," *Elsevier Ecological Economics*, vol. 54, pp. 9–21, 2005. [Online]. Available: https://www.sciencedirect.com/science/article/pii/S0921800905001084

[24] Anders Sandberg, Eric Drexler and Toby Ord, ""On the Likelihood of Observing Extragalactic Civilizations: Predictions from the Self-Indication Assumption"," *Future of Humanity Institute, Oxford University*, Jun 2018.

[25] B. C. Lacki., ""The Log Log Prior for the Frequency of Extraterrestrial Intelligences"," *ArXiv e-prints*, September 2016.

[26] B. Gladman, ""Rotation histories of the natural satellites"," *University of Arizona Press*, pp. 169–170, 1977.

[27] B. Gladman. et al, ""Synchronous locking of tidally evolving satellites"," *Icarus*, vol. 122, pp. 166–192. [Online]. Available: https://www.sciencedirect.com/science/article/pii/S0019103596901177

[28] B. Handwerk. (2014, September) How climate change may have shaped human evolution. [Online]. Available: https://www.smithsonianmag.com/science-nature/how-climate-changemay-have-shaped-human-evolution-180952885/?no-ist

[29] B. Harder. (2002, March) Inner earth may hold more water than the seas. [Online]. Available: https://news.nationalgeographic.com/news/2002/03/03070307waterworld.html

[30] B. I., Pisciotta JM, Zou Y, *Light-Dependent Electrogenic Activity of Cyanobacteria*, 2010, vol. 5.

[31] B. K. G., Charles H. Lineweaver, Yeshe Fenner, ""The galactic habitable zone and the age distribution of complex life in the milky way"," *Science*, January 2004. [Online]. Available: https://arxiv.org/ftp/astro-ph/papers/0401/0401024.pdf

[32] B. Vukotic and M. M. Cirkovic, ""On the timescale forcing in astrobiology"," *Serb. Astron. Journal*, no. 175, p. 7, May 2007. [Online]. Available: https://arxiv.org/pdf/0712.1508.pdf

[33] Britannica. Principle of mediocrity. [Online]. Available: https://www.britannica.com/topic/principle-of-mediocrity

[34] C. A., H.-M., S Seager, M Kuchner, ""Mass-radius relationships for solid exoplanets"," *The Astrophysical Journal*, vol. 669, no. 1279-1297, 2007. [Online]. Available: http://seagerexoplanets.mit.edu/ftp/Papers/Seager2007.pdf

[35] C. D. David, T. Johnstona, Simon W. Poultonc, ""An emerging picture of neoproterozoic ocean chemistry: Insights from the chuar group, grand canyon, usa"," *JOURNAL OF MAGNETIC RESONANCE IMAGING*, 2009. [Online]. Available: http://nrs.harvard.edu/urn3:HUL.InstRepos:10059265

[36] C. D., Drake JW, Charlesworth B, ""Rates of spontaneous mutation"," *Genetics*, vol. 148, p. 166786, April 1998.

[37] C. H. LINEWEAVER, ""Paleontological tests: Human-like intelligence is not a convergent feature of evolution"," *65th International Astronautical Congress, Toronto, Canada*, N/A. [Online]. Available: https://arxiv.org/ftp/arxiv/papers/0711/0711.1751.pdf

[38] C. H. Lineweaver, ""An estimate of the age distribution of terrestrial planets in the universe: Quantifying metallicity as a selection effect"," *Icarus*, vol. 187, p. 13, 2001. [Online]. Available: https://arxiv.org/pdf/astro-ph/0012399.pdf

[39] C. S. Nachman MW, ""Estimate of the mutation rate per nucleotide in humans"," *Genetics*, vol. 156, pp. 297–304, September 2000.

[40] C. T. C. Adami C, Ofria C, ""Evolution of biological complexity"," *PNAS*, vol. 97, pp. 4463–8.

[41] C. Yukna. Mysteries of the dinosaur epoch cases solved? [Online]. Available: http://www.emse.fr/yukna/dinosaurs/mysterydinosaurs.php

[42] Carroll. SB, ""Chance and necessity: the evolution of morphological complexity and diversity"," *Nature*, vol. 409, pp. 1102–9.

[43] Charles H. Lineweaver, D. Grether, ""What fraction of sun-like stars have planets?"," *University of New South Wales*. [Online]. Available: http://www.mso.anu.edu.au/charley/papers/LineweaverGrether03.pdf

[44] Chas Egan, ""Dark Energy, Anthropic Selection Effects, Entropy and Life"," *The University of New South Wales*, 2003.

[45] ckersch. (2015, February) Is there a theoretical maximum size for rocky planets? [Online]. Available: https://worldbuilding.stackexchange.com/questions/9948/is-there-atheoretical-maximum-size-for-rocky-planets

[46] D. C., John W. Drake, Brian Charlesworth, ""Rates of spontaneous mutation"," *GENETICS*, no. 4, pp. 1667–1686, April 1998. [Online]. Available: http://www.genetics.org/content/148/4/1667

[47] D. J. B., Peter J. Franks, Dana L. Royer, *New constraints on atmospheric CO2 concentration for the Phanerozoic*, July 2014, vol. 31.

[48] D. J, S. Benjamin, C. Bartlett, ""Analysis of a precambrian resonance-stabilized day length"," *Geophysical Research Letters*, 2015.

[49] D. R., P. M., G. Gowanlock and S. M. McConnell, ""A model of habitability within the milky way galaxy"," p. 40, 2011. [Online]. Available: https://arxiv.org/pdf/1107.1286.pdf

[50] David Esker. (2009) Dinosaurtheory - the blue planet-the fallacy of an unchanging world. [Online]. Available: http://dinosaurtheory.com/thickatmosphere.html

[51] ——. (2009) Dinosaurtheory - the thick mesozoic atmosphere. [Online]. Available: http://dinosaurtheory.com/thickatmosphere.html

[52] E. L. Schneider S, ""Estimation of past demographic parameters from the distribution of pairwise differences when the mutation rates vary among sites: application to human mitochondrial dna"," *Genetics*, vol. 152, pp. 1079–89, July 1999.

[53] E. L. Wright, ""A cosmology calculator for the world wide web"," *The Publications of the Astronomical Society of the Pacific*, vol. 118, pp. 1711–1715, 2006.

[54] Frank Drake, Carl Sagan, ""The search for extraterrestrial intelligence"," *Scientific American*, 1975.

[55] G. Constable, *Grasslands and Tundra*, 1985.

[56] Gilles Chabrier, ""Galactic stellar and substellar initial mass function"," 2003. [Online]. Available: https://arxiv.org/pdf/astro-ph/0304382.pdf

[57] H. D. Holland, ""The oxygenation of the atmosphere and oceans"," *Transactions of the Royal Society B*, 2006.

[58] H. J. Schloss P, ""Status of the microbial census"," *Microbiol Mol Biol Rev.*, vol. 68, pp. 68–69–1, 2004.

[59] H. K., Suzan Bongers, Pauline Slottje, ""P277long-term exposure to static magnetic fields in mri manufacturing and risk of developing hypertension"," *BMJ Journal*, September 2016.

[60] I. D., Milan M. Cirkovic, Branislav Vukotic, ""Galactic punctuated equilibrium: How to undermine carter's anthropic argument in astrobiology"," *Astrobiology*, 2009. [Online]. Available: https://arxiv.org/ftp/arxiv/papers/0912/0912.4980.pdf

[61] I. Ribas, ""The sun and stars as the primary energy input in planetary atmospheres"," *Solar and Stellar Variability: Impact on Earth and Planets Proceedings IAU Symposium*, p. 631, 2009.

[62] J. R. G. III, ""Implications of the Copernican principle for our future prospects"," *NATURE)*, May 1993.

[63] J. Catanzarite and M. Shao, ""The occurrence rate of earth analog planets orbiting sunlike stars"," p. 19, 2011. [Online]. Available: https://arxiv.org/ftp/arxiv/papers/1103/1103.1443.pdf

[64] J. D. Haqq-Misra and S. D. Baum, ""The sustainability solution to the fermi paradox"," *Elsevier Ecological Economics*, 2009. [Online]. Available: https://arxiv.org/ftp/arxiv/papers/0906/0906.0568.pdf

[65] J. G., Raven P.H, Carol J. Mills, ed., *Understanding Biology (3rd ed.)*, 1995.

[66] J. Korenaga, ""On the likelihood of plate tectonics on super-earths: Does size matter?"," *The Astrophysical Journal Letters*, pp. L43–L46, April 2010. [Online]. Available: https://people.earth.yale.edu/sites/default/files/korenaga10b.pdf

[67] J. P. Vallee, ""Observations of the magnetic fields inside and outside the solar system: From meteorites (10 attoparsecs), asteroids, planets, stars, pulsars, masers, to protostellar cloudlets (< 1 parsec)"," January 1998. [Online]. Available: https://ned.ipac.caltech.edu/level5/March03/Vallee/paper.pdf

[68] J. Rennie. (2014, 18) how does the hubble parameter change with the age of the universe? general relativity - how does the hubble parameter change with the age of the universe? - physics stack exchange. [Online]. Available: http://physics.stackexchange.com/questions/136056/ howdoes-the-hubble-parameter-change-with-the-age-of-the-universe

[69] J. Tainter, *The Collapse of Complex Societies*, 1988.

[70] James H. Kunstler, *The Long Emergency: Surviving the End of Oil, Climate Change, and Other Converging Catastrophes of the Twenty-First Century*, March 2006.

[71] Jim2B. (2015, April) What is the minimum planetary mass to hold an atmosphere over geologic time scales? [Online]. Available: https://worldbuilding.stackexchange.com/questions/13583/ what-is-the-minimumplanetary-mass-to-hold-an-atmosphere-over-geologic-time-scal

[72] K. K. Furusawa C, ""Origin of complexity in multicellular organisms"," *Phys. Rev. Lett*, vol. 84, pp. 6130–3.

[73] K. K., R. David, Oakley Hal, *Biochemistry (2nd ed.)*, 1999.

[74] K. Lewis, ""Moon formation and orbital evolution in extrasolar planetary systems - a literature review"," vol. 11, no. 04003. EPJ Web of Conferences, 2011. [Online]. Available: https://www.epj-conferences.org/articles/ epjconf/pdf/2011/01/epjconfohp201004003.pdf

[75] K. S. G. Schubert, ""Planetary magnetic fields: Observations and models"," *Elsevier -Physics of the Earth and Planetary Interiors*, vol. 187, pp. 92–108, 2011. [Online]. Available: http://www.maths.gla.ac.uk/rs/res/B/ PlanetDyn/Schubert2011.pdf

[76] Keiko. Atobe and Shigeru. Ida, ""Obliquity evolution of extrasolar terrestrial planets"," *Icarus*, 2006. [Online]. Available: https://arxiv.org/pdf/astro-ph/0611669.pdf

[77] L. N. Trefethen, ""Predictions for scientific computing fifty years from now"," *Oxford University Computing Laboratory*, p. 10, June 1998. [Online]. Available: http://eprints.maths.ox.ac.uk/1304/1/NA-98-12.pdf

[78] L. Stryer, ""Biochemistry (2nd ed.)"," *Science*, vol. 328, p. 448, 1981.

[79] M. Cavalli-Sforza, L. Luca, *The History and Geography of Human Genes*, 1996.

[80] M. E., M. M., A. Khamehchi, Khalid Hossain, ""Negative-mass hydrodynamics in a spin-orbitcoupled bose-einstein condensate"," *The Astrophysical Journal*, April 2017. [Online]. Available: https://arxiv.org/pdf/1612.04055.pdf

[81] M. I. Y Takashima, J Miyakoshi, ""Genotoxic effects of strong static magnetic fields in dna-repair defective mutants of drosophila melanogaster"," *Journal of radiation*, vol. 45, pp. 393–397, 2004.

[82] M. J., Longo Giuseppe, Montévil Maël. Dinneen, ""Computation, physics and beyond"," *Lecture Notes in Computer Science*, vol. 97, pp. 289–308.

[83] Marion Hubbert, ""Energy resources: a report to the committee on natural resources of the national academy of sciences-national research council"," Tech. Rep., December 1962.

[84] Micheal. Ruse, *Monad to man: the Concept of Progress in Evolutionary Biology.*, 1996.

[85] Milan M. Ćirković, ""IS MANY LIKELIER THAN FEW? A CRITICAL ASSESSMENT OF THE SELF-INDICATION ASSUMPTION"," *ArXiv e-prints*, 2003.

[86] N. R., M. Cin-Ty, A. Lee, ""Rise of the continents"," *GEOCHEMISTRY*, 2015.

[87] N/A, ""Properties of the pluto-charon binary"," February N/A. [Online]. Available: https://authors.library. caltech.edu/51983/7/Canup.SOM.pdf

[88] Nicholson-W. (2000, March) Setting the scientific record straight on humanity's evolutionary prehistoric diet and ape diets. [Online]. Available: http://www.beyondveg.com/nicholsonw/hb/hb-interview1f.shtml

[89] Nick Bostrom, ""Existential risks: analyzing human extinction scenarios and related hazards"," *Journal of Evolution and Technology*, vol. 9, 2002. [Online]. Available: https://ora.ox.ac.uk/objects/uuid: 827452c3-fcba-41b8-86b0-407293e6617c

[90] O. B. E, D.Kovalevaa, P.Kaygorodova, ""Binary star database bdb development: Structure, algorithms, and vo standards implementation"," *Elsevier*, vol. 11, pp. 119–125, 2015.

[91] P. Bell, ""Viral eukaryogenesis: was the ancestor of the nucleus a complex dna virus?"," *J Molec Biol.*, vol. 53, p. 2516, 2001.

[92] ——, ""Sex and the eukaryotic cell cycle is consistent with a viral ancestry for the eukaryotic nucleus"," *J Molec Biol.*, vol. 243, pp. 54–63, 2006.

[93] P. F. R. Nicolas Flament, Nicolas Coltice, ""A case for late-archaean continental emergence from thermal evolution models and hypsometry"," *Earth and Planetary Science Letters*, vol. 275, November 2008. [Online]. Available: https://www.researchgate.net/publication/222840084Acaseforlate-Archaeancontinentalemergencefromthermalevolutionmodelsandhypsometry

[94] Pavel Kroupa, ""The initial mass function of stars:evidence for uniformity in variable systems"," *SCIENCE*, vol. 295, pp. 82–91, 2002. [Online]. Available: https://arxiv.org/pdf/astroph/0201098.pdf

[95] P.Z.Myers. The mediocrity principle. [Online]. Available: https://www.edge.org/q2011/q1112.htmlmyerspz

[96] R. A. Berner, *Atmospheric oxygen over Phanerozoic time*, 1999.

[97] R. J. Tsvi Piran, ""On the role of grbs on life extinction in the universe"," November 2014. [Online]. Available: https://arxiv.org/pdf/1409.2506.pdf

[98] R. Kopp, *The Paleoproterozoic snowball Earth: A climate disaster triggered by the evolution of oxygenic photosynthesis*, June 2005.

[99] R. L. ARMSTRONG, ""The persistent myth of crustal growth"," *Australian Journal of Earth Sciences*, vol. 38, pp. 613–630, 1991. [Online]. Available: http://www.mantleplumes.org/WebDocuments/Armstrong1991.pdf

[100] R. L. Rudnick, ""Making a continental crust"," *Review Article*, N/A. [Online]. Available: http://www.depts.ttu.edu/gesc/Facpages/Yoshinobu/5362-Tectonics-Web/pdfs2012Tectonics/Rudnick1995Nature.pdf

[101] R. Lovett, ""Early earth may have had two moons"," *Nature*, 2011.

[102] R. M. Canup, ""On a giant impact origin of charon, nix, and hydra"," *The Astronomical Journal*, vol. 141, no. 2, 2010. [Online]. Available: http://iopscience.iop.org/article/10.1088/0004-6256/141/2/35/meta

[103] R. M.Canup, ""Simulations of a late lunar-forming impact"," *Icarus*, 2004.

[104] R. W. Moir, Charles J. Call, ""A novel fusion power concept based on molten-salt technology: Pacer revisited"," *Nuclear Science and Engineering*, 1990. [Online]. Available: http://www.ralphmoir.com/wp-content/uploads/2012/10/novFus90.pdf

[105] Ray Kurzweil. (2001, March) The law of accelerating returns. [Online]. Available: http://www.kurzweilai.net/the-law-of-accelerating-returns

[106] Richard Gordon, Sharov Alexei A, ""Life before earth"," vol. [1304.3381], March 2013. [Online]. Available: https://arxiv.org/abs/1304.3381

[107] Richard Heinberg, *The Party's Over: Oil, War and the Fate of Industrial Societies*, June 2005.

[108] S. A., Roach JC, Glusman G, ""Analysis of genetic inheritance in a family quartet by wholegenome sequencing"," *Science*, vol. 328, pp. 63–69, April 2010.

[109] S. G. Engle and E. F. Guinan, ""Red dwarf stars: Ages, rotation, magnetic dynamo activity and the habitability of hosted planets"," *the Pacific Rim Conference on Stellar Astrophysics ASP Conference Series*, 2011. [Online]. Available: http://www.astronomy.villanova.edu/lward/prcsa2011LWARD.pdf

[110] S Jay Olson, ""On the visible size and geometry of aggressively expanding civilizations at cosmological distances"," *ArXiv e-prints*, Apr 2016.

[111] ——, ""On the Likelihood of Observing Extragalactic Civilizations: Predictions from the Self-Indication Assumption"," *ArXiv e-prints*, Feb 2020.

[112] ——, ""Estimates for the number of visible galaxy-spanning civilizations and the cosmological expansion of life"," *ArXiv e-prints*, Jul 2015.

[113] S. Krasnikov, ""The quantum inequalities do not forbid spacetime shortcuts"," May 2003. [Online]. Available: https://arxiv.org/pdf/gr-qc/0207057.pdf

[114] S. S. D, J. T. Wright, R. Griffith, ""The g infrared search for extraterrestrial civilizations with large energy supplies. ii. framework, strategy, and first result"," *The Astrophysical Journal*, June 2014. [Online]. Available: https://arxiv.org/pdf/1408.1134.pdf

[115] Steven Suan Zhu, ""Gravitational effect on the final stellar to planetary mass ratio"," February 2018.

[116] T. M. D., Charles H. Lineweaver, ""Does the rapid appearance of life on earth suggest that life is common in the universe?"," *ASTROBIOLOGY*, vol. 2, no. 3, 2002. [Online]. Available: http://www.mso.anu.edu.au/charley/papers/LineweaverDavis.pdf

[117] T. Physicist. (2013, April) What kind of telescope would be needed to see a person on a planet in a different solar system? [Online]. Available: http://www.askamathematician.com/2013/04/q-what-kind-of-telescope-would-beneeded-to-see-a-person-on-a-planet-in-a-different-solar-system/

[118] U. Y., MS Morris, KS Thorne, ""Wormholes, time machines, and the weak energy condition"," *Physical Review Letters*, 1988.

[119] V. G. Anastassia Makarieva, ""On the dependence of speciation rates on species abundance and characteristic population size"," *Journal of Biosciences*, vol. 29, March 2004. [Online]. Available: http://www.ias.ac.in/article/fulltext/jbsc/029/01/0119-0128

[120] V.G.Gurzadyan, ""Kolmogorov complexity, string information, panspermia and the fermi paradox"," p. 5, 2005. [Online]. Available: https://arxiv.org/pdf/physics/0508010.pdf

[121] Y. Alibert, W. Benz, ""Formation and composition of planets around very low mass stars"," *Astronomy Astrophysics manuscript*, vol. msrevv3, 2016. [Online]. Available: https://arxiv.org/pdf/1610.03460v1.pdf

[122] Yefei. Yu, ""China's historical average crop yields per acre survey"," *China Academic Journal Publishing House*, p. 13, 2013. [Online]. Available: https://wenku.baidu.com/view/dfee90bc6bec0975f465e2d3.html

[123] S. S. Zhu, ""On the Origin of Extraterrestrial Industrial Civilizations"," *Researchgate*, Feb 2018.

Appendix A

Proof for $P_{df}(t, x)$ represents the cumulative emergence chance of all previous periods

$P_{df}(t, x)$ represents the cumulative emergence chance of all previous periods up to time t can be demonstrated as the follows:

Knowing that $P_{df}(t, x)$ is a close approximation of distribution based on multinomial distribution, the original biocomplexity is represented as the frequency distribution of species possessed different number of traits. Those with a high number of combined traits yields lower relative frequency compares to those with a low number of combined traits since each trait corresponds to a habitat adaptation has < 1 chance among all species, and attaining each trait is largely an independent event, as well as a very low combination value yields from all possible traits.

Assuming at the very beginning, there are only 3 traits available for all species. As a result, there can be only 1 species possessed all 3 traits. So we represent such species as the triplet:

$$(1, 2, 3) \tag{A.0.1}$$

As the next round of evolution proceeds, an extra trait is added to the pool, and now species which possessed 3 traits numbered 4, which are:

$(1, 2, 3)^*$	$(1, 2, 4)$
$(1, 3, 4)$	$(2, 3, 4)$

Which included the one we had from the last round.

Then, we move to the next round with 5 traits. There are 10 possible combinations for species possessed 3 traits. Which included $(1, 2, 3)$ from the 1st round, as well as those from the 2nd round.

$(1, 2, 3)^{**}$	$(1, 2, 4)^*$	$(1, 2, 5)$	$(1, 3, 4)^*$	$(1, 3, 5)$
$(1, 4, 5)$	$(2, 3, 4)^*$	$(2, 3, 5)$	$(2, 4, 5)$	$(3, 4, 5)$

In general, each round adds extra number of traits, and the possible number of combination for a species possessed a given number of traits grows. Each new rounds includes all previous rounds possible combination. Hence, we have shown that $P_{df}(t, x)$ represents the cumulative emergence chance of all previous periods up to time t.

Appendix B

Review and Response to Milan Cirkovic's The Great Silence: Science and Philosophy of Fermi's Paradox

B.1 Intro

Hello Dr Cirkovic,

I contacted you last time in early March, since then, lots have happened. Serbia declared the state of emergence. I planned to leave the country before the airport was closed, but local friends recommended me to stay. I settled myself in Kumodraz, south of Belgrade for the last two months. After following and consuming months of news on the virus and also dedicating myself in formulating mathematical modeling for the pandemic, I spent the recent week thoroughly reading your book *The Great Silence: Science and Philosophy of Fermi's Paradox*. [32]

First of all, I praise your dedication and work to organize all existing solutions to the puzzle with a taxonomic system as a way to organize our ignorance on the matter and giving each a critical philosophical analysis.

Beyond being an excellent researcher, You have a tremendous wealth of knowledge in arts, music, literature, science fiction, science history. By reading your work, one can learn and appreciate the giants who led us to this point in time so far.

I am deeply intrigued and impressed by some of the points you have raised:

1. You raised the concept of complexity, which is also one of the ground concept my model was based on.

2. I totally agree that, without trying to get a deeper theoretical insight, the drake equation is becoming constraining, impractical and supply ammunition to the opponents, and incompatible with type 3 civilizations.

3. I agree that a general need for more theoretical work which unites disparate astrobiological fields is required.

4. Just like you, I am an optimist regarding the existence of extraterrestrial intelligence.

5. I am impressed by your arguments regarding non-exclusivity, and concluding that a zoo hypothesis is a better candidate hypothesis than the hermit hypothesis, since the uniformity of behaviors over long distances and unimaginable diversity is much harder than an agreement between presumably a small number of independent agents required to maintain a zoo.

6. The comparison between Fermi Paradox and Olber's Paradox due to the finite age of sources and marginally by the expansion of the universe is great, if those two paradoxes indeed follow the same line of reasoning, then, philosophically suggesting that all extraterrestrials are young.

7. You mentioned that the conventional estimate is that galaxy contains about 3×10^{11} stars, I pretty much agree though using a list of steps I did come up with a slight higher estimates of 5.1942×10^{11} stars.

8. For self destruction hypothesis, you mentioned it now includes the intentional or accidental misuse of biotechnology, and look at the Covid-19 pandemic.

9. I love how you analyzed introvert big brother requires the suppression of the detection cross section over much longer times than stop worrying and love the bomb, decreasing its credence.

10. You mentioned utilization of gravitational collapse, bulk annihilation of CDM particles and anti particles, hawking evaporation of black holes can be put to use for energy generation, I was unaware of those potential energy sources.

11. The deadly probes example is strikingly analogous to biological pathogen on earth, except it spreads at the cosmic level.

B.2 My Theory

Having fully grasped your thoughts, I will paraphrase my theory [123] in terms of your vocabularies. Based on your taxonomic classification of the solutions, my model only draws a subset of assumptions you have presented, but as you wrote, every SETI project contributes to the reduction of the overall parameter space for cosmic civilization in general, It enables a bit better focus on the regions of interest of the astrobiological landscape. My model is, in your words, an attempt at grounding in detailed numerical modeling which brings together astronomy, social sciences, future studies, and AI. It is essentially a statistical or large number problem. Thanks to my computer science background, I did use the latest platforms and tools attempting to move the discussion from a pure philosophy into real mathematical instantiation. As you have suggested, through more precision, more numerical models, more simulations, more specific scenarios subject to quantification.

I have known Fermi paradox for 16 years, and had many thoughtful discussions during college years with peers regarding this issue, and after many years of searching for an answer, for many years I gradually settled on the conclusion that industrial civilizations are all transient, non-sustainable, which is a solution under your logistic paradigm. You mentioned that we still know too little about the dynamics of culture evolution to be able to put forward a cogent alternative model of technological development vs. time. However, after I have read the works of geologist Marion King Hubbert and energy economists Joseph Tainter, as well as Richard Heinberg, James Kunstler, Micheal Ruppert..etc. I become convinced that all human social evolution is predictable and is based on the extraction of resources and energy, and cultural evolution revolves around the concept of EROEI (energy return over energy invested), which is the central thesis of Joseph Tainter's the collapse of the Roman empire. The exponential extraction of finite, slow renewing geologic resources such as crude, shale, and natural gas will eventually render a peak of industrial output. Every cultural transformation including electronics, aviation, satellites, AI, biotechnology which based on the abundant cheap fossil fuel will one day come to an end.

It was depressing to come to this conclusion, and the whole mainstream in western society is quite silent on this issue, but I was desperately searching for an answer, a way out. I am well-aware Hubbert's recommendation on nuclear breeder and ultimately fusion, but he wrote in 1962 that he is skeptical whether it will ever be possible. 60 years after his writing, the laser driven approach was abandoned at Lawrence Livermore and ITER is not yet in sight. Without fusion, all cosmic travel can be not maintained perpetually. Even if they were carried once, it will be more or less like Ming's Zhenghe voyages. Of course, I found the solution, ultimately, in summer of 2014, when I was in LA, the idea dawned on me, that since hydrogen bomb is the only demonstrable fusion device known, and no container can stand the stress brought by such an explosion, then why not use earth itself as the container? At first, I thought I was crazy but soon a search online I realized nuclear physicists Ralph Moire have already written on this topic in 1990, and its EROEI is guaranteed to be socially and economically viable. All the sudden, I realized sustainability is possible for industrial civilization, humanity's ticket to the future is right at our door step but so far so many have ignored. I soon realized, in order guide people to action is not direct telling them to build such a device. It will not work. It is analogous to writing a beggar a million dollar check. He does not even have the slightest clue to liquidate a check, not even mentioning using the money for investment. For him, his entire life was dedicated to a low level survival strategy. We need to show the beggar the potential new life path available to him and the new rules for life. Therefore, I realized that Fermi paradox needs to be solved one way or another, and all the potential and landscape that can become accessible to human mentioned within the paradox only if fusion power is implemented. With that in mind, I embarked on the mission impossible, determined to write a book as important as Marco Polo's travel, which inspired Columbus's voyage.

So starting in 2015, I did step by step, as what you have suggested, created an overarching explanatory structure, drawing outline of high complexity, building a unified framework for resolving strongFP by joining social science and humanities with the rest of sciences. Much like you, early into my research, I realized that to generate plausible explanatory hypothesis, one has to suppress realism, copernicanism, gradualism or economic assumptions. Since my solution is grounded in realism and naturalism, building on the premise that there are no extraterrestrial civilizations either present on earth or detectable in the solar system or Milky way so far, I have to suppress Copernicanism and propose upfront a version of rare earth hypothesis. My conclusion is that anthropocentrism and Copernicanism does not need to be completely antitheses. You seem to agree with it as well to an extent. As you mentioned, Copernicanism is just a principle, it can not on its own do the explanatory work for us. It needs to be coupled to correct empirical knowledge and theoretical ideas about the world.

In my opinion, the application of Copernicanism is more of an art than science. For example, if we blindly follow Copernicanism, we would expect the sun as the most typical star in the galaxy, but in fact the sun is more massive than 90% of all stars. Likewise, blindly follow Copernicanism implies the peak of terrestrial planet formation occurs at the time of earth's birth, which is refuted by Lineweaver's timescale. But we probably could use Copernicanism to argue that sun is the typical star size for the emergence of intelligent life. In a sense, as long as the concept of mediocrity or typicality is applicable in a some scope regarding the solar system, the principle holds. I tried to downplay the antithesis between rare earth and typicality as much as possible by introducing a list of astronomical and geologic early filters. Where each filter indicates the typicality of earth or the solar system, and each filter is indeed not extremely atypical, thus Copernicanism is conserved yet their multiplicative result does bring a qualitative difference and makes earth rare. Thus, bridging the gap between anthropocentrism and Copernicanism. I have presented this idea early on in my introduction chapter with the thought experiment of Copernicus under imprisonment, and I concluded that, to refine our hypothesis as Many-earth to fewer habitable to fewer hunter-gatherer to rarer agricultural to industrial to an exceptionally rare cosmic expanding human hypothesis. which truly captures the essence of the Fermi Paradox. You seem to agree with me on this as you mentioned rare earth hypothesis needs to be somewhat augmented in order to cope with the possibility of a rare early spreading of intelligence.

I dedicated chapter 2, 3 entirely to these astronomical and geological filters, since they are more likely falsifiable, and being sufficiently early, requiring probably the least amount of fine-tuning. Besides the most common filters, I added the filter of habitable galaxy, the chance of earth getting wet, the total water budget of earth, and the right ocean to land proportions, Gaian window and their particular bottlenecks are not considered but potentially can reduce the contact section even further. I tried my best to clarify the entire list of requirements and requisites for the emergence and evolution of the observers similar to us, and separating the proportion of luck and law leading to us. I have shown that exomoons are not habitable and red dwarfs systems such as Trappist -1 must not be habitable due to extremely strong magnetic fields.

I also realized that filters alone can only decrease the likelihood of peers but there potentially can be civilization evolved on planets older than earth accordingly to Lineweaver's time scale. Just like you mentioned, that additional help could be forthcoming from galactic regulation mechanisms like astrobiological phase transition. Especially if tFH around 10^9 yrs or more, further work on getting a better numerical hold on tFH, which is essential for progress to be made.In my theoretical model, a concept of the EARLIEST WINDOW is introduced, so that the time scale of tFH can be altered to a range of values to fit the given observational constraints on earth. Since we currently is in an empty, unoccupied region of the universe, then, tFH can be lengthened up to a 100% occupation of all space.

I did not apply any of the late filters into my model, first, all those filters are based on social evolution of society, which are less reliable and more speculative than hard science. Secondly, these filters, as you have mentioned, are non-exclusive solutions to the problem. I do acknowledge, that by including these filters, the emergence of technological civilizations and their contact section can be further reduced. However, even with existing early filters it is possible to reduce the emergence of civilization within the galaxy to less than a dozen. Finally, all late filter suffers a major shortcoming. Although all late filter assumes that civilization eventually becoming difficult to detect, there must exist, unless it is extremely transient, a transitional period in which they manifest themselves from the natural environment, and they should most likely be detectable from extra-galactic setting.

Beyond astronomical and geological factors, there are life and evolutionary factors. You mentioned that there is no unbridgeable gap between inorganic matter and living system, under suitable conditions emergence of life is highly probable. Abiogenesis as a modular process with high likelihood whenever relaxed physical and chemical preconditions are met. This is indeed my assumption as well. You mentioned that multi-cellularity is not astronomically improbable, indeed, based on my model, it is when oxygen as a high metabolic fuel becomes available, energy consuming eukaryotes appeared. Therefore, the difficult step was not difficult, rather, building up of free oxygen takes time, and every planet's buildup of free oxygen takes similar amount of time.

Then comes to the core of my model, I attempted, for the first time, a mathematical model for describing the macro-evolution. The model is conceptually similar to a map with three zooming levels. At the greatest level, it sketches the biodiversity of a typical planet given different ratio of land mass to total surface area. Interestingly enough, 29% land surface area does correspond to the maximal biodiversity of a planet given every factors are equal. In terms of habitability, it is as you called a continuous, fine grained, spectrum of habitability.

For each point on the biodiversity curve, a dimensionality expansion renders a multivariate time-dependent exponential lognormal distribution to model biological evolution from the perspective of man. This model is obtained by specifying species as a combination and permutation of traits acquired through evolutionary time, multi-nominal distribution profile of species can be constructed. Those with fewer traits are the most common. A particular multi-nominal distribution is build to model the emergence of civilization by specifying homo sapiens as an outlier. The deviation is calculated based on known cranial capacity of homo sapiens and the explosive growth of angiosperm. The multi-nominal distribution is then transformed/approximated into a more manipulative, generalized lognormal distribution. You have mentioned that existing law describes physics and chemistry does not tell frequency of noogenesis, based on my model, this problem is solved. With this model, I have created concepts in evolutionary biology analogous to classical physics' speed, acceleration, and distance. You have questioned the exact definition of biocomplexity as which might elude us for some time to come, and expecting ever higher peaks of complexity. I defined it mathematically as the area under the lognormal distribution, representing all possible solutions and their frequency distribution by nature's experiment in any given time. Evolutionary speed is governed by stabilizing selection and directional selection. Directional selection correspond to the some killing and displacement of species you mentioned. Indeed, it is highly desirable in order to achieve greater complexity. But the definition of complexity should be clarified further, a greater complexity can be achieved by more species around a static mode/mean value, or greater complexity can be achieved by species over time moving to a greater mode/mean value. Directional selection implies the latter. The general history of evolution of earth indicates a greater mode/mean value with greater number of species around such mode/mean. My model is based on a neocatastrophic paradigm which unites successful features of gradualism with new and dramatic data on mass extinction episodes. Sudden, punctuated changes which present a major ingredient in shaping both earth and galaxy's astrobiological history. In agreement with your words, that Fermi paradox is essentially an evolutionary problem, with punctuated equilibrium when gradualism is considered over long time scales.

Finally, each point along the lognormal distribution can be represented as number of genes/bases of DNA instead of traits. This is the lowest level the map, and correspond to molecular biology.

Once I have formulated the general model for evolution, I assumed that a transcendence through singularity + fusion, super intelligent, post-biological civilization. Once such status is attained, the civilization expands near the speed light in all directions. You have emphasized post biological civilization may not have all the evolutionary instinct to expansion and territorial occupation, and may even obviates expansion drive all together. At the same time, you mentioned natural selection may give a huge advantage to non-biological actors on galactic stage, and whatever possible according to the laws of physics would have occurred at least once. Over time, all non exclusive path may lead to the most economic paths. I assumed all post biological civilization expand based on game theory and economic principle Jevon's paradox (first locally optimize followed by a later expansion) to argue that expansion is still likely, though you may object that economic incentive is based on psychological hard wiring of evolutionary nature. It is possible at least some if not all follow such path, so one can argue this as a possibly late filter. Moreover, I want to present the case as fundamental as possible. By as fundamental as possible I meant the signatures recorded is based on the limits known in pure physics by assuming maximal engineering limits achievable such as near light speed expansion. At the same time, for earthbound observers, maximal detection achievable by assuming there are no instrumental detection threshold limits. Only as such hard limits are proposed, and using the strongest rules possible mathematically, we can formulate the upper bound on detectability and can best verify or falsify these assumptions based on empirical observation.

Even with these most extreme assumptions in place, simulation shows that it is possible that expansion can already be well underway yet we could not possibly make any detection. This is possible due to the wall of semi-invisibility constrained by the known physical limit on the speed of light. That is, in order to detect the next appearing extra-terrestrial industrial civilization, one has to look further out into more distant regions of sky where the snapshot taken occurs at the time when such extra-terrestrial industrial civilization has not yet been evolved. By calculating and locating the distance between earth-bound observers for even earlier arisen industrial civilizations, the distance involved in guaranteeing their appearance, measured in light years, eventually always grows faster than their first arising date measured in years compares to the current time measured in light years. This is the strong case for observational Fermi Paradox. You seem to agree with me on this as well since you mentioned observers in M31, even if they were in possession of instrumentation with infinite sensitivity, would not have been able to detect Homo sapiens.

You also asked, for even crazy leaps on both theoretical and empirical stage, on the most general logical level, what can minds do in the fullness of space and time, free from any constraints except logical coherence and the basic physical law? My answer your question is the construction of worm hole network across the universe. If wormhole indeed prove to shorten cosmic distances, then, within such network, the farthermost distances traversable from earth can be either infinite or much much larger than the observable universe if it is finitely bounded. An expanding cosmic civilization will eventually have a chance to meet every other alien civilization within the universe in a finite amount of time, and every other alien civilization can also meet each other. I call this the Principle of Universal Contacts. In such a cosmically engineered universe, one should able to traverse into neighbor's network and witness the birth of one's own civilization.

In conclusion, my model is based on rare earth and neocatastrophic solution, in agreement with your intuition that most well-rounded solutions comes from them. You mentioned that it is likely a single non-exclusive hypothesis accounts for the most of the explanation in strongFP, and others play a auxiliary role, recommending the smallest number of highly non-exclusive criterion/hypothesis as possible. In my case, it is the relativistic time delay of signal arrival at cosmological distance coupled with the exponentially decreasing emergence chance of civilization into the past accounts for the most of the explanation, and rare earth early filters play a auxiliary role in helping decreasing the likelihood of local emergence. In essence, locally within the galaxy, Lineweaver timescale is reduced significantly due to astrobiological phase transition, then peer planets are reduced by list of early filters, so no patchwork quilt is required.

Since I created the first draft of my work in 2018 and successive refinement up to now, I have received positive encouragement and feedback from many, including Alexei Sharov, senior biological researcher at NIH, Guiseppe Longo, Research Director at CNRS, Nicolas Flament, senior lecturer in geology at university of Wollongong, and David Schwartzman, Professor Emeritus, Howard University. The research as a book was featured by Bookauthority.com as one of the best mathematical paradoxes books.

Hopefully, I am the first but not the last, as a response to your suggestion of huge strides in answering this puzzle on the timescale of years or decades, and ongoing astrobiological revolution has potential to lead to reappraisal of our position in terms of universal complexity, and indeed, there is a sense of excitement, drama, adventure, so uncommon these days in any human endeavor.

www.ingramcontent.com/pod-product-compliance
Lightning Source LLC
Chambersburg PA
CBHW081713220526
45468CB00008B/1825